International Series of Monographs on Physics

Series Editors
J. Birman — City University of New York
S. F. Edwards — University of Cambridge
R. Friend — University of Cambridge
M. Rees — University of Cambridge
D. Sherrington — University of Oxford
G. Veneziano — CERN, Geneva

International Series of Monographs on Physics

142. V. M. Agranovich: *Excitations in organic solids*
141. W. T. Grandy: *Entropy and the time evolution of macroscopic systems*
140. M. Alcubierre: *Introduction to 3 + 1 numerical relativity*
139. A. L. Ivanov, S. G. Tikhodeev: *Problems of condensed matter physics—quantum coherence phenomena in electron—hole and coupled matter-light systems*
138. I. M. Vardavas, F. W. Taylor: *Radiation and climate*
137. A. F. Borghesani: *Ions and electrons in liquid helium*
136. C. Kiefer: *Quantum gravity, Second edition*
135. V. Fortov, I. Iakubov, A. Khrapak: *Physics of strongly coupled plasma*
134. G. Fredrickson: *The equilibrium theory of inhomogeneous polymers*
133. H. Suhl: *Relaxation processes in micromagnetics*
132. J. Terning: *Modern supersymmetry*
131. M. Mariño: *Chern–Simons theory, matrix models, and topological strings*
130. V. Gantmakher: *Electrons and disorder in solids*
129. W. Barford: *Electronic and optical properties of conjugated polymers*
128. R. E. Raab, O. L. de Lange: *Multipole theory in electromagnetism*
127. A. Larkin, A. Varlamov: *Theory of fluctuations in superconductors*
126. P. Goldbart, N. Goldenfeld, D. Sherrington: *Stealing the gold*
125. S. Atzeni, J. Meyer-ter-Vehn: *The physics of inertial fusion*
123. T. Fujimoto: *Plasma spectroscopy*
122. K. Fujikawa, H. Suzuki: *Path integrals and quantum anomalies*
121. T. Giamarchi: *Quantum physics in one dimension*
120. M. Warner, E. Terentjev: *Liquid crystal elastomers*
119. L. Jacak, P. Sitko, K. Wieczorek, A. Wojs: *Quantum Hall systems*
118. J. Wesson: *Tokamaks, Third edition*
117. G. Volovik: *The Universe in a helium droplet*
116. L. Pitaevskii, S. Stringari: *Bose–Einstein condensation*
115. G. Dissertori, I.G. Knowles, M. Schmelling: *Quantum chromodynamics*
114. B. DeWitt: *The global approach to quantum field theory*
113. J. Zinn-Justin: *Quantum field theory and critical phenomena, Fourth edition*
112. R.M. Mazo: *Brownian motion—fluctuations, dynamics, and applications*
111. H. Nishimori: *Statistical physics of spin glasses and information processing— an introduction*
110. N.B. Kopnin: *Theory of nonequilibrium superconductivity*
109. A. Aharoni: *Introduction to the theory of ferromagnetism, Second edition*
108. R. Dobbs: *Helium three*
107. R. Wigmans: *Calorimetry*
106. J. Kübler: *Theory of itinerant electron magnetism*
105. Y. Kuramoto, Y. Kitaoka: *Dynamics of heavy electrons*
104. D. Bardin, G. Passarino: *The Standard Model in the making*
103. G.C. Branco, L. Lavoura, J.P. Silva: *CP Violation*
102. T.C. Choy: *Effective medium theory*
101. H. Araki: *Mathematical theory of quantum fields*
100. L. M. Pismen: *Vortices in nonlinear fields*
99. L. Mestel: *Stellar magnetism*
98. K. H. Bennemann: *Nonlinear optics in metals*
96. M. Brambilla: *Kinetic theory of plasma waves*
94. S. Chikazumi: *Physics of ferromagnetism*
91. R. A. Bertlmann: *Anomalies in quantum field theory*
90. P. K. Gosh: *Ion traps*
88. S. L. Adler: *Quaternionic quantum mechanics and quantum fields*
87. P. S. Joshi: *Global aspects in gravitation and cosmology*
86. E. R. Pike, S. Sarkar: *The quantum theory of radiation*
83. P. G. de Gennes, J. Prost: *The physics of liquid crystals*
73. M. Doi, S. F. Edwards: *The theory of polymer dynamics*
69. S. Chandrasekhar: *The mathematical theory of black holes*
51. C. Møller: *The theory of relativity*
46. H. E. Stanley: *Introduction to phase transitions and critical phenomena*
32. A. Abragam: *Principles of nuclear magnetism*
27. P. A. M. Dirac: *Principles of quantum mechanics*
23. R. E. Peierls: *Quantum theory of solids*

Excitations in Organic Solids

Vladimir M. Agranovich
The University of Texas at Dallas

OXFORD
UNIVERSITY PRESS

Great Clarendon Street, Oxford OX2 6DP

Oxford University Press is a department of the University of Oxford.
It furthers the University's objective of excellence in research, scholarship,
and education by publishing worldwide in

Oxford New York

Auckland Cape Town Dar es Salaam Hong Kong Karachi
Kuala Lumpur Madrid Melbourne Mexico City Nairobi
New Delhi Shanghai Taipei Toronto

With offices in

Argentina Austria Brazil Chile Czech Republic France Greece
Guatemala Hungary Italy Japan Poland Portugal Singapore
South Korea Switzerland Thailand Turkey Ukraine Vietnam

Oxford is a registered trade mark of Oxford University Press
in the UK and in certain other countries

Published in the United States
by Oxford University Press Inc., New York

© V. M. Agranovich, 2009

The moral rights of the author have been asserted
Database right Oxford University Press (maker)

First Published 2009

All rights reserved. No part of this publication may be reproduced,
stored in a retrieval system, or transmitted, in any form or by any means,
without the prior permission in writing of Oxford University Press,
or as expressly permitted by law, or under terms agreed with the appropriate
reprographics rights organization. Enquiries concerning reproduction
outside the scope of the above should be sent to the Rights Department,
Oxford University Press, at the address above

You must not circulate this book in any other binding or cover
and you must impose the same condition on any acquirer

British Library Cataloguing in Publication Data

Data available

Library of Congress Cataloging in Publication Data

Data available

Translated and Typeset in LaTeX by Gerard Czajkowski
Printed in Great Britain
on acid-free paper by
CPI Antony Rowe, Chippenham, Wiltshire

ISBN 978–0–19–923441–7 (Hbk)

1 3 5 7 9 10 8 6 4 2

PREFACE

The fundamental concept of an exciton as an uncharged electronic excitation is an essential part of the modern physics of solids and is now widely used for the explanation of many optical properties such as the absorption and dispersion of light and luminescence. The exciton is also important as a participant in phenomena such as photoemission, photoconductivity, and the photovoltaic effect. There is more than one type of exciton, and in crystals, the type of chemical bonding determines the model of the exciton to be used for the explanation of optical properties. The most widely used are the Frenkel exciton for organic crystals in which the intermolecular bonding is mainly of the weak van der Waals type, and the Wannier–Mott exciton, which is important in inorganic semiconductors, where the bonding is ionic or covalent, and is strong. A descriptive picture of the Frenkel exciton would be that of a electron–hole pair confined to one molecule, constituting an electronically excited molecular state that can transfer between the crystal lattice cells from one molecule to another by virtue of the weak intermolecular interactions; the Wannier–Mott exciton can be depicted as a mobile Coulomb–bonded electron–hole pair of wider separation. In molecular crystals there are also the important charge–transfer excitons, in which the hole–electron pairs are located at neighboring molecules of the crystal. These excited states play an important role in processes of charge separation.

It is worthy to mention that low–temperature spectroscopy of organic molecular crystals originated in the classical experiments carried out in the late 1920s by Pringsheim and Kronenberger[1] (1), also Obreimov and de Haas (2). At that time only the Bloch band scheme for electronic states in crystals was known. This concept predicted very broad absorption bands, in contradiction to the narrow lines observed in the cited experiments. It is known that I. V. Obreimov called this problem to the attention of Ya. J. Frenkel and it was in the 1931 paper of Frenkel (3), where the concept of the exciton in molecular crystals was formulated for the first time. The name "exciton" was introduced by Frenkel in 1936 (4). Following the introduction of the exciton concept, a series of important and puzzling spectroscopical observations using polarized light were reported by Obreimov and Prikhotko (5) the interpretation of which was made by Davydov (6). He rationalized the polarization of the lines in these absorption and fluorescence spectra, which involved what is now referred to as " Davydov splitting". A classical review of experiments on excitons in organic crystals is given in the book by Broude, Rashba and Sheka (7). We have to mention also the well known earlier published books by Hochstrasser(8), Craig and Walmsley (9), Davydov(10), Pope and Swenberg (11), and also in addition the many excellent review papers

---

[1]See the references at the end of Preface.

which I used in this book and which constitute a vast reservoir of information on the properties of molecular crystals.

During the last fifteen to twenty years there has been developed considerable expertise in nanotechnology. Efforts have been devoted to the growth not only bulk but also of crystalline organic layered structures. Improvement in the technique of molecular beam deposition has led to a variety of good quality organic thin films, multilayered structures and heterostructures based on molecular solids, as well as on the combination of organic and inorganic semiconductors. The possibility of growing specially designed systems incorporating different organic crystalline materials with even more flexibility than for multiple quantum wells based on inorganic semiconductors alone, opened a promising field of research from the point of view of fundamental as well as applied physics. The advent of such a new class of organic and inorganic based crystalline materials with potentially important electro–optical linear and nonlinear properties, represents a challenge to scientists to understand their possible advantages in comparison with the usual organic and inorganic materials. And in the recently discovered new types of organic opto–electronic devices such as organic light–emitting diodes (OLEDs) and organic solar cells and in the photogeneration of carriers, the excitons play a decisive role. This has greatly increased interest in the study of properties of excitations in organic materials among scientists working not only in the field of physics, but also in chemistry and biology. For this reason in this book I describe not only the fundamentals of Frenkel exciton and polariton theory in organic crystals but also I consider the electronic excitations and electronic energy transfer in different nanostructures (quantum wells, quantum wires and quantum dots), at interfaces, in multilayers and in microcavities.

The proposed book is an updated English translation of the author's book "Theory of Excitons", published in 1968 (Publishing House Nauka, Moscow, in Russian). It was widely used in USSR and in other countries where students and scientists were able to read Russian. In this new English edition 5 old chapters have been included and updated, and 10 new chapters that contain discussions of new phenomena were added. The title of the book has been changed to reflect the new subject matter that has been added. Among the new chapters are those devoted to hybrid Frenkel–Wannier–Mott excitons in nanostructures, polaritons in organic microcavities including hybrid microcavities, new concept for LED, the mixing of Frenkel and charge-transfer excitons in organic quasi one-dimensional crystals, excitons and polaritons in one and two-dimensional crystals, surface electronic excitations, optical biphonons and Fermi resonance by polaritons.

It is known that excitons and polaritons in general have properties that can be considered within the framework of electrodynamics of continuous media. These properties were described in a monograph by V. M. Agranovich and V. L. Ginzburg ("Crystal Optics with Spatial Dispersion, and Excitons", Springer, 1984) where we used a mainly phenomenological approach. In the present book I provide a microscopical theory of excitations in organic solids and nanostructures, and only in some cases resort to the use of a phenomenological approach. As I have mentioned above, I concentrate attention on the Frenkel and charge–transfer excitons. However, many results presented in this book as, for example, the polariton theory, are quite general and can be used for all types of excitons. I hope that this book will be useful for scientists working on the field of photophysics and photochemistry of organic solids, and for students who are entering this field.

In the book I have used theoretical and experimental results developed by many authors. However, here I would like to thank my former students and collaborators who worked with me in the studies of organic solids in the Institute of Physics and Power Engineering, Obninsk, and in the Institute of Spectroscopy, Troitsk. These include Yuri Konobeev, Oleg Dubovsky, Vratislav Toshich, Vladimir Kravtsov, Valery Rupasov, Vladimir Yudson, Tamara Leskova, Irina Talanina, Anatoly Malshukov, Ivan Lalov, Anatoly Kamchatnov, Yuri Lozovik, Boris Antonyuk, Anvar Zakhidov, Yuri Gartstein, Sergei Kiselev, Denis Basko and Masha Litinskaia.

After Perestroika I worked abroad and enjoyed the collaboration and support of my friends Franco Bassani and Giuseppe La Rocca (Pisa), Peter Reineker (Ulm), Karl Leo and Michael Hoffmann (Dresden), Claude Weisbuch and Henri Benisty (Paris), Steve Forrest (Princeton), Shaul Mukamel (Rochester), and particularly Hobson Wildenthal (Dallas), who made it possible for me to discuss many new and interesting condensed matter problems. I thank my long time friends Joseph Birman (New York City) who persuaded me to prepare an updated translation of my Russian book and Martin Pope for valuable discussions. I am deeply grateful to Michael Philpott who read the manuscript and provided comments, corrections and valuable advices on the material discussed here and to Gerard Czajkowski who translated the book and helped me with the editorial preparation of the text, giving me much valuable advice.

V. M. Agranovich
Dallas, Moscow

References

1. Kronenberger, A. and Pringsheim, P. (1926). *Z. Phys.* **40**, 75.
2. Obreimov, I. V. and de Haas, W. J. (1928). *Proc. Akad. Sci. Amsterdam* **31**, 353.
3. Frenkel, Ya. J. (1931). *Phys. Rev.* **37**, 17.
4. Frenkel, Ya. J. (1936). *Sov. Phys.* **9**, 158.
5. Obreimov, I. V. and Prikhotko, A. F. (1932). *Sov. Phys.* **1**, 203; (1936) **9**, 34; (1936) **9**, 48.
6. Davydov, A. S. (1951). *Theory of Light Absorption in Molecular Crystals*.

UKSSR Akad. Nauk, Kiev; (1964). *Usp. Fiz. Nauk* **82**, 393.
7. Broude, V. L., Rashba, E. I., and Sheka, E. F. (1984). *Spectroscopy of Molecular Excitons*. Springer, Heidelberg.
8. Hochstrasser, R. M. (1966). *Molecular Aspect of Symmetry*. Benjamin, New York.
9. Craig, D. P. and Walmsley, S. M. (1968). *Excitons in Molecular Crystals, Theory and Applications*. Benjamin, New York.
10. Davydov, A. S. (1971). *Theory of Molecular Excitons*. Plenum Press, New York.
11. Pope, M. and Swenberg, C. E. (1999). *Electronic Processes in Organic Crystals and Polymers*. Oxford University Press, Oxford.

CONTENTS

1	**Introduction**	1
2	**Frenkel excitonic states in the Heitler–London approximation**	10
	2.1 Excitons in a molecular crystal with fixed molecules. Splitting of molecular terms in a crystal	10
	2.2 Frenkel, Coulomb, and mechanical excitons	15
	2.3 Application of group theory for the determination of polarization and selection rules for excitonic light absorption. Degeneracy of excitonic levels	23
	2.3.1 Winston theory	23
	2.3.2 Applications to crystals of naphthalene type	26
	2.3.3 Symmetry properties of Coulomb excitons	28
	2.4 Triplet excitons	30
3	**The second-quantized theory of Frenkel excitons**	36
	3.1 Energy operator for a molecular crystal with fixed molecules in the second-quantization representation. Paulions and Bosons	36
	3.2 Excitonic states in the two-level model. Transition to the Heitler–London approximation	39
	3.2.1 Crystals with one molecule per unit cell	42
	3.2.2 Crystals with several molecules per unit cell	43
	3.3 Exciton states beyond the Heitler–London approximation	46
	3.3.1 Small corrections to Heitler–London approximation	49
	3.4 Exciton states in the presence of several molecular states (mixing of molecular configurations)	53
	3.4.1 One molecule per unit cell	55
	3.4.2 Several molecules per unit cell	56
	3.5 Perturbation theory. A comparison with results obtained in the Heitler–London approximation	60
	3.6 Sum rules for the oscillator strengths of excitonic transitions and the hypochromatic effect	62
	3.7 Exciton–phonon interaction	67
	3.8 Spectra and mobility of self-trapped (ST) excitons	71
	3.8.1 Mechanism of self-trapping of Frenkel excitons	73
	3.8.2 Spectra and transport of self-trapped excitons	75
	3.8.3 Self-trapping barrier	75

		3.8.4 Self-trapping of charge-transfer excitons	76
		3.8.5 Self-trapping in one-dimensional structures	76
	3.9	Electron–vibrational excited states in molecular crystals	77
	3.10	Calculation of the exciton states in molecular crystals	85
		3.10.1 Anthracene and naphthalene	91
		3.10.2 Tetracene and pentacene	93
	3.11	Exact transformation from paulions to bosons	94
	3.12	Kinematic biexcitons	98
4	**Polaritons: Excitonic States Taking Account Of Retardation.**		103
	4.1	The crystal energy operator in the presence of a retarded interaction	103
	4.2	Dispersion of polaritons and refraction index of electromagnetic waves	108
		4.2.1 Operators of electric and magnetic fields	116
	4.3	Polariton mechanism of exciton luminescence	118
	4.4	The dielectric tensor and the phenomenology of long-wavelength excitons	121
	4.5	Giant radiative width of small wavevector polaritons in one- and two-dimensional structures ("polariton superradiance")	128
	4.6	Effective radiative lifetime	136
	4.7	Concluding remarks	137
5	**Dielectric theory of Frenkel excitons: local field effects**		140
	5.1	Introduction: the local field method	140
	5.2	Dielectric tensor of cubic crystals	144
	5.3	Effects of impurities	145
	5.4	Dielectric tensor of organic anisotropic crystals	148
	5.5	Dielectric constant of mixed crystalline solutions and polarization of impurity absorption bands	151
	5.6	Optical properties of mixed crystalline solutions	156
	5.7	Energy of the resonance interaction of the impurity molecules	159
	5.8	The higher multipoles in the local field method	162
6	**Biphonons and Fermi Resonance in Vibrational Spectra of Crystals**		166
	6.1	Effects of strong anharmonicity in vibrational spectra of crystals	166
	6.2	Biphonon theory	169
		6.2.1 Biphonons in the overtone frequency region of	

		an intramolecular vibrations: qualitative consideration	169

      6.2.2 Biphonon states   174

      6.2.3 Biphonons in the sum frequency region of the spectrum – the Van Kranendonk model   178

6.3 Green's functions in biphonon theory and Fermi resonance in crystals   180

      6.3.1 Fermi resonance   188

6.4 Fermi resonance with polaritons   189

      6.4.1 Microscopic theory   189

      6.4.2 Macroscopic theory – Transverse, longitudinal, and surface biphonons   194

6.5 Dielectric tensor of a crystal in the spectral region of two-particle phonon states: microscopic theory   196

6.6 Biphonons and biexcitons and the gigantic nonlinear polarizability effect   199

6.7 Experimental investigations of biphonons and Fermi resonance with polariton   202

6.8 Local biphonons in crystals with isotopic substitution impurities   208

      6.8.1 Where does the formation of local states begin in a spectrum of optical vibrations? Effects of strong anharmonicity   208

      6.8.2 Local biphonon in an $^{14}NH_4Br$ crystal containing the isotopic substitution impurity ^{15}N 210

6.9 Conclusion and prospects for further investigations   211

7 The dielectric tensor of crystals in the region of excitonic resonances   215

7.1 On the calculation of the dielectric tensor   215

7.2 The Pitaevsky–Dzyaloshinsky formula for the dielectric tensor   217

7.3 Polariton states in the calculation of the dielectric tensor in the region of Frenkel exciton resonances   220

7.4 The transverse dielectric tensor and dissipation of light waves   224

      7.4.1 The transverse dielectric tensor   224

      7.4.2 The dissipation of polaritons in the vicinity of exciton resonances   227

7.5 Macroscopic and microscopic theories of optical nonlinearity in the region of exciton resonances   229

      7.5.1 On polariton anharmonicity in the nonlinear optical response   231

8 Dielectric tensor of superlattices — 233
8.1 Long-period superlattices — 233
8.2 Spatial dispersion in superlattices — 234
8.2.1 Spatial dispersion in the vicinity of an excitonic resonance ($\ell_{1,2} > a_B$) — 234
8.2.2 Spatial dispersion in the vicinity of an excitonic resonance ($\ell_1 < a_B, \ell_2 > a_B$) — 237
8.2.3 Gyrotropy in superlattices — 238
8.3 Dielectric tensor of superlattices with anisotropic layers — 239
8.3.1 Dielectric tensor of a superlattice — 239
8.3.2 Magnetooptical effects in superlattices — 241
8.3.3 Influence of a static electric field — 243
8.4 Optical nonlinearities in organic multilayers — 244
8.4.1 $\chi^{(2)}$ optical nonlinearities in superlattices — 244

9 Excitations in organic multilayers — 246
9.1 Gas–condensed matter shift and the possibility of governing spectra of Frenkel excitons in thin layers — 246
9.2 Fermi resonance interface modes in organic superlattices — 251
9.2.1 Fermi resonance in molecules — 251
9.2.2 Fermi-resonance wave in a two-layer system — 257
9.2.3 Fermi resonance interface waves — 259
9.2.4 Bistable energy transmission through the interface with Fermi resonance interaction — 262

10 Cavity polaritons in organic microcavities — 266
10.1 Giant Rabi splitting in organic microcavities — 266
10.2 Microcavities with crystalline organics — 269
10.2.1 Introduction — 269
10.2.2 Cavity photons and Coulomb excitons — 270
10.2.3 Cavity polaritons — 272
10.2.4 One molecule in the unit cell — 274
10.2.5 Two molecules in the unit cell — 275
10.2.6 Conclusions — 278
10.3 Microcavities with disordered organics — 278
10.3.1 Qualitative picture — 278
10.4 Macroscopic classical theory — 280
10.4.1 General expressions — 280
10.4.2 The case of vanishing q — 284
10.4.3 The case of large q — 286
10.4.4 Microscopic quantum theory — 288
10.5 The localized end-point polariton states — 290

		10.5.1 Dynamics of a low-energy wavepacket in a perfect microcavity	290

        10.5.1 Dynamics of a low-energy wavepacket in a perfect microcavity   290
        10.5.2 Time evolution of the lowest wavepacket   293
        10.5.3 Concluding remarks   300

11 Charge transfer excitons   301

  11.1 Introduction   301
  11.2 Stark effect and electroabsorption of CTEs   303
  11.3 Phase transition from the dielectric to the conducting state (cold photoconductivity)   305
        11.3.1 Analytical approach   306
        11.3.2 Numerical simulations   309
        11.3.3 Results of numerical simulations   310
        11.3.4 Concluding remarks   313
  11.4 Cumulative photovoltage in asymmetrical donor–acceptor organic superlattices   315
        11.4.1 Introduction   315
        11.4.2 On the mechanisms of the photovoltaic effects in organics   317
        11.4.3 Cumulative photovoltage in an asymmetrical stack of D–A interfaces   318
  11.5 Nonlinear optical response of charge-transfer excitons at donor–acceptor interface   321
        11.5.1 Resonant optical nonlinearity of CTEs: the role of the exciton–exciton repulsion   321
        11.5.2 Photogenerated static electric field: influence on the nonresonant optical response   323

12 Surface excitons   325

  12.1 Introduction   325
        12.1.1 Surface excitons and polaritons   325
        12.1.2 Coulomb surface excitons   327
        12.1.3 The exciton-phonon interaction and the role of surface defects   328
  12.2 Phenomenological theory of surface Coulomb excitons and polaritons   329
        12.2.1 Surface polaritons at the sharp interface between media   329
        12.2.2 Observation of exciton surface polaritons at room temperature   332
        12.2.3 Surface excitons in the presence of a transition layer   333
        12.2.4 Nonlinear surface polaritons   335
  12.3 Site shift surface excitons in molecular crystals   337
        12.3.1 Site shift surface excitons (SSSE) in anthracene   337

	12.3.2 On the radiative width of site shift surface excitons	341
12.4	Edge exciton states in molecular chains	345
	12.4.1 Introduction. Tamm states	345
	12.4.2 Mixing of Frenkel and charge-transfer excitons in a finite molecular chain	346
	12.4.3 Edge and bulk states in a finite molecular chain with mixing of Frenkel and charge-transfer excitons	351
12.5	Application to PTCDA and MePTCDI crystals	354
12.6	Frenkel surface excitons in disordered media	356
	12.6.1 Macroscopic surface excitons and polaritons in isotopically mixed crystalline solutions	356
12.7	Conclusion	359

13 Excitons in Organic-Based Nanostructures — 360

13.1 Introduction — 360
13.2 Hybrid 2D Frenkel–Wannier–Mott excitons at the interface of organic and inorganic quantum wells. Strong coupling regime — 362
    13.2.1 Configuration of heterostructure and general relations — 362
    13.2.2 The coupling matrix element — 364
    13.2.3 Dispersion relations of hybrid states — 368
    13.2.4 Linear optical response of hybrid states — 370
13.3 Hybrid excitons in parallel organic and inorganic semiconductor quantum wires — 372
13.4 On the hybridization of "zero-dimensional" Frenkel and Wannier–Mott excitons — 376
13.5 Nonlinear optics of 2D Hybrid Frenkel–Wannier–Mott excitons — 377
    13.5.1 The resonant $\chi^{(3)}$ nonlinearity — 377
    13.5.2 Second-order susceptibility $\chi^{(2)}$ — 383
13.6 Weak coupling regime in hybrid nanostructures — 384
    13.6.1 The Förster energy transfer — 384
    13.6.2 Förster energy transfer in a planar geometry — 385
    13.6.3 Noncontact pumping of light emitters via nonradiative energy transfer: new concept for light-emitting devices — 387
    13.6.4 First experiments — 388
    13.6.5 Förster energy transfer from quantum dots to organics — 392
    13.6.6 Exciton energy transfer from a quantum dot to its surface states — 395

13.6.7 Exciton energy transfer from organics to semiconductor nanocrystals and carrier multiplication   396
13.7 Hybridization of Frenkel and Wannier–Mott excitons in a 2D microcavity in the regime of strong coupling   398
    13.7.1 Microcavity embedded resonant organic and inorganic quantum wells   398
    13.7.2 Dispersion and relaxation of a polariton in a hybrid microcavity containing a crystalline organic layer and a resonant inorganic quantum well   400
    13.7.3 Classical formalism   400
    13.7.4 QED approach   402
    13.7.5 Exciton–phonon scattering in a microcavity   404
    13.7.6 Estimation of transfer rates   404
    13.7.7 Conclusion   408

14 Mobility of Frenkel Excitons   410
14.1 Diffusion of Frenkel excitons   410
14.2 The diffusion tensor   412
14.3 Weak exciton–phonon coupling: coherent excitons   413
    14.3.1 General expressions   413
    14.3.2 Isotropic exciton effective mass and scattering by acoustic phonons   415
    14.3.3 Temperature dependence of the diffusion constant   417
14.4 Strong exciton–phonon coupling: incoherent excitons   418
14.5 Transport measurements and diffusion of polaritons   420

15 Statistics and collective properties of Frenkel excitons   423
15.1 Approximate second quantization and kinematic interaction   423
15.2 Collective properties of an ideal gas of paulions   425
15.3 Collective properties of Frenkel excitons in the presence of a dynamic interaction   429
    15.3.1 On biexcitons in organic crystals   432
15.4 Kinematic interaction of exciton-polaritons in crystalline organic microcavities   433
15.5 Fermionic character of Frenkel excitons in one-dimensional molecular crystals   434

A Diagonalization of a Hamiltonian quadratic in the Bose-amplitudes   437

B	**Calculations of Polarization in Inorganic Quantum Wells and in Organics**	439
	B.1 General relations	439
	B.2 Polarization in a semiconductor	440
	B.3 Polarization in organics	442
C	**Microscopic Quantum-Mechanical Calculations of the Energy Transfer Rate**	444
D	**Energy transfer in the planar geometry**	446
	D.1 Free excitons	446
	D.2 Localized excitons	451

References	455
Tables	484
Figures	485
Index	489

1

INTRODUCTION

Molecules, atoms, ions, and electrons in condensed medias are constitutive units of the material. Because of the strong interaction between those particles they lose their individual properties. They are replaced by quasiparticles, the quantum of collective motions of condensed materials, that play a fundamental role in modern solid state physics.

At the present time we know a number of quasiparticles in crystals, each of them playing role in explaining specific physical properties of crystals. Excitons, the main topic of the present book, are examples of such particles, and they appear as a result of the quantum mechanical treatment of the collective properties of electrons in crystals.

The foundation of excitons theory was formulated by Frenkel, Peierls and Wannier (1)–(5) more than 70 years ago. After that time the theory has been enriched by many new aspects. The theory has also been exposed to continuous experimental verification, which has confirmed the role of excitons in such processes as absorption of light, luminescence and energy transfer, photochemical processes, etc. Before we examine the experiments, which illustrate the presence and the role of excitons in crystals, we will briefly describe the basic models of excitons, which are mostly used in the interpretation of experimental results.

The first exciton model was proposed by Frenkel (1), (2) in 1930. Frenkel excitons (FE) are mostly observed in molecular crystals.

As is known, crystals are defined as molecular when the interaction between different molecules is much smaller than the interaction between atoms and electrons within the molecule. In consequence, the molecules in such crystals preserve some individual properties. Therefore, in the zeroth approximation, the lowest electronic excited state of the crystal can be considered as a state in which one molecule is excited, and the others in their ground state. As a consequence of the translational symmetry and of the intermolecular interaction, the localization of the excited molecule is not stable, and the excitation energy will be transferred from one molecule to another, propagating like a wave through the crystal.[2]

Let Ψ be the wavefunction of the above described excited state of the crystal, satisfying the stationary Schrödinger equation

$$\hat{H}\Psi = E\Psi, \qquad (1.1)$$

where $\hat{H}$ is the crystal Hamiltonian and E the corresponding eigenvalue. Since the Hamilton operator is invariant under application of the translation sub-

---

[2]The theory of Frenkel excitons will be given in Chapt. 2 and Chapt. 3.

group of the crystal, including all translations which transform the crystal lattice into itself, the eigenfunction Ψ can be chosen as an eigenfunction of any translation operator. In this case for the eigenfunction we have (see, for example, Lyubarsky (6))

$$T_\mathbf{m}\Psi = e^{-i\mathbf{k}\mathbf{m}}\Psi, \tag{1.2}$$

where $T_\mathbf{m}$ is the translation operator, $\mathbf{m}$ an integer–valued lattice vector, and $\mathbf{k}$ the wavevector from the first Brillouin zone. Different eigenfunctions Ψ correspond to different eigenvalues of the operator $T_\mathbf{m}$. Therefore they can be characterized by the wavevector $\mathbf{k}$. Consequently, the exciton energy will also depend on the wavevector $\mathbf{k}$ and the above conclusion does not need the use of a specific model of the exciton; it results in the translational symmetry of the crystal only. One of the fundamental problems of an accurate microscopic theory of excitons consists in the determination of the energies $E(\mathbf{k})$ and the wavefunctions $\Psi_\mathbf{k}$.

For the considered electronic states the wavevector $\mathbf{k}$ is a continuous quantum number characterizing the eigenfunction Ψ. Together with this quantum number the state Ψ can be also characterized by a set s of discrete quantum numbers, so that $\Psi \equiv \Psi_{\mathbf{k}s}$. Making use of the notion of quasiparticles we can say that in the crystal in the state $\Psi_{\mathbf{k}s}$ there is a one quasiparticle of type s, having quasimomentum $\mathbf{k}$ and the energy $E_s(\mathbf{k}) - E_0$, E_0 being the crystal ground state energy.

Just such a type of quasiparticle was named by Frenkel an exciton. For a given and fixed value of the quantum number s the exciton energy $E_s(\mathbf{k})$ becomes a continuum-valued function of $\mathbf{k}$ within the sth energy band, and, in general, $E_s(\mathbf{k}) \neq E_{s'}(\mathbf{k})$ for $s \neq s'$.

Note that in an ideal crystal any stationary state is characterized by a value $\mathbf{k}$, which determines the change of the wavefunction by the translation operation (see eqn 1.2). However, this $\mathbf{k}$ is not always the one continuous quantum number characterizing the state of the crystal. If we consider a crystal with free electrons and holes, the state of the crystal is characterized by two continuous quantum numbers $\mathbf{k}_1$ and $\mathbf{k}_2$, where $\mathbf{k}_1$ is the quasimomentum of the electron, and $\mathbf{k}_2$ is the quasimomentum of the hole. Thus, the two continuous quantum numbers correspond to the state when we have not one, but two quasiparticles in the crystal.

The electron and the hole in the crystal attract themselves and can create a bound state. Obviously, the Frenkel exciton corresponds to the situation when the electron and the hole in a bound state are localized in the same lattice cell (the same molecule). Therefore the Frenkel excitons are also called small–radius excitons. When the radius of the electron–hole bound state is much larger than the lattice constant, the corresponding quasiparticle is called a Wannier–Mott exciton, or a large–radius exciton. Let us consider the latter in more detail.

Assume that a dielectric crystal has one electron in the conduction band with effective mass m_e, and one hole in the valence band with effective mass m_h. If we assume that the interaction between the electron and the hole is given by the

Coulomb potential $V(r) = -e^2/\epsilon r$, where ϵ is the static dielectric constant,[3] and r is the distance between the electron and the hole, the Schrödinger equation for the electron–hole system will have the form[4]

$$\left(-\frac{\hbar^2}{2m_e}\Delta_e - \frac{\hbar^2}{2m_h}\Delta_h - \frac{e^2}{\epsilon r}\right)\Psi = E\Psi. \tag{1.3}$$

This is the Schrödinger equation for a hydrogen-like atom. The solution for $E < 0$, i.e. for the bound states, has the form

$$\Psi_{n\ell m,\mathbf{k}} = e^{i\mathbf{k}\mathbf{R}}F_{n\ell m}(r), \tag{1.4}$$

where $\mathbf{R}$ is the electron–hole pair center-of-mass coordinate, $F_{n\ell m}(r)$ is the hydrogen–like atom eigenfunction with effective charge $Ze = e/\epsilon$, n, ℓ, m are the principal, the angular and the magnetic quantum numbers, respectively, and $\hbar\mathbf{k}$ is the quasimomentum which describes the motion of the electron–hole pair as a whole entity. The energy corresponding to the state (1.4) is given by the expression

$$E_n(\mathbf{k}) = -\frac{\mu e^4}{2\hbar^2\epsilon^2 n^2} + \frac{\hbar^2 k^2}{2(m_e + m_h)}, \quad \text{where} \quad \mu = \left(\frac{1}{m_e} + \frac{1}{m_h}\right)^{-1}. \tag{1.5}$$

The zero value of the energy (1.5) coincides with the bottom of the conduction band and its absolute value for $\mathbf{k} = 0$ is equal to the binding energy of the electron and the hole in the exciton. The relatively simple relation between the Wannier–Mott exciton energy does not hold when the electron–hole interaction is treated more accurately (see (8)), and becomes a relation of a more general type.

Expanding the exciton energy $E_s(\mathbf{k})$ in powers of $\mathbf{k}$ around the point $\mathbf{k} = 0$, we obtain in crystals with inversion symmetry

$$E_s(\mathbf{k}) = E_s(0) + \frac{\hbar^2}{2m_{ij}^*}k_i k_j + \ldots, \tag{1.6}$$

where m_{ij}^* is the tensor of the effective mass of exciton in the sth energy band:

$$m_{ij}^* = \hbar^2 \left\{\frac{\partial^2 E_s(k)}{\partial k_i \partial k_j}\right\}_{k=0}^{-1}. \tag{1.7}$$

If $E_s(\mathbf{k})$ only depends on the absolute value of $\mathbf{k}$, then $m_{ij}^* = m^*\delta_{ij}$, where

$$m^* = \hbar^2 \left\{\frac{d^2 E_s(k)}{dk^2}\right\}_{k=0}^{-1}. \tag{1.8}$$

Depending on the shape of $E_s(k)$ the effective mass can be positive or negative.

---

[3]This type of electron–hole interaction is possible only for values r large compared to the lattice constant.

[4]The theory of Wannier–Mott excitons and, in particular, the limits of validity of eqn (1.3), can be found, for instance, in the textbook by Knox (8) and the review article by Haken (19).

Theory of excitons often uses the so-called effective mass approximation, assuming the energy dispersion as in eqn (1.6), where only the terms displayed on the r.h.s. are taken into account. In this approximation the effective mass also determines the width of the exciton energy band, which for isotropic bands is of order

$$\Delta E_s = \frac{\hbar^2 \pi^2}{2\,|m^*|\,a^2}, \tag{1.9}$$

a being the lattice constant. In this approximation small effective masses correspond to wide energy bands, and, inversely, small energy bands correspond to large effective masses.

As we have seen, the Frenkel exciton and the Wannier–Mott exciton correspond to two limiting situations related to the electron–hole binding process. In the first case one visualizes the electrons and the holes as localized on a given molecule, and their interaction with electrons of other molecules plays a secondary role. In this case the wavefunctions of exciton can be constructed from the wavefunctions of isolated molecules. In the case of Wannier–Mott excitons the mean electron–hole distance is much greater than the crystal lattice constant. It is clear that in this case the interaction energy of electrons and holes strongly depends on the properties of the medium. As can be seen from eqn (1.3), the electron–hole interaction in this simple model is determined by the dielectric permittivity of the medium.

The process of creation of an exciton needs an energy which is less than the energy needed to create free electrons and holes. The difference between these energies, which is equal to the electron–hole binding energy, strongly depends on the crystal type. In semiconductor crystals with a large dielectric constant this binding energy (see eqn 1.5) is of the order of 0.01 eV. A quite different situation occurs in molecular crystals where the binding energy can be of the order of one eV. Therefore, the role of excitons in both types of crystals can be very different. In the first case (for semiconductors with large enough value of the dielectric constant ϵ) excitons quite easily (for example, interacting with phonons) decay into free electrons and holes even at not very low temperatures. As a result we have to discuss how an equilibrium in the excitons–electrons–holes system can be achieved. In the case of high exciton binding energy the process of exciton dissociation requires a high amount of activation energy and is less probable. In this case free charge carriers can be created only by ionization of impurities, as a result of the interaction between excitons and impurities, or in collisions of excitons by excitons. The exciton, being a neutral quasiparticle, does not contribute to the electric conductivity of the crystal.

One of the simplest methods to create excitons is to use electromagnetic radiation. Below a physical picture will be used which existed before the polariton concept had been formulated (see Ch. 4). In this picture the crystal photon with wavevector $\mathbf{q}$ and energy $\hbar\omega = \hbar q c$ propagates in the crystal and can interact with crystal electrons and can decay, creating an exciton. Due to energy and momentum conservation, the energy of an exciton created by the photon decay equals $E_s(\mathbf{q}) = \hbar\omega$. For optical waves the quantity $q \approx 10^5$ cm^{-1} is small compared to the inverse lattice constant so the energy of the exciton is almost equal to the excitonic energy $E_s(0)$ (here we ignore the dissipation broadening). Thus, reasoning in the spirit of the above simplified picture we can say that the energies of exciton absorption lines have to correspond to the values of the energies $E_s(0)$.

In the case of Wannier–Mott excitons this means that the absorption lines are given by the Rydberg formula (1.5). The above theoretical description has been submitted to conscientious experimental verification, which was been initiated by Gross and coworkers (see (7); for the interpretation of experimental results see, for example, Knox (8)). The experimental results have shown that in many crystals, at low temperatures, optical absorption lines can be observed at energies with values which agree quite well with the Rydberg formula (1.5). For example, in the cubic crystal Cu_2O this agreement has been observed for $n \geq 2$, since the terms $n = 1$ and $n = 2$, having a small exciton radius, cannot be described by eqn (1.3) (see (8)).

As has been emphasized above, one of the fundamental tasks of the exciton theory is to determine the function $E_s(\mathbf{k})$. It is also evident that the function $E_s(\mathbf{k})$ can be defined only when the vector $\mathbf{k}$ is a "good" quantum number. We have to keep in mind, however, that even in ideal crystals the exciton can be scattered, for example by the lattice vibrations, changing the value of the exciton $\mathbf{k}$ vector. Thus the exciton has to be considered as a wavepacket made of the wavefunctions $\Psi_{\mathbf{k}s}$ with the vector uncertainty $\delta\mathbf{k}$, and consequently with size of order $\delta x \propto 1/\delta k_x$, etc. When the scattering of excitons is weak, so that the quantity $\delta\mathbf{k}$ is small compared to the vector absolute value $|\mathbf{k}|$, the vector $\mathbf{k}$ can be considered as a good quantum number, and the exciton in this case propagates as a wavepacket with an extension which is much larger than the wavelength and, at the same time, large compared to the lattice constant. Such an exciton propagates through the crystal with group velocity

$$\mathbf{v} = \frac{1}{\hbar} \boldsymbol{\nabla} E_s(\mathbf{k}).$$

When the wavevector uncertainty $\delta\mathbf{k}$ is of the order of $|\mathbf{k}|$ we cannot characterize the exciton by means of a wavevector. In this situation the exciton is localized. If $\delta\mathbf{k}$ is order of $1/d$, d being lattice constant, the exciton is strongly localized at one or two molecules and propagates by hopping from one molecule to another.

Note that an examination of only serial properties of Wannier–Mott excitons (and also Frenkel excitons) can give the values of $E_s(0)$, but will not give the picture of the exciton propagation through the crystal. This information is not enough to determine if the exciton propagates as a wavepacket which envelope many molecules of the crystal, or created by light so that the excitation is strongly localized and cannot be characterized by a wavevector.

To be convinced of the correctness of what was said let us assume, for example, that the effective mass of the hole is very large. In this case the Wannier–Mott exciton practically "stands" at a fixed position, despite the fact that its energy levels follow the Rydberg formula. For such exciton ("localized" in the terminology of Frenkel) the width of the excitonic band (see eqn 1.9) is small so that the interactions generating this band can be neglected. Taking this interaction into account in the next approximations causes hops of the localized exciton from one site of the crystal to another one. This results in the propagation of the electronic excitation energy through the crystal but with a mechanism different from the mechanism of propagation by wavepackets. It is clear that the "localized" exciton is defined not by a wavevector, but by a vector related to the crystal cell, where it is localized.

With the purpose of elucidation the nature of the state which is created by light absorbed by a CdS crystal, Thomas and Hopfield (9) made experiments proving the influence of static electric and magnetic fields $\mathbf{E}$ and $\mathbf{H}$ on positions of a series of exciton absorption lines.

If v is the group velocity of the exciton, the acting electric field in the coordinate system connected with the propagating exciton, is equal to $\mathbf{E} + \frac{1}{c}[\mathbf{v} \times \mathbf{H}]$. Consequently, if the given exciton state has the quadratic Stark effect, in the presence of the magnetic field the exciton line will follow, as a function of $|\mathbf{E}|$, a parabolic dependence on the electric field strength shifted with respect to $\mathbf{E} = 0$. Such a dependence indeed has been observed by Thomas and Hopfield (9). In the effective mass approximation the velocity of the exciton is $\mathbf{v} = \hbar \mathbf{k}/m^*$, where $\mathbf{k}$ is the exciton wavevector, equal to the wavevector of the photon. Thus, since the quantities $\mathbf{E}$, $\mathbf{H}$, and $\mathbf{k}$ turned out to be known, the experiments of Hopfield and Thomas allowed one to estimate the effective mass of the exciton in CdS. They obtained the value of 1.1 m_0, m_0 being the free electron mass, in good agreement with the mass resulting as a sum of the electron and of the hole effective masses (0.2 m_0 and 0.7 m_0, respectively), which have been measured independently in other experiments.

In the case of molecular crystals there are also specific methods, sometimes more indirect, to obtain information on the excitonic states. One of the methods, proposed by Broude, Rashba and Scheka (12), consists of studying the character of impurity states (their intensity and polarization), which occur when the impurity level approaches the excitonic band of the crystal. It follows then from the observed increase of the total absorption that when these energies are closer, the excited state induced by light, being localized at the impurity, catches molecules from the surroundings and in consequence the effective number of light absorbing molecules increases. If the impurity state is located below the exciton band such an effect of flaring up arises only if at the bottom of the exciton band the wavevector $\mathbf{k}_0$, $|\mathbf{k}_0| \ll 1/a$, where a is the crystal lattice constant and also if the exciton band has an enough large width. For narrow excitonic bands this effect is small, owing to the finite linewidths of the impurity- and proper absorption (a more detailed discussion on the impurity absorption in molecular crystals will be given in Section 5.3).

Other information on the nature of excitonic states can be obtained by studying the kinetic parameters characterizing the energy transfer of an electronic excitation through the crystal (see (13), (14)).

It is clear even from a very qualitative consideration, that an exciton with a wavevector **k** will be scattered by phonons, and an increase of temperature will decrease its mean free path. If an exciton is "localized" and changes its local position by hopping from one molecule to another, the thermal motions will intensify the exciton diffusion (for more details about this subject see Ch. 14). Studies of the mobility of excitons are very important for biophysics.

Interesting information about the excitonic states can be obtained from studying the Davydov splitting (15). This problem is discussed in Ch. 2 and Ch. 3. Here we only note that the discovery of this phenomenon has stimulated the investigation of molecular crystal spectra and led to many important conclusions on the nature of intermolecular interactions.

The exciton lifetime depends on the velocity of the transformation of the exciton energy into energy corresponding to other degrees of freedom of the crystal, for example, into the energy of lattice vibrations, to impurities and so on. If the radiative lifetime of the exciton is τ_r and if the nonradiative lifetime is τ_{nr} then the quantum fluorescence yield is given by

$$\eta = \frac{\frac{1}{\tau_r}}{\frac{1}{\tau_r} + \frac{1}{\tau_{nr}}}$$

or

$$\eta = \frac{\tau}{\tau_r}.$$

This exciton lifetime, determining, for instance, the decay of fluorescence, equals

$$\tau = \frac{\tau_r \tau_{nr}}{\tau_r + \tau_{nr}}.$$

In crystals where the probabilities of nonradiative decay processes are smaller (the latter takes place in luminescent crystals with a large quantum luminescence yield), the lifetime for singlet–excitons in pure crystals can be of the order of 10^{-9} s. For triplet-excitons this time can be a few orders of magnitude larger (for example, the lifetime of a triplet exciton in anthracene is of order 10^{-4} s). The characteristic time of exciton scattering by phonons is of the order of picoseconds and thus usually is much less than its radiative lifetime. This means that generally one may assume that during the exciton's lifetime thermodynamic equilibrium of excitons and phonons is established.

These conditions lead to simple and rather general relations between the excitonic absorption spectrum and the spectrum of excitonic luminescence.

Let Q be the vibrational energy of the crystal without an exciton, and Q' be the sum of the crystal vibrational energy and the exciton kinetic energy when the exciton is present in the crystal. If $\omega_0 = E_s(\mathbf{k}_0)/\hbar$, where $\mathbf{k}_0$ is the excitonic wavevector, corresponding to the minimum energy in the excitonic band, then by absorbing or emitting a photon of energy $\hbar\omega$ the following energy conservation relation holds

$$Q' - Q = \hbar(\omega - \omega_0). \qquad (1.10)$$

The number of photons of the excitonic luminescence, created in the crystal per second, depends on the frequency ω as follows

$$W^{\text{lum}}(\omega) = \int \rho^*(Q') A(Q', \omega) \mathrm{d}Q', \qquad (1.11)$$

where $A(Q', \omega)$ is the probability of a spontaneous transition in which the exciton disappears, the photon $\hbar\omega$ is created, and the vibrational energy of the crystal is equal Q' (see eqn 1.10).

The distribution function for energy Q' is given by the Boltzmann-like formula

$$\rho^*(Q') = C^* g^*(Q') e^{-Q'/k_B T}, \qquad (1.12)$$

where T is the temperature of the medium, $g^*(Q')$ is the statistical weight of the energy level Q', and C^* the normalization constant. Formula (1.12) remains valid only when during the lifetime of the exciton the thermodynamic equilibrium of excitons and photons is established (which is assumed). At the same time, the number per unit time of the absorbed photons of frequency ω is given by the expression

$$W^{\text{abs}}(\omega) = \int \rho(Q) B(Q, \omega) u(\omega) \mathrm{d}Q, \qquad (1.13)$$

where

$$\rho(Q) = C\, g(Q) e^{-Q/k_B T},$$

is the distribution function of the vibrational levels for the case without excitons, $g(Q)$ is the statistical weight of the level with energy Q, $u(\omega)$ is the density of the incident radiation, and $B(Q, \omega) u(\omega)$ is the probability of a transition with the creation of an exciton, when the initial vibrational energy of the crystal is equal Q. In what follows we neglect the crystal anisotropy. Thus, making use of the relation (1.10) and of the properties of Einstein coefficients (see (16), § 12),

$$\frac{A(Q', \omega)}{B(Q, \omega)} = \frac{g(Q)}{g^*(Q')} \frac{2\hbar\omega^3}{\pi c^3} n^2 \left| \frac{\mathrm{d}}{\mathrm{d}\omega}(n\omega) \right|, \qquad (1.14)$$

n being the refraction index of the crystal without the excitonic contribution, we finally obtain the relation

$$\frac{W^{\text{lum}}(\omega)}{W^{\text{abs}}(\omega)/u(\omega)} = D \frac{2\hbar\omega^3}{\pi c^3} e^{-\hbar\omega/k_B T} n^2 \left| \frac{\mathrm{d}}{\mathrm{d}\omega}(n\omega) \right|, \qquad (1.15)$$

where D is a factor which weakly depends on ω. The ratio of the quantities $W^{\text{abs}}(\omega)/u(\omega)$ is proportional to the absorption coefficient $\kappa(\omega)$.[5] If in the energetic region under discussion we can neglect the dependence of the refraction index n on ω, the relation (1.14) becomes

$$\frac{W^{\text{lum}}(\omega)}{\kappa(\omega)} = A\omega^3 e^{-\hbar\omega/k_B T}, \qquad (1.16)$$

where A is a factor which depends on T but not on ω. Thus the expression (1.16) gives the relation between the luminescence lineshape and the absorption lineshape. This is a rather general law, reminiscent of the Kirchhoff law, and was first obtained by Stepanov (17) for the case of complex molecules. This relation applies for excitons, when during the lifetime of excitons a thermodynamic equilibrium of excitons and the lattice vibrations is established. It is also clear that the relation (1.16) holds for impurities in the crystal, for electronic states in the vicinity of defects, etc.

We should have in mind two conditions which limit the applicability of the relation (1.16). First, the absorption of a photon by a molecule, crystal, solution, etc. is not necessarily related to an electronic excitation for some spectral region. In such a spectral region the relation between the Einstein coefficients is changed, i.e. the relation (1.14), which makes the relation (1.16) invalid.[6]

The second limitation of the relation (1.16) appears for excitons with a sufficiently large oscillator strength. In this case, for frequencies near to $E_s(0)/\hbar$, the excitons and photons cannot be treated separately (and this is assumed by the derivation of (1.16)). Due to the strong exciton–photon interaction in this frequency region instead of excitons and photons we have to consider polaritons, the new "mixed states" of excitons and photons, having properties different from both those of photons and those of excitons (a more detailed discussion is given in Ch. 4). In this case the relation (1.16) for $\omega \simeq E_s(0)/\hbar$ is not satisfied and needs a new, more rigorous formulation in the framework of polariton theory. The internal structure of polaritons as mixed exciton–photon states is important particularly in the region of strong exciton resonances. In the present book, using classical and quantum theory, we discuss properties of polaritons in bulk organic materials, at surfaces, in superlattices, in low-dimensional structures, in microcavities with crystalline and amorphous organics, and also in resonant hybrid organic–inorganic nanostructures. Very interesting are the problems related to the collective properties of the excitons and polaritons (see Ch. 15).

---

[5] More exactly, the ratio $W^{\text{abs}}(\omega)/u(\omega)$ is equal to the inverse time of the free path of the photon in the crystal.

[6] Calculations concerning this problem made by Ketskemety and coworkers (18) have shown that on the r.h.s. of eqn (1.15), a factor $\eta(\omega)$ occurs, being the relative quantum luminescence yield by excitation with light of frequency ω.

2

FRENKEL EXCITONIC STATES IN THE HEITLER–LONDON APPROXIMATION

2.1 Excitons in a molecular crystal with fixed molecules. Splitting of molecular terms in a crystal

Molecular crystals consist of molecules of noble gases (He, Ne, Ar, Kr, Xe) and of molecules with saturated bonds. An important property of molecular crystals is that they are quite common in organic compounds (benzene, naphthalene, anthracene, and so on). The compounds building the molecular crystals have a relatively low boiling point and sublimation heat. As we mentioned in the Preface the spectroscopy of organic crystals originated in the classical experiments carried out in the late 1920s by Pringsheim and Kronenberger and also by Obreimov and de Haas (1)–(4). The subsequent experimental investigations, which have been performed in all parts of the world (their results can be found in review papers by McClure (5), Wolf (6), Broude et al. (7), Broude, Rashba and Prikhotko (8), and in books by Davydov (9), (10)), and in many more recent review papers which we will mention below in this book, have allowed a comparison of the molecular crystal spectra with those of the same molecules in vapors and solutions. It has been observed that in many cases the spectra of molecular crystals differ qualitatively from those of isolated molecules. The difference originates from the intermolecular interactions, and its theoretical analysis is important in two aspects. One of them is the determination of the symmetry properties of complex molecules. By comparing the absorption of polarized light by a molecular monocrystal with the geometry of the crystal one can establish the direction of the transition dipole moment in the molecule which is responsible for the light absorption. This is correct if the direction of the transition dipole moment in the molecule in the crystal is the same as in the isolate molecule. This assumption corresponds to the so-called "oriented-gas" model and is valid as we will show only in the case of small intermolecular interactions.

Let us note that the above assumption was used by Frenkel and Davydov (9)–(11) by constructing the wavefunctions in the Heitler–London approximation in its simple version. In their theory only the part of the intermolecular interaction was taken into account, which causes the excitation transfer from one molecule to another, and not those which gives the "mixing" of molecular electronic and vibronic configurations.

Thus the second aspect of the analysis of molecular crystal spectra gives an insight into the character of intermolecular electronic interactions in crystals. In those cases when the interaction is sufficiently strong, the crystals do not follow the "oriented gas" of the two-level molecular model and the spectral characteristics obtained in such an approximation do not agree with experimental observations.

In our book we present methods of computation of Frenkel exciton states in molecular crystals, which are not based on the molecular two-level model and Heitler–London approximation (Ch. 3). The methods allow us, in particular, to obtain the Frenkel exciton spectra for arbitrary strength of the intermolecular interaction, assuming that the interaction does not violate the charge neutrality. However, in this section we use the simplest form of the Heitler–London method to construct the wavefunctions and to obtain some qualitative results on the properties of the spectra which occur by the aggregation of molecules into a crystal.

In what follows we assume that the wavefunctions and the correponding eigenvalues of isolated molecules are known. Let the index f label the molecular terms. If a given term has r_0-fold degeneracy, then the corresponding state of energy ε_f will be denoted by $(fr), r = 1, \ldots, r_0$. We denote by $\hat{H}_0$ the Hamiltonian of a molecule fixed at the origin, and by $\varphi_0^{(fr)}$ the corresponding eigenfunction in the state (fr).[7] We have

$$\hat{H}_0 \varphi_0^{(fr)} = \varepsilon_f \varphi_0^{(fr)}. \tag{2.1}$$

We assume that an elementary cell of the crystal contains σ identical but, in general, differently oriented molecules (the molecules inside the crystal cell are labeled by the index $\alpha = 1, \ldots, \sigma$). If $\mathbf{n}$ is an integer-valued crystal lattice vector

$$\mathbf{n} = n_1 \mathbf{a}_1 + n_2 \mathbf{a}_2 + n_3 \mathbf{a}_3,$$

where n_i are integers ($i = 1, 2, 3$), and $\mathbf{a}_i$ are the vectors of the unit cell, the crystal Hamiltonian can be written in the form

$$\hat{H} = \sum_{\mathbf{n}\alpha} \hat{H}_{\mathbf{n}\alpha} + \frac{1}{2} \sum_{\mathbf{n}\alpha,\mathbf{m}\beta}{}' \hat{V}_{\mathbf{n}\alpha,\mathbf{m}\beta}, \tag{2.2}$$

where $\hat{V}_{\mathbf{n}\alpha,\mathbf{m}\beta}$ is the operator of the Coulomb interaction of molecules $\mathbf{n}\alpha$ and $\mathbf{m}\beta$ which depends on the coordinates of their electrons and nuclei, and the prime in (2.2) denotes that the summation excludes the term $\mathbf{n}\alpha \equiv \mathbf{m}\beta$.

The Hamiltonian (2.2) contains only the instantaneous Coulomb interaction between the carriers forming the crystal. Therefore, as in (12), the excitons corresponding to the operator (2.2) will be called Coulomb excitons.

As the molecules have no extra charge, in the first approximation the operator $\hat{V}_{\mathbf{n}\alpha,\mathbf{m}\beta}$ is determined by the dipole-dipole interaction between the molecules. In this case

---

[7] The eigenfunctions of isolated molecules will be assumed real.

$$\hat{V}_{\mathbf{n}\alpha,\mathbf{m}\beta} = \frac{(\hat{\mathbf{p}}_{\mathbf{n}\alpha}\hat{\mathbf{p}}_{\mathbf{m}\beta})|\mathbf{r}_{\mathbf{n}\alpha,\mathbf{m}\beta}|^2 - 3(\hat{\mathbf{p}}_{\mathbf{n}\alpha}\mathbf{r}_{\mathbf{n}\alpha,\mathbf{m}\beta})(\hat{\mathbf{p}}_{\mathbf{m}\beta}\mathbf{r}_{\mathbf{n}\alpha,\mathbf{m}\beta})}{|\mathbf{r}_{\mathbf{n}\alpha,\mathbf{m}\beta}|^5}, \qquad (2.3)$$

where $\mathbf{r}_{\mathbf{n}\alpha,\mathbf{m}\beta}$ is the lattice vector connecting the molecules $\mathbf{n}\alpha$ and $\mathbf{m}\beta$, and $\hat{\mathbf{p}}_{\mathbf{n}\alpha}$ is the operator of the electric dipole moment of the molecule $\mathbf{n}\alpha$.

Higher order multipole terms can also be computed and included in the expression for $\hat{V}_{\mathbf{n}\alpha,\mathbf{m}\beta}$ (see, for example, §§ 41, 42 in (13)). An analogous computation of multipole interactions for small-radius excitons can be found in (41).

First, we assume that the molecules are fixed at their equilibrium positions. In this case the wavefunction of the crystal ground state can be represented as an antisymmetrized product of single molecule ground state wavefunctions. Exchange terms of intermolecular interactions at low excitation states, and, in particular, in the ground state, are small. Therefore we neglect the exchange terms in the following discussion: they will be taken into account in Section 2.4 where the triplet excitons will be discussed and the exchange terms can be important.

In the above approximation the ground state wavefunction of the crystal can be defined in the following way ($f = 0$ represents the single molecule ground state)

$$\Psi_0 = \prod_{\mathbf{n}\alpha} \varphi_{\mathbf{n}\alpha}^0, \qquad (2.4)$$

so that the crystal energy in the state Ψ_0 has the form

$$\mathcal{E}_0 = \left(\Psi_0, \hat{H}\Psi_0\right) = N\sigma\epsilon_0 + \frac{1}{2}\sum_{\mathbf{n}\alpha,\mathbf{m}\beta}{}' \langle 00|\hat{V}_{\mathbf{n}\alpha,\mathbf{m}\beta}|00\rangle, \qquad (2.5)$$

N being the total number of elementary cells in the crystal. The notation for the matrix element in (2.5) follows from the more general expression

$$\langle f_1 f_2 | \hat{V}_{\mathbf{n}\alpha,\mathbf{m}\beta} | f_1' f_2' \rangle = \int \varphi_{\mathbf{n}\alpha}^{*f_1} \varphi_{\mathbf{m}\beta}^{*f_2} \hat{V}_{\mathbf{n}\alpha,\mathbf{m}\beta} \varphi_{\mathbf{n}\alpha}^{f_1'} \varphi_{\mathbf{m}\beta}^{f_2'} d\tau, \qquad (2.6)$$

which will be used in the following.

If we neglect the intermolecular interactions, the crystal wavefunction corresponding to the situation when only one molecule, say $\mathbf{n}\alpha$, is excited, has the form

$$\chi_{\mathbf{n}\alpha}^{(fr)} = \varphi_{\mathbf{n}\alpha}^{(fr)} \prod_{\mathbf{m}\beta \neq \mathbf{n}\alpha} \varphi_{\mathbf{m}\beta}^{(0)}. \qquad (2.7)$$

The energy corresponding to the wavefunction (2.7) is equal to $(N\sigma - 1)\varepsilon_0 + \varepsilon_f$ and is $N\sigma$ degenerate (translational degeneracy) since the crystal energy does not depend on the choice of which of the molecules $\mathbf{n}\alpha$ is excited.

The intermolecular interactions remove the above described degeneracy. For computing the crystal energy in the first approximation, taking into account the intermolecular interaction, we have to use appropriate linear combinations of the wavefunctions (2.7), which can diagonalize the matrix of the intermolecular interaction operator.

Since the required linear combinations should also be the eigenfunctions of the translation operator $T_{\mathbf{m}}$, they can be expressed in the form

$$\Psi_{\mathbf{k}} = \frac{1}{\sqrt{N}} \sum_{\mathbf{n}\alpha,r} u_\alpha^{(fr)} e^{i\mathbf{k}\mathbf{r}_{\mathbf{n}\alpha}} \chi_{\mathbf{n}\alpha}^{(fr)}, \qquad (2.8)$$

where $\mathbf{r}_{\mathbf{n}\alpha} = \mathbf{n} + \boldsymbol{\rho}_\alpha$, and the vector $\boldsymbol{\rho}_\alpha$ determines the position of a molecule inside the crystal cell. It can be easily proved that the function (2.8) satisfies the condition

$$T_{\mathbf{m}} \Psi_{\mathbf{k}} = e^{-i\mathbf{k}\mathbf{m}} \Psi_{\mathbf{k}}, \qquad (2.9)$$

i.e. it reveals the eigenfunction of the translation operator $T_{\mathbf{m}}$ with an integer-valued lattice vector $\mathbf{m}$.

By requiring the normalization of the function (2.8) we obtain the condition

$$\sum_\alpha |u_\alpha|^2 = 1. \qquad (2.10)$$

If the unit cell of the crystal contains only one molecule, and the molecular term is nondegenerate, the wavefunction (2.8) takes the form

$$\Psi_{f\mathbf{k}} = \frac{1}{\sqrt{N}} \sum_{\mathbf{n}} e^{i\mathbf{k}\mathbf{n}} \chi_{\mathbf{n}}^f, \qquad (2.11)$$

and the excitation energy of crystal with one excited molecule is equal to the value

$$E_f(\mathbf{k}) = \langle \Psi_{f\mathbf{k}} | \Delta \hat{H} | \Psi_{f\mathbf{k}} \rangle = (\varepsilon_f - \varepsilon_0) + \mathcal{D}_f + L^f(\mathbf{k}), \qquad (2.12)$$

where $\Delta \hat{H} = \hat{H} - \mathcal{E}_0$. The quantity

$$\mathcal{D} = {\sum_{\mathbf{m}}}' \left\{ \langle f0 | \hat{V}_{\mathbf{nm}} | f0 \rangle - \langle 00 | \hat{V}_{\mathbf{nm}} | 00 \rangle \right\}, \qquad (2.13)$$

is the change of the energy of the interaction of a molecule with its surroundings arising due to the transition of the molecule from a ground state to an excited state. The quantity $L^f(\mathbf{k})$ is given by

$$L^f(\mathbf{k}) = \sum_{\mathbf{m}} M_{0\mathbf{m}}^f e^{i\mathbf{k}\mathbf{m}}; \qquad (2.14)$$

it is determined by the matrix element $M_{\mathbf{nm}}^f$ of the excitation transfer between molecules $\mathbf{n}$ and $\mathbf{m}$

$$M_{\mathbf{nm}}^f = \langle f0 | \hat{V}_{\mathbf{nm}} | 0f \rangle. \qquad (2.15)$$

Thus we found that in the above simple case the intermolecular interaction shifts the excitation energy level and splits the molecular term into an excitonic band.

In what follows we will use the cyclic boundary conditions for the wavefunctions. By this condition a wavefunction Ψ remains invariant with respect to a translation by any of the integer-valued vectors $N_1\mathbf{a}_1, N_2\mathbf{a}_2, N_3\mathbf{a}_3$, where $N_1N_2N_3 = N$ and N is the total number of unit cells in the crystal, which thus becomes a parallelepiped with edges sizes equal to $N_1\mathbf{a}_1, N_2\mathbf{a}_2, N_3\mathbf{a}_3$.

Having in mind the relation (2.9), we now obtain for the wavevector $\mathbf{k}$

$$\mathbf{k} = \sum_{i=1}^{3} \frac{2\pi}{N_i}\nu_i \mathbf{b}_i, \qquad -\frac{N_1}{2} \leq \nu_1 \leq \frac{N_i}{2} \quad (i=1,2,3), \tag{2.16}$$

where $\mathbf{b}_i$ are the basis vectors in the reciprocal lattice, which are related to the primitive basis vectors by the equation $\mathbf{a}_i \mathbf{b}_j = \delta_{ij}$. The vector $\mathbf{k}$ assumes N values (which corresponds to the first Brillouin zone), and the same number N gives the number of excitonic states.

Let us now consider the case of a more complex crystal lattice composed of elementary cells containing a few molecules, with a molecular term f remaining nondegenerate.

Substituting the function (2.8) into the Schrödinger equation

$$\Delta \hat{H}\Psi = E\Psi, \tag{2.17}$$

where the Hamiltonian $\hat{H}$ is given by eqn (2.2), multiplying both sides of the above equation by $\chi_{\mathbf{n}\alpha}$ and integrating over the internal degrees of freedom of the cell molecules, we obtain a system of σ equations for the coefficients u_α^f.

The equations are of the form

$$\sum_{\beta=1}^{\sigma} \mathcal{L}_{\beta\alpha}(\mathbf{k}) = E u_\alpha, \tag{2.18}$$

where

$$\mathcal{L}_{\alpha\beta}(\mathbf{k}) = L_{\alpha\beta}(\mathbf{k}) + \delta_{\alpha\beta}\left(\mathcal{D}_f^\alpha + \Delta\varepsilon_f\right), \qquad \Delta\varepsilon_f = \varepsilon_f - \varepsilon_0,$$
$$L_{\alpha\beta}(\mathbf{k}) = \sum_{\mathbf{m}} \langle f0|\hat{V}_{\mathbf{n}\alpha,\mathbf{m}\beta}|0f\rangle e^{i\mathbf{k}(\mathbf{r}_{\mathbf{m}\beta} - \mathbf{r}_{\mathbf{n}\alpha})}, \tag{2.19}$$

$$\mathcal{D}_f^\alpha = {\sum_{\mathbf{m}\beta}}' \left\{ \langle f0|\hat{V}_{\mathbf{n}\alpha,\mathbf{m}\beta}|f0\rangle - \langle 00|\hat{V}_{\mathbf{n}\alpha,\mathbf{m}\beta}|00\rangle \right\}. \tag{2.20}$$

In crystals which obey symmetry operations, transforming the molecules of different orientations into each other the quantity $\mathcal{D}_f^\alpha$ does not depend on α.

The condition of the existence of nontrivial solutions of the equation system (2.19) leads to a secular equation of the order σ for the quantity E. Since the matrix $\mathcal{L}_{\alpha\beta}(\mathbf{k})$ is hermitian, all roots of the secular equation, $E_\mu^f(\mathbf{k}), \mu = 1, \ldots, \sigma$, are real. They determine σ excitonic bands. To any value μ there corresponds a wavefunction of a crystal excitonic state

$$\Psi_{\mu(f),\mathbf{k}} = \frac{1}{\sqrt{N}} \sum_{n\alpha} u_{\alpha\mu(f)}(\mathbf{k}) e^{i\mathbf{k}\mathbf{r}_{n\alpha}} \chi_{n\alpha}^{f}. \tag{2.21}$$

Thus in crystals where unit cells contain σ molecules, to any single nondegenerate excited state of a free molecule in the crystal corresponds not one, but σ bands of excited states and correspondingly several absorbtion lines. Such a splitting was first discussed by Davydov ((9)–(11)) and is usually called the Davydov splitting,[8] to distinguish it from the Bethe splitting (14).

The degeneracy of a molecular term is usually related to a sufficiently high symmetry of molecules. In crystals, however, the symmetry of fields in the place where the molecule α is located depends on the intermolecular interactions and can be lower than the symmetry of isolated molecules. In this case the degeneracy can be removed and a splitting can appear, which is just the Bethe splitting.

To demonstrate the mechanism of Bethe splitting in molecular crystals we assume that a unit cell contains only one molecule but the term f under consideration is r_0-fold degenerate. In the case being considered, quite analogous to that of the equation system (2.19), we obtain a system of r_0 equations for coefficients $u^{(fr)}$. The solution of the secular equation shows that in crystals with one molecule in the unit cell r_0 excitonic bands appear which correspond to an r_0-fold degenerate molecular term.

If an impurity molecule is characterized by a degenerate term, the Bethe splitting removes partially or totally the degeneracy and instead of one term a multiplicity of terms appears. In crystals, due to their translation symmetry single degenerate terms expand to several excitonic bands.

Let us note that both Bethe and Davydov splittings are not consequences of specific quantum-mechanical properties of molecules. They appear also in the crystals where the molecules are modeled, for example, by classical harmonic oscillators.

In crystals having unit cells with σ molecules with an r_0-fold degenerate term σr_0 excitonic bands appear. In this case the splitting results from the simultaneous influence of many factors leading both to Davydov and Bethe splittings. The analysis of the splitting in this case becomes more cumbersome, although it remains quite elementary.

2.2 Frenkel, Coulomb, and mechanical excitons

Let us consider, as examples, crystals like naphthalene, anthracene, etc. containing two identical molecules in the unit cell. In these crystals molecular states are nondegenerate. We have to find the eigenvectors $u_1(\mu)$, $u_2(\mu)$ and the eigenvalues

---

[8]In his work on this topic Frenkel considered crystals with one molecule in the unit cell. Crystals with molecular cells containing a few molecules, in the context of excitonic properties, were first investigated by Davydov (11).

E_μ of the matrix $\mathcal{L}_{\alpha\beta}(\mathbf{k})$ which result from eqns (2.19)–(2.20). This matrix has the form

$$\begin{pmatrix} L_{11}(\mathbf{k}) + \mathcal{D}_f + \Delta\varepsilon_f & L_{12}(\mathbf{k}) \\ L_{21}(\mathbf{k}) & L_{22}(\mathbf{k}) + \mathcal{D}_f + \Delta\varepsilon_f \end{pmatrix}. \tag{2.22}$$

The eigenvalues and the eigenvectors of the above matrix are given by the equations

$$E_\mu(\mathbf{k}) = \Delta\varepsilon_f + \mathcal{D}_f + \frac{L_{11}(\mathbf{k}) + L_{22}(\mathbf{k})}{2}$$
$$- \frac{(-1)^\mu}{2} \sqrt{[L_{11}(\mathbf{k}) - L_{22}(\mathbf{k})]^2 + 4L_{12}^2(\mathbf{k})}, \tag{2.23}$$

$$\{u_{\alpha\mu}(\mathbf{k})\} = \begin{pmatrix} \cos\frac{\gamma}{2} & \sin\frac{\gamma}{2} \\ \sin\frac{\gamma}{2} & -\cos\frac{\gamma}{2} \end{pmatrix}, \quad \mu = 1, 2, \tag{2.24}$$

where γ is defined by the relation

$$\tan\gamma = \frac{2L_{12}(\mathbf{k})}{|L_{11}(\mathbf{k} - L_{22}(\mathbf{k})|}. \tag{2.25}$$

Making use of the relations (2.21) and (2.24) we obtain the wavefunctions of the excitonic states $\mu = 1, 2$

$$\Psi_{1,\mathbf{k}} = \frac{1}{\sqrt{N}} \sum_\mathbf{n} e^{i\mathbf{k}\mathbf{n}} \left(\cos\frac{\gamma}{2} \chi_{\mathbf{n}1} + \sin\frac{\gamma}{2} \chi_{\mathbf{n}2} \right),$$
$$\Psi_{2,\mathbf{k}} = \frac{1}{\sqrt{N}} \sum_\mathbf{n} e^{i\mathbf{k}\mathbf{n}} \left(\sin\frac{\gamma}{2} \chi_{\mathbf{n}1} - \cos\frac{\gamma}{2} \chi_{\mathbf{n}2} \right). \tag{2.26}$$

If for all values $\mathbf{k}$ the following inequality holds

$$|L_{12}(\mathbf{k})| \ll |L_{11}(\mathbf{k}) - L_{22}(\mathbf{k})|, \tag{2.27}$$

then $\gamma \simeq 0$. In this case molecules with orientation $\alpha = 1$ form excitonic states independently of molecules with orientation $\alpha = 2$. If, in contrast

$$|L_{12}(\mathbf{k})| \gg |L_{11}(\mathbf{k}) - L_{22}(\mathbf{k})|,$$

then the approximate values of the eigenenergies are given by

$$E_\mu(\mathbf{k}) = \Delta\varepsilon_f + \mathcal{D}_f + L_{11}(\mathbf{k}) - (-1)^\mu |L_{12}(\mathbf{k})|. \tag{2.28}$$

Making use of (2.23) we can compute the Davydov splitting

$$E_1(\mathbf{k}) - E_2(\mathbf{k}) = \sqrt{(L_{11} - L_{22})^2 + 4L_{12}^2(\mathbf{k})}. \tag{2.29}$$

Consider now one more interesting limiting case. Let us assume that the distance between molecules inside the unit cell is small compared to the lattice constant and, in addition, the orientation of these molecules is such that the interaction between the molecules inside the unit cell is large compared with the interaction between molecules belonging to different elementary cells. In this case the width of excitonic bands, which is determined by the interaction between molecules from different cells, can be much smaller than the Davydov splitting. If the widths of exciton bands in the considered case is very small, the scattering of excitons by phonons or defects even in an ideal crystal structure can make the exciton wavevector a "bad" quantum number. Therefore even at not very high crystal temperatures the excitons demonstrating the Davydov splitting can be, in the same time, localized within the unit cell. They can propagate in the crystal by jumping from one cell to another and not as wavepackets covering many elementary cells.

Let us continue to consider the properties of Coulomb excitons. When computing the exciton energy the whole Coulomb interaction between crystal molecules is taken into account, the exciton energy $E_\mu(\mathbf{k})$ appears as a nonanalytic function of $\mathbf{k}$ for small $\mathbf{k}$ values, even in the case of a nondegenerate excitonic band. This singularity of Coulomb exciton results from the fact that by computation of the exciton energy a long-range electric field contribution of the interaction between molecules is involved.

To demonstrate this effect we first consider an exciton in a crystal with one molecule in the unit cell having a nondegenerate molecular term f. In this case the exciton energy, in the Heitler–London approximation, is given by the relation (2.12).

We assume that an isolated molecule has a nonzero transition dipole matrix element

$$\langle 0|\hat{\mathbf{p}}_{n\alpha}|f\rangle \equiv \mathbf{p}_\alpha^{0f} \neq 0. \tag{2.30}$$

If we now take into account in the operator $V_{n\alpha,m\beta}$ only the dipole-dipole interaction, and will assume that this operator is determined by relation (2.3), the matrix element (2.15) becomes

$$M_{0\mathbf{m}}^f = \frac{\left|\mathbf{p}_0^{0f}\right|^2 |\mathbf{m}|^2 - 3\left(\mathbf{p}_0^{0f}\mathbf{m}\right)^2}{|\mathbf{m}|^5}. \tag{2.31}$$

In consequence, the quantity $L(\mathbf{k})$ can be put into the form

$$L(\mathbf{k}) = -\mathbf{p}^{0f}\boldsymbol{\mathcal{E}}_0^{f0}, \tag{2.32}$$

where

$$\boldsymbol{\mathcal{E}}^{f0} = -\sum_{\mathbf{m}}{}' \frac{e^{i\mathbf{k}\mathbf{m}}}{|\mathbf{m}|^3}\left\{\mathbf{p}^{0f} - 3\frac{\mathbf{m}}{|\mathbf{m}|^2}\left(\mathbf{p}^{0f}\mathbf{m}\right)\right\}. \tag{2.33}$$

It can be seen from eqn (2.33) that $\boldsymbol{\mathcal{E}}^{f0}$ is the electric field at the point $\mathbf{m} = 0$, created by dipoles located at the lattice points $\mathbf{m} \neq 0$ and with value depending on the lattice cell position by the following relation

$$\mathbf{p(n)} = \mathbf{p}^{0f} e^{i\mathbf{kn}}. \tag{2.34}$$

The total field (2.33) can be represented, as was first shown by Ewald (15), as a sum of two contributions. One of them for small $\mathbf{k}$ corresponds to a long-wavelength (macroscopic) field which will be present in a medium where the dipoles are distributed not in discrete lattice points, but continuously according to the distribution

$$\mathbf{P(r)} = \frac{\mathbf{p}^{0f}}{\Delta} e^{i\mathbf{kr}}, \tag{2.35}$$

where Δ is the volume of a unit cell, and $\mathbf{P(r)}$ represents in this case the polarization of an unit volume. The second contribution to the total field is usually called the internal (local, or short-wavelength) field.

The long-wavelength field can be easily found if we take into account that in a medium without external charges the longitudinal component of the induction vector $\mathcal{D}$ vanishes, and the macroscopic electric field is longitudinal, if the retardation, as assumed in the theory of Coulomb excitons, is not taken into account. From this considerations we obtain

$$\mathcal{D}_\| = \mathbf{E}_\| + 4\pi \mathbf{P}_\| = \mathbf{E} + 4\pi \mathbf{k} \frac{(\mathbf{kP})}{k^2} = 0, \tag{2.36}$$

which gives, according to (2.35),

$$\mathbf{E(r)} = -4\pi \mathbf{k} \frac{(\mathbf{kP})}{k^2} = -\frac{4\pi}{\Delta} \frac{\mathbf{k}\left(\mathbf{kp}^{0f}\right)}{k^2} e^{i\mathbf{kr}}. \tag{2.37}$$

The expression (2.37) is a nonanalytic function of $\mathbf{k}$ and for small $\mathbf{k}$ depends on $\mathbf{s} = \mathbf{k}/k$. The local field contribution of the total field (2.33) does not depend on $\mathbf{s}$ when $\mathbf{k} \to 0$. Thus the separation of the long-wavelength field part corresponds to the separation of the part nonanalytic in $\mathbf{k}$ of the total field $\mathcal{E}^{f0}$.

Let us now represent (2.33) as a sum (15; 16)

$$\mathcal{E}_j^{f0} = E_{j0} + \sum_{j'} Q_{jj'}(\mathbf{k}) p_{j'}^{f0}, \tag{2.38}$$

where the second term corresponds to the internal field. Substituting (2.38) into (2.32) and making use of (2.12) we obtain the following relation

$$E_f(\mathbf{k}) = \Delta\epsilon_f + \mathcal{D}_f - \sum_{j'} Q_{jj'}(\mathbf{k}) p_j^{f0} p_{j'}^{f0} + \frac{4\pi \left|\mathbf{p}^{0f}\right|^2}{\Delta} \cos^2\theta, \tag{2.39}$$

where θ is the angle between the vectors $\mathbf{k}$ and $\mathbf{p}^{0f}$. The last term in (2.39) being determined by the long-wavelength field, is a nonanalytic function of $\mathbf{k}$.

In crystals with several molecules in the unit cell the quantity (2.19) can also be put into the form

$$L_{\alpha\beta}(\mathbf{k}) = -\mathbf{p}_\alpha^{0f} e^{-i\rho_\alpha} \mathcal{E}_{0\alpha,\beta}^{f0}, \qquad (2.40)$$

where $\mathcal{E}_{0\alpha,\beta}^{f0}$ is the electric field at the point $\mathbf{r} = \mathbf{r}_{n\alpha}$ resulting from the dipoles located at the lattice points of the lattice β with the following dependence on the lattice vectors

$$\mathbf{P}_{\mathbf{m}\beta} = \mathbf{p}_\beta^{f0} e^{i\mathbf{k}\mathbf{r}_{m\beta}}. \qquad (2.41)$$

In this case the field $\mathcal{E}_{0\alpha,\beta}^{f0}$ can be represented as a sum

$$\left[\mathcal{E}_{0\alpha,\beta}^{f0}(\mathbf{k})\right]_j = E_j e^{i\mathbf{k}\boldsymbol{\rho}_\alpha} + e^{i\mathbf{k}\boldsymbol{\rho}_\alpha} \sum_{j'} Q_{jj'}\binom{\mathbf{k}}{\alpha\beta} p_{\beta,j'}^{0f}, \qquad (2.42)$$

where

$$E_j = -\frac{4\pi}{\Delta} \frac{k_j}{|\mathbf{k}|} \left(\frac{\mathbf{k}}{|\mathbf{k}|} \sum_\alpha \mathbf{p}_\alpha^{0f}\right) \qquad (2.43)$$

is a nonanalytic function of $\mathbf{k}$.

The coefficients $Q_{jj'}\binom{\mathbf{k}}{\alpha\beta}$ for given $\mathbf{k}$ only depend on the lattice structure (their explicit form, which is not needed here, can be found in the book by Born and Huang (16), eqn (30.31); a detailed description of the Ewald method of separation of the field is also given there). These coefficients for $\mathbf{k} \to 0$ tend to values which do not depend on the direction $\mathbf{s} = \mathbf{k}/k$. It follows from formulas (2.40) and (2.42) that the quantity $L_{\alpha\beta}(\mathbf{k})$ and, in consequence, the eigenvalues of the matrix $\mathcal{L}_{\alpha\beta}$ (see (2.19)]) are, in general, nonanalytic functions of $\mathbf{k}$.

If, however, we omit in the expression for $L_{\alpha\beta}(\mathbf{k})$ the terms, which correspond to the first term on the r.h.s. of eqn (2.42), i.e. if in the interaction matrix the terms corresponding to the long-wavelength field are neglected, the matrix $L_{\alpha\beta}(\mathbf{k})$ becomes a new matrix, which in the following will be denoted by $\tilde{L}_{\alpha\beta}(\mathbf{k})$ and which results in an analytic function of $\mathbf{k}$. The elementary excitations corresponding to the matrix $\tilde{L}_{\alpha\beta}(\mathbf{k})$ will be called, as in (12), mechanical excitons. Their energies $\tilde{E}_\mu(\mathbf{k})$, different from the case of Coulomb excitons, are analytic functions of $\mathbf{k}$. This property, as will be shown below, makes the group-theoretical classification of mechanical excitons much simpler than the corresponding classification for Coulomb excitons.

This is, however, not the only reason to consider mechanical excitons because mechanical excitons can be used as states of zeroth approximation in calculation of the crystal dielectric tensor (see Sections 7.1, 7.3 and the monographs (12), (16) for this problem). In addition, as will be shown below, using the states of mechanical excitons and also the dielectric tensor, one can establish the energies of Coulomb excitons. Here we do not repeat the calculations of the dielectric tensor by making use of mechanical excitons states, since these calculations can be found in Ch. IV of the monograph (12). We will use only the results of these calculations.

According to the formula (12.14b) from (12), the dielectric tensor of a crystal with dipole-allowed excitonic states and with the spatial dispersion neglected, has the form

$$\epsilon_{ij}(\omega) = \delta_{ij} - \frac{8\pi}{\Delta\hbar} \sum_{\mu} \frac{P^i_{0;0\mu}(0) P^j_{0;0\mu}(0) \tilde{\omega}_\mu(0)}{\omega^2 - \tilde{\omega}_\mu^2(0)}. \tag{2.44}$$

In the above expression $\tilde{\omega}_\mu(0) = \tilde{E}_\mu(0)/\hbar$, where $\tilde{E}_\mu(0)$ is the energy of a mechanical exciton μ for $\mathbf{k} = 0$, $\mathbf{P}_{0;0\mu}(0)$ is the transition dipole matrix element of a unit cell, which is obtained by using the crystal ground state wavefunctions and the wavefunctions of a crystal with the mechanical exciton of type μ for $\mathbf{k} = 0$; Δ is the unit cell volume.

Applying this general expression (2.44) to the case of a crystal with one molecule per unit cell and taking into account only one molecular excited state f, one obtains

$$\epsilon_{ij}(\omega) = \delta_{ij} - \frac{8\pi}{\Delta\hbar} \frac{P^i_{0;0f}(0) P^j_{0;0f}(0) \tilde{\omega}_\mu(0)}{\omega^2 - \tilde{\omega}_f^2(0)}, \tag{2.45}$$

the frequency $\tilde{\omega}_f(0) = \tilde{E}_f(0)/\hbar$ being given by the formula (2.39) where the last term has been omitted, and the vector $\mathbf{P}_{0;0f}$ is given by the relation

$$\mathbf{P}_{0;0f} = \mathbf{p}^{0f}. \tag{2.46}$$

The frequencies of Coulomb excitons can be found from the dielectric tensor (see Section 7.4 and (12), § 2). If, for a given direction, $\mathbf{k}$ mechanical excitons have only the vectors $\mathbf{P}_{0;0\mu}$ which are not perpendicular to the direction of $\mathbf{k}$, then the frequencies of corresponding Coulomb excitons for this direction of the wavevector can be found from the equations[9]

$$\epsilon_{ij}(\omega) s_i s_j = 0. \tag{2.47}$$

Making use of the above equation in the case of a simple model crystal, where the formula (2.45) is valid, we obtain for the Coulomb exciton energy $E = \hbar\omega$ the expression

$$E^2 = \hbar^2 \tilde{\omega}_f^2(0) + \frac{8\pi}{\Delta} \left|\mathbf{p}^{0f}\right|^2 \hbar \tilde{\omega}_f(0) \cos^2\theta, \tag{2.48}$$

θ being the angle between the vectors $\mathbf{p}^{0f}$ and $\mathbf{s}$. Extracting the square root of both sides of the above equation and assuming that the quantity

$$\frac{4\pi \left|\mathbf{p}^{0f}\right|^2 \cos^2\theta}{\hbar \tilde{\omega}_f(0) \Delta} \tag{2.49}$$

is small compared to unity, we obtain for the excitonic energy the same expression as given above in (2.39).

---

[9] A reverse situation is considered in Section 7.4.

In reality the molecules usually have a number of excited states. The existence of such states creates screening of the intermolecular interaction and even neighboring molecules do not interact as they would in a vacuum. Interactions between distant molecules are lowered by the action of the polarization of the medium, for which not only the bound electronic states, but also free charges can contribute. Such screening effects of the obviously depend on the distance between molecules, which makes a quantitative discussion of this effect rather difficult.

A similar situation appears in the theory of Wannier–Mott excitons (see, for example, (17), § 4), where the interaction between the electron and the hole is given by $-e^2/\epsilon r$ only at large enough distances.

If, however, the frequencies of mechanical excitons and the dielectric tensor are assumed to be known, this screening effect of for the case of excitons can be obtained in a rather simple way. The expression (2.45), by taking into account the polarization of the medium depending on the remaining excited states of the molecule, can be put into the form

$$\epsilon_{ij}(\omega) = \epsilon_{ij}^0 - \frac{8\pi}{\Delta \hbar} \frac{P^i_{0;0f}(0)P^j_{0;0f}(0)\omega_\mu(0)}{\omega^2 - \tilde{\omega}_f^2(0)}, \qquad (2.50)$$

where ϵ_{ij}^0 for given frequency region $\omega \simeq \omega_f$ depends weekly on ω and is determined by the contribution to $\epsilon_{ij}(\omega)$ of distant resonances. Thus, by making use of the equation (2.47), we obtain the relation

$$E_f(0) = \tilde{E}_f(0) + \frac{4\pi \left|\mathbf{p}^{0f}\right|^2}{\Delta \left(\epsilon_{ij}^0 s_i s_j\right)} \cos^2\theta, \qquad (2.51)$$

instead of (2.39). The difference between the above expression and (2.39) consists of the fact that the quantity $\left|\mathbf{p}^{0f}\right|^2$ is replaced by $\left|\mathbf{p}^{0f}\right|^2/\epsilon_{ij}^0 s_i s_j$ just in the energetic region of the Coulomb exciton, where the contribution of the long-wavelength field is relevant.

A quite analogous role of the medium polarization can be discussed in the case of more complex crystals, where the unit cell contains more than one molecule. We consider, as an example, the case of two molecules in the unit cell. In the vicinity of a nondegenerate molecular term, however, taking into account the Davydov splitting eqn (2.50) now becomes

$$\epsilon_{ij}(\omega) = \epsilon_{ij}^0 - \frac{8\pi}{\Delta \hbar} \left[\frac{P^i_{0;01}(0)P^j_{01;0}(0)\tilde{\omega}_1(0)}{\omega^2 - \tilde{\omega}_1^2(0)} + \frac{P^i_{0;02}(0)P^j_{02;0}(0)\tilde{\omega}_2(0)}{\omega^2 - \tilde{\omega}_2^2(0)} \right], \qquad (2.52)$$

where $\mu = 1, 2$ correspond to the states of the mechanical exciton, and $\tilde{\omega}_{1,2}$ are the corresponding frequencies, i.e. the frequencies arising due to the Davydov splitting.

Substituting the expression (2.52) into eqn (2.47) we obtain the frequencies of Coulomb excitons in terms of that of mechanical excitons

$$\omega_{1,2}^2(\mathbf{s}) = \frac{\tilde{\omega}_1^2(0) + \tilde{\omega}_2^2(0) + a + b}{2} \\ \pm \frac{1}{2}\sqrt{[\tilde{\omega}_1^2(0) - \tilde{\omega}_2^2(0) + a - b]^2 + 4ab}, \quad (2.53)$$

where

$$a = \frac{8\pi}{\Delta\hbar} \frac{(\mathbf{sP}_{0;01})^2}{(\epsilon_{ij}^0 s_i s_j)} \tilde{\omega}_1(0), \qquad b = \frac{8\pi}{\Delta\hbar} \frac{(\mathbf{sP}_{0;02})^2}{(\epsilon_{ij}^0 s_i s_j)} \tilde{\omega}_2(0). \quad (2.54)$$

Equations (2.53) give, in particular, the Davydov splitting for Coulomb excitons $\omega_1(\mathbf{s}) - \omega_2(\mathbf{s})$

$$\omega_1(\mathbf{s}) - \omega_2(\mathbf{s}) = \frac{1}{2\omega_f}\sqrt{[\tilde{\omega}_1^2(0) - \tilde{\omega}_2^2(0) + a - b]^2 + 4ab}. \quad (2.55)$$

It is seen that this splitting is different from the splitting of mechanical excitons (by the derivation of (2.55) we make use of the relation

$$\omega_1(\mathbf{s}) + \omega_2(\mathbf{s}) \simeq 2\omega_f,$$

where ω_f is the frequency of the corresponding transition in an isolated molecule).

As we have shown above, the exciton energy depends not only on the characteristics of a single molecular term f, as it would follow from an elementary approach based on the Heitler–London approximation, but, in general, depends on all excited states of the molecule. This property is reflected by the fact that, as we have shown, the energies of Coulomb excitons can be expressed in terms of the crystal dielectric tensor, which includes contributions of all resonances.

Using, in the calculation of the exciton energies, certain phenomenological data derived from the dielectric tensor ϵ_{ij}^0 makes this calculations easier. However, we should have in mind that the existence of many excited states or in other words taking into account the mixing of molecular configurations under the influence of intermolecular interactions, leads not only to the appearance in the calculation the value $\epsilon_{ij}^0 \neq 0$, but also influences the local fields. Therefore, by using the tensor $\epsilon_{ij}^0 \neq 0$ the energies and the transition dipole moments of mechanical excitons must also be computed by taking into account the distant excited states. The more rigorous microscopical theory of Coulomb excitons taking into account the mixing of molecular configurations will be presented in Ch. 3. The different and in some sense more transparent semiphenomenological approach will be presented in Ch. 5.

2.3 Application of group theory for the determination of polarization and selection rules for excitonic light absorption. Degeneracy of excitonic levels

2.3.1 Winston theory

As can be seen from formula (2.44) which defines the dielectric tensor without spatial dispersion, nonzero contributions to the tensor ϵ_{ij} come from those mechanical excitons, for which the dipole matrix elements $\mathbf{P}_{0;0\mu}$ of an elementary cell are nonzero.

The dipole moment operator of the total crystal is given by

$$\hat{\mathbf{P}}^0 = \sum_{n\alpha} \hat{\mathbf{p}}_{n\alpha}. \tag{2.56}$$

Thus, by making use of the formula (2.4) which gives the crystal wavefunction for the ground state, then using the formula (2.8), which for $\mathbf{k} = 0$ gives the wavefunction of a Coulomb exciton with $\mathbf{k} = 0$ in the excitonic band μ, we obtain

$$\mathbf{P}_{0;0\mu} = \frac{1}{N}\langle\Psi_0|\mathbf{P}^0|\Psi_{\mu 0}\rangle = \frac{1}{\sqrt{N}}\sum_{\alpha,r} u^{fr}_{\alpha\mu}(0)\mathbf{p}^{0r}_\alpha. \tag{2.57}$$

The expression for $\mathbf{P}_{0;0\mu}$ in the case of mechanical excitons has the same form (2.57), but the functions $u^{fr}_{\alpha\mu}(0)$ must be replaced by $\tilde{u}^{fr}_{\alpha\mu}(0)$, obtained by neglecting the effects of the long-wavelength field. Since the operator $\hat{\mathbf{P}}^0$ is transformed like a polar vector, and the wavefunction Ψ_0 is invariant under all crystal symmetry transformations, the matrix element (2.57) will be nonzero only for those excitonic states whose wavefunctions are transformed like the components of a polar vector. If, for example, the function $\Psi_{0\mu}$ transforms like the x-component of a polar vector, the vector $\mathbf{P}_{0;0\mu}$ will be parallel to the x-axis. Thus the symmetry properties of the excitonic wavefunctions determine the polarization of a light wave which can create a given type of exciton. In the above example only a light wave polarized in the x-direction will be absorbed, obviously, if we restrict the consideration to dipole-type absorption. In a similar way, for example, the quadrupole absorption in the excitonic region of the spectrum can be discussed (for details see, for example, § 8 in (12)).

As we have shown above, the energies of mechanical excitons for $\mathbf{k} = 0$ do not depend on the direction of $\mathbf{k}$. This property simplifies the group-theoretical classification of the states of mechanical excitons and will be used below.

The first application of group-theoretical methods for the qualitative analysis of excitonic absorption spectra in molecular crystals was given by Davydov (9)–(11). He succeeded, for a number of crystals, explaining the polarization of light absorption, which appears when molecules agglomerate into crystals. An exhaustive presentation of results obtained by Davydov can be found in the books (10), (11). In these works, in order to obtain the wavefunctions, the coefficients $u^{fr}_{\alpha\mu}$ were first computed and then used for determining symmetry properties of excitonic states.

Winston (18) (see also (19)) proposed an alternative method of a qualitative study of the structure of excitonic states with $\mathbf{k} = 0$, where the determination of coefficients $u_{\alpha\mu}$ is not needed.

In the method of Winston three groups are used: the symmetry group of the molecule, the site groups $\alpha = 1, 2, \ldots, \sigma$ and the crystal factor group. The symmetry group of the molecule contains all symmetry operations, which leave the molecule unchanged. The site groups α contain only those elements of the molecule symmetry, which leave unchanged not only the molecule α, but also the whole crystal. Clearly each site group is a subgroup of the molecule symmetry group. Both are point groups.

For explaining the notion of a factor group we notice that any space group contains the subgroup of translations T comprising an infinite set of all parallel displacements which leave the crystal lattice unchanged. The full space group can be obtained from this subgroup by adding H elements ("rotational elements") involving rotations and reflections with H equal to the number of elements of the point group of the considered crystal class. Every element from the space group can be considered as a product of one of translations from the subgroup T and one of the rotation elements. If the space group of the crystal does not contain screw rotations or glide reflections, then the set of "rotational elements" forms a point group, which is equivalent to the point group of a given crystal class.

Let $T_\mathbf{m}$ be an element of the translation subgroup T, and h_i be one of the "rotational elements". A set of elements of type $T_\mathbf{m} h_i$ denoted by H_i (where $\mathbf{m}$ is an integer-valued lattice vector, which for given i takes all possible values $m_{1,2,3} = 0, \pm 1, \pm 2, \ldots$), will be named a class H_i. Using all rotational elements we can construct a number of affine classes equal to the number of symmetry elements in the point group of the crystal class. One can show that any element of the space group is contained in one and only one class and these classes, considered as a new object (with the product of classes defined as class containing all products of elements of two classes) also form a group. This group is named a factor group of the crystal space group. It can be shown that the factor group is isomorphic to the point group of the crystal class (see, for example, (12), Appendix).

According to the method of Winston, we choose the wavefunctions in the form

$$\Psi_\alpha^f(\mathbf{k}) = N^{-1/2} \sum_\mathbf{n} e^{i\mathbf{k}\mathbf{r}_{n\alpha}} \chi_{\mathbf{n}\alpha}^f. \tag{2.58}$$

These functions form the irreducible representations of the translational subgroup since

$$T_\mathbf{m} \Psi_\alpha^f(\mathbf{k}) = e^{i\mathbf{k}\mathbf{m}} \Psi_\alpha^f(\mathbf{k}). \tag{2.59}$$

It follows from the above equation that the functions $\Psi_\alpha^f(\mathbf{k})$, where $\alpha = 1, 2, \ldots, \sigma$, and with $\mathbf{k} = 0$, remain invariant under translations. Therefore any element from the adjacent class H_i and acting on the function $\Psi_\alpha^f(0)$, always gives the same result

$$h_i T_{\mathbf{m}} \Psi_\alpha^f(0) = h_i \Psi_\alpha^f(0) = \sum_\beta u_{\alpha\beta} \Psi_\beta^f. \tag{2.60}$$

Equation (2.60) defines how an element of the factor group transforms the functions $\Psi_\alpha^f(0)$. We see that the representation of the crystal factor group can be given by the set of functions $\Psi_\alpha^f(0), \alpha = 1, 2, \ldots, \sigma$. This means that the excitonic states with $\mathbf{k} = 0$ and with wavefunctions being correct linear combinations of the functions $\Psi_\alpha^f(0)$, can be classified in terms of irreducible representations of the crystal factor group, or with respect to the crystal class group being isomorphic to the factor group.

We first assume that the functions $\varphi_{n\alpha}^f$ have the symmetry of a local site α and thus the wavefunctions $\chi_{n\alpha}^f$ realize the irreducible representation of the αth site group, where the index f labels this representation. Let $\chi_{\text{s.g}}^f(R)$ be the character of a matrix which corresponds to an element R in the fth irreducible representation of the site group, and $\chi_{\text{f.g.}}^i(R)$ be the character of a matrix corresponding to an element R of a factor group in its ith irreducible representation. The character of an reducible representation of a factor group, spanned on functions $\Psi_\alpha^f(0)$ and corresponding to an element R, is given by

$$\chi^{\text{f.g.}}(R) = \sum_\alpha \delta_\alpha(R) \chi_{\text{s.g}}^f(R), \tag{2.61}$$

where

$$\delta_\alpha(R) = \begin{cases} 1, & \text{when } R \text{ is an element of a site group } \alpha, \\ 0, & \text{when } R \text{ is not an element of a site group } \alpha. \end{cases} \tag{2.62}$$

For establishing how many irreducible representations of type i factor group are contained in the above described factor group reducible representation, we have to find the number

$$\begin{aligned} n_i^f &= \frac{1}{H} \sum_R \chi_{\text{f.g.}}^i(R)^* \chi^{\text{f.g.}}(R) \\ &= \frac{1}{H} \sum_{R,\alpha} \delta_\alpha(R) \chi_{\text{s.g.}}^{*f}(R) \chi_{\text{f.g.}}^i(R), \end{aligned} \tag{2.63}$$

where H is the number of elements in the factor group. Since the site group is a subgroup of the factor group, the ith irreducible representation of the factor group can be reducible in the site group. Thus to each ith representation of the factor group there corresponds a set of numbers a_{if} such that

$$\chi_{\text{f.g.}}^i(R) = \sum_{f_1} a_{if_1} \chi_{\text{s.g.}}^{f_1}(R), \tag{2.64}$$

R being an element of the site group. By the usual group theoretical equation we have also

$$a_{if} = (1/n) \sum_{R_{\text{s.g.}}} \chi^{f*}_{\text{s.g.}}(R)\chi^{i}_{\text{f.g.}}(R), \qquad (2.65)$$

where n indicates the number of elements in the site group. Substituting (2.64) into (2.63) we get

$$n_i^f = \frac{1}{H} \sum_{R,\alpha,f_1} \delta_\alpha(R) a_{if_1} \chi^{f_1*}_{\text{s.g.}}(R) \chi^{f}_{\text{s.g.}}(R). \qquad (2.66)$$

The characters of the irreducible representation satisfy the normalization condition

$$\sum_R \delta_\alpha(R) \chi^{*f}_{\text{s.g}}(R) \chi^{f_1}_{\text{s.g}}(R) = n\delta_{ff_1}. \qquad (2.67)$$

Since $H = n\sigma$, where σ indicates the number of molecules in the unit cell, we obtain

$$n_i^f = a_{if} = \frac{1}{n} \sum_{R_{\text{s.g.}}} \chi^{i*}_{\text{f.g.}}(R) \chi^{f}_{\text{s.g.}}(R). \qquad (2.68)$$

The relation (2.68) was derived by Winston. It enables us to establish the symmetry properties of excitonic states which occur in a crystal where one molecule is in an excited state corresponding to the fth irreducible representation of the site group.

2.3.2 Applications to crystals of naphthalene type

Consider, as an example, crystals of naphthalene type, C_{2h}^5 being the factor group and C_i the site group. Their characters are collected in Tables 2.1 and 2.2.

Consider, for example, what happens with the state transforming according to the irreducible representation $f = A_u^{s.g.}$ of site group in the crystal. Using the formula (2.68) we obtain

$$n_i^f := \begin{cases} 0, & \text{if } i = A_g; \\ 0, & \text{if } i = B_g; \\ 1, & \text{if } i = A_u; \\ 1, & \text{if } i = B_u. \end{cases}$$

Table 2.1 *Characters of the irreducible representations of group C_{2h}.*

i	Irreducible representation	Components of the polar vector	Symmetry operations			
			E	C_2	i	σ
1	A_g		1	1	1	1
2	A_u	$r_{\mathbf{b}}$	1	1	-1	-1
3	B_g		1	-1	1	-1
4	B_u	$r_{\mathbf{a}}, r_{\mathbf{c}}$	1	-1	-1	1

Table 2.2 *Characters of the irreducible representations of the group C_i.*

C_i	E	i
A_g	1	1
A_u	1	−1

Table 2.3 *Character table for the irreducible representations of group D_{2h}.*

	E	C_2^a	C_2^b	C_2^c	i	σ^a	σ^b	σ^c
A_{1g}	1	1	1	1	1	1	1	1
B_{1g}	1	−1	−1	1	1	−1	−1	1
A_{2g}	1	1	−1	−1	1	1	−1	−1
B_{2g}	1	−1	1	−1	1	−1	1	−1
A_{1u}	1	1	1	1	−1	−1	−1	−1
B_{1u}	1	−1	−1	1	−1	1	1	−1
A_{2u}	1	1	−1	−1	−1	−1	1	1
B_{2u}	1	−1	1	−1	−1	1	−1	1

This means that each state which is transformed according to the irreducible representation A_u of the site group is split in the crystal into two exciton states of symmetry corresponding to the factor group irreducible representations A_u and B_u, respectively. Sometimes such splitting is named a factor group splitting.

It follows also from the data of Table 2.1 that transitions to these exciton states are allowed: transition to the state $A_u^{\text{f.g.}}$ for the polarization along the **b**-axis of the crystal, and transition to the state $B_u^{\text{f.g.}}$ for the in-plane polarization, where the plane is determined by vectors **a** and **c**.

Now it is important to establish the relation between states in a site group and states in an isolated molecule. An isolated naphthalene molecule has the symmetry properties of group D_{2h}, which contains the group C_i as its subgroup.

Let us compare the characters of group D_{2h} (Table 2.3) with the characters of group C_i. We can see that some irreducible representations of the molecular group inside the site group (i.e. for elements of the site group symmetry) have the same characters and thus inside the site group they have the same symmetry. For example, each of the wavefunctions which transform according to the one-dimensional irreducible representations $A_{1u}, B_{1u}, A_{2u}, B_{2u}$ of the isolated molecule group transform according to the irreducible representations A_u of the site group C_i. As was shown, the state with such a site group symmetry splits in the crystal into two exciton states. This means that to each frequency of molecular transitions to the state of symmetry A_{1u}, or B_{1u} or A_{2u} or B_{2u} will correspond a pair of exciton states and that the transitions to these states have to be dipole allowed as explained above.

In a similar way crystals with different symmetries can be considered. If an isolated molecule has a number of degenerate states (as, for example, in a benzene molecule), first we have to establish according to what irreducible representations of the site group the wavefunctions of this degenerate molecular term are transformed (it may happen that the degeneracy is removed within the site group), and then make use of the formula (2.68), which enables the determination of the symmetries of the exciton states, corresponding to any of the above-mentioned irreducible representations of the site group.

2.3.3 Symmetry properties of Coulomb excitons

Finally we notice that the wavefunctions of Coulomb excitons $(\mu, \mathbf{k})$, even for $\mathbf{k} = 0$ do not transform according to any of the irreducible representations of the factor group. This results from the fact that, in general, elements of the factor group change the direction of the exciton wavevector characterizing a given wavefunction. At the same time the energy of a Coulomb exciton $(\mu, \mathbf{k})$ depends on the direction $\mathbf{s} = \mathbf{k}/k$ even for $\mathbf{k} = 0$. Therefore a wavefunction with a different direction of vector $\mathbf{s}$ corresponds, in general, to a different exciton energy value. A different situation occurs for mechanical excitons, where the energy for $\mathbf{k} \to 0$ does not depend on the direction $\mathbf{s}$. Thus by classification of states of mechanical excitons the direction $\mathbf{s} = \mathbf{k}/k$ for $\mathbf{k} \to 0$ is not important, and this classification appears quite analogous to the classification of excited states of atoms and molecules.

In the case of Coulomb excitons their wavefunctions can be classified not with regard to the irreducible representations of the crystal factor group, but with regard to the symmetry group of the wavevector, being a subgroup of the crystal space group containing only those elements which leave the wavevector unchanged (or transforms it to an equivalent vector, i.e. differing by an integer-valued reciprocal lattice vector).

Even though, in the work by Winston this restriction was not stressed, his formula (2.68) is valid only for mechanical excitons. Correct linear combinations of functions $\Psi_\alpha^f(0)$ for mechanical excitons transform according to irreducible representations of the crystal factor group, therefore the calculations of the corresponding coefficients defining this linear combinations can be made by group-theoretical methods, as it was mentioned by Winston (18). They consist of a decomposition of a reducible representation, spanned on functions $\Psi_\alpha^f(0), \alpha = 1, 2, \ldots, \sigma$, into irreducible representations (see, for example, (21), § 26, and also (20)). The corresponding coefficients $\tilde{u}_{\alpha\mu}(0)$, as mentioned previously, determine for $\mathbf{k} \to 0$ the s-independent wavefunctions of mechanical excitons and not those of Coulomb excitons. The coefficients $\tilde{u}_{\alpha\mu}(0)$ will coincide with the coefficients $u_{\alpha\mu}(0)$ appropriate for wavefunctions of Coulomb excitons only for a special choice of the directions $\mathbf{k}$, namely for the case when the group of the wavevector $\mathbf{k}$ for $\mathbf{k} \to 0$ coincides with the crystal class group.

Application of group theory can also explain the important problem of degeneracy of excitonic states. This degeneracy can be conditioned by invariance of the crystal Hamiltonian upon elements of its space group (in this case it is sometimes called "compulsory" degeneracy, see (21), § 40) or, as shown by Herring (23), it can result from the crystal invariance upon the time-reversal operation.

Let us first consider the case of "compulsory" degeneracy. To this end we assume that an exciton state $E_\mu(\mathbf{k})$ is degenerated for $\mathbf{k} = \mathbf{k}_0$, which means there are p excitonic states $\Psi_{\mathbf{k}_0\mu\ell}$ associated with the wavevector $\mathbf{k}_0$, $\ell = 0, 1, \ldots, p$, being the index of the corresponding excitonic band. It can be shown (see (21), §30) that the ensemble of functions $\Psi_{\mathbf{k}_0\mu\ell}$ is invariant with respect to the elements of the wavevector group $G_{\mathbf{k}_0}$. In consequence, those functions transform upon one of its irreducible representations, having dimension p. So, if the group $G_{\mathbf{k}_0}$ has only one-dimensional representations, we cannot expect that for $\mathbf{k} = \mathbf{k}_0$ a band degeneracy appears. Inversely, if the group $G_{\mathbf{k}_0}$ does not possess one-dimensional representations, then at $\mathbf{k} = \mathbf{k}_0$ a degeneracy occurs, i.e. each excitonic band for $\mathbf{k} = \mathbf{k}_0$ will coincide with one or more neighboring excitonic bands.

If the wavevector group $G_{\mathbf{k}_0}$ does not contain screw rotations or glide reflections, then an ensemble of "rotation elements" being the wavevector group, constitutes a point group $\hat{G}_{\mathbf{k}_0}$, which forms a subgroup of the crystal class group. In this case, to each irreducible representation of the wavevector group there corresponds an irreducible representation of the group $\hat{G}_{\mathbf{k}_0}$ of the same dimension, so the question of the "compulsory" degeneracy can be discussed in terms of the irreducible representations of the group $\hat{G}_{\mathbf{k}_0}$. It can be shown (see (21), §30) that this correspondence occurs for all vectors $\mathbf{k}_0$ inside the first Brillouin zone. For these $\mathbf{k}_0$ vectors the correspondence holds also in the general case only if $\hat{G}_{\mathbf{k}_0}$ is understood as a point group, in which the above mentioned ensemble of "rotation elements" converts, and if one does not make the distinction between essential and nonessential screw axes and slip planes (i.e. if the "rotation elements" are taken without possible translations by noninteger lattice vectors).

As an example we consider the case of the naphthalene crystal, C_{2h}^5 being its space group. The point group C_{2h} (see Table 2.1) has only one-dimensional representations. Since any subgroup of the group C_{2h} can also have only one-dimensional representations, it is clear that in crystals of naphthalene type the "compulsory" degeneracy for excitonic states inside the first Brillouin zone is not possible.

As a second example consider the case of a quartz crystal, with the space group D_3. One of the representations of the point group D_3 has dimension two. Thus, if the vector $\mathbf{k}_0$ is parallel to the three-fold symmetry axis when the point group $\hat{G}_{\mathbf{k}_0}$ coincides with the point group D_3, a double degeneracy of excitonic terms is, in general, possible.

There is another reason for degeneracy of excitonic states, being a consequence of the structure of the Schrödinger equation. Indeed, since the Hamiltonian is a self-conjugated operator, wavefunctions $\Psi_{\mathbf{k}_0\mu\ell}^*$ ($\ell = 1, 2, \ldots, p$), where the star means complex conjugate, as well as wavefunctions $\Psi_{\mathbf{k}_0\mu\ell}$ ($\ell = 1, 2, \ldots, p$), correspond to the same energy value $E_\mu(\mathbf{k})$. But the functions $\Psi^*_{\mathbf{k}_0\mu\ell}$ transform under the irreducible representation of the translational subgroup corresponding to the wavevector $-\mathbf{k}_0$. Thus, if vectors $\mathbf{k}_0$ and $-\mathbf{k}_0$ are equivalent, then $E_\mu(\mathbf{k}_0) = E_\mu(-\mathbf{k}_0)$ and at the $\mathbf{k}_0$ point we find a contact of not p, but of $2p$ excitonic bands, if the linear superposition of functions $\Psi_{\mathbf{k}_0\mu\ell}$ ($\mathbf{k}_0, \mu$ fixed, $\ell = 1, 2, \ldots, p$) does not coincide with the linear superposition of functions $\Psi^*_{\mathbf{k}_0\mu\ell}$. Such a coincidence will not appear when the irreducible representation $(\mathbf{k}_0, \mu)$ of the wavevector $\mathbf{k}_0$ group, spanned on functions $\Psi_{\mathbf{k}_0\mu\ell}$, is not real.[10]

Here we will not report the relations, which allow the determination of when a representation $(\mathbf{k}_0, \mu)$ is real, since this problem is discussed both in the work by Herring (23), and in books on group theory (see, for example, (21), § 30, see also (24)). We only present one of the results obtained by Herring, which says that energy bands for quasiparticles, when its spin[11] is not taken into account, should be in contact for wavevectors ending at the Brillouin surface and perpendicular to a second-order screw axis of the reciprocal lattice (if such an axis exists). Applying this result to the case of the naphthalene crystal, whose space group contains a screw axis of second-order we obtain that the degeneracy of the excitonic bands occurs in the space of wavevectors for planes limiting the Brillouin zone, which are perpendicular to the screw axis of the reciprocal lattice of the naphthalene crystal.

Above we had in mind that the wavevector $\mathbf{k}$ is a good quantum number and thus that all exciton states are coherent. In the opposite case, which can occur, for example, as a result of exciton-phonon scattering or scattering by lattice defects, the exciton energy bands are not characterized by the $\mathbf{k}$ value. In this case incoherent localized states can appear, for which the translation symmetry of the crystal is not important and which are similar, for example, to excitations in amorphous materials. In some solids the coexistence of coherent and incoherent excitations can also be possible.

2.4 Triplet excitons

In previous sections a hidden assumption was made that an excitation of a molecule does not change its spin. Since the majority of molecules have spin zero in the ground state, this assumption is valid only for singlet excitons, i.e. excitons which are created by transitions of crystal molecules in one of their excited singlet states.

At the same time we must have in mind that for the majority of aromatic molecular crystals such as, for example, naphthalene, the lowest molecular electron excited state is a triplet state. In these circumstances triplet excitons exhibit a number of specific properties, which has been used for their study.

We notice that singlet–triplet transitions in molecules are forbidden by spin selection rules and the lifetime of triplet states in crystals exceeds by many orders

---

[10] A representation is called real when, in a linear function space corresponding to it, a basis can be chosen such that matrices of all operators of this representation are real.

[11] In the case of triplet excitons this problem was discussed by Sternlicht and McConnell (26).

of magnitude the lifetime for singlet states. This circumstance makes triplet excitons important by transfer of electronic excitation energy in crystals (see also Ch. 14), and also facilitates the creation in crystals of high concentrations of excitons, which allow experimental observation of exciton–exciton interactions (see, for example, (28)). The theory of triplet excitons was formulated mainly by McConnel and coworkers (see (26)–(30)), who then used their results for analysis of electronic paramagnetic resonance (EPR) spectra in various crystals, where triplet excitons exist.

As shown by Sternlicht and McConnell (26), the specific property of the theory of triplet excitons is that the usual perturbation theory, which is used for computation of wavefunctions and excitonic energies of singlet Frenkel excitons (see, for example, § 1), must be improved in the case of triplet excitons. As has been shown by Merrifield (31), this improvement is related to the fact that matrix elements of transfer of triplet excitation from one crystal molecule to another, computed by antysymmetrized crystal wavefunctions, where one molecule is in an excited triplet state and all others in the ground state, can be reduced to two-electron exchange integrals between orbits of different molecules. These integrals are small, which indicates the necesssity of taking into account intermolecular transfer via intermediate states by computing the matrix elements of triplet excitation transfer.

Crystal states, where one molecule is ionized, and the conduction band contains one electron, can be used as those intermediate states, as has been shown in (31) (the role of such intermediate states in theory of photoconductivity of molecular crystals has been discussed by Lyons (32)). The use of intermediate states becomes indispensable when the second-order perturbation theory is applied in the case of a degenerate term. According to (33), correct linear combinations of crystal states, containing one molecule in a triplet state and all remaining in the ground state, can be found by perturbation theory when in the corresponding secular equation the following effective Hamiltonian is used

$$\hat{H} = \hat{H}_0 + \sum_{n\alpha, m\beta} R_{n\alpha, m\beta} |\chi_{n\alpha}^{*f}\rangle\langle\chi_{m\beta}^{f}|\,;$$

$$R_{n\alpha, m\beta} = -\sum_{kq} \left(E_{kq} - E^0\right)^{-1} V_{kq, m\beta} V_{n\alpha, kq}\,, \qquad (2.69)$$

$$V_{kq, n\alpha} = \langle\psi_{kq}|\hat{H}_0|\chi_{n\alpha}^{f}\rangle,$$

where ψ_{kq} is the wavefunction of a crystal having one electron in the conduction band (**k** being the electron wavevector and q the band index); $\chi_{n\alpha}^{f}$ is the wavefunction of a crystal where the molecule $n\alpha$ is at excited triplet state f and all other molecules are at the ground state. The Hamilton operator $\hat{H}_0$ in eqn (2.69) is defined in eqn (2.2).

It is evident that ionized states become important when for neighboring molecules the following inequality is satisfied[12]

$$|R_{n\alpha,m\beta}| > \left|\left(\hat{H}_0\right)_{n\alpha,m\beta}\right|. \qquad (2.70)$$

If the part of the interaction which depends on electronic spins is not taken into account, the molecule energy level in a triplet state is three-fold degenerate with respect to the molecular spin orientation ($M = 0, \pm 1$). In this case there are σ excitonic bands corresponding to each value of M (σ being the number of molecules in unit cell), and the respective wavefunctions have the form

$$\Psi_{\mu\mathbf{k}}^{M} = \frac{1}{\sqrt{N}} \sum_{n\alpha} u_{\alpha\mu}^{M}(\mathbf{k}) e^{i\mathbf{k}\mathbf{r}_{n\alpha}} \chi_{n\alpha}^{fM}. \qquad (2.71)$$

The coefficients $u_{\alpha\mu}^{M}$ in (2.71), similar to the case of singlet excitons, are obtained from eqns (2.18), but for triplet excitons the coefficients $L_{\alpha\beta}$ are given in the form

$$L_{\alpha\beta}(\mathbf{k}) = \sum_{m} \langle \chi_{n\alpha}^{fM} | \hat{H} | \chi_{m\beta}^{fM} \rangle e^{i\mathbf{k}(\mathbf{r}_{n\alpha} - \mathbf{r}_{m\beta})}. \qquad (2.72)$$

Let us remark that in crystals consisting of aromatic molecules, to which the theory of Sternlicht and McConnell (26) was applied, the excited triplet states are not three-fold degenerate even when an external magnetic field is absent. Due to the dipole spin–spin interaction between electrons the degeneracy is totally or partially removed, depending on the symmetry of the excited state wavefunction. By a phenomenological description of this splitting the so-called Spin-Hamiltonian is usually applied

$$\hat{H}_s = D\left(\hat{S}_z^2 - \frac{1}{3}\hat{S}^2\right) + E\left(\hat{S}_x^2 - \hat{S}_y^2\right), \qquad (2.73)$$

$\hat{S}_x, \hat{S}_y, \hat{S}_z$ being the components of the molecular spin, and D, E are the spin–spin interaction constants, which describe the splitting of the triplet term. Comparing the splitting with, for example, EPR data for $\mathcal{H} \to 0$, $\mathcal{H}$ being the magnetic field intensity, we can establish the values of D and E. Taking, as an example, the naphthalene molecule, and having in mind the results of Hutchison and Mangum (34), we obtain

$$D = \pm 0.1006 \text{ cm}^{-1}, \quad E = \mp 0.0138 \text{ cm}^{-1}. \qquad (2.74)$$

As follows from calculations given, for example, in Ref. (39), also for triplet excitons the energy of the resonant interaction in molecular crystals is always

---

[12] As shown by Choi et al. (40) contributions of ionized states and of the exchange interaction to the width of triplet excitonic band are of the same order of magnitude. When ionized states are taken into account, the Davydov splitting in anthracene and naphthalene increases by about 50%.

high compared to D and E, and the spin–spin interaction can be computed as a first-order perturbation, where the functions (2.71), obtained without taking into account this interaction, serve as zeroth-order approximation.

Calculations of triplet excitonic bands for a number of crystals have been performed by Sternlicht and McConnell (26) and they have shown that the resonant interaction, proper for a crystal, can substantially change the EPR spectra when molecules aggregate to a crystal. An interesting effect related to the resonant interaction has been predicted for the naphthalene crystal (26). The EPR spectra of naphthalene molecules embedded as a solid solution into a durene crystal show four lines (34). This crystal, similar to the naphthalene crystal, contains two molecules in its unit cell. Thus, inside the crystal a naphthalene molecule can have two orientations. Each of these orientations provides two lines in the EPR spectra,[13] and only in such cases, when the direction of the applied magnetic field is invariant with respect to symmetry operations of the durene crystal, four EPR lines conglomerate into two lines.

At the same time in the naphthalene crystal, where the molecules can also have two different orientations independent of the magnetic field direction, the EPR spectrum should exhibit, according to Sternlicht and McConnell (26), only two lines, because due to the resonant interaction and movement of the exciton, the presence of two nonequivalent orientations of molecules inside the unit cell cannot be revealed.

A second specific property of EPR triplet excitonic spectra, which was pointed out by Deigen and Pekar (36) and Sternlicht and McConnell (26), is related to the hyperfine structure resulting from the interaction of electrons of excited molecules with magnetic moments of crystal nuclei. According to (26) because of the motion of triplet excitons a hyperfine structure in their EPR spectra vanishes. As shown by Deigen and Pekar (36), in excitons (and also in polarons) the energy correction resulting from the hyperfine interaction vanishes in a first approximation. In consequence, in this approximation the EPR linewidth related to the hyperfine interaction also vanishes and the corresponding hyperfine structure also should be absent. Since in local centers in many cases it is the hyperfine interaction which determines both the linewidth and hyperfine structure of EPR spectra (see, for example, (34)) the study of the linewidth and hyperfine structure of EPR spectra can be used as an experimental method to distinguish excitons from excitations of local centers.

All these considerations are widely used for the interpretation of experimental results obtained from radiospectroscopic measurements performed on organic crystals, exhibiting triplet excitations. We mention here Ref. (38), where EPR spectra of naphthalene doped by deuterated naphthalene were examined. In this work, using the data on the EPR linewidth and assuming that the dominant

---

[13] When degeneracy is totally removed, two lines in the EPR spectra correspond to transitions between triplet states, for which the selection rule $\Delta M = \pm 1$ is satisfied. Only these transitions are allowed in high magnetic fields. However, as shown in (35), in some cases transitions with $\Delta M = \pm 2$ have been observed.

spin-relaxation mechanism results from hopping, the frequency of hopping of triplet excitons from one molecule to another inside the naphthalene unit cell was estimated. It was found from this estimation that the diffusion coefficient D for triplet excitons has the value $D = 5 \cdot 10^{-4} \text{cm}^2/\text{s}$ which is in agreement with results obtained in other experiments.

In the papers by Berk et al. (42) the EPR linewidths of triplet excitons in single-crystal pyrene at room temperature have been measured in experiments performed at 24 GHz. The data are fitted to a formula first presented by Reineker (43) in a theory based on the Haken–Strobl–Reineker model of exciton motion (the Haken–Strobl–Reineker model can be applied for triplet excitons because they have an exciton bandwidth small in comparison with the thermal energy $k_B T$; for more details see the review paper by Reineker (44)). This formula was rederived in the paper by Berk et al. (42) more directly from Blume's stochastic Liouville formalism (45). The agreement was excellent. This result again implied that the dominant spin-relaxation mechanism in pyrene, as in anthracene and presumably in similar molecular crystals, results from hopping between differently oriented molecules in the unit cell.

While the Sternlicht and McConnel paper (26) was originally devoted to resonance phenomenon, by justifying the exciton spin Hamiltonian, its main influence has been on the large number of experiments on the magnetic field dependence of delayed fluorescence from molecular crystals. A review of fluorescence studies can be found in the paper by Swenberg and Giacintov (46).

The transfer energy by triplet excitons, as with singlets, depends strongly on the structure of the exciton band. The structure of these bands for triplet excitons most probably arises from short-range intermolecular interactions. If in some crystal this interaction is largest along one of the crystallographic axes, the triplet exciton states should be mostly one-dimensional. Such an interesting conclusion resulted from the studies of triplet excitons in 1,4-dibromonaphthalene crystal performed by Hochstrasser and Whiteman (47).

It was shown in this paper that the band triplet states are essentially those of a linear chain characterized by nearest neighbor interactions along the **c** crystallographic axis. Although the crystal is topologically a three-dimensional network, for the energy transfer by triplet excitons and trapping, the crystal behaves as a set of linear chains. It was shown that a heavy doping of DBN-h_6 (host) with up to 18% DBN-d_6 (guest) yielded the expected discrete spectra of a random linear array corresponding to a nearest neighbor interaction of 7.4 ± 0.1 cm^{-1} and total bandwidth of 29.6 ± 0.4 cm^{-1} for the zero–zero transition. A symmetric mode at 0+520 cm^{-1} was shown to have an exciton bandwidth of 15 cm^{-1}, and an asymmetric vibrational level at 0+250 cm^{-1} was shown to have a vanishing bandwidth. To become more acquainted with the experimental and the theoretical results on the triplet excitons one can suggest the papers by McConnell and coworkers (26)–(30) and the review articles by Grescishkin and Ajbinder (37). Also an excellent review by Hochstrasser of the theory and experiments with triplet exciton states of molecular crystal can be found in (48). In this paper the author analyzes the results of the studies of triplet exciton states in crystals of aromatic hydrocarbons (benzene, naphthalene, anthracene, perylene), in some substituted benzene, for small polyatomics and ions and also on triplet excitons in N-heterocyclics, pyrazine, p-benzoquinones, and benzophenone.

We also mention Ref. (38), where EPR spectra of naphthalene doped by deuterated naphthalene were examined. In this work, using the EPR linewidth the frequency of transitions of triplet excitations from one molecule to another inside the naphthalene unit cell was estimated and from them the diffusion coefficient D for triplet excitons, obtaining the value $D = 5 \cdot 10^{-4}$ cm^2/s. This value is in agreement with results obtained in other experiments.

Our last remark is on the charged triplet excitons in tetracene. In molecular crystals a charge carrier and a molecular (Frenkel) exciton are attracted to each other. This attraction arises due to an increase of the molecular static polarizability upon electronic excitation, and may be responsible for the formation of a bound state of the free exciton and the free carrier charge. Such charged Frenkel excitons are analogous to trions (bound states of a Wannier–Mott exciton and a charge carrier) in inorganic semiconductors. The theory of charged excitons in molecular crystals has been developed by Agranovich *et al.* (49). It was shown that the binding energy of the charged Frenkel excitons can be of the order of several hundred cm^{-1}. Particularly interesting are charged triplet excitons in tetracene. In this crystal the energy of the lowest charged triplet exciton is much smaller than the energy of the lowest excitation in ions. This circumstance prevents the transformation of a charged triplet exciton into an ion excitation and promotes the formation of rather stable charged triplet excitation. Such states under the influence of the static electric field can transfer not only the charge but also the triplet exciton energy.

3

THE SECOND-QUANTIZED THEORY OF FRENKEL EXCITONS

3.1 Energy operator for a molecular crystal with fixed molecules in the second-quantization representation. Paulions and Bosons

In Ch. 2 we have presented an elementary theory of excitonic states in crystals, which was based on perturbation theory in the form proposed by Heitler and London. This type of approximation will be called the Heitler–London approximation. As is well known, studying systems which consist of a large number of identical interacting subsystems (atoms, particles, molecules, etc.) can be done by making use of the second-quantization formalism. In this chapter we will study excitonic states by means of the second-quantization formalism and show how the results presented in Ch. 2 can be obtained and how a more accurate approximations can be handled (1), (2), see also (3), (4). In particular, we discuss the mixing of molecular configurations which is due to the intermolecular interaction. Moreover, by computing energies of Frenkel excitons we take into account crystal states where not only one (as in the Heitler–London approximation) but several molecules are in an excited state.

In the Schrödinger picture the energy operator for a crystal with fixed molecules with regard to the relation (2.2) is given by

$$\hat{H} = \sum_n \hat{H}_n + \frac{1}{2} {\sum_{n,m}}' \hat{V}_{nm}, \qquad (3.1)$$

where $n \equiv \{\mathbf{n}, \alpha\}$, $m \equiv \{\mathbf{m}, \beta\}$; $\mathbf{n}, \mathbf{m}$ are lattice vectors, and the prime by the summation means that the terms $\mathbf{n}\alpha \equiv \mathbf{m}\beta$ are omitted.

The transition to the second quantization representation requires a choice of a complete orthonormal set of functions, characterizing a certain isolated subsystem. We can choose the eigenfunctions[14] φ_n^f of isolated molecule Hamiltonians $\hat{H}_n$ as such functions. The eigenfunctions correspond to eigenvalues ε_f.

In the second-quantization representation the crystal states are characterized by occupation numbers N_{fn} which indicate the state f of the nth molecule.

Each molecule can occupy only one stationary state. Therefore the occupation numbers satisfy the relation

---

[14] As we have seen in Ch. 2, the eigenfunctions φ_n^f are orthonormalized: $\left(\varphi_n^f, \varphi_m^{f'}\right) =$ $= \delta_{nm}\delta_{ff'}$. In some cases, see, for example, (12), the eigenfunctions are not orthogonal for $n \neq m$. In this situation, instead of eigenfunctions of isolated molecules, linear combinations of eigenfunctions can be used, which are orthogonal.

$$\sum_f N_{nf} = 1. \tag{3.2}$$

As a consequence of the above relation we obtain

$$\sum_{n,f} N_{nf} = \sigma N, \tag{3.3}$$

where σN is the total number of molecules in the crystal.

The crystal wavefunction depends on all occupation numbers and will be denoted as

$$|\ldots N_{nf} \ldots\rangle. \tag{3.4}$$

All operators in the second-quantization representation act on functions depending on occupation numbers. The occupation number operator will be denoted by $\hat{N}_{nf}$. It is diagonal in the occupation number states representation

$$\hat{N}_{nf}|\ldots N_{nf} \ldots\rangle = N_{nf}|\ldots N_{nf} \ldots\rangle. \tag{3.5}$$

The numbers N_{nf} are equal to 0 or 1. A hermitian number state operator $\hat{N}_{nf}$ can be expressed in terms of two nonhermitian operators $b^\dagger_{nf}$ and b_{nf} by the following relation

$$\hat{N}_{nf} = b^\dagger_{nf} b_{nf}. \tag{3.6}$$

Since N_{nf} are equal to zero or one, $N_{nf} = (N_{nf})^2$, and by (3.5) it is enough to assume that the operators $b^\dagger_{nf}$, b_{nf} fulfill the relations

$$b^\dagger_{nf}|\ldots N_{nf} \ldots\rangle = (1 - N_{nf})|\ldots (N_{nf} + 1) \ldots\rangle,$$
$$b_{nf}|\ldots N_{nf} \ldots\rangle = N_{nf}|\ldots (1 - N_{nf}) \ldots\rangle. \tag{3.7}$$

It follows from the above relations that the operator $b^\dagger_{nf}$ can be called the creation operator of the states nf, and b_{nf} the annihilation operator for the same states. It also follows from (3.7) that the operators obey the anticommutation rules

$$b_{nf} b^\dagger_{nf} + b^\dagger_{nf} b_{nf} = 1,$$
$$b_{nf} b_{nf} = b^\dagger_{nf} b^\dagger_{nf} = 0. \tag{3.8}$$

The operators b_{nf} and $b^\dagger_{nf}$ corresponding to different values n and f, act on diffefent variables in the wavefunction (3.4) and thus must commute.

Introducing operator functions

$$\hat{\psi}(\ldots \xi_n \ldots) = \sum_{nf} b_{nf} \varphi^f_n(\xi_n),$$

$$\hat{\psi}^\dagger(\ldots \xi_n \ldots) = \sum_{nf} b^\dagger_{nf} \varphi^{*f}_n(\xi_n), \tag{3.9}$$

we obtain the transition from operators in the Schrödinger representation into operators in the second-quantization representation. For example, the operator of the total number of crystal molecules can be put into the form

$$\hat{N} = \int \hat{\psi}^\dagger(\ldots \xi_n \ldots) \hat{\psi}(\ldots \xi_n \ldots) \mathrm{d}\xi = \sum_{nf} b^\dagger_{nf} b_{nf}. \tag{3.10}$$

The crystal energy operator without intermolecular interactions can be obtained from the transformation

$$\sum_n \hat{H}_n(\xi_n) \rightarrow \hat{H}_0,$$

$$H_0 \equiv \int \hat{\psi}^\dagger(\ldots \xi_n \ldots) \sum_n \hat{H}_n(\xi_n) \hat{\psi}(\ldots \xi_n \ldots) \, \mathrm{d}\xi \tag{3.11}$$

$$= \sum_{nf} b^\dagger_{nf} b_{nf} \varepsilon_f.$$

Any operator which in the Schrödinger representation can be represented by a sum of operators $\hat{V}_n(\xi_n)$, each of them acting only on the internal coordinates of a single molecule, transforms according to the rule

$$\sum_n \hat{V}_n(\xi_n) \rightarrow \hat{V} \equiv \int \hat{\psi}^\dagger(\ldots \xi_n \ldots) \sum_n \hat{V}_n(\xi_n) \hat{\psi}(\ldots \xi_n \ldots) \, \mathrm{d}\xi$$

$$= \sum_{n,f,g} b^\dagger_{ng} b_{nf} \langle g|\hat{V}_n|f\rangle, \tag{3.12}$$

where

$$\langle g|\hat{V}_n|f\rangle \equiv \int \varphi^{*g}_n \hat{V}_n(\xi_n) \varphi^f_n \, \mathrm{d}\xi_n. \tag{3.13}$$

Any operator which in the Schrödinger representation consists of a sum of operators $\hat{U}_{nm}(\xi_n, \xi_m)$, each of them acting on internal coordinates of two molecules, transforms according to the rule

$$\sideset{}{'}\sum_{n,m} \hat{U}_{nm} \rightarrow \hat{U} \equiv \int \hat{\psi}^\dagger(\ldots \xi \ldots) \hat{\psi}^\dagger(\ldots \xi' \ldots) \sum_{n,m} \hat{U}_{nm} \hat{\psi}(\ldots \xi' \ldots)$$

$$\times \hat{\psi}(\ldots \xi \ldots) \, \mathrm{d}\xi \, \mathrm{d}\xi' \tag{3.14}$$

$$= \sum_{n,m,f,g,f',g'} b^\dagger_{nf'} b^\dagger_{mg'} b_{mg} b_{nf} \langle f'g'|\hat{U}_{nm}|gf\rangle,$$

where

$$\langle f'g'|\hat{U}_{nm}|fg\rangle$$

$$\equiv \int \varphi_n^{*f'}(\xi_n)\,\varphi_m^{*g'}(\xi_m)\,\hat{U}_{nm}\,\varphi_m^{g}(\xi_m)\,\varphi_n^{f}(\xi_n)\,\mathrm{d}\xi_n\,\mathrm{d}\xi_m. \qquad (3.15)$$

Making use of the relations (3.11) and (3.14) we obtain the crystal energy operator (3.1) in the second-quantization representation

$$\hat{H} = \sum \varepsilon_f b_{nf}^\dagger b_{nf} + \frac{1}{2}{\sum}' b_{nf'}^\dagger b_{mg'}^\dagger b_{mg} b_{nf} \langle f'g'|\hat{V}_{nm}|fg\rangle, \qquad (3.16)$$

where f', f, g', g are quantum numbers characterizing the stationary molecular states. The sum in the first term of the r.h.s. of (3.16) goes over all values of n and f. In the second term, the summation contains all values of n, m, f, f', g and g' where $n \neq m$.

3.2 Excitonic states in the two-level model. Transition to the Heitler–London approximation

We now consider the lowest crystal excited state, when the molecules remain neutral. In some cases by computing the crystal energy and the properties of corresponding excitonic states it is enough to consider only one molecular excited state. In such cases the summation in (3.16) reduces to the terms where f and g are equal 0 or f.

Molecular eigenfunctions φ^0 and φ^f can always be chosen as real. In this case the matrix elements in the operator (3.16) are real and symmetric. We first consider such crystal excited states, which correspond to the excitation of only one crystal molecule. According to this assumptions the operator (3.16) must contain:
(1) interactions between molecules which are in the ground state; such interactions are characterized by matrix elements of type $\langle 00|\hat{V}_{nm}|00\rangle$;
(2) interactions of the excited molecule with remaining nonexcited; in this case interactions are characterized by matrix elements of type $\langle 0f|\hat{V}_{nm}|0f\rangle$;
(3) matrix elements of type $\langle 0f|\hat{V}_{nm}|f0\rangle \equiv M_{nm}^f$, characterizing excitation transfer between molecules, and matrix elements (which are equal to M_{nm}^f up to effects of electron exchange between molecules, not considered here)

$$\langle 00|\hat{V}_{nm}|ff\rangle = \langle ff|\hat{V}_{nm}|00\rangle \simeq M_{nm}^f. \qquad (3.17)$$

Making use of the property (3.2) we have

$$\hat{N}_{n0} = 1 - \hat{N}_{nf}, \qquad (3.18)$$

and the operator (3.16), when taking into account only the terms linear in occupation numbers, has the from

$$\hat{H} = \mathcal{E}_0 + \hat{H}_1 + \hat{H}_2 + \hat{H}_3, \qquad (3.19)$$

where

$$\mathcal{E}_0 = N\sigma\varepsilon_0 + \frac{1}{2}\sum_{N,m}{}' \langle 00|\hat{V}_{nm}|00\rangle \qquad (3.20)$$

is a constant factor;

$$\hat{H}_1 = \sum_n (\Delta\varepsilon_f + \mathcal{D}_f)\hat{N}_{nf}, \qquad \Delta\varepsilon_f = \varepsilon_f - \varepsilon_0 \qquad (3.21)$$

is the excitation energy of an isolated molecule;

$$\mathcal{D}_f = \sum_m \left\{\langle 0f|\hat{V}_{nm}|0f\rangle - \langle 00|\hat{V}_{nm}|00\rangle\right\} \qquad (3.22)$$

being the difference of the interactions of the excited and a nonexcited molecule with all remaining crystal molecules in ground state;

$$\hat{H}_2 = \sum_{n,m}{}' M_{nm}^f b_{n0}^\dagger b_{mf}^\dagger b_{m0} b_{nf}, \qquad (3.23)$$

$$\hat{H}_3 = \frac{1}{2}\sum_{n,m}{}' M_{nm}^f \left\{b_{n0}^\dagger b_{m0}^\dagger b_{mf} b_{nf} + b_{nf}^\dagger b_{mf}^\dagger b_{m0} b_{n0}\right\}. \qquad (3.24)$$

It is convenient to introduce new operators defined by the relations

$$P_{nf} = b_{n0}^\dagger b_{nf}, \qquad P_{nf}^\dagger = b_{nf}^\dagger b_{n0}. \qquad (3.25)$$

It follows from the above definitions that the operator P_{nf} annihilates the state f, and $P_{nf}^\dagger$ creates the state f of the molecule n. By (3.18) and (3.25), taking into account the equality $\hat{N}_{nf}\hat{N}_{f0} = 0$, we obtain

$$P_{nf}^\dagger P_{nf} = \hat{N}_{nf}\left(1 - \hat{N}_{n0}\right) = \hat{N}_{nf}, \qquad (3.26)$$

$$P_{nf}P_{nf}^\dagger = \hat{N}_{n0} = 1 - \hat{N}_{nf}. \qquad (3.27)$$

By the relations (3.26) and (3.27) we obtain the following permutation rules

$$P_{nf}P_{nf}^\dagger + P_{nf}^\dagger P_{nf} = 1,$$
$$P_{nf}P_{nf}^\dagger - P_{nf}^\dagger P_{nf} = 1 - 2\hat{N}_{nf}. \qquad (3.28)$$

An analogous procedure can be used in consideration of the terms in the total Hamiltonian (3.16) which arise due to the presence of matrix elements $\langle f0|\hat{V}_{nm}|00\rangle$, $\langle f0|\hat{V}_{nm}|ff\rangle$ and $\langle f0|\hat{V}_{nm}|f0\rangle$. As result the total Hamiltonian can be presented by the relation:

$$\hat{H} = \hat{H}_1 + \hat{H}_2 + \hat{H}_3 + \hat{H}_4 + \hat{H}_5 + \hat{H}_6, \qquad (3.29)$$

where

$$\hat{H}_1 = \sum_n (\Delta\varepsilon_f + \mathcal{D}_f)P_{nf}^\dagger P_{nf},$$

$$\hat{H}_2 = {\sum_{n,m}}' M^f_{nm} P^\dagger_{mf} P_{nf},$$

$$\hat{H}_3 = \frac{1}{2}{\sum_{n,m}}' M^f_{nm}\left(P^\dagger_{mf}P^\dagger_{nf} + P_{mf}P_{nf}\right),$$

$$\hat{H}_4 = \frac{1}{2}\sum_{n} \Phi_{nm} P^\dagger_{nf} P_{nf} P^\dagger_{mf} P_{mf},$$

$$\Phi_{nm} = \langle ff|\hat{V}_{nm}|ff\rangle + \langle 00|\hat{V}_{nm}|00\rangle - 2\langle f0|\hat{V}_{nm}|f0\rangle,$$

$$\hat{H}_5 = {\sum_{n,m}}' \langle f0|\hat{V}_{nm}|00\rangle \left(P^\dagger_{nf} + P_{nf}\right),$$

$$\hat{H}_6 = {\sum_{n,m}}' \left(\langle 0f|\hat{V}_{nm}|ff\rangle - \langle f0|\hat{V}_{nm}|00\rangle\right)(P^\dagger_{nf} + P_{nf})P^\dagger_{mf}P_{mf}.$$

In the crystal containing molecules with a center of inversion and only in the dipole approximation for the intramolecular interaction, the parts of the total Hamiltonian $\hat{H}_4 = \hat{H}_5 = \hat{H}_6 = 0$, and at the same time the value $\mathcal{D}_f$, is equal to zero. However, it is known that the value D_f determining the so-called gas–condensed matter shift even exists in crystals with center of inversion (and thus due to the contribution of the nondipole intermolecular interaction) and can be of the order of 0.1 eV (in anthracene crystals the value of the gas–condensed matter shift is about 2000 cm^{-1}). This indicates that even in centrosymmetric crystals the terms in the total Hamiltonian $\hat{H}_4, \hat{H}_5, \hat{H}_6$ may be neglected only with some care.

Below we take into account in the total Hamiltonian only terms $\hat{H}_i$, $i = 1, 2, 3$ which are quadratic in the operators $P^\dagger$, P_{mf} and which with high accuracy determine the properties of one-exciton states. The role of the exciton–exciton interaction $\hat{H}_4$ can be taken into account in a consideration of biexciton states and collective properties of Frenkel excitons in Ch. 15.

For typical values of the intensity of electromagnetic radiation only a small number of crystal molecules is excited, i.e. the following inequality

$$\langle N_{nf}\rangle \ll 1, \tag{3.30}$$

holds. By this assumption the operators (3.25) fulfill the common commutation rules for Bose operators[15]

$$P_{nf} \equiv B_{nf}, \quad P^\dagger_{nf} = B^\dagger_{nf},$$
$$B_{n'f'}B^\dagger_{nf} - B^\dagger_{nf}B_{n'f'} = \delta_{nn'}\delta_{ff'}, \tag{3.31}$$

whereas the remaining combinations of the operators B_{nf} commute.

Making use of relations (3.28) we put the operator (3.21) into the form

$$\hat{H}_1 = \sum_n (\Delta\varepsilon_f + \mathcal{D}_f) B^\dagger_{nf} B_{nf}. \tag{3.32}$$

---

[15] A more accurate transition from operators $P_{nf}, P^\dagger_{nf}$ to Bose operators $B_{nf}, B^\dagger_{nf}$ is given in Section 3.11.

The operators (3.23) and (3.24) can be transformed in a similar way

$$\hat{H}_2 = {\sum_{n,m}}' M_{nm}^f B_{mf}^\dagger B_{nf}, \tag{3.33}$$

$$\hat{H}_3 = \frac{1}{2} {\sum_{n,m}}' M_{n,m}^f (B_{mf}^\dagger B_{nf}^\dagger + B_{mf} B_{nf}). \tag{3.34}$$

The operator $\hat{H}_2$ is a sum of operators, describing the excitation transfer from one molecule to another. The operator $\hat{H}_3$ describes the creation and annihilation of two excitons which simultaneously occupy two crystal molecules.

Let us first consider the expression (3.19) without the operator $\hat{H}_3$. Such an approximation corresponds to the Heitler–London approximation. In this approximation the excitation energy operator has the form

$$\Delta \hat{H} = \hat{H} - \mathcal{E}_0 = \sum_n (\Delta \varepsilon_f + \mathcal{D}_f) B_{nf}^\dagger B_{nf} + {\sum_{n,m}}' M_{nm}^f B_{mf}^\dagger B_{nf}. \tag{3.35}$$

We will compute the eigenvalues of $\Delta \hat{H}$ for the cases of a crystal with one molecule per unit cell and in the general case separately.

3.2.1 Crystals with one molecule per unit cell

In crystals with one molecule per unit cell the indices n, m become the crystal latttice vectors $\mathbf{n}$ and $\mathbf{m}$. Diagonalization of the excitation energy operator

$$\Delta \hat{H} = \hat{H} - \mathcal{E}_0 = \sum_{\mathbf{n}} (\Delta \varepsilon_f + \mathcal{D}_f) B_{\mathbf{n}f}^\dagger B_{\mathbf{n}f} + {\sum_{\mathbf{n},\mathbf{m}}}' M_{\mathbf{n}\mathbf{m}}^f B_{\mathbf{m}f}^\dagger B_{\mathbf{n}f} \tag{3.36}$$

can be performed by replacement of the operators $B_{\mathbf{n}f}$ by new operators $B_f(\mathbf{k})$ obtained from the following unitary transformation

$$B_{\mathbf{n}f} = \frac{1}{\sqrt{N}} \sum_{\mathbf{k}} B_f(\mathbf{k}) \exp(i\mathbf{k}\,\mathbf{n}), \tag{3.37}$$

where $\mathbf{k}$ is the wavevector defined by the relations (2.16). Substituting the expansion (3.37) into eqn (3.31) we can see that the operators $B_f(\mathbf{k})$ fulfill the following commutation rules

$$B_{f'}(\mathbf{k}') B_f(\mathbf{k}) - B_f^\dagger(\mathbf{k}) B_{f'}(\mathbf{k}') = \delta_{\mathbf{k}\mathbf{k}'} \delta_{ff'}. \tag{3.38}$$

In consequence, the operators $B_f^\dagger(\mathbf{k})$ appear as Bose operators of creation of states with quantum numbers f and wavevectors $\mathbf{k}$. The operators $B_f(\mathbf{k})$ are annihilation operators of those states.

EXCITONIC STATES IN THE TWO-LEVEL APPROXIMATION 43

Substituting (3.37) into the expression (3.36) we obtain

$$\Delta \hat{H} = \sum_{\mathbf{k}} \{\Delta \varepsilon_f + \mathcal{D}_f + L^f(\mathbf{k})\} B_f^\dagger(\mathbf{k}) B_f(\mathbf{k}), \tag{3.39}$$

where

$$L^f(\mathbf{k}) \equiv {\sum_{\mathbf{m}}}' M_{\mathbf{nm}}^f \exp\left[i\mathbf{k}(\mathbf{n} - \mathbf{m})\right]. \tag{3.40}$$

The operator (3.39) is diagonal with respect to the operators $\hat{N}_f = B_f^\dagger(\mathbf{k}) B_f(\mathbf{k})$ being the occupation numbers of excitonic states $\mathbf{k}f$. Therefore the eigenfunction of (3.39) has the form

$$|\ldots N_f(\mathbf{k}) \ldots\rangle,$$

where the quantum numbers $N_f(\mathbf{k}) = 0, 1, \ldots$ indicate the number of excitonic excitations of the given type. The crystal states with one excitonic excitation $\mathbf{k}f$ are obtained when the quantum numbers are chosen as $N_{f'}(\mathbf{k'}) = \delta_{\mathbf{kk'}}\delta_{ff'}$. According to (3.39), the excitation energy of those states is given as

$$E_f(\mathbf{k}) = \delta \varepsilon_f + \mathcal{D}_f + L^f(\mathbf{k}). \tag{3.41}$$

The above result is identical with (2.12) of Ch. 2.

3.2.2 Crystals with several molecules per unit cell

If a crystal contains σ molecules per unit cell, the indices n and m in (3.35) are given in the form

$$n \equiv (\mathbf{n}, \alpha), \qquad m \equiv (\mathbf{m}, \beta),$$

where $\mathbf{n}$ and $\mathbf{m}$ indicate the position of the unit cell; the numbers $\alpha, \beta = 1, 2, \ldots, \sigma$ describe the position and the orientation of molecules within the unit cell.

Diagonalization of the operator (3.35) can be carried out in two steps. First we use the canonical transformation

$$B_{nf} = \frac{1}{\sqrt{N}} \sum_{\mathbf{k}} A_{\alpha f}(\mathbf{k}) \exp\left(i\mathbf{k}\mathbf{r}_{\mathbf{n}\alpha}\right), \tag{3.42}$$

where the new operators $A_{\alpha f}(\mathbf{k})$ obey the following commutation rules

$$A_{\alpha' f'}(\mathbf{k'}) A_{\alpha f}^\dagger(\mathbf{k}) - A_{\alpha f}^\dagger(\mathbf{k}) A_{\alpha' f'}(\mathbf{k'}) = \delta_{ff'}\delta_{\alpha\alpha'}\delta_{\mathbf{kk'}}. \tag{3.43}$$

Substituting (3.42) into (3.35) we obtain the following expression for the excitation energy operator

$$\Delta \hat{H} = \sum_\alpha \sum_{\mathbf{k}} (\Delta \varepsilon_f + \mathcal{D}_f) A_{\alpha f}^\dagger(\mathbf{k}) A_{\alpha f}(\mathbf{k})$$

$$+ \sum_{\mathbf{k},\alpha,\beta} L^f_{\alpha\beta}(\mathbf{k}) A^{\dagger}_{\beta f}(\mathbf{k}) A_{\alpha f}(\mathbf{k}), \tag{3.44}$$

where

$$L^f_{\alpha\beta}(\mathbf{k}) \equiv \sum_{\mathbf{m}} M^f_{\mathbf{n}\alpha,\mathbf{m}\beta} \exp\left[i\mathbf{k}\left(\mathbf{r}_{\mathbf{n}\alpha} - \mathbf{r}_{\mathbf{m}\beta}\right)\right]. \tag{3.45}$$

For simplicity, we introduce the shorthand notation

$$\mathcal{L}^f_{\alpha\beta}(\mathbf{k}) = (\Delta\varepsilon_f + \mathcal{D}_f)\,\delta_{\alpha\beta} + L^f_{\alpha\beta}(\mathbf{k}). \tag{3.46}$$

Since we consider only one fth excited state, the index f will be omitted. Making use of this notation, we put the operator (3.44) into the form

$$\Delta\hat{H} = \sum_{\mathbf{k},\alpha,\beta} \mathcal{L}_{\alpha\beta}(\mathbf{k}) A^{\dagger}_{\beta}(\mathbf{k}) A_{\alpha}(\mathbf{k}). \tag{3.47}$$

Diagonalization of the operator (3.47) can be performed by making use of the following canonical transformation[16]

$$A_{\mu}(\mathbf{k}) = \sum_{\mu=1}^{\sigma} u_{\alpha\mu}(\mathbf{k}) B_{\mu}(\mathbf{k}), \tag{3.48}$$

where the functions $u_{\alpha\mu}$ obey the relations

$$\sum_{\mu} u^*_{\alpha\mu} u_{\beta\mu} = \delta_{\alpha\beta}, \qquad \sum_{\alpha} u^*_{\alpha\mu} u_{\alpha\nu} = \delta_{\mu\nu}. \tag{3.49}$$

As can be verified by substitution of the operators (3.48) into the commutation relations (3.43), the new operators satisfy the commutation rules

$$B_{\mu}(\mathbf{k}) B^{\dagger}_{\nu}(\mathbf{k}') - B^{\dagger}_{\nu}(\mathbf{k}') B_{\mu}(\mathbf{k}) = \delta_{\mathbf{k}\mathbf{k}'}\delta_{\mu\nu}. \tag{3.50}$$

As the canonical transformation (3.48) diagonalizes the energy operator (3.47), the following equality must be fulfilled

$$\Delta\hat{H} = \sum_{\mathbf{k},\alpha,\beta,\mu,\nu} u_{\alpha\nu}(\mathbf{k}) \mathcal{L}_{\alpha\beta}(\mathbf{k}) u^*_{\beta\mu}(\mathbf{k}) B^{\dagger}_{\mu}(\mathbf{k}) B_{\nu}(\mathbf{k})$$
$$= \sum_{\mu\mathbf{k}} E_{\mu}(\mathbf{k}) B^{\dagger}_{\mu}(\mathbf{k}) B_{\mu}(\mathbf{k}). \tag{3.51}$$

The above relation holds when the coefficients $u_{\alpha\mu}$ satisfy the following system of equations

$$\sum_{\beta} \mathcal{L}_{\beta\alpha}(\mathbf{k}) u_{\beta\mu}(\mathbf{k}) = E_{\mu}(\mathbf{k}) u_{\alpha\mu}(\mathbf{k}). \tag{3.52}$$

The above system of equations is identical with those given by eqn (2.18).

---

[16] See Appendix A.

EXCITONIC STATES IN THE TWO-LEVEL APPROXIMATION 45

It follows from the relation (3.52) that to any fixed value of the wavevector $\mathbf{k}$ there correspond σ excitonic crystal states with distinct values μ. The wavefunction

$$|\ldots N_\mu(\mathbf{k})\ldots\rangle$$

with $N_\mu(\mathbf{k})$ being the number of excitations of type $\mu, \mathbf{k}$, is therefore an eigenfunction of the operator (3.51) and of the number state operator

$$\hat{N}_\mu(\mathbf{k}) = B_\mu^\dagger(\mathbf{k}) B_\mu(\mathbf{k}). \tag{3.53}$$

The relation

$$E_\mu(\mathbf{k}) = \sum_{\beta,\alpha} u_{\alpha\mu}^*(\mathbf{k}) \mathcal{L}_{\alpha\beta}(\mathbf{k}) u_{\beta\mu}(\mathbf{k}) \tag{3.54}$$

gives the energy of the exciton $\mu\mathbf{k}$. The wavefunction of an excited state with one exciton $\mu\mathbf{k}$ in the second-quantization representation is obtained from the ground state wavefunction by the formula

$$|1\mu(\mathbf{k})\rangle = B_\mu^\dagger(\mathbf{k})|0\rangle. \tag{3.55}$$

The ground state function $|0\rangle$ corresponds to the wavefunction $\Psi_0^0 = \prod_{\mathbf{n}\alpha} \varphi_{\mathbf{n}\alpha}^0$ in the coordinate representation (see eqn 2.4). We put the function (3.55) into the coordinate representation. The following is obtained by the relations (3.48) and (3.42)

$$B_\mu^\dagger(\mathbf{k}) = \sum_\alpha u_{\alpha\mu}^*(\mathbf{k}) A_\alpha^\dagger(\mathbf{k}),$$

$$A_\alpha^\dagger(\mathbf{k}) = \frac{1}{\sqrt{N}} \sum_{\mathbf{n}} e^{i\mathbf{k}\mathbf{r}} B_{\mathbf{n}\alpha}^\dagger.$$

With respect to (3.25) we can write

$$B_\mu^\dagger(\mathbf{k}) = \frac{1}{\sqrt{N}} \sum_{\mathbf{n},\alpha} u_{\alpha\mu}^*(\mathbf{k}) e^{i\mathbf{k}\mathbf{r}_{\mathbf{n}\alpha}} b_{\mathbf{n}\alpha,f}^\dagger b_{\mathbf{n}\alpha,0}. \tag{3.56}$$

As a final result we obtain the excitonic excitation function in the coordinate representation

$$\Psi_\mu^f(\mathbf{k}) = \frac{1}{\sqrt{N}} \sum_{\mathbf{n},\alpha} u_{\alpha\mu}^* e^{i\mathbf{k}\mathbf{r}_{\mathbf{n}\alpha}} b_{\mathbf{n}\alpha}^\dagger b_{\mathbf{n}\alpha,0} \Psi_0$$

$$= \frac{1}{\sqrt{N}} \sum_{\mathbf{n},\alpha} u_{\alpha\mu}^* e^{i\mathbf{k}\mathbf{r}_{\mathbf{n}\alpha}} \varphi_{\mathbf{n}\alpha}^f \prod_{\mathbf{m}\beta} \varphi_{\mathbf{m}\beta}^0, \quad \mathbf{m}\beta \neq \mathbf{n}\alpha. \tag{3.57}$$

The above result coincides with the previously obtained expression (2.8).

3.3 Exciton states beyond the Heitler–London approximation

In the previous section we have obtained the crystal excitation energy in the Heitler–London approximation, i.e. neglecting the term $\hat{H}_3$ in the crystal energy operator (3.19). We now wish to determine the crystal energy without using this simplification.

Making use of (3.25) we can put the operator (3.19) into the form (the fixed index f is not indicated)

$$\Delta \hat{H} = \sum_n (\Delta \varepsilon + \mathcal{D}) B_n^\dagger B_n$$
$$+ \sum_{n,m} M_{nm} \left(B_m^\dagger B_n + \frac{1}{2} B_m^\dagger B_n^\dagger + \frac{1}{2} B_m B_n \right). \quad (3.58)$$

Diagonalization of the operator (3.58) can be performed by means of the unitary transformation[17]

$$B_n \equiv B_{n\alpha} \quad (3.59)$$
$$= \frac{1}{\sqrt{N}} \sum_{\mathbf{k},\mu} \left\{ u_{\alpha\mu}(\mathbf{k}) e^{i\mathbf{k}\mathbf{r}_{n\alpha}} B_\mu(\mathbf{k}) + v_{\alpha\mu}^*(\mathbf{k}) e^{-i\mathbf{k}\mathbf{r}_{n\alpha}} B_\mu^\dagger(\mathbf{k}) \right\},$$

where the new operators $B_\mu(\mathbf{k})$ satisfy the commutation relations (3.50). The transformation (3.59) is unitary if the following conditions are satisfied

$$\sum_\mu \left\{ u_{\alpha\mu}^*(\mathbf{k}) u_{\beta\mu}^*(\mathbf{k}) - v_{\beta\mu}^*(\mathbf{k}) v_{\alpha\mu}(\mathbf{k}) \right\} = \delta_{\alpha\beta},$$
$$\sum_\mathbf{k} \exp[i\mathbf{k}(\mathbf{m} - \mathbf{n})] = N\delta_{\mathbf{mn}}. \quad (3.60)$$

We can prove by substituting the expressions (3.59) into (3.58) that the operator (3.58) can be put into the diagonal form

$$\Delta \hat{H} = -\sum_{\mathbf{k},\mu,\alpha} E_\mu(\mathbf{k}) v_{\alpha\mu}(\mathbf{k}) v_{\alpha\mu}^*(\mathbf{k}) + \sum_{\mu,\mathbf{k}} E_\mu(\mathbf{k}) B_\mu^\dagger(\mathbf{k}) B_\mu(\mathbf{k}), \quad (3.61)$$

when the coefficients of the unitary transformation (3.59) satisfy the equations

$$[\Delta \varepsilon + \mathcal{D} - E_\mu(\mathbf{k})] u_{\alpha\mu}(\mathbf{k}) + \sum_\beta L_{\alpha\beta}(\mathbf{k}) [u_{\beta\mu}(\mathbf{k}) + v_{\beta\mu}(\mathbf{k})] = 0,$$

$$\quad (3.62)$$

$$[\Delta \varepsilon + \mathcal{D} - E_\mu(\mathbf{k})] v_{\alpha\mu}(\mathbf{k}) + \sum_\beta L_{\alpha\beta}(\mathbf{k}) [u_{\beta\mu}(\mathbf{k}) + v_{\beta\mu}(\mathbf{k})] = 0,$$

where $L_{\alpha\beta}$ are given by eqn (3.45). The first term in (3.61) contributes to the crystal ground state.

[17] For the diaganolization of quadratic forms of Bose operators, see Appendix A.

If we determine the crystal excitation energy referred to the renormalized vacuum state $|0\rangle$, i.e. the state without excitons, then the excitation energy operator (3.58) takes the form

$$\Delta \hat{H} = \sum_{\mu,\mathbf{k}} E_\mu(\mathbf{k}) B_\mu^\dagger(\mathbf{k}) B_\mu(\mathbf{k}). \tag{3.63}$$

In consequence, a state with an exciton $\mu\mathbf{k}$ has energy $E_\mu(\mathbf{k})$ which can be computed by solving eqns (3.62) with regard to the conditions (3.60). The function $E_\mu(\mathbf{k})$ characterizes the excitonic spectrum, i.e. the dependence of the exciton energy on the excitonic quasimomentum $\hbar\mathbf{k}$.

In the simpler case, when the crystal contains a single molecule in its unit cell, only one excitonic band exists ($\alpha = \beta = \mu = 1$), and eqns (3.60), (3.62) become

$$\begin{aligned} u^2 - v^2 &= 1, \\ [\Delta\varepsilon + \mathcal{D} - E(\mathbf{k})]\,u + L(\mathbf{k})[u+v] &= 0, \\ [\Delta\varepsilon + \mathcal{D} - E(\mathbf{k})]\,v + L(\mathbf{k})[u+v] &= 0. \end{aligned} \tag{3.64}$$

Solving the above system of equations, we obtain

$$\begin{aligned} E(\mathbf{k}) &= \left\{ [\Delta\varepsilon + \mathcal{D} + L(\mathbf{k})]^2 - L^2(\mathbf{k}) \right\}^{1/2}, \\ [\Delta\varepsilon + \mathcal{D} + E(\mathbf{k})]\,v &= [\Delta\varepsilon + \mathcal{D} - E(\mathbf{k})]\,u. \end{aligned} \tag{3.65}$$

If

$$L(\mathbf{k}) \ll \Delta\varepsilon + \mathcal{D}, \tag{3.66}$$

then the results (3.65) can be approximated by

$$\begin{aligned} E(\mathbf{k}) &\simeq \Delta\varepsilon + \mathcal{D} + L(\mathbf{k}) - \frac{1}{2}L^2(\mathbf{k})(\Delta\varepsilon + \mathcal{D})^{-1} \\ &\simeq \Delta\varepsilon + \mathcal{D} + L(\mathbf{k}), \\ v &\simeq \frac{L(\mathbf{k})u}{2(\Delta\varepsilon + \mathcal{D}) + L(\mathbf{k})} \simeq 0, \quad u \simeq 1. \end{aligned} \tag{3.67}$$

The results (3.67) coincide with those obtained by the Heitler–London approximation (3.41). Thus the inequality (3.66) indicates the limits of applicability of the Heitler–London approximation. In molecular crystals for excitations f, which correspond to the lowest electronic excitations of free molecules, the quantity $L(\mathbf{k})$ is mostly less than 10^3 cm^{-1}, when the quantity $\Delta\varepsilon + \mathcal{D}$ is of order $3 \cdot 10^4$ cm^{-1}. Thus the inequality (3.66) is satisfied and the Heitler–London approximation can be used for those states. In some crystals (for example, in anthracene) the second molecular excitation corresponds to the value $|L^f(\mathbf{k})| \sim 10^4$cm^{-1}. In such cases corrections to the Heitler–London approximation can be of importance (see also below Section 3.10).

With the aim of obtaining general expressions for corrections to the Heitler–London approximation in crystals with several molecules per unit cell, we put eqns (3.62) into the form

$$v_{\alpha\mu}(\mathbf{k}) = \frac{\Delta\varepsilon + \mathcal{D} - E_\mu(\mathbf{k})}{\Delta\varepsilon + \mathcal{D} + E_\mu(\mathbf{k})} u_{\alpha\mu}(\mathbf{k}), \qquad (3.68)$$

$$[E_\mu(\mathbf{k}) - \Delta\varepsilon - \mathcal{D}] u_{\alpha\mu} = 2\left(1 + \frac{E_\mu(\mathbf{k})}{\Delta\varepsilon + \mathcal{D}}\right)^{-1} \sum_\beta L_{\alpha\beta}(\mathbf{k}) u_{\beta\mu}(\mathbf{k}).$$

The Heitler–London approximation results from (3.68) if we substitute $E_\mu^0(\mathbf{k}) = \Delta\varepsilon + \mathcal{D}$ on the r.h.s. In this case

$$v_{\alpha\mu}^{\mathrm{H-L}} = 0, \qquad (3.69)$$

$$\sum_\beta \left\{ [E_\mu^{\mathrm{H-L}}(\mathbf{k}) - \Delta\varepsilon - \mathcal{D}] \delta_{\alpha\beta} - L_{\alpha\beta}(\mathbf{k}) \right\} u_{\beta\mu}^{\mathrm{H-L}}(\mathbf{k}) = 0. \qquad (3.70)$$

Equations (3.69), with regard to (3.60), yield for the Heitler–London approximation

$$\sum_\mu u_{\alpha\mu}^{*\mathrm{H-L}}(\mathbf{k}) u_{\beta\mu}^{\mathrm{H-L}}(\mathbf{k}) = \delta_{\alpha\beta}. \qquad (3.71)$$

Equations (3.70) and (3.71) coincide, as one might expect, with eqns (3.52) and (3.49) giving the energy and the wavefunctions in the Heitler–London approximation. For computing the excitonic energies within this approximation we notice that the quantities

$$u_{\alpha\mu}(\mathbf{k}), \qquad \alpha = 1, 2, \ldots, \sigma \quad \text{and} \quad u_{\alpha\mu}^{\mathrm{H-L}}(\mathbf{k}), \qquad \alpha = 1, 2, \ldots, \sigma, \qquad (3.72)$$

are eigenvectors of the same matrix $L_{\alpha\beta}(\mathbf{k})$ (see eqns 3.68 and 3.70). Therefore we can assume that they differ only by a normalization factor. The corresponding eigenvalues will coincide. In consequence,

$$E_\mu^{\mathrm{H-L}}(\mathbf{k}) - \Delta\varepsilon - \mathcal{D} = \frac{E_\mu^2(\mathbf{k}) - (\Delta\varepsilon + \mathcal{D})^2}{2(\Delta\varepsilon + \mathcal{D})} \qquad (3.73)$$

or

$$E_\mu(\mathbf{k}) = \left\{ (\Delta\varepsilon + \mathcal{D}) \left[2 E_\mu^{\mathrm{H-L}}(\mathbf{k}) - (\Delta\varepsilon + \mathcal{D}) \right] \right\}^{1/2}. \qquad (3.74)$$

The equality (3.74) allows us to determine improved values of the excitonic energies, when the values $E_\mu^{\mathrm{H-L}}(\mathbf{k})$ resulting from the Heitler–London approximation are known.

In the following we wish to establish the relations between the quantities $u_{\alpha\mu}(\mathbf{k})$ and $u_{\alpha\mu}^{\mathrm{H-L}}(\mathbf{k})$. Those quantities for all α and fixed μ can differ by a multiplicative factor which can be found from the normalization conditions of $u_{\alpha\mu}(\mathbf{k})$

and $u^{\mathrm{H-L}}_{\alpha\mu}(\mathbf{k})$. Making use of the relations (3.60) and of the first of relations (3.68) we obtain that the normalization condition for $u_{\alpha\mu}(\mathbf{k})$ reads

$$\sum_{\mu}\left[1-\left(\frac{\Delta\varepsilon+\mathcal{D}-E_{\mu}(\mathbf{k})}{\Delta\varepsilon+\mathcal{D}+E_{\mu}(\mathbf{k})}\right)^{2}\right]u^{*}_{\alpha\mu}(\mathbf{k})u_{\beta\mu}(\mathbf{k})=1. \qquad (3.75)$$

By comparing (3.75) and (3.71) we obtain the relation between $u^{\mathrm{H-L}}_{\alpha\mu}$ and $u_{\alpha\mu}(\mathbf{k})$ in the form

$$u^{\mathrm{H-L}}_{\alpha\mu}(\mathbf{k})=2\frac{\sqrt{E_{\mu}(\mathbf{k})(\Delta\varepsilon+\mathcal{D})}}{\Delta\varepsilon+\mathcal{D}+E_{\mu}(\mathbf{k})}u_{\alpha\mu}(\mathbf{k}). \qquad (3.76)$$

3.3.1 Small corrections to Heitler–London approximation

Beyond the Heitler–London approximation (HLA), when we take into account in the Hamiltonian of the crystal all the terms quadratic in Bose operators (including terms which do not conserve the number of quasiparticles), the crystal ground state wavefunction and the wavefunction of a state with one exciton contains also the states with several excited molecules. The contributions of such nonresonant terms to the energy of one-exciton states and to its wavefunctions are usually small because the ratio of resonant intermolecular interactions to the value of the exciton energy is small. For this reason in such cases the mentioned terms in the Hamiltonian can be omitted. However, sometimes these contributions can be interesting. For a more exact estimation of these small contributions of multiexciton states to the HLA the correct statistics of the molecular excitations become important. It was shown in the paper (7) how these corrections to the HLA can be calculated more rigorously taking into account the exact paulion statistics of molecular excitations. Using this approach one can estimate the correction to the HLA for crystals of general structure and beyond the two-level model, thus taking into account also the mixing of molecular configurations. For the model of a one-dimensional chain of two-level molecules and in the nearest-neighbor approximation our results for contribution from multiexciton states coincide with the results obtained earlier by Bakalis and Knoester (8) who used the Jordan–Wigner transformation from paulions to fermions (see Ch. 15).

To demonstrate the idea of the calculations we consider below the case of a molecular crystal with one two-level molecule in the unit cell (Subsection 3.4.1). For this model the second-quantized Hamiltonian of the crystal has the form

$$\hat{H}=E_{0}\sum_{n}\hat{P}^{\dagger}_{nf}\hat{P}_{nf}+\sum_{n,m}{}^{\prime}\hat{V}^{f}_{n-m}\hat{P}^{\dagger}_{mf}\hat{P}_{nf}$$
$$+\frac{1}{2}\sum_{n,m}\hat{V}^{f}_{n-m}\left(\hat{P}^{\dagger}_{mf}\hat{P}^{\dagger}_{nf}+\hat{P}_{mf}\hat{P}_{nf}\right). \qquad (3.77)$$

Here E_0 is the energy of the molecular excitation where also the gas–condensed matter shift is taken into account, and $\hat{P}^{\dagger}_n, \hat{P}_n$ are the exciton creation and annihilation Pauli operators on the molecule n, where n labels the unit cells of the crystal. $V(n-m)$ is the translationaly invariant matrix element of the intermolecular interaction between the molecules n and m. The molecular wavefunctions may be chosen in such a way that they are real and symmetric. If in this Hamiltonian we omit the nonresonant terms of type $\hat{P}\hat{P}, \hat{P}^\dagger \hat{P}^\dagger$ which do not conserve the number of quasiparticles, the rest of the Hamiltonian will describe the crystal in the HLA which we already considered. In this approximation the states of the crystal are characterized by a definite number of quasiparticles (excitons). The lowest energy states are those with one exciton, for which the statistics of the quasiparticles is unimportant. The one-exciton energy and its wavefunction in this approximation coincide with what was calculated in subsection 3.4.1 for crystals with one molecule per unit cell. If one wants to go beyond the HLA (to consider the full paulion Hamiltonian), both the ground and lowest excited states become a superposition of different many-particle states, and the question of the quasiparticle statistics arises or, equivalently, of the commutation relations between the operators $P^\dagger, P$. The operators, corresponding to different molecules, commute analogously to those for bosons. At the same time, since two or more excitations cannot reside on the same molecule n, the operators for the same molecule satisfy the relations (3.39) and $\hat{P}_n^\dagger \hat{P}_n^\dagger = \hat{P}_n \hat{P}_n = 0$, analogous to those for fermion operators. Since the terms $\hat{P}\hat{P}, \hat{P}^\dagger \hat{P}^\dagger$ lead to mixing of the multiexciton bands separated from each other by the energy of at least $2E_0$ with the interaction constant $V \ll E_0$, the corrections to the one-particle energy, as follows from eqn (3.65), is of the order of V^2/E_0. Indeed, according to (3.65), the lowest order correction $\delta E_\mu(k)$ to the energy of the one-particle excitation energy in the HLA is given by

$$\delta E_\mu(k) = -\frac{\left[E_\mu^{\text{HLA}}(k) - E_0\right]^2}{2E_0}. \quad (3.78)$$

To estimate the importance of such corrections it is reasonable to compare them with the width of the exciton states. For the lowest energy exciton in anthracene this correction is of the order of a few cm^{-1}, while the width of the exciton level even at low temperatures is of the order of several tens of cm^{-1}. Thus, for the lowest energy in anthracene the correction may be neglected. However, for example, for the lowest energy exciton in α-quaterthienyl (T4) this correction is already a few hundred cm^{-1}, and for the second (intense) transitions in anthracene and in many other molecular solids this correction can be of the order of a thousand cm^{-1}. And it is clear that the corrections to the HLA can be particularly important in the theory which takes into account the mixing of molecular configurations with the distance between mixed excited states smaller or of the order of the intermolecular interaction.

The contribution to the ground state energy (equal to zero in the HLA) in the boson approximation, when all terms quadratic in Bose amplitudes are taken into accout, to the lowest order is equal to

$$E(g) = -\frac{N}{4}\sum_n (V(n)^2). \tag{3.79}$$

To calculate the corrections to the HLA we may take advantage of the smallness of V/E_0 explicitly, using perturbation theory with respect to the terms that do not conserve the number of particles. All the first-order corrections to the energies vanish since the perturbation has no diagonal matrix elements in the HLA eigenstate basis. The correction to the energy E_s of the state $|s\rangle$ due to the perturbation V is given by

$$\delta E_s = \sum_{s'} \frac{\langle s|\hat{V}|s'\rangle \langle s'|\hat{V}|s\rangle}{E_s - E_{s'}}. \tag{3.80}$$

In our case the perturbation is represented by the nonresonant terms in the paulion Hamiltonian

$$\hat{V} = \frac{1}{2}\sum_n \sum_m V(n-m)\hat{P}_n \hat{P}_m + \text{h.c.} \equiv \hat{V}^- + \hat{V}^+, \tag{3.81}$$

where we denote by $\hat{V}^-$ and $\hat{V}^+$ the destruction and creation parts of $\hat{V}$, respectively. The states $|s\rangle$ of interest are either the vacuum $|0\rangle$ or a HLA one-particle state $\hat{P}^\dagger_{k\mu}|0\rangle$. Since the operator $\hat{V}$ changes the number of particles by 2, the intermediate states $|s'\rangle$ are only two- or three-particle manifolds, respectively. To our precision we may set $E_s - E_{s'} = -2E_0$ and put the denominator out of the sum. Then, using the completeness of the intermediate states, we obtain simple formulas for the ground state energy and the one-particle excitation energy correction,

$$E^g = -\frac{\langle 0|\hat{V}^-\hat{V}^+|0\rangle}{2E_0} \tag{3.82}$$

and

$$\delta E_\mu(\mathbf{k}) = -\frac{\langle 0|\hat{P}_{\mathbf{k}\mu}\hat{V}^-\hat{V}^+\hat{P}^\dagger_{\mathbf{k}\mu}|0\rangle}{2E_0} \tag{3.83}$$

(we remind the reader that the quantity $\delta E_\mu(\mathbf{k})$ is the correction to the difference between the one-particle and the ground state energies, which is measured in experiments). Thus, the calculation of these corrections is reduced to evaluating the averages of the type $\langle 0|\hat{P}\hat{P}\hat{P}^\dagger \hat{P}^\dagger|0\rangle$ for $E^{(g)}$ and $\langle 0|\hat{P}\hat{P}\hat{P}\hat{P}^\dagger\hat{P}^\dagger\hat{P}^\dagger|0\rangle$ for $\delta E_\mu(\mathbf{k})$ over the vacuum state. These averages are calculated in the site representation n using the commutation relations for paulion operators. As a result, we obtain that the correction to the ground state energy is given by the same expression as for bosons. For the excitation energy correction we obtain

$$\delta E_\mu(\mathbf{k}) = -\frac{\left(E_\mu^{\text{HLA}}(k) - E_0\right)^2}{2E_0} + \frac{1}{E_0}\sum_{n\neq 0}(V(n))^2. \tag{3.84}$$

The first term coincides with that calculated in the boson approximation, the second one thus may be considered as an additional contribution arising due to paulion statistics of Frenkel excitons. We see that this contribution is of the same order and has the opposite sign to the first term. It is interesting that for crystals with one molecule in the unit cell it does not depend on the wavevector. We may also calculate the correction to the transition dipole moment, which determines the excitonic contribution to the optical properties of the crystal. These calculations demonstrate that this correction coincides with the result of the bosonic approach, as in the case of the ground state energy (7).

It is instructive to apply our results for a 1D crystal and compare them with the results obtained for J-aggregates by Bakalis and Knoester (8). They considered the particular case of the simplest one-dimensional linear chain of two-level molecules and the intermolecular coupling only between the nearest neighbors. In this case the diagonalization of exciton Hamiltonian may be performed exactly, using for J-aggregates the Jordan–Wigner transformation from paulion to fermion operators (9). This model corresponds to one molecule in the unit cell, $\hat{V}(1) = \hat{V}(-1) = J$, all the rest of the interaction being zero. The results for the one-particle excitation and the ground state energies are

$$E(k) = \sqrt{E_0^2 + 4E_0 J \cos k + 4J^2},$$
$$E^{(g)} = \frac{N}{2}\left[E_0 - \int_0^\pi \frac{dk}{\pi} E(k)\right]. \qquad (3.85)$$

Expanding these expressions to second order in $2J/E_0$, we obtain the same results as those following from our expressions (3.84) and (3.78). In the nearest-neighbor approximation the HLA exciton energy is $E^{HLA}(k) = J \cos k$, and the correction is given by

$$\delta E(k) = -(2J^2/E_0)\cos^2 k + 2J^2/E_0.$$

Thus, in this particular case the correction to the result of the boson approximation is important and for $k = 0$ it even compensates the contribution obtained in the boson approximation. Let us mention also that the results of our calculations applied to a one-dimensional crystal give the possibility of taking into account or estimating the contribution of the resonance intermolecular interaction beyond the nearest-neighbor approximation and for several molecules in the unit cell. In higher dimensionality one might expect the paulionic effects to be less pronounced than in 1D, since the excitons are allowed "to escape each other", not being constrained to move on the same straight line. However, this turns out not to be the case. The demonstration of this conclusion can also be found in (7), the same as the consideration of crystals with multilevel molecules.

Before we finish the discussion on the corrections to the HLA it will be useful to say a few words on classical oscillator dipole theory. Mahan (10) was the first to do a self-consistent strong coupling theory for molecular solids (see also his review paper (11)). In this approximation the molecule is considered as a harmonic oscillator and not as a two-level model. This approach is correct for high-frequency intramolecular optical vibrations. However, Mahan calculated the

Davydov splitting for exciton resonance in anthracene, and for the first time obtained reasonable agreement with available experimental data. He used a dipole approximation for the intermolecular interaction and the only ingredients in his theory were the resonance frequencies and oscillator strength. In contrast to quantum theory described in this chapter the classical dipole theory does not take into account the contribution of the nondipole interaction, which are important in the majority of solids. It is clear that also multiexciton states including states with few quantum of excitations on the same molecule (what is forbidden for the two-level model) in classical harmonic oscillator theory contribute to the energy of excitons. However, in the framework of the classical theory it is impossible to develop the estimation of corrections which we discussed here.

Below we do not define more precisely the contribution of multiexciton states. Their importance in each concrete case can be estimated with the use of the approach developed in (7).

3.4 Exciton states in the presence of several molecular states (mixing of molecular configurations)

In previous sections we discussed crystal excitonic states starting from the assumption that by creation of those states only two stationary molecular states are included: the ground state (0), and the excited state (f). This assumption is valid when the excited state (f) is sufficiently far from the remaining excited states. If we deal with a group of near states, then all those states can participate in the creation of excitonic states ("mixing" of molecular terms). Theoretical discussion of the above case was given by Craig (12); (13) in the Heitler–London approximation, and by the author (2) by means of second quantization. Below we present the results obtained by second quantization, which permit us to go beyond the HLA.

If we retain in (3.16) only the matrix elements of type

$$\langle 00|\hat{V}_{nm}|00\rangle, \quad \langle 0g|\hat{V}_{nm}|0f\rangle,$$

and

$$M_{nm}^{gf} = \langle 0g|\hat{V}_{nm}|f0\rangle = \langle 0f|\hat{V}_{nm}|g0\rangle, \tag{3.86}$$

where f and g label molecular excited states, then, according to the normalization condition

$$\sum_{f} \hat{N}_{nf} + \hat{N}_{n0} = 1, \tag{3.87}$$

we obtain

$$\hat{H} = \mathcal{E}_0 + \hat{H}_1 + \hat{H}_2 + \hat{H}_3, \tag{3.88}$$

$\mathcal{E}_0$ being a constant factor

$$\mathcal{E}_0 = N\varepsilon_0 + \frac{1}{2}{\sum_{n,m}}' \langle 00|\hat{V}_{nm}|00\rangle,$$

and

$$\hat{H}_1 = \sum_{n,f} (\Delta\varepsilon_f + \mathcal{D}_f)\,\hat{N}_{nf},$$

$$\hat{H}_2 = \frac{1}{2}{\sum_{n,m,f,g}}' M_{nm}^{gf} \left\{ b_{n0}^\dagger b_{mg}^\dagger b_{m0} b_{nf} + b_{n0} b_{mg} b_{m0}^\dagger b_{nf}^\dagger \right\}, \qquad (3.89)$$

$$\hat{H}_3 = \frac{1}{2}{\sum_{n,m,f,g}}' M_{nm}^{gf} \left\{ b_{n0}^\dagger b_{mg} b_{m0}^\dagger b_{nf} + b_{n0} b_{mg}^\dagger b_{m0} b_{nf}^\dagger \right\}.$$

Making use of formulas (3.25) and (3.31) we pass to the operators B_{nf} and transform the energy operator (3.88) to the form

$$\Delta\hat{H} = \hat{H} - \mathcal{E}_0 = \sum_{nf} (\Delta\varepsilon_f + \mathcal{D}_d)\, B_{nf}^\dagger B_{nf}$$

$$+ \frac{1}{2}{\sum_{n,m,g,f}}' M_{nm}^{gf} \left(B_{mg}^\dagger + B_{mg} \right)\left(B_{nf}^\dagger + B_{nf} \right). \qquad (3.90)$$

If we neglect the operator $\hat{H}_3$ in (3.88) and thus go to the Heitler–London approximation, then

$$\Delta\hat{H} = \sum_{\mathbf{n}f} (\Delta\varepsilon_f + \mathcal{D}_d)\, B_{\mathbf{n}f}^\dagger B_{\mathbf{n}f} + {\sum_{\mathbf{n,m},g,f}}' M_{\mathbf{nm}}^{fg} B_{\mathbf{m}g}^\dagger B_{\mathbf{n}f}. \qquad (3.91)$$

Here f and g are integers $1, 2, \ldots, \ell$, ℓ being the number of molecular excited states taken into account.

Diagonalization of the operator (3.91) can be done in two steps. In the first step we perform the canonical transformation (3.42) obtaining

$$\Delta\hat{H} = \sum_{\mathbf{k},\alpha,g,\beta,f} \mathcal{L}_{\alpha\beta}^{fg}(\mathbf{k}) A_{\beta g}^\dagger(\mathbf{k}) A_{\alpha f}(\mathbf{k}), \qquad (3.92)$$

where

$$\mathcal{L}_{\alpha\beta}^{fg} \equiv (\Delta\varepsilon_f + \mathcal{D}_f)\,\delta_{\alpha\beta}\delta_{fg} + L_{\alpha\beta}^{fg}(\mathbf{k}), \qquad (3.93)$$

$$L_{\alpha\beta}^{fg}(\mathbf{k}) \equiv \sum_{\mathbf{m}} M_{\mathbf{n}\alpha,\mathbf{m}\beta}^{fg} \exp\{i\mathbf{k}(\mathbf{n} - \mathbf{m})\}. \qquad (3.94)$$

The second step of the diagonalization consists of a canonical transformation

MIXING OF MOLECULAR CONFIGURATIONS

$$A_{\alpha f}(\mathbf{k}) = \sum_{\mu,g} u^*_{\alpha f, \mu(g)}(\mathbf{k}) B_\mu(\mathbf{k}) \tag{3.95}$$

to new operators $B_\mu(\mathbf{k})$ which satisfy the following commutation rules

$$B_\mu(\mathbf{k}) B^\dagger_\nu(\mathbf{k}') - B^\dagger_\nu(\mathbf{k}') B_\mu(\mathbf{k}) = \delta_{\mathbf{k}\mathbf{k}'} \delta_{\mu\nu}. \tag{3.96}$$

The matrix elements of the transformation matrix u satisfy the system of equations

$$\sum_{\beta,g} \mathcal{L}^{gf}_{\beta\alpha}(\mathbf{k}) u_{\beta,g,\mu}(\mathbf{k}) = E_\mu(\mathbf{k}) u_{\alpha f,\mu}(\mathbf{k}), \tag{3.97}$$

where the indices α, β are integers $1, 2, \ldots, \sigma$, and the indices f, g run over the set $1, 2, \ldots, \ell$.

Solving eqns (3.97) for a fixed value of $\mathbf{k}$ we obtain $\sigma\ell$ roots $E_\mu(\mathbf{k})$ with the corresponding coefficients $u_{\alpha f,\mu}(\mathbf{k})$.

3.4.1 One molecule per unit cell

In the special case of crystals with one molecule per unit cell eqns (3.97) have the simpler form

$$\sum_g \mathcal{L}^{fg}(\mathbf{k}) u_{g,\mu}(\mathbf{k}) = E_\mu(\mathbf{k}) u_{f,\mu}(\mathbf{k}), \tag{3.98}$$

where

$$\mathcal{L}^{fg}(\mathbf{k}) = (\Delta\varepsilon_f + \mathcal{D}_f)\delta_{fg} + L^{fg}(\mathbf{k}),$$
$$L^{fg}(\mathbf{k}) = \sum_{\mathbf{m}} M^{fg}_{\mathbf{nm}} \exp\{i\mathbf{k}(\mathbf{n} - \mathbf{m})\}.$$

In this case the Davydov splitting is absent, and when molecular terms are degenerate, a splitting occurs which is analogous to the Bethe splitting (see Section 2.1).

If only two excited states contribute, the quantities $E_\mu(\mathbf{k})$ can be given explicitly:

$$E_\mu(\mathbf{k}) = \frac{1}{2}\left\{ A_1 + A_2 + (-1)^\mu \left[(A_1 - A_2)^2 + |L^{12}(\mathbf{k})|^2 \right]^{1/2} \right\}, \tag{3.99}$$

with

$$A_s \equiv \Delta\varepsilon_s + \mathcal{D}_s + L^{ss}(\mathbf{k}), \qquad \mu, s = 1, 2.$$

The coefficients $u_{f\mu}$, up to a normalization constant, are given by the equations

$$u_{1\mu}(\mathbf{k}) = \frac{2L^{12}(\mathbf{k}) u_{2\mu}(\mathbf{k})}{A_2 - A_1 - (-1)^\mu \sqrt{(A_1 - A_2)^2 + |L^{12}(\mathbf{k})|^2}}. \tag{3.100}$$

The wavefunctions corresponding to crystal excited states have the form

$$\Psi_\mu(\mathbf{k}) = \frac{1}{\sqrt{N}} \sum_{nf} u_{f\mu}(\mathbf{k}) e^{i\mathbf{k}\mathbf{n}} \varphi_\mathbf{n}^f \prod_{\mathbf{m}\neq\mathbf{n}} \varphi_\mathbf{m}^0. \qquad (3.101)$$

Comparing (3.99) and (3.100) we can see that the contribution of the second excited intramolecular state to the excitonic energy, which is near to the energy of the first excited intramolecular state, becomes relevant only if the following inequality holds

$$(A_2 - A_1)^2 \leq \left|L^{12}(\mathbf{k})\right|^2. \qquad (3.102)$$

3.4.2 *Several molecules per unit cell*

The role of mixing of molecular states can be particularly important for excitonic bands with a small Davydov splitting in crystals with unit cells containing several molecules.

Qualitatively this phenomenon can be explained as follows. Even if for each mixing term the contribution of mixing is small, the corresponding energy differences (i.e. the splitting energies) can, in general, change quite significantly. Besides, the mixing influences also the transition intensities, corresponding to all components of the splitting (see Section 3.6 of this chapter).

Since both the splitting energies and intensities of excitonic transitions can be established experimentally with a high degree of accuracy, theoretical determination of those quantities is important.

As we have shown above, for the calculation of the $\sigma\ell$ quantities $E_\mu(\mathbf{k})$ from eqn (3.97) we must solve a secular equation of order $\sigma\ell$. It is clear that this procedure is quite complicated even for $\ell = 2$ and $\sigma > 2$.

We mention here that taking into account the crystal symmetry properties simplifies the computation of excitonic states in the case of mixing. The basic simplification is based on the fact that the crystal energy operator can be transformed to a new representation where the wavefunctions correspond to the irreducible representation of the crystal space group and in this representation the crystal energy operator (without operators describing the term mixing) is diagonal. In this new representation the secular equation separates, and we have to compute a determinant of order ℓ instead of a determinant of order $\sigma\ell$.

We wish to discuss the problem more exactly. Now the operator $\hat{H}_3$ which appears in eqn (3.88) will be taken into account, since its contribution for the crystal Hamiltonian is relevant, and will not complicate the calculations.

With regard to eqn (3.90) the operator $\Delta\hat{H}$ can be put into the form

$$\Delta\hat{H} = \sum_f \Delta\hat{H}^f + \frac{1}{2} \sum_{\substack{f\neq g \\ n,m}}' M_{nm}^{fg} \left(B_{mg}^\dagger + B_{mg}\right)\left(B_{nf}^\dagger + B_{nf}\right), \qquad (3.103)$$

where the operator

$$\Delta\hat{H}^f = \sum_n (\Delta\varepsilon_f + \mathcal{D}_f) B_{nf}^\dagger B_{nf}$$

$$+ \frac{1}{2} {\sum_{n,m}}' M_{nm}^f \left(B_{mf}^\dagger + B_{mf} \right) \left(B_{nf}^\dagger + B_{nf} \right), \qquad (3.104)$$

has the form (3.58) and gives the excitonic states which correspond to the fth molecular term when the contribution of mixing of molecular terms is neglected.

As we have shown in Section 3.3, the Bose operators B_{nf} and $B_{nf}^\dagger$ can be replaced by operators $B_{\mu(f)}(\mathbf{k})$ and $B_{\mu(f)}^\dagger(\mathbf{k})$, linearly connected with B_{nf} and $B_{nf}^\dagger$, which diagonalize the operator $\Delta \hat{H}^f$. In particular, with respect to (3.63),

$$\Delta \hat{H}^f = - \sum_{\mu(f),\mathbf{k},\alpha} E_{\mu(f)}(\mathbf{k}) \left| v_{\alpha\mu(f)} \right|^2$$
$$+ \sum_{\mu(f),\mathbf{k}} E_{\mu(f)}^f(\mathbf{k}) B_{\mu(f)}^\dagger(\mathbf{k}) B_{\mu(f)}(\mathbf{k}), \qquad (3.105)$$

where

$$B_{\mathbf{n}\alpha,f} = \frac{1}{\sqrt{N}} \sum_{\mu(f),\mathbf{k}} \left\{ u_{\alpha\mu(f)}(\mathbf{k}) e^{i\mathbf{k}\mathbf{r}_{\mathbf{n}\alpha}} B_{\mu(f)}(\mathbf{k}) \right.$$
$$\left. + v_{\alpha\mu(f)}^*(\mathbf{k}) e^{-i\mathbf{k}\mathbf{r}_{\mathbf{n}\alpha}} B_{\mu(f)}^\dagger(\mathbf{k}) \right\}. \qquad (3.106)$$

The index $\mu(f)$ in eqns (3.105) and (3.106) denotes one of the excitonic states which corresponds to the fth molecular term without mixing effects, so that $\sum_{\mu(f)}$ denotes the summation over all excitonic states which correspond to the molecular term f. It is evident that in the new coordinates the operator $\sum_f \Delta \hat{H}^f$, being equal to $\Delta \hat{H}$ (see eqn 3.103) only when the mixing contributions are neglected (which means the contributions proportional to the matrix elements $M_{nm}^{fg}, f \neq g$), is also diagonal. Taking into account these matrix elements, we obtain for the operator $\Delta \hat{H}$ the following expression

$$\Delta \hat{H} = - \sum_{f,\mu(f),\mathbf{k},\alpha} E_{\mu(f)}(\mathbf{k}) \left| v_{\alpha\mu(f)}(\mathbf{k}) \right|^2$$
$$+ \sum_{f,\mu(f),\mathbf{k}} E_{\mu(f)}(\mathbf{k}) B_{\mu(f)}^\dagger(\mathbf{k}) B_{\mu(f)}(\mathbf{k})$$
$$+ \frac{1}{2} \sum_{f \neq g,\mu(f),\mu(g)} \left\{ D_{\mu(f),\mu(g)}^{(1)}(\mathbf{k}) B_{\mu(f)}(\mathbf{k}) B_{\mu(g)}(-\mathbf{k}) \right. \qquad (3.107)$$
$$+ D_{\mu(f)\mu(g)}^{(1)*}(\mathbf{k}) B_{\mu(f)}^\dagger(\mathbf{k}) B_{\mu(g)}^\dagger(-\mathbf{k})$$
$$\left. + 2 D_{\mu(f)\mu(g)}^{(2)}(\mathbf{k}) B_{\mu(f)}(\mathbf{k}) B_{\mu(g)}^\dagger(\mathbf{k}) \right\},$$

where

$$D^{(1)}_{\mu(f)\mu(g)}(\mathbf{k}) = \sum_{\alpha,\beta} L^{fg}_{\alpha\beta}(\mathbf{k}) t_{\alpha\mu(f)}(\mathbf{k}) t_{\beta\mu(g)}(-\mathbf{k}),$$

$$D^{(2)}_{\mu(f)\mu(g)}(\mathbf{k}) = \sum_{\alpha,\beta} L^{fg}_{\alpha\beta}(\mathbf{k}) t_{\alpha\mu(f)}(\mathbf{k}) t^*_{\beta\mu(g)}(\mathbf{k}), \qquad (3.108)$$

$$t_{\alpha\mu(f)}(\mathbf{k}) = u_{\alpha\mu(f)}(\mathbf{k}) + v_{\alpha\mu(f)}(\mathbf{k}).$$

In crystals with inversion symmetry

$$t^*_{\alpha\mu(f)}(\mathbf{k}) = t_{\alpha\mu(f)}(-\mathbf{k}), \qquad (3.109)$$

(see Section 2.1). In consequence, in such crystals

$$D^{(1)}_{\mu(f),\mu(g)}(\mathbf{k}) = D^{(2)}_{\mu(f)\mu(g)}(\mathbf{k}). \qquad (3.110)$$

Making use of relations (3.68) and (3.76) we obtain

$$\begin{aligned} t_{\alpha\mu(f)}(\mathbf{k}) &= \frac{2\left(\Delta\varepsilon_f + \mathcal{D}_f\right)}{\Delta\varepsilon_f + \mathcal{D}_f + E_{\mu(f)}(\mathbf{k})} u_{\alpha\mu(f)}(\mathbf{k}) \\ &= \sqrt{\frac{\Delta\varepsilon_f + \mathcal{D}_f}{E_{\mu(f)}}}(\mathbf{k}) u^{\text{H}-\text{L}}_{\alpha\mu(f)}(\mathbf{k}). \end{aligned} \qquad (3.111)$$

Therefore

$$D^{(1)}_{\mu(f),\mu(g)}(\mathbf{k}) = \frac{\sqrt{(\Delta\varepsilon_f + \mathcal{D}_f)(\Delta\varepsilon_g + \mathcal{D}_g)} D^{(1)\text{H}-\text{L}}_{\mu(f)\mu(g)}(\mathbf{k})}{\sqrt{E_{\mu(f)}(\mathbf{k}) E_{\mu(g)}(-\mathbf{k})}}, \qquad (3.112)$$

where

$$\begin{aligned} D^{(1)\text{H}-\text{L}}_{\mu(f)\mu(g)}(\mathbf{k} &= \sum_{\alpha\beta} L^{fg}_{\alpha\beta}(\mathbf{k}) u^{\text{H}-\text{L}}_{\alpha\mu(f)}(\mathbf{k}) u^{*\text{H}-\text{L}}_{\beta\mu(g)}(\mathbf{k}) \\ &\equiv \langle \Psi_{\mu(g)}(\mathbf{k}) | {\sum_{n,m}}' \hat{V}_{nm} | \Psi_{\mu(f)}(\mathbf{k}) \rangle \end{aligned} \qquad (3.113)$$

is the matrix element of the operator $\sum_{n,m}' \hat{V}_{nm}$ computed with the excitonic wavefunctions of the states $(\mu(f), \mathbf{k})$ and $(\mu(g), \mathbf{k})$ which are taken in the Heitler–London approximation.

Since the operator $\sum_{n,m}' \hat{V}_{nm}$ is invariant upon the crystal space group, and the functions $\Psi_{\mu(f)}(\mathbf{k})$ and $\Psi_{\mu(g)}(\mathbf{k})$ transform upon irreducible representation of the space group, the matrix element (3.113) will not be zero only for excitonic states $(\mu(f), \mathbf{k})$ and $(\mu(g), \mathbf{k})$ which transform upon the same irreducible representation of the crystal group.

In order to diagonalize the quadratic form (3.107) we introduce Bose amplitudes $B_\rho(\mathbf{k})$ (see eqn (A.2) in Appendix A)

$$B_{\mu(f)}(\mathbf{k}) = \sum_\rho \left[u_{\mu(f),\rho}(\mathbf{k}) B_\rho(\mathbf{k}) + v^*_{\mu(f),\rho}(\mathbf{k}) B^\dagger_\rho(-\mathbf{k}) \right], \qquad (3.114)$$

with coefficients u and v obeying the equations

MIXING OF MOLECULAR CONFIGURATIONS 59

$$\left[E_\rho - E_{\mu(f)}(\mathbf{k})\right] u_{\mu(f),\rho}(\mathbf{k})$$
$$= \sum_{\mu(g), g \neq f} \left\{ D^{(2)}_{\mu(f),\mu(g)}(\mathbf{k}) u_{\mu(g),\rho}(\mathbf{k}) + D^{*(1)}_{\mu(f),\mu(g)}(\mathbf{k}) v_{\mu(g),\rho}(\mathbf{k}) \right\},$$
(3.115)

$$- \left[E_\rho + E_{\mu(f)}(\mathbf{k})\right] v_{\mu(f),\rho}(\mathbf{k})$$
$$= \sum_{\mu(g), g \neq f} \left\{ D^{*(2)}_{\mu(f),\mu(g)}(\mathbf{k}) v_{\mu(g),\rho}(\mathbf{k}) + D^{(1)}_{\mu(f),\mu(g)}(\mathbf{k}) u_{\mu(g),\rho}(\mathbf{k}) \right\}.$$

In these new coordinates the operator $\Delta \hat{H}$ can be expressed in the form

$$\Delta \hat{H} = - \sum_{f,\mu(f),\mathbf{k},\alpha} E_{\mu(f)}(\mathbf{k}) \left| v_{\alpha\mu(f)}(\mathbf{k}) \right|^2 - \sum_{f,\mu(f),\mathbf{k},\rho} E_\rho(\mathbf{k}) \left| v_{\mu(f),\rho}(\mathbf{k}) \right|^2$$
$$+ \sum_{\rho,\mathbf{k}} E_\rho(\mathbf{k}) B^\dagger_\rho(\mathbf{k}) B_\rho(\mathbf{k}).$$
(3.116)

The new excitation energies – the quantities $E_\rho(\mathbf{k})$ – can be obtained by the condition of vanishing of the determinant of the equation system (3.115). The coefficients $u_{\mu(f),\rho}(\mathbf{k})$ and $v_{\mu(f),\rho}(\mathbf{k})$ are normalized by the relation

$$\sum_{f,\mu(f)} \left\{ \left| u_{\mu(f),\rho}(\mathbf{k}) \right|^2 - \left| v_{\mu(f),\rho}(\mathbf{k}) \right|^2 \right\} = 1.$$
(3.117)

In crystals with inversion center, where the condition (3.110) is satisfied, we obtain the relation

$$v_{\mu(f),\rho}(\mathbf{k}) = - \frac{E_\rho - E_{\mu(f)}(\mathbf{k})}{E_\rho + E_{\mu(f)}(\mathbf{k})} u_{\mu(f),\rho}(\mathbf{k}).$$
(3.118)

Substituting (3.118) into the first of equations (3.115), we obtain a system of equations for the coefficients $u_{\mu(f),\rho}(\mathbf{k})$, which is of the form

$$\left[E_\rho - E_{\mu(f)}(\mathbf{k})\right] u_{\mu(f),\rho}(\mathbf{k})$$
$$= \sum_{\mu(g), g \neq f} u_{\mu(g),\rho}(\mathbf{k}) \left\{ D^{(1)}_{\mu(f),\mu(g)}(\mathbf{k}) - D^{*(1)}_{\mu(f),\mu(g)}(\mathbf{k}) \frac{E_\rho - E_{\mu(g)}(\mathbf{k})}{E_\rho + E_{\mu(g)}(\mathbf{k})} \right\}.$$
(3.119)

The above equations have a simpler form for crystals of type naphthalene, anthracene, and similar, where for $\mathbf{k} = 0$ the states are not degenerate. In such crystals the quantities $D^{(1)}$ and $D^{(2)}$ (3.108) are not only equal, but also real (because $t_{\alpha\mu(f)}$ are real; see eqns 3.108 and 3.68). Therefore the equation system (3.119) can be put into the form

$$\left[E_\rho^2 - E_{\mu(f)}^2(\mathbf{k})\right] \tilde{u}_{\mu(f),\rho}(\mathbf{k})$$

$$= 2 \sum_{\mu(g), g \neq f} D^{(1)}_{\mu(f),\mu(g)}(\mathbf{k}) E_{\mu(g)}(\mathbf{k}) \tilde{u}_{\mu(g),\rho}(\mathbf{k}), \tag{3.120}$$

where we have used the notation

$$\tilde{u}_{\mu(f),\rho}(\mathbf{k}) = \frac{u_{\mu(f),\rho}(\mathbf{k})}{E_\rho + E_{\mu(f)}(\mathbf{k})}. \tag{3.121}$$

As we have mentioned previously, the quantities $D^{(1)}_{\mu(f),\mu(g)}(\mathbf{k})$ are not zero only for such excitonic states $\mu(f)$ and $\mu(g)$, which transform upon the same irreducible representation of crystal space group. Thus the order of the equation which determines the quantities E_ρ^2 is equal to (or smaller than) the number of molecular terms which were taken into account by the mixing of terms, and does not depend on the number of molecules per unit cell.

Hence, if we consider only two molecular excited states f and g, the excitonic energy is given by the formula

$$E_{1,2}^2(\mathbf{k}) = \frac{1}{2} \left[E_{\mu(f)}^2(\mathbf{k}) + E_{\mu(g)}^2(\mathbf{k}) \right]$$
$$\tag{3.122}$$
$$\pm \frac{1}{2} \sqrt{\left[E_{\mu(f)}^2(\mathbf{k}) - E_{\mu(g)}^2(\mathbf{k}) \right]^2 + 16 E_{\mu(f)}(\mathbf{k}) E_{\mu(g)}(\mathbf{k}) \left[D^{(1)}_{\mu(f),\mu(g)}(\mathbf{k}) \right]^2}.$$

3.5 Perturbation theory. A comparison with results obtained in the Heitler–London approximation

If the modulus of the matrix elements $D^{(1),(2)}_{\mu(f),\mu(g)}(\mathbf{k}), f \neq g$ is small compared with the quantities $|E_{\mu(f)} - E_{\mu(g)}|$, then eqns (3.115) can be solved by taking an expansion in terms of a small matrix $D^{(1),(2)}(\mathbf{k})$. For simplicity, we consider here the case when the matrices $D^{(1)}(\mathbf{k})$ and $D^{(2)}(\mathbf{k})$ are real and equal. In this case eqns (3.115) take the form

$$\left[E - E_{\mu(f)}(\mathbf{k}) \right] u_{\mu(f)}(\mathbf{k})$$
$$= \sum_{\mu(g), g \neq f} D^{(1)}_{\mu(f),\mu(g)}(\mathbf{k}) \left[u_{\mu(g)}(\mathbf{k}) + v_{\mu(g)}(\mathbf{k}) \right],$$
$$\tag{3.123}$$
$$- \left[E + E_{\mu(f)}(\mathbf{k}) \right] v_{\mu(f)}(\mathbf{k})$$
$$= \sum_{\mu(g), g \neq f} D^{(1)}_{\mu(f),\mu(g)}(\mathbf{k}) \left[u_{\mu(g)}(\mathbf{k}) + v_{\mu(g),\rho}(\mathbf{k}) \right].$$

Now we estimate corrections to the energy of the exciton $(\mu(h), \mathbf{k})$ which are due to the mixing of molecular terms. In the zeroth order of perturbation theory $E =$
$= E_{\mu(h)}(\mathbf{k})$ and, with respect to the normalization condition (3.117), $u_{\mu(g)}(\mathbf{k}) =$

$= \delta_{\mu(h),\mu(g)}$, $v_{\mu(g)} = 0$ for all values of g. In the first order of perturbation theory we put

$$u_{\mu(g)}(\mathbf{k}) = \delta_{\mu(h),\mu(g)} + u^{(1)}_{\mu(g)}(\mathbf{k}), \qquad v_{\mu(g)}(\mathbf{k}) = v^{(1)}_{\mu(g)}(\mathbf{k}),$$
$$E_\mu(\mathbf{k}) = E_{\mu(h)}(\mathbf{k}) + E^{(1)}_{\mu(h)}(\mathbf{k}). \tag{3.124}$$

Substituting the quantities (3.124) into eqns (3.123) for $f \neq g$ and comparing expressions of the same order in the expansion parameter we obtain

$$u^{(1)}_{\mu(f), f \neq h} = \frac{D^{(1)}_{\mu(f),\mu(h)}(\mathbf{k})}{E_{\mu(h)} - E_{\mu(f)}}, \qquad v^{(1)}_{\mu(f), f \neq h} = -\frac{D^{(1)}_{\mu(f),\mu(h)}(\mathbf{k})}{E_{\mu(h)} + E_{\mu(f)}}. \tag{3.125}$$

Now we substitute the expressions (3.124) and (3.125) into eqns (3.123) putting $f = h$ and thus obtain the energy correction as

$$E^{(1)}_{\mu(h)}(\mathbf{k}) = \sum_{\mu(g),\, g \neq h} \frac{\left[D^{(1)}_{\mu(h),\mu(g)}(\mathbf{k})\right]^2}{E_{\mu(h)}(\mathbf{k}) - E_{\mu(g)}(\mathbf{k})}$$
$$- \sum_{\mu(g),\, g \neq h} \frac{\left[D^{(1)}_{\mu(h),\mu(g)}(\mathbf{k})\right]^2}{E_{\mu(h)}(\mathbf{k}) + E_{\mu(g)}(\mathbf{k})}. \tag{3.126}$$

We now compare this expression for the energy $E^{(1)}_{\mu(h)}(\mathbf{k})$ with the corresponding expression obtained in the Heitler–London approximation. We have in mind the result by Craig (12); (13):

$$E^{(1)\mathrm{H-L}}_{\mu(h)}(\mathbf{k}) = \sum_{\mu(g), g \neq h} \frac{\left[D^{(1)\mathrm{H-L}}_{\mu(h),\mu(g)}(\mathbf{k})\right]^2}{E^{\mathrm{H-L}}_{\mu(h)}(\mathbf{k}) - E^{\mathrm{H-L}}_{\mu(g)}(\mathbf{k})}. \tag{3.127}$$

The above expression is similar to the first term on the r.h.s. of eqn (3.126), but not identical, since $E^{(1)\mathrm{H-L}}_{\mu(h)} \neq E^{(1)}_{\mu(h)}(\mathbf{k})$ and, with respect to (3.112),

$$\left[D^{(1)}_{\mu(h),\mu(g)}(\mathbf{k})\right]^2 = \frac{(\Delta \varepsilon_h + \mathcal{D}_h)(\Delta \varepsilon_g + \mathcal{D}_g)}{E_{\mu(f)}(\mathbf{k}) E_{\mu(g)}(\mathbf{k})} \left[D^{(1)\mathrm{H-L}}_{\mu(h)\mu(g)}(\mathbf{k})\right]^2. \tag{3.128}$$

The second term on the the r.h.s. of eqn (3.126) has no analogous counterpart in (3.127).

In order to exemplify of the formulas (3.126) and (3.127) we consider crystals of anthracene type with two molecules per unit cell. In such crystals one usually considers a mixing of the first excited state (1) with the second excited state (2). Since the oscillator strength of the first excited state is much smaller than those of the second excited state, the role of mixing is more relevant for the lower electronic transition (for details. see Section 3.10 of this chapter).

Let us denote by $\mu = a$ and $\mu = b$ the components of the Davydov splitting. If higher excited states are not considered, than we obtain from eqn (3.126) for the lower Davydov doublet upon mixing

$$E_a(\mathbf{k}) = E_{a(1)}(\mathbf{k}) + \frac{\left[D^{(1)}_{a(1),a(2)}(\mathbf{k})\right]^2}{E_{a(1)} - E_{a(2)}} - \frac{\left[D^{(1)}_{a(1),a(2)}(\mathbf{k})\right]^2}{E_{a(1)} + E_{a(2)}},$$

$$E_b(\mathbf{k}) = E_{b(1)}(\mathbf{k}) + \frac{\left[D^{(1)}_{b(1),b(2)}(\mathbf{k})\right]^2}{E_{b(1)} - E_{b(2)}} - \frac{\left[D^{(1)}_{b(1),b(2)}(\mathbf{k})\right]^2}{E_{b(1)} + E_{b(2)}}. \quad (3.129)$$

Simultaneously, for the same case [see (3.127)]

$$E^{(1)H-L}_{a(1)}(\mathbf{k}) = \frac{\left[D^{(1)H-L}_{a(1),a(2)}(\mathbf{k})\right]^2}{E^{H-L}_{a(1)}(\mathbf{k}) - E^{H-L}_{a(2)}(\mathbf{k})},$$

$$E^{(1)H-L}_{b(1)}(\mathbf{k}) = \frac{\left[D^{(1)H-L}_{b(1),b(2)}(\mathbf{k})\right]^2}{E^{H-L}_{b(1)}(\mathbf{k}) - E^{H-L}_{b(2)}(\mathbf{k})}. \quad (3.130)$$

3.6 Sum rules for the oscillator strengths of excitonic transitions and the hypochromatic effect

In the following section we derive the sum rule for excitonic transitions in a crystal which can be considered as analogous to the sum rule for dipole transitions in atoms.

For this purpose we introduce two operators. The first is

$$\hat{\Pi}(\mathbf{k}) = \sum_{n\alpha} e^{i\mathbf{k}\mathbf{r}_{n\alpha}} \hat{\mathbf{p}}_{n\alpha}, \quad (3.131)$$

where $\hat{\mathbf{p}}_{n\alpha} = \sum_\nu e_\nu \mathbf{r}^\alpha_\nu$ is the dipole moment operator of the molecule $n\alpha$, e_ν is the charge of the νth particle, being part of the molecule $n\alpha$, and $\mathbf{r}^\alpha_\nu$ is a vector which indicates the position of the νth particle. The second operator is of the form

$$\frac{\mathrm{d}}{\mathrm{d}t}\hat{\Pi}(\mathbf{k}) = \sum_{n\alpha} e^{i\mathbf{k}\mathbf{r}_{n\alpha}} \frac{\mathrm{d}}{\mathrm{d}t}\hat{\mathbf{p}}_{n\alpha}. \quad (3.132)$$

Since the operator (3.131) does not depend explicitly on time, the operator (3.132) can be put into the form (see, for example, (14))

$$\frac{\mathrm{d}}{\mathrm{d}t}\hat{\Pi}(\mathbf{k}) = \frac{1}{i\hbar}\left(\hat{H}\hat{\Pi}(\mathbf{k}) - \hat{\Pi}(\mathbf{k})\hat{H}\right), \quad (3.133)$$

$\hat{H}$ being the Hamilton operator, describing elementary excitations in crystals in the excitonic energy region. With regard to the formula (3.116) the operator $\hat{H}$ takes the form

SUM RULES FOR EXCITONIC TRANSITIONS

$$\hat{H} = \mathcal{E}_0 + \sum_{\rho,\mathbf{k}} E_\rho(\mathbf{k}) B_\rho^\dagger(\mathbf{k}) B_\rho(\mathbf{k}). \tag{3.134}$$

As concerning the operator $\mathbf{\Pi}(\mathbf{k})$, its form in the second-quantization representation can be found by making use of the formulas (3.12), (3.25), (3.106), and (3.114).

Crystals of anthracene type have an inversion center. Therefore in each molecular state the mean value of the dipole moment $\langle g | \mathbf{p}_\alpha | g \rangle$ is zero. With regard to this property and taking in the expression for $\mathbf{\Pi}(\mathbf{k})$ only contributions linear with respect to Bose amplitudes $B_\rho(\mathbf{k})$ and $B_\rho^\dagger(\mathbf{k})$, we obtain

$$\hat{\mathbf{\Pi}} = \sqrt{N} \sum_\rho \left\{ \mathbf{P}^{(1)}(\mathbf{k},\rho) B_\rho(-\mathbf{k}) + \mathbf{P}^{(2)}(\mathbf{k},\rho) B_\rho^\dagger(\mathbf{k}) \right\}, \tag{3.135}$$

where

$$\mathbf{P}^{(1)}(\mathbf{k},\rho) = \sum_{\alpha,f,\mu(f)} \langle 0|\mathbf{p}_\alpha|f\rangle \left[t_{\alpha\mu(f)}(-\mathbf{k}) u_{\mu(f),\rho}(-\mathbf{k}) \right.$$
$$\left. + t^*_{\alpha\mu(f)}(\mathbf{k}) v_{\mu(f),\rho}(\mathbf{k}) \right], \tag{3.136}$$
$$\mathbf{P}^{(2)}(\mathbf{k},\rho) = \sum_{\alpha,f,\mu(f)} \langle 0|\mathbf{p}_\alpha|f\rangle \left[t_{\alpha\mu(f)}(-\mathbf{k}) v^*_{\mu(f),\rho}(-\mathbf{k}) \right.$$
$$\left. + t^*_{\alpha\mu(f)}(\mathbf{k}) u^*_{\mu(f),\rho}(\mathbf{k}) \right].$$

In crystals with an inversion center

$$\mathbf{P}^{(1)}(\mathbf{k},\rho) = \mathbf{P}^{(2)}(\mathbf{k},\rho). \tag{3.137}$$

If not only the whole crystal has an inversion center, but also the constituent molecules have the same property, then the quantities, which occur in eqn (3.136), are real and symmetric with respect to the tranformation $\mathbf{k} \to -\mathbf{k}$. In this case, making use of the formulas (3.111) and (3.118), we obtain

$$\mathbf{P}^{(1)}(\mathbf{k},\rho) = \mathbf{P}^{(2)}(\mathbf{k},\rho) \tag{3.138}$$
$$= \sum_{\alpha,f,\mu(f)} \langle 0|\mathbf{p}_\alpha|f\rangle \frac{4\left(\Delta\varepsilon_f + \mathcal{D}_f\right) E_{\mu(f)}(\mathbf{k}) u_{\alpha\mu f}(\mathbf{k}) u_{\mu(f),\rho}(\mathbf{k})}{\left[\Delta\varepsilon_f + \mathcal{D}_f + E_{\mu(f)}(\mathbf{k})\right] \left[\Delta\varepsilon_f + \mathcal{D}_f + E_\rho(\mathbf{k})\right]}.$$

Making use of the relations (3.133), (3.134) and (3.135) we obtain the operator $\frac{d}{dt}\hat{\mathbf{\Pi}}(\mathbf{k})$ in the second-quantization representation. It is given by the expression

$$\hat{\dot{\mathbf{\Pi}}}(\mathbf{k}) \equiv \frac{d}{dt}\hat{\mathbf{\Pi}}(\mathbf{k}) = -\frac{i\sqrt{N}}{\hbar} \left\{ \mathbf{P}^{(1)}(\mathbf{k},\rho) E_\rho(-\mathbf{k}) B_\rho(-\mathbf{k}) \right.$$

$$-\mathbf{P}^{(2)}(\mathbf{k},\rho)E_\rho(\mathbf{k})B_\rho^\dagger(\mathbf{k})\right\}. \tag{3.139}$$

The operators $\hat{\Pi}_x(\mathbf{k})$ and $\hat{\Pi}_y(\mathbf{k})$ for any $x,y = 1,2,3$ satisfy simple commutation relations

$$\hat{\Pi}_x(\mathbf{k})\hat{\Pi}_y(-\mathbf{k}) - \hat{\Pi}_y(-\mathbf{k})\hat{\Pi}_x(\mathbf{k}) = -i\hbar N\sigma\left(\sum_\nu \frac{e_\nu^2}{m_\nu}\right)\delta_{xy}. \tag{3.140}$$

The above relations can be obtained by making use of the expressions for the operators $\hat{\Pi}$ and $\hat{\Pi}$ in the coordinate representation (see eqns 3.131 and 3.132). The commutation relations cannot depend on the choice of representation and will also be valid for the operators in the second-quantization representation. Thus, substituting (3.135) and (3.139) into the relations (3.140) we obtain the desired sum rule

$$\sum_\rho \left\{ P_x^{(1)}(\mathbf{k},\rho)P_y^{(1)}(-\mathbf{k},\rho)E_\rho(-\mathbf{k}) + P_y^{(1)}(-\mathbf{k},\rho)P_x^{(2)}(\mathbf{k},\rho)E_\rho(\mathbf{k})\right\}$$

$$= \hbar^2\sigma\left(\sum_\nu \frac{e_\nu^2}{m_\nu}\right)\delta_{xy}. \tag{3.141}$$

The sum rule (3.141), which is valid for all values of $\mathbf{k}$, will be used in the following section. The relation has a simpler form when we neglect the state mixing in the Heitler–London approximation, and the index ρ corresponds to $(f,\mu(f))$. In this case for crystals having an inversion center, the vector

$$\mathbf{P}^{(1)}(\mathbf{k},\mu(f)) = \mathbf{P}^{(2)}(\mathbf{k},\mu(f))$$
$$= \sum_\alpha \langle 0|\mathbf{p}_\alpha|f\rangle u_{\alpha\mu}^{\mathrm{H-L}}(\mathbf{k}) = \mathbf{P}_{0,\mu(f)}^{\mathrm{H-L}}(\mathbf{k}), \tag{3.142}$$

up to the multiplicative factor $\sqrt{N}$, is equal to the matrix element of the unit cell dipole moment computed with the ground state and the excited state crystal wavefunctions taken in the Heitler–London approximation. Performing in (3.141) summation over x and y we obtain in this approximation

$$\sum_{f,\mu(f)} F_{\mu(f)}^{\mathrm{H-L}} = 1, \tag{3.143}$$

where

$$F_{\mu(f)}^{\mathrm{H-L}}(\mathbf{k}) = \frac{2\left|\mathbf{P}_{0,\mu(f)}^{\mathrm{H-L}}\right|^2}{3\hbar^2\sigma\left(\sum_\nu \frac{e_\nu^2}{m_\nu}\right)} E_{\mu(f)}^{\mathrm{H-L}}(\mathbf{k}) \tag{3.144}$$

is the oscillator strength of the transition between the ground state and the state $|\mu(f),\mathbf{k}\rangle$. This quantity describes the absorption intensity of light with frequency $\omega = E_\mu(\mathbf{k})/\hbar$ in the Heitler–London approximation.

If we do not use the Heitler–London approximation, the expression (3.144) must be replaced by

$$F_\rho(\mathbf{k}) = \frac{2\,|\mathbf{P}(\mathbf{k},\rho)|^2}{3\hbar^2 \sigma \left(\sum_\nu \frac{e_\nu^2}{m_\nu}\right)} E_\rho(\mathbf{k}). \tag{3.145}$$

Problems related to light absorption in the excitonic region of the spectrum will be discussed in the following chapters. At this point we only notice that the formulas (3.136), (3.137), and (3.145) allow one to establish the oscillator strength of an excitonic transition which are more exact than that by use of the Heitler–London approximation, and the differences can be substantial.

For demonstration of the above statement let us consider crystals with an inversion center, for which the relation (3.137) holds. Taking into account the relation (3.76) we obtain instead of (3.138)

$$\mathbf{P}^{(1)}(\mathbf{k},\rho) \tag{3.146}$$

$$= \sum_{f,\mu(f)} \frac{2 E_{\mu(f)}(\mathbf{k}) u_{\mu(f),\rho}(\mathbf{k})}{E_{\mu(f)}(\mathbf{k}) + E_\rho(\mathbf{k})} \sqrt{\frac{\Delta \varepsilon_f + \mathcal{D}_f}{E_{\mu(f)}(\mathbf{k})}}\, \mathbf{P}^{\text{H}-\text{L}}_{0,\mu(f)}(\mathbf{k}).$$

When neglecting the mixing, then

$$\rho \equiv \mu(g), \qquad u_{\mu(f),\mu(g)} = \delta_{\mu(f),\mu(g)}, \tag{3.147}$$

so that

$$\mathbf{P}^{(1)}(\mathbf{k},\mu(f)) = \sqrt{\frac{\Delta \varepsilon_f + \mathcal{D}_f}{E_{\mu(f)}(\mathbf{k})}}\, \mathbf{P}^{\text{H}-\text{L}}_{0,\mu(f)}(\mathbf{k}). \tag{3.148}$$

The above result differs from those obtained by the Heitler–London approximation by taking into account the crystal states where not only one but several crystal molecules are excited. The relation between oscillator strengths of components of the Davydov splitting taken in the above approximation can differ substantially from the corresponding relation obtained by the Heitler–London approximation. Making use of (3.144), (3.145), and (3.148) we obtain for two components μ and μ' of the Davydov splitting

$$\frac{F_{\mu(f)}(\mathbf{k})}{F_{\mu'(f)}(\mathbf{k})} = \frac{\left|\mathbf{P}^{\text{H}-\text{L}}_{0,\mu(f)}(\mathbf{k})\right|^2}{\left|\mathbf{P}^{\text{H}-\text{L}}_{0,\mu'(f)}(\mathbf{k})\right|^2} = \left(\frac{E^{\text{H}-\text{L}}_{\mu'(f)}(\mathbf{k})}{E^{\text{H}-\text{L}}_{\mu(f)}(\mathbf{k})}\right) \frac{F^{\text{H}-\text{L}}_{\mu(f)}(\mathbf{k})}{F^{\text{H}-\text{L}}_{\mu'(f)}(\mathbf{k})}. \tag{3.149}$$

Therefore, for example, in anthracene crystals, where the second molecular transition splits into two lines, for which[18] $E^{\text{H}-\text{L}}_a(k \simeq 0) = 53 \cdot 10^3$ cm^{-1} and

---

[18] The energy values $E^{\text{H}-\text{L}}_a$ and $E^{\text{H}-\text{L}}_b$ result from (3.74) by inserting for E_a and E_b the experimental values $37 \cdot 10^3$ cm^{-1}, $53 \cdot 10^3$ cm^{-1}, and $\Delta \varepsilon_f + \mathcal{D}_f \simeq 40 \cdot 10^3$ cm^{-1}.

$E_b^{\text{H-L}}(k \simeq 0) = 37 \cdot 10^3$ cm^{-1}, the ratio of oscillator strengths when taking into account multimolecular excited states[19] differs from the corresponding relation obtained by the Heitler–London approximation by the factor $E_a^{\text{H-L}}/E_b^{\text{H-L}} \simeq$ $\simeq 1.4$. An analogous situation occurs for intensive transitions with a large splitting in other crystals, as, for example, naphthalene, etc.

For transitions with small intensity and by neglecting the mixing of molecular configurations, the contribution of multimolecular excited states is negligible.

A quite different situation arises when we consider the mixing of excitonic states having weak oscillator strength with excitonic states having large oscillator strength. This effect can be discussed by making use of the relation (3.146) which, when contributions of higher excited states are taken into account, yields results quite different from those obtained by the Heitler–London approximation.

For verifying the above statement we start with the expression (3.146) for the case when the distance between excitonic states is much larger than the matrix elements of the matrix $D(\mathbf{k})$ (see Sections 3.4 and 3.5), which determines the intensity of mixing of excitonic states. Assuming in (3.146) that $\rho = \mu(h)$ where h indicates a molecular state with small oscillator strength, and making use of the formulas (3.124)–(3.126) and (3.128) up to terms linear in components of $D(\mathbf{k})$ we obtain[20]

$$\mathbf{P}^{(1)}(\mathbf{k}, \mu(h)) = \mathbf{P}^{\text{H-L}}_{0,\mu(h)}(\mathbf{k}) \qquad (3.150)$$
$$- \sum_{f,\mu(f), f \neq h} 2\left(\Delta\varepsilon_f + \mathcal{D}_f\right) \frac{D^{(1)\text{H-L}}_{\mu(h),\mu(f)}(\mathbf{k})}{E^2_{\mu(f)}(\mathbf{k}) - E^2_{\mu(h)}(\mathbf{k})} \mathbf{P}^{\text{H-L}}_{0,\mu(f)}(\mathbf{k}).$$

At the same time the corresponding expression, obtained in the Heitler–London approximation, has the form

$$\mathbf{P}^{(1)}(\mathbf{k}, \mu(h)) = \mathbf{P}^{\text{H-L}}_{0,\mu(h)}(\mathbf{k})$$
$$- \sum_{f,\mu(f), f \neq h} 2 \frac{D^{(1)\text{H-L}}_{\mu(h),\mu(f)}(\mathbf{k})}{E_{\mu(f)}(\mathbf{k}) - E_{\mu(h)}(\mathbf{k})} \mathbf{P}^{\text{H-L}}_{0,\mu(f)}(\mathbf{k}), \qquad (3.151)$$

which coincides with (3.150) if we take

$$\frac{2\left(\Delta\varepsilon_f + \mathcal{D}_f\right)}{E_{\mu(f)} + E_{\mu(h)}} = 1. \qquad (3.152)$$

If the mixing of molecular configurations is important, so that the second term in (3.150) gives a substantial contribution to the value $\mathbf{P}^{(1)}(\mathbf{k}, \mu(h))$, then the

---

[19]This means crystal states when several molecules are simultaneously excited.

[20]We also used the relation $\sqrt{\frac{\Delta\varepsilon_h + \mathcal{D}_h}{E_{\mu(h)}}} \simeq 1$, resulting with a high accuracy for the states with small oscillator strength.

approximation (3.151) may not be exact enough. For example, in the case of anthracene the expression (3.152) for the two lowest excited states takes the value $(2 \cdot 40)/(27+37) \simeq 1.3$ and strongly influences the corresponding oscillator strength. A similar situation may arise for other crystals also.

Thus, even in such cases, when the conditions of applicability of perturbation theory are satisfied, i.e. when the matrix elements $D^{(1),(2)}_{\mu(f),\mu(g)}(\mathbf{k}), f \neq g$, are small compared to the quantities $|E_{\mu(f)} - E_{\mu(g)}|$, the contribution of multimolecular states, in general, is important.

Finally, we wish to present here just for illustration, the results of Hoffmann (4), which permit the discussion of the role of different factors in the computation of excitonic energies by taking into account the mixing of molecular configurations. Using the above presented theory for computing excitonic energies in biopolymers, Hoffmann has considered a one-dimensional model crystal with one molecule per unit cell. For the molecule three energy states were taken into account: the ground state 0, the first excited state B, and the second excited state A. He also assumed that the transitions between the ground state and the states A and B are dipole-allowed (with oscillator strengths f_A and f_B, respectively) and the transition dipole moments and the polymer axis are placed in the same plane.

In Figs 3.1 and 3.2 we display the results of calculations of the hypochromatic effect (i.e. the quantity of the relative change of oscillator strength) for the lower excited state, which occurs when molecules aggregate to the crystal, as a function of the angle θ_A between the dipole moment of the transition to the state A with the polymer axis. In Fig. 3.1 the calculations were performed using the values

$$\theta_B = \theta_A + \frac{\pi}{3}, \quad f_A = f_B = 1, \quad E_A = 6.2 \text{ eV}, \quad E_B = 4.77 \text{ eV},$$

and the lattice constant $d = 0.6$ nm. The data used in Fig. 3.2 were the following

$$\theta_B = \theta_A, \quad f_A = 1, \quad f_B = 0.01, \quad E_A = 6.2 \text{ eV}, \quad E_B = 4.77 \text{ eV},$$

and the lattice constant $d = 0.8$ nm. In both figures the dotted lines correspond to results obtained by perturbation theory, and the solid lines are obtained by the more accurate theory presented in the previous section. As follows from the displayed curves, for some values of the angle θ_A and for the parameters chosen by Hoffmann, the inaccuracy of results obtained by perturbation theory can attain values of the order of 50%. It is evident that for another choice of parameters this inaccuracy can be larger or smaller than those obtained by Hoffmann.

3.7 Exciton–phonon interaction

In the previous sections we assumed that the crystal molecules rest at their equilibrium positions at the lattice sites. Such an assumption makes it impossible to discuss several properties of excitons as, for example, the absorption lineshape and others, which result from the exciton–phonon interaction.

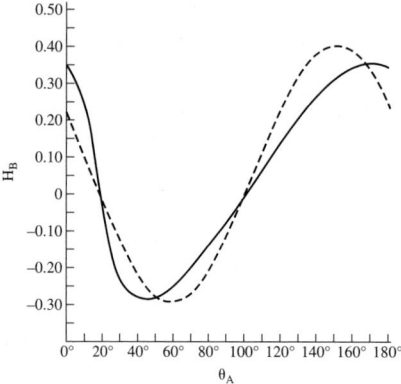

FIG. 3.1. The hypochromatic effect for the lower excited state, which occurs when molecules aggregate to the crystal, as a function of the angle θ_A between the dipole moment of the transition to the state A with the polymer axis. The parameters used are $\theta_B = \theta_A + \frac{\pi}{3}, f_A = f_B = 1, E_A = 6.2 \text{eV}, E_B = 4.77$ eV, and the lattice constant $d = 0.6$ nm. Reprinted with permission from Hoffmann (4). Copyright 1963, Radiation Research.

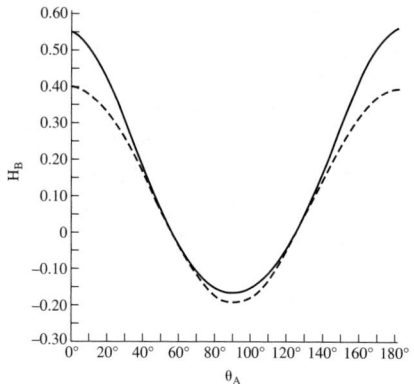

FIG. 3.2. The same as in Figure 3.1, for a different set of parameters: $\theta_B = \theta_A, f_A = 1, f_B = 0.01, E_A = 6.2 \text{eV}, E_B = 4.77 \text{eV}$, and the lattice constant $d = 0.8$ nm. Reprinted with permission from Hoffmann (4). Copyright 1963, Radiation Research.

With the aim of obtaining an operator, describing this interaction, we assume that the crystal molecules are displaced from their equilibrium positions and that their orientations also differ from that in equilibrium.

In this case the crystal Hamiltonian in the excitonic energy region can be taken in the form (3.90), where now the quantities $\mathcal{E}_0, \mathcal{D}_f, M_{nm}^{fg}$ depend on small displacements characterizing molecules deviations from their equilibrium positions and orientations. Expanding those quantities with respect to deviations we obtain a Hamiltonian which depends on the exciton creation and annihilation operators and on variables characterizing molecular displacements. Including also the kinetic energy operator of displacements of atomic nuclei, we obtain the full crystal energy operator in the excitonic energy region in the form

$$\hat{H} = \hat{H}_e + \hat{H}_L + \hat{H}_{eL}, \quad (3.153)$$

where $\hat{H}_e$ is the Hamilton operator in the excitonic energy region for an ideal crystal lattice, $\hat{H}_L$ is the Hamilton operator describing the lattice vibrations when excitons are absent, $\hat{H}_{eL}$ is the operator characterizing the exciton–phonon interaction, depending on the operators of creation and annihilation of excited states in molecules and on the deviations of molecules from the equilibrium positions.

Expanding the quantity $\mathcal{E}_0$ in (3.90) with respect to deviations from equilibrium up to quadratic terms and introducing normal coordinates the Hamiltonian H_L can be written as a sum of Hamiltonians which correspond to harmonic oscillators in their normal coordinates. Then we use the phonon creation and annihilation operators, i.e. the operators $b_{\mathbf{q}r}^\dagger$ and $b_{\mathbf{q}r}$ ($\mathbf{q}$ is the phonon wavevector and r indicates the corresponding frequency branch) and obtain the Hamiltonian H_L in the form

$$\hat{H}_L = \sum_{\mathbf{q}r} \hbar\omega(\mathbf{q},r)\left(b_{\mathbf{q}r}^\dagger b_{\mathbf{q}r} + \frac{1}{2}\right). \qquad (3.154)$$

If we now pass from the operators $B_{nf}, B_{nf}^\dagger$ to the exciton creation and annihilation operators, i.e. to the operators $B_\mu(\mathbf{k}), B_\mu^\dagger(\mathbf{k})$ which diagonalize the operator $\hat{H}_e$, we obtain the exciton–phonon interaction operator, linear with respect to displacements of molecules, in the form

$$\hat{H}_{eL} = \sum_{\mathbf{k}\mu,\mu'}\sum_{\mathbf{q}r} F(\mathbf{k}+\mathbf{q},\mu;\mathbf{k},\mu';\mathbf{q}r) B_\mu^\dagger(\mathbf{k}+\mathbf{q}) B_{\mu'}(\mathbf{k}) \left(b_{\mathbf{q}r} + b_{-\mathbf{q}r}^\dagger\right). \qquad (3.155)$$

In the derivation of (3.155) we have neglected terms proportional to $B^\dagger B^\dagger$ and BB which describe the simultaneous creation and annihilation of two excitons.

The coefficients F which occur in (3.155) satisfy the condition

$$F(\mathbf{k}+\mathbf{q},\mu;\mathbf{k},\mu';\mathbf{q}r) = F^*(\mathbf{k},\mu';\mathbf{k}+\mathbf{q},\mu;-\mathbf{q}r), \qquad (3.156)$$

which results from the hermicity of the operator $\hat{H}_{eL}$. If the excitonic bands are sufficiently separated and the effects of "mixing" of excitonic bands, resulting from the exciton–phonon interaction, are weak, we can neglect in (3.155) the terms where $\mu \neq \mu'$. Then in the region of an isolated μth excitonic band we have

$$\hat{H}_{eL} = \sum_{\mathbf{k},\mathbf{q}r} F(\mathbf{k}+\mathbf{q},\mu;\mathbf{k},\mu;\mathbf{q}r) B_\mu^\dagger(\mathbf{k}+\mathbf{q}) B_\mu(\mathbf{k}) \left(b_{\mathbf{q}r} + b_{-\mathbf{q}r}^\dagger\right). \qquad (3.157)$$

The exciton–phonon interaction operator in the form (3.155) was used by several authors (see, for example, (16)–(18)).

The coefficients F in (3.155) can be expressed as first derivatives of the quantities $\mathcal{D}_f$ and M_{nm}^f with respect to the displacements of molecules from equilibrium. The operator (3.155), obtained by this approximation, was used in Ref. (18) by analysis of the conditions of applicability of a weak exciton–phonon interaction when the operator (3.155) can be considered as a small perturbation.

At the same time Frenkel (see for this argument also the monograph by Davydov (19)) discussed another limiting case, which is that of a strong exciton–phonon interaction. In the case of a strong exciton–phonon interaction we assume that the time of the excitation transfer from one molecule to an adjacent molecule is large compared to the time when displacements of molecules to new equilibrium positions take place. The displacements result from changes of interaction forces induced by excitation of a molecule. Therefore in the crystal there occurs a local deformation, which propagates through the crystal together with the excitation. The dependence of the equilibrium positions of molecules on the position of the excited molecule is such that the excitonic bandwidth diminishes. For such excitons the wavevector is no longer a "good" quantum number (a similar situation occurs in the theory of polarons (20)). In this case the motion of the exciton ("localized" in Frenkel terminology), similar to the motion of carriers in semiconductors with small carrier mobility (21); (22), takes the form of jumps.

For computing the wavefunction of a "localized" exciton the adiabatic approximation (see (20), §§ 28,29) can be used. The first step in this approximation consists of establishing the wavefunction χ and the corresponding eigenenergy U for the electronic subsystem assuming that the positions of atomic nuclei are fixed. Thus, denoting by r the set of electronic coordinates and by R the set of nuclear coordinates, we have $\chi \equiv \chi(R)$, $U \equiv U(R)$, i.e. the wavefunction χ and the energy U depend on R treated as parameters in this approximation.

The function $U(R)$ acts as a potential energy in the nuclei subsystem and the wavefunction for this subsystem is obtained from the equation

$$\left[\hat{T}_R + U(R)\right] \Phi(R) = E\, \Phi(R), \tag{3.158}$$

$\hat{T}_R$ being the nuclear kinetic energy operator.

Assuming that effects of mixing of molecular states, which can be caused by intermolecular interactions, are negligible, the wavefunction χ for the electronic subsystem and for a state, when only one molecule, say $\mathbf{n}\alpha$, is excited, can be approximately given in the form

$$\chi^f_{\mathbf{n}\alpha}(r) = \varphi^f_{\mathbf{n}\alpha} \prod_{\mathbf{m}\beta \neq \mathbf{n}\alpha} \varphi^0_{\mathbf{m}\beta}. \tag{3.159}$$

When one of the crystal molecules is excited, the equilibrium positions of crystal molecules are displaced. Thus the minimum of the function $U(R)$ does not correspond to equilibrium positions of crystal molecules, when all molecules are in the ground state. Thus, expanding the function $U(R)$ with respect to deviations from the new equilibrium and neglecting the terms of the order higher than two and introducing normal coordinates, corresponding to the vibrations of nuclei with respect to new equilibrium positions, we obtain for the eigenfunctions and eigenvalues of the operator $\hat{T}_R + U(R)$ the following system of equations

$$\Phi_N(R) \equiv \Phi_{\ldots N_\kappa \ldots}(R) = \prod_\kappa \Phi_{N_\kappa}\left(q_\kappa - q_{\kappa,\mathbf{n}\alpha}\right), \tag{3.160}$$

$$E^f_{\mathbf{n}\alpha,N} \equiv E_{\mathbf{n}\alpha,\ldots N_\kappa \ldots} = E^f_{\mathbf{n}\alpha,0} + \sum_\kappa \hbar\omega_\kappa \left(N_\kappa + \frac{1}{2}\right). \tag{3.161}$$

The function $\Phi_N(R)$ in (3.160) represents the wavefunction of a harmonic oscillator with the normal coordinate q_κ; $q_{\kappa,\mathbf{n}\alpha}$ is its equilibrium position corresponding to such a crystal state, when one molecule $\mathbf{n}\alpha$ is excited, and the quantum number $N_\kappa = 0, 1, 2, \ldots$ indicates the state of the oscillator κ.

The quantities $E^f_{\mathbf{n}\alpha,0}$ in (3.161), due to the translational symmetry, do not depend on the vector $\mathbf{n}$. In crystals where their symmetry operations contain an element which transforms the molecules of a unit cell into each other, the quantities $E^f_{\mathbf{n}\alpha,0}$ also do not depend on the index α. The wavefunction in the considered simple factorized approximation has the form

$$\Psi^f_{\mathbf{n}\alpha,N}(r, R) = \chi^f_{\mathbf{n}\alpha}(r) \Phi_{\ldots N_\kappa \ldots}(R). \tag{3.162}$$

The states (3.162) were obtained under the assumption that the resonant intermolecular interaction is small. Here we notice that attempts to improve the above method of computation of crystal states in the case of a strong exciton–phonon interaction has been made in a series of works (see (23)–(25)).

In particular, Rashba (23) has considered such crystal excited states, where the excitation is not centered at one molecule, as was assumed in (3.162), but is smeared out about a certain finite crystal region. In this case, when the time of the resonant excitation transfer from one molecule to another is small compared to the time needed by the molecule to achieve a new equilibrium position[21] a local deformation can occur within some excitation region; similar behavior is observed in the case of polarons (20). The shape of the deformation is consistent with the shape of the excitation distribution inside the same crystal region. If, in particular, the resonant interaction tends to zero, the states, obtained in Ref. (23), are identical with that given by formula (3.162).

Finally we remark that within the same excitonic band, if its width is sufficiently large, one can find both states, for which the exciton–phonon interaction is strong, and states where it is weak so that these states can be computed by means of perturbation theory (for a more detailed discussion of this problem see Ref. (18)). Only in the limit of very narrow excitonic bands do the excitonic states show the character of "localized" excitons, on which we have concentrated our attention. In all references which we have mentioned above the variational method was used, which gives only the lowest states in the excitonic band.

3.8 Spectra and mobility of self-trapped (ST) excitons

We return to the self-trappping (ST) of Frenkel excitons in 3D organic structures. As a result of the Franck–Condon principle, photo-exciting a crystal from its ground state, with a regular lattice, leads to the initial creation of a coherent

---

[21] As shown above, in such cases we can deal with free excitons, not connected with a local crystal deformation. Since all the states were obtained by the variational method, the physically realizable states are those which minimize the crystal energy. As shown in Ref. (23), in one-dimensional crystals with strong resonant intermolecular interaction the exciton–phonon interaction leads to a situation when the minimal energy of a crystal with exciton corresponds to a state with local deformation.

exciton. In order to pass into a ST state, the latter has to overcome a barrier (in a 3D lattice), which may be done either by thermoactivated tunneling or by a thermoactivated transition. A review of the relevant theory is given in Ref. (27). It was shown that the rate of ST of thermalized excitons can be represented as

$$W(T) = \omega_v B(T) \exp[-S(T)], \tag{3.163}$$

where ω_v is the characteristic phonon frequency, the pre-exponential factor $B(T)$ is always large in comparison with unity, and $S(T)$ is the temperature-dependent Hamiltonian action. At temperatures with $k_B T > \hbar \omega_v$, we have

$$S(T) = \frac{U}{k_B T}, \tag{3.164}$$

where U is the height of the barrier. Thus, as could be expected, the ST rate then follows an Arrhenius law.

The experimental investigations of ST in organic crystals mainly concern optical spectra, in particular time-resolved spectra. Good examples are the experiments by Matsui and coworkers reported in Refs. (28). To discuss these spectra, let us consider the energy $E_c(\eta)$ of the crystal in its ground and excited states as functions of the coordinate η that undergoes a strong displacement upon ST. We assume that the ground state energy has its minimum at $\eta = 0$ and that this value of η also represents the local minimum in the excited state, as is shown in Fig. 3.3. The absolute minimum corresponds to the ST state, so that the dependence of the total energy on η is described by an asymmetric double-well potential. Thus the coherent states are protected by a barrier; their optical spectra have to be analogous to the spectra in a regular crystal with weak exciton–phonon coupling. We can expect the existence of narrow zero-phonon lines in the absorption and fluorescence spectra and a Davydov splitting may be observed. The existence of ST states leads to the appearance of additional broad and red-shifted bands in the fluorescence spectra. Upon ST the movement of wavepackets is replaced by jumps. At high temperature these jumps are thermoactivated leading to the Arrhenius law equation for the diffusion constant. At low temperature quantum tunnelling occurs, in analogy to quantum diffusion of impurities in solids (see Ref. (29) for a review).

Numerous picosecond experiments have been performed on exciton ST in organic crystals (28). Such experiments give the possibility of studying the dynamics of the ST process and allow one to determine the height of the ST barrier and the rate of the transient free-exciton luminescence. For some crystals these investigations also gave the possibility of tracing the pathways to self-trapping (cf. Fig. 3.3).

An example is pyrene, which is a crystal with a rather strong exciton–phonon interaction and a barrier height $U \approx 262$ cm^{-1}. It was demonstrated that upon photogeneration of excitons in this crystal, even at low temperature, the process of self-trapping not always requires relaxation to the bottom of the free-exciton band (with $\mathbf{k} = \mathbf{0}$; black arrows in Fig. 3.3), but sometimes takes place directly

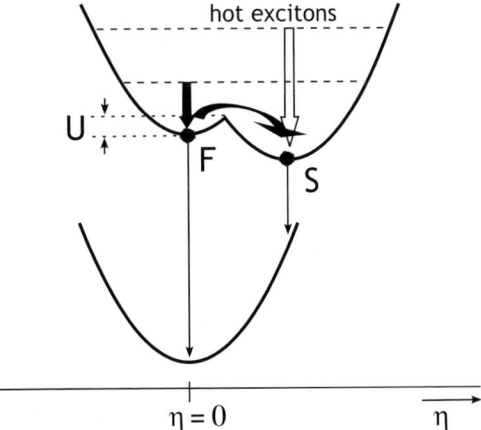

FIG. 3.3. Ground state potential and asymmetric double-well potential associated with the phenomenon of exciton self-trapping, as a function of the coordinate η that undergoes a strong displacement upon self-trapping. F is the bottom of the free-exciton band, in which the lattice is not distorted ($\eta = 0$), S denotes the lowest self-trapped exciton state, and U is the barrier height. The luminescence from the self-trapped state is red-shifted relative to the free-exciton luminescence. Upon photoexcitation of the system, two pathways towards the self-trapped state occur. The first possibility is that the created excitons first relax towards the bottom of the free-exciton well, after which they may further relax to the self-trapped state through tunneling or a thermoactivated process. This pathway is indicated by the filled arrows. The second possibility is that high-energy (hot) excitons relax directly to the self-trapped state, as indicated by the open arrow. Reprinted with permission from Knoester et al. (47). Copyright Elsevier (2003).

from the states with large **k** (hot excitons) avoiding the ST barrier, as shown by the open arrow in Fig. 3.3.

3.8.1 Mechanism of self-trapping of Frenkel excitons

First of all, let us explain why the ST of Frenkel excitons, which in contrast to electrons (holes) have no charge, may arise. It is useful to start from the limit of a narrow exciton band and strong exciton–phonon coupling. If Δ denotes the exciton bandwidth in the regular lattice, then $\tau_1 \approx \hbar/\Delta$ is the time-scale for the transfer of the excitation from one molecule to another. The second relevant time-scale is τ_d, which denotes the time necessary for molecules to be displaced to new equilibrium positions upon a change in the electronic state. Such changes into a new, locally deformed, lattice configuration arise from the fact that the intermolecular interactions (like the van der Waals interaction) are changed upon excitation of one of the molecules. In the limit of a narrow exciton band and strong exciton–lattice interaction, we have $\tau_1 \gg \tau_d$. If E_d is the energy of the local deformation (the analog of the polaron energy shift), then $\tau_d \approx \hbar/E_d$. It is also clear that τ_d is longer than the characteristic period of lattice vibration $2\pi/\omega_v$, where ω_v is a typical phonon frequency (in organic solids, $\omega_p \approx 100$ cm^{-1}). We thus have in the limit considered the condition $\tau_1 \gg \tau_d \geq 2\pi/\omega_v$. It follows that in this limit ($\Delta \ll \hbar\omega_v$) the phonons (displacements) are relatively fast, which allows them to follow the slowly moving excitons.

Thus, in the limit of narrow exciton bandwidth, the local deformation travels through the crystal following the molecular excitation. In other words, as has been metaphorically described by Frenkel, the exciton while moving through the lattice "drags with itself the entire load of atomic displacements". Thus, the already narrow exciton band is transformed into an even narrower band of "dressed" excitons. However, since the dissipative width of such states in most cases is large compared to the narrow bandwidth, the wavevector cannot be considered a "good" quantum number. The band picture of "dressed" excitons is destroyed and localized molecular excited states (dressed with a deformation) propagate by hops from one molecule to another. It is clear that here we meet the case of incoherent excitons, with energies that, in contrast to the case of coherent excitons, do not depend on the wavevector. (For the diffusion constant of such excitons see Ch. 14) It is useful to note that triplet excitons in organic solids have a rather small bandwidth (of the order of 10 cm^{-1}), implying that the occurrence of low-energy ST triplet states is very likely.

We mention here also an excimer (originally short for excited dimer) which is a type of small radius self-trapped exciton state. In organic solids this excited state can be considered as a dimeric molecule formed from two molecules, when one of the molecules is in an electronic excited state. Excimers are usually formed between two molecules that would not bond if both were in the ground state and the molecule pyrene is one of the canonical examples of an excimer. The excimer states play an important role in applications and in photochemistry (more information can be found in (30)).

We now turn to crystals with a wide exciton band, $\Delta \gg \hbar\omega_v$. Under this condition, the exciton system is fast and, thus, the usual adiabatic approximation may be used in the ST theory. This situation was first analyzed by Deigen and Pekar (31), who showed that in this case ST states of small radius $R_{ST} \approx a$ may be formed (a is the lattice constant). For localization on a scale $\Delta x \approx a$, the uncertainty of the wavevector is $\Delta k \approx 1/a$, which is of the order of magnitude of the Brillouin zone. If the localization (deformation) energy $E_d > \Delta$ (strong exciton–phonon coupling), a ST will be formed and its energy will lie below the bottom of exciton band. On the other hand, if $E_d < \Delta$, the excitons in the lowest-energy states remain coherent and ST states (if they arise at all) may appear only in the region of higher energies. More information on the ST of wide-band excitons may be found in the review paper by Rashba (32).

3.8.2 *Spectra and transport of self-trapped excitons*

The interaction of excitons with phonons governs the nature of the exciton motion. Coherent motion occurs for weak exciton–phonon coupling, while incoherent transport takes place for strong coupling. In the limit of strong interaction, the phonons not only affect the motion of the excitons, but they may also alter the exciton state itself considerably, to a so-called self-trapped exciton. The self-trapping (ST) of excitons is analogous to the ST of electrons and holes in ionic crystals, which also arises from the interaction of the quasiparticles with lattice vibrations. For electrons in ionic crystals the possibility of ST was first pointed out by Landau in 1933 (33). He showed that due to the electron–ion interaction the states of the electron "trapped by the lattice" (i.e. the states in which the lattice around an electron is strongly deformed) have an energy smaller than that of the Bloch states in the regular lattice. The word "polaron" was later introduced by Pekar to denote such trapped states; he also developed the first consistent theoretical treatment of the self-trapped electron state by considering the model of a large-radius local state (34).

The physics of exciton ST has many features in common with electron (hole) ST. Therefore, the methods and results of the theory of electron ST have been widely used in the development of the theory of exciton ST. The effects of strong exciton–phonon coupling in organic crystals and the possibility of exciton ST were discussed in the papers by Peierls (35), Frenkel (36), and Davydov (37).

More recent discussion of exciton ST and the synthesis of almost all available approaches may be found in the reviews published as Refs. (32), (38)–(40). The reader is referred to this literature for a detailed account of exciton ST. In the following subsections, we will restrict ourselves to a qualitative description of the most general and characteristic features of this phenomenon.

3.8.3 *Self-trapping barrier*

In his seminal paper on self-trapping (33), Landau already noticed that to make the transition from the coherent (free) state to the ST state, the particle has to overcome a barrier. In the case of small-radius excitons (Frenkel excitons) the existence of a barrier in crystals composed of large molecules may be stipulated by purely spatial limitations. The excited molecule tries to pass into a ST state with low energy, but the surrounding molecules, having no "wish" to be displaced or change their orientation, may prevent such passage. Thus, the coherent states are protected by a barrier, which makes them meta-stable and gives them the opportunity to exist during a finite time, even if the passage into the ST state is accompanied by a considerable energy gain. One then speaks of the coexistence of coherent and ST excitons. It should be stressed that this coexistence is not a general phenomenon and occurs only for three-dimensional (3D) systems. As was shown by Rashba (32), no barrier exists for excitons in 1D systems. In this case after initial excitation of a coherent exciton, a monotonic lowering of the total energy takes place towards the energy of the ST state. Thus, the coherent exciton states are absolutely unstable and the creation of ST states occurs without any restrictions. For polarons in 3D ionic crystals, the same physical picture was found earlier by Pekar (34).

3.8.4 Self-trapping of charge-transfer excitons

As a brief intermezzo, we mention some peculiarities of ST of charge-transfer excitons. The electron–hole interaction energy in a charge-transfer exciton is of the order of e^2/a, where a is the distance between electron and hole. This energy may be a few eV. The change of this energy under the influence of lattice vibrations may compete with the interaction of the free electron and hole with phonons and may even be the dominant mechanism of the exciton–phonon interaction. Such a situation is expected to take place for many organic crystals (see Refs. (41) and (42)). The theory of ST of charge-transfer excitons in such crystals can be found in Refs. (41), (43).

Another interesting situation arises in the ST of excitons in which the electron and hole are spatially separated and are localized on different filaments (polymers or quantum wires) or different planes (or quantum wells). In such structures the electron–hole Coulomb interaction changes when these filaments or planes are deformed. As a result a strong exciton–phonon interaction may exist, even if the individual quasiparticles (electron and hole) have very small interaction with the phonons. The theory of ST of this type of excitations may be found in Ref. (44).

3.8.5 Self-trapping in one-dimensional structures

To conclude the discussion of self-trapping Frenkel excitons we will make here a few remarks on the ST in one-dimensional structures. This problem for infinite chains was investigated for the first time by Rashba (26). This paper considered the model of a 1D crystal with the excitonic bands being much wider than the typical energy of phonons. Such a limiting case takes place, as is known, for many J-aggregates. Just for this case of "light" excitons and weak exciton–phonon coupling one can expect a large size of the self-trapped exciton states. For rather short chains, when the length of localization ℓ_0 is of order of the chain's length L, one can expect the quantum confinement of self-trapping states. The ST in linear chains of finite length has been considered in (45) and applied to the case of J-aggregates. The properties of J-aggregates have been investigated for a long time, but recently this interest has increased considerably in connection with new experimental results on the superradiance of excitonic states, its temperature dependence, and numerous applications in nanostructures (OLEDs, solar cells, and so on). It is known (see, for instance, the review papers by Knoester (46), (47)) that in J-aggregates with length $L \ll \lambda$ (λ is the optical wavelength) an overwhelming part of the total oscillator strength is concentrated in the lowest excitonic state and equals $F \cong 0.85\, f_0(L/a)$, where f_0 is the oscillator strength of a monomer and a is the lattice constant. This leads to superradiance from the lowest excitonic state and its domination in the absorption spectrum of the crystal and is correct only if disorder and exciton–phonon interaction are neglected. The self-trapping of excitons may destroy this simple picture so that it may take place only for chains with length L small compared to the length ℓ_0 of self-trapping. It was shown in (45) that for long enough chains the value of F does not increase with growth of L, but tends to the saturation limit $F_1 \cong 5f_0(\ell_0/2a)$. The oscillator strength of the next bright state also tends to the same limit with growth of L, and analogous relations hold true for the following bright states. Contrary to the case of an infinite chain, where only one self-trapped state exists, in chains of finite length several self-trapped states can arise, the number of which depends on the length of the chain. The discussion of the influence of quantum confinement and self-trapping on the superradiance of 1D molecular crystals can be found in the paper (45). The quantum confinement effect in this case can be expected for a chain length of the order (or less) of the length of self-trapping (see also (48), ((49)).

3.9 Electron–vibrational excited states in molecular crystals

In the previous considerations we have assumed that the Frenkel exciton is a pure electronic intramolecular excitation, which can propagate through the crystal, and considered phonons as elementary lattice excitations, corresponding to molecular degrees of freedom (displacements and rotations). The frequencies of this vibrations are of the order 10–100 cm^{-1} and are characteristic only for crystals. Therefore these are often named lattice vibrations, to distinguish them from oscillations of atoms inside the molecule, when the molecular center-of-mass is at rest and there is no rotation of the whole molecule. The frequencies of these intramolecular vibrations can exceed by an order of magnitude the phonon frequencies and achieve values of 500–2000 cm^{-1}. They show a weak dependence on the aggregation type. Thus absorption spectra of molecular crystals, as also absorption spectra of individual crystal molecules, contain both lines corresponding to pure electronic excitations and quite clear absorption lines corresponding to various combinations of electronic excitations with intramolecular excitations. The frequencies of these lines in the visible and ultraviolet parts of the spectrum are given approximately by the formula

$$\nu = \nu_{el}^f + n \cdot \nu_1 + m \cdot \nu_2 + \ldots; \qquad n, m = 0, 1, 2, \ldots, \qquad (3.165)$$

ν_{el}^f being the frequency of the pure electronic transition $0 \to f$ and $\nu_1, \nu_2, \ldots$ being the frequencies of intramolecular vibrations.

The inaccuracy of the approximation (3.165) is related to the fact that the frequencies of intramolecular vibrations depend strongly on the state of the electronic subsystem (which means that the frequencies ν_i depend on the state f; in particular, by electronic excitation of molecules of the aromatic series the frequencies ν_i change by a quantity of order 50 cm^{-1}). In addition, in (3.165) anharmonicity is neglected, being important for high values of the numbers $n, m, \ldots$.

Due to the large differences in the masses of electrons and nuclei it is convenient to use the adiabatic approximation by analysis of molecular spectra. In particular, this approximation enables one to explain specific properties of the relation (3.165). It also allows one to examine the intensity distribution of the above electronic–vibrational (vibronic) molecular states. As was first noticed by Simpson and Peterson (50) (see also the papers by McClure (51)), when molecules aggregate into crystals, their electronic–vibrational spectra change and new properties arise. We discuss the changes first for the case of an isolated molecule and then pass to the case of a crystal.

Considering an isolated molecule we can distinguish two subsystems: the slow subsystem, consisting of heavy nuclei, and the fast subsystem, consisting of light electrons. In the adiabatic approximation the fast subsystem follows the instantaneous configuration of the slow subsystem, which "feels" only an average configuration of the electronic cloud. Therefore in the adiabatic approximation the molecular wavefunction (r is the electron coordinate, and R the nucleus coordinates) can be taken in the form

$$\Psi_{f,n}(r,R) = \psi_f(r,R)\varphi_n^f(R), \tag{3.166}$$

where $\psi_f(r,R)$ is the electronic wavefunction of fth state, depending on the nuclei configuration R treated as the parameter, and $\varphi_n^f(R)$ is the wavefunction of the nucleus being in the state n, which does not depend on the instantaneous electronic configuration, but depends on electronic state.

Taking the molecular Hamiltonian in the form

$$\hat{H} = \hat{T}_R + \hat{V}(r,R), \tag{3.167}$$

where $\hat{T}_R$ is the nuclear kinetic energy, and $\hat{V}(r,R)$ is the nuclei interaction energy and the energy of electrons attracted by fixed nuclei, one can obtain the following equations for the functions displayed in (3.166)

$$\hat{V}(r,R)\psi_f(r,R) = V_f(R)\psi_f(r,R),$$
$$\left[\hat{T}_R + V_f(R)\right]\varphi_n^f(R) = E_{f,n}\varphi_n^f(R). \tag{3.168}$$

Here $V_f(R)$ is the eigenvalue of the operator $\hat{V}(r,R)$ and $E_{f,n}$ is the total energy of the molecule in the vibronic state (f,n). As follows from (3.168), $V_f(R)$ acts as the potential energy for the nuclei motion. So we obtain that the nuclei equilibrium positions depend on the state f of the electronic subsystem. It can be shown that the adiabatic approximation gives reasonable results when for given values R the distance between surfaces $V_f(R)$ (for simplicity, the state f is considered as non-degenerate) and $V_{f'}(R), f \neq f'$ is much larger than the frequencies of intramolecular vibrations. Only in this case can we assume that excitation of small intramolecular vibrations does not induce the transition $f \to$
$\to f'$.

The intensity of a dipole transition from the ground state ($f=0, n=0$) to an arbitrary vibronic state is given by the square of the dipole matrix element of the molecular dipole moment $\hat{\mathbf{p}}$. The dipole moment $\hat{\mathbf{p}}$ consists of two contributions,

$\hat{\mathbf{p}} = \hat{\mathbf{p}}_e + \hat{\mathbf{p}}_N$, where $\hat{\mathbf{p}}_e$ depends on the electronic coordinates, and $\hat{\mathbf{p}}_N$ on nuclear coordinates. By the orthogonality of the electronic wavefunctions we have

$$(\Psi_{f,n}, \hat{\mathbf{p}}_N \Psi_{00}) \equiv \int \Psi_{f,n}^* \hat{\mathbf{p}}_N \Psi_{00} dr dR = 0. \qquad (3.169)$$

In consequence

$$\hat{\mathbf{p}}_{0n}^{0f} \equiv (\Psi_{f,n}, \hat{\mathbf{p}} \Psi_{00}) = \int \varphi_n^{f*}(R) \varphi_0^0(R)\, \mathbf{p}_e^{0f}(R)\, dR, \qquad (3.170)$$

with

$$\mathbf{p}_e^{0f} = \int \psi_f^*(r, R)\, \hat{\mathbf{p}}_e \psi_0(r, R)\, dr. \qquad (3.171)$$

If the quantity (3.171) weakly depends on R, which is the case for sufficiently intense dipole transitions, it can be considered as a constant taken at a certain value $R = R_0$, R_0 being the molecular nuclei configuration at the ground state, and removed from integration in (3.170). Thus we obtain

$$(\Psi_{f,n}, \hat{\mathbf{p}} \Psi_{00}) = \mathbf{p}_e^{0f} \left(\varphi_n^f, \varphi_0^0 \right). \qquad (3.172)$$

The wavefunctions φ_n^f, φ_0^0 are, in general, not orthogonal, being eigenfunctions of different operators (for $f \neq 0$). But the functions φ_0^0 are normalized, so that

$$\sum_n \left| (\varphi_n^f, \varphi_0^0) \right|^2 = 1. \qquad (3.173)$$

In consequence, according to (3.170) and (3.172),

$$\sum_n \left| \mathbf{p}_{0n}^{0f} \right|^2 = \left| \mathbf{p}_e^{0f} \right|^2 \sum_n \left| (\varphi_n^f, \varphi_0^0) \right|^2 = \left| \mathbf{p}_e^{0f} \right|^2. \qquad (3.174)$$

The frequencies of intramolecular vibrations are approximately by order smaller than the frequencies of pure electronic transitions in the considered spectral region, and the quantities $\left| (\varphi_n^f, \varphi_0^0) \right|^2$ decrease rapidly with increasing n. Therefore one can read off from (3.174)[22] that the sum of oscillator strengths for vibronic transitions in this approximation equals the oscillator strength for pure electronic transitions by the assumption, that nuclear vibrations are neglected. Now the question arises, to what extent is this approximation valid for vibronic transitions in crystals. Further examination of this question requires a detailed study of the nature of excitonic states in molecular crystals, which appear when electronic states of the crystal interact with intramolecular vibrations.

It was shown in the above mentioned paper by Simpson and Peterson (50) (mainly by qualitative arguments) that by examination of the interactions between excitons and intramolecular vibrations in molecular crystals two limiting

---

[22]Note that the oscillator strength is proportional not only to the square of the dipole matrix element but also to the transition frequency.

cases can be considered. One of them, which in (50) was termed the case of weak coupling, appears when the time of resonant electronic excitation transfer to an adjacent molecule is much larger than the relaxation time of the molecule nuclei, which is needed by nuclei for transition to new equilibrium positions when molecular electrons are excited.

Let V be the width of an excitonic band, and $\delta \leq \hbar\nu$ be the width of the absorption line corresponding to a pure electronic transition in a molecule.[23] The time of electronic excitation transfer can be estimated as $\tau_e \simeq \hbar/V$, and the nuclear relaxation time $\tau_N \simeq \hbar/\delta$ so that the case of weak coupling (or weak intermolecular resonant interaction) occurs if

$$\frac{V}{\hbar\nu} \leq \frac{V}{\delta} \ll 1. \tag{3.175}$$

In this case, or at least in the zeroth approximation, the approximation of an oriented-gas is valid. In spite of the interaction of excitons with intramolecular vibrations the wavefunction of a crystal molecule in a vibronic state is given by eqn (3.166), i.e. it is the same as in the case of an isolated molecule. Therefore in this limiting case the sum of the oscillator strengths of vibronic transitions, as in the case of an isolated molecule, is equal to the oscillator strength of an electronic transition when molecular vibrations are taken into account. By including, for example, by means of the Heitler–London approximation, the exciton–intermolecular interactions (see Ch. 2), we can find excitonic bands corresponding to a given vibronic transition, the corresponding Davydov splitting, etc. Excitonic states, which now correspond to electronic–vibrational waves, similar to the case of neglecting the interaction of excitons with intramolecular vibrations, are determined inside every excitonic band by a wavevector and have proper effective mass. The main difference by computation of excitonic energies in this case consists of the fact that now the transition dipole matrix element is not given by $\mathbf{p}_e^{0f}$, but by the expression (3.172), which even for a transition with $n = 0$ differs from $\mathbf{p}_e^{0f}$ by the factor $\left(\varphi_0^f, \varphi_0^0\right) \neq 1$. The values of the quantities $\mathbf{p}_e^{0f}$ and $\left(\varphi_n^f, \varphi_0^0\right)$ are mostly determined by experiments on optical spectra of molecules in the gas phase.

Let us now assume that in a crystal an inequality opposite to (3.175) holds which means that the time of intermolecular excitonic transfer is smaller than the time of intramolecular configuration and frequency rearrangement. Using also in this case the adiabatic approximation we can consider not only the intramolecular electron motion, but also the energy transfer along the crystal as fast processes.[24] In this limiting case (which can be called the case of strong resonant intermolecular coupling) and in the adiabatic approximation the crystal wavefunction has the form

---

[23] Here ν corresponds to those intramolecular vibrations which most strongly interact with the electronic subsystem in the resonance region.

[24] This assumption is not valid for excitonic states with small excitonic group velocities.

$$\Psi_{in}(r, R) = \psi_i(r, R)\varphi_n^i(R), \tag{3.176}$$

where $\psi_i(r, R)$ are the wavefunctions of crystal electrons in state i, obtained by fixed nuclei positions R, and $\varphi_n^i(R)$ are the nuclear wavefunctions, describing the nuclear oscillations with respect to their equilibrium positions when the electronic subsystem is in state i. If the electronic wavefunction $\psi_i(r, R)$ is smoothed out over a crystal region which is much larger than the elementary cell (note that for $R = R_0$, R_0 being the nuclear positions of nuclei in an ideal lattice, the states $\psi_i(r, R)$ go over into excitonic steady states which are equally distributed over the whole crystal volume), the nuclei positions inside each crystal molecule are almost the same as their positions in an ideal lattice. If we neglect these small differences, i.e. if we assume that exciton–vibrational waves do not affect nuclear equilibrium positions and frequencies, we arrive at the model of independent electronic and vibronic oscillations.

Due to the translational symmetry to each intramolecular vibration there corresponds at least one optical branch in the lattice vibration spectrum (in practice, the number of branches is given by σr, where σ is the number of molecules per unit cell, and r is the multiplicity of degeneracy of intramolecular vibrations). Therefore in this approximation the excited states of the crystal are described by the wavefunction

$$\Psi_{\mathbf{k}\mu,\dots N_\kappa\dots}(r, R) = \Psi_{\mu\mathbf{k}}(r) \prod_\kappa \Phi_{N_\kappa}(q_\kappa), \tag{3.177}$$

where $\kappa \equiv (\mathbf{q}, \nu)$, $\mathbf{q}$ being the wavevector of an optical phonon, ν is the branch index; $\Phi_N(q)$ is the wavefunction of a harmonic oscillator in state N, ($N = 0, 1, 2, \dots$) (see also eqn (3.160, 3.161). The state (3.177) corresponds to the energy

$$E_{\mathbf{k}\mu,\dots N_\kappa\dots} = E_\mu(\mathbf{k}) + \sum_{\mathbf{q}\nu} \hbar\omega_\nu(\mathbf{q}) N_{\mathbf{q}\nu}, \tag{3.178}$$

so that if we neglect the dependence on the wavevector, the above expression coincides with the semiempirical formula (3.165).

In the above approximation excitons and optical phonons act as independent and uncorrelated quasiparticles. In contrast to the case of weak resonant coupling, the vibronic states are characterized by a set of wavevectors, corresponding to the excitonic wavevector and to the phonon wavevectors, and not by one value of the wavevector. Thus by excitation by light of these multiparticle states the conservation laws for the wavevector and for the energy have the form (note that in the ground state excitons and molecule oscillations are absent because of the low temperature T)

$$\mathbf{Q} = \mathbf{k} + \sum_\nu \mathbf{q}_\nu N_{\mathbf{q}\nu}, \qquad \hbar\omega = E_{\mathbf{k}\mu,\dots N_\kappa\dots} \tag{3.179}$$

and describe the decay of a photon into a number of quasiparticles, whereas the same laws in the case of the weak coupling and strong exciton–phonon interaction have the form

$$\mathbf{Q} = \mathbf{k}, \qquad \hbar\omega = E_\mu(\mathbf{Q}). \tag{3.180}$$

It follows from the above discussion that considering excitons and intramolecular phonons as independent particles is approximate. It is clear, in particular, that a sufficiently strong exciton–phonon interaction can create propagating bound states, when electronic and vibronic excitations are centered on the same molecule. These states correspond to the previously discussed weak resonant interaction case. But the existence of such states in the vibronic spectrum does not exclude the existence of free excitons and intramolecular phonons states. Both types of states can usually coexist in the vibronic spectrum, in analogy to the case of two interacting particles, where continuum states, corresponding to free particles, coexist with bound states.

Quantitative discussion of vibronic spectra can be found in the papers by Merrifield (52), Suna (53), and Rashba (54).

Merrifield (52) has shown that the vibronic spectra can be studied by using the following Hamiltonian

$$\hat{H} = \sum_{\mathbf{k}\mu} E_\mu(\mathbf{k}) B_\mu^\dagger(\mathbf{k}) B_\mu(\mathbf{k}) \tag{3.181}$$
$$+ \sum_{\mathbf{q}\nu} \hbar\omega_\nu(\mathbf{q}) a_\nu^\dagger(\mathbf{q}) a_\nu(\mathbf{q}) + \sum_{\mu\mathbf{k},\mu'\mathbf{k}'} \hat{V}_{\mathbf{k}\mathbf{k}'}^{\mu\mu'} B_\mu^\dagger(\mathbf{k}) B_{\mu'}(\mathbf{k}'),$$

where $E_\mu(\mathbf{k})$ is the energy of the exciton ($\mu\mathbf{k}$), $\hbar\omega_\nu(\mathbf{q})$ is the energy of the optical phonon ($\nu,\mathbf{q}$), $a_\nu^\dagger(\mathbf{q}), a_\nu(\mathbf{q})$ are Bose creation and annihilation operators of the phonon ($\nu\mathbf{k}$), $B_\mu^\dagger(\mathbf{k}), B_{\mu'}(\mathbf{k}')$ are Bose creation and annihilation operators of the exciton ($\mu\mathbf{k}$), and $\hat{V}_{\mathbf{k}\mathbf{k}'}^{\mu\mu'}$ is an operator of the exciton–phonon interaction containing contributions linear and quadratic in the operators $a^\dagger, a$.

It has also been shown by Merrifield (52) that linear terms in $\hat{V}_{\mathbf{k}\mathbf{k}'}^{\mu\mu'}$ appear due to displacements of nuclear equilibrium positions caused by electronic molecular excitation, whereas changes in vibration frequencies give rise to quadratic terms. By derivation of (3.181) the adiabatic approximation for isolated molecules has been applied so that (3.181) can be used for examination of vibronic states by an arbitrary intensity of resonant intermolecular interaction.

Including in (3.181) only terms linear with respect to the operators $a^\dagger, a$, as representative of the exciton–phonon interaction, Merrifield (52) has discussed the process of "dressing" of excitons into a "coat" of virtual optical phonons. Although Merrifield (52), by means of the variational method, has applied his approach to a one-dimensional chain, his results are qualitatively valid in the general case. In particular, he has shown that by increasing the exciton–phonon interaction the excitonic bandwidth decreases. It decreases even for the lowest vibronic state, where real phonons are absent.

In contrast to Merrifield (52), Suna (53) has considered not the lowest vibronic state, but also a state with one quantum of intramolecular vibration. For this state the Dyson equation for the excitonic Green's function has been obtained. However, in this equation the exact formula for the mass operator was replaced by an approximate expression, corresponding to a Feynmann graph for a one-phonon line. Having the Green's function, Suna has obtained new spectral branches.

The discussion, initiated by Simpson and Peterson (50), Merrifield (52), and Suna (53) has been continued by Rashba (54) (see also the paper by Philpott (55)). Rashba assumed that the main mechanism of the exciton–phonon interaction is the changing Δ_ν of the frequency of intramolecular vibrations, when the molecule is excited.

The frequencies of intramolecular vibrations in an electronically excited molecule are smaller than the corresponding frequencies when the molecule is in its ground state. Thus a transition of an electronic–vibrational excitation (when the exciton and the quantum of vibration are on the same molecule) into the state, when the electronic and vibrational excitations appear in different molecules, can be considered as a change of the potential energy of their relative motion.

Let us assume, for simplicity, that the excitonic bandwidth when the exciton–phonon interaction is not taken into account, is large compared to the νth phonon bandwidth. In this case the intermolecular phonon is slow and practically "rests" in the site, and the exciton moves in an attractive field induced by the phonon. Thus, if the quantity Δ_ν is not small compared to the excitonic bandwidth, the lower part of the energy spectrum is revealed as a band of one-particle states, corresponding to the joint motion of electro- and vibrational excitation with a quasimomentum $\mathbf{k}$.[25] In this case the electronic and the vibrational excitation comprise a bound state, supporting one another in a finite crystal region and resting a certain time at the same crystal lattice site.[26] Higher in the energy scale we find dissociated states, i.e. two-particle excitations corresponding to the independent motion of the electronic and the vibronic excitation. If the quantity Δ_ν is small compared with the excitonic bandwidth, the bound states may not appear, and the two-particle excitations comprise the whole spectrum.

As shown in Ref. (54), in all cases the approximation of a weak resonant coupling of Simpson and Peterson (50) allows us to establish correctly the position of the "center-of-mass" of the vibronic spectrum and, in addition, its integral intensity (see also Merrifield (56)). A comparison of the theoretical predictions from Ref. (50) and experimental results can be found in Ref. (57).

To conclude our qualitative discussion of properties of vibronic states in organic crystals we would like again to bring to the attention of the readers the book by Broude *et al.* (57) where the spectral properties of the simplest vibronic states (one exciton + one quantum of vibration localized on the same or on different molecules) were investigated theoretically as well as experimentally. The next step was done recently by Hoffmann and Soos (58) in a paper

---

[25] Creation of a propagating bound state is related to the fact that phonons can also propagate. However, the effective mass corresponding to the phonon motion is quite large so that the bands of one-particle states are narrow, and the respective Davydov splitting is small.

[26] If the depth of the "well", i.e. the quantity Δ_ν, is large compared to the excitonic bandwidth, then in such states excitons and phonons are located at the same site (50).

where the problem of vibronic states has been investigated theoretically within the Holstein model (59). The Holstein model gives the possibility to take into account also multiparticle states; however, it assumes a one-dimensional molecular chain. Each molecule has one vibrational and one electronic degree of freedom. Vibrationally, each molecule n has one effective configuration coordinate λ_n. The vibrational potential is $V_n^{\text{gr}} = \lambda_n^2$ in the electronic ground and $V_n^{\text{ex}} = (\lambda_n - g)^2$ in the excited state. Thus, the shift of equilibrium value of the vibrational coordinate is here taken into account. Below all energies are measured in units of the vibrational quantum $\hbar\omega$. The dimensionless exciton–phonon coupling constant g is related to the vibrational relaxation energy (Franck–Condon energy) of the excited molecule by $E_{\text{FC}} = g^2$. The creation and annihilation operators for vibrations in the potential V_n^{gr} are denoted by $b_n^\dagger$ and b_n. Electronically, molecule n can be either in the ground state or in the first excited state. Operators $a_n^\dagger$ are introduced to create an excitation at site n from the electronic ground state of the chain $|0^{el}\rangle$. In the model, used in the above paper, the quasiparticle $a_n^\dagger$ is an exciton localized in site n. The hopping integral J (in units of $\hbar\omega$) describes the nearest-neighbor transfer of the exciton. Using these definitions, the complete Holstein Hamiltonian for a Frenkel exciton (FE) can be written as

$$H = H_{\text{elec}}^{\text{FE}} + H^{\text{ph}} + H^{\text{FE-ph}},$$
$$H_{\text{elec}}^{\text{FE}} = J \sum_n (a_n^\dagger a_{n+1} + a_{n+1}^\dagger a_n),$$
$$H^{\text{ph}} = \sum_n b_n^\dagger b_n, \qquad (3.182)$$
$$H^{\text{FE-ph}} = \sum_n a_n^\dagger a_n \left[-g \left(b_n^\dagger + b_n \right) + g^2 \right].$$

Here, the last term $H^{\text{FE-ph}}$ couples linearly to the otherwise independent exciton and phonon systems. Thus, the Holstein Hamiltonian operates on states that generally consist of both exciton and phonon excitations. Just such states, if they contain at least one exciton, are called, as we explained in this section, vibronic states. The major goal of this article was to investigate the structure of phonon clouds for molecular crystals of current interest, in which the exciton–phonon coupling constant typically is in the order of 1. It was shown that the molecular vibron model with joint exciton–phonon configurations is justified only for strong exciton–phonon coupling $g \leq 1$ and the same as predicted by the weak resonance intermolecular interaction $J \ll 1$. This result coincides with the prediction of the so-called model of weak resonance coupling by Simpson and Peterson (50). As shown in Ref. (58), this regime is approximately realized for PTCDA and MePTCDI spectra ($J \approx 0.27$). For larger values of J, the effects of delocalized phonon clouds become significant. Thus, in these cases, an extended phonon cloud basis should be used for the calculation of the vibronic states.

3.10 Calculation of the exciton states in molecular crystals

In the experimental and theoretical investigation of the optical and electronic properties of molecular crystals, the linear polyacenes naphthalene, anthracene, tetracene, and pentacene form an important series because of the similarities of their molecular and crystal structures. Information which may be obtained experimentally from the crystals includes the energies and oscillator strengths of the exciton transitions, the splitting in energy of the exciton bands due to crystal fields, i.e. the factor group splitting (Davydov splitting), and the directional dispersion in the above quantities. The intensity of the lowest medium intensity transition (oscillator strength $f = 0.1$) is approximately the same for all members of the series. However, the observed factor group splitting of this transition on the (001) face increase with the size of the molecule. We describe below the methods of calculations of exciton bands in organic crystals and some results published by Schlosser and Philpott (60), (61). In these papers the reader can find the most complete results of calculations for the above mentioned crystals of linear polyacenes. To explain the main idea of the approach used in (60), (61) we have to recall that in resonance not only the dipole–dipole interaction, but also higher multipole interactions contribute strongly to the short-range intermolecular resonance interaction. However, when additional transitions and thus mixing of molecular configurations are included in the calculations, the problem quickly becomes intractable (60), (61). For this reason the authors use the molecular orbital (MO) method. In the MO technique, the interacting transition charge densities are expanded in terms of molecular wavefunctions which have to be known and the appropriate interatomic integrals must be evaluated. The calculated interactions include both dipole–dipole point interactions and other nondipole interactions. It was demonstrated in the quoted papers that, by appropriate scaling, the interactions calculated by the MO method may be separated into dipolar and nondipolar parts. Due to the short-range nature of the nondipolar interaction, the number of interactions necessary to calculate lattice sums is small. The convergence problem of the lattice sum of long-range dipolar interactions may be treated separately with a mathematical technique known as the Ewald method (62). Since in this case Schlosser and Philpott discussed systems which contain several coupled transitions of medium of strong intensity and there are different interaction sums associated with each transition, the second-quantized theory of molecular excitons described in this chapter was used by them to calculate exciton energies and oscillator strengths. Vibrational structure was included in the calculation provided the electronic and vibrational parts of the molecular wavefunction are separable, i.e. the Born–Oppenheimer approximation is valid.

Experimentally, crystals of polyacenes have been studied by absorption and reflection spectroscopy. The pioneering work by Clark and his students, using a microtome to deliberately cut specific crystal faces of organic crystals which were then studied by reflection spectroscopy, was very important. The corresponding absorption spectra have been derived by Kramers–Kronig analysis (63) (equations for the Kramers–Kronig transformation can be found in (64)). Their study of the second singlet of anthracene showed that a Davydov/factor group splitting could be of the order of eV, which is in good agreement with the expected value if we take into account the very large oscillator strength of the second singlet excited state in the anthracene molecule. Figures 3.4, 3.5 demonstrate typical

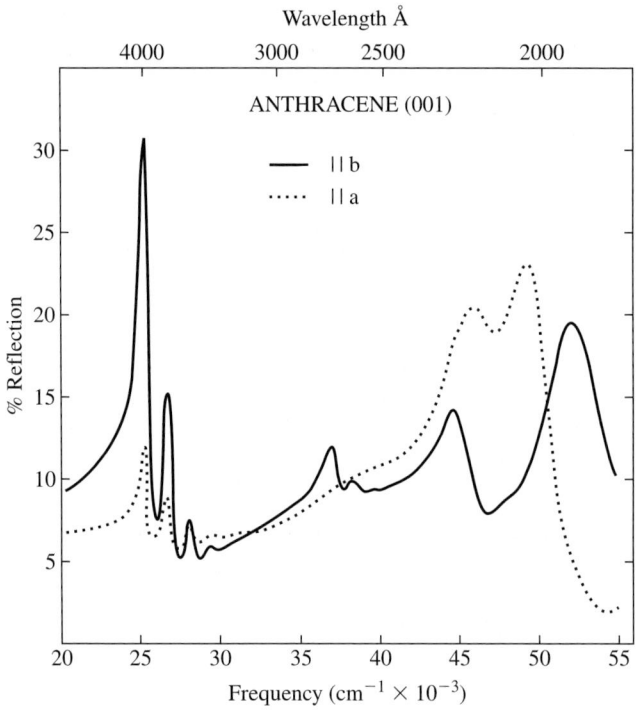

FIG. 3.4. Reflection spectra of the (001) face. The solid and dotted curves were obtained with the incident light polarized parallel to the **b** and **a** crystallographic axes, respectively. Reprinted with permission from Clark *et al.* (63). Copyright 1970, American Institute of Physics.

examples of such reflection and absorption spectra. Because of the large oscillator strengths of many of the transitions, absorption studies have been limited to thin crystals on the (001) face and weak or medium intensity transitions. The use of reflection spectroscopy and Kramers–Kronig analysis removes these limitations and, because crystals are generally available with several well-developed faces, allows the probing of the Brillouin zone in different directions. Reflection spectroscopy has been used to examine several faces of anthracene and the first singlet of tetracene on the (001) face (numerous references can be found in (60), (61)). Schlosser and Philpott (60), (61) carried out the examination of the role

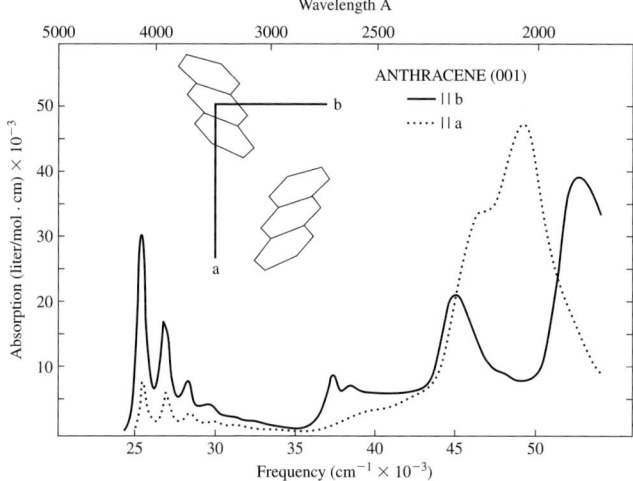

FIG. 3.5. Derived absorption spectra of the (001) face. The solid and dotted curves were obtained from the corresponding curves of Fig. 3.4 by Kramers–Kronig analyses. The projection of the two molecules of the unit cell onto the (001) plane is shown. Reprinted with permission from Clark *et al.* (63). Copyright 1970, American Institute of Physics.

of nondipolar interactions, crystal induced mixing of molecular configurations, and crystal structure on the exciton energies. They have also computed oscillator strengths and Davydov splittings of naphthalene, anthracene, tetracene, and pentacene. Table 3.1 contains crystallographic data.

Table 3.1 *Crystallographic data.*

	Naphthalene	Anthracene	Tetracene	Pentacene
system	monoclinic	monoclinic	triclinic	triclinic
a (Å)	8.235	8.561	7.90	7.90
b (Å)	6.003	6.036	6.03	6.06
c (Å)	8.658	11.163	13.53	16.01
α (°)	90.0	90.0	100.3	101.9
β (°)	122.92	124.7	113.2	112.6
γ (°)	90.0	90.0	86.3	85.8
space group	P2$_1$/a	P2$_1$/a	P$\bar{1}$	P$\bar{1}$
Ref.	(65)	(66)	(67)	(67)

Table 3.2 contains the summary of available free molecule data which were

used in calculations (60), (61) and which can be interesting in comparing with data for crystals. We have no possibility of going into details of the calculations,

Table 3.2 *Summary of free molecule data for naphthalene, anthracene, tetracene, and pentacene,* [a] *Vapor phase data;* [b] *No data available, the Franck–Condon factors of tetracene were used.*

Molecular transition	Naphthalene	Anthracene	Tetracene	Pentacene
I	35000+ +n1430 cm^{-1} ξ_n^2=0.168,0.280,0.249 0.187,0.118 [a] n=0,1,2,3,4 M axis,$\mu_{\rm I}$=0.54 Å	26000+ +n1400 cm^{-1} ξ_n^2=0.324,0.316,0.218 0.093,0.050 [a] n=0,1,2,3,4 M axis,$\mu_{\rm I}$=0.61 Å	21100+ +n1430 cm^{-1} ξ_n^2=0.271,0.327,0.209 0.134,0.060 [a] n=0,1,2,3,4 M axis,$\mu_{\rm I}$=0.69 Å	17100+ +n1350 cm^{-1} ξ_n^2=0.271,0.327,0.209 0.134,0.060 [b] n=0,1,2,3,4 M axis,$\mu_{\rm I}$=0.73 Å
II	45 350 cm^{-1} L axis, $\mu_{\rm II}$=1.56 Å	39 900 cm^{-1} L axis, $\mu_{\rm II}$=1.87 Å	34 000 cm^{-1} L or M axis, $\mu_{\rm II}$=0.52 Å	28 900 cm^{-1} L or M axis, $\mu_{\rm II}$=0.56 Å
III	52 500 cm^{-1} M axis,$\mu_{\rm III}$=0.41 Å	44 500 cm^{-1} M axis,$\mu_{\rm III}$=0.64 Å	36 500 cm^{-1} L axis,$\mu_{\rm III}$=2.1 Å	33 000 cm^{-1} L axis,$\mu_{\rm III}$=2.6 Å
IV	62 000 cm^{-1} M axis,$\mu_{\rm IV}$=0.38 Å	54 000 cm^{-1} M axis,$\mu_{\rm IV}$=0.83 Å	44 000 cm^{-1} L axis,$\mu_{\rm IV}$=0.57 Å	35 000 cm^{-1} L axis,$\mu_{\rm IV}$=0.8 Å
V	66 450 cm^{-1} M axis, $\mu_{\rm V}$=0.83 Å	64 500 cm^{-1} M axis, $\mu_{\rm V}$=0.83 Å	47 500 cm^{-1} L axis, $\mu_{\rm V}$=0.91 Å	44 000 cm^{-1} L axis, $\mu_{\rm V}$=0.7 Å
VI	62 000 cm^{-1}	54 000 cm^{-1}	44 000 cm^{-1} M axis,$\mu_{\rm VI}$=0.68 Å	35 000 cm^{-1}
Ref.	(68; 69)	(70)	(71; 72)	(73)

and below we will demonstrate only their results. First, however, it is useful to say a few words about the structure of the polyacenes consider as the most typical crystal, the crystal of anthracene.

The anthracene crystal is of monoclinic type. The elementary cell of this crystal is spanned by the primitive lattice vectors **a**, **b**, and **c**, where the vector **b** is oriented along the monoclinic axis, containing two molecules (cf. Fig. 3.6). The point group of the anthracene crystal is C_{2h}^5. The vectors **a** and **b** are orthogonal, and the vectors **a** and **c** form an angle such that $\mathbf{ac} = ac\cos 125°$. The lengths of the primitive vectors are: $a = 0.856$ nm, $b = 0.604$ nm, and $c = 1.116$ nm.

As given in Ch. 2, the position of the molecular center-of-mass within the elementary cell is determined by the vectors $\mathbf{r}_{n\alpha} = \mathbf{n} + \boldsymbol{\rho}_\alpha$, where **n** is an integer-valued lattice vector. In the case of anthracene

$$\rho = \begin{cases} 0, & \text{if } \alpha = 1; \\ (\mathbf{a}+\mathbf{b})/2, & \text{if } \alpha = 2. \end{cases} \quad (3.183)$$

The most interesting in anthracene are excitonic states, which appear in the

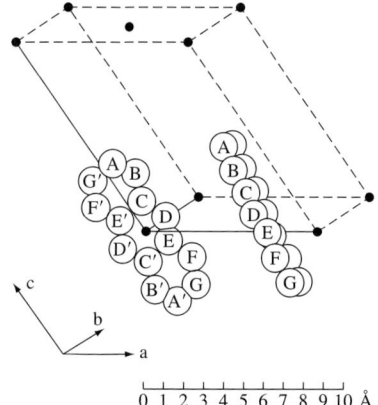

FIG. 3.6. The structure of the primitive cell in the anthracene crystal.

Table 3.3 *The angles formed by the mean and the longitudinal axis with the primitive cell vectors* $\mathbf{a}, \mathbf{b}$ *and* $\mathbf{c}' = \mathbf{a} \times \mathbf{b}$ *for the case of anthracene molecule.*

	M	L
a	71.3	119.7
b	26.6	97.0
c′	71.8	30.6

region of the two lowest singlet molecular excitations f_1 and f_2, $\varepsilon_{f_1} \simeq 26\,500$ cm^{-1}, $\varepsilon_{f_2} \simeq 40\,000$ cm^{-1}; transitions to them from the ground state are dipole allowed. The matrix element of the dipole moment $\mathbf{p}^{0f_1}$ is located in the plane of the anthracene molecules and is oriented along the mean axis M. The matrix element $\mathbf{p}^{0f_2}$ is located in the same plane, but is oriented along the longitudinal axis L. The angles formed by the mean and the longitudinal axis with the primitive cell vectors $\mathbf{a}, \mathbf{b}$ and $\mathbf{c}' = \mathbf{a} \times \mathbf{b}$ are, for the case of molecule $\alpha = 1$, displayed in Table 3.3. Orientations in the molecule $\alpha = 2$ are obtained from that of the molecule $\alpha = 1$ by reflection with respect to the plane formed by the vectors $\mathbf{a}$ and $\mathbf{c}$.

Calculations of the lowest excitonic states in anthracene were performed by Silbey *et al.* (74), Craig (12), Fox and Yatsiv (75), Davydov and Sheka (76), and Claxton (77) within the Heitler–London approximation. In the papers by Silbey *et al.* (74), Craig (12), Fox and Yatsiv (75), and Claxton (77), only the states $\mathbf{k} \simeq 0$ were considered, and the mixing of molecular configurations has been taken into account. Davydov and Sheka (76) have neglected the mixing of molecular configurations, but for some directions of the wavevector they obtained the form of excitonic bands by means of the Heitler–London approximation.

Silbey et al. (74) and Claxton (77) have computed the contribution from higher multipoles into short-range intermolecular interactions and, in addition, they discussed the role of retardation. According to this we notice that, concerning the positions of resonance excitonic absorption lines, it is not correct to include the retardation by computing the resonance frequencies. By definition the excitonic absorption lines correspond to frequencies for which, when the dissipation is neglected, the refraction index $n(\omega)$ becomes infinite. We will show that the poles in $n(\omega)$ (see also (1) and (78), §12 and §14) appear when $\omega = E_\mu(\mathbf{Q})/\hbar$, and where $E_\mu(\mathbf{Q})$ is the excitonic energy obtained by neglecting the retardation ($\mathbf{Q}$ is here the light wavevector).

As explained previously, the so-called polarization ratio, being the relation of the oscillator strengths of different bands of the Davydov splitting, is an important characteristic of excitonic spectra. In the case of anthracene, where one of the multiplets is always polarized along the **b**-axis (representation A_u) and the other perpendicular to it (representation B_u), the following expression describes the polarization ratio[27]

$$P\left(\frac{b}{a}\right) = \left[\frac{P_b(0, \rho \equiv A_u)}{P_a(0, \rho \equiv B_u)}\right]^2, \tag{3.184}$$

$\mathbf{P}(0, \rho \equiv A_u)$ and $\mathbf{P}(0, \rho \equiv B_u)$ being the dipole moment matrix elements defined by (3.138), and

$$P_b(0, \rho) = \frac{\mathbf{P}(0,\rho)\mathbf{b}}{|\mathbf{b}|}, \qquad P_a(0, \rho) = \frac{\mathbf{P}(0,\rho)\mathbf{a}}{|\mathbf{a}|}, \tag{3.185}$$

where **a** and **b** are the primitive vectors of the anthracene crystal lattice (see Fig. 3.6). In the Heitler–London approximation, when the relation (3.142) holds,

$$\mathbf{P}(0, \rho \equiv A_u) = \frac{1}{\sqrt{2}}\left(\mathbf{p}_1^{0f} - \mathbf{p}_2^{0f}\right),$$
$$\mathbf{P}(0, \rho \equiv B_u) = \frac{1}{\sqrt{2}}\left(\mathbf{p}_1^{0f} + \mathbf{p}_2^{0f}\right). \tag{3.186}$$

In this approximation, making use of the data from Table 3.3, we find for $f = f_1$

$$P\left(\frac{b}{a}\right) = \left(\frac{\cos 26.6^0}{\cos 71.3^0}\right)^2 \simeq 7.8. \tag{3.187}$$

Note also that when the mixing of molecular configurations is not taken into account, the above polarization ratio should be valid for all Davydov doublets, corresponding to vibronic molecular states. However, as was found experimentally by Brodin and Marisova (79), these ratios are different for different doublets

---

[27]Because of the small value of the Davydov splitting, the quantity $P(b/a)$ for lower excitonic bands practically coincides with the oscillator strength ratio.

and, moreover, their values are different from the value 7.8, which is appropriate for the oriented-gas model. We present in Table 3.4 the results of experiments by Brodin and Marisova and Wolf (79), (80) for the polarization ratio $P(b/a)$ and the Davydov splitting Δ_D in anthracene for the lower excitonic bands and for $\mathbf{k} \parallel \mathbf{c'}$.

Table 3.4 *The results of experiments by Brodin and Marisova and Wolf (79), (80) for the polarization ratio $P(b/a)$ and the Davydov splitting Δ_D in anthracene for the lower excitonic bands.*

Bands	Δ_D	$P(b/a)$
0–0	220 (190)	5
0–1	141 (150)	4.5
0–2	58 (80)	3
0–3		

The mixing of molecular configurations changes substantially both the Davydov splitting and the polarization ratio. Let us now go on to discuss the results of the calculation performed in (60), (61).

3.10.1 Anthracene and naphthalene

The crystal structures of naphthalene and anthracene are identical except for the length of the unit cell in the **c**-direction. Both crystals are monoclinic and the two molecules in the unit cell are related by a glide plane perpendicular to the **b**-axis. The energies and oscillator strengths are calculated for directions of **k** which are perpendicular to the observed crystal faces and are to be compared with values obtained from the normal incidence absorption or reflection spectra from those faces. The most prominent crystal face is (001); other -naturally occurring faces are (201), (101) and (110). On faces which include the **b** crystal axis, the principal polarization directions of the exciton branches are determined by symmetry to be either parallel to or perpendicular to the **b**-axis and the analysis of the experimental spectra is simplified. The (010) face is chosen because it includes all exciton branches polarized perpendicular to **b**. The face is unique in that the exciton branches polarized parallel to **b** are impossible to excite with photons at normal incidence and therefore have zero intensity. The calculated exciton energies and oscillator strengths taking account of dipolar as well as nondipolar interactions of anthracene are collected in Table 3.5. Analogous results for naphthalene can be found in (60), (61).

Results are given for $k = 0$ and several different directions of **k**. For purposes of comparison, the values calculated using the dipole approximation for all interactions with **k** perpendicular to the face (001) are also given in Table 3.5. Below, to give an impression of the state of art, we mention in some detail, for

Table 3.5 *Exciton energies* (cm^{-1}) *and oscillator strengths for the* $(0,0,1)$, $(2,0,\bar{1})$, *and* (010) *faces of anthracene.* [a] *Dipole approximation.* [b] *MO method.*

Molecular transition		(001) [a]		(001) [b]		$(\bar{1}10)$ [b]	(010) [b]	
		ν	f	ν	f	ν	ν	f
I_{0-0}	$\perp b$	25625	0.064	25691	0.049	26093	25681	0.042
	$\|b$	25266	0.276	25456	0.248	25648	26185	0.032
	$\Delta\nu$	359		235		445	503	
I_{0-1}	$\perp b$	27139	0.033	27174	0.028	27497	27171	0.023
	$\|b$	27014	0.080	27074	0.094	27154	27619	0.048
	$\Delta\nu$	126		99		343	488	
I_{0-2}	$\perp b$	28639	0.019	28660	0.017	28869	28659	0.013
	$\|b$	28573	0.045	28607	0.053	28650	28976	0.050
	$\Delta\nu$	66		53		219	317	
I_{0-3}	$\perp b$	30130	0.009	30140	0.008	30228	30140	0.002
	$\|b$	30102	0.021	30117	0.024	30136	30277	0.025
	$\Delta\nu$	20		22		92	137	
I_{0-4}	$\perp b$	31560	0.006	31566	0.005	31613	31565	0.004
	$\|b$	31542	0.013	31553	0.015	31563	31638	0.013
	$\Delta\nu$	17		13		50	72	
II	$\perp b$	49047	0.665	48338	0.733	37143	37142	4.320
	$\|b$	36760	0.008	36594	0.018	36630	36650	0.006
	$\Delta\nu$	12887		11744		513	492	
III	$\perp b$	42621	0.485	43290	0.507	45047	44068	0.085
	$\|b$	43164	0.812	43145	0.819	43877	45407	0.142
	$\Delta\nu$	-542		146		1170	1339	
IV	$\perp b$	56745	0.186	56157	0.013	54799	53264	0.185
	$\|b$	52137	0.946	52158	0.933	52922	55507	0.225
	$\Delta\nu$	4608		3999		1806	2243	
V	$\perp b$	65174	0.002	65045	0.006	65305	63733	0.081
	$\|b$	63008	0.719	63011	0.718	63546	66383	0.221
	$\Delta\nu$	2166		2034		1759	2659	

crystal faces (001) and (201), the results of calculations and their comparison with experimental data only for anthracene and for the lowest energy exciton state. Other information can be obtained directly from (60), (61). In the dipole approximation, the calculated factor group splitting of the transition I_{0-0} is 359 cm^{-1} on face (001). Using the MO method, the splitting is reduced to 235 cm^{-1}. The observed room temperature splitting equals 190 cm^{-1}. Inclusion of the nondipolar part of the Coulomb interaction brings the calculated factor group splittings of the transition I_{0-0} on face (001) into better agreement with experiment. On the face (001) the experimental polarization ratio with the value 4.7

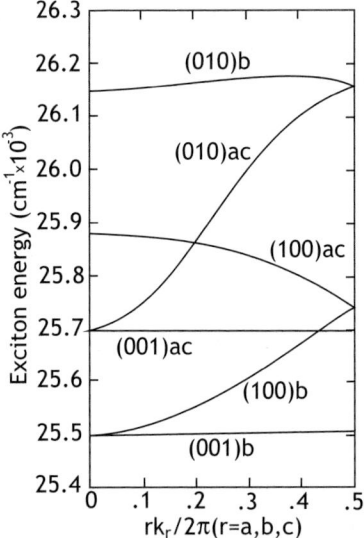

FIG. 3.7. Energy of the two exciton branches of the 0-0 component of the 3800 Å transition for **k** normal to faces (001), (010), and (100). Reprinted with permission from Philpott (81). Copyright 1971, American Institute of Physics.

for the transition I_{0-0} is in good agreement with the predicted polarization ratios of 5.1 using the MO method and 4.3 in the dipole approximation. As follows from comparison with experiment, the oscillator strengths of transition I are overestimated by about a factor of 2 using the MO method.

The results of calculations of exciton bands in the anthracene crystal in the dipole–dipole approximation are shown on Fig. 3.7. The letters a, b, c on the Fig. 3.7 indicate the orientation of the transition dipole moment. The results of calculations of exciton bands for the 2500-Å electronic transition can also be found in the quoted paper.

3.10.2 *Tetracene and pentacene*

Crystals of tetracene and pentacene are triclinic, with two molecules per unit cell. The dimensions of the unit cell in the directions of the **a** and **b** axes are similar to those of naphthalene and anthracene; the dimension in the direction of the **c**-axis increases proportionally to the number of benzene rings in the molecule. In all four crystal structures the long axes of both molecules in the unit cell are approximately parallel to the **c** crystal axis. However, in the triclinic structures of tetracene and pentacene, the only crystal symmetry operation is inversion. The results of calculation for tetracene crystals are shown in Table 3.6. Analogous results for pentacene can be found in (60), (61). The experimental data of the splitting cm^{-1} for the I_{0-0} transition in the tetracene crystal are in the interval 600–700 cm^{-1}. For the I_{0-0} transition in a crystal of pentacene the measured value of the splitting is 1030 cm^{-1} which means poor agreement with results of calculations (the references and experimental data for higher transitions including polarization ratios can be found in (60), (61)). The comparison of the results of calculations with experiments demonstrates the importance of mixing of molecular configurations and also that the inclusion of higher transitions tends to increase the splitting of the lowest transitions. However, as mentioned in the quoted papers, the lack of reliable experimental free molecule data serving as a basis of the calculations creates difficulties, particularly for the pentacene crystal.

Finally, some remarks to conclude this section. The first one is on the role of charge transfer states in calculations of energies and wavefunction of Frenkel excitons. Considering strong excited crystal states as, for example, those related to the second electronic transition in the anthracene molecule, the mixing effects with crystal excited states, where the neutrality of some molecules is violated, can be, in general, relevant. The importance of such states for molecular crystals was first emphasized by Lyons (82) (see also the papers (83)–(86) in this context). However, as it has been shown in (74) for the case of singlet excitons which can be excited in the dipole approximation, that the role of these states (i.e. states with charge transfer; see also Section 3.4) is not very significant.

The next remark concerns the structure of wavefunctions which have been used in calculations. In the above discussion the crystal was considered as infinite, so that by computing the exciton energies the interactions with molecules placed at arbitrary distance (long-range interaction) have been taken formally into account. However, in real crystals such interactions are not very important, since excitons, by virtue of scattering processes, must be considered as wavepackets of dimensions, resulting from the uncertainty principle, i.e. proportional to the inverse uncertainty of the excitonic wavevector. This circumstance excludes interactions of very distant molecules and makes excitonic properties in some degree independent of the crystal dimension; it also removes the nonanalytic dependence of the exciton energy on the wavevector in the limit $\mathbf{k} \to 0$. A quantitative theory of all these effects is a future task. However, it is quite clear that if the intensity of scattering processes depends on temperature, so the dimensions of the wavepackets, and in consequence the related energies, must also depend on temperature.

3.11 Exact transformation from paulions to bosons

First let us remark that the replacement of Pauli operators by Bose operators applied in this chapter is only approximate since the occupation numbers of paulions can be either 0 or 1, whereas the occupation numbers for bosons can take all nonnegative integer values 0,1,2,....[28] Therefore the replacement of the operators P_s and $P_s^\dagger$ by Bose operators can provoke uncontrolled errors in all

---

[28]The results of this section can also be applied in the theory of magnetism (see, for example, the papers by Goldhirsh et al. (87), by Goldhirsh and Yakhot (88) and the supplement by Yu. Rudoy and Yu. Tserkovnikov in Tyablikov's book (89)).

Table 3.6 *Exciton energies* (cm^{-1}) *and oscillator strengths for the (0,0,1) face of tetracene.*

Molecular transition		Dipole approx. 6-state model II L			MO method 9-state model II M		
		ν	$f_{\|b}$	$f_{\perp b}$	ν	$f_{\|b}$	$f_{\perp b}$
I_{0-0}		19877	0.394	0.004	20116	0.538	0.002
		20736	0.003	0.040	20701	0.000	0.052
	$\Delta\nu$	859			584		
I_{0-1}		21939	0.070	0.001	21997	0.147	0.001
		22229	0.002	0.024	22208	0.000	0.029
	$\Delta\nu$	290			211		
I_{0-2}		23625	0.036	0.000	23673	0.084	0.000
		23795	0.001	0.012	23784	0.000	0.015
	$\Delta\nu$	170			111		
I_{0-3}		25100	0.014	0.000	25202	0.062	0.000
		25287	0.001	0.009	25279	0.000	0.010
	$\Delta\nu$	187			77		
I_{0-4}		26773	0.002	0.003	26727	0.036	0.000
		26792	0.020	0.001	26769	0.000	0.005
	$\Delta\nu$	19			42		
	$\Delta\nu$	187			77		
II		34129	0.001	0.000	33330	0.012	0.104
		34184	0.000	0.000	33664	0.212	0.005
	$\Delta\nu$	55			334		
III		25924	0.015	0.000	31167	0.556	0.023
		63740	0.015	0.230	54420	0.033	0.006
	$\Delta\nu$	37816			23253		
IV		42364	1.420	0.054	40954	1.760	0.038
		43009	0.050	0.211	42520	0.021	0.340
V		45531	0.147	0.286	43996	0.028	0.024
		46191	0.047	0.017	45979	0.130	0.163
VI		52365	0.194	0.074	50930	0.096	0.067
		52859	0.044	0.025	52691	0.120	0.090

the situations where the number of bosons is larger than one. These errors are called "contributions of nonphysical states" (see, for example, (89)).

We can, however, improve the transition from Pauli operators to Bose operators B_s and $B_s^\dagger$ requiring that for an arbitrary number of bosons the number of paulions will equal 0 or 1 (90). To this aim we write the Pauli operators in the form

$$P_s = \left(\sum_{\nu=0}^{\infty} a_\nu B_s^{\dagger\nu} B_s^\nu\right)^{1/2} B_s, \qquad P_s^\dagger = B_s^\dagger \left(\sum_{\nu=0}^{\infty} a_\nu B_s^{\dagger\nu} B_s^\nu\right)^{1/2}, \qquad (3.188)$$

a_ν being real coefficients. We require that the operators $P_s, P_s^\dagger$ satisfy the relation

$$P_s P_s^\dagger + P_s^\dagger P_s = 1, \qquad (3.189)$$

if B_s and $B_s^\dagger$ in (3.188) are Bose operators. Substituting (3.188) into (3.189) and using the identity

$$B_s^{\dagger(\nu+1)} B_s^{(\nu+1)} = \left(\hat{N}_s - \nu\right) B_s^{\dagger\nu} B_s^\nu,$$

with $\hat{N}_s = B_s^\dagger B_s$, we obtain from (3.189) that

$$P_s P_s^\dagger + P_s^\dagger P_s = \sum_{\nu=0}^{\infty} a_\nu \left[2 B_s^{\dagger(\nu+1)} B_s^{(\nu+1)} + (\nu+1) B_s^{\dagger\nu} B_s^\nu\right] = 1,$$

if

$$a_\nu = \frac{-2}{\nu+1} a_{\nu-1}, \qquad a_0 = 1, \qquad (3.190)$$

$$a_\nu = \frac{(-2)^\nu}{(1+\nu)!}. \qquad (3.191)$$

So we obtained the exact transition from Pauli to Bose operators in the form

$$P_s = \left[\sum_{\nu=0}^{\infty} \frac{(-2)^\nu}{(1+\nu)!} B_s^{\dagger\nu} B_s^\nu\right]^{1/2} B_s,$$

$$P_s^\dagger = B_s^\dagger \left[\sum_{\nu=0}^{\infty} \frac{(-2)^\nu}{(1+\nu)!} B_s^{\dagger\nu} B_s^\nu\right]^{1/2}. \qquad (3.192)$$

The occupation number operator for paulions $\hat{L}_s = P_s^\dagger P^s$ is thereby expressed by the occupation number operator for bosons by the relation

$$\hat{L}_s = P_s^\dagger P_s = \hat{N}_s + \sum_{\nu=1}^{\infty} \frac{(-2)^\nu}{(1+\nu)!} \hat{N}_s \left(\hat{N}_s - 1\right) \ldots \left(\hat{N}_s - \nu\right). \qquad (3.193)$$

We can easily verify that $L_s = 0$ corresponds to states with an even number of bosons, whereas $L_s = 1$ corresponds to states with an odd number of bosons. Thus the transformations (3.192) and (3.193) do not create states with "unphysical" number of paulions (i.e. numbers $L_s > 1$).

We can also verify that from (3.192) it follows that

$$P_s^2 = \left(P_s^\dagger\right)^2 = 0.$$

To obtain this we note that the operator

$$\left[\sum_{\nu=0}^\infty \frac{(-2)^\nu}{(1+\nu)!} B_s^{\dagger\nu} B_s^\nu\right]^{1/2}$$

acting on states corresponding to odd numbers of bosons gives the value zero. Therefore the result of applying the operator

$$P_s^2 = \left[\sum_{\nu=0}^\infty \frac{(-2)^\nu}{(1+\nu)!} B_s^{\dagger\nu} B_s^\nu\right]^{1/2} B_s \left[\sum_{\nu=0}^\infty \frac{(-2)^\nu}{(1+\nu)!} B_s^{\dagger\nu} B_s^\nu\right]^{1/2} B_s$$

to an arbitrary boson state also gives zero, which follows from the structure of the operator P_s^2. In a similar way we can prove that $\left(P_s^\dagger\right)^2 = 0$.

If in (3.192) we restrict the summation to the first term (with $\nu = 0$), then we obtain $P_s = B_s$, $\hat{L}_s = \hat{N}_s$, i.e. the approximation used in the previous subsection.

Taking into account also the term with $\nu = 1$, we have

$$P_s = \sqrt{1-\hat{N}_s}\, B_s, \qquad P_s^\dagger = B_s^\dagger \sqrt{1-\hat{N}_s}, \qquad (3.194)$$

and hence obtain the well-known Holstein–Primakoff transformation (91). Here we have

$$P_s P_s^\dagger + P_s^\dagger P_s = 1 - \hat{N}_s \left(\hat{N}_s - 1\right), \qquad (3.195)$$

so that the r.h.s. of (3.195) is equal to unity if the numbers of bosons are restricted to the values 0,1. Thus if in crystals many elementary excitations can occur, these uncontrolled errors are possible.

By application of the exact representation (3.192) the terms in the sum on the r.h.s. with $\nu \geq 1$ can be considered as small operators because the smallness increases with increasing ν. Indeed, for Bose operators we have

$$B_s^{\dagger\nu} B_s^\nu = \hat{N}_s \left(\hat{N}_s - 1\right) \ldots \left(\hat{N}_s - \nu + 1\right). \qquad (3.196)$$

Consequently, the operator (3.196) vanishes identically for the class of functions which correspond to boson numbers $N_s < \nu$. By increasing ν this class of functions becomes larger. Therefore the square root of $\sum_\nu$ in (3.192) can be rewritten as a power series

$$\sum_\nu b_\nu B_s^{\dagger\nu} B_s^\nu.$$

To obtain the coefficients b_ν we make use of the fact that in the representation of occupation numbers for bosons for arbitrary integers $N_s \geq 0$ the following relation should be fulfilled (see also eqn 3.196)

$$\left[\sum_\nu a_\nu N_s (N_s - 1) \ldots (N_s - \nu + 1)\right]^{1/2}$$
$$= \sum_\nu b_\nu N_s (N_s - 1) \ldots (N_s - \nu + 1).$$

Taking, in the above relation, $N_s = 0$ and making use of (3.191) we obtain $b_0 = 1$. Inserting $N_s = 1$ we obtain $b_1 = -1$. In a similar fashion, inserting $N_s = 2$, we obtain $b_2 = \frac{1}{2}\left(1 + \frac{\sqrt{3}}{3}\right)$, etc.

Having determined the coefficients b_ν we can put the transformation (3.192) into the form

$$P_s = \left[\sum_{\nu=0}^\infty b_\nu B_s^{\dagger\nu} B_s^\nu\right] B_s, \qquad P_s^\dagger = B_s^\dagger \left[\sum_{\nu=0}^\infty b_\nu B_s^{\dagger\nu} B_s^\nu\right]. \qquad (3.197)$$

It is interesting to note that by taking in the above expansions only the terms with $\nu = 0$ and $\nu = 1$, we obtain the relation

$$P_s = \left(1 - \hat{N}_s\right) B_s, \qquad P_s^\dagger = B_s^\dagger \left(1 - \hat{N}_s\right), \qquad (3.198)$$

which differs from the expansion of the Holstein–Primakoff transformation (3.194) into powers with respect to $\hat{N}_s$:

$$P_s = \left(1 - \frac{1}{2}\hat{N}_s\right) B_s, \qquad P_s^\dagger = B_s^\dagger \left(1 - \frac{1}{2}\hat{N}_s\right). \qquad (3.199)$$

The difference is a result of inaccuracy of the latter expansion, where the neglected terms in the state $N_s = 1$ are nonzero, whereas all ignored terms in (3.198) for $N_s = 1$ are identically equal to zero.

Substituting the expansion (3.197) into (3.19) and (3.29) we obtained the desired expansion of the Hamiltonian in powers with respect to Bose operators, when not only the dynamic, but also the kinetic interaction is taken into account. The new anharmonicity terms do not contain kinematic corrections. The role of this anharmonicity in the theory of third order nonlinear optical effects has been discussed in the article by Ovander (92).

The kinematic interaction appears first in terms of fourth order. It can be considered in the theory of nonlinear optical effects in a quite similar fashion as has been done in the article by Toshich (93), who considered nonlinear optical effects of fourth order in the excitonic spectral region. The kinematic interaction of polaritons in organic microcavities will be mentioned in Ch. 10.

3.12 Kinematic biexcitons

The problem of electronic biexcitons (excitonic molecules) in molecular crystals has been repeatedly discussed (see Ref. (97) and references therein). Such attention is motivated by the important role which, as in the case of semiconductors

(39), Frenkel biexcitons could act in the processes of photoluminescence, hyper-Raman scattering, two-photon absorption and four-wave mixing. For molecular crystals, similar to anthracene crystals, however, we are not aware of any experiments where biexcitons were definitely observed. In this situation, further experimental as well as theoretical studies certainly present interest, particularly when the increasing significance of organic materials in photonics is considered. It is usually supposed that the attraction between excitons is necessary for biexciton states to appear. That is true for the low-energy biexciton levels splitting off the band of two-particle states. But if excitons repel each other, then biexciton levels may split off the band in the region of higher energy. It is clear that we may expect the existence of such repulsion-induced states only for molecular crystals since in semiconductors the higher part of the exciton band usually overlaps with the electron–hole continuum. Of course, it makes sense to speak of such split-off states only if their broadening is smaller than the value of the splitting, i.e. in crystals with relatively weak exciton–phonon coupling. Note also that in molecular crystals with several molecules in the unit cell the number of two-particle bands rapidly increases. Depending on the exciton band structure the two-particle bands may either overlap or be separated by gaps. In this situation the question arises as to whether gap Frenkel biexcitons are formed, similar to the gap local Frenkel excitons or local optical phonons appearing in crystals with impurities or vacancies (98).

The biexciton states described above have energies outside the free-exciton two-particle continuum and can decay only due to exciton–photon or exciton–phonon interactions. However, if the biexciton energy lies in the continuum, such a biexciton can decay spontaneously into states with two free (unbound) excitons. These quasibound states display themselves as resonant peaks in the two-particle continuum and can be observed in linear and nonlinear optical processes if these resonances are sufficiently narrow, e.g. if they are situated in the region of low density of two-particle states.

Let us assume that every molecule in the crystal has only two states, ground and excited. Then a general Hamiltonian for Frenkel excitons in the Heitler–London approximation (neglecting the particle nonconserving terms) including pair interaction may be written as (1), (6)

$$\hat{H} = \sum_n E_0 \hat{P}_n^\dagger \hat{P}_n + \sum_{n \neq m} M_{n,m} \hat{P}_n^\dagger \hat{P}_m. \qquad (3.200)$$

Here $n = (n, \alpha)$ labels the molecules in the crystal, where n is the lattice vector and α enumerates the molecules in a unit cell; $\hat{P}_n^\dagger$ and $\hat{P}_n$ are the creation and annihilation operators of an exciton on molecule n, obeying Pauli commutation relations. E_0 is the renormalized excitation energy in the monomer (see Section 3.2), and M_{nm} is the matrix element of the excitation energy transfer from the molecule m to the molecule n.

As long as one is interested in one-particle states of a system, the statistics does not matter. But if states with the number of particles larger than one are considered, the situation is different. Particles obeying Pauli statistics are extremely inconvenient to treat by standard many-body methods, since Pauli commutation relations are not preserved under unitary transformations. For this reason one usually applies some transformation to a standard statistics (see, for example, the previous subsection). The Jordan–Wigner transformation from paulion to fermion operators preserves the state space, but it makes the system practically untreatable in more than one dimension. Another way to treat a paulion system is to introduce boson operators (see subsection 3.11). However, the state spaces for paulions and bosons are completely different, since an arbitrary number of bosons can reside on the same molecule. To correct this point, one has to introduce an additional interaction term (as we have already mentioned in this chapter) to the boson Hamiltonian for bosons to have the same properties as paulions. This additional interaction is referred to as a kinematic interaction because it is not related to the fourth-power term in the paulion Hamiltonian, but inevitably appears merely due to the mixed statistics of paulions. In the following, we employ neither the fermion nor the boson picture but we prefer to keep the term kinematic interaction for paulions to express the simple fact that even if the Hamiltonian is bilinear in paulion operators and therefore may be diagonalized exactly by a unitary transformation to plane waves, the state with several plane waves is not an eigenstate of the quadratic paulion Hamiltonian. These plane waves are subject to scattering, i.e. some interaction between quasiparticles is still present even for a quadratic Hamiltonian. It was shown in (99), (100) that in some cases this residual repulsive kinematic interaction is sufficient to form bound states. To demonstrate this effect we consider a molecular crystal of reduced dimensionality, e.g. such as 1D J–aggregates. The situation turns out to depend significantly on the number of exciton bands, which is, in turn, determined by the number of molecules in the unit cell. The contribution of the obtained bound state to the nonlinear polarizability $\chi^{(3)}$ was calculated in (100).

The problem of the kinematic interaction between two paulions is similar to the problem of localized states of an exciton in the presence of a vacancy (98). Indeed, the kinematic interaction governing the relative motion of two Pauli particles is formally analogous to the one-particle potential created by a vacancy, which cannot be occupied by an exciton. In this case the equation determining the localized exciton state energy E is

$$\sum_{k,\mu} \frac{|u_{\alpha\mu}(k)|^2}{E - E_\mu(k)} = 0, \qquad (3.201)$$

where k is the exciton wavevector, $E_\mu(k)$ is the unperturbed dispersion of the μth exciton band, and $u_{\alpha\mu}(k)$ is the unperturbed exciton wavefunction on the αth molecule in the unit cell – the molecule which is lacking. It is easy to see that in the case of the one-exciton band ($u_{\alpha\mu}(k) = 1$) no local state can split off the continuum since either $E < E_\mu(k)$ and the sum is strictly negative or $E > E_\mu(k)$ and the sum is strictly positive. However, for two bands separated by a gap and

E lying in the gap, the two sums corresponding to the bands have different signs and the equation may be satisfied. Inspired by this analogy, a simple example of a perfect infinite one-dimensional molecular chain with two identical molecules in the unit cell has been considered (99), (100). The Schrödinger equation

$$\hat{H}|2\rangle = E|2\rangle, \tag{3.202}$$

for two-exciton state

$$|2\rangle = \sum_{n\alpha, m\beta} \Psi_{n,m}^{\alpha,\beta} \hat{P}_{n,\alpha}^{\dagger} \hat{P}_{m,\beta}^{\dagger} |0\rangle \tag{3.203}$$

leads to the following equations for the wavefunctions

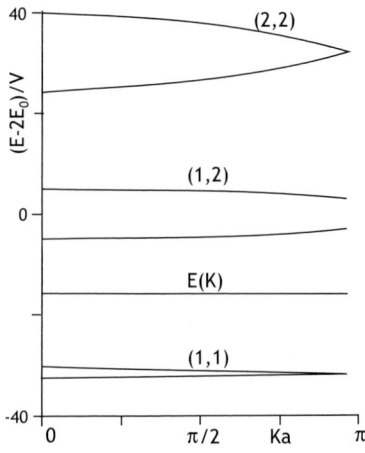

FIG. 3.8. Spectrum of excitonic states in 1D molecular crystal with two molecules per unit cell: bands of unbound two-particle excitonic states (1,1), (1,2) and (2,2), and gap biexciton level $E(k)$. Reprinted with permission from Agranovich et al. (99). Copyright Elsevier (2001).

$$(E - 2E_0)\,\Psi_{n,m}^{\alpha,\beta} = \sum_{\ell,\gamma} \left(M_{n,\ell}^{\alpha,\gamma} \Psi_{l,m}^{\gamma,\beta} + M_{m,\ell}^{\beta,\gamma} \Psi_{n\ell}^{\alpha,\gamma} \right), \tag{3.204}$$

with $\Psi_{\ell\ell}^{\gamma,\gamma} = 0$. It can be shown that the energies of a kinematic biexciton can be found as solutions of the equation:

$$\det |T_{\alpha,\beta}(E,\mathbf{K})| = 0, \tag{3.205}$$

where

$$T_{\alpha,\beta}(E,\mathbf{K}) = \frac{1}{N} \sum_{\mu\nu, \mathbf{p+q}=\mathbf{K}} \frac{u_{\mu,\mathbf{p}}^{\alpha} u_{\nu,\mathbf{q}}^{\beta} u_{\mu,\mathbf{p}}^{\beta*} u_{\nu,\mathbf{q}}^{\gamma*}}{E - 2E_0 - \varepsilon_{\mu,\mathbf{p}} - \varepsilon_{\nu,\mathbf{q}}},$$

and where $\epsilon_{\mu,\mathbf{p}}$ is the one-exciton energy in the band μ with wavevector $\mathbf{p}$, $u^\alpha_{\mu,\mathbf{p}}$ being the corresponding eigenfunction of the one-exciton problem. The performed calculations showed that a biexciton resonance indeed appears between the highest and the lowest two-exciton bands of the structure. We have no space to go into details of the calculations and present only the result. It was found that for the model of a 1D chain investigated in (99) the kinematic biexciton has a positive effective mass at $K = 0$. The dispersion of the kinematic biexciton is very small, as one can see in Fig. 3.8, where the biexciton branch is shown together with free two-exciton bands (1,1), (1,2) and (2,2). In the paper (100) was also shown that a kinematic biexciton state should manifest itself, e.g. as a resonant peak in the two-photon absorption spectrum at a frequency lying in the gap between the two one-exciton bands. The kinematic biexcitons might be observed experimentally, e.g. in J–aggregates with several molecules per unit cell.

4

POLARITONS: EXCITONIC STATES TAKING ACCOUNT OF RETARDATION.

4.1 The crystal energy operator in the presence of a retarded interaction

In previous chapters we considered elementary crystal excitation taking into account only the Coulomb interaction between carriers. From the point of view of quantum electrodynamics (see, for example, (1)) such an interaction is conditioned by an exchange of virtual scalar and longitudinal photons, so that the potential energy, corresponding to this interaction, depends on the carrier positions and not on their velocity distribution. As is well-known, the exchange of virtual transverse photons leads to the so-called retarded interaction between charges.

In what follows we find it convenient to choose the Coulomb gauge for the vector potential. In this gauge the vector potential satisfies the relation (see (1), § 6)

$$\text{div } \mathbf{A} = 0, \tag{4.1}$$

so that the total Hamiltonian of a system consisting of carriers and the electromagnetic field can be written in the form

$$\hat{H} = \hat{H}_1 + \hat{H}_2 + \hat{H}_{\text{int}}, \tag{4.2}$$

$\hat{H}_1$ being the crystal energy operator when only the Coulomb interaction is accounted for, $\hat{H}_2$ is the energy operator of transverse photons of the electromagnetic field in vacuum, and $\hat{H}_{\text{int}}$ is the operator of the interaction between all carriers with the field of the transverse photons.

The operator $\hat{H}_1$ for a molecular crystal with fixed molecules is defined by the expression (3.1) and the elementary excitation spectrum in the excitonic energy region corresponding to this Hamiltonian was discussed in Chs. 2 and 3.

The Hamiltonian $\hat{H}_2$ of the transverse photon field has the form (see, for example, (1))

$$\hat{H}_2 = \sum_{\mathbf{k}, j=1,2} \hbar k c \, a_{\mathbf{k}j}^\dagger a_{\mathbf{k}j}. \tag{4.3}$$

In the above expression c is the light velocity in vacuum, $\mathbf{k}$ is the photon wavevector, $\hbar k c$ is the photon energy, j labels two possible transverse polarizations, and $a_{\mathbf{k}j}^\dagger, a_{\mathbf{k}j}$ are the Bose creation and annihilation operators of a photon $\mathbf{k}, j$. We notice that the summation over $\mathbf{k}$ in (4.3) runs over the whole infinite reciprocal lattice, and is different from the summation over $\mathbf{k}$ carried out in previous chapters.

The operator describing the interaction of carriers with the transverse photon field in the nonrelativistic (as concerns the carrier motion) approximation is given by

$$\hat{H}_{\text{int}} = -\frac{1}{c}\sum_{\mathbf{n}\alpha}\sum_{\nu}^{(\mathbf{n}\alpha)} \frac{e_\nu}{m_\nu}\hat{\mathbf{A}}\left(\boldsymbol{\rho}_\nu^{\mathbf{n}\alpha}\right)\hat{\mathbf{j}}_\nu^{(\mathbf{n}\alpha)}$$
$$+\frac{1}{2c^2}\sum_{\mathbf{n}\alpha}\sum_{\nu}^{(\mathbf{n}\alpha)} \frac{e_\nu}{m_\nu}\hat{\mathbf{A}}^2\left(\boldsymbol{\rho}_\nu^{\mathbf{n}\alpha}\right). \quad (4.4)$$

Here $\sum_\nu^{\mathbf{n}\alpha}$ denotes summation over all carriers in the molecule $\mathbf{n}\alpha$, $\boldsymbol{\rho}_\nu^{\mathbf{n}\alpha}$ indicates the position of the νth carrier in the molecule $(\mathbf{n}\alpha)$, and $\hat{\mathbf{j}}_\nu^{(\mathbf{n}\alpha)}$ is the corresponding momentum operator. If we assume that the molecules reside at their equilibrium positions, the ion masses in (4.4) must be considered as infinite.

The vector potential operator is defined as (1)

$$\hat{\mathbf{A}}(\mathbf{r}) = \sum_{\mathbf{k},j=1,2}\left(\frac{2\pi\hbar c^2}{Vkc}\right)^{1/2}\mathbf{e}_{\mathbf{k}j}\left(a_{\mathbf{k}j}e^{i\mathbf{k}\mathbf{r}} + a_{\mathbf{k}j}^\dagger e^{-i\mathbf{k}\mathbf{r}}\right), \quad (4.5)$$

where V is the volume where the cyclic boundary conditions are applied, and $\mathbf{e}_{\mathbf{k}j}$ are unit vectors such that the three vectors $\mathbf{e}_{\mathbf{k}j}, \mathbf{k}$ are mutually orthogonal.

If the retarded interaction is ignored and the operator $\hat{H}_{\text{int}}$ is removed from the total Hamiltonian (4.2), then the operator $\hat{H}$ becomes a sum of two independent Hamiltonians, one of them $(\hat{H}_1)$ describing the crystal elementary excitations – those which occur when the retardation effects are ignored and the second $(\hat{H}_2)$ giving the elementary excitations – transverse photons in vacuum. The presence of the operator $\hat{H}_3$ leads to the interaction between carriers with the transverse electromagnetic field. In the case of an atomic gas this interaction causes, in particular, the so-called radiative width of energetic levels of excited states. In the case of an infinite crystal possessing translational symmetry the radiative width of excitonic states vanishes.[29]

The above property can be explained as follows. An excited electron in an isolated atom interacts with an infinite number of degrees of freedom of the transverse electromagnetic field, to which elementary excitation energies near to the energy of the excited electron correspond. On the other hand, in an infinite crystal, where the conservation of quasimomentum holds, the exciton can interact only with those photons which have the same wavevector as the exciton.[30] Consequently, the exciton in an infinite crystal interacts at most with two resonating degrees of freedom of the transverse electromagnetic field (two polarization orientations), therefore the field acts not as a dissipative subsystem, but as a dynamical subsystem. As we show below the interaction between

---

[29] Excluding situations when the width of an excitonic band is small, so that we deal with "localized excitations" rather than with excitons. This problem is discussed in the following.

[30] Processes of an "Umklapp" type can be ignored here.

the excitonic subsystem with transverse electromagnetic field leads in the linear approximation to the creation of new collective states, which in 3D structures can be dissipated only by the action of other interactions, for example, by the exciton–phonon interaction. In modern literature these new one-quasiparticle excited states (the coherent superposition of one exciton state and one photon state) are named polaritons.

First of all we have to mention that the above described situation of resonance is not related to any quantum effects. Moreover, the role of the transverse electromagnetic field in crystal oscillations in the infrared part of the spectrum was discussed by means of the classical dynamics of crystal lattices a long time ago by Born and Ewald (2) (see also (3) and (4)), and later by a semiphenomenological approach in (5), (6). It is evident, however, that a quantum theory of polaritons in the region of electronic transitions can also be important particularly for the discussion of quantum effects.

The first attempt to construct such a theory was done by Fano (7). In contrast to earlier work (8), Fano assumed that the oscillator strengths of electromagnetic transitions are not small and did not consider the electromagnetic field as a perturbation. He investigated an isotropic solid medium, characterized by an energy operator taken as the sum of energy operators of harmonic oscillators. It follows from the discussion in Ch. 3 (see also (9)) that such a model in the excitonic energy region can be justified by the method of approximated second-quantization (see Ch. 3). However, in (7) the Coulomb interaction between oscillators has been taken into account in the random phase approximation. In this approximation the short-range interaction is neglected and as a result the local field corrections were lost.[31]

Later this problem has been considered independently and more consistently in the work by Hopfield (12) and the author (9). They performed quantization of the electromagnetic field in the region of exciton resonances and demonstrated how new quasiparticles – polaritons, the coherent superpositions of excitons and transverse photons arise. In these papers, in contrast to (7), the Coulomb gauge for the vector potential has been used. As shown above, in this gauge the total Hamiltonian $\hat{H}$ of the system consisting of charge carriers and the electromagnetic field can be put into the form (4.2) where it is assumed that in the exciton Hamiltonian the exciton state are found taking account of short-range as well as long-range Coulomb interactions. Hopfield (12) has considered elementary excitations characterized by the Hamiltonian (4.2) making use of the mathematical methods given in (7), taking the operator $\hat{H}_1$ in the form (3.134) and considering, as did Fano, only cubic crystals. However, the majority of molecular crystals are very anisotropic. Therefore in the following in the discussion of the polariton spectrum we use the results of (9), which remain valid for crystals of arbitrary symmetry and allow us easily to investigate the effects of exciton state mixing,

---

[31]The contribution of the short-range Coulomb interaction was later taken into account by Fano in (10) for the same classical model. A generalization for the case of liquids has been given in (11).

low-dimensional structures, and different nonlinear optical effects (see below and Ch. 7, Section 7.5).

A few words on the history of the term "polariton". This term was first used by Hopfield (12). In his paper on page 1558 Hopfield wrote "The polarization field 'particles' analogous to photons will be called 'polaritons'. (Excitons will be shown to be one kind of polariton in Section VI. Optical phonons are another example of polaritons.)". Thus, this definition did not refer to new quasiparticles arising when taking into account the strong interaction of excitons and transverse photons. The author of this book had not read the paper by Hopfield attentively enough. In his talk at the All Union Conference on Luminescence in Leningrad in 1959, he discussed the results of his polariton theory (9), the polariton mechanism of exciton luminescence and the role of defects (this talk was published next year in *Uspekhi Fiz. Nauk* (13)). Believing that it is only appropriate to introduce new terms to denote new phenomena, he began to use the term "polaritons" for the new quasiparticles arising when taking into account the strong exciton–photon interaction. Two years later, S.I. Pekar pointed out to the present author on his redefining of Hopfield's definition but it was already too late. It is precisely the present author's definition of the term that has turned out to be generally accepted.

Let us continue to consider in detail the operator characterizing the interaction of electrons with the field of transverse photons, given by the expression (4.4). The radius-vector of the νth electron of the molecule $\mathbf{n}\alpha$ contained in this expression is given by

$$\boldsymbol{\rho}_\nu^{\mathbf{n}\alpha} = \mathbf{r}_{\mathbf{n}\alpha} + \boldsymbol{\rho}'_{\nu,\mathbf{n}\alpha}, \tag{4.6}$$

where $\boldsymbol{\rho}'_{\nu,\mathbf{n}\alpha}$ indicates the position of the νth electron with respect to a lattice node $(\mathbf{n}\alpha)$. It will be proved below that only the long-wavelength photons, for which $ka \ll 1$, a being the lattice constant, substantially interact with excitons. Therefore the vector potential $\hat{\mathbf{A}}(\boldsymbol{\rho}_\nu^{\mathbf{n}\alpha})$, with regard to (4.5), can be written in the form

$$\hat{\mathbf{A}}(\boldsymbol{\rho}_\nu^{\mathbf{n}\alpha}) = \sum_{\mathbf{k},j=1,2} \left(\frac{2\pi\hbar c^2}{Vkc}\right)^{1/2} \mathbf{e}_{\mathbf{k}j}\left[a_{\mathbf{k}j}\left(1 + i\mathbf{k}\boldsymbol{\rho}'_{\nu,\mathbf{n}\alpha}\right)e^{i\mathbf{k}\mathbf{r}_{\mathbf{n}\alpha}} + \text{h.c.}\right]. \tag{4.7}$$

By substituting (4.7) into (4.4) and denoting the first term by $\hat{H}_{\text{int}}^{(1)}$ we obtain

$$\hat{H}_{\text{int}}^{(1)} = -\sum_{\mathbf{k},j=1,2} \left(\frac{2\pi\hbar}{Vkc}\right)^{1/2} \left[a_{\mathbf{k}j}\mathbf{e}_{\mathbf{k}j}\hat{\boldsymbol{\Pi}}(\mathbf{k}) + \text{h.c.}\right], \tag{4.8}$$

with

$$\hat{\boldsymbol{\Pi}}(\mathbf{k}) = \sum_{\mathbf{n}\alpha} e^{i\mathbf{k}\mathbf{r}_{\mathbf{n}\alpha}} \left\{\sum_\nu^{(\mathbf{n}\alpha)} \frac{e_\nu}{m_\nu}\hat{\mathbf{j}}_\nu^{\mathbf{n}\alpha}\left(1 + i\mathbf{k}\boldsymbol{\rho}'_{\nu,\mathbf{n}\alpha}\right)\right\}. \tag{4.9}$$

If we neglect in (4.9) the terms $\mathbf{k}\boldsymbol{\rho}'_{\nu,\mathbf{n}\alpha}$, the operator (4.9) goes over into the operator defined by expressions (3.132) and (3.139). In what follows we limit the discussion to this approximation. As will be shown in the following, it allows one to determine elementary excitation spectra in the region of dipole-allowed molecular transitions.

Substituting the expression (3.139) into (4.8) we obtain for the case of crystals posessing an inversion center

$$\hat{H}_{\text{int}}^{(1)} = -\sum_{\mu,\mathbf{k}j} T(j,\mathbf{k},\mu) \Big\{ a_{\mathbf{k}j} [B_\mu(-\mathbf{k}) - B_\mu^\dagger(\mathbf{k})]$$
$$+ a_{\mathbf{k}j}^\dagger [B_\mu(\mathbf{k}) - B_\mu^\dagger(-\mathbf{k})] \Big\}, \quad (4.10)$$

where

$$T(j,\mathbf{k},\mu) = i\left(\frac{2\pi N}{Vkc\hbar}\right)^{1/2} E_\mu(\mathbf{k})\mathbf{P}(\mathbf{k},\mu)\mathbf{e}_{\mathbf{k}j}, \quad (4.11)$$

for vectors $\mathbf{k}$ from the first Brillouin zone; otherwise $T = 0$.[32] It can be easily seen that the second contribution to the interaction Hamiltonian, up to a certain constant term, is given by

$$\hat{H}_{\text{int}}^{(2)} = \frac{\hbar\omega_0^2}{4} \sum_{\mathbf{k}j} \frac{1}{kc} \left(2a_{\mathbf{k}j}^\dagger a_{\mathbf{k}j} + a_{\mathbf{k}j} a_{-\mathbf{k}j} + a_{\mathbf{k}j}^\dagger a_{-\mathbf{k}j}^\dagger\right). \quad (4.12)$$

Here ω_0 denotes the so-called plasma frequency

$$\omega_0^2 = \frac{4\pi e^2 N\sigma}{mV} S, \quad (4.13)$$

S being the number of electrons in the molecule. Using (4.3), (4.10), (4.12) and taking $\hat{H}_1$ in the form (3.134), we obtain for the total Hamiltonian (4.2) the expression

$$\hat{H} = \mathcal{E}_0 + \sum_{\mu\mathbf{k}} E_\mu(\mathbf{k}) B_\mu^\dagger(\mathbf{k}) B_\mu(\mathbf{k}) + \sum_{\mathbf{k}j} \hbar kc\, a_{\mathbf{k}j}^\dagger a_{\mathbf{k}j} \left(1 + \frac{\omega_0^2}{2k^2c^2}\right)$$
$$- \sum_{\mu\mathbf{k}j} T(j,\mathbf{k},\mu) \Big\{ a_{\mathbf{k}j} [B_\mu(-\mathbf{k}) - B_\mu^\dagger(\mathbf{k})] + a_{\mathbf{k}j}^\dagger [B_\mu(\mathbf{k}) - B_\mu^\dagger(-\mathbf{k})] \Big\}$$
$$+ \frac{\hbar\omega_0^2}{4} \sum_{\mathbf{k}j} \frac{1}{kc} \left(a_{\mathbf{k}j} a_{-\mathbf{k}j} + a_{\mathbf{k}j}^\dagger a_{-\mathbf{k}j}^\dagger\right). \quad (4.14)$$

To obtain the expressions (4.10) and (4.12) we neglected, in the operator $\hat{H}_{\text{int}}$, terms of type $a_{\mathbf{k}+\mathbf{b},j} B_\mu^\dagger(\mathbf{k})$ and similar, $\mathbf{b}$ being an integer-valued reciprocal lattice vector. Such terms relate excitonic states with the wavevectors $\mathbf{k}$ from the

---

[32] In (4.11) $\mathbf{P}(\mathbf{k},\mu) \equiv \mathbf{P}_{0;\mathbf{k}\mu}$ is the dipole moment matrix element of the crystal elementary cell $\langle 0|\hat{\mathbf{P}}|\mathbf{k}\mu\rangle$.

first Brillouin zone with short-wavelength (Roentgen) photons. In spite of an infinite number of such states, the interaction with them does not create a radiative width, since the corresponding photon energy is more than a thousand times larger than the exciton energy. Those states can be regarded as a perturbation by computing the elementary excitations energies. In the lowest order of perturbation theory they do not contribute. In the second and higher orders of perturbation for elementary excitations with $ka \ll 1$ we obtain a correction to the energy of the order $E_\mu (ka)^2$. When concerning the short-wavelength elementary excitations with $ka \propto 1$, these corrections and, in general, the retarded interaction, are also not important. We therefore proceed by taking into account the interaction of excitons and photons with their wavevectors lying only in the first Brillouin zone.

4.2 Dispersion of polaritons and refraction index of electromagnetic waves

The operator $\hat{H}$, given by eqn (4.14), is quadratic with respect to the Bose operators $a_{\mathbf{k}j}, a_{\mathbf{k}j}^\dagger, B_\mu(\mathbf{k})$, and $B_\mu(\mathbf{k})^\dagger$. Thus the computation of exciton spectra when the retardation is taken into account is equivalent to the diagonalization of the expression (4.14). In this expression only those amplitudes are connected, for which the wavevectors are equal to $\mathbf{k}$ or $-\mathbf{k}$, respectively. Therefore instead of (4.14) it is sufficient to diagonalize the quadratic form

$$\begin{aligned}
\hat{H}_{\mathbf{k}} = &\sum_{\mu \mathbf{k}} E_\mu(\mathbf{k}) \left[B_\mu^\dagger(\mathbf{k}) B_\mu(\mathbf{k}) + B_\mu^\dagger(-\mathbf{k}) B_\mu(-\mathbf{k}) \right] \\
&+ \sum_{j=1,2} \hbar k c \left(a_{\mathbf{k}j}^\dagger a_{\mathbf{k}j} + a_{-\mathbf{k}j}^\dagger a_{-\mathbf{k}j} \right) \left(1 + \frac{\omega_0^2}{2k^2 c^2} \right) \\
&+ \frac{\hbar \omega_0^2}{2kc} \sum_{j=1,2} \left(a_{\mathbf{k}j} a_{-\mathbf{k}j} + a_{\mathbf{k}j}^\dagger a_{-\mathbf{k}j}^\dagger \right) \\
&- \sum_{\mu, j=1,2} T(j, \mathbf{k}, \mu) \Big\{ \left(a_{\mathbf{k}j} + a_{-\mathbf{k}j}^\dagger \right) \left[B_\mu(-\mathbf{k}) - B_\mu^\dagger(\mathbf{k}) \right] \\
&+ \left(a_{-\mathbf{k}j} + a_{\mathbf{k}j}^\dagger \right) \left[B_\mu(\mathbf{k}) - B_\mu^\dagger(-\mathbf{k}) \right] \Big\}.
\end{aligned} \quad (4.15)$$

The diagonalization of the above quadratic form can be carried out by the transition to new Bose operators ξ_ρ and $\xi_\rho^\dagger$ (see Appendix A):

$$\begin{aligned}
B_\mu(\mathbf{k}) &= \sum_\rho \left[\xi_\rho(\mathbf{k}) u_{\mathbf{k}\mu}(\rho) + \xi_\rho^\dagger(-\mathbf{k}) v_{-\mathbf{k}\mu}^*(\rho) \right], \\
a_{\mathbf{k}j} &= \sum_\rho \left[\xi_\rho(\mathbf{k}) u_{\mathbf{k}j}(\rho) + \xi_\rho^\dagger(-\mathbf{k}) v_{-\mathbf{k}j}^*(\rho) \right].
\end{aligned} \quad (4.16)$$

In these relations the quantities u and v, according to the formulas (A.5), satisfy the following system of equations [33]

$$[E_\mu(\mathbf{k}) - \mathcal{E}] u_{\mathbf{k}\mu} + \sum_{j=1,2} T(j, \mathbf{k}, \mu)(u_{\mathbf{k}j} + v_{-\mathbf{k}j}) = 0,$$

$$[E_\mu(\mathbf{k}) + \mathcal{E}] v_{-\mathbf{k}\mu} - \sum_{j=1,2} T(j, \mathbf{k}, \mu)(u_{\mathbf{k}j} + v_{-\mathbf{k}j}) = 0,$$

$$(\hbar k c - \mathcal{E}) u_{\mathbf{k}j} - \sum_\mu T(j, \mathbf{k}, \mu)(u_{\mathbf{k}\mu} - v_{-\mathbf{k}\mu}) + \frac{\hbar \omega_0^2}{2kc}(u_{\mathbf{k}j} + v_{-\mathbf{k}j}) = 0,$$

$$(\hbar k c + \mathcal{E}) v_{-\mathbf{k}j} - \sum_\mu T(j, \mathbf{k}, \mu)(u_{\mathbf{k}\mu} - v_{-\mathbf{k}\mu}) + \frac{\hbar \omega_0^2}{2kc}(u_{\mathbf{k}j} + v_{-\mathbf{k}j}) = 0,$$

(4.17)

and the normalization condition

$$\sum_\mu \left(|u_{\mathbf{k}\mu}(\rho)|^2 - |v_{-\mathbf{k}\mu}(\rho)|^2\right) + \sum_{j=1,2}\left(|u_{\mathbf{k}j}(\rho)|^2 - |v_{-\mathbf{k}j}(\rho)|^2\right) = 1. \quad (4.18)$$

In these new variables the crystal Hamiltonian becomes [34]

$$\hat{H} = \mathcal{E}_0 - \sum_{\mathbf{k}j\rho} \mathcal{E}_\rho(\mathbf{k})|v_{\mathbf{k}j}(\rho)|^2 - \sum_{\mathbf{k}\mu\rho}\mathcal{E}_\rho(\mathbf{k})|v_{\mathbf{k}\mu}(\rho)|^2$$

$$+ \sum_{\mathbf{k}\rho} \mathcal{E}_\rho(\mathbf{k}) \xi_\rho^\dagger(\mathbf{k}) \xi_\rho(\mathbf{k}). \quad (4.19)$$

To each value ρ there corresponds a branch of new elementary excitations (polaritons). Comparing, in the system (4.17), the first and the second equation and the third and the fourth equation, we obtain

$$v_{-\mathbf{k}\mu}(\rho) = -\frac{E_\mu(\mathbf{k}) - \mathcal{E}_\rho}{E_\mu(\mathbf{k}) + \mathcal{E}_\rho} u_{\mathbf{k}\mu}(\rho), \qquad v_{-\mathbf{k}j}(\rho) = \frac{\hbar k c - \mathcal{E}_\rho}{\hbar k c + \mathcal{E}_\rho} u_{\mathbf{k}j}. \quad (4.20)$$

Making use of eqn (4.20) we can eliminate the quantities v from eqn (4.17) thus obtaining

$$[E_\mu^2(\mathbf{k}) - \mathcal{E}^2] u_{\mathbf{k}\mu} + \sum_{j=1,2} 2T(j, \mathbf{k}, \mu) \hbar k c \frac{E_\mu(\mathbf{k}) + \mathcal{E}}{\hbar k c + \mathcal{E}} u_{\mathbf{k}j} = 0,$$

---

[33] The same equations have been obtained in a different way by Ball and McLachlan (14).

[34] In what follows we omit the constant factor in the Hamiltonian $\hat{H}$, which is equal to the ground state energy and is obtained by taking into account the retardation. Note the paper by Mavroyannis (15) where a detailed discussion of the influence of the retardation on the crystal ground state energy has been carried out. Later the same problem was discussed by Bassani and Quattropani (16).

$$\left(\hbar^2 k^2 c^2 + \hbar^2 \omega_0^2 - \mathcal{E}^2\right) u_{\mathbf{k}j} \tag{4.21}$$
$$- \sum_\mu 2T(j,\mathbf{k},\mu) E_\mu(\mathbf{k}) \frac{\hbar k c + \mathcal{E}}{E_\mu(\mathbf{k}) + \mathcal{E}} u_{\mathbf{k}\mu} = 0.$$

For the longitudinal excitons, $\mu = \mu_\parallel$ and $T(j,\mathbf{k},\mu) = 0$ as follows from (4.11). In consequence, those excitons do not interact with the transverse photons and retardation effects for them are not essential. From (4.17) and (4.18) we obtain for the longitudinal excitons[35]

$$\mathcal{E}_\rho = E_{\mu_\parallel}(\mathbf{k}); \quad \rho \equiv (\mu_\parallel, \mathbf{k}); \quad u_{\mathbf{k}\mu_\parallel}(\rho) = 1,$$
$$v_{\mathbf{k}\mu}(\rho) = u_{\mathbf{k}j}(\rho) = v_{\mathbf{k}j}(\rho) = 0. \tag{4.22}$$

Now consider the excitons for which eqn (4.22) are not satisfied. In this case we can use the first of eqn. (4.21) to express the quantities $u_{\mathbf{k}\mu}$ in terms of $u_{\mathbf{k}j}$ and substitute the resulting expressions into the second of eqn. (4.21), thus obtaining a system of two linear homogeneous equations for the quantities $u_{\mathbf{k}j}, j = 1, 2$. The equations have the form

$$\left(\hbar^2 k^2 c^2 + \hbar^2 \omega_0^2 - \mathcal{E}^2\right) u_{\mathbf{k}j} - \sum_{j'=1,2} u_{\mathbf{k}j'} A_{jj'} = 0, \tag{4.23}$$

where

$$A_{jj'} = \sum_\mu 4\hbar k c \, E_\mu(\mathbf{k}) \frac{T(j,\mathbf{k},\mu) T(j',\mathbf{k},\mu)}{\mathcal{E}^2 - E_\mu^2(\mathbf{k})}. \tag{4.24}$$

The condition of the vanishing of the determinant of the above system of equations gives the elementary excitations energy of the total system (electrons plus field) when the retardation is taken into account. The equation obtained for the quantity $\mathcal{E}(\mathbf{k})$ has the form

$$\left(\hbar^2 k^2 c^2 + \hbar^2 \omega_0^2 - \mathcal{E}^2\right)^2 + \left(\hbar^2 k^2 c^2 + \hbar^2 \omega_0^2 - \mathcal{E}^2\right)(A_{11} + A_{22})$$
$$+ A_{11} A_{22} - A_{12} A_{21} = 0. \tag{4.25}$$

In (4.25), according to its derivation, the vector $\mathbf{k}$ is real. Consequently the eqn (4.25) gives a spectrum of polaritons consisting, in general, of allowed and forbidden energy bands.[36] We can also proceed in a different way and express the refraction index $n(\omega, \mathbf{s})$ defined by the relations

$$\mathbf{k} = \frac{\omega}{c} n(\omega, \mathbf{s}) \mathbf{s}, \quad \mathbf{s} = \frac{\mathbf{k}}{k}, \tag{4.26}$$

as a function of the frequency $\omega = \mathcal{E}/\hbar$. It is clear that the dependences $\mathcal{E}_\rho(\mathbf{k})$ and $n(\omega, \mathbf{s})$ are equivalent.

---

[35]Analogous relations are always valid when $T(j,\mathbf{k}\mu) = 0$. We have $T(j,\mathbf{k}\mu) = 0$ for $\mathbf{k} = 0$, when the exciton band cannot be excited by light in the dipole approximation (i.e. when the matrix element $\langle 0|\mathbf{P}|0\mu\rangle$ is 0).

[36]The index ρ labels the allowed energy bands. The number of values which ρ can achieve is given by a sum of the considered exciton bands + two ($j = 1, 2$ – two branches of the transverse photons).

In what follows we first consider the quantity $n(\omega, \mathbf{s})$. Using (4.25) and (4.26) we find

$$n^2(\omega, \mathbf{s}) = 1 - \frac{\hbar^2 \omega_0^2}{\mathcal{E}^2} + \frac{A_{11} + A_{22}}{2\mathcal{E}^2}$$

$$\pm \left[\left(\frac{A_{11} - A_{22}}{\mathcal{E}^2} \right)^2 + 4 \left(\frac{A_{12}}{\mathcal{E}^2} \right)^2 \right]^{1/2}. \tag{4.27}$$

In (4.27) the contributions of the order $\mathcal{E}^{-2}$ cancel identically. Indeed, since

$$\frac{E_\mu^2}{\mathcal{E}^2(\mathcal{E}^2 - E_\mu^2)} \equiv \frac{1}{\mathcal{E}^2 - E_\mu^2} - \frac{1}{\mathcal{E}^2},$$

then, from (4.11) and (4.24) we obtain

$$\frac{1}{\mathcal{E}^2} A_{jj'} = 8\pi \left(\frac{N}{V} \right) \sum_\mu E_\mu(\mathbf{k}) |\mathbf{P}(\mathbf{k}, \mu)|^2 \cos \varphi_j(\mu \mathbf{k}) \cos \varphi_{j'}(\mu \mathbf{k})$$

$$\times \left(\frac{1}{\mathcal{E}^2} - \frac{1}{\mathcal{E}^2 - E_\mu^2(\mathbf{k})} \right),$$

where $\varphi_j(\mu \mathbf{k})$ is the angle formed by the vectors $\mathbf{P}(\mu, \mathbf{k})$ and $\mathbf{e}_{\mathbf{k}j}$. Making use of the sum rule (3.141), which for nongyrotropic crystals takes the form

$$2 \sum_\mu E_\mu(\mathbf{k}) |\mathbf{P}(\mathbf{k}, \mu)|^2 \cos \varphi_j(\mu \mathbf{k}) \cos \varphi_{j'}(\mu \mathbf{k}) = \frac{\hbar^2 e^2}{m} \sigma S \delta_{jj'}, \tag{4.28}$$

S being the number of electrons in the molecule, we finally obtain

$$n^2(\omega, \mathbf{s}) = 1 - \frac{3}{2} \sum_\mu \frac{\omega_0^2 F_\mu(\mathbf{k}) \sin^2 \varphi(\mu, \mathbf{k})}{\omega^2 - \Omega_\mu^2(\mathbf{k})}$$

$$\pm \frac{3}{2} \left\{ \left[\sum_\mu \frac{\omega_0^2 F_\mu(\mathbf{k}) \left(\cos^2 \varphi_1(\mu, \mathbf{k}) - \cos^2 \varphi_2(\mu, \mathbf{k}) \right)}{\omega^2 - \Omega_\mu^2(\mathbf{k})} \right]^2 \right. \tag{4.29}$$

$$\left. + 4 \left[\sum_\mu \frac{\omega_0^2 F_\mu(\mathbf{k}) \cos \varphi_1(\mu, \mathbf{k}) \cos \varphi_2(\mu, \mathbf{k})}{\omega^2 - \Omega_\mu^2(\mathbf{k})} \right]^2 \right\}^{1/2}.$$

Here $\varphi(\mu, \mathbf{k}) \equiv \varphi_3(\mu, \mathbf{k})$ is the angle formed by the vectors $\mathbf{k}$ and $\mathbf{P}(\mathbf{k}\mu)$, and

$$\sin^2 \varphi(\mu, \mathbf{k}) = \cos^2 \varphi_1(\mu, \mathbf{k}) + \cos^2 \varphi_2(\mu, \mathbf{k}),$$

$$F_\mu(\mathbf{k}) = \frac{2m\,|\mathbf{P}(\mu\mathbf{k})|^2}{3\sigma S e^2 \hbar^2} E_\mu(\mathbf{k}), \qquad \Omega_\mu(\mathbf{k}) = \frac{E_\mu(\mathbf{k})}{\hbar}. \tag{4.30}$$

According to (4.28)

$$\sum_\mu F_\mu(\mathbf{k}) = 1. \tag{4.31}$$

When the frequency ω is large compared to all excitons frequencies $\Omega_\mu(\mathbf{k})$, the crystal electrons can be considered as free. In this case, as is known (see, for example, (17), §59)

$$n^2(\omega, \mathbf{s}) = 1 - \frac{\omega_0^2}{\omega^2}. \tag{4.32}$$

The same relation, as it should be, results from the formula (4.29), if we neglect the quantities Ω_μ as small compared to ω and use the sum rule (4.28).

If, on the r.h.s. of (4.29) in the long-wave limit ($ka \ll 1$), the dependence on $\mathbf{k}$ is neglected (i.e. the quantities Ω_μ and F_μ are taken at $\mathbf{k} = 0$), then for each value of ω eqn (4.29) gives two values $n_1(\omega, \mathbf{s}), n_2(\omega, \mathbf{s})$ of the refractive index. If $\omega \simeq \Omega_\mu$, i.e. if it is near to one of the Coulomb frequencies, then from the sum over μ in (4.29) we can extract the resonant terms, which will be dominant.

With the purpose of illustrating the relation (4.29) we consider some special cases.

1. The frequency ω is, in vicinity of nondegenerate exciton bands, $\omega \simeq \Omega(\mathbf{k})$. In this case we obtain from (4.29)

$$n_1^2(\omega, \mathbf{s}) = \epsilon_1(\omega, \mathbf{s}) - \frac{3\omega_0^2 F(\mathbf{k}) \sin^2 \varphi(\mathbf{s})}{\omega^2 - \Omega^2(\mathbf{k})},$$
$$n_2^2(\omega, \mathbf{s}) = \epsilon_2(\omega, \mathbf{s}), \tag{4.33}$$

where ϵ_1 and ϵ_2 weakly depend on the frequency ω in the considered part of the spectrum and $\varphi(\mathbf{s})$ is the angle between the wavevector $\mathbf{k}$ and the transition dipole moment.

2. The crystal possesses cubic symmetry. In this case exciton states, which correspond to nonvanishing values $F_\mu(\mathbf{s})$, are either longitudinal exciton states $(\mu = \mu_\|, \mathbf{P}(\mathbf{k}, \mu_\|) \parallel \mathbf{s})$, or transverse exciton states, which for $\mathbf{k} \to 0$ corresponds to the two-fold degenerate exciton band

$$\mu = \mu_\perp^{(1,2)}, \qquad \mathbf{P}\left(\mathbf{k}, \mu_\perp^{(1)}\right) \perp \mathbf{s}, \qquad \mathbf{P}\left(\mathbf{k}, \mu_\perp^{(2)}\right) \perp \mathbf{s},$$
$$\mathbf{P}\left(\mathbf{k}, \mu_\perp^{(1)}\right) \perp \mathbf{P}\left(\mathbf{k}, \mu_\perp^{(2)}\right), \qquad \left|\mathbf{P}\left(\mathbf{k}, \mu_\perp^{(1)}\right)\right| = \left|\mathbf{P}\left(\mathbf{k}, \mu_\perp^{(2)}\right)\right|.$$

In cubic crystals, and also in crystals of arbitrary symmetry, longitudinal exciton states do not contribute to (4.29). It can be easily verified if we have regard to the fact that for the longitudinal waves, $\varphi(\mu_\|, \mathbf{k}) = 0$, and $\varphi_j(\mu_\|, \mathbf{k}) = \pi/2$, $j = 1, 2$. In the case of transverse excitons the expression in brackets in (4.29) is identically equal to zero. Indeed, choosing the vectors $\mathbf{P}\left(\mu_\perp^{(1,2)}, \mathbf{k}\right)$ to be parallel to the unit vectors of the transverse polarization $\mathbf{e}_j, j = 1, 2$, we find that $\cos\varphi_j\left(\mu_\perp^{(j')}, \mathbf{k}\right) = \delta_{jj'}, j', j = 1, 2$. Therefore

$$\cos\varphi_1\left(\mu_\perp^{(j)}, \mathbf{k}\right) \cos\varphi_2\left(\mu_\perp^{(j)}, \mathbf{k}\right) = 0$$

for each value $j = 1, 2$, and, moreover

$$\sum_{i=1,2} \frac{\omega_0^2 F_{\mu_\perp}(\mathbf{k}) \left[\cos^2\varphi_1\left(\mu_\perp^{(i)}, \mathbf{k}\right) - \cos^2\varphi_2\left(\mu_\perp^{(i)}, \mathbf{k}\right)\right]}{\omega^2 - \Omega_\mu^2(\mathbf{k})} = 0.$$

In consequence, in cubic crystals

$$n_1(\omega, \mathbf{s}) = n_2(\omega, \mathbf{s}) \equiv n(\omega, \mathbf{s}),$$

and, in particular, in the limit $\mathbf{k} \to 0$

$$n^2(\omega, \mathbf{s}) = 1 - \frac{3}{2}\sum_{\mu_\perp} \frac{\omega_0^2 F_\mu(0)}{\omega^2 - \Omega_\mu^2(0)} \equiv n^2(\omega). \tag{4.34}$$

In the vicinity of one of the resonances

$$n^2(\omega) = \epsilon_0(\omega) - \frac{3\omega_0^2 F_{\mu_\perp}(0)}{\omega^2 - \Omega_{\mu_\perp}^2}.$$

We note that if on the r.h.s. of (4.29) we do not put $\mathbf{k} = 0$ then, by virtue of $\mathbf{k} = \omega n(\omega, \mathbf{s})\mathbf{s}/c$ the equation (4.29) becomes an equation for $n^2(\omega, \mathbf{s})$ which can have more than two solutions. This effect,[37] as one of the possible effects of the spatial dispersion, was discussed in detail by Agranovich and Ginzburg (19). It is important to note that the new solutions for $n(\omega, \mathbf{s})$ can occur only near the frequencies $\Omega_\mu(0)$ where, however, the absorption not accounted in (4.29), can be essential and has to be taken into account.

In some cases it is useful to know the dependence of the exciton frequencies on the wavevector. The inverse dependence, i.e. the dependence of $\mathbf{k}$ on ω, is, as we have seen before, given by the relation $\mathbf{k} = \omega n(\omega, \mathbf{s})\mathbf{s}/c$. The calculation of the function $\omega(\mathbf{k})$ can be carried out by using the previously obtained expressions for $n(\omega, \mathbf{s})$. So, for example, for transverse waves in cubic crystals we obtain from (4.34) the following equation for $\omega(\mathbf{k})$:

$$\frac{k^2 c^2}{\omega^2} = \epsilon_0(\omega) - \frac{3\omega_0^2 F_{\mu_\perp}(0)}{\omega^2 - \Omega_{\mu_\perp}(0)}, \tag{4.35}$$

with solutions

$$\omega_{1,2}^2(k) = \frac{1}{2}\left[\Omega_{\mu_\perp}^2(0) + \frac{k^2 c^2}{\epsilon_0} + \frac{3\omega_0^2 F_{\mu_\perp}(0)}{\epsilon_0}\right]$$

---

[37] This effect for excitons was first pointed by Pekar (18).

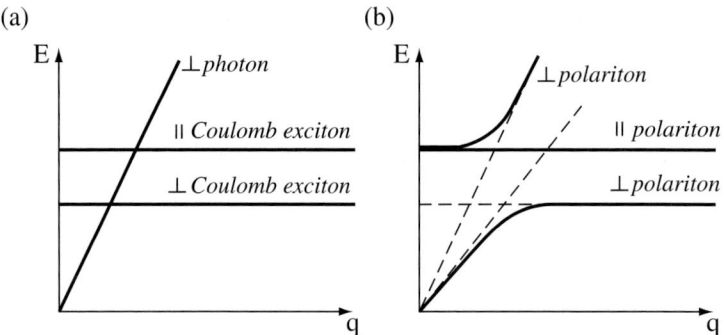

FIG. 4.1. Dispersion relation for Coulomb excitons and polaritons.

$$\pm \frac{1}{2} \left\{ \left[\Omega_{\mu_\perp}^2(0) + \frac{k^2 c^2}{\epsilon_0} + \frac{3\omega_0^2 F_{\mu_\perp}(0)}{\epsilon_0} \right]^2 - \frac{4k^2 c^2}{\epsilon_0} \Omega_{\mu_\perp}^2(0) \right\}^{1/2}. \quad (4.36)$$

It follows from the above relation that the retarded interaction is important only in the vicinity of wavevectors $k \approx \sqrt{\epsilon_0}\Omega/c$, i.e. in that part of the spectrum, where the frequencies of the Coulomb excitons are near to those of the transverse photons. When the retardation is ignored, the branches of the Coulomb excitons and the transverse photons intersect (Fig. 4.1a). This intersection is removed when the retardation is taken into account (Fig. 4.1b). In a similar way the dependence $\omega(\mathbf{k})$ for polaritons can be found for crystals with different symmetries.

As a second example of the application of the above described theory we consider crystals with one molecule per unit cell. We wish to consider polaritonic states with frequencies near to the nondegenerate electronic excitation of the molecule, where the transition from the ground to the excited state is dipole allowed.

If the retardation is ignored, then two coinciding branches of the transverse photons (polarization $j = 1, 2$) and one exciton branch, polarized along the molecule transition dipole, are the lowest elementary excitations of the crystal.

For each orientation of the photon wavevector we choose the polarization in such way that one of them ($j = 1$) is placed in the plane, formed by the molecule transition dipole moment and the photon wavevector, and the second ($j = 2$) is perpendicular to this plane.

With regard to (4.11), by this choice of photon polarization the quantity $T(j = 2, \mathbf{k})$ vanishes, whereas

$$T(j = 1, \mathbf{k}) = i \left(\frac{2\pi N}{\hbar k c V} \right)^{1/2} \left(\mathbf{p}^{0f} \mathbf{e}_1 \right) E(\mathbf{k}) \neq 0.$$

Compared with the expression (4.11), in the above expression we have omitted

the index μ, since we deal with an isolated exciton band, and have regard to the fact that
$$\mathbf{P}(\mathbf{k}\rho) = \mathbf{p}^{0f},$$
where f is the index indicating the considered molecular electronic excited state.

Within an exciton band, where the polariton energy $\mathcal{E}_\rho$ satisfies the inequality
$$\frac{|\mathcal{E}_\rho - E(\mathbf{k})|}{E(\mathbf{k}) + \mathcal{E}} \ll 1,$$
and with regard to eqn (4.20), we have
$$|u_{\mathbf{k}\mu}(\rho)| \gg |v_{-\mathbf{k}\mu}(\rho)|.$$

It also follows from (4.20) that
$$v_{-\mathbf{k}j}(\rho) = \frac{n_\rho - 1}{n_\rho + 1} u_{\mathbf{k}j}(\rho),$$
$n_\rho \equiv \hbar k c / \mathcal{E}_\rho(\mathbf{k})$. Thus, taking into account the equality $T(2,\mathbf{k}) = 0$, we obtain from the first of equations (4.17)
$$u_{\mathbf{k}1} = -\frac{(n_\rho + 1)[E(\mathbf{k}) - \mathcal{E}_\rho(\mathbf{k})]}{2T(1,\mathbf{k})n_\rho} u_{\mathbf{k}\mu}(\rho). \tag{4.37}$$

The above relation, together with the normalization (4.18), allows one to compute the absolute values of the quantities $u_{\mathbf{k}1}(\rho)$ and $u_{\mathbf{k}\mu}(\rho)$. It should be noted that, because of (4.17), $u_{\mathbf{k}j} = v_{\mathbf{k}j} = 0$ for the $j = 2$ polarization, provided that the polariton state ρ appeared, as was supposed, as a result of mixing of an exciton state with the transverse photon having the polarization $j = 1$ (only for $j = 1$ is the quantity $T(j, \mathbf{k}) \neq 0$).

Neglecting the terms $|v_{\mathbf{k}\mu}|$ as small compared to $|u_{\mathbf{k}\mu}|$, we finally obtain
$$|u_{\mathbf{k}\mu}(\rho)|^2 = \left\{ 1 + \frac{\mathcal{E}_\rho(\mathbf{k})[E(\mathbf{k}) - \mathcal{E}_\rho(\mathbf{k})]^2}{\hbar k c |T(1,\mathbf{k})|^2} \right\}^{-1}.$$

In the above case of a nondegenerate exciton band the dispersion for polaritons, resulting from mixing of photon states $j = 1$ with exciton states, is given by the first of equations (4.33), where we put $\Omega(\mathbf{k}) = = E(\mathbf{k})/\hbar$, $\mathcal{E} = \hbar \omega$. Solving this equation with respect to $\mathcal{E}^2(\mathbf{k})$, we obtain[38]

$$\mathcal{E}_{1,2}^2(\mathbf{k}) = \frac{E^2(\mathbf{k})}{2} + \frac{\hbar^2 c^2 k^2 + 3\hbar^2 \omega_o^2 F \sin^2 \varphi}{2\epsilon_1} \tag{4.38}$$

[38]It follows from (4.37) that the energy splitting $\mathcal{E}_1 - \mathcal{E}_2$ depends on the direction of $\mathbf{k}$ and vanishes when $\varphi = 0$, i.e. when the vector $\mathbf{k}$ is parallel to the vector $\mathbf{p}^{0f}$.

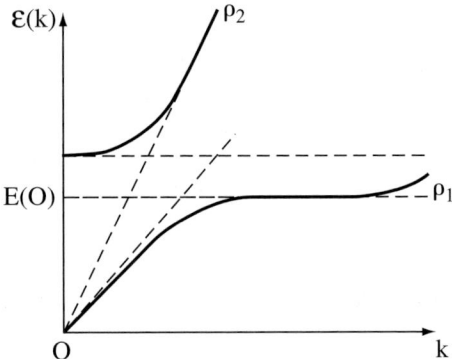

FIG. 4.2. Polariton dispersion.

$$\pm \frac{1}{2} \left\{ \left[\frac{E^2(\mathbf{k})}{2} + \frac{\hbar^2 c^2 k^2 + 3\hbar^2 \omega_o^2 F \sin^2 \varphi}{2\epsilon_1} \right]^2 - \frac{4c^2 k^2 E^2(\mathbf{k}) \hbar^2}{\epsilon_1} \right\}^{1/2}.$$

This relation is schematically sketched in Fig. 4.2. At the lower branch ($\rho = \rho_1$) the energy $\mathcal{E}_\rho(\mathbf{k}) \to E(\mathbf{k})$ with increasing $|\mathbf{k}|$. We also have

$$u_{\mathbf{k}\mu}(\rho_1) \to 1, \qquad v_{\mathbf{k}\mu}(\rho_1) \to 0, \qquad u_{\mathbf{k}j}(\rho_1), \; v_{\mathbf{k}j}(\rho_1) \to 0.$$

On the other hand $\mathcal{E}_{\rho_2}(\mathbf{k}) \to \hbar kc/\epsilon_1$, so that $u_{\mathbf{k}\mu}(\rho_2) \to 0, v_{\mathbf{k}\mu}(\rho_2) \to 0$. In this way, by virtue of (4.16), the operator $\xi_{\rho_1}(\mathbf{k}) \to B_\mu(\mathbf{k})$, i.e. in this spectral region the creation or annihilation of a polariton ($\rho_1 \mathbf{k}$) practically coincides with the creation or annihilation of a Coulomb exciton with wavevector $\mathbf{k}$.

In a similar way we can verify that by increasing $|\mathbf{k}|$ the quantity $|u_{\mathbf{k}1}(\rho_2)| \to$
$\to 1$ so that the operator $\xi_{\rho_2}(\mathbf{k}) \to a_{\mathbf{k}1}$. In this spectral region the polariton states ρ_2 do not differ essentially from the transverse photon states (the dispersion rule for this photons is $\mathcal{E}_{\rho_2} = \hbar kc/\epsilon_1$, where the quantity ϵ_1 which appears in (4.33), accounts for the contribution of far resonances). In the spectral region, where $\mathcal{E}_\rho(\mathbf{k}) \simeq E(\mathbf{k})$, the polariton states are neither Coulomb excitons, nor transverse photons.

The above property is very important in the theory of nonlinear optical effects, in a number of problems related to exciton kinetics at low temperatures, etc.

4.2.1 Operators of electric and magnetic fields

In this subsection we derive formulas for the operators of the electric and the magnetic field in crystals, expressing them in terms of the Bose amplitudes $\xi_\rho^\dagger$ and ξ_ρ.

Using the Coulomb gauge for the vector potential we can decompose the electric field operator into two parts:

$$\hat{\mathbf{E}}(\mathbf{r}) = \hat{\mathbf{E}}^\perp(\mathbf{r}) + \hat{\mathbf{E}}^\parallel(\mathbf{r}), \tag{4.39}$$

where $\hat{\mathbf{E}}^\perp(\mathbf{r})$ and $\hat{\mathbf{E}}^\parallel(\mathbf{r})$ are the transverse and the longitudinal parts, respectively. The first of the above operators is given by

$$\hat{\mathbf{E}}^\perp(\mathbf{r}) = -\frac{1}{c}\frac{\partial \hat{\mathbf{A}}}{\partial t} = -\frac{i}{\hbar c}\left(\hat{H}\hat{\mathbf{A}} - \hat{\mathbf{A}}\hat{H}\right). \tag{4.40}$$

The crystal Hamiltonian, which appears in (4.40), in both cases of the retarded and Coulomb interaction, is expressed in terms of $\xi_\rho^\dagger$ and ξ_ρ by the formula (4.19). As concerning the operator $\hat{\mathbf{A}}(\mathbf{r})$, making use of (4.5) and (4.16) we find

$$\hat{\mathbf{A}}(\mathbf{r}) = \sum_{\mathbf{k}j,\rho}\left(\frac{2\pi\hbar c^2}{Vkc}\right)^{1/2}\mathbf{e}_{\mathbf{k}j}\left[u_{\mathbf{k}j}(\rho) + v_{\mathbf{k}j}(\rho)\right]\xi_\rho(\mathbf{k})e^{i\mathbf{k}\mathbf{r}} + \text{h.c.} \tag{4.41}$$

Inserting (4.19) into (4.41) and (4.40) and making use of the commutation rules for the Bose amplitudes ξ_ρ and $\xi_\rho^\dagger$ we obtain the following expression for the operator $\hat{\mathbf{E}}^\perp$

$$\hat{\mathbf{E}}^\perp(\mathbf{r}) = \frac{i}{\hbar}\sum_{\mathbf{k}j,\rho}\left(\frac{2\pi\hbar c^2}{Vkc}\right)^{1/2}\mathcal{E}_\rho(\mathbf{k})\mathbf{e}_{\mathbf{k}j}\left[u_{\mathbf{k}j}(\rho) + v_{\mathbf{k}j}(\rho)\right]\xi_\rho(\mathbf{k})e^{i\mathbf{k}\mathbf{r}} + \text{h.c.} \tag{4.42}$$

To find the operator $\hat{\mathbf{E}}^\parallel(\mathbf{r})$, we consider only the long-wavelength limit of the longitudinal field, where some well-known relations from phenomenological theory can be applied. In particular, in virtue of the solenoidal character of the induction vector (div $\mathbf{D} = 0$), we have $\mathbf{k}\,\mathbf{D}(\omega,\mathbf{k}) = 0$ for plane waves. Simultaneously $\mathbf{D} = \mathbf{E} + 4\pi\,\mathbf{P}$ (see also Section 4.4) so that the longitudinal parts of the vectors $\mathbf{E}$ and $\mathbf{P}$ are related by

$$\hat{\mathbf{E}}^\parallel = -4\pi\,\hat{\mathbf{P}}^\parallel, \tag{4.43}$$

$\hat{\mathbf{P}}$ being the polarization per unit volume.

In the exciton region of the spectrum this operator can be written as[39]

$$\hat{\mathbf{P}} = \sum_{\mu,\mathbf{k}}\mathbf{P}(\mathbf{k},\mu)B_\mu(\mathbf{k})e^{i\mathbf{k}\mathbf{r}} + \text{h.c.} \tag{4.44}$$

where $\mathbf{P}(\mathbf{k},\mu)$ is the amplitude of the dipole moment operator per unit volume corresponding to the transition from the ground state to the state with a Coulomb exciton $(\mathbf{k}\mu)$.

---

[39] Clearly (4.44) is an approximation since it contains only matrix elements corresponding to transitions from the ground state to excited states, whereas transitions between excited crystal states are not accounted for. However, this inadequacy is not very essential if we consider only situations where the number of excitons is small compared to the number of molecules. When considering situations caused by high intensity radiation, a more precise expression for $\hat{\mathbf{P}}$ is needed.

Now we can use (4.43) and (4.44) obtaining

$$\hat{\mathbf{E}}^{\|} = -4\pi \sum_{\mu,\mathbf{k},\rho} \mathbf{k}\frac{(\mathbf{k}\mathbf{P}(\mathbf{k}\mu))}{k^2} B_\mu(\mathbf{k})e^{i\mathbf{k}\mathbf{r}} + \text{h.c.}, \qquad (4.45)$$

and, in consequence, in terms of the variables ξ_ρ and $\xi_\rho^\dagger$ (see eqn 4.16)

$$\hat{\mathbf{E}}^{\|} = -4\pi \sum_{\mu,\mathbf{k},\rho} \mathbf{k}\frac{(\mathbf{k}\mathbf{P}(\mathbf{k}\mu))}{k^2} \left[u_{\mathbf{k}\mu}(\rho) + v_{\mathbf{k}\mu}(\rho)\right]\xi_\rho(\mathbf{k})e^{i\mathbf{k}\mathbf{r}} + \text{h.c.} \qquad (4.46)$$

In this manner, with regard to (4.39), we obtain for the total electric field

$$\hat{\mathbf{E}}(\mathbf{r}) = \sum_{\rho,\mathbf{k}} \mathbf{S}_\rho(\mathbf{k})\xi_\rho(\mathbf{k})e^{i\mathbf{k}\mathbf{r}} + \text{h.c.}, \qquad (4.47)$$

where

$$\mathbf{S}_\rho(\mathbf{k}) = -4\pi \sum_{\mu,\mathbf{k},\rho} \mathbf{k}\frac{(\mathbf{k}\mathbf{P}(\mathbf{k}\mu))}{k^2} \left[u_{\mathbf{k}\mu}(\rho) + v_{\mathbf{k}\mu}(\rho)\right]$$
$$+\frac{i}{\hbar}\sum_j \left(\frac{2\pi\hbar}{Vkc}\right)^{1/2} \mathcal{E}_\rho(\mathbf{k})\mathbf{e}_{\mathbf{k}j}\left[u_{\mathbf{k}j}(\rho) + v_{\mathbf{k}j}(\rho)\right]. \qquad (4.48)$$

The magnetic field operator $\mathcal{B}(\mathbf{r})$ can be obtained in an analogous manner. Since $\hat{\mathcal{B}}(\mathbf{r}) = \text{rot}\,\hat{\mathbf{A}}$, we obtain from (4.41)

$$\hat{\mathcal{B}}(\mathbf{r}) = i\sum_{\mathbf{k}j,\rho}\left(\frac{2\pi\hbar c^2}{Vkc}\right)^{1/2}[\mathbf{e}_{\mathbf{k}j}\times\mathbf{k}]\left[u_{\mathbf{k}j}(\rho) + v_{\mathbf{k}j}(\rho)\right]\xi_\rho(\mathbf{k})e^{i\mathbf{k}\mathbf{r}} + \text{h.c.} \qquad (4.49)$$

Using (4.44), (4.47), and (4.49) we can prove that the operators $\hat{\mathbf{E}}, \hat{\mathbf{P}}$, and $\hat{\mathcal{B}}$ obey the macroscopic Maxwell equations for crystals.

4.3 Polariton mechanism of exciton luminescence

As was discussed previously, the interaction between excitons and transverse photons (the retarded interaction) in crystals in the exciton spectral region leads to new hybrid elementary excitations (polaritons). These quasiparticles remain undamped if scattering processes such as, for example, the exciton–phonon interaction, are not taken into account. The retarded interaction results in a significant renormalization of the exciton and transverse photon spectra in the region of small wavevector k where the density of states is also small. At large k values ($k > E_{\text{exc}}/\hbar c$), the polariton gradually transforms into an exciton with growth of the wavevector. In this region of the exciton band the exciton–polaritons strongly interact with vibrations and defects of the lattice, which restricts their mobility. Nevertheless, during their lifetime, these polaritons can reach the region of the polariton band with small wavevectors.[40] For the polariton with small k values ($k < E_{\text{exc}}/\hbar c$) the exciton part of the energy is relatively small and the essential part is that of the energy of the transverse photon field. These polaritons, weakly interacting with the lattice vibrations, are capable of reaching the crystal surface and yielding the luminescence light. Therefore, the transitions in which a polariton with large momentum k is converted into polariton with small k can result in transformation of the electronic excitation energy into observable light. The above qualitative model of the so-called polariton mechanism of luminescence was first treated in the papers (13), (20). The direct calculation by Toyozawa (20) has shown that, despite the existence of the lower polariton branch, the time of "sliding" of the polariton from the range of large k values to the range of small k values is always of the order of the "old" radiation exciton lifetime, which can be calculated for the exciton using Fermi's golden rule. The region of the lower branch polariton where the dispersion curves of the Coulomb exciton and transverse photon meet, is known as the "bottle neck".

Defects of crystal structures or impurities can also participate in the processes of polariton luminescence. Of course, they are known, generally speaking, to give rise to local electronic excitation levels and their contribution to luminescence can be significant owing to possible transfer of the energy of excitons (or polaritons) to electrons of the defect with its subsequent radiation. However, as mentioned in (13), defects can give rise to exciton luminescence also without localization of the energy at the defect. This process of scattering and energy relaxation can be readily treated in terms of polaritons since the only perturbation in this case is the perturbing effect of the defect and not the exciton–photon interaction. For any coupling strength the exciton–transverse photon interaction is already included in the zeroth approximation. In this case the above mechanism corresponds to the transition within the polariton band from the region of large **k** values to the region of small **k** values below the region of the "bottle neck" at a constant energy. If the exciton band minimum is at $k = k_0$ and $k_0 \gg \Omega_{\text{exc}}/c$, then at low temperature, when only a small part of the band with $k \approx k_0$ is populated, the defect can give rise to a high-intensity narrow (the width $\delta \approx k_B T$) line $\hbar\omega = E(k_0)$ in the luminescence spectrum. If the effective mass of the exciton is negative the bottom of an exciton energy band is not always in the center of the Brillouin zone (Fig. 4.3 d). In gyrotropic crystals (see Fig. 4.3 e,f) the bottom of the exciton band is always in this position for polaritons with right-hand or left-hand circular polarization and this can lead to circular polarization of the long-wavelength lines of exciton luminescence (13).

Above we have assumed that the exciton wavevector is a good quantum number. However, it is clear that even in ideal crystals, in some cases, this assumption may not be valid. Consider, for example, exciton states in a very narrow exciton band. Since there always exist processes leading to scattering

---

[40]Here we assume for simplicity that the exciton has a positive effective mass and for this reason the mentioned process of energy relaxation goes in the direction of small polariton energies.

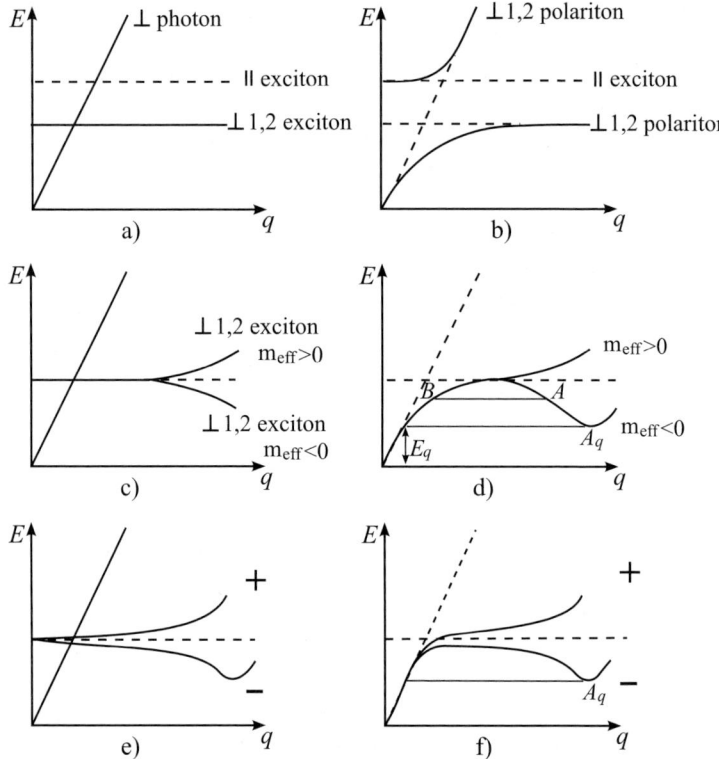

FIG. 4.3. The dispersion of polariton in cubic crystals. Nongyrotropic crystals: (a) The dependences of exciton and photon energy on wavevector, the retardation neglected; (b) the same but with retardation taken into account. The symbols ∥ and ⊥ indicate longitudinal and transverse polarization of excitons; (c) retardation neglected but dependence of the exciton energy on the wavevector taken into account; here and in (d), (e), and (f) only the lower branch of the polaritons shown; (d) the retardation and dependence of exciton energy on wavevector are taken into account. Gyrotropic crystals: (e) Dispersion of excitons in the cubic gyrotropic crystals if retardation is neglected; (f) the same when retardation is also taken into account; A_q denotes the position of the bottom of the polariton energy.

of an exciton accompanied with variations of its energy and momentum, the corresponding energy uncertainty can be of the order of the exciton bandwidth or even larger. Let M be the exciton bandwidth and $\tau(\mathbf{k})$ the exciton scattering time, $\mathbf{k}$ being the exciton wavevector. The quantity $\hbar/M$ describes approximately the time needed for excitation transfer from one molecule in the crystal to another one. If the inequality

$$\frac{\hbar}{\tau} \geq M$$

is satisfied, the uncertainty of the exciton energy is larger than the exciton bandwidth. In these circumstances the wavevector of the exciton is not a good quantum number; the band picture for excitons is not a correct one. The exciton wavefunctions must be chosen to describe the molecular excitations localized in different crystal lattice nodes and this has to be taken also into account in consideration of the exciton energy transfer in lattice. For such "localized" excitons the interaction of excitations with the transverse photon field is quite analogous to that which concerns molecules in a solid solution. However, this picture is correct if the oscillator strength of the electronic transition is not very large. In the opposite situation due to strong resonance of excitations in the medium with light, a small region of coherent polariton states can appear in the spectrum of excitations in the region of resonance, similar to the spectrum of cavity polaritons in a microcavity with disordered organics (see Ch. 10).

Note that in some cases, even for sufficiently wide exciton bands, when the wavevector of the exciton is a good quantum number, the polaritonic picture in the spectrum of quasiparticles does not appear. This can take place if the excitonic transition has a small transition dipole moment and as a consequence a small oscillator strength and a small value of the transverse–longitudinal splitting of the exciton, but at the same time there is a large broadening of the exciton resonance. In the region of such weak resonances the effect of the retardation on the excitation spectrum can be ignored. In this case we can treat Coulomb excitons and transverse photons, and not polaritons, as elementary excitations of the zeroth approximation and consider the exciton–photon interaction in the framework of Fermi's golden rule as a weak perturbation giving rise to transitions between states of the zeroth approximation. The remarks made above demonstrated (see also Fig. 4.3) that the effects of retardation and the polariton picture exist even if the spatial dispersion is absent or is negligible. They are different physical effects. But, of course, the spatial dispersion in some cases (in the vicinity of intensive exciton resonances) exerts an important influence on the dispersion of polaritons. As will be shown in Section 4.5, the radiative width of exciton (polariton) states appears in ideal one- and two-dimensional structures. In such cases the crystal translational symmetry, when considering the interaction of excitons with photons, preserves the conservation, not of the total photon and exciton wavevectors, but only of some its projections (one or two, respectively), which yields some interesting peculiarities (see (21) and also Section 4.5).

4.4 The dielectric tensor and the phenomenology of long-wavelength excitons

The analysis of excitons in the presence of retardation given in the previous sections of this chapter was based on making use of the Hamiltonian (4.2), (4.4). This Hamiltonian (see (1)) corresponds to the microscopic Maxwell equations. Since we are interested in long-wavelengths electromagnetic waves in crystals, the analysis can be carried out in the framework of macroscopic electrodynamics (17).

In the case of nonmagnetic media, and without distinguishing the magnetic induction **B** and the magnetic field intensity $\mathcal{H}$, the macroscopic Maxwell equations take the form

$$\text{rot } \mathbf{B} = \frac{1}{c}\frac{\partial \mathcal{D}}{\partial t} + \frac{4\pi}{c}\mathbf{j}_{\text{ext}}, \qquad \text{rot } \mathbf{E} = -\frac{1}{c}\frac{\partial \mathbf{B}}{\partial t},$$
$$\text{div } \mathcal{D} = 4\pi\rho_{\text{ext}}, \qquad \text{div } \mathbf{B} = 0. \qquad (4.50)$$

Here $\mathcal{D}$ and **E** are the electric induction and the electric field strength, respectively, and $\mathbf{j}_{\text{ext}}, \rho_{\text{ext}}$ are the external electric current and the charge densities. As is known, the Maxwell equations (4.50) must be complemented by the so-called material equations, which enable the expression of $\mathcal{D}$ as a function of **E**. In the linear approximation the relation between $\mathcal{D}$ and **E** can be taken as

$$\mathcal{D}_i(\mathbf{r},t) = \int_{-\infty}^{t} dt' \int d^3 r' \hat{\epsilon}_{ij}(\mathbf{r},\mathbf{r}',t,t') E_j(\mathbf{r}',t'). \qquad (4.51)$$

All properties of the medium are incorporated in the integral kernel in (4.51). For example, in vacuum

$$\hat{\epsilon}_{ij}(\mathbf{r},\mathbf{r}',t,t') = \delta_{ij}\delta(\mathbf{r}-\mathbf{r}')\delta(t-t') \quad \text{and} \quad \mathcal{D} = \mathbf{E}. \qquad (4.52)$$

In condensed matter the electric induction is given by

$$\mathcal{D} = \mathbf{E} + 4\pi\mathbf{P}, \qquad (4.53)$$

where **P** is the crystal unit volume polarization induced by the fields **E** and **B**.

Equation (4.51) refers to the causality principle owing to which the induction at time t is determined only by the field at times $t' \leq t$.

If the system is in a stationary state and in consequence its properties do not change in time, then the kernel $\hat{\epsilon}_{ij}$ can depend only on the difference $t - t'$. If the medium is uniform in space (which means it is invariant upon displacement by any vector), then $\hat{\epsilon}_{ij}$ depends only on the difference $\mathbf{r} - \mathbf{r}'$.

Under such conditions

$$\mathcal{D}_i(\mathbf{r},t) = \int_{-\infty}^{t} dt' \int d^3 r' \hat{\epsilon}_{ij}(\mathbf{r}-\mathbf{r}',t-t') E_j(\mathbf{r}',t'), \qquad (4.54)$$

and for fields which are plane waves

$$\mathbf{E}(\mathbf{r},t) = \mathbf{E}(\omega,\mathbf{k})e^{i(\mathbf{kr}-\omega t)},$$
$$\mathcal{D}(\mathbf{r},t) = \mathcal{D}(\omega,\mathbf{k})e^{i(\mathbf{kr}-\omega t)}, \qquad (4.55)$$

we have

$$\mathcal{D}_i(\omega,\mathbf{k}) = \epsilon_{ij}(\omega,\mathbf{k})E_j(\omega,\mathbf{k}), \qquad (4.56)$$

with
$$\epsilon_{ij}(\omega, \mathbf{k}) = \int_0^\infty d\tau \int d^3 R \, e^{-i(\mathbf{kR} - \omega\tau)} \hat{\epsilon}_{ij}(\mathbf{R}, \tau). \qquad (4.57)$$

Sometimes instead of the tensor ϵ_{ij} it proves convenient to use the inverse tensor ϵ_{ij}^{-1}. In this case, obviously

$$E_i(\omega, \mathbf{k}) = \epsilon_{ij}^{-1} \mathcal{D}_j(\omega, \mathbf{k}). \qquad (4.58)$$

The quantity ϵ_{ij} is called the complex dielectric tensor. In the framework of the phenomenological theory this tensor is supposed to be known.

From (4.54) and (4.57) it follows that the dependence of the tensor $\epsilon_{ij}(\omega, \mathbf{k})$ on $\mathbf{k}$ (this dependence is called the spatial dispersion; for details see (19)) is related to the fact that the magnitude of the electric induction at point $\mathbf{r}$ is determined by the magnitude of the electric field not only at the point $\mathbf{r}$, but also by its values in a certain region around $\mathbf{r}$. Thus the spatial dispersion originates from a nonlocal relation of $\mathcal{D}$ and $\mathbf{E}$. A similar dependence on ω (time or frequency dispersion) is related to the nonlocality in time of the relation (4.54). The eigenfrequencies of the medium ω_i can be located in the considered frequency interval ω, so that the parameter characterizing the frequency dispersion – the ratio ω_i/ω – can be of the order 1. Thus, the frequency dispersion in the region of resonances can be large.

A quite different situation arises in the case of spatial dispersion. Wavelengths λ in the optical spectral region are much larger than the extension of the region centered at $\mathbf{r}$, from which essential contributions to the space integral in (4.54) originate ($\lambda = \lambda_0/n$, λ_0 being the wavelengths of normal waves with frequency ω in vacuum, $\lambda_0 \equiv 2\pi c/\omega$, n is the refractive index, increasing in the resonance region). In dielectric media this region of nonlocality is usually of the order of one or few lattice constants a, so that the ratio a/λ is small: $a/\lambda \sim 10^{-3}$. Therefore in most cases the spatial dispersion is small and its role is important only when it yields new phenomena as, for example, natural optical activity. If ω approaches a resonance value, then the wavelength λ in the medium decreases, so that the ratio a/λ increases. In this case new phenomena can occur (see, for details (19)).

The determination of the tensor $\epsilon_{ij}(\omega, \mathbf{k})$ for crystals, as for other condensed media, can be done by means of a microscopic theory (see Ch. 7). If the tensor $\epsilon_{ij}(\omega, \mathbf{k})$ is assumed to be known, we can solve the field equations (4.50) for small $\mathbf{k}$, obtaining all normal electromagnetic waves and their dispersion.

The normal waves result from eqn (4.50) if external sources are absent (i.e. for $\mathbf{j}_{\text{ext}} = \rho_{\text{ext}} = 0$). In this case we obtain from eqn (4.50) the following wave equation

$$\text{rot rot } \mathbf{E} + \frac{1}{c^2} \frac{\partial^2 \mathcal{D}}{\partial t^2} = 0, \qquad (4.59)$$

which for plane waves takes the form

$$\mathcal{D} = \frac{c^2}{\omega^2} \left[k^2 \mathbf{E} - \mathbf{k}(\mathbf{kE}) \right]. \qquad (4.60)$$

Substituting (4.57) into (4.60) we find

$$\left[\frac{\omega^2}{c^2}\epsilon_{ij}(\omega,\mathbf{k}) - k^2\delta_{ij} + k_ik_j\right]E_j(\omega,\mathbf{k}) = 0. \tag{4.61}$$

The above system of algebraic equations has a nontrivial solution when its determinant vanishes, i.e. when

$$\left|\frac{\omega^2}{c^2}\epsilon_{ij}(\omega,\mathbf{k}) - k^2\delta_{ij} + k_ik_j\right| = 0. \tag{4.62}$$

The solution gives the dispersion relation, i.e. the dependence of ω on $\mathbf{k}$, in the form

$$\omega_\ell = \omega_\ell(\mathbf{k}), \quad \ell = 1, 2, \ldots, \tag{4.63}$$

where the index ℓ refers to the normal waves – polaritons.

Instead of (4.63) we can use the inverse relation

$$n^2 = \frac{k^2 c^2}{\omega^2}, \tag{4.64}$$

where the refraction index n becomes a function of ω and $\mathbf{s} = \mathbf{k}/k$. In this case the determinantal equation (4.62) becomes an equation for $n^2(\omega,\mathbf{s})$

$$n^4\left(\epsilon_{ij}s_is_j\right) - n^2\left\{(s_i\epsilon_{ij}s_j)\operatorname{Tr}\hat{\epsilon}_{ij} - s_is_j\epsilon_{i\ell}\epsilon_{\ell j}\right\} + |\epsilon_{ij}| = 0. \tag{4.65}$$

Note that eqns (4.62) and (4.65) determine the dispersion relation only for those normal waves for which the electric vector is not longitudinal. Otherwise $\mathbf{E}(\omega,\mathbf{k}) = = (\mathbf{k}/k)E(\omega,\mathbf{k})$ and $\mathcal{D} = 0$ by virtue by eqn (4.60). On the other hand, it follows from (4.56) that for $\mathbf{E} \neq 0$ and $\mathcal{D} = 0$ the dispersion of longitudinal waves can be found from the equations

$$|\epsilon_{ij}(\omega,\mathbf{k})| = 0. \tag{4.66}$$

For example, in the case of isotropic media and when the spatial dispersion is absent, the relation (4.66) simplifies to

$$\epsilon(\omega) = 0. \tag{4.67}$$

If a normal wave is not longitudinal, then the longitudinal part of the field $\mathbf{E}$ can be excluded from (4.56) (see also (19), §2). Such an operation is possible since for normal waves div $\mathcal{D} = 0$, i.e. $\mathcal{D}\mathbf{s} = 0$. Using (4.56) and decomposing the electric field into transverse and longitudinal part $\mathbf{E}^\perp$ and $\mathbf{E}^\|$, respectively

$$\mathbf{E}(\omega,\mathbf{k}) = \mathbf{E}^\perp(\omega,\mathbf{k}) + \mathbf{E}^\|(\omega,\mathbf{k}), \tag{4.68}$$

we obtain

$$\epsilon_{ij}s_i\left(E_j^\perp + s_jE^\|\right) = 0, \tag{4.69}$$

and thus
$$\mathbf{E}^{\|} = -\mathbf{s}\frac{\epsilon_{ij}s_i E_j^\perp}{\epsilon_{rt}s_r s_t}. \tag{4.70}$$

Thus, we obtain instead of (4.56) the relation [41]

$$\mathcal{D}_i(\omega,\mathbf{k}) = \epsilon_{\perp,ij}(\omega,\mathbf{k})E_j^\perp(\omega,\mathbf{k}), \tag{4.71}$$

which for normal waves is equivalent to (4.56). However, this relation for fields with external sources when $\text{div}\,\mathcal{D} \neq 0$, opposite to (4.56), cannot be applied. The tensor $\epsilon_{\perp,ij}$ can be expressed in terms of the tensor ϵ_{ij} by the following relation

$$\epsilon_{\perp,ij} = \eta_{ii_1}\left(\epsilon_{i_1 j_1} - \frac{\epsilon_{i_1 \ell}\epsilon_{mj_1}s_\ell s_m}{\epsilon_{rt}s_r s_t}\right)\eta_{j_1 j}, \qquad \eta_{ij} = \delta_{ij} - s_i s_j. \tag{4.72}$$

This relation was obtained in (19). If, for example,

$$\epsilon_{ij} = \delta_{ij} - \frac{A p_i p_j}{\omega^2 - \omega_0^2},$$

then from (4.72) we find

$$\epsilon_{\perp,ij} = \eta_{ii_1}\left(\delta_{i_1 j_1} - \frac{A p_{i_1} p_{j_1}}{\omega^2 - \omega_0^2 + A(\mathbf{ps})^2}\right)\eta_{j_1 j}.$$

Because of
$$s_i \epsilon_{\perp,ij} = \epsilon_{\perp,ij} s_j = 0,$$

one can see that the tensor $\epsilon_{\perp,ij}$ can be considered as a two-dimensional tensor acting in a plane perpendicular to the vector $\mathbf{s}$.

For this reason let us choose a new coordinate system with the z-axis parallel to $\mathbf{s}$. In this new coordinate system the components of the tensor $\epsilon_\perp$ will be denoted by $\epsilon_{\perp,\alpha\beta}$, $\alpha,\beta = 1,2$. The field equation (4.60), with regard to (4.71), is of the form

$$\epsilon_{\perp,\alpha\beta}E_\beta^\perp = n^2 E_\alpha^\perp. \tag{4.73}$$

This evidently leads to a determinant, the vanishing of which gives the following dispersion relation

$$n^4 - n^2\left(\epsilon_{\perp,11} + \epsilon_{\perp,22}\right) + \epsilon_{\perp,11}\epsilon_{\perp,22} - \epsilon_{\perp,12}\epsilon_{\perp,21} = 0, \tag{4.74}$$

which for waves with $\mathbf{E}^\perp \neq 0$ is equivalent to the relation (4.62).

Concluding this section we consider the relation between the resonance energies of the dielectric susceptibility (corresponding to mechanical excitons, i.e. excitons calculated if the retardation is ignored) and the energies of Coulomb

---

[41] The tensor $\epsilon_{\perp,ij}$ for the region of exciton resonances was first calculated using microscopic theory by Pekar (18).

excitons. These relations are nontrivial and useful particularly for anisotropic crystals because just the frequencies of Coulomb excitons are resonance frequencies of light absorption and their dependence on the direction of the wavevector reflects a nonanalytical dependence of Coulomb excitons on the wavevector which we already discussed in Chs. 2 and 3.

The existence of this relation should be no surprise since, as we have demonstrated, the tensor $\epsilon_{ij}(\omega, \mathbf{k})$ determines the frequencies of all normal electromagnetic waves in a condensed medium. But this relation, as we show below, permits a simplified consideration of some properties of Coulomb excitons including the dependence of Coulomb exciton energies on $\mathbf{s}$ for $\mathbf{k} \to 0$, which by using microscopic theories is quite tedious.

When retardation is ignored, we can omit, in eqn (4.50), the terms containing time derivatives, vanishing in the limit $c \to \infty$. In this case for $\mathbf{j}_{\text{ext}} = \rho_{\text{ext}} = 0$, i.e. for normal waves, the equations for the electric field have the form

$$\operatorname{rot} \mathbf{E} = 0, \quad \operatorname{div} \mathcal{D} = 0,$$

which for plane waves give the relations

$$\mathbf{s} \times \mathbf{E}(\omega, \mathbf{k}) = 0, \quad \mathbf{s}\mathcal{D}(\omega, \mathbf{k}) = 0. \tag{4.75}$$

The first of these equations predicts that in the case of no retardation the electric field of normal waves is either longitudinal or equal to zero. Then, using the material equation (4.56) and the second of equations (4.75), we obtain for such waves ($\mathbf{E} = \mathbf{s}E$)

$$(s_i s_j \epsilon_{ij}(\omega, \mathbf{k})) E = 0$$

and for $E \neq 0$

$$s_i s_j \epsilon_{ij}(\omega, \mathbf{k}) = 0. \tag{4.76}$$

By this relation we can express the Coulomb exciton frequencies ω (or energies $\hbar\omega$) for $\mathbf{E} \neq 0$ as functions of $\mathbf{k}$, and for $k \to 0$ as functions of $\mathbf{s} = \mathbf{k}/k$.

With the purpose of illustrating eqn (4.76) we consider the energy region of a nondegenerate linear polarized dipole-allowed transition in a uniaxial crystal. When neglecting the spatial dispersion, the tensor $\epsilon_{ij}(\omega)$ has the following nonzero components (the z-axis is oriented along the optical axis):

$$\epsilon_{11} = \epsilon_{22} = \epsilon_0, \quad \epsilon_{33} = \epsilon_0 - \frac{A^{\|}}{\omega^2 - \omega_\mu^2}, \tag{4.77}$$

where we can assume that the quantity ϵ_0 in the considered frequency region does not depend on ω, ω_μ is the resonance frequency, and $A^{\|}$ gives the oscillator strength of the transition. In eqn (4.77) we have in mind that in the case of a nondegenerate dipole-allowed transition this transition must be polarized parallel to the optical axis.

Substituting (4.77) into (4.76) we find for the Coulomb exciton energy the following expression

$$E_\mu(0) = \hbar\omega \simeq \hbar\omega_\mu + \frac{1}{2}\frac{\hbar A^\|}{\omega_\mu \epsilon_0}\cos^2\theta, \qquad (4.78)$$

θ being the angle between the optical axis and the vector **s**. In a similar way we can discuss the dependence on **s** in the limit $\mathbf{k} \to 0$ for crystals obeying different symmetries.

If in a normal wave the electric field is $\mathbf{E} = 0$, instead of (4.76) we must use another relation resulting directly from (4.58). Since the vector $\mathcal{D}$ may differ from zero even for $\mathbf{E} = 0$, the determinant of the equation system

$$\epsilon_{ij}^{-1}\mathcal{D}_j = 0 \qquad (4.79)$$

should vanish. Therefore the dispersion law, i.e. the dependence $\omega = \omega(\mathbf{k})$, for Coulomb excitons with $\mathbf{E} = 0$ and $\mathcal{D} \neq 0$, has the form

$$\left|\epsilon_{ij}^{-1}(\omega, \mathbf{k})\right| = 0. \qquad (4.80)$$

Equation (4.80), for given **k**, may have multiple roots. Among them we choose only those solutions which obey the relation (4.79) and are transverse with respect to $\mathcal{D}$

$$\mathcal{D}\mathbf{k} = 0. \qquad (4.81)$$

The above condition is quite essential, since (4.81) is one of the field equations.

Consider, as an example, a crystal having a dielectric tensor with components given by the relations (4.77). One can easily verify that in this case eqn (4.80) takes the form

$$\frac{1}{\epsilon_{33}} = 0, \qquad (4.82)$$

which inserted into (4.79) gives, for the components of $\mathcal{D}$, the results $\mathcal{D}_1 = \mathcal{D}_2 = 0$, $\mathcal{D}_3 \neq 0$. Therefore for a vector **k** which is not perpendicular to the optical axis, we have no solutions with $\mathbf{E} = 0$ and $\mathcal{D} \neq 0$ (the condition (4.81) is not satisfied). If the vector **s** is perpendicular to the optical axis, then a solution exists and, by virtue of (4.82) and (4.77), we find for them $\omega = \omega_\mu$.

The same result can be obtained from (4.78), if we put $\theta \to \pi/2$.

In the vicinity of a dipole allowed degenerate transition, polarized perpendicular to the optical axis, instead of (4.77) we have

$$\epsilon_{11}(\omega) = \epsilon_{22}(\omega) = \epsilon_0 - \frac{A^\perp}{\omega^2 - \omega_\mu^2}, \qquad \epsilon_{33}(\omega) = \epsilon_0. \qquad (4.83)$$

In this case eqn (4.76) for excitons with $\mathbf{E} \neq 0$ yields

$$E_\mu(0) = \hbar\omega \simeq \hbar\omega_\mu + \frac{1}{2}\frac{\hbar A^\perp}{\omega_\mu \epsilon_0}\sin^2\theta. \qquad (4.84)$$

At the same time, for excitons with $\mathbf{E} = 0$ we obtain from (4.79) and (4.80)

$$\frac{1}{\epsilon_{11}(\omega)} = 0, \qquad \mathcal{D}_1 \neq 0, \qquad \mathcal{D}_2 \neq 0, \qquad \mathcal{D}_3 = 0. \tag{4.85}$$

The condition (4.81) now takes the form

$$\mathcal{D}_1 k_1 + \mathcal{D}_2 k_2 = 0$$

and can be satisfied for $\mathcal{D} \neq 0$. Equation (4.85) gives $\omega = \omega_\mu$.

Consequently in the vicinity of an allowed transition, polarized perpendicular to the optical axis in a uniaxial crystal, with Coulomb interaction fully taken into account, for a given value of $\mathbf{k}$ we obtain not one, but two exciton states. This distinction from the case when the tensor ϵ_{ij} is given by the relations (4.77), results from properties of states determined by the poles of ϵ_{ij}.

The poles of the tensor ϵ_{ij} correspond to exciton states obtained by solving the Coulomb problem, when the long-range counterpart of the Coulomb interaction is ignored or absent.[42] This problem will be discussed in greater depth in Ch. 7. Here we only notice that the tensor (4.80) corresponds to the frequency region of nondegenerate, and the tensor (4.83) to double-degenerate mechanical excitons.

When the long-range part of the Coulomb interaction is taken into account, in the second case the degeneracy is lifted, while in the first case we observe only a frequency shift. It is appropriate to note here that when the Coulomb interaction is fully accounted for, the exciton energy, in accordance with the results of the microscopic theory (see Chs. 2 and 3), depends on $\mathbf{s}$ when $\mathbf{k} \to 0$, i.e. reveals a nonanalytic function of $\mathbf{k}$ for $\mathbf{k} \to 0$.

In the determination of the form of the nonanalycity in the framework of the microscopic theory dipole sums must be considered. The above analysis seems to be simpler.

4.5 Giant radiative width of small wavevector polaritons in one- and two-dimensional structures ("polariton superradiance")

Up to this point we have considered the influence of retardation on the form of exciton bands in three-dimensional crystals. Examination of excitons in two- and one-dimensional crystals, to which we devote this section, is also of interest, since we observe a number of important and specific phenomena. As an aside we notice that the examination of exciton spectra from one- and two-dimensional crystals is important from the point of view of possible applications (quantum wells, quantum wires, polymers, etc.). Below we follow the paper (21).[43]

---

[42]See (19), §2, and Section 2.2, where those excitons were called "mechanical excitons".

[43]The results of this old paper later were re-created in a long list of publications. Some of them can be found, for example, in (22).

First consider the case of one-dimensional molecular crystals. In these crystals

$$\mathbf{n} = n\mathbf{d} \qquad \left(n = 0, \pm 1, \pm 2, \ldots, \pm \frac{N}{2} \right),$$

where $\mathbf{d}$ is the elementary cell vector of the chain, and $N+1$ is the number of molecules in chain. The exciton wavevector in such crystals is oriented along $\mathbf{d}$ and has the property

$$-\frac{\pi}{d} \leq k \leq \frac{\pi}{d}. \tag{4.86}$$

At the same time the wavevector of a transversal photon can be presented as the sum $\mathbf{q} = \mathbf{k}_{\|} + \mathbf{q}_{\perp}$, where $\mathbf{k}_{\|} = k_{\|}\frac{\mathbf{d}}{d}$; the vector $\mathbf{q}_{\perp}$ is the transverse with respect to $\mathbf{d}$ component of $\mathbf{q}$. It is unlimited with respect to its length and orientation. Therefore, making use of the relation

$$\sum_{\mathbf{n}} e^{i\mathbf{q}\mathbf{n}} = \begin{cases} N, & \frac{\mathbf{q}\mathbf{d}}{d} = \frac{2\pi}{d}m, & m = 0, \pm 1, \ldots, \\ 0, & \frac{\mathbf{q}\mathbf{d}}{d} \neq \frac{2\pi}{d}m, & m = 0, \pm 1, \ldots, \end{cases} \tag{4.87}$$

we obtain in a quite similar way as was done in Section 4.1 for three-dimensional crystals, the following expression for the operator of the exciton–phonon interaction [44]

$$\hat{H}_{\text{int}} = -\sum_{\mu,j,\mathbf{q}} T(j,\mathbf{q},\mu) \Big\{ a_{\mathbf{q}j} \left[B_\mu(-\mathbf{k}_\|) - B_\mu^\dagger(\mathbf{k}_\|) \right]$$
$$+ a_{\mathbf{q}j}^\dagger \left[B_\mu(\mathbf{k}_\|) - B_\mu^\dagger(-\mathbf{k}_\|) \right] \Big\}, \tag{4.88}$$

where

$$T(j,\mathbf{q},\mu) = i \left(\frac{2\pi N}{V\hbar q c} \right)^{1/2} E_\mu(\mathbf{k}_\|) \, \mathbf{P}(\mathbf{k}_\|,\mu) \, \mathbf{e}_{\mathbf{q}j}. \tag{4.89}$$

The quantity V in (4.89) is the fundamental volume for photons: $V = L_1 L_2 L_3$. If the z-axis is chosen parallel to the vector $\mathbf{d}$, then $L_3 = Nd$ is the fundamental length related to cyclic boundary conditions for excitons.[45]

Let us assume that in an initial state the exciton $(\mu, \mathbf{k})$ propagates along the chain. If the operator (4.88) is considered as the cause of transitions, than in the framework of perturbation theory the probability per unit time that an exciton decays into a photon, is given by the relation

$$W_\mu = \frac{2\pi}{\hbar} \sum_{\mathbf{q}j} |\langle \text{init}| \hat{H}_{\text{int}} |\text{fin}\rangle|^2 \, \delta\left(E_\mu(\mathbf{k}) - \hbar q c\right). \tag{4.90}$$

---

[44] In $\hat{H}_{\text{int}}$ we omit the terms proportional to $\mathbf{A}^2$. Therefore we focuss attention only on those spectral region where $\mathcal{E} \simeq E_\mu(\mathbf{k})$.

[45] Only by this condition are the relations (4.87) satisfied.

Since the exciton ($\mu\mathbf{k}$) interacts only with those photons for which $\mathbf{k}_\| = \mathbf{k}$ (in (4.88) we consider only processes in the optical part of the spectrum and Umklapp type processes are ignored) the quantity (4.90) can be expressed as

$$W_\mu = \frac{2\pi}{\hbar} \sum_{\mathbf{q}_\perp j} |T(j, \mathbf{q}_\perp, \mu)|^2 \, \delta\left(E_\mu(\mathbf{k}) - \hbar c \sqrt{k^2 + q_\perp^2}\right). \tag{4.91}$$

According to (4.89)

$$\sum_{\mathbf{q}j} |T(j, \mathbf{q}, \mu)|^2 = \frac{2\pi N}{V \hbar q c} E_\mu^2(\mathbf{k}_\|) \left\{ |\mathbf{P}(\mathbf{k}_\|\mu)|^2 - \frac{(\mathbf{qP}(\mathbf{k}_\|\mu))^2}{q^2} \right\}. \tag{4.92}$$

Hence, multiplying (4.91) by the density of states of photons with given $\mathbf{k}_\| = \mathbf{k}$, i.e. by the quantity

$$\frac{L_1 L_2 q_\perp \, dq_\perp \, d\varphi}{(2\pi)^2},$$

and performing the integration we finally obtain that the radiative width $\Gamma_\mu(\mathbf{k}) \equiv \hbar W_\mu(\mathbf{k})$ is equal to

$$\Gamma_\mu(\mathbf{k}) = \frac{\pi E_\mu^2(\mathbf{k}) |\mathbf{P}(\mathbf{k}\mu)|^2}{\hbar^2 c^2 d} \left[1 + \cos^2\theta - \frac{\hbar^2 c^2 k^2}{E_\mu^2(\mathbf{k})} (2\cos^2\theta - \sin^2\theta) \right], \tag{4.93}$$

if $k < k_0 \equiv E_\mu(k_0)/\hbar c$, and is equal to zero if $k > k_0$ (in (4.93) θ is the angle formed by the vector $\mathbf{P}(\mathbf{k}_\|\mu)$ and the chain axis).

The radiative width of an excited isolated molecule is proportional to (see (1))

$$\frac{E_\mu^3(\mathbf{k})|\mathbf{P}|^2}{\hbar^2 c^3}.$$

Hence the above calculated radiative width in one-dimensional crystals with $k < k_0$, as follows from (4.93), is greater than the corresponding quantity in an isolated molecule by $1/k_0 d$ times. In the optical spectrum region $1/k_0 d = \lambda/(2\pi d) \simeq 10^2$–$10^3$. In consequence, the lifetime of exciton states in one-dimensional ideal molecular crystals may be several orders of magnitude smaller than the corresponding lifetime in isolated molecules.

An entirely analogous procedure can be applied to determine the radiative width of exciton states in two-dimensional molecular crystals. In this case an exciton can decay only into a photon, which has an in-plane component of the wavevector equal to the exciton wavevector. Denote this component by $\mathbf{k}_\|$, and the component perpendicular to the crystal plane by $\mathbf{q}_\perp$ ($\mathbf{q} = \mathbf{k}_\| + \mathbf{q}_\perp$). Then the probability of decay of an exciton with given wavevector $\mathbf{k}$ is equal to

$$W_\mu(\mathbf{k}) = \frac{2\pi}{\hbar} \sum_{j, \mathbf{q}_\perp} |T(j, \mathbf{k} + \mathbf{q}_\perp, \mu)|^2 \, \delta\left(E_\mu(\mathbf{k}) - \hbar c \sqrt{k^2 + q_\perp^2}\right).$$

Since in this case
$$\frac{(\mathbf{qP})^2}{q^2} = \frac{k_\parallel^2 P^2 \cos^2\theta + q_\perp^2 P^2 \cos^2\varphi + 2k_\parallel q_\perp P^2 \cos\theta\cos\varphi}{k_\parallel^2 + q_\perp^2},$$

where θ is the angle formed by the vectors $\mathbf{k}_\parallel$ and $\mathbf{P}$, and φ the angle formed by the vector $\mathbf{P}$ and the normal to the crystal surface,

$$\Gamma_\mu(\mathbf{k}) = \frac{4\pi E_\mu^3(\mathbf{k})}{\hbar^3 c^3} |\mathbf{P}(\mathbf{k},\mu)|^2 \left\{ 1 - \cos^2\varphi - \frac{k^2}{k_0^2}\left(\cos^2\theta - \cos^2\varphi\right) \right.$$

$$\left. - \frac{2k\sqrt{k_0^2 - k^2}}{k_0^2} \cos\theta\cos\varphi \right\} \frac{1}{d^2 k_0 \sqrt{k_0^2 - k^2}} \quad (4.94)$$

if $|\mathbf{k}| < k_0$ and $\Gamma_\mu(\mathbf{k}) = 0$ if $|\mathbf{k}| > k_0$.

The function $\Gamma_\mu(\mathbf{k})$ has a singularity for $|\mathbf{k}| \to k_0$. However, if we put $\sqrt{k_0^2 - k^2} \sim k_0$, then the radiative width (4.94) is larger than the corresponding quantity for an isolated level by $1/(d^2 k_0^2) \sim 10^4$ times. In these conditions the exciton lifetime τ can be of the order $\tau_0 (dk_0)^2 \sim 10^{-12}$–$10^{-13}$ ($\tau_0 \simeq 10^{-8}$–10^{-9} s is the lifetime of a free molecule excitation).

In order to discuss more correctly the question of radiative width we keep in mind the fact that the operator (4.88) provokes not only a radiative damping of exciton states, but also changes their dispersion rule. To obtain this dispersion, we add to (4.88) operators of the free exciton and transverse photon fields, as was done in Sections 4.1 and 4.2, and diagonalize the total Hamiltonian so obtained. Recall for comparison that in the case of an ideal 3D crystal after such diagonalization of the total Hamiltonian the radiative width of new excitations (polaritons) disappeared. We show below that in 1D and 2D the results are completely different.

It can be easily verified that the corresponding transition coefficients u and v satisfy the relations

$$[E_\mu(\mathbf{k}) - \mathcal{E}] u_\mu(\mathbf{k}) + \sum_{j,\mathbf{q}_\perp} T(j,\mathbf{q}_\perp + \mathbf{k}, \mu) \{u_j(\mathbf{q}_\perp + \mathbf{k}) + v_j(-\mathbf{k} + \mathbf{q}_\perp)\} = 0,$$

$$(\hbar qc - \mathcal{E}) u_j(\mathbf{q}_\perp + \mathbf{k}) - T(j,\mathbf{q}_\perp + \mathbf{k}, \mu)[u_\mu(\mathbf{k}) - v_\mu(-\mathbf{k})] = 0,$$

$$v_j(-\mathbf{k} + \mathbf{q}_\perp) = \frac{\hbar qc - \mathcal{E}}{\hbar qc + \mathcal{E}} u_j(\mathbf{q}_\perp + \mathbf{k}), \quad (4.95)$$

$$v_\mu(-\mathbf{k}) = -\frac{E_\mu(\mathbf{k}) - \mathcal{E}}{E_\mu(\mathbf{k}) + \mathcal{E}} u_\mu(\mathbf{k}).$$

Eliminating $u_j(\mathbf{q}), v_j(\mathbf{q})$ and $v_\mu(-\mathbf{k})$ from eqns (4.95) we obtain the following equation determining the dependence of the exciton energy on $\mathbf{k}$, when the retardation is accounted for

$$E_\mu^2(\mathbf{k}) - \mathcal{E}^2 = \sum_{j,\mathbf{q}_\perp} \frac{4\hbar c E_\mu \sqrt{k^2 + q_\perp^2}}{\hbar^2 (k^2 + q_\perp^2) c^2 - \mathcal{E}^2} |T(j,\mathbf{q}_\perp + \mathbf{k}, \mu)|^2. \quad (4.96)$$

By examination of the above relation we must keep in mind that the quantities are defined by (4.89) only in the case of long waves, when $qd \ll 1$. Only in this limiting case are the quantities T expressed in terms of matrix elements of dipole, quadrupole, etc. unit cell operators. For $qd > 1$ the expression (4.89) is inappropriate and must be replaced by a more accurate one. Since for $qd > 1$ the quantities T decay quite rapidly, we can use the expression (4.89), but the summation over $\mathbf{q}_\perp$ in (4.96) will be limited to terms

$$|\mathbf{q}_\perp| < q_0, \qquad \text{where} \qquad q_0 \simeq \frac{1}{d}.$$

First we consider the relation (4.96) for $\mathcal{E} \simeq E_\mu$ in the case of a one-dimensional chain. Making use of the relations (4.92),(4.89), replacing in (4.96) the summation by integration and performing the integration over angles in the $\mathbf{q}_\perp$ plane, we obtain instead of (4.96), the following equation

$$E_\mu^2(\mathbf{k}) - \mathcal{E}^2 = A \int_0^{q_0^2} \frac{dz}{z-a} \left[1 - \frac{k^2 \cos^2\theta + \frac{z}{2}\sin^2\theta}{k^2 + z} \right], \qquad (4.97)$$

where

$$a = \frac{\mathcal{E}^2 - \hbar^2 k^2 c^2}{\hbar^2 c^2}, \qquad A = \frac{2E_\mu^3(\mathbf{k}) |\mathbf{P}(\mathbf{k},\mu)|^2}{d\,\hbar^2 c^2}. \qquad (4.98)$$

Since the constant A is proportional to the square of the dipole matrix element, it is clear that the retardation can be essential only for sufficiently strong dipole transitions. If, for example, we take $|\mathbf{P}| \simeq 2e \times 10^{-8}$ cm, $E_\mu \simeq 3$ eV, $d \simeq 5 \times 10^{-8}$ cm, then we obtain for A the value $A \simeq 2 \times 10^{-4}$ (eV)^2. If $|\mathbf{P}|$ is equal to $(1/2)e \times 10^{-8}$ cm, then $A \simeq 10^{-5}$ (eV)^2, etc.

Solutions of (4.97) are, in general, complex. For a field $F \sim \exp(-i\omega t)$: $\mathcal{E} = \mathcal{E}' - i\mathcal{E}''$. In those spectral region, where $|\mathcal{E}''| \ll |\mathcal{E}'|$, we can put the photon Green's function expression $1/(z-a)$ into the form

$$\frac{1}{z-a-i\epsilon} = \mathcal{P}\frac{1}{z-a} + i\pi\delta(z-a), \qquad \epsilon \to +0, \qquad (4.99)$$

where now

$$a = \frac{(\mathcal{E}')^2 - \hbar^2 k^2 c^2}{\hbar^2 c^2}, \qquad (4.100)$$

and $\mathcal{P}$ indicates the principal part. In a first approximation, wave damping can be neglected. Then the equation for $\mathcal{E}'(k)$ takes the form

$$E_\mu^2(\mathbf{k}) - (\mathcal{E}')^2 = A\,\mathcal{P} \int_0^{q_0^2} \frac{dz}{z-a} \left[1 - \frac{k^2 \cos^2\theta + \frac{z}{2}\sin^2\theta}{k^2 + z} \right]. \qquad (4.101)$$

The imaginary part $\mathcal{E}''(k)$ (half the total radiative width) can be estimated approximately as

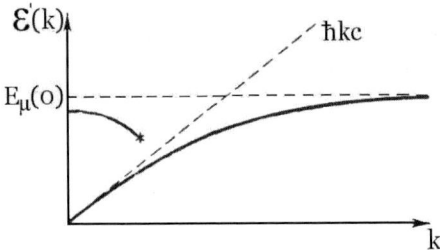

FIG. 4.4. The dispersion relation $\mathcal{E}'(k)$ for angle $\theta = \pi/2$.

$$\mathcal{E}'' = \frac{\pi A}{2\mathcal{E}'} \int_0^{q_0^2} \left[1 - \frac{k^2 \cos^2 \theta + \frac{z}{2} \sin^2 \theta}{k^2 + z}\right] \delta\left[z - \frac{(\mathcal{E}')^2 - \hbar^2 c^2 k^2}{\hbar^2 c^2}\right] dz. \quad (4.102)$$

Performing the integration in (4.101) and keeping in mind that in the spectral region under discussion $k^2 \ll q_0^2$, $|a| \ll q_0^2$, we obtain

$$E_\mu^2(\mathbf{k}) - (\mathcal{E}')^2 = \frac{A}{2}(1 + \cos^2 \theta) \ln\left|\frac{\hbar^2 q_0^2 c^2}{(\mathcal{E}')^2 - \hbar^2 c^2 k^2}\right|$$
$$+ \frac{A(1 - 3\cos^2 \theta)}{2} \frac{\hbar^2 k^2 c^2}{(\mathcal{E}')^2} \ln\left|\frac{k^2}{a}\right|. \quad (4.103)$$

We note again that the equation determining $\mathcal{E}'(\mathbf{k})$ applies only in the spectral region, where $|\mathcal{E}''| < |\mathcal{E}'|$. To establish this spectral region we must calculate the quantity $\mathcal{E}'(\mathbf{k})$. From (4.102) we find that if $\mathcal{E}'(k) > \hbar k c$,

$$\mathcal{E}'' = \frac{\pi A}{2\mathcal{E}'}\left[1 - \frac{k^2 \cos^2 \theta + \frac{a}{2}\sin^2 \theta}{k^2 + a}\right] \quad (4.104)$$

and $\mathcal{E}'' = 0$ if $\mathcal{E}'(k) < \hbar k c$.

This dependence of the dressed exciton dispersion $\mathcal{E}'(k)$ for angle $\theta = 0$ when the transition dipole moment is perpendicular to chain is displayed in Fig. 4.4. For another orientation of the exciton transition dipole moment the dependence of $\mathcal{E}'(k)$ can be very different. For excitons with small transition dipole moment the renormalization of the exciton dispersion due to account of retardation is usually small and can be important only at low temperature of order of 1–2 K or less because of the smallness of the parameter A/E_μ. In the same situation the radiative width of exciton states with small wavevectors determined by the same parameter A/E_μ can be a hundred-fold larger than the radiative width of a molecule in solution. Very interesting is the problem of the temperature dependence of the radiative lifetime and we come back to the discussion of this problem later.

The states which have radiative width equal to zero cannot be excited by light.[46] Therefore only the states with energy $\mathcal{E}'(k) > \hbar c k$ can occur in the absorption spectrum.

Quite analogously we can consider retardation effects in two–dimensional crystals. In this case the excitonic dispersion rule has the form

$$E_\mu^2(\mathbf{k}) - \mathcal{E}^2 = \frac{B}{2} \int_{-q_0}^{q_0} \frac{dq_\perp}{q_\perp^2 - a} \left\{ 1 - \frac{k^2 \cos^2 \theta + q_\perp^2 \cos^2 \varphi + 2kq_\perp \cos\theta \cos\varphi}{k^2 + q_\perp^2} \right\}, \qquad (4.105)$$

where

$$a = \frac{\mathcal{E}^2 - \hbar^2 k^2 c^2}{\hbar^2 c^2}, \qquad B = \frac{8 E_\mu^3(\mathbf{k}) |\mathbf{P}(\mu, \mathbf{k})|^2}{d^2 \hbar^2 c^2}. \qquad (4.106)$$

The angles θ and φ were defined above by discussion of eqn (4.93). For the sake of simplicity, we take $\theta = \pi/2$, i.e. assume that the vectors $\mathbf{P}(\mu, \mathbf{k})$ are perpendicular to the crystal plane (hence automatically $\varphi = 0$). In this case the equation for $\mathcal{E}'(k)$ (by the assumption that $|\mathcal{E}'| \gg |\mathcal{E}''|$ and that $q_0 \gg k$) takes the form

$$E_\mu^2(k) - (\mathcal{E}')^2 = B \frac{\hbar^2 k^2 c^2}{(\mathcal{E}')^2} \mathcal{P} \int_0^{q_0} \frac{dq_\perp}{q_\perp^2 - a} - B \frac{\pi}{2} \frac{\hbar^2 k^2 c^2}{(\mathcal{E}')^2}. \qquad (4.107)$$

On the other hand

$$\mathcal{E}''(k) = \begin{cases} \frac{\pi B}{4 \mathcal{E}'(k)} \frac{\hbar^2 k^2 c^2}{(\mathcal{E}')^2} \frac{1}{\sqrt{a}} \frac{\pi}{2}, & \text{if } \mathcal{E}'(\mathbf{k}) > \hbar k c; \\ 0 & , \text{if } \mathcal{E}'(\mathbf{k}) < \hbar k c. \end{cases} \qquad (4.108)$$

If $a < 0$, which occurs for $\mathcal{E}'(\mathbf{k}) < \hbar k c$,

$$\mathcal{P} \int_0^{q_0} \frac{dq_\perp}{q_\perp^2 - a} \simeq \frac{1}{\sqrt{|a|}} \frac{\pi}{2},$$

then eqn (4.107) can be rewritten in the form

$$E_\mu^2(\mathbf{k}) - (\mathcal{E}')^2 (\mathbf{k}) = \frac{B \pi \hbar^2 c^2 k}{2 (\mathcal{E}')^2} \left(\frac{k}{\sqrt{|a|}} - 1 \right). \qquad (4.109)$$

For large k, when $\hbar k c \gg \mathcal{E}'$, $\sqrt{|a|} \sim k$, from (4.109) we find $\mathcal{E}'(k) \simeq E_\mu(\mathbf{k})$.

---

[46]When the interaction with phonons is ignored.

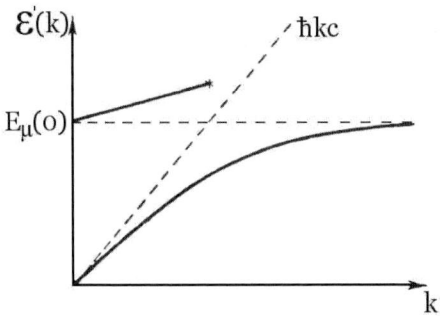

FIG. 4.5. Excitonic dispersion in two-dimensional crystals with retardation.

Now consider those spectral regions, where $\mathcal{E}'(\mathbf{k}) \ll E_\mu(\mathbf{k})$. In this case solutions of eqn (4.109) satisfying the condition $\mathcal{E}' < \hbar k c$ exist only if $k > \tilde{k}$, where $\tilde{k}$ is defined as

$$\tilde{k} = \frac{2\pi B}{E_\mu^2(0)}. \tag{4.110}$$

If $a > 0$, which happens when $\mathcal{E}'(\mathbf{k}) > \hbar k c$, then

$$\mathcal{P} \int_{-q_0}^{q_0} \frac{dq_\perp}{\mathbf{q}_\perp^2 - a} \simeq -\frac{2}{q_0}.$$

Since in the spectral region under consideration $k \ll q_0$, $\sqrt{|a|} \ll q_0$, $\hbar q_0 c \gg E_\mu(\mathbf{k})$, the first term on the r.h.s. of eqn (4.107) can be neglected. Hence, in this approximation

$$(\mathcal{E}'(\mathbf{k}))^2 = \frac{E_\mu^2(\mathbf{k})}{2}\left[1 + \sqrt{1 + \frac{k}{\bar{k}}}\right], \tag{4.111}$$

where

$$\bar{k} = \frac{E_\mu^4(\mathbf{k})}{2\pi B \hbar^2 c^2}. \tag{4.112}$$

Estimated values of $\bar{k}$ are of the order $\bar{k} \geq 10^6\text{--}10^8$ cm^{-1}. Therefore in those spectral regions where $k \simeq E_\mu/\hbar c \simeq 3 \cdot 10^5$ cm^{-1}, we have $k/\bar{k} \ll 1$ so that

$$\mathcal{E}'(\mathbf{k}) = E_\mu(\mathbf{k})\left(1 + \frac{k}{8\bar{k}}\right). \tag{4.113}$$

We note that excitons of this branch, even for $k \to 0$, possess a large group velocity

$$v = \frac{E_\mu(0)}{8\hbar \bar{k}} \simeq 10^7\text{--}10^9 \text{ cm/s}$$

and hence should weakly interact with lattice vibrations.

Note, however, that the dependence of the polariton energy on the wavevector, which arose when only retardation is taken into account, is correct only if we can neglect the dependence of the energy of the Coulomb exciton E_μ on $\mathbf{k}$, arising from instantaneous Coulomb interaction. For example, if we apply this theory for 2D quantum well polaritons, the linear term in the dispersion of polaritons will be cancelled because in this case the linear term as a function of the energy of the quantum well exciton on the wavevector has the same value with opposite sign.

Equation (4.109) applies if $|\mathcal{E}'| < |\mathcal{E}''|$. The lower branch ($a < 0$) satisfies this condition since there $\mathcal{E}'' = 0$. At the upper branch ($a > 0$) eqn (4.113) applies only for those values of $\mathbf{k}$ for which the quantity defined in (4.108) is much smaller than $\mathcal{E}'(\mathbf{k})$. This condition determines the limit of the upper branch. In Fig. 4.5, where the excitonic dispersion in two-dimensional crystals with retardation accounted for is schematically sketched, the limit is indicated by an asterisk.

Above we discussed the dispersion rule and radiative width of states in the special case of vectors $\mathbf{P}(\mathbf{k}, \mu)$ perpendicular to the crystal plane. Other situations can be treated in the same manner. The dispersion for this case, as compared with those displayed in Fig. 4.5, will change. To calculate the dispersion of a polariton in a wide region of the it is necessary to take into account the terms proportional to $\mathbf{A}^2$ in the operator of the exciton–photon interaction.

The above results are valid for crystal dimensions which are large compared to the wavelength

$$\lambda_0 = \frac{2\pi}{k_0}, \quad \text{where} \quad k_0 = \frac{E_\mu(\mathbf{k}_0)}{\hbar c}.$$

If, for example, L_3 denotes the chain length, then the uncertainty of the exciton wavevector is equal to $\Delta k \sim 1/L_3$, and if $k_0 \leq 1/L_3$, then in the spectral region, which is relevant from the point of view of retardation effects $k \sim k_0$, and the wavevector is not a good quantum number. In consequence, the above results are not valid. In other words, the results are valid if the chain length $L_3 \geq \lambda_0$ or, in the case of two-dimensional crystals, both dimensions L_1 and L_2 also satisfy the condition $L_1, L_2 \geq \lambda_0$.

Note that disorder, which was neglected above, is also able to destroy coherent "superradiant" emission of 1D and 2D polaritons. This can happen when the uncertainty of the exciton wavevector arising due to scattering of an exciton by disorder will be of the order of value $E_{\text{exc}}/\hbar c$. A quantitative theory of this effect has been developed in the paper by Orrit et al. (23).

4.6 Effective radiative lifetime

It was demonstrated above that in low-dimensional structures only exciton–polaritons from states with small wavevectors $k < k_0$ can spontaneously undergo radiative decay while conserving energy and momentum. A distribution of excitons in the energy band is expected to have a longer effective lifetime since only a small fraction of the occupied states have a radiative width. Studies of such a distribution need to consider the competition between radiative decay and scattering by phonons or defects. Let us continue the discussion of this problem having in mind a 1D structure in some matrix. It is easier to calculate the radiative decay time of a distribution of excitons in chain assuming that exciton–polaritons are in thermal equilibrium with the matrix. In this approximation the temperature-dependent decay rate is

$$\Gamma(T) = \frac{\int_0^\infty dE' \exp\left(-E'/k_B T\right) \Gamma(E')}{\int_0^\infty dE' \exp\left(-E'/k_B T\right)}$$

where $\Gamma(E') = 2\mathcal{E}''$ is the radiative width of the exciton–polariton with energy E' and k_B is the Boltzmann constant. If the constant A determining the intensity of the retardation effect (see Section 4.5) is small and we can neglect renormalization of 1D polariton dispersion, then if the temperature $T \gg T_0$ with $T_0 = (1/k_B)\hbar^2 k_0^2/2m_{\text{exc}}$ and m_{exc} being the positive effective mass of the exciton, and using the relation (4.104) for $\mathcal{E}''$ for $\theta = 0$ (in this case the transition dipole moment directed parallel to the chain), we obtain for the effective radiative width $\Gamma(T)$

$$\Gamma(T) \sim \frac{1}{\sqrt{T}}$$

or for radiative lifetime the relation (24)

$$\tau_{\text{eff}}(T) \sim \sqrt{T}.$$

However, the assumption that the thermal distribution of the excitons is maintained also in the region of small wavevectors which mainly contributes to the value of $\Gamma(T)$ can be justified only if the quasielastic scattering of large-wavelength excitons by acoustic or optical phonons is much faster than the radiative process. It is clear that such a relation between the effective radiative rate and the rate of elastic scattering can be expected only for high enough temperatures. A similar calculation for the effective radiative lifetime has been performed in (25). It is clear that the generalization of theory to the region of small temperature needs consideration of the competition of the exciton radiative decay in the framework of the Bolzman kinetic equation.

4.7 Concluding remarks

The appearance of an enhanced radiative width and renormalization of the exciton dispersion are the main effects arising in one- and two-dimensional structures under the influence of retardation (21). Qualitatively these effects are valid for Frenkel as well as for Wannier–Mott excitons. In contrast to 3D structure where in the exciton–photon interaction all three components of the momentum have to be conserved and as a result a picture with 3D polaritons arises, for structures of lower dimensionality only the in-plane momentum for 2D structures is conserved and only one component for the 1D structure. An exciton in both cases is coupled to a continuum of photon states. There is no possibility of reversible strong coupling. Exciton absorption and luminescence become allowed in first order. This fact played an important role in the understanding of optical properties of low-dimensional structures (26). In both cases and particularly for 2D structures the theory predicts a large enhancement of the radiative width and consequently a large decrease of the radiative lifetime in comparison with these characteristics for a monomer or for excitations in samples with size small in comparison with the wavelength. Sometimes this effect is named superradiance. In 2D structures the enhancement factor for the radiative width is $\sim (\lambda/2\pi d) \simeq 10^4 - 10^5$ and thus can be particularly large. Hanamura (27) calculated the radiative decay rate Γ_0 for an exciton with zero momentum and found the relation

$$\Gamma_0 = 1/\tau_0 = 24\pi \left(\frac{\lambda}{a_B}\right)^2 \gamma_s$$

where $\gamma_s = 45|\mathbf{d}_{cv}|^2/3\hbar\lambda^3$, with $\mathbf{d}_{cv}$ being the transition dipole matrix element, representing the one-electron radiative decay rate per unit cell. However, the radiative rate of the Wannier–Mott exciton γ is proportional not to $|\mathbf{d}_{cv}|^2$ but to $|\mathbf{d}_{cv}|^2 (d/a_B)^2$, where d is the lattice constant, and where the new factor $(d/a_B)^2$ determines the probability for the electron and hole forming a Wannier–Mott exciton to be in the same 2D cell. For this reason it is evident that the enhancement of the radiative rate due to retardation for a Wannier–Mott exciton in a semiconductor quantum well is proportional to the same factor $(\lambda/d)^2$ and thus is the same as in organic 2D structure. We have already mentioned the observation of fast electron–hole recombination in organic monolayers and in inorganic quantum wells and briefly mentioned the investigation of the same processes in 1D structures.

An ultrafast decay of Frenkel exciton was observed for anthracene monolayers in picosecond measurements performed by Aaviksoo et al. (28). We will mention these experiments also in Ch. 9. Similar data for Wannier–Mott excitons in inorganic materials have been obtained for the first time by Deveaud et al. (29) in examining the radiative decay of excitons in GaAs quantum wells. Recently this effect was also investigated for different inorganic semiconductor quantum wells. The optical properties, including temperature dependence of the radiative lifetime in the region of exciton resonances in the case of 1D structures – semiconductor quantum wires – with length larger than the wavelength, have been studied (see a review paper by Bassani (30)) for carbon nanotubes (see (31; 32) and references herein), and in isolated chains of polydiacetilene dispersed in their monomer crystals (33).

The scattering of polaritons by disorder in 1D structures is very important for the interpretation of the results. The disorder strongly influences the states in 1D structures making all states weakly or strongly localized (34). Any imperfections in the chain replace the 1D wire by a collection of finite boxes. The sizes of the boxes depend on the disorder and can be different. The strong localization by definition is localization of polaritons with mean free path small in comparison with the polariton wavelength. For such localized states the wavevector is no longer a good quantum number characterizing the state of the polariton, and the corresponding radiative lifetime will be the same as for a monomer or box smaller than the wavelength. In the case of weak localization of polaritons the wavelength of the polariton is small in comparison with its mean free path and the wavevector is a good quantum number. The length of localization in this case can be large but in all cases it is restricted by the length of the boxes in the 1D chain or by the length of the chain. Thus, the influence of disorder will not only decrease the enhancement factor but can also create a strongly non-one-exponent decay of fluorescence if the quantum yield of the fluorescence is high and thus if the nonradiative processes are be important.

As shown, the theory predicts a very definite dependence of the effective radiative lifetime on temperature. A $T^{1/2}$ variation of τ has, in semiconductor quantum wires, been observed only in a narrow T range (35). In the case of organic chains the absolute values of τ and of the fluorescence quantum yield η have been determined in macroscopic experiments on ensembles of chains and the values obtained range between 10 and 70 K for poly-3BCMU isolated red chains. Knowledge of both η and τ allows one to calculate the exciton radiative τ_{rad} and nonradiative τ_{nonrad} lifetimes. It was found in (36) that the calculated values of τ_{rad} follow the law $\tau_{\text{rad}} = (80)\, T^{1/2}$ in picoseconds, verifying the expected 1D law in the investigated T range. Such a comparison with the theoretical formula for τ_{rad} used above has some grounds because for a chain in a monomer matrix the rate of 3D acoustic phonon absorption or emission is about a few picoseconds and this provides an efficient process for thermalization of excitons in a chain with the lattice assumed above in calculating $\Gamma(T)$.

We have no possibility to discuss in more details these experiments on 1D structures. Readers interested in the experiments and the corresponding theory are referred to the original publications.

5

DIELECTRIC THEORY OF FRENKEL EXCITONS: LOCAL FIELD EFFECTS

5.1 Introduction: the local field method

This chapter presents a consistent description of the local field method and some of its results. First, we should make a brief note.

It is well-known that the simplest approach to the study of the optical properties of condensed media is the macroscopic electrodynamics approach, making use of the concept of the dielectric constant tensor $\epsilon_{ij}(\omega, k)$ where ω and k are the frequency and the wavevector of the light wave. Calculation of this tensor for a specific medium is, however, a problem of microscopic theory. For instance, the procedures for calculating the tensor $\epsilon_{ij}(\omega, k)$ for the excitation region of the spectrum in crystals are discussed in (1) and (2). These procedures are based on using the different types of exciton states of the crystal (Coulomb or mechanical excitons, see Ch. 3) which are treated as the zeroth-order states when we attempt to find a linear response to an external electromagnetic perturbation. These procedures are fairly general. However, this does not mean that we must know the exciton states of the crystal in order to calculate the dielectric constant tensor. We shall expand below on these points when we consider the spectra of the lowest singlet excited states of molecular crystals (pure crystals, crystals containing impurities, and crystalline solutions) where the mutual interaction between molecules does not result in charge transfer. The intermolecular interaction here is purely classical in character and is determined by the van der Waals forces which only result in electronic energy transfer and in mixing of the molecular configurations (2).

Numerous theoretical and experimental studies have been carried out in this field so that a whole branch of molecular optics – the optics of molecular crystals and molecular liquids – has been established. Even before Frenkel put forward his exciton concept, workers in this branch of optics had developed a variety of exact and approximate methods for the theoretical description of optical phenomena; many of these methods were also substantiated in experimental studies. However, after the discovery of excitons the use of these methods became increasingly rare and many of the results obtained with them have not been sufficiently understood in the framework of exciton theory. Therefore, further development and generalization of these methods were impeded. On the other hand, since the results of pre-excitonic molecular optics were underestimated, the optical properties of crystals were treated in terms of only exciton theory even in those cases when this could be done much more easily by using the earlier, simpler and no less clear physical concepts. These circumstances could not, of course, fail to influence the development of the theory of the optics of crystals. The resulting situation is discussed in (3) and this chapter is largely based on the results presented therein. This paper was concerned mainly with the calculation of the dielectric constant tensor for crystals consisting of identical or different molecules, a discussion of their optical properties (dispersion and absorption of light), and a determination of the energy of the resonance dipole–dipole interaction between the impurity molecules resulting in excitation transfer as a function of the dielectric properties of the host crystal.

We can use the local field method to find the dielectric constant tensor for systems of this type. This method goes back to Lorentz who used it to derive the well-known formula for the optical refractive index in isotropic media (the Lorenz–Lorentz formula). Let us recall the derivation of this formula.

According to Lorentz, the electric field $\mathbf{E}'$ acting on a molecule in the isotropic medium (local field) and causing its polarization is not equal to the mean (macroscopic) field $\mathbf{E}$ which satisfies the phenomenological Maxwell's equations, but is determined by

$$\mathbf{E}' = \mathbf{E} + \frac{4\pi}{3}\mathbf{P} = \frac{\epsilon+2}{3}\mathbf{E}, \qquad \mathbf{P} = \frac{(\epsilon-1)\mathbf{E}}{4\pi},$$

where $\mathbf{P}$ is the polarization per unit volume, and ϵ is the dielectric constant of the medium. On the other hand, since the polarization per unit volume is $\mathbf{P} = N_0 a \mathbf{E}'$ (where a is the polarizability of the molecule, and N_0 is the number of molecules per unit volume) the of electric induction vector $\mathbf{D}$ is given by the relationship,

$$\mathbf{D} = \mathbf{E} + 4\pi\mathbf{P} = \left[1 + \frac{4\pi}{3}N_0 a(\epsilon+2)\right]\mathbf{E} = \epsilon\mathbf{E},$$

which directly yields the Lorenz–Lorentz formula

$$\frac{\epsilon-1}{\epsilon+2} = \frac{4\pi}{3}N_0 a.$$

However, this formula, which expresses the dielectric constant of the medium in terms of the polarizability of an individual molecule, is only a rough approximation even for cubic crystals with the van der Waals forces acting between the isotropic molecules. For instance, the formula takes no account of spatial dispersion. Moreover, it does not take into account the contribution of the higher multipoles to the energy of the intermolecular interaction which is important at distances of the order of lattice constant.

We show below how these inaccuracies in the Lorenz–Lorentz formula can be eliminated. We employ the local field method to modify this formula and apply it to anisotropic organic crystals of complex structure. We discuss below also a variety of their optical properties which were earlier analyzed in a less general form only in the framework of exciton theory. We explain how it has to be applied to the local field method to calculate the dielectric tensor of the anisotropic molecular crystal taking into account the mixing of molecular configurations, spatial dispersion, and short-range as well as long-range intermolecular interactions. The results obtained give the possibility to discuss the effect of the medium on the energy of the resonance interaction between the impurity molecules that are important for the estimation of the probability of electronic energy transfer.

Let the unit cell of the crystal contain σ identical molecules ($\alpha = 1, 2, \ldots, \sigma$) that are differently oriented with respect to crystallographic axes. According to Born and Huang ((4), Section 30), when a plane electromagnetic wave of amplitude $\mathbf{E}(\omega, \mathbf{k})$ propagates in the crystal the electric field $\mathbf{E}^\alpha$ acting on the molecule α is not equal to the mean field but is given by:

$$E_i^\alpha = E_i + \sum_{\beta j} Q_{ij}^{\alpha\beta}(\mathbf{k}) p_j^\beta, \tag{5.1}$$

where $\mathbf{p}^\beta$ is the amplitude of the dipole moment induced in the molecules of the species α, and the coefficients $Q_{ij}^{\alpha\beta}(\mathbf{k})$ (the internal field coefficients) are determined by the lattice structure only. If $a_{ij}^\alpha(\omega)$ is the polarizability of the molecule with orientation α (for simplicity, we assume here and below that the molecules have no static dipole moments) then we have

$$p_i^\alpha = a_{ij}^\alpha(\omega) E_j^\alpha. \tag{5.2}$$

At the same time, of course, the molecules can have static moments of higher multipolarity both in the ground and in the excited state. Generally speaking, these moments are different in the ground and in the excited states. The molecules in these states also have different energies of interaction with the environment so that the frequencies of the intramolecular transitions are somewhat shifted with respect to the transition frequencies in the vacuum. We shall assume below that the only difference between the tensor $a_{ij}(\omega)$ and the respective tensor for a molecule in the vacuum is determined by this frequency shift.

Substitution of eqn (5.2) into eqn (5.1) yields:

$$E_i^\alpha = E_i + \sum_{\beta j_1} Q_{ij_1}^{\alpha\beta}(\mathbf{k}) a_{j_1 j}^\alpha(\omega) E_j^\alpha. \tag{5.3}$$

Now we can express the local fields $\mathbf{E}^\alpha$ in terms of $\mathbf{E}$:

$$E_i^\alpha(\omega, \mathbf{k}) = A_{ij}^\alpha(\omega, \mathbf{k}) E_j(\omega, \mathbf{k}). \tag{5.4}$$

If we know the tensor A_{ij} we can directly find the dielectric tensor for the crystal Indeed, since the polarization per unit volume is

$$P_i = \frac{1}{v} \sum_\alpha p_i^\alpha, \tag{5.5}$$

where v is the volume of the unit cell, then the dielectric constant tensor is

$$\epsilon_{ij}(\omega, \mathbf{k}) = \delta_{ij} + \frac{4\pi}{v} \sum_{\alpha j_1} a_{ij_1}^\alpha A_{j_1 j}^\alpha. \tag{5.6}$$

Equation (5.6) in the usual dipole approximation makes possible a very simple analysis of the problem of how the internal field corrections affect the optical properties of crystals. A number of results for pure crystals in the dipole approximation are reported by Dunmur (5) and Cummins et al. (6) . For instance, Dunmur (5) calculated the internal field coefficients for $\mathbf{k} = 0$ for the crystals of anthracene, naphthalene, benzene, phenanthrene, biphenyl and iodoform. Wünsche (7) analyzed the case of uniaxial crystals with one or two molecules in the unit cell and (8) the problem of the local field in the framework of nonlinear crystal optics (see also Fulton (9)). Apart from these problems we discuss below the optical properties of mixed crystalline solutions, take into account some effects of spatial dispersion in the medium, and find the energy of the resonance interaction between the impurity molecules. Here we shall only note that the use of the tensor a_{ij}^α independent of $\mathbf{k}$ for describing the polarizability of an individual molecule means that we take into account only the dipole polarization of the molecule. This approximation is valid if we deal with the optical properties in the region of resonances of the molecule having high enough oscillator strengths. However, this approximation is by no means essential for the following analysis because the more general and more exact expression of polarizability can be used. The consideration which we develop below, introducing as known the tensor $a_{ij}(\omega)$ is no more phenomenological than the theory of small-radius Frenkel excitons which assumes the wavefunctions and energies of the states of isolated molecule, rather than the tensor $a_{ij}(\omega, \mathbf{k})$ to be known. Note also that the theory of point groups can be used to analyze the structure of this tensor and this analysis is particularly simple in the regions of certain resonances of the molecule.

Since we assume below that the polarizability of the molecule in vacuum is known we can use the more general expression $a_{ij}(\omega, \mathbf{k})$ for it in eqn (5.2). Of course, the expression for the internal field should include in this case the terms corresponding to the internal field of quadrupoles,, octupoles, etc., apart from the term given in eqn (5.1) and corresponding to the internal field in the lattice of dipoles. The electric field produced by these multipoles decreases with distances faster than the dipole field so that in calculation of the corresponding internal field one does not encounter the well-known difficulties with summing the series as were so successfully overcome by Ewald for a lattice of dipoles (4). Thus, finding the internal field for the higher multipoles is a more or less routine procedure and we shall not discuss it below (these results in the most general form are presented by Khokhlov (10) who also gives references to earlier studies; see also Philpott and Mahan (11) and Cummins et. al. (6).

The tensor $a_{ij}^\alpha(\omega)$, rather than the tensor $a_{ij}(\omega, \mathbf{k})$, in eqn (5.2) has to be used for the investigation of optical properties of crystal only in the region of dipole-forbidden transitions. Below we will discuss the optical properties of nongyrotropic pure and impure crystals only in the spectral region of dipole-allowed transitions. However, a more general analysis can be carried out in a similar way (see Section 5.8).

5.2 Dielectric tensor of cubic crystals

As was shown by Born and Huang (4) in cubic crystals with one molecule per unit cell the tensor $Q_{ij}^{\alpha\beta}$ is reduced to the scalar $Q_{ij} = (4\pi/3v)\delta_{ij}$ where v is the volume of the unit cell, if we ignore spatial dispersion. Moreover, $a_{ij} = a\delta_{ij}$ so that eqn (5.3) yields the tensor $A_{ij} = A\delta_{ij}$ where $A = [1 - (4\pi a/3v)]^{-1}$. Substituting this expression into eqn (5.6) we obtain $\epsilon_{ij} = \epsilon\delta_{ij}$ where

$$\epsilon(\omega) = 1 + \frac{4\pi}{v}a\left(1 - \frac{4\pi}{3v}a\right)^{-1}, \tag{5.7}$$

so that $A = (\epsilon+2)/3$. Equation (5.7) directly yields the Lorenz–Lorentz formula:

$$\frac{\epsilon - 1}{\epsilon + 2} = \frac{4\pi}{3v}a. \tag{5.8}$$

Let us now consider the dispersion of ϵ taking into account only one of the resonances of $a(\omega)$. In this approximation we obtain:

$$a(\omega) = \frac{F_1}{\omega_1^2 - \omega^2}, \quad F_1 = \frac{2p_1^2\omega_1}{\hbar} = \frac{e^2}{m}f_1, \tag{5.9}$$

where f_1 is the oscillator strength, ω_1 is the frequency of the $0 \to 1$ transition in the isolated molecule, and p_1 the corresponding dipole moment of the transition. Substitution of eqn (5.9) into eqn (5.7) yields:

$$\epsilon(\omega) = 1 + \frac{(4\pi/v)F_1}{\omega_1^2 - \omega^2 - (8\pi/3v\hbar)p_1^2\omega_1}. \tag{5.10}$$

Equation (5.10) shows that if we take into account the internal field correction, that is, the fact that $A \neq 1$, the oscillator strength for the transition is not changed and only the resonance frequency is shifted – the $\epsilon(\omega)$ resonance is shifted by $\Delta\omega \ll \omega_1$ towards lower frequencies from the frequency of the transition in the isolated molecule:

$$\Delta\omega = \frac{4\pi}{3v\hbar}p_1^2.$$

Decay is taken into account by adding the imaginary terms $-i\delta$ to the denominator of eqn (5.10). Since $\epsilon = (n + i\kappa)^2$ where n and κ are the refractive and absorption indexes, for $\delta \to 0$ we obtain,

$$\int_{-\infty}^{+\infty} 2n(\omega)\kappa(\omega)d\omega = 2\pi^2 F_1/\omega_1 v. \tag{5.11}$$

In this case we have,

$$\kappa(\omega) = \frac{(2\pi/v)F_1\delta(\omega)}{(\omega_\perp^2 - \omega^2)^2 + \delta^2(\omega)} \frac{1}{n(\omega)},$$

where $\omega_\perp^2 = \omega_1^2 - (4\pi/3v)F_1$.

5.3 Effects of impurities

Let us now consider a crystal containing a certain number of substitutional impurity molecules (the treatment can be readily modified for the case of interstitial impurities). Now the internal field depends on the spatial distribution of the impurities. If we ignore fluctuations of this distribution and replace the internal field with its mean value we obtain

$$\epsilon(\omega) = 1 + 4\pi N_0 a(\omega)\frac{\epsilon+2}{3} + 4\pi N_1 \left[\tilde{a}(\omega) - a(\omega)\right]\frac{\epsilon+2}{3}, \tag{5.12}$$

where $\tilde{a}(\omega)$ is the polarizability of the impurity molecule, $N_0 = 1/v$, and N_1 is the concentration of impurity molecules (the accuracy of this approximation is discussed in Section 5.6). If $\delta\epsilon$ is a small variation of ϵ due to the presence of impurities ($\delta\epsilon \sim N_1$ for $N_1 \ll N_0$) then eqn (5.12) yields

$$\delta\epsilon = \left(1 - \frac{4\pi N_0 a}{3}\right)^{-1} 4\pi N_1 (\tilde{a} - a) \frac{\epsilon_0+2}{3},$$

where $\epsilon_0 = \epsilon$ for $N_1 = 0$. Hence (see eqn 5.8) we have

$$\epsilon(\omega) = \epsilon_0(\omega) + 4\pi N_1 \left[\tilde{a}(\omega) - a(\omega)\right]\left(\frac{\epsilon_0+2}{3}\right)^2. \tag{5.13}$$

Using eqn (5.8) we can rewrite eqn (5.13) as

$$\epsilon(\omega) = \epsilon_0 - \frac{c}{3}(\epsilon_0+2)(\epsilon_0-1) + 4\pi N_1\tilde{a}(\omega)\left(\frac{\epsilon_0+2}{3}\right)^2, \tag{5.14}$$

where $c = N_1 v$. Assuming that for the impurity molecule we can write, instead of eqn (5.9), the relationship (taking into account decay)

$$\tilde{a}(\omega) = \frac{2\tilde{p}_1^2\tilde{\omega}_1/\hbar}{\tilde{\omega}_1^2 - \omega^2 - i\tilde{\delta}},$$

and that the frequency $\tilde{\omega}_1$ is in the region of transparency of the solvent, we obtain for the frequencies $\omega \sim \tilde{\omega}_1$

$$2n\kappa = 8\pi N_1 \left(\frac{\epsilon_0+2}{3}\right)^2 \tilde{p}_1^2 \frac{\tilde{\omega}_1\tilde{\delta}/\hbar}{\left[(\tilde{\omega}_1)^2 - \omega^2\right]^2 + \tilde{\delta}^2}.$$

Hence, integration in the region of the absorption band yields:

$$\int 2n\kappa \, d\omega = 4\pi^2 N_1 \left[\frac{\epsilon_0(\tilde{\omega}) + 2}{3} \right]^2 \frac{\tilde{p}_1^2}{\hbar}. \tag{5.15}$$

Thus, we have found that the coefficient of absorption of light by the impurity in a medium with dielectric constant $\epsilon_0(\omega)$ and the integral on the left-hand side of eqn (5.15), known as the Kravets integral, are proportional to the squared Lorentz factor. The change of the oscillator strength formally means that under the effect of the medium the dipole moment $\tilde{\mathbf{p}}_1$ of the transition in the impurity is replaced by the effective dipole moment $(\tilde{\mathbf{p}}_1)_{\text{eff}} = \tilde{\mathbf{p}}_1(\epsilon_0 + 2)/3$. (Evidently, formally any crystal defect can be considered as an impurity placed where an excited state is localized.) If we have, in the vicinity of the impurity, $\epsilon_0(\tilde{\omega}_1) \gg 1$ the intensity I of impurity absorption at the impurity resonance increases as $(\tilde{\omega}_1 - \omega_0)^{-2}$ where ω_0 is the resonance frequency for the dielectric tensor of the pure crystal $\epsilon_0(\omega)$. If the resonance frequency of the pure crystal corresponds to an exciton transition with positive effective mass and if the oscillator strength of the exciton transition is large enough and its width is small in comparison with the difference $\omega_0 - \omega_1$, then the dependence on the frequency of the intensity of impurity absorption changes. In this case $I \sim (\tilde{\omega}_1 - \omega_0)^{-3/2}$ as shown by Rashba (12). In the molecular crystal the effects of spatial dispersion even in the region of resonances usually are small and the width of the exciton transition is usually rather large even at low temperature. One can expect for this reason that the effect discussed of "the flaring up" of impurity absorption in the vicinity of the host crystal exciton resonance can be understood in the framework of more simple theory predicted dependence $I \sim (\tilde{\omega}_1 - \omega_0)^{-2}$. A similar effect should take place for such "impurity" centers as biexcitons (in luminescence) and various other luminescence centers (for instance, exciton localized at the impurity, etc.). Since the experimentally measured quantity in eqn (5.15) is its left-hand side, a correct introduction of the internal field correction (the Lorentz factor in this case) makes it possible to find the oscillator strength for the molecules in solution. Of course, this can be done only if no chemical bonds arise between the molecules of the solute and the solvent, no aggregates of the impurity molecules are formed, and so on. Note also that the effect analogous to "the flaring up" of impurity absorption can be expected for impurity-induced natural and magnetic optical activity (a rotation of the polarization plane (13)).

In the above discussion we have used a two-level model for the molecules of the impurity and the solvent. A natural question to consider now is how the existence of many resonances of the polarizability of the molecules affects the above results. When we derived eqn (5.13) we did not use the two-level model for describing the molecules of the solvent or the impurity. Therefore, this relation remains valid in the case of many resonances and only the right-hand side of eqn (5.15) is changed. Namely, when we have many levels $i = 1, 2, \ldots$ this term is transformed into a sum of contributions on the individual resonances:

$$\sum_i 4\pi^2 N_1 \left[\frac{\epsilon_0(\tilde{\omega}_i) + 2}{3}\right]^2 \frac{\tilde{p}_i^2}{\hbar}. \tag{5.16}$$

Now let us consider the impurity-free crystal and assume that the isolated molecules of the crystal have polarizability,

$$a(\omega) = a_0 + a_1(\omega), \tag{5.17}$$

where the function $a_1(\omega)$ is given by eqn (5.9) while a_0 is determined by the contributions of the far resonances to the polarizability of the molecule and so a_0 can be assumed to be a constant independent of ω in the frequency range $\omega \approx \omega_1$. Substituting eqn (5.17) into eqn (5.7) we now obtain

$$\epsilon(\omega) = \epsilon_b + \frac{(4\pi/v)F_1\left[(\epsilon_b + 2)/3\right]^2}{\omega_\perp^2 - \omega^2}, \tag{5.18}$$

where

$$\epsilon_b = 1 + \frac{4\pi}{v}a_0\left(1 - \frac{4\pi}{3v}a_0\right)^{-1} \tag{5.19}$$

is a quantity (see eqn 5.7) corresponding to the dielectric constant of the crystal under the assumption that the polarizability of the crystal molecule lacks the resonance term for the frequency ω_1 (that is, $a_1(\omega) \equiv 0$). The resonance frequency $\omega_\perp$ is given by

$$\omega_\perp^2 = \omega_1^2 - \frac{4\pi}{3v}F_1\left(\frac{\epsilon_b + 2}{3}\right).$$

If we introduce weak decay into eqn (5.18) then eqn (5.11) is replaced by

$$2\int n(\omega)\kappa(\omega)\mathrm{d}\omega = \frac{2\pi^2}{v\omega_\perp}F_1\left(\frac{\epsilon_b + 2}{3}\right)^2. \tag{5.20}$$

We might call the quantity ϵ_b the background dielectric constant ϵ_b with respect to the resonance at the frequency $\omega_\perp$. Since $\epsilon_b \neq 1$ the existence of the background dielectric constant alters the oscillator strength of the transition. This is especially significant for the spectral range of low-intensity molecular transitions (see Ch. 3).

If the molecule has two or more close resonances in the given frequency range then we must also distinguish the terms $a_2(\omega)$, $a_3(\omega)$, etc. in eqn (5.17), apart from the terms indicated there. These terms can be readily taken into account though the resulting formulas for $\epsilon(\omega)$ become somewhat cumbersome.

Equations (5.18) and (5.20) are special cases of the more general formulas derived in Ch. 3; they demonstrate for the given crystal model how the mixing of molecular configurations due to intermolecular interaction in the crystal affects the oscillator strengths of the dipole transitions. We shall only stress here the significant difference between these equations and similar relationships (5.15) and (5.16) for the impurity molecules in solutions. When we consider impurity molecules in a solution the Lorentz factor on the r.h.s. of eqn (5.15) includes the dielectric constant of the solvent at the transition frequency of the impurity. In the case of pure substances we see from eqn (5.20) that this quantity in the Lorentz factor is replaced by the background dielectric constant which is not at all equal to the squared index of refraction of light in the crystal at the transition frequency. In the two-level model we have $\epsilon_b = 1$ and eqn (5.18) is transformed into eqn (5.10), and eqn (5.20) into eqn (5.11).

Finally, note that eqn (5.8) can be used to obtain resonances of the molecular polarizability from the data on dispersion and absorption of light in the crystal (14). Since $\epsilon = (n + i\kappa)^2$ where n and κ are the index of refraction and absorption coefficient in the crystal, eqn (5.8) yields:

$$\mathrm{Im}\, a(\omega) = \frac{9n\,\kappa/2\pi N_0}{\left(n^2 - \kappa^2 + 2\right)^2 + 4n^2\kappa^2},$$

where N_0 is the number of molecules per unit volume. This formula has been successfully used by Bakhshiev (14) for a number of examples. He found that the maxima of the function $\mathrm{Im}\, a(\omega)$ which determines the molecular absorption can be considerably shifted with respect to the maxima of the quantity $\kappa(\omega)$ which determines the absorption of the crystal.

5.4 Dielectric tensor of organic anisotropic crystals

As an example of an anisotropic crystal, let us consider a molecular crystal with $\sigma > 1$ molecules per unit cell. Let us investigate the optical properties of the crystal in the frequency range $\omega \approx \omega_1$ where ω_1 is the transition frequency to that of a nondegenerate excited state of the isolated molecule. The polarizability of the molecule in the vicinity of the dipole-allowed transition can be presented by the relation

$$a_{ij}(\omega) = \frac{F_1 \ell_i \ell_j}{\omega_1^2 - \omega^2}, \tag{5.21}$$

where F_1 is a quantity proportional to the oscillator strength of the $0 \to 1$ transition and $\boldsymbol{\ell}$ is the unit vector parallel to the vector of the dipole moment of this transition. Since the different molecules in the unit cell have different orientations their polarizability tensor is

$$a_{ij}^\alpha(\omega) = \frac{F_1 \ell_i^\alpha \ell_j^\alpha}{\omega_1^2 - \omega^2}, \qquad \alpha = 1, 2, \ldots, \sigma. \tag{5.22}$$

When we substitute eqn (5.22) into eqn (5.3) and find the scalar product on the left with $\boldsymbol{\ell}^\alpha$ we obtain the following system of equations for the quantities $(\mathbf{E}^\alpha \boldsymbol{\ell}^\alpha)$ where $\alpha = 1, 2, \ldots, \sigma$:

$$(\mathbf{E}^\alpha \boldsymbol{\ell}^\alpha) - \sum_\beta M_{\alpha\beta}(\omega, \mathbf{k})(\mathbf{E}^\beta \boldsymbol{\ell}^\beta) = (\mathbf{E}\boldsymbol{\ell}^\alpha). \tag{5.23}$$

Here,
$$M_{\alpha\beta}(\omega, \mathbf{k}) = \frac{F_1}{\omega_1^2 - \omega^2} \sum_{ij} Q_{ij}^{\alpha\beta}(\mathbf{k}) \ell_i^\alpha \ell_j^\beta. \tag{5.24}$$

For a crystal with one molecule per unit cell eqn (5.23) directly yields

$$\left(\mathbf{E}^1 \boldsymbol{\ell}^1\right) = [1 - M_{11}(\omega, \mathbf{k})]^{-1} \left(\mathbf{E}\boldsymbol{\ell}^1\right),$$

so that the polarization per unit volume is

$$P_i = \frac{1}{v} a_{ij}^1(\omega) E_j^{(1)} = \frac{F_1 \ell_i^{(1)} \ell_j^{(1)}}{v(\omega_1^2 - \omega^2)} \frac{E_j}{1 - M_{11}(\omega, \mathbf{k})}.$$

Taking this relationship and eqn (5.24) we find the following expression for the dielectric constant tensor:

$$\epsilon_{ij}^0(\omega, \mathbf{k}) = \delta_{ij} + \frac{(4\pi F_1/v) \ell_i \ell_j}{\omega_1^2 - \omega^2 - F_1 \sum_{i_1 j_1} Q_{i_1 j_1}(\mathbf{k}) \ell_{i_1} \ell_{j_1}}. \tag{5.25}$$

The resonance of this expression occurs at $\omega = \Omega_1(\mathbf{k})$ where

$$\Omega_1^2(\mathbf{k}) = \omega_1^2 - F_1 \sum_{i_1 j_1} Q_{i_1 j_1}(\mathbf{k}) \ell_{i_1} \ell_{j_1}. \tag{5.26}$$

The tensor (5.25) has the symmetry of a uniaxial crystal. If one of the coordinate axes, for instance, the x-axis, is taken parallel to $\boldsymbol{\ell}$ then the tensor (5.25) is reduced to diagonal form with the following nonzero components:

$$\epsilon_2 = \epsilon_3 = 1, \qquad \epsilon_1 = 1 + \frac{4\pi F_1/v}{\Omega_1^2 - \omega^2}. \tag{5.27}$$

Let us now consider crystals with two molecules per unit cell. This crystal group includes anthracene, naphthalene, and other crystals of the aromatic series which were the objects of many experimental studies.

For crystals of the anthracene type there are symmetry operations that transform molecules with $\alpha = 1$ into molecules with $\alpha = 2$. Therefore, when $\mathbf{k} = 0$ or when vectors $\mathbf{k} \neq 0$ are parallel or perpendicular to the monoclinic axis the relationships $M_{11}(\omega, \mathbf{k}) = M_{22}(\omega, \mathbf{k})$ and $M_{12}(\omega, \mathbf{k}) = M_{21}(\omega, \mathbf{k})$ are satisfied. For such vectors $\mathbf{k}$ the solution of the system of equations (5.23) is less cumbersome. It can be easily seen that in this case we have

$$\left(\mathbf{E}^{(\alpha)} \boldsymbol{\ell}^{(\alpha)}\right) = \frac{1}{2} \left[\frac{L_j^{(1)}}{1 - M_{11}(\omega, \mathbf{k}) - M_{12}(\omega, \mathbf{k})} \right.$$

$$-\frac{(-1)^\alpha L_j^{(2)}}{1 - M_{11}(\omega, \mathbf{k}) + M_{12}(\omega, \mathbf{k})}\Bigg] E_j, \qquad \alpha = 1, 2, \qquad (5.28)$$

where

$$\mathbf{L}^{(1)} = \boldsymbol{\ell}^{(1)} + \boldsymbol{\ell}^{(2)}, \qquad \mathbf{L}^{(2)} = \boldsymbol{\ell}^{(1)} - \boldsymbol{\ell}^{(2)}. \qquad (5.29)$$

Substituting these relationships into eqn (5.5) we also can find the dielectric tensor

$$\epsilon_{ij}(\omega, \mathbf{k}) = \delta_{ij} + \frac{2\pi}{v} F_1 \left[\frac{L_i^{(1)} L_j^{(1)}}{\Omega_1^2(\mathbf{k}) - \omega^2} + \frac{L_i^{(2)} L_j^{(2)}}{\Omega_2^2(\mathbf{k}) - \omega^2} \right]. \qquad (5.30)$$

Here,

$$\Omega_1^2(\mathbf{k}) = \omega_1^2 - F_1 \sum_{ij} Q_{ij}^{11}(\mathbf{k}) \ell_i^{(1)} \ell_j^{(1)} - F_1 \sum_{ij} Q_{ij}^{12}(\mathbf{k}) \ell_i^{(1)} \ell_j^{(2)},$$

$$\Omega_2^2(\mathbf{k}) = \omega_1^2 - F_1 \sum_{ij} Q_{ij}^{11}(\mathbf{k}) \ell_i^{(1)} \ell_j^{(1)} + F_1 \sum_{ij} Q_{ij}^{12}(\mathbf{k}) \ell_i^{(1)} \ell_j^{(2)}. \qquad (5.31)$$

Since the quantities $Q_{ij}(\mathbf{k})$ are analytic functions of $\mathbf{k}$ the same is true for the frequencies $\Omega_{1,2}(\mathbf{k})$ as can be seen also from eqn (5.26). This, of course, is not surprising since (see also Chs. 3 and 4) the resonances of the tensor $\epsilon_{ij}(\omega, \mathbf{k})$ occur at the so-called mechanical exciton frequencies which are analytic functions of $\mathbf{k}$, regardless of the model being used. The vectors $\mathbf{L}^{(1)}$ and $\mathbf{L}^{(2)}$ are orthogonal. Therefore, if the coordinate axes x and y are parallel to the vectors $\mathbf{L}^{(1)}$ and $\mathbf{L}^{(2)}$ the tensor ϵ_{ij} is transformed to diagonal form with nonzero components:

$$\epsilon_{xx}(\omega, \mathbf{k}) = 1 + \frac{2\pi}{v} \frac{F_{1,xx}}{\Omega_1^2(\mathbf{k}) - \omega^2},$$

$$\epsilon_{yy}(\omega, \mathbf{k}) = 1 + \frac{2\pi}{v} \frac{F_{1,yy}}{\Omega_2^2(\mathbf{k}) - \omega^2}, \qquad (5.32)$$

$$\epsilon_{zz}(\omega, \mathbf{k}) = 1, \quad F_{1,xx} = F_1 \left|\mathbf{L}^{(1)}\right|^2, \quad F_{1,yy} = F_1 \left|\mathbf{L}^{(2)}\right|^2,$$

$$F_{1,xx}/F_{1,yy} = \cot^2(\theta/2), \qquad \cos\theta = \ell_1 \ell_2.$$

These relationships show that absorption of light with the electric vector $\mathbf{E} \parallel \mathbf{L}^{(1)}$ propagating along the z–axis must occur at the frequency $\Omega_1(\mathbf{k})$. If the electric vector $\mathbf{E} \parallel \mathbf{L}^{(2)}$ light absorption occurs at $\omega = \Omega_2(\mathbf{k})$. Thus, though we assumed that the excited state in the isolated molecule (the frequency ω_1) is nondegenerate, the absorption spectrum of a crystal with two molecules per unit cell should have two differently polarized lines of light absorption. Since we know the tensor ϵ_{ij} (see eqn 5.30 and 5.32) we can find the position of the absorption line for arbitrary polarizations and directions of propagation of the light and also take into account the spatial dispersion effects in crystals of arbitrary shape.

This phenomenon (Davydov splitting) has been studied in many systems. Although it was found from the application of Frenkel's exciton theory, actually the theory of small-radius excitons was not needed to understand it. As was shown above, it was sufficient to generalize the Lorenz–Lorentz formula for the case of anisotropic crystals to describe this phenomenon. This cannot be said about the exciton spectra of semiconductors. The theory of large-radius excitons had to be developed to describe them. This, of course, does not diminish the significance of the development of the small-radius exciton theory as stimulated by the studies of Frenkel, Peierls, Davydov, and other authors. As we know, it is only this theory that makes it possible to scent for such phenomena as transfer of electronic excitation energy in crystals, the optical properties of crystals at high exciton pumping, nonlinear optical effects, the fine details of the light absorption, and other matters.

Let us now continue the discussion of the optical properties of crystals in the region of impurity light absorption. In Section 5.2 we treated only cubic crystals. Here we shall consider anisotropic crystals using the equations establishing the relation between the local field and the mean field. It should be noted that, although the optics of impurity centers in crystals has a rich history, the interest in this branch of solid state spectroscopy is still high. One of the reasons is the use of crystals with impurities in new optical devices; other reasons are the discovery of the optical analog of the Mössbauer effect, studies of the capture of excitons by impurities, energy transfer via donors, and many other interesting optical phenomena occurring, for instance, in anisotropic matrices.

In this connection, let us discuss how the dielectric properties of the matrix affect the optical properties of the impurity in the anisotropic medium.

5.5 Dielectric constant of mixed crystalline solutions and polarization of impurity absorption bands

Let us start with the simplest model of a crystal in which the polarizability of the molecules of the main substance in vacuum is given by eqn (5.22). For the sake of simplicity, we shall assume that the impurity is of the substitutional type for which the polarizability tensor $\tilde{a}_{ij}^{\alpha}(\omega)$ differs from eqn (5.21) only in the resonance frequency and the oscillator strength so that we have,

$$\tilde{a}_{ij}^{\alpha}(\omega) = \frac{\tilde{F}_1 \ell_i^{(\alpha)} \ell_j^{(\alpha)}}{(\tilde{\omega}_1)^2 - \omega^2}, \qquad \tilde{F}_1 \neq F_1, \qquad \omega_1 \neq \tilde{\omega}_1. \tag{5.33}$$

If the impurity distribution is assumed to be uniform the local field, on the average, can be approximated (in the mean polarizability approximation) by

$$E_i^\alpha = E_i + \sum_{\beta j} Q_{ij}^{\alpha\beta}(\mathbf{k}) \bar{p}_j^\beta, \tag{5.34}$$

where $\bar{p}^\beta$ is the mean polarization of the site β,

$$\bar{p}_i^\beta = (1-c) a_{ij}^\beta E_j^\beta + c \tilde{a}_{ij}^\beta E_j^\beta, \tag{5.35}$$

and c is the ratio of the number of the impurity molecules to the total number of molecules in the crystal. Now, using eqns (5.34), (5.35), (5.22), and (5.33) we can find an equation for the quantities $\mathbf{E}^\alpha \boldsymbol{\ell}^\alpha$, $\alpha = 1, 2, \ldots, \sigma$, similar to eqn (5.23). As can be readily seen, this equation has the form:

$$\mathbf{E}^\alpha \boldsymbol{\ell}^\alpha - \sum_\beta \tilde{M}_{\alpha\beta}(\omega, \mathbf{k})(\mathbf{E}^\beta \boldsymbol{\ell}^\beta) - (\boldsymbol{\ell}^\alpha \mathbf{E}^\alpha), \tag{5.36}$$

where

$$\tilde{M}_{\alpha\beta}(\omega, \mathbf{k}) = \sum_{ij} Q_{ij}^{\alpha\beta}(\mathbf{k}) \ell_i^\alpha \ell_j^\beta \left[\frac{(1-c)F_1}{\omega_1^2 - \omega^2} + \frac{c\tilde{F}_1}{\tilde{\omega}_1^2 - \omega^2} \right]. \tag{5.37}$$

For crystals having one molecule per unit cell the polarization per unit volume is

$$p_i = \frac{1}{v} \left[(1-c) a_{ij}^{(1)}(\omega) E_j^{(1)} + c \tilde{a}_{ij}^{(1)}(\omega) E_j^{(1)} \right],$$

so that we obtain:

$$\epsilon_{ij}(\omega, \mathbf{k}) = \delta_{ij} + \frac{4\pi}{v} \left[\frac{(1-c)F_1}{\omega_1^2 - \omega^2} + \frac{c\tilde{F}_1}{\tilde{\omega}_1^2 - \omega^2} \right] \left[1 - \tilde{M}_{11}(\omega, \mathbf{k}) \right]^{-1} \ell_i \ell_j. \tag{5.38}$$

When $c \to 0$, eqn (5.38) is transformed into eqn (5.25). Here we are interested in the case $c \ll 1$ when we can expand eqn (5.38) in powers of c and retain only the term linear in c along with ϵ_{ij}^0. It can be easily seen that in this case:

$$\epsilon_{ij}(\omega, \mathbf{k}) = \epsilon_{ij}^0(\omega, \mathbf{k}) + 4\pi N_1 \left[\tilde{a}_{ij}(\omega) - a_{ij}(\omega) \right] \left[1 + \xi \left(\epsilon_1^0 - 1 \right) \right]^2, \tag{5.39}$$

where $\epsilon_{ij}^0(\omega, \mathbf{k})$ is the dielectric constant tensor for the pure crystal as defined by eqn (5.25),

$$\xi = \sum_{ij} v \, Q_{ij}(\mathbf{k}) \ell_i \ell_j, \qquad N_1 = c/v,$$

and ϵ_1^0 is given by eqn (5.27). Thus, eqn (5.39) is similar to eqn (5.14) derived above for cubic crystals and serves as its generalization for anisotropic crystals of the given type.

Before we proceed to the case of crystals with two molecules per unit cell we should note that the so-called mean polarizability approximation which makes it possible to idealize the crystal as an array of molecules having mean polarizability given by eqn (5.35) is, actually, a very old approach. In former times (see, for instance, Section 6 of (15)), this approximation when applied to molecular systems with the van der Waals interaction was referred to as the "additive refraction approximation". It should be stressed that, although eqn (5.38) provides just a very convenient extrapolation procedure, its accuracy increases with decreasing c so that the term linear in c in eqn (5.39) proves to be exact (but it does not take into account the effect of excitation delocalization discussed below).

Now let us consider crystals of the anthracene type having two molecules per unit cell. Since the system of equations (5.36) differs from eqn (5.23) only in the replacement $M_{\alpha\beta}(\omega,\mathbf{k}) \to \tilde{M}_{\alpha\beta}(\omega,\mathbf{k})$ we can use eqn (5.28) for the quantities $(\mathbf{E}^\alpha \boldsymbol{\ell}^\alpha)$ satisfying the system of equations (5.36) and write directly,

$$(\mathbf{E}^\alpha \boldsymbol{\ell}^\alpha) = \frac{1}{2}\left[\frac{L_j^{(1)}}{1-\tilde{M}_{11}-\tilde{M}_{12}} - \frac{(-1)^\alpha L_j^{(2)}}{1-\tilde{M}_{11}+\tilde{M}_{12}}\right]E_j, \qquad \alpha = 1, 2. \quad (5.40)$$

Furthermore, since the dielectric constant tensor of the given medium is found from the relationship:

$$\epsilon_{ij}(\omega,\mathbf{k})E_j = E_i + \frac{4\pi}{v}(1-c)\sum_\alpha a_{ij}^\alpha E_j^\alpha + \frac{4\pi}{v}c\sum_\alpha \tilde{a}_{ij}^\alpha E_j^\alpha, \quad (5.41)$$

we can substitute eqn (5.40) into eqn (5.41) and finally obtain:

$$\epsilon_{ij}(\omega,\mathbf{k}) = \delta_{ij} + \frac{2\pi}{v}\left[\frac{(1-c)F_1}{\omega_1^2-\omega^2} + \frac{c\tilde{F}_1}{\tilde{\omega}_1^2-\omega^2}\right] \quad (5.42)$$

$$\times \left[\frac{L_i^{(1)}L_j^{(1)}}{1-\tilde{M}_{11}(\omega,\mathbf{k})-\tilde{M}_{12}(\omega,\mathbf{k})} + \frac{L_i^{(2)}L_j^{(2)}}{1-\tilde{M}_{11}(\omega,\mathbf{k})+\tilde{M}_{12}(\omega,\mathbf{k})}\right].$$

In this approximation (the additive refraction or mean polarizability approximation) eqn (5.42) with the addition of eqn (5.37) fully determines the dependence of the dielectric constant tensor on the impurity concentration c. The optical properties of mixed crystals with large impurity concentrations are discussed in Section 5.6. As we did above for crystals with one molecule per unit cell, we shall discuss here the case of small values of c when we can ignore terms of the order of c^2, c^3, etc. in the expansion of the tensor (5.42) in powers of c. Then we obtain:

$$\epsilon_{ij}(\omega,\mathbf{k}) = \epsilon_{ij}^0(\omega,\mathbf{k}) + \frac{2\pi}{v}cF_1\left(\frac{1}{\tilde{\omega}_1^2-\omega^2} - \frac{1}{\omega_1^2-\omega^2}\right)$$

$$\times \left[\frac{L_i^{(1)}L_j^{(1)}}{(1-M_{11}-M_{12})^2} + \frac{L_i^{(2)}L_j^{(2)}}{(1-M_{11}+M_{12})^2}\right] \quad (5.43)$$

(for the sake of simplicity, here and below we assume $\tilde{F}_1 = F_1$ which is justified for an isotopic mixture). Using eqn (5.24) we can easily see that the term proportional to $c/(\omega_1^2-\omega^2)$ in eqn (5.43) identically vanishes for $\omega \to \omega_1$. (A similar situation is found for cubic crystals and anisotropic crystals with one molecule per unit cell discussed above as can be seen from eqns (5.14) and (5.39), respectively.)

Thus, in the frequency range of impurity absorption $\tilde{\omega}_1 \approx \omega$, the only resonance term in eqn (5.43) provides a contribution with the form

$$\delta\epsilon_{ij}(\omega,\mathbf{k}) = \frac{2\pi}{v}c\frac{F_1}{\tilde{\omega}_1^2-\omega^2}\left[\frac{L_i^{(1)}L_j^{(1)}}{(1-M_{11}-M_{12})^2} + \frac{L_i^{(2)}L_j^{(2)}}{(1-M_{11}+M_{12})^2}\right]. \quad (5.44)$$

However, if we recall what was written in Section 5.2 about the role of the local field correction on impurity spectra, we can write eqn (5.44) directly without expanding eqn (5.42) in powers of c.

Indeed, in the isolated molecule of an isotopic impurity having orientation α the transition dipole is $\mathbf{p}^\alpha = p\boldsymbol{\ell}^\alpha$. The effective value corresponding to this dipole moment should be found from the condition

$$-\mathbf{p}^\alpha \mathbf{E}^\alpha = -\mathbf{p}^\alpha_{\text{eff}} \mathbf{E}. \tag{5.45}$$

Making use of eqn (5.28) we obtain

$$\mathbf{p}^\alpha_{\text{eff}} = \frac{p}{2}\left(\frac{\mathbf{L}^{(1)}}{1 - M_{11} - M_{12}} - \frac{(-1)^\alpha \mathbf{L}^{(2)}}{1 - M_{11} + M_{12}}\right). \tag{5.46}$$

On the other hand, we have by definition,

$$\delta\epsilon_{ij}(\omega, \mathbf{k}) = \frac{4\pi}{v}c\frac{F_1}{p^2}\sum_\alpha \frac{(\mathbf{p}^\alpha_{\text{eff}})_i (\mathbf{p}^\alpha_{\text{eff}})_j}{\tilde{\omega}_1^2 - \omega^2}, \tag{5.47}$$

and substitution of eqn (5.46) into eqn (5.47) directly yields eqn (5.44).

Equation (5.44) makes it possible to analyze the dependence of the intensity of impurity absorption on the polarization of the incident light. Rashba has treated this problem within the framework of the theory of small-radius excitons (2), (12) for the isotopic substitutional impurities.

For the given crystal model the tensor $\epsilon_{ij}(\omega, \mathbf{k})$ is reduced to the diagonal form if the coordinate axes x and y are parallel to the vectors $\mathbf{L}^{(1)}$ and $\mathbf{L}^{(2)}$. Then we have

$$\epsilon_{11}(\omega, \mathbf{k}) = \epsilon^0_{11}(\omega, \mathbf{k}) + \frac{2\pi}{v}c\frac{F_1}{\tilde{\omega}_1^2 - \omega^2 - i\eta}\frac{\left|\mathbf{L}^{(1)}\right|^2}{(1 - M_{11} - M_{12})^2},$$

$$\epsilon_{22}(\omega, \mathbf{k}) = \epsilon^0_{22}(\omega, \mathbf{k}) + \frac{2\pi}{v}c\frac{F_1}{\tilde{\omega}_1^2 - \omega^2 - i\eta}\frac{\left|\mathbf{L}^{(2)}\right|^2}{(1 - M_{11} + M_{12})^2}, \tag{5.48}$$

$$\epsilon_{22}(\omega, \mathbf{k}) = 1,$$

where a possible weak decay has been taken into account ($\eta > 0$).

For instance, assume that the light is polarized along the x-axis (parallel to $\mathbf{L}^{(1)}$). Assuming that the matrix is transparent at the frequency $\omega \approx \tilde{\omega}_1$ we derive the following expression for the absorption coefficient of the impurity

$$\kappa_{\text{I}}(\omega) = \frac{\pi c}{2v\tilde{\omega}_1}\frac{1}{n^0_{\text{I}}(\omega)}\frac{F_1 \left|\mathbf{L}^{(1)}\right|^2 \delta(\omega - \tilde{\omega}_1)}{(1 - M_{11} - M_{12})^2}, \tag{5.49}$$

Here $n^0_{\text{I}}(\omega) = \sqrt{\epsilon^0_{xx}(\omega, \mathbf{k})}$ is the refractive index of the matrix for light polarized along the x-axis. Similarly, we obtain for light polarized along the y-axis (parallel to $\mathbf{L}^{(2)}$):

$$\kappa_{\mathrm{II}}(\omega) = \frac{\pi c}{2v\tilde{\omega}_1} \frac{1}{n_{\mathrm{II}}^0(\omega)} \frac{F_1 \left|\mathbf{L}^{(2)}\right|^2 \delta(\omega - \tilde{\omega}_1)}{(1 - M_{11} + M_{12})^2}, \qquad (5.50)$$

where $n_{\mathrm{II}}^0(\omega) = \sqrt{\epsilon_{yy}^0(\omega, \mathbf{k})}$ can be found from eqn (5.32). The ratio of the integral absorption intensities $\bar{\kappa}_{\mathrm{I,II}}$ (given by $\int \kappa(\omega)d\omega$) corresponding to the absorption coefficients κ_{I} and κ_{II} is, evidently, equal to

$$\frac{\bar{\kappa}_{\mathrm{I}}(\tilde{\omega}_1)}{\bar{\kappa}_{\mathrm{II}}(\tilde{\omega}_1)} = \frac{n_{\mathrm{II}}^0(\tilde{\omega}_1)}{n_{\mathrm{I}}^0(\tilde{\omega}_1)} \left. \frac{\left|\mathbf{L}^{(1)}\right|^2 (1 - M_{11} + M_{12})^2}{\left|\mathbf{L}^{(2)}\right|^2 (1 - M_{11} - M_{12})^2} \right|_{\omega=\tilde{\omega}_1}. \qquad (5.51)$$

In order to rewrite this relationship in a more explicit form let us turn back to the dielectric constant tensor (5.32) for the pure crystal. Using eqns (5.24) and (5.31) we can rewrite eqn (5.51) in the form:

$$\frac{\bar{\kappa}_{\mathrm{I}}(\tilde{\omega}_1)}{\bar{\kappa}_{\mathrm{II}}(\tilde{\omega}_1)} = \frac{n_{\mathrm{II}}^0(\tilde{\omega}_1)}{n_{\mathrm{I}}^0(\tilde{\omega}_1)} \frac{F_{1,xx} \left[\tilde{\omega}_1^2 - \Omega_2^2(\mathbf{k})\right]^2}{F_{1,yy} \left[\tilde{\omega}_1^2 - \Omega_1^2(\mathbf{k})\right]^2}. \qquad (5.52)$$

If the frequency $\tilde{\omega}_1$ is close to the lattice absorption frequencies so that the difference $|\tilde{\omega}_1 - \Omega_{1,2}|$ is suffiently small, eqn (5.52) can also be written as:

$$\frac{\bar{\kappa}_{\mathrm{I}}}{\bar{\kappa}_{\mathrm{II}}} = \frac{n_{\mathrm{II}}^0(\tilde{\omega}_1)}{n_{\mathrm{I}}^0(\tilde{\omega}_1)} \frac{F_{1,xx} (\tilde{\omega}_1 - \Omega_2(\mathbf{k}))^2}{F_{1,yy} (\tilde{\omega}_1 - \Omega_1(\mathbf{k}))^2}. \qquad (5.53)$$

This relationship shows that when the frequency $\tilde{\omega}_1$ tends, for instance, to the frequency Ω_1 the quantity $\bar{\kappa}_{\mathrm{I}}$ can become anomalously high as compared with $\bar{\kappa}_{\mathrm{II}}$. Thus, absorption at the impurity frequency becomes sharply polarized even though sharp polarization of the impurity absorption may be absent far from the frequencies Ω_1 and Ω_2. Equation (5.53) for $\mathbf{k} = 0$ has been derived by Rashba (12). The importance of taking into account spatial dispersion in eqn (5.53) will be discussed below. Now we shall make a few remarks on the nature of the effect described by eqn (5.53). As eqn (5.46) implies, this effect is due entirely to the internal field correction which increases when the frequency $\tilde{\omega}_1$ approaches the resonances of the dielectric constant tensor for the pure crystal. Of course, this mechanism determining the effect of the matrix occurs also in crystals of any structure and also for interstitial impurities. However, the structure and the effect of the internal field can differ in different cases. Now let us continue the analysis of eqn (5.53). Assume that the frequency $\tilde{\omega}_1$ is in the region of a high-intensity dipole transition in the matrix and it is just this transition that determines the dispersion of the refractive indexes $n_{\mathrm{I,II}}^0(\omega)$ in the given frequency range. Then, if $\tilde{\omega}_1 < \Omega_1(0), \Omega_2(0)$ and the differences $|\Omega_{1,2} - \tilde{\omega}_1|$ are small but yet large enough that we can neglect the spatial dispersion of the medium, we obtain:

$$n_{\rm I}^0(\omega) \sim \sqrt{\frac{F_{1,xx}}{|\omega - \Omega_1(0)|}}, \qquad n_{\rm II}^0(\omega) \sim \sqrt{\frac{F_{1,yy}}{|\omega - \Omega_2(0)|}}.$$

Under these conditions eqn (5.53) is transformed into:

$$\frac{\bar{\kappa}_{\rm I}}{\bar{\kappa}_{\rm II}} = \sqrt{\frac{F_{1,xx}}{F_{1,yy}}} \left|\frac{\tilde{\omega}_1 - \Omega_2(0)}{\tilde{\omega}_1 - \Omega_1(0)}\right|^{3/2}. \tag{5.54}$$

If the spatial dispersion must be taken into account then, since for small **k** we have:

$$\Omega_1(\mathbf{k}) = \Omega_1(0) + \mu_1 \left(n_{\rm I}^0\right)^2, \qquad \Omega_2(\mathbf{k}) = \Omega_2(0) + \mu_2 \left(n_{\rm II}^0\right)^2, \tag{5.55}$$

eqn (5.53) can be rewritten as,

$$\frac{\bar{\kappa}_{\rm I}}{\bar{\kappa}_{\rm II}} = \frac{F_{1,xx}}{F_{1,yy}} \left\{ \frac{\sqrt{n_{\rm II}^0(\tilde{\omega}_1)}\left[\tilde{\omega}_1 - \Omega_2(0) - \mu_2\left(n_{\rm II}^0\right)^2\right]}{\sqrt{n_{\rm I}^0(\tilde{\omega}_1)}\left[\tilde{\omega}_1 - \Omega_1(0) - \mu_1\left(n_{\rm I}^0\right)^2\right]} \right\}^2, \tag{5.56}$$

where $n_{\rm I,II}^0$ satisfy the equations:

$$n_{\rm I}^2(\tilde{\omega}_1) = \epsilon_{xx}^0(\tilde{\omega}_1, \mathbf{k}) = 1 + \frac{2\pi}{v} \frac{F_{1,xx}}{\Omega_1^2(0) + \mu_1 n_{\rm I}^2 - \tilde{\omega}_1^2},$$

$$n_{\rm II}^2(\tilde{\omega}_1) = \epsilon_{yy}^0(\tilde{\omega}_1, \mathbf{k}) = 1 + \frac{2\pi}{v} \frac{F_{1,yy}}{\Omega_2^2(0) + \mu_2 n_{\rm II}^2 - \tilde{\omega}_1^2}. \tag{5.57}$$

The nature of the dependences $n_{\rm I}(\omega)$ and $n_{\rm II}(\omega)$ are known to be dependent on the signs of the coefficients μ_1 and μ_2, respectively (1). Therefore, the experimental results for the ratio (5.53) can, in principle, be used to identify the spatial dispersion effects of the crystal matrix. However, the decay of the excited states of both the impurity and the matrix makes this difficult. If the levels of these states are wide enough then it becomes practically impossible to "sneak up" on the frequency $\Omega_1(0)$ and to distinguish the impurity absorption from the matrix absorption.

5.6 Optical properties of mixed crystalline solutions

Now we shall discuss some features of the electromagnetic wave spectrum in mixed crystalline solutions using the equations for the dielectric constant tensor of the solution derived in the mean polarizability approximation. We know that the effects of concentration broadening of absorption spectra are lost in this approximation. However, Onodera and Toyozawa (16), Dubovsky and Konobeev (17), Hoschen and Jortner (18), and Hong and Robinson (19), who have actually studied the corrections to the mean polarizability approximation, have shown that this approximation is more or less suitable for treating such relatively coarse features of a spectrum as, for instance, the dependence of the position of the center of an absorption band on the composition of the solution and some others.

Moreover, the mean polarizability approximation can yield highly accurate results for dispersion and optical anisotropy of crystalline solutions outside the absorption band. This is due to the fact that the concentration broadening in crystals of this type (with only van der Waals interactions between the molecules) does not affect the integral oscillator strength of a transition. The mean polarizability approximation served as the basis for the procedure developed by Obreimov for the analysis of the composition of multicomponent systems as applied to a wide variety of isotopic mixtures, both liquid and crystalline (for the details, see (20)).

Let us illustrate the mean polarizability approximation by applying it to the spectrum of a binary mixture of isotopically related molecules in a cubic crystal. In this case eqn (5.12) gives

$$\epsilon(\omega) = 1 + \frac{4\pi N_0 \bar{a}}{1 - (4\pi/3)N_0 \bar{a}}, \qquad \bar{a} = (1-c)a + c\tilde{a}, \qquad (5.58)$$

where c is the relative concentration of the impurity. Thus, the resonances of $\epsilon(\omega)$ correspond to the frequencies that depend on c and satisfy the equation

$$\frac{1}{A} = \frac{1-c}{\omega_1^2 - \omega^2} + \frac{c}{\tilde{\omega}_1^2 - \omega^2}, \qquad (5.59)$$

where $A = (4\pi/3)N_0 F_1$.

In deriving eqn (5.59) we have used the fact that the polarizabilities of the different isotopic molecules are given by eqn (5.9) and differ from one another only in the values of the resonance frequencies. Equation (5.59) enables us to determine the shift of the absorption lines of the mixture as a function of its composition in the given approximation. The dependence of the frequencies of the longitudinal waves on the composition of the mixture can be found from the equation $\epsilon(\omega) = 0$. Using eqn (5.58) we find that these frequencies satisfy the equation

$$-\frac{1}{2A} = \frac{1-c}{\omega_1^2 - \omega^2} + \frac{c}{\tilde{\omega}_1^2 - \omega^2}.$$

In a similar way we can find the dependence, for instance, of the frequency of the surface waves, on the composition of the mixture (from the condition $\epsilon(\omega) = -1$) and many other parameters of the mixture that are determines by its dielectric constant.

The treatment of anisotropic crystalline solutions can also be carried out very simply. For such solutions consisting of two types of isotopic molecules and having two molecules per unit cell the dielectric constant tensor has been derived above in the additive refraction approximation (see eqn 5.42). This equation shows that in the polarization $\mathbf{L}^{(1)}$ the resonances of the tensor ϵ_{ij} correspond to the frequencies ω satisfying the equation

$$1 - \tilde{M}_{11}(\omega, \mathbf{k}) - \tilde{M}_{12}(\omega, \mathbf{k}) = 0, \tag{5.60}$$

and in the polarization $\mathbf{L}^{(2)}$ the respective frequencies satisfy the equation

$$1 - \tilde{M}_{11}(\omega, \mathbf{k}) + \tilde{M}_{12}(\omega, \mathbf{k}) = 0. \tag{5.61}$$

Using eqn (5.37) we can transform eqns (5.60) and (5.61) into

$$\frac{1}{A_{11} - A_{12}(-1)^\rho} = \frac{1-c}{\omega_1^2 - \omega^2} + \frac{c}{\tilde{\omega}_1^2 - \omega^2}, \qquad \rho = 1, 2, \tag{5.62}$$

where

$$A_{11} = F_1 \sum_{ij} Q_{ij}^{11} \ell_i^1 \ell_j^1, \qquad A_{12} = F_1 \sum_{ij} Q_{ij}^{12} \ell_i^1 \ell_j^2.$$

Equation (5.62) implies that a doublet of lines should correspond to each of the polarizations $\rho = 1, 2$ in the light absorption spectrum. If we denote the solutions of eqn (5.62) for each of the polarizations ρ by $\Omega_{\rho\lambda}^2$ ($\lambda = 1, 2$) then after expansion into elementary fractions we can write the tensor $\epsilon_{ij}(\omega, \mathbf{k})$ as

$$\epsilon_{ij}(\omega, \mathbf{k}) = \delta_{ij} + \sum_{\rho, \lambda} \frac{\Pi_{\rho\lambda}}{\Omega_{\rho\lambda}^2 - \omega^2} L_i^{(\rho)} L_j^{(\rho)}, \tag{5.63}$$

where

$$\Pi_{\rho\lambda} = \frac{2\pi}{v} \frac{F_1}{[A_{11} - (-1)^\rho A_{12}]^2} \left[\frac{1-c}{(\omega_1^2 - \Omega_{\rho\lambda}^2)^2} + \frac{c}{(\tilde{\omega}_1^2 - \Omega_{\rho\lambda}^2)^2} \right]^{-1},$$

$$\rho = 1, 2, \qquad \lambda = 1, 2. \tag{5.64}$$

The solutions of eqns (5.62)–(5.64) in this approximation fully determine the dependences on c of not only the intensities and positions of the light absorption bands for a crystal but also the dispersion of the refractive indexes. The above equations can be readily extended to the case of multicomponent isotopic solutions. Then eqns (5.36) and (5.63) retain their form but eqns (5.37), (5.62), and (5.64), respectively, are replaced by the more general equations:

$$\tilde{M}_{\alpha\beta}(\omega, \mathbf{k}) = \sum_k \frac{F_1 c_k}{\omega_k^2 - \omega^2} \sum_{ij} Q_{ij}^{\alpha\beta}(\mathbf{k}) \ell_i^\alpha \ell_j^\beta,$$

(here c_k are the relative concentrations, $k = 1, 2, \ldots, s$, so that $\sum_k c_k = 1$)

$$\frac{1}{A_{11} - (-1)^\rho A_{12}} = \sum_k \frac{c_k}{\omega_k^2 - \omega^2}, \tag{5.65}$$

$$\Pi_{\rho\lambda} = \frac{2\pi}{v} \frac{F_1}{[A_{11} - (-1)^\rho A_{12}]^2} \left[\sum_k \frac{c_k}{(\omega_k^2 - \Omega_{\rho\lambda}^2)^2} \right]^{-1}, \tag{5.66}$$

where $\Omega_{\rho\lambda}^2$ are the solutions of eqn (5.62) for fixed $\rho = 1, 2$ and $\lambda = 1, 2, \ldots, s$.

It should be stressed that the results obtained here for the positions of the lines and the intensities of absorption in such solutions coincide with the results of Broude and Rashba (21) which were derived in the framework of the theory of small-radius excitons using an approximation identical with the mean polarizability approximation (the additive refraction approximation; the entire tensor $\epsilon_{ij}(\omega, \mathbf{k})$ was not found in this study).

Apart from isotopic mixtures, the acting field method described here can be similarly applied to molecular mixtures of quite different molecules. Here we should only bear in mind the fact that in this more general case the resonance frequencies ω_k of individual molecules in solution become functions of the composition, in contrast to the case of isotopic molecules for which all the static multipoles can be considered independent of the isotopic composition to a high degree of accuracy. This dependence on composition can be easily understood since it is due to the shift of the resonance frequency of the individual molecule mentioned in Section 5.1. As shown there, this shift is caused by the change upon excitation of the energy of interaction of the molecule with the environment. This environment differs for different compositions of the mixture and this is the cause of the above dependence. We can take it into account by using the following notation: ω_k^0 is the resonance of the molecule in vacuum, and $D_{k\ell}$ is the shift of this resonance in the environment consisting only of the molecules of the species ℓ. Clearly, in the mean polarizability approximation we obtain for the mixture,

$$\omega_k \equiv \omega_k(c_1, c_2, \ldots) = \omega_k^0 + \sum_\ell D_{k\ell} c_\ell.$$

5.7 Energy of the resonance interaction of the impurity molecules

In the preceding sections of this chapter the local field method was applied mainly to the theory of optical properties of crystals and crystalline solutions. In this section we will investigate the influence of the local field corrections on the resonance interaction between the impurity molecules. Just this interaction determines the transfer of intramolecular (electronic or vibrational) excitation energy from one impurity molecule to another.

We assume that the main role in this interaction is played by the dipole–dipole interaction as in the Förster mechanism. The operator of this interaction in vacuum is

$$\hat{V}_{ab}^0 = \frac{R^2 \hat{\mathbf{p}}_a \hat{\mathbf{p}}_b - 3(\hat{\mathbf{p}}_a \mathbf{R})(\hat{\mathbf{p}}_b \mathbf{R})}{R^5},$$

where $\hat{\mathbf{p}}_a$ and $\hat{\mathbf{p}}_b$ are the operators of the dipole elements of the impurity molecules a and b, and $\mathbf{R}$ is the vector between the centers of these molecules. If initially the molecule a was in the excited state f and the identical molecule b was in the ground state and in the final state, the situation is reversed, and the matrix element of the operator $\hat{V}_{ab}^0$ corresponding to this transition would be:

$$M_{ab} \equiv V_{ab}^{f0,0f} = \frac{R^2 \mathbf{p}_a^{f0} \mathbf{p}_b^{0f} - 3\left(\mathbf{p}_a^{f0}\mathbf{R}\right)\left(\mathbf{p}_b^{0f}\mathbf{R}\right)}{R^5}. \tag{5.67}$$

If the impurity molecules are in a polarizing medium, then the appropriate corrections should be made in the interaction energy operator and its matrix element. For simplicity, we consider now the case of an isotropic medium. Assuming that the distance between molecules is enough large, $c/\omega_{0f} \gg R \gg d$ (d is the crystal lattice parameter) at first sight, it might seem that this can be done just by dividing the matrix element M_{ab} by $\epsilon(\omega_{0f})$ where $\epsilon(\omega_{0f})$ is the dielectric constant of the solvent at frequency ω_{0f}. However, the result would not be correct even though it corresponds to the well-known result obtained within the framework of the phenomenological Maxwell equations. Indeed, if we introduce a source dipole μ_a at a point a into these equations the electric field produced by this dipole is of the order of $\mu_a/\epsilon R^3$ so that the resulting interaction energy is indeed of the order of

$$U \sim \mu_a \mu_b / R^3 \epsilon(\omega_{0f}). \tag{5.68}$$

However, we should bear in mind that all the quantities entering into the phenomenological Maxwell equations should be regarded as phenomenological quantities which can be considerably different from the respective microscopic quantities. This fact is well known, for instance, for the dielectric constant $\epsilon(\omega)$. Here it is expressed as the difference between the polarizability of individual molecules and the polarizability of the crystal. The same is true for diagonal and off-diagonal elements of the dipole operator. The phenomenological (effective) value of this dipole in the medium differs considerably from the corresponding value in vacuum. For the model of two dipoles in an isotropic medium eqn (5.67) is replaced by

$$(M_{ab})_{\text{eff}} \equiv \left(V_{ab}^{f0,0f}\right)_{\text{eff}} = \frac{1}{\epsilon}\left(\frac{\epsilon+2}{3}\right)^2 V_{ab}^{f0,0f}, \tag{5.69}$$

where ϵ is the dielectric constant of the medium at the transition frequency.

Equation (5.69) is derived within the framework of the microscopic theory (22), (23) where anisotropic crystals are also treated. Mahan (24) and Mahan and Mazo (25) derived eqn (5.69) for static dipoles ($\omega_{0f} = 0$) in cubic crystals. Nienhuis and Deutch (26) also reported calculations of the energy of interaction between charges and dipoles in a medium.

A quite elementary and general analysis is sufficient to understand the new factors appearing in eqn (5.69) as compared with eqn (5.67) since these factors have a macroscopic nature.

Indeed, the dipole moment appearing in eqn (5.45) is not the dipole moment in vacuum but an effective dipole moment found by taking into account the internal field correction. In the isotropic medium this is $\mu_a = [(\epsilon+2)/3]p_a^{0f}$ which yields precisely eqn (5.69) if we take into account eqn (5.68). This conclusion is valid for the local centers of any nature in the nonconducting medium. We mean here media in which the intermolecular interaction does not violate the neutrality of molecules. The specific effects found in ionic crystals are discussed by Smith and Dexter (27).

Proceeding from the above discussion, let us now find the general form of the energy of the dipole–dipole resonance radiationless interaction in anisotropic crystals. To do this we must generalize the relations of Section 5.4. Assume that $a_\nu(\omega)$ are the principal values of the polarizability tensor of the individual molecule ($\nu = 1, 2, 3$) and ℓ^ν are the directions of its principal axes. Then we can write this tensor as

$$a_{ij}^\alpha(\omega) = \sum_\nu a_\nu(\omega) \ell_i^{\alpha\nu} \ell_j^{\alpha\nu}.$$

If we take into account only one of the excited states in the molecule this equation is replaced by eqn (5.21). In the region of transparency many of the excited molecular states make comparable contributions to the polarizability of the molecule so that the approximation (5.21) becomes insufficient. A similar situation occurs also in crystalline solutions. In this case, in the mean polarizability approximation (the additive refraction approximation) we should write, for instance, for isotopic mixtures,

$$a_{ij}^\alpha(\omega) = \sum_\nu \bar{a}_\nu(\omega) \ell_i^{\alpha\nu} \ell_j^{\alpha\nu}, \qquad \bar{a}(\omega) = \sum_\rho c_\rho a_\nu^\rho(\omega),$$

where c_ρ is the relative concentration of isotopic component ρ, and $a_\nu^\rho(\omega)$ are the respective principal values of the polarizability tensor of the molecule ($\nu = 1, 2, 3$). Using eqns (5.1) and (5.2) we find that the projections of the internal field $\mathbf{E}^\alpha$ on the directions $\ell^{\alpha\nu}$, that is, the quantities $\mathbf{E}^\alpha \ell^{\alpha\nu}$, satisfy a system of 3σ equations ($\alpha = 1, 2, \ldots, \sigma;\ \nu = 1, 2, 3$):

$$\left(\mathbf{E}^\alpha \ell^{\alpha\nu} \right) = \left(\mathbf{E}\ell^{\alpha\nu} \right) + \sum_{\beta\nu'} M_{\alpha\beta}^{\nu\nu'}(\omega, \mathbf{k}) \left(\mathbf{E}^\beta \ell^{\beta\nu'} \right),$$

where

$$M_{\alpha\beta}^{\nu\nu'}(\omega, \mathbf{k}) = a^\nu(\omega) \sum_{ij} Q_{ij}^{\alpha\beta}(\mathbf{k}) \ell_i^{\alpha\nu} \ell_j^{\beta\nu'}.$$

For fixed α the unit vectors $\ell^{\alpha\nu}$ make up a triad of mutually orthogonal vectors and, therefore, we can use the solution of the above system of inhomogeneous equations to find the form of the tensor $A_{ij}^\alpha(\omega, \mathbf{k})$ in eqn (5.4). Then, if $\mathbf{p}_a^{0f,\alpha}$ is the value in vacuum for the matrix element of the dipole moment of the substitutional impurity at the site α, its effective value is

$$\mu_{ai}^{0f,\alpha} = \left(\mathbf{p}_a^{0f,\alpha} \right)_j A_{ji}^\alpha(\omega, \mathbf{k}).$$

Therefore, if we ignore spatial dispersion the energy of the dipole–dipole interaction is

$$V_{ab}(\mathbf{R}) = -\left(\mathbf{p}_a^{0f,\alpha} \right)_{j_1} A_{j_1 i}^\gamma(\omega^{0f}) A_{j_2 j}^{\gamma'}(\omega^{0f}) \left(\mathbf{p}_b^{0f,\gamma'} \right)_{j_2}.$$

$$\times \frac{\partial^2}{\partial x_i \partial x_j} \frac{1}{\sqrt{\epsilon_{\alpha\beta}^{-1} x_\alpha x_\beta}} \left(\det \hat{\epsilon}^{-1} \right)^{1/2}, \tag{5.70}$$

where $\epsilon_{\alpha\beta}(\omega^{0f})$ is the dielectric tensor of the solvent at the frequency ω^{0f} (here γ and γ' denote the sites of the impurities a and b).

If we take into consideration spatial dispersion (2) we obtain in $V_{ab}(\mathbf{R})$ additional terms that decrease more quickly with increasing R. However, these terms are significant only for the frequencies ω^{0f} that are close enough to the intrinsic absorption frequencies of the medium. In the case of isotropic media, eqn (5.70) is replaced by eqn (5.69).

Finally, note that both for pure crystals and for isotopically mixed crystals (for which $M_{\alpha\beta}^{\nu\nu'} \to \tilde{M}_{\alpha\beta}^{\nu\nu'}$, $a^\nu(\omega) \to \bar{a}(\omega)$) the solution of the system of equations for the quantities $\mathbf{E}^\alpha \ell^{\alpha\nu}$ makes it possible to generalize the above results and to take into account many resonances in the molecule (that is, to take into account the mixing of molecular configurations under the effect of the intermolecular interaction: these problems are discussed also in the excellent review by Mahan (28).)

5.8 The higher multipoles in the local field method

To illustrate how we can take into account the higher multipoles in the framework of the local field method let us consider, along with the dipole polarization, also the quadrupole q_{ij} and the octupole $q_{ij\ell}$ polarizations of the molecule. In this approximation, the operator for the energy of interaction of the molecule with the external monochromatic electric field $E(\mathbf{r},t)$ is

$$\hat{V} = \sum_i \hat{p}_i E_i + \sum_{ij} \hat{q}_{ij} E_{ij} + \sum_{ij\ell} \hat{q}_{ij\ell} E_{ij\ell}, \tag{5.71}$$

where $\hat{p}_i, \hat{q}_{ij}$ and $\hat{q}_{ij\ell}$ are the operators for the dipole, quadrupole, and octupole moments of the molecule, $E_{ij} = \partial E_i / \partial x_j$, and $E_{ij\ell} = \partial^2 E_i / \partial x_j \partial x_\ell$.

Using eqn (5.71) and time-dependent perturbation theory((29), Section 40) we can find the multipoles induced by the external field. However, as was noted in Section 5.1, the local field serves as the perturbing field for the molecules in a crystal. Therefore, for the αth molecule ($\hat{q}^\alpha$ differ from the conventional tensors (see (30)) in numerical factors) the multipoles can be written as:

$$\begin{aligned}
p_i^\alpha &= \sum_j a_{ij}^\alpha E_j^\alpha + \sum_{j\ell} a_{ij\ell}^\alpha E_{j\ell}^\alpha + \sum_{j\ell m} a_{ij\ell m}^\alpha E_{j\ell m}^\alpha, \\
q_{ij}^\alpha &= \sum_\ell b_{ij,\ell}^\alpha E_\ell^\alpha + \sum_{\ell m} b_{ij,\ell m}^\alpha E_{\ell m}^\alpha + \sum_{\ell m n} b_{ij,\ell m n}^\alpha E_{\ell m n}^\alpha, \\
q_{ij\ell}^\alpha &= \sum_m c_{ij\ell,m}^\alpha E_m^\alpha + \sum_{mn} c_{ij\ell,mn}^\alpha E_{mn}^\alpha + \sum_{mnp} c_{ij\ell,mnp}^\alpha E_{mnp}^\alpha.
\end{aligned} \tag{5.72}$$

The tensors a^α, b^α and c^α in eqn (5.72), which depend on ω, have the form of the sum of resonance terms; each of the terms corresponds to a transition from the ground state to one of the excited states of the molecule. If, for the sake of simplicity, we take into consideration only one of the resonances corresponding to the transition from the ground state to the nondegenerate excited state of the molecule having excitation energy $\hbar\omega_1$, the tensors a^α, b^α and c^α can be factorized so that the treatment becomes considerably simpler. In this approximation, which corresponds to ignoring the mixing of the molecular configurations, the tensor $b^\alpha_{ij,\ell mn}$, for instance, has the form

$$b^\alpha_{ij,\ell mn} = \frac{2\omega_1}{\omega_1^2 - \omega^2} (q^\alpha_{ij})^{01} (q^\alpha_{\ell mn})^{10},$$

so that eqn (5.72) can be rewritten as:

$$p^\alpha_i = \frac{2\omega_1/\hbar}{\omega_1^2 - \omega^2} (p^\alpha_i)^{01} V^{10}_\alpha,$$

$$q^\alpha_{ij} = \frac{2\omega_1/\hbar}{\omega_1^2 - \omega^2} (q^\alpha_{ij})^{01} V^{10}_\alpha, \qquad (5.73)$$

$$q^\alpha_{ij\ell} = \frac{2\omega_1/\hbar}{\omega_1^2 - \omega^2} (q^\alpha_{ij\ell})^{01} V^{10}_\alpha,$$

where

$$V^{10}_\alpha = \sum_i (p^\alpha_i)^{10} E^\alpha_i + \sum_{ij} (q^\alpha_{ij})^{10} E^\alpha_{ij} + \sum_{ij\ell} (q^\alpha_{ij\ell})^{10} E^\alpha_{ij\ell}. \qquad (5.74)$$

Equations (5.72), which are a generalization of eqn (5.2), show that when we take into account higher multipoles within the framework of the local field method we must find not only the local field amplitude but also the amplitudes of its derivatives. Bearing in mind the above discussion and the results reported by Born and Huang (4) and Khokhlov (10), we can write the local field and its derivatives in the following form:

$$E^\alpha_i = E_i + \sum_{\beta j} Q^{\alpha\beta}_{ij} p^\beta_j + \sum_{\beta j\ell} Q^{\alpha\beta}_{ij\ell} q^\beta_{j\ell} + \sum_{\beta j\ell m} Q^{\alpha\beta}_{ij\ell m} q^\beta_{j\ell m},$$

$$E^\alpha_{ij} = E_{ij} + \sum_{\beta\ell} \tilde{Q}^{\alpha\beta}_{ij\ell} p^\beta_\ell + \sum_{\beta\ell m} \tilde{Q}^{\alpha\beta}_{ij\ell m} q^\beta_{\ell m} + \sum_{\beta\ell mn} \tilde{Q}^{\alpha\beta}_{ij\ell mn} q^\beta_{\ell mn}, \qquad (5.75)$$

$$E^\alpha_{ij\ell} = E_{ij\ell} + \sum_{\beta m} \tilde{\tilde{Q}}^{\alpha\beta}_{ij\ell m} p^\beta_m + \sum_{\beta mn} \tilde{\tilde{Q}}^{\alpha\beta}_{ij\ell mn} q^\beta_{mn} + \sum_{\beta mnp} \tilde{\tilde{Q}}^{\alpha\beta}_{ij\ell mnp} q^\beta_{mnp}.$$

The internal field coefficients $Q^{\alpha\beta}$, $\tilde{Q}^{\alpha\beta}$ and $\tilde{\tilde{Q}}^{\alpha\beta}$ depend only on the lattice structure; the explicit expressions for them are not given here; they can be found from the results of Born and Huang (4) and Khokhlov (10). It is clear that eqns (5.75) are a generalization of eqn (5.1) and are reduced to it for $q^\alpha_{ij} = q^\alpha_{ij\ell} = 0$.

Now, multiply the first of eqn (5.75) by $(p^\alpha_i)^{10}$, the second by $(q^\alpha_{ij})^{10}$, and the third by $(q^\alpha_{ij\ell})^{10}$, and take a sum over the subscripts i, j, and ℓ. Using eqns (5.73) and (5.74) we can easily see that the quantities V_α^{10} (where $\alpha = 1, 2, \ldots, \sigma$) satisfy the following system of σ equations:

$$V_\alpha^{10} = K_\alpha + \sum_\beta M_{\alpha\beta}^{\text{eff}}(\mathbf{k}) V_\beta^{10}. \tag{5.76}$$

Here

$$M_{\alpha\beta}^{\text{eff}}(\mathbf{k}) = \frac{2\omega_1/\hbar}{\omega_1^2 - \omega^2} \Bigg\{ \sum_{ij} Q_{ij}^{\alpha\beta}(\mathbf{k}) \left(p_i^\alpha\right)^{01} \left(p_j^\beta\right)^{10}$$

$$+ \sum_{ij\ell} \left[Q_{ij\ell}^{\alpha\beta}(\mathbf{k}) \left(p_i^\alpha\right)^{01} \left(q_{j\ell}^\beta\right)^{10} + \tilde{Q}_{ij\ell}^{\alpha\beta}(\mathbf{k}) \left(p_\ell^\beta\right)^{10} \right] \tag{5.77}$$

$$+ \sum_{ij\ell m} \left[Q_{ij\ell m}^{\alpha\beta}(\mathbf{k}) \left(p_i^\alpha\right)^{01} \left(q_{j\ell m}^\beta\right)^{10} + \tilde{Q}_{ij\ell m}^{\alpha\beta}(\mathbf{k}) \left(q_{ij}^\alpha\right)^{10} \left(q_{\ell m}^\beta\right)^{10} \right.$$

$$\left. + \tilde{\tilde{Q}}_{ij\ell m}^{\alpha\beta}(\mathbf{k}) \left(q_{ij\ell}^\alpha\right)^{01} \left(p_{mn}^\beta\right)^{10} \right] + \sum_{ij\ell mn} \left[\tilde{Q}_{ij\ell mn}^{\alpha\beta}(\mathbf{k}) \left(q_{ij}^\alpha\right)^{01} \left(q_{\ell mn}^\alpha\right)^{10} \right.$$

$$\left. + \tilde{\tilde{Q}}_{ij\ell mn}^{\alpha\beta}(\mathbf{k}) \left(q_{ij\ell}^\alpha\right)^{01} \left(q_{mn}^\beta\right)^{10} \right]$$

$$+ \sum_{ij\ell mnp} \tilde{\tilde{Q}}_{ij\ell mnp}^{\alpha\beta}(\mathbf{k}) \left(q_{ij\ell}^\alpha\right)^{01} \left(q_{mnp}^\beta\right)^{10} \Bigg\},$$

$$K_\alpha = \sum_i \left(p_i^\alpha\right)^{10} E_i + \sum_{ij} \left(q_{ij}^\alpha\right)^{10} E_{ij} + \sum_{ij\ell} \left(q_{ij\ell}^\alpha\right)^{10} E_{ij\ell}. \tag{5.78}$$

In contrast to the tensor $M_{\alpha\beta}$ (5.24), the tensor $M_{\alpha\beta}^{\text{eff}}$ takes into account the dipole–quadrupole, dipole–octupole, quadrupole–quadrupole, octupole–octupole, and quadrupole–octupole interactions, in addition to the dipole–dipole interaction. However, if we are dealing with crystals of the type of anthracene or naphthalene that consist of molecules having an inversion center, then the matrix elements $(q_{ij}^\alpha)^{01}$ of the quadrupole moment operator vanish for the dipole-allowed $0 \to 1$ transitions. In this case only the dipole–dipole, dipole–octupole, and octupole–octupole interactions make contributions to eqn (5.77). Moreover, for these transitions we can omit in eqn (5.78) the second and the third terms since for macroscopic fields $E_i(\mathbf{r}) = E_i \exp(i\mathbf{k}\mathbf{r})$ their derivatives with respect to the coordinates are small (these terms are significant only when we analyze the gyrotropy effects in which case eqn (5.71) should take into account the interaction of the molecule with the magnetic field, and also in the frequency range of the dipole-forbidden transitions). Proceeding from the above discussion we come to the conclusion that the system of equations (5.76) is quite analogous to the system of equations (5.23) with the only difference that the matrix (5.77) gives a more accurate description of the intermolecular interaction. Therefore, in the region of the dipole-allowed transitions the taking into account of the higher multipoles (see eqns 5.5 and 5.73) only shifts somewhat the resonances of the dielectric constant tensor. All other results derived above remain unchanged. However, in the region of dipole-forbidden transitions the terms with E_{ij} and $E_{ij\ell}$ should be retained and we have here also to use the relation

$$P_\ell = \frac{1}{v} \sum_\alpha \left(p_\ell^\alpha + iq_{\ell j}^\alpha k_j - q_{\ell jm}^\alpha k_j k_m \right),$$

where the term quadratic in the wavevector components describes the contribution of the quadrupole and the magnetic dipole. The second term in the expression for the dipole moment is absent for molecules with an inversion center but will play an important role in the calculation of the effects of the gyrotropy of the crystal. It is interesting to apply the local field method for the study of gyrotropy of molecular crystals (both pure crystals and crystals with impurities) and, in particular, gyrotropy of crystalline solutions as a function of their composition. The mean polarizability approximation can, apparently, be used in the region of transparency for calculating the frequency dependence of the rotatory power of a crystalline solution. It would be no less interesting to study the optical properties of molecular crystals as functions of magnetic fields (the Faraday effect) and static electric fields (the Stark effect). (The theory of the Stark effect for impurities is developed by Dunmur and Munn (31).) These results are used for interpreting the experimental results for interstitial azulene molecules in naphthalene crystals reported by Hochstrasser and Noe (32). We mention here these effects because their treatment in the framework of the local field method would probably be the most suitable and simplest approach available.

6

BIPHONONS AND FERMI RESONANCE IN VIBRATIONAL SPECTRA OF CRYSTALS

6.1 Effects of strong anharmonicity in vibrational spectra of crystals

The application of lasers in optical experimental techniques has led to a rapid development of research into the properties of elementary excitations in solids. In addition to the conventional methods of linear crystal optics, Raman scattering of light (RSL) has become one of the principal research methods, as have its various modifications, such as coherent active Raman spectroscopy and others.

Up-to-date optical methods enable one to obtain sufficiently accurate and comprehensive information on various processes, in particular on those accompanied by the simultaneous creation or annihilation of several quasiparticles (two in the simplest case). These processes are of special interest because in them a "residual" interaction between quasiparticles, which need not necessarily be weak in all cases, should be manifest to a greater or lesser degree.

For phonons such a "residual" interaction is anharmonicity, which is commonly ignored in the calculation of the frequencies and amplitudes of the normal vibrations of the crystal lattice. In this harmonic approximation, advanced at the very beginning of the development of present-day solid-state theory (1)–(3), the excited states of the lattice are associated with sets of various numbers of phonons of one kind or another. The energy, for example, of the excited state of a lattice with two phonons equals

$$E_{\ell_1\ell_2}(\mathbf{k}_1, \mathbf{k}_2) = \hbar\Omega_{\ell_1}(\mathbf{k}_1) + \hbar\Omega_{\ell_2}(\mathbf{k}_2), \qquad (6.1)$$

where ℓ_1 and ℓ_2 indicate the branches of the phonon spectrum, and $\mathbf{k}_1$ and $\mathbf{k}_2$ are the wavevectors of the phonons. In contrast to a state with a single phonon, the state being considered is characterized by the values of quasimomenta, $\hbar\mathbf{k}_1$ and $\hbar\mathbf{k}_2$, and, consequently, is a two-particle state. The energy of the states of a lattice with a large number of phonons can be written in a similar manner. Since in such many-particle states the phonons $(\mathbf{k}_1\ell_1)$, $(\mathbf{k}_2\ell_2)$, etc. do not interact when anharmonicity is neglected, the bandwidth of a many-particle state is found to be equal to the sum of the widths of the energy bands of the isolated phonons (for simplicity, we consider the case of nonoverlapping bands).

Taking anharmonicity of the lattice vibrations into account leads to interaction of the phonons with one another. When this interaction turns out to be sufficiently strong, the formation of bound states of quasiparticles becomes possible in addition to the above-mentioned multiparticle states. Such states are absent in the description of the crystal in the framework of the harmonic approximation. In such states the quasiparticles move in the crystal as a whole and, therefore, like isolated phonons, can be characterized by a single value of the wavevector.

The bound state of two phonons is usually called a biphonon. Quite a comprehensive theory of biphonons has been developed and, what is of prime significance, convincing evidence has been obtained of their existence in various kinds of crystals.

The study of bound states of optical phonons or, stated more generally, the study of the effects of strong anharmonicity in the vibrational spectra of crystals, has grown out of the development of modern solid-state physics and to no less extent, it was motivated by the requirements of experiment.

Biphonons, as well as other more complicated phonon complexes, should appear in the spectra of inelastically scattered neutrons. Nevertheless, up to the present, the most vital experimental data were obtaining by analyzing the spectra of RSL by polaritons.

As is well known, the selection rules allow RSL by polaritons only in crystals without a center of inversion. This is precisely the kind of crystal in which Fermi resonance with polaritons (to be discussed below) was found to be the physical phenomenon in which the special features of the biphonon spectrum were most evident.

Besides the region of basic (fundamental) frequencies of lattice vibrations, the polariton (light) branch in crystals also intersects the region of two-particle, three-particle, etc. states. Resonance with these states influences the dispersion law of the polariton and the result of this influence can be expediently investigated by the observation of the spectra of RSL by polaritons. What actually occurs here is a resonance, similar to the Fermi resonance, since one of the normal waves in the crystal (the polariton) resonates with states that are analogous to overtones or to combination tones of intramolecular vibrations.

The most useful experimental data were obtained in investigations of the Fermi resonance of the polariton with two-particle states. This led to the discovery of biphonons in many crystals. Before beginning a discussion of the results obtained in these investigattions, we have several comments to make on the development of this research from the historical point of view.

It should be noted, to begin with, that biphonons are quite analogous to bound states of two magnons (4)–(6).

As a matter of fact, states of this kind were investigated in crystalline hydrogen by Van Kranendonk (7), (8) a long time ago, and soon afterward they were observed experimentally (9).

In the papers by Van Kranendonk (7), (8) only the bound states of two different quasiparticles were considered under the condition that the motion of one of them can be ignored in a first approximation (the Van Kranendonk model, see Subsection 6.2.3). This made the Van Kranendonk model inapplicable for analysis of the biphonon spectrum in the frequency region of overtones, as well as combination tones, corresponding to phonons with comparable bandwidths. But this model was found to be suitable for analysis of the biphonon spectrum in the region of combination tone frequencies for many crystals (CO_2, NO_2, OCS, etc.; see the paper by Bogani (10)) and also for analysis of the spectrum of vibronic states in molecular crystals (see the review by Sheka (11)). Useful information can be found also in the book by Califano et al. (12) and in the paper by Gardini (13), who studied the biphonons in N_2O crystals.

A generalization of biphonon theory beyond the Van Kranendonk model was made later (14)–(17). Subsequently, the effect of biphonons on polariton dispersion in the spectral region of two-particle states was investigated in a number of papers (18)–(22), and the contribution of biphonons to the nonlinear polarizability of a crystal was discussed in (23)–(25). Problems of the theory of local and quasilocal biphonons in disordered media were discussed in a number of papers (14), (26) –(28). The influence of anharmonicity in crystals on the spectra of inelastically scattered neutrons was considered by Krauzman et al. (29), Prevot et al. (30), and in Ref. (31).

In the following we shall again touch upon the results of these investigations to a greater or lesser extent. Here we only note that the investigations by Ruvalds and Zawadowski (15), (16) were undertaken in connection with the attempts to interpret the second-order RSL spectra in diamond. The interest provoked by this crystal was due to the fact that as far back as 1946 Krishnan (32) observed a sharp peak in the RSL spectrum of diamond at a frequency exceeding, as it was considered at that time, twice the maximum frequency of a phonon ($\Omega_0 \approx 1332.5 \pm 0.5$ cm^{-1}) by about 1.9 ± 1.5 cm^{-1}. Since the nature of this peak remained unclear, Ruvalds and Zawadowski (15), (16) advanced the hypothesis that the peak was due to the excitation of a biphonon. The phonon frequency Ω_0 corresponds to the value $\mathbf{k} \approx 0$, optical phonons have negative effective mass at low $\mathbf{k}$ values and, to form biphonons with an energy exceeding that of two-particle states, repulsion and not attraction is required.

Subsequently, a peak in the RSL spectra, similar to the one observed by Krishnan, was not found in some crystals, such as silicon and germanium, which have the same type of structure as diamond and have even stronger anharmonicity than diamond. This encouraged Tubino and Birman (33) to improve the accuracy of the calculations of the structure of the phonon bands in crystals with a diamond-type structure. It was shown as a result of comprehensive investigations that the dispersion curve of the above-mentioned optical phonon in diamond has its highest maximum not at $\mathbf{k} = 0$, but at $\mathbf{k} \neq 0$. The result of these calculations indicates that the peak experimentally observed in the RSL spectra of diamond falls within the region of the two-phonon continuum. It cannot correspond to a biphonon and is most likely related to features of the density of two-particle (dissociated) states.

The bound state of two phonons for the overtone frequency region was evidently first identified by Ron and Hornig (34). In this investigation they measured the absorption spectrum of the HCl crystal in the region of overtone frequencies of the fundamental vibration (i.e. at $\omega \approx 2\Omega$, where $\Omega = 2725$ cm^{-1}). It was found that along with the wide absorption band corresponding to the excitation of two free phonons (this bandwidth equals 2Δ, where $\Delta \approx 90$ cm^{-1} is the phonon bandwidth), there was also an absorption peak in the region of lower frequencies with the maximum at $\omega = 5313$ cm^{-1} and a half-width approximately equal to 20 cm^{-1}. Later, biphonons were detected in the region of overtone vibrations in many crystals.

In the subsequent sections of this chapter we discuss the fundamentals of biphonon theory, consider the special features of the Fermi resonance, including Fermi resonance with polaritons, and also analyze the data obtained in the infrared (IR) absorption and RSL spectra (see also the review (18)).

6.2 Biphonon theory

6.2.1 *Biphonons in the overtone frequency region of an intramolecular vibrations: qualitative consideration*

The conclusion that the anharmonicity of optical vibrations in crystals can be very strong in the region of overtone frequencies of intramolecular vibrations follows even from purely qualitative considerations. Indeed, in isolated molecules the anharmonicity energy A for intramolecular vibration usually amounts to 1––3% of the energy $\hbar\Omega$ of a quantum of the fundamental vibrations. Here the anharmonicity energy A is understood to be the quantity $A = (2\hbar\Omega - E_2)/2$, where E_2 is the energy of the excited state with quantum number $n = 2$. For $\hbar\Omega = 1000$ cm^{-1}, for example, A is usually found to be approximately 10–30 cm^{-1}. At the same time, the energy of the intermolecular interaction in crystals, determining the phonon energy bandwidth Δ for the indicated frequency region, can also have a value of the order of magnitude of tens of reciprocal centimeters, as follows, for instance, from measurements of the second-order RSL spectra (37). This is precisely why the dimensionless ratio A/Δ, as previously mentioned in Section 6.1, is not generally a small quantity. Therefore, when the phonon bandwidth is of the order of A, the optical vibrational spectrum in the region of overtone or combination tone frequencies may have an extremely complex structure.

To find a model Hamiltonian (14), (15) that would be sufficiently simple for analysis and, at the same time, would allow a discussion of the most interesting physical effects, we shall consider the problem of the occurrence of biphonons in more detail, using, as an example, a molecular crystal in the frequency region corresponding to the first overtone (i.e. at $\omega \approx 2\Omega$). If there are two quanta of molecular vibrations in the crystal, localized on different molecules, the energy of the crystal, intermolecular interaction being neglected, is $E = 2\hbar\Omega$. If both quanta are localized on a single molecule, $E = 2\hbar\Omega - 2A$ owing to the intramolecular anharmonicity. Hence, in the language of quasiparticles, we can contend that in the case being discussed the intramolecular anharmonicity ($A > 0$) leads to decreasing of the energy of the crystal as the particles approach one another and, consequently, corresponds to their attraction. But the localization of quasiparticles (intramolecular phonons) on a single molecule leads to an increase in the kinetic energy of their relative motion. Since, by order of magnitude, this energy is equal to the phonon energy bandwidth, the state with two phonons bound to each other (a biphonon) certainly occurs when $A \gg \Delta$. Then, besides the energy band of two-particle states, eqn (6.1), corresponding to the independent motion of two phonons, there are, in the molecular overtone frequency region of the spectral spectrum, biphonon states with a lower energy (for $A > 0$). The number of such states, even for $A > \Delta$, as will be shown below (see also (14)), depends upon the structure of the unit cell and is not equal, in general, to the number of molecules per unit cell, as in the case with the number of optical branches corresponding to the fundamental tone.

In the limiting case of weak intermolecular interaction ($\Delta \to 0$ and $|A| \gg \gg \Delta$), biphonons have an extremely simple structure; they go over into the states of molecules excited to the second vibrational level. Here the spectrum of the crystal in the frequency region being considered consists of two lines, with the corresponding crystal energies $E = 2\hbar\Omega - 2A$ (both quanta "sit" on the molecule and the state of the crystal is N-fold degenerate, where N is the number of molecules in the crystal) and $E = 2\hbar\Omega$ (the quanta "sit" on different molecules and the state of the crystal is $[N(N-1)/2]$-fold degenerate).

If, on the contrary, anharmonicity is weak ($|A| \ll \Delta$), biphonons are not created outside the band of two-particle states. But inside the band of two-particle states, as shown by Pitaevsky (38), only weakly bound states of biphonons are formed (even for the smallest value of $|A|$; it is necessary, of course, that the value of the binding energy of the biphonon be greater than the width δ of the phonon level; regarding the feasibility of observing the states discussed by Pitaevsky(38), see below).

A model Hamiltonian that describes the excitation spectrum of the crystal in the energy region $E = 2\hbar\Omega$ can be readily constructed on the basis of the qualitative considerations presented above. As a matter of fact, the Hamiltonian of the crystal, describing the effect of intermolecular interaction on the spectrum, for example, of nondegenerate molecular vibrations can be written in the harmonic approximation as follows:

$$\hat{H}_0 = \sum_n \hbar\Omega B_n^\dagger B_n + \sum_{n,m}{}' V_{nm} B_n^\dagger B_m, \qquad (6.2)$$

where $B_n^\dagger$ and B_n are the Bose creation and annihilation operators of a quantum of intramolecular vibrations with energy $\hbar\Omega$ in molecule n, and V_{nm} is the matrix element of the interaction of molecules n and m corresponding to the transfer of one quantum from molecule m to molecule n. If a unit cell of the crystal contains several molecules, then the subindex n is composite, $n \equiv (\mathbf{n}, \alpha)$, where $\mathbf{n}$ is an integer-valued lattice vector, and α labels the molecules in the unit cell: $\alpha = 1, 2, \ldots, \sigma$.

To determine the energy of the phonons it is necessary to diagonalize the Hamiltonian (6.2). In a crystal with a single molecule per unit cell, for instance, this leads to the relation

$$E \equiv E(\mathbf{k}) = \hbar\Omega + V(\mathbf{k}) \tag{6.3}$$

for an optical phonon with wavevector $\mathbf{k}$, where $V(\mathbf{k})$ is given by the expression

$$V(\mathbf{k}) = \sum_m V_{nm} \exp\left[i\mathbf{k}\left(\mathbf{m}-\mathbf{n}\right)\right]. \tag{6.4}$$

A degenerate intramolecular vibration in a crystal with even a single molecule per unit cell corresponds to a number of phonon bands equal to the degeneracy multiplicity ν. If, moreover, the number of molecules per unit cell $\sigma > 1$, then the number of phonon bands that are related to the given intramolecular vibration of frequency Ω becomes equal to $\sigma\nu$. Hence the possible energies of the phonon are determined by the quantities $E_\ell(\mathbf{k})$, where the index ℓ assumes the values $\ell = 1, 2, \ldots, \sigma\nu$.

The afore-said concerning the structure of the optical phonon spectrum is, of course, well known. We dwelt on this question in such detail because in the spectral region with $\omega \approx 2\Omega$ being discussed, the number of bands of two-particle states of the form of eqn (6.1) (equal to $\sigma\nu(\nu\sigma+1)/2$) can be very large in the general case ($\nu \neq 1$ and $\sigma \neq 1$). Since the distance between the bands $E_\ell(\mathbf{k})$ is of the order of the bandwidth, the bands may overlap, forming a quite complex spectrum of two-particle states in the energy region $E = 2\hbar\Omega$. An experimental investigation of two-particle states presents formidable difficulties, requiring the application of not only the most effective optical techniques, but also careful theoretical analysis.

To take intramolecular anharmonicity into consideration, it is necessary to add the operator

$$\hat{H}_\text{A} = -A \sum_n \left(B_n^\dagger\right)^2 B_n^2, \tag{6.5}$$

to the Hamiltonian (6.2). Hence, the total Hamiltonian is

$$\hat{H}' = \hat{H}_0 + \hat{H}_\text{A}. \tag{6.6}$$

Having been derived on the basis of purely qualitative considerations, the Hamiltonian (6.6) naturally requires some substantiation. It can be shown (see (10) and references therein) that when the cubic terms of the intramolecular anharmonicity, which do not conserve the number of quasiparticles, are taken into account, as well as anharmonicity of the fourth order, we obtain a Hamiltonian of the form (6.6), provided that the natural frequencies of the intramolecular vibrations appearing in the Hamiltonian $\hat{H}_0$ are considered to have been found taking anharmonicity into account. This provides a correction of the order of $A/\hbar\Omega$.

We shall not present here the relations between the phenomenological quantity A and the anharmonicity constants appearing in the potential energy of the molecule. Such relations are not required here because these anharmonicity constants can be found only by making use of some model, whereas the quantity A can be determined directly from a comparison of the first- and the second-order molecular spectra. However, this type of relations exists for some crystals also. For example, Bogani (10) considering one- and two-phonon renormalized Green's functions, calculated nonlinear terms in biphonon Hamiltonian. He found that these terms correspond to phonon–phonon attraction in good agreements with his experimental results. More recent calculations can be found in the papers by Proville (39).

When we take many intramolecular vibrations into account, the Hamiltonian $\hat{H}'$ is of the following form

$$\hat{H}' = \sum_{n,j} \hbar\Omega_j \left(B_n^j\right)^\dagger B_n^j + \sum_{\substack{n,m \\ j \leq j'}} V_{nm}^{jj'} \left(B_n^j\right)^\dagger B_m^{j'}$$

$$- \sum_{n,j \leq j'} A(jj') \left(B_n^j\right)^\dagger \left(B_n^{j'}\right)^\dagger B_n^j B_n^{j'}. \quad (6.7)$$

It becomes necessary to make use of this Hamiltonian (see also Section 6.2.3) when discussing the spectra of crystals containing molecules with degenerate or close frequencies. Assuming, however, that the frequency Ω of the intramolecular vibration is nondegenerate, we shall continue our discussion of the expression (6.6) (or 6.7, if necessary).

An important feature of the operator (6.6) is that it commutes with the operator of the total number of vibration quanta $\hat{N} = \sum_n B_n^\dagger B_n$. Hence, in the steady state of the crystal the number of such quanta is a conserved quantity. In particular, in a state with a single ℓ quantum $|1\rangle_\ell$, the anharmonicity, eqn (6.5), is not important:

$$\hat{H}_A |1\rangle_\ell = 0,$$

so that

$$\hat{H}' |1\rangle_\ell = \hat{H}_0 |1\rangle_\ell = E_\ell(\mathbf{k}) |1\rangle_\ell.$$

The operator (6.5) is chosen so as to provide the correct energy values for a crystal with two vibration quanta when intermolecular interaction is neglected. If, for instance, these quanta are located on different molecules, n and m, the corresponding wavefunction of the crystal is

$$\psi_{nm} = B_n^\dagger B_m^\dagger |0\rangle,$$

where $|0\rangle$ is the ground state of the crystal. Since in this case $\hat{H}_A \psi_{nm} = 0$, we have $\hat{H}' \psi_{nm} = 2\hbar\Omega \psi_{nm}$.

If, however, both quanta are located on a single molecule, i.e. $n = m$, then, using the properties of a Bose operator, we find that $\hat{H}_A \psi_{nn} = -2A\psi_{nn}$. Hence $\hat{H}' \psi_{nn} = (2\hbar\Omega - 2A)\psi_{nn}$.

Thus, strictly speaking, the operator (6.6) can be employed for investigating the states of a crystal with two vibration quanta only when the intramolecular anharmonicity of the form given in (6.5) dominates, and the part of the anharmonicity that is associated with the presence of intermolecular interaction can be neglected.[47] Since, by assumption, $A/\hbar\Omega \ll 1$ and $\Delta/\hbar\Omega \ll 1$, the total Hamiltonian (6.6) also neglects terms that do not conserve the number of quasiparticles (inessential corrections are introduced if they are taken into account).

We underline, however, that even in the limit of large A values, intermolecular anharmonicity may turn out to be important in calculating biphonon bandwidths. Here the Hamiltonian $\hat{H}'$ should include the term

$$\hat{H}_T = \frac{1}{2}\sum_{n,m}{}' W_{nm} \left(B_n^\dagger\right)^2 B_m^2, \tag{6.8}$$

which leads to the simultaneous transfer of two vibration quanta from molecule m to molecule n and back again ($n \neq m$). This transfer $n \to m$ of two vibration quanta is allowed, of course, when relation (6.7) is used as well. But in this approximation, the corresponding matrix element is nonzero only in the second order of perturbation theory with respect to the intermolecular interaction V_{nm}. It is readily evident that this matrix element equals V_{nm}^2/A, so that terms with W_{nm} may be omitted under the condition that

$$\frac{V_{nm}^2}{A} \gg |W_{nm}|. \tag{6.9}$$

Even if this inequality is satisfied for small values of $|\mathbf{n} - \mathbf{m}|$, it may, in general, be violated for large $|\mathbf{n} - \mathbf{m}|$ values because for dipole-active overtones $|W_{nm}| \sim |\mathbf{n} - \mathbf{m}|^{-3}$, whereas the quantity V_{nm}^2 can decrease with increasing $|\mathbf{n} - \mathbf{m}|$ proportionally to $|\mathbf{n} - \mathbf{m}|^{-6}$ or more rapidly. It should also be noted that for dipole-active overtones, it is important to take operator (6.8) into account also because this corresponds to Coulomb long-range interactions and in cubic crystals, for instance, leads to longitudinal–transverse splitting of the biphonon. Fortunately, the inclusion of the operator (6.8) in the total Hamiltonian (see also (40) and (41)) only slightly modifies the calculation procedure, as will be illustrated below.

When the anharmonicity is so large that the inverse inequality holds instead of (6.9), then terms with W_{nm} precisely make the main contribution to the energy bandwidth of the biphonon. In this case the energy of the biphonon is

$$E_b(\mathbf{k}) = 2\hbar\Omega - 2A + \sum_m W_{nm} \exp\left[i\mathbf{k}\left(\mathbf{m} - \mathbf{n}\right)\right] + 0\left(\Delta^2/A\right). \tag{6.10}$$

Note that this relation, as previously mentioned, is exact in the limit of large A values. In the more general case both types of the anharmonicity have to be taken into account.

---

[47] The opposite situation has been investigated by Lalov (40).

In connection with the afore-said, for further analysis of the biphonon states we shall make use of a Hamiltonian of the form

$$\hat{H} = \hat{H}_0 + \hat{H}_A + \hat{H}_T, \qquad (6.11)$$

which is more general than (6.6).

To obtain Hamiltonian (6.11) we proceeded above from the model of a molecular crystal. Actually, its range of application includes nonmolecular crystals as well, provided we are concerned with optical phonons whose bandwidth is much narrower than the phonon frequency. In these spectral regions the vibrations of the atoms inside the unit cell are similar to intramolecular vibrations in molecular crystals, since the comparatively narrow phonon bandwidth is indicative of the weakness of the interaction between the vibrations of atoms located in different unit cells.

6.2.2 Biphonon states

The wavefunction of states of the crystal with two vibration quanta, of interest to us, can be written as

$$|2\rangle = \sum_{n,m} \psi(n,m) B_n^\dagger B_m^\dagger |0\rangle, \qquad \psi(n,m) = \psi(m,n), \qquad (6.12)$$

where $\psi(n,m)$ has the meaning of the wavefunction of two phonons in the Schrödinger representation. The function $|2\rangle$ should satisfy the Schrödinger equation

$$\hat{H}|2\rangle = E_2|2\rangle, \qquad (6.13)$$

where E_2 is the required excitation energy of the crystal with two quanta. Since

$$\hat{H}_A|2\rangle = -A \sum_n \left(B_n^\dagger\right)^2 B_n^2 \sum_{\ell,m} \psi(\ell,m) B_\ell^\dagger B_m^\dagger |0\rangle$$

$$= -2A \sum_{\ell,m} \psi(\ell,m) \delta_{\ell m} B_\ell^\dagger B_m^\dagger |0\rangle,$$

$$\hat{H}_0|2\rangle = 2\hbar\Omega|2\rangle + \sum_{n,m} \left(\sum_\ell [V_{n\ell}\psi(\ell,m) + V_{m\ell}\psi(\ell,n)] \right) B_n^\dagger B_m^\dagger |0\rangle,$$

$$\hat{H}_T|2\rangle = \frac{1}{2} \sum_{n,m} \sum_p \left(W_{np}^A + W_{mp}^A\right) \psi(p,p) \delta_{nm} B_n^\dagger B_m^\dagger |0\rangle,$$

we find that $\psi(n,m)$ satisfies the following system of equations:

$$(E_2 - 2\hbar\Omega)\psi(n,m) - \sum_\ell [V_{n\ell}\psi(\ell,m) + V_{m\ell}\psi(\ell,n)]$$

$$= \left(-2A\psi(n,n) + {\sum_p}' W_{np}^A \psi(p,p) \right) \delta_{nm}. \qquad (6.14)$$

Before solving this system of equations we note that biphonon states, like other crystal states, transform according to the irreducible representations of the crystal space group. Therefore, the range of allowed values of its wavevector $\mathbf{K}$, as for other elementary quasiparticles (for a phonon, for instance), is determined within the first Brillouin zone. Consequently, the wavefunction of the biphonon, corresponding to the wavevector $\mathbf{K}$, can be searched for in the form

$$\psi(n,m) = \exp\left[\frac{1}{2}i\mathbf{K}(\mathbf{n}+\mathbf{m})\right]\varphi_{\alpha\beta}(\mathbf{n}-\mathbf{m}), \qquad (6.15)$$

where the function φ, determining the internal structure of the biphonon, satisfies, in accordance with (6.12), the symmetry condition

$$\varphi_{\alpha\beta}(\mathbf{n}-\mathbf{m}) = \varphi_{\alpha\beta}(\mathbf{m}-\mathbf{n}). \qquad (6.16)$$

To find the biphonon energy levels, we introduce the two-particle Green's function in the zeroth approximation $G_E^0(n,m|\ell,\ell')$, $n=\mathbf{n},\alpha; m=\mathbf{m},\beta$ and so on, satisfying the equation

$$(E-2\hbar\Omega)G_E^0(n,m|\ell,\ell')$$
$$-\sum_p\left[V_{np}G_E^0(p,m|\ell,\ell') + V_{mp}G_E^0(p,n|\ell,\ell')\right] = \delta_{n\ell}\delta_{m\ell'}. \qquad (6.17)$$

As is known, the Green's function of an arbitrary self-conjugated operator $\hat{L}$ has the form

$$G_E(r|r') = \sum_\lambda \frac{\varphi_\lambda(r)\varphi_\lambda^*(r')}{E-E_\lambda}, \qquad (6.18)$$

where E_λ and φ_λ are its λth eigenvalue and eigenfunction. In our case, the role of this operator is played by the energy operator $\hat{H}_0$, whereas the states λ correspond to the eigenstates of a crystal with two phonons when the anharmonicity is neglected. If μ is the number of the phonon band and $\mathbf{k}$ is the wavevector of the phonon, then the wavefunction of a state with two free phonons is determined by the equation

$$\Psi_{\mu\mathbf{k},\mu'\mathbf{k}'}(\mathbf{n}\alpha,\mathbf{m}\beta) = \frac{a_\alpha^\mu(\mathbf{k})a_\beta^{\mu'}(\mathbf{k}')}{N}\exp\left[i(\mathbf{k}\mathbf{n}+\mathbf{k}'\mathbf{m})\right], \qquad (6.19)$$

i.e. it is equal to the product of the wavefunction of the separate phonons $\psi_{\mu\mathbf{k}}(\mathbf{n}\alpha) = a_\alpha^\mu(\mathbf{k})\exp(i\mathbf{k}\mathbf{n})/\sqrt{N}$, where N is the number of unit cells in the volume of the crystal. Making use of (6.18) we now obtain

$$G_E^0(n,m|\ell,\ell') = \sum_{\mu,\mu',\mathbf{k}\mathbf{k}'} \frac{\psi_{\mu\mathbf{k}}(n)\psi_{\mu'\mathbf{k}'}(m)\psi_{\mu\mathbf{k}}^*(\ell)\psi_{\mu'\mathbf{k}'}(\ell')}{E-\varepsilon_\mu(\mathbf{k})-\varepsilon_{\mu'}(\mathbf{k}')}. \qquad (6.20)$$

Hence, as follows from eqn (6.14),

$$\psi(m,n) = \sum_\ell G_E^0(n,m|\ell,\ell)\left(-2A\psi(\ell,\ell) + {\sum_p}' W_{\ell p}^A \psi(p,p)\right). \qquad (6.21)$$

Assuming $n = m$ in this equation, and making use of relation (6.15), we obtain a system of σ equations for the quantities $\varphi_{\alpha\alpha}(0)$. This system is of the following form

$$\varphi_{\alpha\alpha}(0) = \sum_{\beta} R_{\alpha\beta}(E, \mathbf{K}) \varphi_{\beta\beta}(0), \quad (6.22)$$

where

$$R_{\alpha\beta}(E, \mathbf{K}) = \sum_{\boldsymbol{\ell}} \left(-2A G_E^0 \left(\mathbf{n}\alpha, \mathbf{n}\alpha | \boldsymbol{\ell}\beta, \boldsymbol{\ell}\beta \right) \right.$$
$$\left. + \sum_{\mathbf{p},\gamma} G_E^0 \left(\mathbf{n}\alpha, \mathbf{n}\alpha | \mathbf{p}\gamma, \mathbf{p}\gamma \right) W_{\mathbf{p}\gamma, \boldsymbol{\ell}\beta}^A \right) \exp\left[i\mathbf{K}\left(\boldsymbol{\ell}-\mathbf{n}\right)\right]. \quad (6.23)$$

The energy values E of the states being investigated can be obtained from the condition that the determinant of the system of equations (6.22) equals zero, i.e. from the equation

$$|\delta_{\alpha\beta} - R_{\alpha\beta}(E, \mathbf{K})| = 0. \quad (6.24)$$

Crystals with one molecule per unit cell are most easily analyzed. Here eqn (6.24) is of the form

$$1 = R(E, \mathbf{K}). \quad (6.25)$$

Moreover, since in this case

$$G_E^0 \left(\mathbf{n}, \mathbf{n} | \boldsymbol{\ell}, \boldsymbol{\ell} \right) = \frac{1}{N^2} \sum_{\mathbf{k}, \mathbf{k}'} \frac{e^{i(\mathbf{n}-\boldsymbol{\ell})(\mathbf{k}+\mathbf{k}')}}{E - \varepsilon(\mathbf{k}) - \varepsilon(\mathbf{k}')}, \quad (6.26)$$

and

$$\varepsilon(\mathbf{k}) = \hbar\Omega + \sum_{m} V_{nm} \exp\left[i\mathbf{k}\left(\mathbf{m}-\mathbf{n}\right)\right] \equiv \hbar\Omega + V(\mathbf{k}),$$

we find that eqn (6.25) can be written as follows:

$$1 = \frac{1}{N} \sum_{\mathbf{q}} \frac{-2A + W^A(\mathbf{K})}{E - \varepsilon\left(\mathbf{K}/2 + \mathbf{q}\right) - \varepsilon\left(\mathbf{K}/2 - \mathbf{q}\right)}, \quad (6.27)$$

where

$$W^A(\mathbf{K}) = \sum_{\boldsymbol{\ell}} W_{p\ell}^A \exp\left[i\mathbf{K}\left(\boldsymbol{\ell}-\mathbf{p}\right)\right].$$

For $\mathbf{K} = 0$ and $W^A = 0$, the form of (6.27) coincides with that of the equation for the energy of a quantum of local vibration in the vicinity of an isotopic impurity, and its analysis is actually known (see, e.g. (42)). Here we only stress that for $|A| \gg |V(\mathbf{k})|$ the root of eqn (6.27), lying outside the energy region of two-particle states $E(\mathbf{k}, \mathbf{k}') = 2\hbar\Omega + V(\mathbf{k}) + V(\mathbf{k}')$, always exists. To terms of the order of $|V(\mathbf{k})|^2/|A|$ it is given by the expression (cf. eqn 6.10)

$$E_2(\mathbf{K}) = 2\hbar\Omega - 2A + W^A(\mathbf{K}) - \frac{1}{2AN}\sum_q \left[V\left(\frac{1}{2}\mathbf{K} + \mathbf{q}\right) + V\left(\frac{1}{2}\mathbf{K} - \mathbf{q}\right)\right]^2.$$

If the anharmonicity is small, so that $|A| < |V(\mathbf{k})|$, eqn (6.27) may not have a solution $E_2(\mathbf{k})$ lying outside the spectral region of two-particle states. In this case no biphonons are formed, but, under the influence of the anharmonicity, new maxima, not directly associated with Van Hove singularities, can sometimes occur in the density of two-particle states. These maxima may appear, in particular, in RSL spectra and in the spectra of inelastically scattered neutrons. They are due to the possibility of the formation of quasistationary ("resonance") quasibound states of two quasiparticles that are capable of dissociating, at the same energy, into two free phonons. We shall dwell upon the ensuing special features in discussing the experimental data.

If a unit cell of the crystal contains several molecules, then, as has already been noted, the shape of the spectrum of two-particle states becomes more complicated even when anharmonicity is ignored. As concerns the number of biphonon bands, it is equal, under conditions of strong anharmonicity $|A| \gg |V_{nm}|$) for nondegenerate vibrational transitions, to the number σ of molecules in the unit cell.

Indeed, as follows from the relation (6.20), the energy E of the biphonon in this case is

$$E - \varepsilon_\mu(\mathbf{k}) - \varepsilon_{\mu'}(\mathbf{k}') \approx E - 2\hbar\Omega,$$

so that

$$G_E^0(n, m|\ell, \ell') \approx \frac{1}{E - 2\hbar\Omega}\delta_{n\ell}\delta_{m\ell'}.$$

In the limit of large $|A|$ values the expression for $R_{\alpha\beta}$ (see eqn 6.23) simplifies:

$$R_{\alpha\beta} = -\frac{2A\delta_{\alpha\beta}}{E - 2\hbar\Omega} + \frac{W_{\alpha\beta}^A(\mathbf{K})}{E - 2\hbar\Omega}.$$

Substituting this relation into eqn (6.22) we find that the quantities $\varphi_{\alpha\alpha}(0)$ satisfy a system of σ equations:

$$(E - 2\hbar\Omega + 2A)\varphi_{\alpha\alpha}(0) = \sum_\beta W_{\alpha\beta}^A(\mathbf{K})\varphi_{\beta\beta}(0)$$

similar to the system of equations for the Frenkel exciton in a crystal with σ molecules per unit cell. This proves the statement made above on the number of biphonon bands in the limiting case of large $|A|$ values.

If, however, anharmonicity is not too strong, then, as has been pointed out (14), the number of biphonon bands may not be equal to σ (for $A = 0$ and $W^A = 0$ there are absolutely no such bands).

Light absorption by biphonons in anisotropic crystals can be strongly polarized, the corresponding polarization of the absorption line being closely associated with the crystal symmetry. This should be kept in mind in discussing experimental investigations.

6.2.3 Biphonons in the sum frequency region of the spectrum – the Van Kranendonk model

We shall now discuss the special features in the spectrum of excited states of a crystal in the sum frequency region of intramolecular vibrations. Assume, for instance, that we are concerned with the frequency region $\omega \approx \Omega_1 + \Omega_2$, where Ω_1 and Ω_2 are the frequencies of two nondegenerate intramolecular vibrations. Here the model Hamiltonian of the crystal, for the spectral region being considered, can be represented in the form (see also eqn 6.7)

$$\hat{H} = \hat{H}_0^{(1)} + \hat{H}_0^{(2)} - 2A \sum_n \left(B_n^{(1)}\right)^\dagger B_n^{(1)} \left(B_n^{(2)}\right)^\dagger B_n^{(2)}$$
$$+ \frac{1}{2} {\sum_{n,m}}' W_{n,m}^A \left(B_n^{(1)}\right)^\dagger \left(B_n^{(2)}\right)^\dagger B_m^{(1)} B_m^{(2)}, \qquad (6.28)$$

where the operators $\hat{H}_0^{(1)}$ and $\hat{H}_0^{(2)}$ are expressed in terms of the creation and annihilation operators $\left(B_n^{(i)}\right)^\dagger$ and $B_n^{(i)}$ (where $i = 1, 2$).

The Hamiltonian (6.28) commutes with the operator

$$\hat{I}_{nm} = \left(B_n^{(1)}\right)^\dagger \left(B_m^{(2)}\right)^\dagger B_n^{(2)} B_m^{(1)} + \left(B_n^{(2)}\right)^\dagger \left(B_m^{(1)}\right)^\dagger B_n^{(1)} B_m^{(2)},$$

which realizes the interchange of the quanta $\hbar\Omega_1$ and $\hbar\Omega_2$, localized on molecules n and m. Consequently, the wavefunctions of a crystal with two quanta of different kinds can be either even or odd with respect to the interchange of the coordinates n and m. Since, in the case being considered,

$$|2\rangle = \sum_{n,m} \psi(n,m) \left(B_n^{(1)}\right)^\dagger \left(B_m^{(2)}\right)^\dagger |0\rangle, \qquad (6.29)$$

the afore-said means that there are even stationary states, for which $\psi(n,m) = \psi(m,n)$, as well as odd ones, for which $\psi(n,m) = -\psi(m,n)$.

Since for odd states $\psi(n,n) = 0$, the Hamiltonian (6.28) cannot lead to the occurrence of odd biphonon states. This conclusion follows directly from the fact that the result of the action of the anharmonicity operator in Hamiltonian (6.28) on the odd wavefunctions of the biphonon is equal to zero. For example,

$$\hat{H}_A^{(2)}|2\rangle = -2A \sum_n \psi(n,n) \left(B_n^{(1)}\right)^\dagger \left(B_n^{(2)}\right)^\dagger |0\rangle = 0,$$

and in a similar way for the operator containing the quantity W^A.

If we include in the Hamiltonian (6.28) not only the intramolecular anharmonicity (proportional to A) and the anharmonicity due to the intermolecular interaction W^A, but also the part of the intermolecular anharmonicity that is of the form $\sum_{nm} L_{nm} \hat{I}_{nm}$, where L_{nm} is the matrix element of the operator of intermolecular interaction between molecules n and m, corresponding to an interchange of quanta, the odd biphonon states can also separate from the band of two-particle states. But since $|L_{nm}| \leq \left|V_{nm}^{(i)}\right|$, where $i = 1, 2$, the odd biphonon levels, in contrast to the even ones, must always be close to the band of two-particle states.

So far the odd biphonon states have not been discovered experimentally. In molecular crystals their contribution to the absorption spectrum or Raman scattering spectrum should be relatively small because in the odd states (6.29) there are no configurations in which both quanta "sit" on one and the same molecule.

Transitions to such configurations are those that usually correspond to relatively high oscillator strengths, because the intramolecular constants of both mechanical and electrical anharmonicity exceed as a rule, the corresponding constants of intermolecular anharmonicity. Later on we shall return again to these problems in finding the dielectric constant of a crystal for the spectral region of two-particle states. We shall now continue our discussion of the Hamiltonian (6.28).

If in expression (6.28) we put the quantity $V_{nm}^{(2)}$ equal to zero, i.e. ignore the possibility of motion of one of the quanta, we arrive at the case that was investigated by Van Kranendonk (7). Assuming that the quantum $\hbar\Omega_2$ is localized on the molecule $n = 0$ and averaging the Hamiltonian (6.28) over the indicated state, we obtain:

$$\hat{H} = \sum_n \hbar\Omega_1 \left(B_n^{(1)}\right)^\dagger B_n^{(1)} + {\sum_{n,m}}' V_{nm}^{(1)} \left(B_n^{(1)}\right)^\dagger B_m^{(1)} - 2A \left(B_0^{(1)}\right)^\dagger B_0^{(1)} + \hbar\Omega_2,$$

which corresponds to the motion of the quantum $\hbar\Omega_1$ in a crystal with an "isotopic" substitutional impurity located at the lattice point $n = 0$ and with the shifted frequency of intramolecular vibration $\Omega_1' = \Omega_1 - 2A/\hbar$. Thus, in the Van Kranendonk model the energy calculation for the biphonon is reduced to calculations of the energy of a local vibration in the region of an isotopic defect. In the given case this energy is determined by the equation

$$1 = -\frac{2A}{N} \sum_k \frac{1}{E_2 - \hbar(\Omega_1 + \Omega_2) - V^{(1)}(\mathbf{k})}$$

$$\equiv -\frac{2A}{N} \int \frac{\rho(\varepsilon) d\varepsilon}{E_2 - \hbar\Omega_2 - \varepsilon},$$

where $V^{(1)}(\mathbf{k})$ is determined by the relation (6.4) with $V_{nm} = V_{nm}^{(1)}$, $\rho(\varepsilon)$ is the density of $B^{(1)}$-phonon states, and $\int \rho(\varepsilon) d\varepsilon = N$. It is clear that the Van Kranendonk model is not suitable for analysis of biphonon spectra in the overtone frequency region or in the region of combination tones, corresponding to phonons with comparable bandwidths. But the application of this model was found to be highly succsessful in analysis of the infrared (IR) spectrum of crystalline hydrogen in the region of its rotation–vibrational band (see the review of theoretical and experimental investigation compiled by Van Kranendonk and Karl (8)).

As a conclusion of this subsection we write down the equation which determines the energy of the biphonon when the motion of both quanta is taken into account. It can readily be seen that within the framework of the model (6.28) this equation is of the form

$$1 = \frac{1}{N} \sum_q \frac{-2A + W^A(\mathbf{K})}{E_2 - \hbar\Omega_1 - \hbar\Omega_2 - V^{(1)}(\mathbf{K}/2 + \mathbf{q}) - V^{(2)}(\mathbf{K}/2 - \mathbf{q})},$$

which is a simple generalization of eqn (6.27).

6.3 Green's functions in biphonon theory and Fermi resonance in crystals

The relations obtained above for the energies of biphonons can also be derived by applying the Green's function method. This method is found to be extremely useful within the framework of the model being discussed, because all the Green's functions required for calculating also the dielectric tensor of the crystal, its nonlinear polarizabilities, its density of states, its RSL cross-section, and other physical properties, can be found exactly, without resorting to perturbation theory. We illustrate the afore-said below considering also the cases when Fermi resonance is present in the crystal.

Upon Fermi resonance in an isolated molecule, the frequency of one of the molecular vibrations turns out to be close to the overtone frequency, or the sum frequency of some other vibrations. We may say that in this case the mentioned resonance occurs between two excited states of the molecule (for simplicity we assume here that all vibrations are nondegenerate). If the anharmonicity is absent these states do not interact and as the overtones usually have a small oscillator strength in the considered approximation we can expect in the spectrum, for example, only one absorption line. However, owing to the anharmonicity of intramolecular vibrations, the resonance can lead to the characteristic doublet of comparable intensity in the absorption spectra or the RSL spectra or (depending upon the symmetry of the molecule and the type of vibration) in both. If degenerate vibrations also participate in the Fermi resonance, the number of lines in these spectra can even be larger (43).

Let us now consider what is the analog of a Fermi resonance in a molecule when we consider the crystals. In going over from an isolated molecule to a crystal, the branches of optical phonons appear. In the region of overtone and sum frequencies, several bands of many-particle states arise and, if anharmonicity is sufficiently strong, bands of states with quasiparticles bound to one another (for instance, biphonons) will also appear. Thus, in crystals a large number of two- or three-particle excited states of the crystal can resonate with some one-particle states substantially complicating the spectra obtained.

In order to analyze these spectra and, in particular, to investigate the effects of Fermi resonance on biphonon spectra, it is necessary to generalize the Hamiltonian (6.11) to some extent.

We shall assume that the conditions for Fermi resonance are satisfied in a free molecule, i.e. that there are two (for the sake of simplicity) nondegenerate vibrations with the frequencies Ω_1 and Ω_2, for which, for instance, $2\Omega_1 \approx \Omega_2$. In this case, when taking the intramolecular anharmonicity (with the constant Γ) into account, it is necessary to add to the Hamiltonian (6.11) the sum of two terms: $\hat{H}_0(C)$ and $\hat{H}_F(B,C)$, where

$$\hat{H}_0(C) = \sum_n \hbar\Omega_2 C_n^\dagger C_n + \sum_{n,m} V_{nm}^{(2)} C_n^\dagger C_m, \tag{6.30}$$

$$\hat{H}_F(B,C) = \Gamma \sum_n \left[\left(B_n^\dagger\right)^2 C_n + C_n^\dagger (B_n)^2 \right], \tag{6.31}$$

so that the total Hamiltonian $\hat{H}$ assumes the form (see eqn 6.8 for $\hat{H}_T(B)$)

$$\hat{H} = \hat{H}_0(B) + \hat{H}_0(C) + \hat{H}_A(B) + \hat{H}_T(B) + \hat{H}_F(B,C). \tag{6.32}$$

We shall assume for simplicity that there is a single molecule per unit cell of the crystal and proceed to a momentum reprepresentation for the operators B and C:

$$B_\mathbf{n} = \frac{1}{\sqrt{N}} \sum_\mathbf{k} B_\mathbf{k} e^{i\mathbf{k}\mathbf{n}},$$

$$C_\mathbf{n} = \frac{1}{\sqrt{N}} \sum_\mathbf{k} C_\mathbf{k} e^{i\mathbf{k}\mathbf{n}}.$$

In this representation, the Hamiltonian (6.32) is of the form

$$\begin{aligned}\hat{H} = & \sum_\mathbf{k} \left[\varepsilon_1(\mathbf{k}) B_\mathbf{k}^\dagger B_\mathbf{k} + \varepsilon_2(\mathbf{k}) C_\mathbf{k}^\dagger C_\mathbf{k} \right] \\ & - \frac{1}{N} \sum_{\mathbf{k},\mathbf{k}',\mathbf{q}} \tilde{A}(\mathbf{k}+\mathbf{k}') B_\mathbf{k}^\dagger B_{\mathbf{k}'}^\dagger B_\mathbf{q} B_{\mathbf{k}+\mathbf{k}'-\mathbf{q}} \\ & + \frac{\Gamma}{\sqrt{N}} \sum_{\mathbf{k},\mathbf{k}'} \left(B_\mathbf{k}^\dagger B_{\mathbf{k}'}^\dagger C_{\mathbf{k}+\mathbf{k}'} + \text{h.c.} \right),\end{aligned} \tag{6.33}$$

where

$$\tilde{A}(\mathbf{k}) = A - \frac{1}{2} W^A(\mathbf{k}).$$

If we introduce the operator

$$\hat{T}(\mathbf{k}) = \frac{1}{\sqrt{N}} \sum_{\mathbf{q}} B_{\mathbf{k}/2-\mathbf{q}} B_{\mathbf{k}/2+\mathbf{q}}, \tag{6.34}$$

then the Hamiltonian (6.33) can be written in the more concise form:

$$\hat{H} = \sum_{\mathbf{k}} \left\{ \varepsilon_1(\mathbf{k}) B_{\mathbf{k}}^\dagger B_{\mathbf{k}} + \varepsilon_2(\mathbf{k}) C_{\mathbf{k}}^\dagger C_{\mathbf{k}} - \tilde{A}(\mathbf{k}) \hat{T}(\mathbf{k})^\dagger \hat{T}(\mathbf{k}) \right.$$
$$\left. + \Gamma \left[\hat{T}(\mathbf{k})^\dagger C_{\mathbf{k}} + C_{\mathbf{k}}^\dagger \hat{T}(\mathbf{k}) \right] \right\}. \tag{6.35}$$

Assuming the temperature of the crystal to be zero, we shall find the retarded Green's function

$$G_{\mathbf{k}}^{(1)}(t) = -\mathrm{i}\theta(t) \langle 0 | C_{\mathbf{k}}(t) C_{\mathbf{k}}^\dagger(0) | 0 \rangle,$$

where $\theta(t) = 1$ for $t > 0$ and $\theta(t) = 0$ for $t < 0$. We derive the equation for this function by making use of the Heisenberg equation of motion:

$$\mathrm{i}\hbar \frac{\mathrm{d}\hat{F}}{\mathrm{d}t} = \hat{F}\hat{H} - \hat{H}\hat{F},$$

which can be applied to the operator $\hat{F}$ not dependent on time explicitly, the relation $\mathrm{d}\theta/\mathrm{d}t = \delta(t)$ for the θ-function and the commutation relations for the Bose operators B and C. Differentiating the expression for $G_{\mathbf{k}}^{(1)}(t)$ with respect to t we obtain

$$-\mathrm{i}\frac{\mathrm{d}G_{\mathbf{k}}^{(1)}(t)}{\mathrm{d}t} = -\delta(t) - \theta(t) \left\langle 0 \left| \frac{\mathrm{d}C_{\mathbf{k}}}{\mathrm{d}t}, C_{\mathbf{k}}^\dagger(0) \right| 0 \right\rangle. \tag{6.36}$$

Since

$$\mathrm{i}\hbar \frac{\mathrm{d}C_{\mathbf{k}}}{\mathrm{d}t} = C_{\mathbf{k}}\hat{H} - \hat{H}C_{\mathbf{k}} = \varepsilon_2(\mathbf{k}) C_{\mathbf{k}} + \Gamma T(\mathbf{k}),$$

eqn (6.36) assumes the form

$$\mathrm{i}\frac{\mathrm{d}G_{\mathbf{k}}^{(1)}}{\mathrm{d}t} = \delta(t) + \frac{1}{\hbar}\varepsilon_2(\mathbf{k}) G_{\mathbf{k}}^{(1)} + \frac{\Gamma}{\hbar} G_{\mathbf{k}}^{(2)}, \tag{6.37}$$

where

$$G_{\mathbf{k}}^{(2)}(t) = -\mathrm{i}\theta(t) \langle 0 | T(\mathbf{k}, t) C_{\mathbf{k}}^\dagger(0) | 0 \rangle. \tag{6.38}$$

Next we introduce the Green's function

$$G_{\mathbf{k},\mathbf{q}}^{(3)}(t) = -\mathrm{i}\theta(t) \frac{1}{\sqrt{N}} \langle 0 | B_{\mathbf{k}/2+\mathbf{q}}(t) B_{\mathbf{k}/2-\mathbf{q}}(t) C_{\mathbf{k}}^\dagger(0) | 0 \rangle, \tag{6.39}$$

such that obviously

$$\sum_{\mathbf{q}} G_{\mathbf{k},\mathbf{q}}^{(3)}(t) = G_{\mathbf{k}}^{(2)}(t). \tag{6.40}$$

Differentiating the relation (6.39) with respect to time, we obtain in a similar way

$$-\mathrm{i}\frac{\mathrm{d}G_{\mathbf{k},\mathbf{q}}^{(3)}}{\mathrm{d}t} = -\frac{1}{\hbar}\left[\varepsilon_1\left(\frac{1}{2}\mathbf{k}+\mathbf{q}\right) + \varepsilon_1\left(\frac{1}{2}\mathbf{k}-\mathbf{q}\right)\right] G_{\mathbf{k},\mathbf{q}}^{(3)} \tag{6.41}$$

$$+ i\theta(t) \frac{\tilde{A}(\mathbf{k})}{\hbar\sqrt{N}} \langle 0 | B_{\mathbf{k}/2-\mathbf{q}}(t) B_{\mathbf{k}/2+\mathbf{q}}(t) T^\dagger(\mathbf{k},t) T(\mathbf{k},t) C_{\mathbf{k}}^\dagger(0) | 0 \rangle$$

$$- i\theta(t) \frac{\Gamma}{\hbar\sqrt{N}} \langle 0 | B_{\mathbf{k}/2-\mathbf{q}}(t) B_{\mathbf{k}/2+\mathbf{q}}(t) T^\dagger(\mathbf{k},t) T(\mathbf{k},t) C_{\mathbf{k}}^\dagger(0) | 0 \rangle.$$

This equation simplifies if we take into account that for arbitrary operators D_1 and D_2 (see the definition 6.34)

$$\langle 0 | B_{\mathbf{k}/2-\mathbf{q}}(t) B_{\mathbf{k}/2+\mathbf{q}}(t) T^\dagger(\mathbf{k},t) D_1 D_2 | 0 \rangle$$
$$= \langle 0 | D_2^\dagger D_1^\dagger T(\mathbf{k},t) B_{\mathbf{k}/2+\mathbf{q}}^\dagger(t) B_{\mathbf{k}/2-\mathbf{q}}^\dagger(t) | 0 \rangle$$
$$= \frac{2}{\sqrt{N}} \langle 0 | D_2^\dagger D_1^\dagger | 0 \rangle = \frac{2}{\sqrt{N}} \langle 0 | D_1 D_2 | 0 \rangle.$$

Hence, eqn (6.41) can also be written as follows:

$$-i \frac{dG^{(3)}_{\mathbf{k},\mathbf{q}}}{dt} = -\frac{1}{\hbar} \left[\varepsilon_1 \left(\frac{1}{2}\mathbf{k} + \mathbf{q} \right) + \varepsilon_1 \left(\frac{1}{2}\mathbf{k} - \mathbf{q} \right) \right] G^{(3)}_{\mathbf{k},\mathbf{q}}$$
$$+ \frac{2\tilde{A}(\mathbf{k})}{\hbar N} G^{(2)}_{\mathbf{k}} - \frac{2\Gamma}{\hbar N} G^{(1)}_{\mathbf{k}}. \tag{6.42}$$

Turning now to the Fourier representation with respect to time, we find that

$$G^{(3)}_{\mathbf{k},\mathbf{q}}(\omega) = -\frac{\left(2\tilde{A}/N\right) G^{(2)}_{\mathbf{k}}(\omega) - (2\Gamma/N) G^{(1)}_{\mathbf{k}}(\omega)}{\hbar\omega - \varepsilon_1\left(\frac{1}{2}\mathbf{k}+\mathbf{q}\right) - \varepsilon_1\left(\frac{1}{2}\mathbf{k}-\mathbf{q}\right)}, \tag{6.43}$$

so that (see eqn 6.40)

$$G^{(2)}_{\mathbf{k}}(\omega) = -\left[2\tilde{A}(\mathbf{k}) G^{(2)}_{\mathbf{k}}(\omega) - 2\Gamma G^{(1)}_{\mathbf{k}}(\omega)\right] R(E, \mathbf{k}), \tag{6.44}$$

where

$$R(E, \mathbf{k}) = \frac{1}{N} \sum_{\mathbf{q}} \frac{1}{\hbar\omega - \varepsilon_1\left(\frac{1}{2}\mathbf{k}+\mathbf{q}\right) - \varepsilon_1\left(\frac{1}{2}\mathbf{k}-\mathbf{q}\right)}$$
$$= 2 \int \frac{\rho_0(\varepsilon, \mathbf{k})}{E - \varepsilon} d\varepsilon, \tag{6.45}$$
$$E = \hbar\omega,$$
$$\rho_0(\varepsilon, \mathbf{k}) = \frac{1}{2N} \sum_{\mathbf{q}} \delta\left[\varepsilon - \varepsilon_1\left(\frac{1}{2}\mathbf{k}+\mathbf{q}\right) - \varepsilon_1\left(\frac{1}{2}\mathbf{k}-\mathbf{q}\right)\right]$$

is the density of two-particle states with the total wavevector $\mathbf{k}$.

In a similar manner, making use of eqn (6.37), we can obtain a second relation linking the functions $G_{\mathbf{k}}^{(1)}(\omega)$ and $G_{\mathbf{k}}^{(2)}(\omega)$:

$$G_{\mathbf{k}}^{(1)}(\omega)\left[\hbar\omega - \varepsilon_2(\mathbf{k})\right] - \Gamma G_{\mathbf{k}}^{(2)}(\omega) = \hbar. \tag{6.46}$$

From relations (6.44) and (6.45) we obtain the required Green's functions:

$$G_{\mathbf{k}}^{(1)}(\omega) = \frac{\hbar\left[1 + 2\tilde{A}(\mathbf{k})R(E,\mathbf{k})\right]}{[E - \varepsilon_2(\mathbf{k})]\,\Delta(E,\mathbf{k})}, \tag{6.47}$$

where

$$\Delta(E,\mathbf{k}) = 1 + 2\left(\tilde{A}(\mathbf{k}) - \frac{\Gamma^2}{E - \varepsilon_2(\mathbf{k})}\right)R(E,\mathbf{k}), \tag{6.48}$$

$$G_{\mathbf{k}}^{(2)}(\omega) = \frac{2\hbar\Gamma R(E,\mathbf{k})}{[E - \varepsilon_2(\mathbf{k})]\,\Delta(E,\mathbf{k})}. \tag{6.49}$$

The function $G_{\mathbf{k},\mathbf{q}}^{(3)}(\omega)$ is also found to be completely determined, in accordance with eqn (6.43).

Along with the functions $G_{\mathbf{k}}^{(1)}$ and $G_{\mathbf{k}}^{(2)}$, a number of other Green's functions must be known in order to calculate the dielectric tensor of a crystal in the overtone frequency region, as well as the RSL cross-section and the cross-section of nonlinear optical processes. Among these others we required the two-particle Green's function $G_{\mathbf{k},\mathbf{k}',\mathbf{q}}^{(4)}(t)$, which is determined by the relation

$$G_{\mathbf{k},\mathbf{k}',\mathbf{q}}^{(4)}(t) = -i\theta(t)\langle 0|B_{\mathbf{k}/2+\mathbf{q}}(t)B_{\mathbf{k}/2-\mathbf{q}}(t)B_{\mathbf{k}/2+\mathbf{q}}^{\dagger}(0)B_{\mathbf{k}/2-\mathbf{q}}^{\dagger}(0)|0\rangle. \tag{6.50}$$

The calculation of this function is perfectly analogous to the foregoing so that only the final result will be given below. It can be shown that

$$G_{\mathbf{k},\mathbf{q},\mathbf{q}'}^{(3)}(\omega) = -\frac{2}{\sqrt{N}} \frac{\tilde{A}(\mathbf{k})G_{\mathbf{k},\mathbf{q}'}^{(5)}(\omega) - \Gamma G_{\mathbf{k},\mathbf{q}'}^{(6)}(\omega)}{E - \varepsilon_1\left(\frac{1}{2}\mathbf{k}+\mathbf{q}\right) - \varepsilon_1\left(\frac{1}{2}\mathbf{k}-\mathbf{q}\right)} + \frac{\hbar(\delta_{\mathbf{q}+\mathbf{q}'} + \delta_{\mathbf{q}-\mathbf{q}'})}{E - \varepsilon_1\left(\frac{1}{2}\mathbf{k}+\mathbf{q}\right) - \varepsilon_1\left(\frac{1}{2}\mathbf{k}-\mathbf{q}\right)}, \tag{6.51}$$

where $G_{\mathbf{k},\mathbf{q}'}^{(5)}(\omega)$ is the Fourier component of the Green's function

$$G_{\mathbf{k},\mathbf{q}'}^{(5)}(\omega) = -i\theta(t)\langle 0|T(\mathbf{k},t)B_{\mathbf{k}/2-\mathbf{q}'}^{\dagger}(0)B_{\mathbf{k}/2+\mathbf{q}'}^{\dagger}(0)|0\rangle.$$

and is determined by the relation

$$G_{\mathbf{k},\mathbf{q}'}^{(5)}(\omega) = \frac{2\hbar}{\sqrt{N}\left[E - \varepsilon_1\left(\frac{1}{2}\mathbf{k}+\mathbf{q}'\right) - \varepsilon_1\left(\frac{1}{2}\mathbf{k}-\mathbf{q}'\right)\right]} \cdot \frac{1}{\Delta(E,\mathbf{k})}, \tag{6.52}$$

whereas

$$G^{(6)}_{\mathbf{k},\mathbf{q}'}(\omega) = \frac{\Gamma}{E - \varepsilon_2(\mathbf{k})} G^{(5)}_{\mathbf{k},\mathbf{q}'}(\omega) \tag{6.53}$$

is the Fourier component of the Green's function

$$G^{(6)}_{\mathbf{k},\mathbf{q}'}(t) = -i\theta(t)\langle 0|C_{\mathbf{k}}(t)B^{\dagger}_{\mathbf{k}/2+\mathbf{q}'}(0)B^{\dagger}_{\mathbf{k}/2-\mathbf{q}'}(0)|0\rangle.$$

It follows from the expressions given above for the Green's functions $G^{(i)}(\omega)$, $i = 1, 2, \ldots, 6$, that, when anharmonicity is characterized by the constants $\tilde{A}$ and Γ, this leads to the appearance, along with poles of the type of eqn (6.2), of a new type of poles for the Green's function. These poles are determined by the equation

$$\Delta(\mathbf{k}, \omega) = 0. \tag{6.54}$$

This equation, a generalization of eqn (6.27), enables one to calculate the energy of biphonons, taking into account the Fermi resonance of the two-particle B-phonon states with the band of C-phonons. A comparison of relations (6.54) and (6.27) indicates that taking the Fermi resonance into account leads to a renormalization of the anharmonicity constant

$$\tilde{A}(\mathbf{k}) \to \tilde{A}(\mathbf{k}) - \frac{\Gamma^2}{E - \varepsilon_2(\mathbf{k})}, \qquad E = \hbar\omega.$$

The new anharmonicity "constant" becomes a function of the energy E and its effective magnitude in the energy region being considered is found to depend substantially on the position of the C-phonon energy with respect to the band of two-particle states. Hence, Fermi resonance, in general, strongly affects the conditions for the formation of biphonons and the positions of their levels. This equation, determining the energy $E = E' + i\gamma$ of the biphonon, can be rewritten, for convenience, in the form

$$\Phi_1(E) = \Phi_2(E), \tag{6.55}$$

where

$$\Phi_1(E) = -1 + \frac{\Gamma^2/\tilde{A}(\mathbf{k})}{\tilde{\varepsilon}_2(\mathbf{k}) - E}, \qquad \tilde{\varepsilon}_2 = \varepsilon_2 + \Gamma^2/\tilde{A}, \tag{6.56}$$

$$\Phi_2(E) = 2\tilde{A}(\mathbf{k})\,\mathcal{P}\int \frac{\rho_0(\varepsilon, \mathbf{k})}{E - \varepsilon}\,d\varepsilon + 2\pi i \tilde{A}(\mathbf{k})\rho_0(E, \mathbf{k}), \tag{6.57}$$

and $\mathcal{P}$ denotes the principal value of the integral. The function $\Phi_1(E)$ for the case $\tilde{A} > 0$ is given schematically in Fig. 6.1 (it is taken into account here that $\Gamma^2/\tilde{A}\varepsilon_2(\mathbf{k}) \ll 1$). In the same figure the dashed line shows the value of $\Phi_1(E)$ for the case when the Fermi resonance is neglected (i.e. for $\Gamma = 0$; in this case $\Phi_1(E) = -1$). Presented in Fig. 6.2 is the relationship Re $\{\Phi_2(E)\}$ for the same case of $\tilde{A} >> 0$ with damping ignored. Within the framework of the

186 BIPHONONS AND FERMI RESONANCE

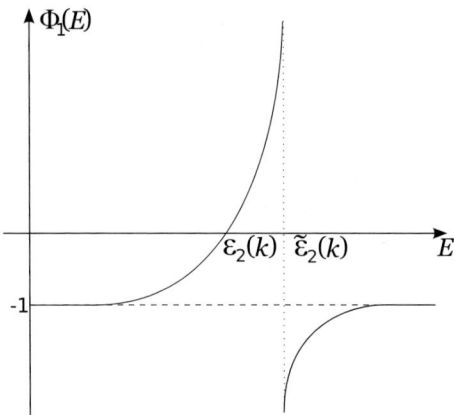

FIG. 6.1. The function $\Phi_1(E)$; $\varepsilon_2(\mathbf{k})$ is the C-phonon energy and $\tilde{\varepsilon}_2 = \varepsilon + \Gamma^2/A$.

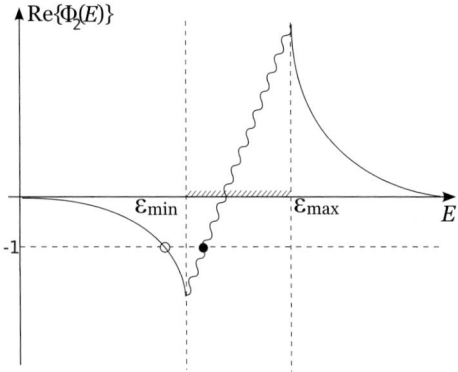

FIG. 6.2. The function Re $\{\Phi_2(E)\}$ is shown by the full line. The dashed line represents the function $\Phi_1(E) = -1$. The open circle represents a biphonon, the filled circle a quasibiphonon.

model being discussed, the damping of the biphonon states may be associated with dissociation into two free phonons. It is clear that such quasistationary (resonance) states – quasibiphonons – can have a physical meaning only in the region of small $\rho_0(E, \mathbf{k})$ values, where their width $\gamma \ll \text{Re}\{E\} = E'$.

The function Re $\{\Phi_2(E)\}$ is everywhere finite in three-dimensional crystals. In particular, for states with small $\mathbf{k}$ values $\rho(\varepsilon, \mathbf{k}) \approx \rho(\varepsilon, 0)$ and in the vicinity of the edge $\bar{\varepsilon}$ (where $\bar{\varepsilon} = \varepsilon_{\min}$ or $\bar{\varepsilon} = \varepsilon_{\max}$) of the band of two-particle states $\rho_0(\varepsilon, 0) = \rho_0 |\bar{\varepsilon} - \varepsilon|^{1/2}$. Therefore, as $E \to \bar{\varepsilon}$, then outside the band of two-particle states the function $\Phi'_2(E) = \text{Re}\{\Phi_2(E)\}$, while remaining finite itself, only has an infinite derivative (see Fig. 6.2). Also shown by a dashed line in Fig. 6.2 is the function $\Phi_1(E) = -1$ for $\Gamma = 0$. It follows from this figure that when the value of $|\Phi_2(E)|$ for $E \to \varepsilon_{\min}$ tends to a value less than unity (i.e. for $|\Phi(\varepsilon_{\min})| < 1$), a bound state of two phonons – biphonon – is formed. Along with the biphonon state, with energy $E_b < \varepsilon_{\min}$ (open circle in Fig. 6.2), a quasibiphonon (solid circle in Fig. 6.2) is also formed with energy E_{qb}, lying within the band of two-particle states. When anharmonicity is not too strong, so that the value of $|\Phi(\varepsilon_{\min})|$ does not exceed unity too much, the quasibiphonon falls in the region of low density of two-particle states, and its width is found to be small compared to the bandwidth. Then a distinct peak, not associated with van Hove singularities, is formed in the density of states found when anharmonicity is taken into account.

As a matter of fact, the density of states with the total wavevector $\mathbf{q} + \mathbf{q}' = \mathbf{k}$ is determined by the relation

$$\rho(E, \mathbf{k}) = -\frac{1}{2\pi N} \sum_{\mathbf{q}} (1 + \delta_{\mathbf{q}0})\, \text{Im}\, \left\{ G^{(4)}_{\mathbf{k},\mathbf{q},\mathbf{q}}(E + i\gamma) \right\}, \qquad \gamma \to +0, \quad (6.58)$$

while for noninteracting phonons (i.e. for $\tilde{A} = \Gamma = 0$), the relation is

$$\rho(E, \mathbf{k}) \equiv \rho_0(E, \mathbf{k}) = \frac{1}{2N} \sum_{\mathbf{q}} (1 + \delta_{\mathbf{q}0})\, \delta\left[E - \varepsilon_1\left(\frac{1}{2}\mathbf{k} + \mathbf{q}\right) - \varepsilon_1\left(\frac{1}{2}\mathbf{k} - \mathbf{q}\right)\right].$$

If $\tilde{A} \neq 0$, but $\Gamma = 0$ (no Fermi resonance), the function $G^{(4)}_{\mathbf{k},\mathbf{q},\mathbf{q}}(E)$, according to eqn (6.50), is determined by the relation

$$G^{(4)}_{\mathbf{k},\mathbf{q},\mathbf{q}}(E) = \left[E - \varepsilon_1\left(\frac{1}{2}\mathbf{k} + \mathbf{q}\right) - \varepsilon_1\left(\frac{1}{2}\mathbf{k} - \mathbf{q}\right)\right]^{-1} - \frac{4\tilde{A}(\mathbf{k})\Delta^{-1}(E, \mathbf{k})}{N\left[E - \varepsilon_1\left(\frac{1}{2}\mathbf{k} + \mathbf{q}\right) - \varepsilon_1\left(\frac{1}{2}\mathbf{k} - \mathbf{q}\right)\right]^2}, \quad (6.59)$$

so that

$$\rho(E, \mathbf{k}) = \rho_0(E, \mathbf{k}) + \frac{16\tilde{A}^2(\mathbf{k}) D(E, \mathbf{k}) \rho_0(E, \mathbf{k})}{\left[1 + 2\tilde{A}(\mathbf{k}) R'(E, \mathbf{k})\right]^2 + 16\pi^2 \tilde{A}^2(\mathbf{k}) \rho_0^2(E, \mathbf{k})}, \quad (6.60)$$

where

$$R'(E, \mathbf{k}) = \text{Re}\, \{R(E, \mathbf{k})\} = 2\, \mathcal{P} \int \frac{\rho_0(\varepsilon, \mathbf{k})}{E - \varepsilon}\, d\varepsilon,$$

$$D(E, \mathbf{k}) = \mathcal{P} \int \frac{\rho_0(\varepsilon, \mathbf{k})}{(E - \varepsilon)^2}\, d\varepsilon. \quad (6.61)$$

In the range of energies E in which $\rho_0(E, \mathbf{k}) = 0$, i.e. outside the band of two-particle states, relation (6.60) reduces to the following

$$\rho(E, \mathbf{k}) = 4D(E, \mathbf{k})\left|\tilde{A}(\mathbf{k})\right| \delta\left[1 + 2\tilde{A}(\mathbf{k})R(E, \mathbf{k})\right], \tag{6.62}$$

or

$$\rho(E, \mathbf{k}) = \delta\left(E - E_\mathrm{b}\right),$$

where the energy E_b of the biphonon is the root of the equation

$$1 + 4\tilde{A}(\mathbf{k}) \int \frac{\rho_0(\varepsilon, \mathbf{k})}{E - \varepsilon} \, d\varepsilon = 0. \tag{6.63}$$

Within the band of two-particle states, in the range of energies $E \approx E_\mathrm{qb}$, where E_qb is the root of the equation

$$1 + 4\tilde{A}(\mathbf{k}) \, \mathcal{P}\!\int \frac{\rho_0(\varepsilon, \mathbf{k})}{E - \varepsilon} \, d\varepsilon = 0,$$

relation (6.60) can be written in the form

$$\rho(E, \mathbf{k}) = \rho_0(E, \mathbf{k}) + \frac{1}{\pi} \frac{\gamma}{(E - E_\mathrm{qb})^2 + \gamma^2}, \tag{6.64}$$

where $\gamma = \pi \rho_0(E_\mathrm{qb}, \mathbf{k})/D(E_\mathrm{qb}, \mathbf{k})$ is the half-width of the quasibiphonon level. From this expression for γ it follows that this half-width can be sufficiently small only when the level of the quasibiphonon is within the region of low density $\rho_0(E, \mathbf{k})$ of the levels of two-particle states.

6.3.1 *Fermi resonance*

Now let us proceed to a discussion of the case with Fermi resonance ($\tilde{A} \neq 0$ and $\Gamma \neq 0$).

In this case the position and number of roots of eqn (6.55) depends substantially on the relation between the quantities $\tilde{A}$ and Γ, and on the position of the energy $\varepsilon_2(\mathbf{k})$ with respect to the band of two-particle states. To illustrate the afore-said, the curves $\Phi_1(E)$ and $\Phi_2(E)$ are shown in Figs. 6.3, 6.4, and 6.5, and the roots of eqn (6.55) are indicated for three limiting situations.

Figure 6.3 corresponds to the case in which the energy $\tilde{\varepsilon}_2(\mathbf{k}) = \varepsilon_2(\mathbf{k}) + \Gamma^2/A(\mathbf{k})$, is located below the band of two-particle states and sufficiently distant from the bottom of the band, ε_min. In this case the number of solutions to eqn (6.55), lying outside the band of two-particle states, is equal to two. One of these (the lower one) is generically associated with the C-phonon state and goes over into this state as the energy of the C-phonon moves away from the band of two-particle states.

In the case depicted in Fig. 6.4, the energy $\tilde{\varepsilon}_2(\mathbf{k})$ is within the band of two-particle states, and the quantity $\tilde{A} \approx \Gamma$. For the energy region $E < \varepsilon_\mathrm{min}$, eqn (6.55) has only a single solution, and no quasibiphonon state is formed.

If the energy of the C-phonon lies above the band of two-particle states and its bandwidth Δ is large compared to Γ, a situation is possible that corresponds

FERMI RESONANCE WITH POLARITONS

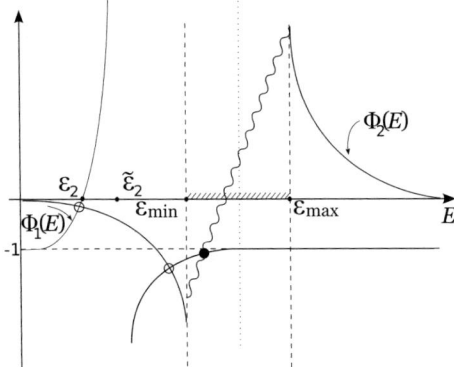

FIG. 6.3. The functions $\Phi_1(E)$ and $\Phi_2(E)$ in the presence of a Fermi resonance: $\varepsilon_{\min}$ and $\varepsilon_{\max}$ are the boundaries of the band of two-particle dissociated states. The open and filled circles represent the biphonon and quasibiphonon, respectively, $\tilde{\varepsilon} < \varepsilon_{\min}$.

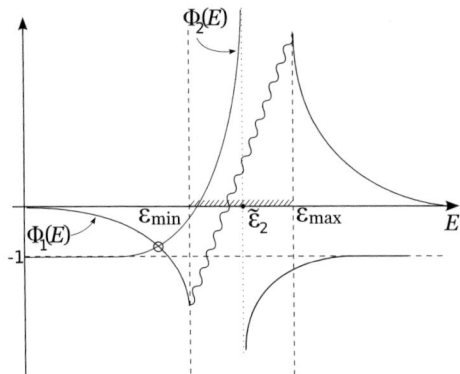

FIG. 6.4. Same as Fig. 6.3, with $\varepsilon_{\min} < \tilde{\varepsilon}_2 < \varepsilon_{\max}$.

to Fig. 6.5. Here the solutions of eqn (6.55) lie on different sides of the band of two-particle states, and the formation of a quasibiphonon state is possible.

Relations (6.50), (6.52), and (6.53) are required to find the density of states in the presence of Fermi resonance, and this can be done in a way similar to the case of $\Gamma = 0$. We shall not give the corresponding calculations here, but shall turn to a discussion of a more general situation arising in the case of Fermi resonance with a polariton.

6.4 Fermi resonance with polaritons

6.4.1 *Microscopic theory*

Let us consider the effects that arise when the branch of C phonons corresponds to dipole-active vibrations. In the region of small values of $k = 2\pi/\lambda$, where

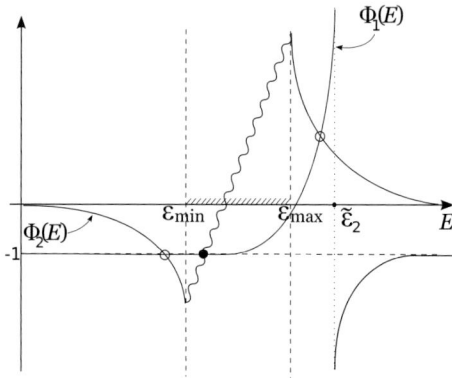

FIG. 6.5. Same as Fig. 6.3, with $\tilde{\varepsilon}_2 > \varepsilon_{\max}$.

λ is the wavelength of light with frequency $\omega = \varepsilon_2/\hbar$, C phonons of this kind strongly interact with the transverse photons. As a result, in the region of long wavelengths, new elementary excitations – phonon-polaritons (see Ch. 4) – are formed instead of C phonons and transverse photons.

The spectrum of polaritons can be found by means of Maxwell's macroscopic equations (see Ch. 4), provided that the dielectric tensor of the medium (44) is assumed to be known. Without going into details, we emphasize here that always a gap appears in the polariton spectrum (here we ignore spatial dispersion) in the region of the fundamental dipole-active vibration (C-phonon, exciton, etc.). At present, there is a sufficiently detailed theory for RSL by phonon-polaritons, taking many phonon bands into consideration. With this theory the RSL cross-section can be calculated for various scattering angles provided that the dielectric tensor of the crystal is known, as well as the dependence of the polarizability of the crystal on the displacement of the lattice sites and the electric field generated by this displacement (45).

It is an essential fact that the above-mentioned gaps in the polariton spectrum, if they arise, as well as the corresponding interaction between the photon and phonon, are nonzero within the framework of linear theory and, in general, do not require that anharmonicity be taken into account. Therefore, it makes sense to denote as a polariton Fermi resonance only such situations where vibrations of overtone or combination tone frequencies resonate with the polariton. We now turn our attention to an analysis of such rather complex situations, requiring that multiparticle excited states of the crystal be taken into consideration. Shown schematically in Fig. 6.6 is a typical polariton spectrum, as well as a band of two-particle states of B phonons. If, under the effect of anharmonicity, biphonons with energy $E = E_b$ are formed, these states also resonate with the polariton, influencing its spectrum.

The RSL by polaritons is extremely intense for many crystals that lack a center of inversion, whereas second-order RSL (i.e. RSL accompanied by the

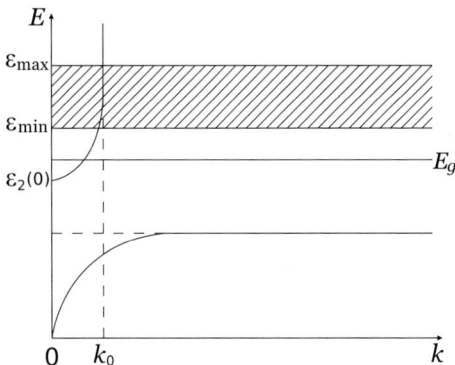

FIG. 6.6. Polariton dispersion in the Fermi resonance region, neglecting anharmonicity: $\varepsilon_2(0)$ is the energy of the longitudinal vibration at $k = 0$; $\varepsilon_{\min}$ and $\varepsilon_{\max}$ are the minimum and maximum energy values in the band of two-particle states.

simultaneous excitation of two quasiparticles) is relatively weak, as a rule. The intersection of the biphonon levels and bands of two-particle states with the polariton branch, shown in Fig. 6.6, under the influence of anharmonicity leads to the partial transfer of RSL intensity from the phonon-polariton to the biphonon and two-particle states. Along with this effect, which is of great significance in the experimental investigation of the above-mentioned states, substantial changes occur in the spectra of polaritons and two-particle states in the region of the resonance (see Fig. 6.6). These changes may also appear in the RSL spectra. Since, as will be shown in the following, the nature of these changes depends on whether or not biphonons are formed, we can come to the conclusion on the existence of biphonons experimentally by investigating the spectra of RSL by polaritons in the region of a polariton Fermi resonance. Extensive experimental data are available at present in this field of investigation and they are reviewed below briefly in Section 6.7. We now return to the results of the preceding section and show how Fermi resonance with polaritons can be investigated within the framework of microscopic theory (see also (19)).

To take the interaction between phonons and photons into consideration, it is necessary to add to the Hamiltonian (6.32), the Hamiltonian $\hat{H}_0(a)$ of the free field of transverse photons and the Hamiltonian $\hat{H}_{\text{int}}$ for the interaction of the field of transverse photons with phonons. The linear transformation from the operators a and C to the polariton creation and annihilation operators, i.e. to the operators $\xi_\rho^\dagger(\mathbf{k})$ and $\xi_\rho(\mathbf{k})$, diagonalizes the quadratic part of the total Hamiltonian. The two-particle states of the crystal, corresponding to the excitation of two B phonons, usually have a small oscillator strength and the retardation for such states can be neglected. In view of the afore-said, the quadratic part of the total Hamiltonian with respect to the Bose operators can be written in the form of the sum $\hat{H}_0(B) + \hat{H}_0(\xi)$, where

$$\hat{H}_0(\xi) = \sum_{\rho,\mathbf{k}} \mathcal{E}_\rho(\mathbf{k}) \xi_\rho^\dagger(\mathbf{k}) \xi_\rho(\mathbf{k}),$$

and $\mathcal{E}_\rho(\mathbf{k}) \equiv \hbar\omega_\rho(\mathbf{k})$ is the energy of the polariton $\rho\mathbf{k}$.

Let us now direct our attention to the Hamiltonian $\hat{H}_{\text{int}}$ and discuss the structure of its terms that are cubic in the quasiparticle creation and annihilation operators.

For a crystal with a one molecule per unit cell, these terms are:

$$\hat{H}_{\text{int}}^{(3)} = \sum_{\mathbf{n}} \sum_{\rho,\mathbf{k}} \left(\Gamma_\rho(\mathbf{k}) \left(B_{\mathbf{n}}^\dagger\right)^2 \xi_\rho(\mathbf{k}) \frac{e^{i\mathbf{n}\mathbf{k}}}{\sqrt{N}} + \text{h.c} \right). \tag{6.65}$$

To find the quantity $\Gamma_\rho(\mathbf{k})$ it is necessary to take into account the expression of the operators $C_\mathbf{n}$ in terms of the polariton creation and annihilation operators (see Ch. 4)

$$C_\mathbf{n} = \frac{1}{\sqrt{N}} \sum_{\rho,\mathbf{k}} e^{i\mathbf{n}\mathbf{k}} \left[u_\rho(\mathbf{k}) \xi_\rho(\mathbf{k}) + v_\rho^*(-\mathbf{k}) \xi_\rho^\dagger(-\mathbf{k}) \right].$$

Hence, taking eqn (6.31) into account, we find that $\Gamma_\rho(\mathbf{k}) = \Gamma u_\rho(\mathbf{k})$. However, now along with the operator (6.31), $\hat{H}_{\text{int}}$ also has to include an operator corresponding to direct interaction between the transverse photons and overtones. This operator corresponds to

$$\hat{H} = \sum_{\mathbf{n},\mathbf{k},j} \left[D(\mathbf{k}j) \left(B_{\mathbf{n}}^\dagger\right)^2 a_{\mathbf{k}j} + \text{h.c} \right], \tag{6.66}$$

where the so-called electrooptic anharmonicity is taken into account (37). Hence the constants $\Gamma_\rho(\mathbf{k})$, appearing in eqn (6.65), are determined by the total contribution of both the mechanical anharmonicity, eqn (6.31), and the electrooptic anharmonicity, eqn (6.66). Consequently, these constants depend, in general, on two independent phenomenological constants, Γ and D.

Now, when we compare the Hamiltonian (6.33) with the Hamiltonian

$$\hat{H} = \hat{H}_0(B) + \hat{H}_0(\xi) + \hat{H}_{\text{int}}^{(3)} + \hat{H}^{(4)}(B), \tag{6.67}$$

where $\hat{H}^{(4)}(B)$ is the third term in (6.33), we come to the conclusion that the general structure of the Hamiltonian is conserved in going over to polaritons. This releases us from the necessity to repeat the calculations, so that we can proceed directly to the formulation of results.

We direct our attention, first of all, to an analysis of the dispersion law for polaritons in the region of the Fermi resonance. For this purpose, by analogy with

eqn (6.47), we write the expression for the Fourier components of the Green's function
$$G^{(1)}_{\mathbf{k}\rho}(t) = -\mathrm{i}\,\theta(t)\langle 0|\xi_\rho(\mathbf{k},t)\xi^\dagger_\rho(\mathbf{k},0)|0\rangle.$$
Taking into account only the polariton branch ρ that intersects the region of two-particle states, we find that when we take anharmonicity into consideration
$$G^{(1)}_{\mathbf{k}\rho}(E) = \frac{1 + 2\tilde{A}(\mathbf{k})R(E,\mathbf{k})}{\tilde{\Delta}(E,\mathbf{k})\left[E - \mathcal{E}_\rho(\mathbf{k})\right]}, \qquad (6.68)$$
where
$$\tilde{\Delta}(E,\mathbf{k}) = 1 + 2\left(\tilde{A}(\mathbf{k}) - \frac{|\Gamma_\rho(\mathbf{k})|^2}{E - \mathcal{E}_\rho(\mathbf{k})}\right)R(E,\mathbf{k}). \qquad (6.69)$$

The poles of the Green's function (6.68) determine the dispersion of the polariton in the region of the Fermi resonance. The energy of the polariton is determined from the condition $\tilde{\Delta}(E,\mathbf{k}) = 0$, which can be written in the form
$$1 + 2\tilde{A}(\mathbf{k})R(E,\mathbf{k}) = \frac{|\Gamma_\rho(\mathbf{k})|^2}{E - \mathcal{E}_\rho(\mathbf{k})}R(E,\mathbf{k}). \qquad (6.70)$$

The l.h.s. of eqn (6.70) may be equal to zero outside the spectrum of two-particle states if a biphonon level exists for $\Gamma_\rho(\mathbf{k}) = 0$. If, in this case, the energy of the biphonon is $E_\mathrm{b}(\mathbf{k})$, the l.h.s. of eqn (6.70) can be written, for $E \approx E_\mathrm{b}(\mathbf{k})$, in the form $\alpha^2(\mathbf{k})\left[E - E_\mathrm{b}(\mathbf{k})\right]$, so that eqn (6.70) assumes the form
$$\left[E - E_\mathrm{b}(\mathbf{k})\right]\left[E - \mathcal{E}_\rho(\mathbf{k})\right] = \frac{|\Gamma_\rho(\mathbf{k})|^2}{\alpha^2(\mathbf{k})\tilde{A}(\mathbf{k})}. \qquad (6.71)$$

It follows from this relation that at a certain $\mathbf{k}_0$, such that $E_\mathrm{b}(\mathbf{k}_0) = \mathcal{E}_\rho(\mathbf{k}_0)$, i.e.

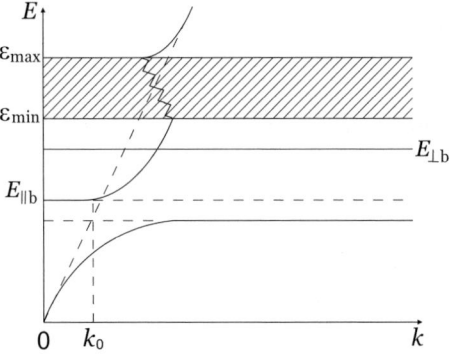

FIG. 6.7. Same as Fig. 6.6, but with anharmonicity of vibrations taken into account.

at the point where the polariton intersects the biphonon level, a gap is formed in the polariton spectrum (see Fig. 6.7) with the half-width

$$\delta = \left| \Gamma_\rho(\mathbf{k}_0)/\alpha(\mathbf{k}_0) \sqrt{\tilde{A}(\mathbf{k}_0)} \right|.$$

If, however, there is no biphonon, no gap appears in the polariton spectrum outside the region of two-particle states. Consequently, the experimental observation of such a gap is, at the same time, experimental proof of the existence of the biphonon state (see Section 6.7).

6.4.2 *Macroscopic theory – Transverse, longitudinal, and surface biphonons*

Note that the microscopic theory of Fermi resonance with polaritons, developed above, cannot be directly applied to cubic crystals, because triply degenerate states correspond to dipole-active transitions in such crystals (for the corresponding generalization of the theory, see (41)). However, as was mentioned previously, the polariton spectrum can also be found within the framework of macroscopic electrodynamics, which requires that the dielectric tensor of the crystal be known. The results of a proper analysis, as could be expected, are equivalent to those obtained in microscopic theory. We shall use the macroscopic theory in the following in application to cubic crystals. Using this approach we shall show additionally how the longitudinal and surface biphonons can also be found (see also (15)).

The dielectric tensor in a cubic crystal is reduced, as is well known, to the scalar dielectric function $\epsilon(\omega)$ when spatial dispersion is neglected. In the region of the band of two-particle states, this function can be presented in the form

$$\epsilon(\omega) = \epsilon_\infty - \frac{F_\mathrm{b} \Omega_{\perp \mathrm{b}}^2}{\omega^2 - \Omega_{\perp \mathrm{b}}^2} - \mathcal{P} \int \frac{F(\omega') \omega'^2 d\omega'}{\omega^2 - \omega'^2}, \qquad (6.72)$$

where ϵ_∞ is a quantity that is determined by the contribution of distant resonances and, in the spectral region being considered, can be assumed independent of ω, $\Omega_{\perp \mathrm{b}} = E_\mathrm{b}^\perp(0)/\hbar$, $E_\mathrm{b}^\perp(0)$ is the energy of a transverse biphonon with wavevector $\mathbf{k} = 0$, F_b is a factor proportional to the biphonon oscillator strength, and $F(\omega')$ is a quantity proportional to the strength of an oscillator corresponding to the excitation of two free B phonons with total energy $\hbar\omega'$. The quantity $F(\omega')$ is also proportional to the density of energy levels with the total wavevector $\mathbf{k} = 0$ in the band of two-particle states, so that $F(\omega') = 0$ if the frequency ω' is outside this band. The integral over frequencies in eqn (6.72) is taken in the sense of the principal value. It is also assumed that resonances $\epsilon(\omega)$, corresponding to the frequencies of other fundamental dipole vibrations of the lattice (i.e. frequencies found in the harmonic approximation), do not fall within the frequency range being considered.

Polaritons in cubic crystals can be transverse or longitudinal and as we neglected the spatial dispersion, the polariton dispersion law, i.e. the dependence

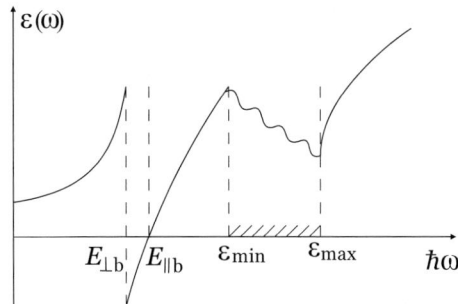

FIG. 6.8. Dependence of the dielectric function on the frequency in the region of overtone frequencies: $E_{\|b}$ and $E_{\perp b}$ are the energies of the longitudinal and transverse biphonons, respectively; $\varepsilon_{\min}$ and $\varepsilon_{\max}$ are the minimum and maximum energy values in the band of two-particle states.

of the frequency on the wavevector, can be determined for transverse polaritons from the relation

$$\epsilon(\omega) = \frac{k^2 c^2}{\omega^2}. \tag{6.73}$$

At the same time the frequencies of longitudinal waves obey the equation

$$\epsilon(\omega) = 0. \tag{6.74}$$

These waves are not observed in the infrared absorption spectra, but, with proper selection of the polarization of the incident and scattered light, they can be observed in RSL spectra (see Section 6.7).

In the frequency region being considered when $\Omega_{\perp b} < \varepsilon_{\min}$, the relation $\epsilon(\omega)$ can be schematically represented in the form shown in Fig. 6.8.

It follows from the vanishing of $\epsilon(\omega)$ at a certain frequency $\omega \equiv \Omega_{\|b}$, where $\epsilon(\Omega_{\|b}) = 0$ and $\Omega_{\|b} > \Omega_{\perp b}$ that a longitudinal biphonon with energy $\hbar\Omega_{\|b}$ is also formed, simultaneously with the transverse biphonon that leads to the appearance of a gap in the polariton spectrum. We point out that longitudinal–transverse splitting of the biphonon in the case being discussed does not comply with the well-known Lyddane–Sachs–Teller formula, which is valid for the region of an isolated frequency of the fundamental vibration. Owing to the contribution of the integral the low-frequency value of $\epsilon(\omega)$ is determined by the relation

$$\epsilon(0) = \epsilon_\infty + \int F(\omega') \, d\omega',$$

so that the Lyddane–Sachs–Teller relation $\Omega_{\|b}^2 = \epsilon(0)\Omega_{\perp b}^2/\epsilon(\infty)$ is not, in general, valid for the frequencies $\Omega_{\|b}$ and $\Omega_{\perp b}$ (see also (41)).

In the region of longitudinal–transverse splitting $\Omega_{\perp b} < \omega < \Omega_{\|b}$, where $\epsilon(\omega) < 0$, surface biphonons should also exist. At the boundary with the vacuum, their dispersion law satisfies the relation

$$k^2 = \frac{\omega^2}{c^2} \frac{\epsilon(\omega)}{\epsilon(\omega) - 1}, \qquad (6.75)$$

where retardation is taken into account. It follows from (6.75) that as $c \to \infty$, i.e. when retardation is neglected, the frequency Ω_{sb} of a surface biphonon satisfies the relation

$$\epsilon(\Omega_{\mathrm{sb}}) = -1.$$

Surface biphonons could be investigated, for example, by the attenuated total reflection (ATR) method. In contrast to RSL by polaritons, this method is effective, as is well known, both for crystals with and without inversion center. In this sense, it is a more universal method. In conclusion we point out that in degenerate semiconductors Fermi resonance with plasmons (47) is also possible along with Fermi resonance with phonons and polaritons. The spectrum of plasmophonons has been measured in many semiconductors by the RSL method (see, e.g. Mooradian and McWhorter (48)).

6.5 Dielectric tensor of a crystal in the spectral region of two-particle phonon states: microscopic theory

In this section we demonstrate how the dielectric tensor ϵ_{ij} can be found in the spectral region of two-particle states, within the framework of the same simplest model of a crystal as used previously in Section 6.2.

According to the theory of linear response the transverse dielectric tensor of a crystal at temperature $T = 0$ is determined by the relation

$$\epsilon_{\perp,ij}(\omega) = \delta_{ij} + 4\pi \chi_{\perp,ij}(\omega), \qquad (6.76)$$

where

$$\chi_{\perp,ij}(\omega) = -\frac{1}{\hbar V} \left[\Phi_{ij}(\omega + i\epsilon) + \Phi_{ij}(-\omega + i\epsilon) \right], \qquad \epsilon \to +0, \qquad (6.77)$$

$$\Phi_{ij}(t) = -\frac{1}{2} \theta(t) \langle 0 | M_i(t) M_j(0) + M_j(t) M_i(0) | 0 \rangle, \qquad (6.78)$$

V is the volume of the crystal, M is its dipole moment,

$$\mathbf{M} = \sum_n \mathbf{M}_n, \qquad (6.79)$$

with $\mathbf{M}_n$ being the dipole moment of the molecule, where $n \equiv (\mathbf{n}\alpha)$.

In determining the quantities $\Phi_{ij}(\omega)$ we shall make use of the Hamiltonian (6.32) in which, as assumed, the Coulomb interaction has been completely taken into account. In this case, linear response theory determines only the so-called (see (44) and the next Ch. 7) transverse dielectric tensor $\epsilon_{\perp,ij}(\omega, \mathbf{k})$. This tensor relates the induction vector $\mathbf{\mathcal{D}}$ to the transverse part of the macrofield $\mathbf{E}$: $D_i(\omega, \mathbf{k}) = \epsilon_{\perp,ij}(\omega, \mathbf{k}) E_j^{\perp}(\omega, \mathbf{k})$, see also eqn (7.3).

Expanding the quantity $\mathbf{M}_n$ in a series in terms of the normal coordinates of the molecule and going over to the excitation creation and annihilation operators, we obtain up to the second-order terms

$$\mathbf{M}_{ni} = d_i^C \left(C_n + C_n^\dagger \right) + d_i^{BB} \left(B_n + B_n^\dagger \right)^2 + \ldots, \qquad i = x, y, z. \qquad (6.80)$$

It is precisely the terms that have been written out here that make, in the spectral region $\omega \approx 2\Omega_1 \approx \Omega_2$ being considered, the main contribution to the transition dipole moment of the molecule. Applying relations (6.79) and (6.34) we find that

$$\frac{1}{\sqrt{N}} \mathbf{M}_i = d_i^C \left(C_{k=0} + C_{k=0}^\dagger \right) + d_i^{BB} \left[T(\mathbf{k}=0) + T^\dagger(\mathbf{k}=0) \right] + \tilde{M}_i, \qquad (6.81)$$

where the operator $\tilde{M}_i$ includes terms of the form of $B_q^\dagger B_q$ making no contribution to the quantity $\Phi_{ij}(\omega)$. Making use of (6.78) we obtain

$$\frac{1}{N} \Phi_{ij}(t) = d_i^C d_j^C G_0^{(1)}(t) + \frac{1}{2} \left(d_i^{BB} d_j^C + d_j^{BB} d_i^C \right) \left[G_0^{(2)}(t) + G_0^{(7)}(t) \right]$$
$$+ d_i^{BB} d_j^{BB} G_0^{(8)}(t),$$

where

$$G_0^{(7)}(t) = \sum_q G_{k,q}^{(6)}(t) = -i\theta(t)\langle 0 | C_k(t) T_k^\dagger(0) | 0 \rangle,$$

$$G_0^{(8)}(t) = \sum_q G_{k,q}^{(5)}(t) = -i\theta(t)\langle 0 | T_k(t) T_k^\dagger(0) | 0 \rangle.$$

From relations (6.52) and (6.53) we obtain

$$G_k^{(8)}(\omega) = \frac{2\hbar R(E, \mathbf{k})}{\Delta(E, \mathbf{k})},$$

$$G_k^{(7)}(\omega) = \frac{\Gamma}{E - \varepsilon_2(\mathbf{k})} G_k^{(8)}(\omega) = \frac{2\hbar \Gamma R(E, \mathbf{k})}{[E - \varepsilon_2(\mathbf{k})] \Delta(E, \mathbf{k})} = G_k^{(2)}(\omega),$$

so that

$$\frac{1}{\hbar N} \Phi_{ij}(t) = \frac{d_i^C d_j^C}{\hbar \omega - \varepsilon_2(0)} \qquad (6.82)$$
$$+ \frac{2R(\omega, 0)}{\Delta(\omega, 0)} \left(\frac{\Gamma d_i^C}{\hbar \omega - \varepsilon_2(0)} + d_i^{BB} \right) \left(\frac{\Gamma d_j^C}{\hbar \omega - \varepsilon_2(0)} + d_j^{BB} \right).$$

Together with relations (6.77) and (6.76), relation (6.82) completely determines the dielectric tensor of the crystal in the spectral region being discussed. Let us now turn to certain limiting cases. In the first place we point out that if anharmonicity is neglected ($\tilde{A} = 0$ and $\Gamma = 0$)

$$\frac{1}{\hbar N}\Phi_{ij}(t) = \frac{d_i^C d_j^C}{\hbar\omega - \varepsilon_2(0)} + 2R(\omega, 0)d_i^{BB}d_j^{BB},$$

so that (see eqn 6.77; in the following $\epsilon \to +0$)

$$\chi_{\perp,ij}(\omega) = -\frac{1}{v}\left[\frac{2\varepsilon_2(0)d_i^C d_j^C}{(\hbar\omega)^2 - \varepsilon_2^2(0) + \mathrm{i}\epsilon} + 4d_i^{BB}d_j^{BB}\left(\mathcal{P}\int\frac{2x\rho_0(x)\mathrm{d}x}{E^2 - x^2} - \mathrm{i}\pi\rho_0(E)\right)\right],$$

where $v = V/N$, $E = \hbar\omega$.

If we ignore the small effect of two-phonon transitions on the dispersion of the refraction index n, then in the region of the energies $E = \hbar\omega$, corresponding to the excitation of two-particle states, the absorption coefficient κ is found to be directly associated with the density of two-particle states. In fact, for light polarized, for instance, along the x-axis, which coincides, by definition, with the direction of the vector $\mathbf{d}^{BB}$, the dielectric tensor has a nonvanishing component

$$\epsilon_{\perp,11} = 1 + 4\pi\chi_{\perp,11} = (n + \mathrm{i}\kappa)^2 \approx n^2 + 2\mathrm{i}n\kappa,$$

so that

$$\kappa(E) = \frac{8\pi^2}{vn_0}\left|\mathbf{d}^{BB}\right|^2 \rho_0(E),$$

where n_0 is the value of the refractive index.

Next we shall show that taking anharmonicity into account leads, in general, to the appearance of new maxima of $\kappa(E)$. We assume, for simplicity, that $\Gamma = 0$, but that $\tilde{A} \neq 0$. Then, for the region of two-particle states

$$\kappa(E) = -\frac{4\pi\left|\mathbf{d}^{BB}\right|^2}{n_0 v}\mathrm{Im}\left\{\frac{R(\omega + \mathrm{i}\epsilon, 0)}{1 + 2\tilde{A}(0)R(\omega + \mathrm{i}\epsilon, 0)}\right\}$$

$$= \frac{\left|\mathbf{d}^{BB}\right|^2}{n_0 v}\frac{8\pi^2\rho_0(E)}{\left[1 + 2\tilde{A}(0)R'(E, 0)\right]^2 + 16\pi^2\left|\tilde{A}(0)\right|^2\rho_0^2(E)}.$$

The analysis of this expression is similar to that for relation (6.60). Specifically, if eqn (6.63) is satisfied for a certain value of $E = E_\mathrm{b}$ outside the region of two-particle states, then, for this region of the spectrum,

$$\kappa(E) = \frac{\pi^2\left|\mathbf{d}^{BB}\right|^2}{2n_0 v\left|\tilde{A}(0)\right|^2 D(E_\mathrm{b}, 0)}\delta\left(E - E_\mathrm{b}\right). \tag{6.83}$$

In the vicinity of the quasibiphonon energy, where $E \approx E_\mathrm{qb}$ (see eqn 6.64),

$$\kappa(E) = \frac{\pi^2 \left|\mathbf{d}^{BB}\right|^2}{2n_\rangle v \left|\tilde{A}(0)\right|^2 D(E_b, 0)} \frac{1}{\pi} \frac{\gamma}{(E - E_{\text{qb}})^2 + \gamma^2}. \qquad (6.84)$$

A discussion of the more general situation with $\Gamma \neq 0$ is more cumbersome and is not given here. We only point out that the appearance of biphonon states leads to a new resonance of $\kappa(E)$ when Fermi resonance is taken into account. But the appearence of quasibiphonon states can substantially alter the density of states inside the band of two-particle states.

6.6 Biphonons and biexcitons and the gigantic nonlinear polarizability effect

Nonlinear optical effects can be described within the framework of macroscopic electrodynamics (see, e.g. Bloembergen (49)), by applying the nonlinear relation between the induction vector $\mathcal{D}$ and the strength $\mathbf{E}$ of the macroscopic electric field. When the value of E in the light wave is small compared to the intra-atomic electric fields, this nonlinear relation can be written in the form of the expansion

$$\begin{aligned}
D_i(\mathbf{r}, t) = &\, \epsilon_{ij}(\omega) \exp(-\mathrm{i}\omega t) \\
&+ 4\pi \chi_{ij\ell}(\omega, \omega') E_j(\omega) E_\ell(\omega') \exp[-\mathrm{i}(\omega + \omega')] \\
&+ 4\pi \chi_{ij\ell m}(\omega, \omega', \omega'') E_j(\omega) E_\ell(\omega') E_m(\omega'') \exp[-\mathrm{i}(\omega + \omega' + \omega'')] \\
&+ \ldots,
\end{aligned}$$

where $\chi_{ij\ell}$, $\chi_{ij\ell m}$, etc. are nonlinear polarizabilities, characterizing the nonlinear optical properties of the medium.

In the preceding section it was shown that the formation of bound states of phonons leads to the appearance of a new type of resonance of the dielectric tensor $\epsilon_{ij}(\omega)$. It is clear, of course (23), that the nonlinear polarizabilities should have analogous resonances, and this also concerns, besides biphonons, other types of bound states of quasiparticles, such as biexcitons, electron–exciton complexes, etc.

An investigation of the contribution of the bound states of quasiparticles to the nonlinear polarizabilities is of interest for many reasons. The main ones are the new opportunities for studying the properties of bound states, as well as the gigantic values of the nonlinear polarizabilities that can be reached, precisely as a result of the new type of resonances.

The very existence of nonlinear polarizabilities is due to the presence of some anharmonicity in the medium. Anharmonicity is usually regarded as a weak perturbation in calculating these polarizabilities. It is clear, however, that when anharmonicity leads to the formation of states of quasiparticles bound to each other, the polarizabilities, in the region of the resonances corresponding to these states, become nonanalytic functions of the anharmonicity constants. For this reason ordinary perturbation theory is found to be inapplicable in their calculation, and more general methods are required. In accordance with two papers,

(23) and (24), we intend to show how the method of Green's functions can be employed for this case.

Let us first consider the dispersion of the nonlinear polarizability tensor of third order. The relation of this tensor to the triple-time correlation function is determined by the following expression (50) (here and in the following $\hbar = 1$):

$$\chi_{ij\ell} = -\frac{\pi}{V}\left[\mathcal{K}_{ij\ell}\left(\omega+\omega',\omega'\right) + \mathcal{K}_{ij\ell}\left(\omega+\omega',\omega\right)\right], \tag{6.85}$$

where

$$\mathcal{K}_{ij\ell}\left(\omega+\omega',\omega'\right) = \frac{1}{2\pi}\int_{-\infty}^{\infty} d\tau_1 \int_{-\infty}^{\infty} d\tau_2 \, \exp\left\{i\left[\left(\omega+\omega'\right)\tau_1+\omega'\tau_2\right]\right\}$$
$$\times \theta\left(\tau_1\right)\theta\left(\tau_2\right)\left\langle 0\left|\left[\left[\hat{\mathbf{M}}_i\left(\tau_1\right),\hat{\mathbf{M}}_j\left(0\right)\right],\hat{\mathbf{M}}_\ell\left(-\tau_2\right)\right]\right|0\right\rangle.$$

Here $[\hat{A},\hat{B}]$ denotes the anticommutator, $[\hat{A},\hat{B}] = \hat{A}\hat{B} + \hat{B}\hat{A}$.

Assuming that the Hamiltonian is of the form (6.32) and that in the dipole moment $\mathbf{M}$ of the crystal (see eqn 6.79) the term $\sqrt{N}d_i^B\left(B_0+B_0^\dagger\right)$ is also included in the expansion (6.81), we reach the conclusion that in order to find the tensor $\chi_{ij\ell}$ is necessary to know double-time Green's functions of the form

$$G^{(1)}\left(\tau_1,\tau_2\right) = \theta\left(\tau_1\right)\theta\left(\tau_2\right)\left\langle 0\left|\left[\left[B_0\left(\tau_1\right)B_0(0)\right],C_0^\dagger\left(-\tau_2\right)\right]\right|0\right\rangle,$$
$$G^{(2)}\left(\tau_1,\tau_2\right) = \theta\left(\tau_1\right)\theta\left(\tau_2\right)\left\langle 0\left|\left[\left[B_0\left(\tau_1\right)C_0^\dagger(0)\right],B_0\left(-\tau_2\right)\right]\right|0\right\rangle,$$
$$G^{(3)}\left(\tau_1,\tau_2\right) = \theta\left(\tau_1\right)\theta\left(\tau_2\right)\left\langle 0\left|\left[\left[B_0\left(\tau_1\right)B_0(0)\right],\hat{\Phi}\left(-\tau_2\right)\right]\right|0\right\rangle,$$

where

$$\hat{\Phi} = \sum_k B_k^\dagger\left(\tau_2\right)B_{-k}^\dagger\left(\tau_2\right), \quad \text{etc.}$$

In applying the model Hamiltonian (6.32) all Green's functions can be exactly determined, although, admittedly, they correspond to the crystal temperature $T=0$. Their calculation is extremely cumbersome, but is similar, to a considerable extent, to that set forth in Section 6.3. Consequently, we shall not give here the relations obtained from them, but only the final expression for the tensor $\chi_{ij\ell}$. We shall assume here that the frequency $\omega \approx \omega'$ is close to the limiting frequency of the B phonon, i.e. that $\hbar\omega \approx \hbar\omega' \approx \varepsilon_1(0)$, whereas the sum $\hbar\left(\omega+\omega'\right)$ is close to the energy of the biphonon. It is convenient, in this case, to separate out terms which have a resonance at the biphonon frequency, putting the remaining terms in $\tilde{\chi}_{ij\ell}$. In view of the afore-said, the expression for $\chi_{ij\ell}$ can be presented as follows:

$$\chi_{ij\ell}\left(\omega,\omega'\right)$$
$$= \tilde{\chi}_{ij\ell} + \frac{\Gamma d_i^C d_j^B d_\ell^B / v}{\left[\hbar\left(\omega+\omega'\right)-\varepsilon_2(0)\right]\left[\hbar\omega'-\varepsilon_1(0)\right]\left[\hbar\omega-\varepsilon_1(0)\right]\Delta\left(\omega+\omega',0\right)}$$

$$+\frac{d_i^{BB} d_j^B d_\ell^B}{[\hbar\omega - \varepsilon_1(0)][\hbar\omega' - \varepsilon_1(0)] \Delta(\omega + \omega', 0)}, \qquad (6.86)$$

where the quantity $\Delta(\omega, 0)$ is determined by relation (6.48). It follows from expression (6.86) that along with the ordinary resonances at $\hbar\omega = \varepsilon_1(0)$ or $\hbar(\omega + \omega') = \varepsilon_2(0)$, the nonlinear polarizability tensor has a resonance at the bound state of quasiparticles, i.e. at $\hbar(\omega + \omega') = E_b$, where E_b is the biphonon energy. If the binding energy of the biphonon substantially exceeds the linewidth, then at $\omega = \omega'$ (in this case the tensor $\chi_{ij\ell}$ corresponds to the generation of a second harmonic) the tensor $\chi_{ij\ell}(\omega, \omega)$ will manifest a resonance if the frequency ω lies within the transparency region.

Note that for $\hbar(\omega + \omega') = E_b$ it is necessary to take the damping of the biphonon into account in the expression for $\Delta(\omega + \omega', 0)$. In this case, along with the real part of the tensor $\chi_{ij\ell}(\omega, \omega)$, an imaginary part is also present. This corresponds, as is well known, to the occurrence of two-photon absorption that is accompanied, in the given case, by the excitation of a biphonon. As applied to excitons this question has been discussed by Hanamura (51) within the framework of a somewhat different approach. We refer to it here (see also Flytzanis (52)) because both the generation of a second harmonic and two-photon absorption are processes that are completely described by the nonlinear polarizability of the crystal found above with bound states taken into account. Actually, these processes can be investigated by a single method.

A relation analogous to (6.86) also obtains for the nonlinear polarizability tensor $\chi_{ij\ell m}$. In this case, for $\omega > 0$, $\omega' > 0$ and $\omega'' > 0$ as well,

$$\chi_{ij\ell m}(\omega, \omega', -\omega'') \qquad (6.87)$$
$$= \tilde{\chi}_{ij\ell m} + \frac{2A d_i^{BB} d_j^{BB} d_\ell^B d_m^B}{[\hbar\omega - \varepsilon_1(0)][\hbar\omega' - \varepsilon_1(0)][\hbar\omega'' - \varepsilon_1(0)][\hbar\omega''' - \varepsilon_1(0)] \Delta(\omega + \omega', 0)}.$$

This tensor corresponds to the four-photon process of scattering of light by light, for which the conservation law

$$\hbar\omega + \hbar\omega' = \hbar\omega'' + \hbar\omega'''$$

holds. If the energy $\hbar(\omega + \omega') \approx E_b$ and the frequencies ω, ω' and ω'' (and consequently ω''' as well) are close to the frequency $\varepsilon_1/\hbar$ of the B phonon, the nonlinear polarizability becomes very large (gigantic). With respect to excitons and biexcitons in a CuCl crystal, a gigantic nonlinear polarizability $\chi_{ij\ell m}$ was experimentally recorded by Grun et al. (53). It follows from the afore-said that the formation of such gigantic nonlinear polarizabilities in the region of biphonon or biexciton resonances, or the occurence of intense two- or three-photon absorption is not a chance effect. Such gigantic resonance effects should be observed in all spectral regions in which dipole-active bound states of quasiparticles exist. In particular, such effects may exist in the infrared region of the spectra as well in all crystals in which dipole-active biphonons are detected.

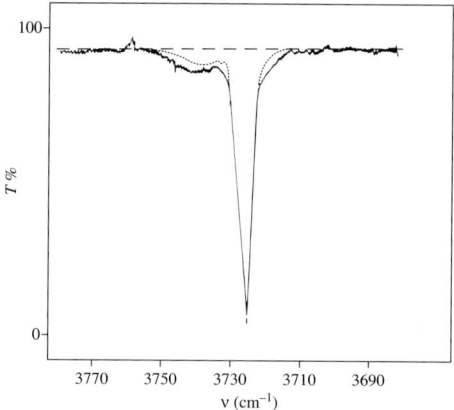

FIG. 6.9. The $\nu_1 + \nu_3$ band of CO_2. Observed and calculated transmission spectrum for a thickness of approximately 1.8 μm. The dotted line represents the calculated absorption spectrum (10).

6.7 Experimental investigations of biphonons and Fermi resonance with polariton

As has been pointed out previously, in many crystals biphonons were observed both in infrared absorption spectra and in RSL spectra.

In discussing infrared absorption spectra, we refer, first of all, to the papers by Dows and Schettino (54) and of Schettino and Salvi (55). Dows and Schettino (54) investigated the CO_2 crystal spectrum in the frequency region corresponding to the combination tone of the intramolecular vibrations ν_1 and ν_3 ($\nu_1 + \nu_3 \approx 3720$ cm^{-1}). Schettino and Salvi (55) measured the infrared (IR) spectra of N_2O and OCS crystals. The CO_2 and N_2O molecules are linear, have no permanent dipole moments, and form a simple cubic lattice upon crystallization. This lattice has four molecules per unit cell, which are oriented along the axes of a tetrahedron. The OCS molecule is also linear, but it forms a crystal of the trigonal system with one molecule per unit cell.

Since the CO_2 molecule is symmetrical, its stretching vibration ν_1 is IR inactive and has practically no dispersion. Hence, the Van Kranendonk model can be employed to interpret the experimenatl results obtained in the region of the combination frequency $\nu_1 + \nu_3$. This has actually been taken into account by Bogani (10). Shown in Fig. 6.9 is the transmission spectrum, measured by Dows and Schettino (54), of a crystal of 1.8 μm thickness. Calculations carried out by Bogani (10) indicate that the sharp absorption peak obtained in this case corresponds to the excitation of a biphonon.

Similar results were obtained (55) for the N_2O crystal in the frequency region that corresponds to the combination frequency $\nu_1 + \nu_3$. The N_2O molecule is not symmetrical and therefore all of its three intramolecular vibrations, ν_1, ν_2 and ν_3, are IR active. Its flexural vibration ν_2 is two-fold degenerate and, owing to

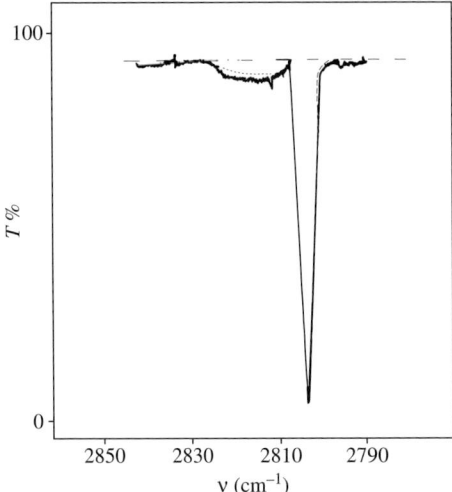

FIG. 6.10. Transmission spectrum of the $\nu_2 + \nu_3$ band of N_2O. The dotted line shows the calculated spectrum for a thickness of approximately 18 μm (10).

the small value of the dipole moment, the dispersion is weak throughout the Brillouin zone (less than 3 cm^{-1}). Hence, the Van Kranendonk model can be applied for the frequency region $\nu_2 + \nu_3$, as well as for the region $\nu_1 + \nu_2$ (see Bogani (10)). A transmission spectrum, measured by Schettino and Salvi (55) of a film of N_2O crystal, 18 μm thick, in the $\nu_2 + \nu_3$ band, is shown in Fig. 6.10.

Extensive experimental data have been obtained in studying the effects of anharmonicity in crystals by the RSL method and, in particular, in the observation of RSL by polaritons.

As a result of this research, and in full agreement with the prediction of theory, it was shown that:

(a) The formation of dipole-active biphonons leads to a discontinuity in the polariton branch outside the band of two-particle states of phonons, and that such a shape of the polariton branch cannot be explained on the basis of the theory in the harmonic approximation (see (56)–(60)).

(b) In cubic crystals, along with the transverse biphonons that lead to a discontinuity in the polariton branch (see (a)), longitudinal biphonons are also formed, see (61).

(c) In many crystals Fermi resonance of the phonon branches very strongly influences the position and other characteristics of biphonons (see, for example, (62)).

(d) In a number of crystals discontinuities of the polariton branch are observed inside the band of two-particle states as well (see, e.g. (63)–(65); (62), see also Fig. 6.7).

The cited experiments stimulated a more comprehensive theoretical analysis of the dispersion of polaritons within the band of two-particle states (21), (22).

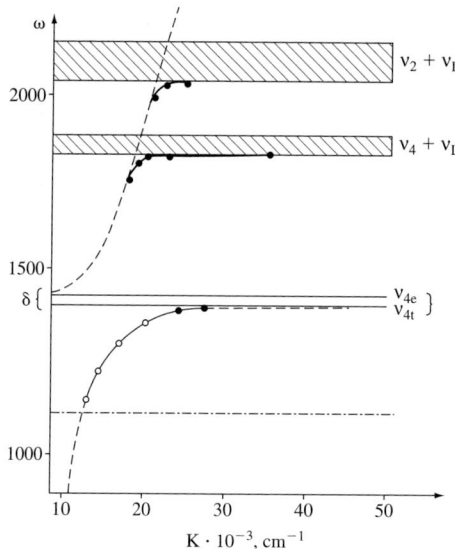

FIG. 6.11. Polariton dispersion in the NH$_4$Cl crystal. Longitudinal–transverse splitting of the biphonon is observed in the frequency region $\nu \approx 1460$ cm^{-1} (from (61)).

Inside the band of two-particle states a channel opens up for polariton decay into two phonons. This process leads to a broadening of the polariton line (see, e.g. (59), (60)), and also a change in the polariton dispersion law. The two most important effects in the latter case are: (a) interference of scattering by a polariton and two-particle states (this can lead to drops in intensity of the Fano antiresonance type; see (22)), and (b) the presence of singularities in the density of two-particle states (those, in particular, that correspond to quasibiphonons).

Experimental investigations of the effects of strong anharmonicity in phonon and polariton spectra are reported in many papers. In order to give some idea of the state-of-the-art concerning experimental research along these lines, as well as in the study of crystals, we shall make certain comments.

We note, first of all, that the first convincing proof of the presence of a gap in the polariton spectrum, appearing in the vicinity of the biphonon level, was obtained by Mavrin and Sterin (56) and by Winter and Claus (57) in investigating lithium niobate (LiNbO$_3$). In this crystal the dispersion branch of an ordinary polariton undergoes splitting (of approximately 5 cm^{-1}) at a frequency of 537 cm^{-1}. There are no fundamental vibrations of the lattice in this spectral region (the closest has a frequency of 582 cm^{-1}) and, consequently, the observed splitting is associated with the presence of a biphonon. This same kind of splitting was subsequently observed in many crystals, in particular in HIO$_3$ (59), (60) and NH$_4$Cl (58), (61), (62).

In the low-temperature phase IV, NH$_4$Cl crystals belong to the cubic class T_d and have one molecule per unit cell. Though these are not molecular crystals, the optical vibrations of the NH_4^+ ions, like the molecules in a molecular crystal, interact relatively weakly with one another. The vibrational spectra of this crystal have been investigated very comprehensively. This, precisely, is the crystal in which the longitudinal biphonon state was first indentified and the magnitude of the longitudinal–transverse splitting (see Fig. 6.11) was determined by Gorelik et al. (58), (61) from the RSL spectra for various light polarizations.

No less interesting results were obtained in previously mentioned papers by Polivanov (59), (60). Polivanov investigated the spectra of RSL by polaritons in the HIO_3 crystal. This crystal is biaxial (symmetry point group 222). The phonon spectrum of this crystal has been much studied (see Krauzman et al. (66)) and is usually subdivided into four groups (latticed vibrations (0 – 220 cm^{-1}), deformation vibrations of the IO_3 group (290 – 400 cm^{-1}), streching vibrations of the IO_3 group (600 – 845 cm^{-1}), and vibrations of the OH group: out-of-plane (torsional) vibrations (560 cm^{-1}), plane (deformation) vibrations (1160 cm^{-1}) and stretching vibrations (2940 cm^{-1})). Thus, for $\omega > 1160$ cm^{-1}, only a single first-order line with frequency $\omega \approx 2940$ cm^{-1} can be observed in the scattering spectra. Nevertheless, a wide band can be observed in the RSL spectra in the frequency region of approximately 2270 cm^{-1} (scattering angle $\varphi \approx 90°$). On the low-frequency edge of this band, i.e. at $\omega = 2270$ cm^{-1}, there is a quite narrow and intense peak. It was pointed out by Polivanov (59) that scattering in the frequency region 2270 $cm^{-1} < \omega < 2940$ cm^{-1} corresponds to the excitation of two- and three-particle states. It was also postulated that the peak at $\omega = 2270$ cm^{-1} indicates the existence of a biphonon that split off the overtone band of the fundamental vibration with frequency $\Omega = 1160$ cm^{-1}. A second paper by Polivanov (60) confirmed this hypothesis. In the investigation of RSL by polaritons (scattering angle $\varphi = 0°$ to $4°$) it was shown (Fig. 6.12) for the same crystal that the polariton branch has a discontinuity at frequency $\omega = 2270$ cm^{-1} and is greatly broadened in the band of two-particle (dissociated) states. Upon cooling the crystal from $T = 300$ K to $T = 80$ K a substantial decrease in the biphonon linewidth was observed (60), as well as the occurrence of polariton branch discontinuities within the band of two-particle states.

The broadening of the polariton branch in the region of dissociated states is due to the opening of a new decay channel, an effect mentioned above. As to the discontinuities of the polariton branch that occur in the same spectral region upon lowering of the temperature of the crystal, their cause is still not quite clear. Only in subsequent investigations can we hope to establish which of these discontinuities are due to critical points in the density of two-particle states, to the formation of quasibiphonons, or to interference of the Fano antiresonance, see (21), (22). The experimental investigations of biphonons in disordered media as a function of impurity concentration have been performed by Belousov and Pogarev (67). We discuss the results of these investigations in Subsection 6.8.1.

New possibility in studies of biphonons and Fermi resonance has appeared due to the development of coherent anti-Stokes Raman spectroscopy (CARS). As was shown by Flytzanis and his group (68)–(70) this technique for noncen-

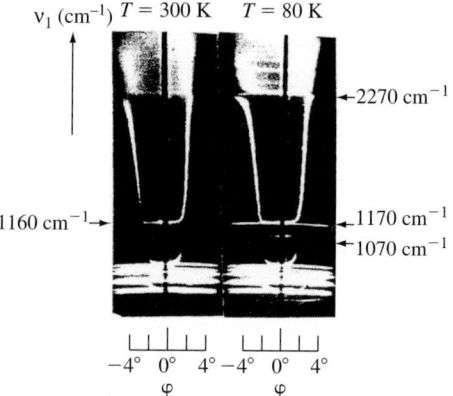

FIG. 6.12. Fragments of phonon spectra in scattering by polaritons in the HIO_3 crystal that were obtained by Polivanov (60) at $T = 300$ K and $T = 80$ K; φ is the scattering angle.

trosymmetric crystals gives direct access to the propagation and loss of coherence of picosecond phonon–polariton pulses of widely varying wavevector and frequency. It was shown in (68) that it is possible to create and follow the spatial and temporal evolution in crystals of a picosecond pulse of propagating Raman-active polaritons by use of a time and space resolved CARS. This method gives direct access to the energy propagation velocity and dephasing time of phonon-polaritons over a large portion of the polariton dispersion curve. A schematic outline of the experiment (68) is given in Fig. 6.13. Coherent excitation of the polariton at time $t = 0$ and position $X = 0$ is provided by two time-coincident optical picosecond pulses of wavevectors $\mathbf{k}_L$ and $\mathbf{k}_S$ and frequencies ω_L and ω_S, respectively. The polariton wavevector is $\mathbf{k}_\pi = \mathbf{k}_L - \mathbf{k}_S$ and, hence, its propagation direction can be changed by changing the angle between the two beams. The resonant excitation of a polariton with frequency $\omega_\pi(\mathbf{k}_\pi)$, the frequency of a polariton of wavevector $\mathbf{k}_\pi$, can be obtained by frequency tuning ω_L such that $\omega_L - \omega_S = \omega_\pi$. The interaction between polariton, laser and Stokes pulses can be described by a phenomenological energy density

$$V = \left(-d_E E_L E_S^+ E_\pi^+ - d_Q E_L E_S^+ Q_\pi^+\right) + \text{c.c.}, \tag{6.88}$$

where E_L and E_S are the electric field at ω_L(laser) and ω_S (Stokes), respectively, E_π and Q_π represent the electric field and transverse optical mechanical vibration associated with the polariton and where d_E, d_Q are parameters. This coupling between fields produces a picosecond duration polariton wavepacket at $t = 0$ and $X = 0$ which then propagates in the crystal in a direction determined by the polariton wavevector $\mathbf{k}_\pi = \mathbf{k}_L - \mathbf{k}_S$, and at a speed given by the polariton group velocity $\mathbf{V}_g(\mathbf{k}_\pi)$. The temporal and spatial evolution of the propagating excitation is followed by phase-matched coherent anti-Stokes scattering at $\omega_a =$

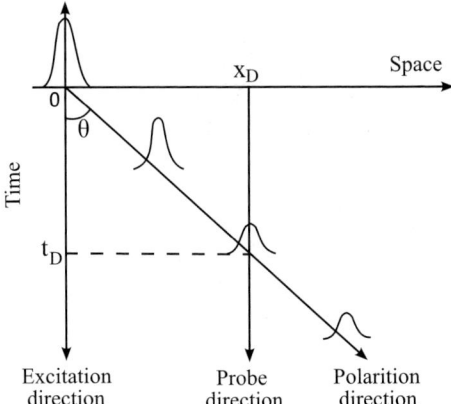

FIG. 6.13. The polariton wavepacket, created by coherent scattering of picosecond laser and Stokes pulses, propagates in the crystal at the angle θ with respect to the excitation direction. This angle is determined by the excitation wavevector geometry (see above). The coherent amplitude of the propagating wavepacket may be measured by phase-matched coherent anti-Stokes scattering of a probe pulse suitably delayed in time (t_D) and displaced in space (by X_D). Reprinted with permission from Gale et al. (68). Copyright (1986), American Physical Society.

$= \omega_p + \omega_\pi$ of the third pulse ω_p displaced with respect of the excitation in time (by t_D) and in space (by x_D) (see Fig. 6.13). The measured dependence of the spatial displacement x_D, where the signal is a maximum, on time delay t_D is directly related to the energy-propagation characteristics of the polariton wavepacket. Using this technique the same authors have performed the first time- and space-resolved study of polariton Fermi resonance in ammonium chloride NH_4Cl, in a frequency region where the polariton is in strong interaction with the polar ($2\nu_4$) two-phonon band, which extends roughly from 2800 to 2900 cm^{-1} and which before have been studied by conventional Raman scattering technique, (71) (Fig. 6.14).

In Fig. 6.15 we demonstrate the polariton dephasing rate vs. polariton frequency measured in (69) at temperature 78 K and the results of calculations done with the help of Fermi resonance theory. Figure 6.15 demonstrates a dramatic frequency dependence of the dephasing rate at polariton frequencies close to the edges of the $2\nu_4$ band. The very rapid increase of the dephasing rate inside the $2\nu_4$ band is a clear indication that the isoenergetic two-phonon states provide a direct and very efficient relaxation channel. In the paper (70) the relaxation dynamics of the longitudinal lower-energy component of the $((\nu_2+\nu_4), \nu_3)$ polar two-phonon Fermi quadruplet in ordered ammonium chloride have been investigated with the use coherent picosecond spectroscopy. The obtained results provide more confirmation that in ordered crystals the Fermi resonance plays a crucial role not only in the creation, but also in the relaxation of bound

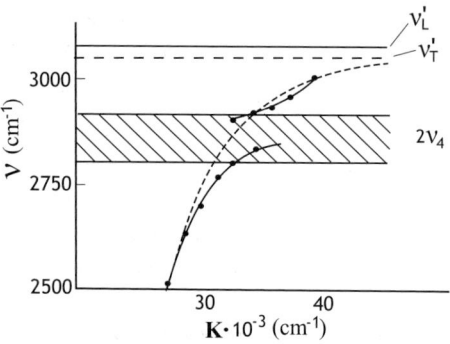

FIG. 6.14. The polariton dispersion curve of ammonium chloride in the vicinity of the $2\nu_4$ two-phonon quasicontinuum (shaded region) obtained by near forward Raman scattering (filled circles). The dashed curve is the calculated dispersion curve in the absence of polariton Fermi resonance. Reprinted with permission from Mitin et al. (71). Copyright (1975), American Physical Society.

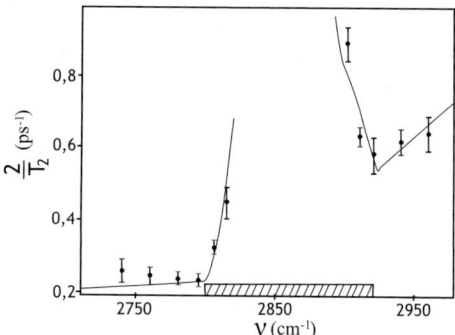

FIG. 6.15. Measured polariton dephasing rate $2/T_2$ ps^{-1} vs polariton frequency, under the conditions of Fermi resonance with the $2\nu_4$ band (shaded region), at a crystal temperature 78 K. The full line is calculated (see text).

multiphonon states.

6.8 Local biphonons in crystals with isotopic substitution impurities

6.8.1 Where does the formation of local states begin in a spectrum of optical vibrations? Effects of strong anharmonicity

We begin, in this section, a discussion of features of the biphonon spectrum in imperfect crystals. For the sake of simplicity we shall assume that the crystal contains only the simplest point defects: isotopic substitution impurities. Before going over to a theory of local biphonons, we shall make several qualitative remarks concerning the effect of anharmonicity on the spectrum of local vibrations in a crystal having an isotopic impurity (here Δ denotes the isotopic shift). We point out, first of all, that in the harmonic approximation, for instance in a crystal with one molecule per unit cell, the frequency ω_ℓ of local vibrations split off the band $\varepsilon(k)$ of nondegenerate optical vibrations satisfies the equation

$$1 = \frac{\Delta}{N} \sum_k \frac{1}{\hbar\omega_\ell - \varepsilon(\mathbf{k})}. \tag{6.89}$$

It is a well-known (see, for example, Lifszitz (72)) that in three-dimensional crystals this equation has a solution for ω_ℓ, outside the frequency band $\varepsilon(\mathbf{k})/\hbar$, only at sufficiently large values of $|\Delta|$, i.e. $|\Delta| \gg \Delta_c$, where $\Delta_c \approx T_1$ and T_1 is the width of the band of optical phonons in a perfect crystal. If the preceding inequality is fulfilled, local states appear, not only in the region $\omega_\ell \approx \Omega$, but also in the region of the second and higher overtones, i.e. at the frequencies $2\omega_\ell, 3\omega_\ell$, etc. The afore-said is, of course, obvious when one takes into account that all of these states correspond to different quantum numbers ($n = 1, 2, \ldots$) of a harmonic oscillator, i.e. the normal local vibrations of a crystal having an isotopic impurity. If, however, the inequality $|\Delta| > \Delta_c$ is not complied with, i.e. if $|\Delta| < \Delta_c$, then eqn (6.89) has no solutions for ω_ℓ that lie outside the band $\varepsilon(\mathbf{k})$. In this latter case, the presence of an isotopic defect does not lead to the formation of local states. It is clear that in the harmonic approximation such states do not appear either in the fundamental tone region (i.e. at $\omega_\ell \approx \Omega$) or in the overtone region.

The anharmonicity can qualitatively change the pattern of the local state spectrum leading to the formation of such states even in cases ($|\Delta| < \Delta_c$) when in the harmonic approximation there are no states localized in the region of the defect (14).

Let us assume that the anharmonicity constant A is large compared to the width $T_1 = 2T$ (where T is the half-width) of the phonon band. In this case, as was shown in Subsection 6.2.2, the width of the biphonon band is of the order of T_1^2/A, i.e. small compared to the width T_1 of the band of optical phonons. Important here is, however, the comparison of the width of the biexciton band with isotope shift. Indeed, in the limiting case of strong anharmonicity, the biphonon energy is $E \approx 2\hbar\omega - 2A$ and the biphonon state $|2\rangle$ is just the coherent superposition of the states of two-fold excited molecules. It is clear then that an elementary generalization of an equation of the type (6.89) can be used to find biphonon local states. Specifically, the equation for the frequency of a local biphonon $\omega_\ell^{(2)}$, i.e. the localized state split off the biphonon zone, can be written as follows:

$$1 = \frac{\tilde{\Delta}}{N} \sum_{\mathbf{K}} \frac{1}{\hbar\omega_\ell^{(2)} - E(\mathbf{K})}, \tag{6.90}$$

where $E(\mathbf{K})$ is the energy of a biphonon with wavevector $\mathbf{K}$ in a perfect crystal, and $\tilde{\Delta}$ is the isotopic shift of the two-fold excited state of an impurity molecule:

$$\tilde{\Delta} = 2\Delta. \tag{6.91}$$

Since the states of the biphonon, like those of the phonon, are characterized by only a value of the wavevector, an analysis of eqn (6.90) is analogous to that of eqn (6.89). On the basis of the result of such an analysis, which we have already used for phonons, it can be contended that the level of a local biphonon is formed if

$$\left|\tilde{\Delta}\right| > \tilde{\Delta}_c, \tag{6.92}$$

where the critical value is $\tilde{\Delta}_c \approx T_2$, T_2 is the width of the biphonon band ($T_2 \sim$ $\sim T_1^2/A$) or if:

$$2|\Delta| > \tilde{\Delta}_c. \tag{6.93}$$

It is clear that inequality (6.92) or (6.93) can be complied with even in the case when

$$|2\Delta| < \Delta_c,$$

i.e. when the following inequalities are simultaneously complied with:

$$T_1^2/2A \ll |\Delta| \ll T_1. \tag{6.94}$$

Complying with the r.h.s. inequality does not lead to the formation of local states in the region of the fundamental tone. However, if we comply with the second (left-hand) inequality, splitting off the level of a local biphonon is provided for. Thus, the main reasons of the influence of anharmonicity is in a strong decrease of the width of the biphonon band if anharmonicity is strong enough, from one side, and in a two-fold increasing of the isotopic shift. This demonstrates the importance of the role of anharmonicity in forming the spectra of local states. In the limiting situation under discussion the spectrum of local states begins at $E = 2\hbar\Omega$ rather than at $E \approx \hbar\Omega$. Clearly, such an effect can also occur for three-phonon vibrations, etc. (on the theory of bound three-phonon complexes see the paper by Dubovsky (73)). For the theory of local biphonons see also (74), (75).

6.8.2 Local biphonon in an $^{14}NH_4Br$ crystal containing the isotopic substitution impurity ^{15}N

We already mentioned the experimental research conducted by Belousov et al. (77). In these investigations they studied the spectrum of local states in the frequency region $\omega \approx 2800$ cm^{-1} in a ^{14}NH$_4$Br crystal (a cubic crystal of group T_d^1) having as an impurity the isotope ^{15}N. It was first shown experimentally in these investigations that, in accordance with predictions (14), anharmonicity actually can led to the formation of a local biphonon under conditions in which no local phonons exist.

In the Raman spectrum of a pure crystal of ^{14}NH$_4$Br, two narrow lines (of width not larger than 0.2 cm^{-1}) are observed in the region $\omega_4(\mathbf{k} \approx 0)$ of one-phonon transitions. The lines correspond to TO (1397.5 cm^{-1}) and LO

(1413.5 cm^{-1}) phonons. Upon introducing the isotope ^{15}N, only neglilible broadening and shift of these lines occur in crystals ^{15}N$_x$^{14}N$_{1-x}$H$_4$Br (where $x = 0.05$), and no local one-phonon vibration is set up. Its absence and the typically one-mode nature of the one-phonon spectrum are explained by the fact that the isotopic shift $\Delta = -6$ cm^{-1} (Price et al. (76)) is small compared to the width of the one-phonon zone ($2T = 36$ cm^{-1}). Such a value of the one-phonon zone width follows from measurements of the Raman spectrum for the region of the $\omega_4(\mathbf{k}) + \omega_4(-\mathbf{k})$ band. The picture of a spectrum in this region for a natural isotope ^{15}N content ($x = 0.0037$) is shown in Fig. 6.16(a). The band 2795–2867 cm^{-1} corresponds to two-particle transitions $\omega_4(\mathbf{k}) + \omega_4(-\mathbf{k})$. Its low-frequency boundary coincides with twice the frequency $\omega_4(\mathbf{k} \approx 0)$ of a TO phonon and its width (according to the selection rules) is equal to twice the width of the phonon zone $\omega_4(\mathbf{k})$. The narrow line at 2792.5 cm^{-1} (see Fig. 6.16a), located below the zone of dissociated two-particle states, corresponds, obviously, to the excitation of a biphonon. Its occurrence in a second-order spectrum is thereby an indication of quite strong anharmonicity with the characteristic constant $A \approx$ $\approx 30 - 35$ cm^{-1}.

An additional line at $\hbar\omega' = 2788$ cm^{-1} was found even in a crystal with a natural content of the isotope ^{15}N by Belousov et al. (77). When the concentration is increased to $x = 0.05$ and $x = 0.3$, the integrated intensity of this line increases (see Fig. 6.16b). It is exactly this fact that enables us to regard this line as being correspondingly coupled to the impurity of the two-phonon state (local biphonon).

Experimental investigations of two-phonon spectra of crystals with defects are continuing. This is what makes the further theoretical analysis of local multiple-particle states in crystals a timely object of research.

A more complete discussion of biphonons in crystals with defects can be found in (74). A review of the spectroscopy of excitons and phonons in disordered molecular crystals can also be found in (75).

6.9 Conclusion and prospects for further investigations

The possibility of the formation of biphonons and other larger phonon complexes substantially enriches the vibrational spectrum of multiparticle states. Though many results have been obtained in this line of research, both on the experimental and theoretical sides, much still remains to be done. Among others, three-phonon and other, more complex bound states of phonons deserve further study (their analysis has only been initiated by Lalov (78), Krivenko et al. (79), and Gotshev and Lalov (80). A more recent review paper on the theory of multiphonon complexes can be found in (74). However, even biphonons and their role in many optical processes have not yet been investigated to a sufficiently comprehensive degree. In connection with the afore-said, we should like to call attention to the timeliness of calculations of biphonons in crystals of various structures for the region of degenerate vibrational transitions.

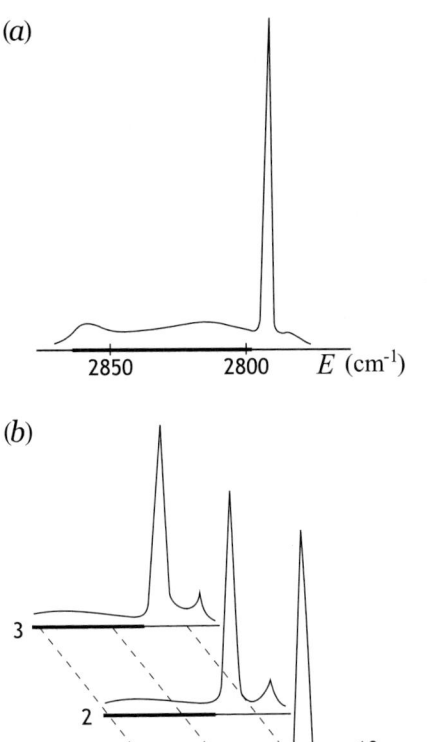

FIG. 6.16. Raman scattering spectra for ^{15}N$_x$^{14}N$_{1-x}$H$_4$Br crystals in the region of $\omega_4(\mathbf{k}) + \omega_4(-\mathbf{k})$ transitions. The heavy line distinguishes the region of two-particle transitions of unbound phonons. The peak at 2788 cm^{-1} corresponds to the excitation of local biphonons. Local phonons do not exist. Concentrations are (a) $x = 0.0037$; (b); (1) $x = 0.0037$ and (3) $x = 0.3$.

The investigations of biphonons in one- and two-dimensional crystals may also be of special interest. In such crystals, as well as in three-dimensional crystals in which certain phonons can be assumed to be quasi-one-dimensional or quasi-two-dimensional, the conditions for the formation of biphonons should be more favorable, all other things being equal, than in ordinary three-dimensional crystals.

Speaking of biphonons in one-dimensional crystals, we should also like to draw attention to their possible role in the transfer of energy along protein molecules (81). In such molecules there are vibrations with energy $\hbar\Omega \approx 0.2$ eV. Hence, the formation of stable and mobile complexes, consisting of two or three vibration quanta, may prove to be an essential factor in the transfer of the

CONCLUSION AND PROSPECTS

energy $E \approx 0.5$ eV (see (81)). It is quite clear, of course, that such biphonon energy transfer over macroscopic distances, which are the only important ones in biology, can be considered possible only under the condition that the lifetime of the vibration quanta is sufficiently long. In this sense, the problem posed here has the same significance as in the assessment of the role of Davydov soliton transfer (82), (83).

It has already been pointed out that the conditions for the formation of local or quasilocal biphonons in disordered crystals can differ substantially from the analogous conditions for the formation of local states in the region of the fundamental frequencies. In view of the development of experimental research of spectra of disordered crystals in the region of overtones and the sum of the fundamental vibrations, what seems to be especially timely and of great theoretical interest is the further analysis of the above-mentioned conditions for the formation of local biphonons. In a more general sense, this means the analysis of the spectra of disordered crystals in the spectral region of multiparticle states. In the theoretical papers cited above (27), (28) only the simplest version of the coherent potential approximation was applied. In these papers the method was employed to investigate the spectra of isotopically disordered crystals. Hence, the analysis of the spectra of mixed crystals, consisting of different molecules, is also timely. The coherent potential method (84) was devised, as is well-known, especially for calculating the spectra of one-particle states (electrons, phonons, and excitons). In dealing with two-particle states, however, a number of theoretical problems arise when anharmonicity is taken into account (see also (85)). These problems require additional analysis and are associated primarily with the need to provide proper asymptotic behavior of the method at low impurity concentrations (going over to the biphonon or quasibiphonon equation).

Also worthy of further development is the theory of surface biphonons. The conditions required for the formation of these states are different from those of the formation of surface states for the spectral region of the fundamental vibrations. It was demonstrated on the model of a one-dimensional crystal (26) that situations may exist, in general, in which the surface state of the phonon is not formed and the spectrum of surface states begins only in the frequency region of the overtones or combination tones of the vibrations.

When the frequency of the surface biphonon lies within the band of the surface polariton, Fermi resonance occurs and the dispersion curve of the polariton is subject to a number of essential changes (gaps appear, etc. (86)). Consequently, experimental research of surface polariton dispersion under these conditions could yield, like similar investigations of bulk polaritons, a great deal of interesting information, not only about the surface biphonons themselves, but about the density of states of surface phonons and the magnitude of their anharmonicity constants as well.

In conclusion, we stress the fact that biphonons and other effects of strong anharmonicity should appear, not only in the absorption and luminescence spectra of pure and impure crystals, in nonlinear processes, in RSL spectra, and in the spectra of inelastically scattered neutrons, but, possibly, also in nonradiative decay processes of electronically excited states of the crystal. An analysis of this problem, as well as of many other manifestations of states of phonons bound to one another, is of exceptional interest and will undoubtedly be the object of future research.

7
THE DIELECTRIC TENSOR OF CRYSTALS IN THE REGION OF EXCITONIC RESONANCES

7.1 On the calculation of the dielectric tensor

In the previous chapters we repeatedly discussed the various methods of calculation of the dielectric tensor $\epsilon_{ij}(\omega, \mathbf{k})$. In this chapter we will demonstrate one more method based on applying the results of microscopic polariton theory presented in the previous chapters and the theory developed by Dzyaloshinsky and Pitayevsky (1).

Before discussing this new method it is useful to recall briefly the methods which we have already discussed. Note, first of all that calculations of the dielectric tensor must be based, as is known, upon a microscopic theory. Such a theory for ionic crystals was first developed by Born and Ewald (2) for the infrared spectral region. The application of this approach for the region of exciton resonances has also been demonstrated in (3). In an approach identical to that of Born and Ewald (2) the mechanical excitons (see Section 2.2) are taken as states of zeroth-approximation. In the calculation of these states the Coulomb interaction between charges has to be taken into consideration without the contribution of the long-range macroscopic part of the longitudinal electric field. If this procedure can be carried out, then the Maxwell total macroscopic fields $\mathbf{E}$ and $\mathbf{H}$ can be taken as perturbations. In the first order of perturbation theory, we find

$$P_i(\omega, \mathbf{k}) = \chi_{ij}(\omega, \mathbf{k}) E_j(\omega, \mathbf{k}), \qquad (7.1)$$

$\chi_{ij}(\omega, \mathbf{k})$ being the dielectric polarizability (susceptibility) tensor, so that

$$\epsilon_{ij}(\omega, \mathbf{k}) = \delta_{ij} + 4\pi \chi_{ij}(\omega, \mathbf{k}). \qquad (7.2)$$

By establishing the relation (7.1) we must use quantum–mechanical perturbation theory, considering the perturbation fields $\mathbf{E}$ and $\mathbf{H}$ as classical (non-quantized) quantities.

The main difficulty, which may arise by applying the above-described scheme with mechanical excitons, consists in the necessity of excluding the macroscopic longitudinal electrical field from the equations of motion (or from the Coulomb Hamiltonian) in the calculation of the unperturbed zeroth-approximation states.

For ionic crystals, within the classical treatment of ion motion (see (2)), and for molecular crystals in the Frenkel exciton spectral region (see (3) and Section 2.2), the Ewald procedure was successfully used to exclude the macroscopic part of longitudinal field. This permitted one to compute the tensor $\epsilon_{ij}(\omega, \mathbf{k})$ using the given above scheme. However, the problem of the extraction of the macroscopic longitudinal field from the Coulomb Hamiltonian in the calculation of zeroth-approximation states can be avoided in calculations of the so-called transverse dielectric tensor. This tensor determines the relation between the induction **D** and the transverse part of macroscopic electric field **E**. This relation can be obtained only when external charges and currents are absent and the relation div **D** = 0 holds. As shown in Section 4.4, instead of the tensor $\epsilon_{ij}(\omega, \mathbf{k})$ the transverse dielectric tensor $\epsilon_{\perp,ij}(\omega, \mathbf{k})$ can be used. Making use of this tensor we must express the induction **D** as a function of the transverse component of the electric field **E**, i.e. by $\mathbf{E}^\perp$ (see eqn 4.71)

$$D_i(\omega, \mathbf{k}) = \epsilon_{\perp,ij}(\omega, \mathbf{k}) E_j^\perp(\omega, \mathbf{k}). \tag{7.3}$$

Linearity of the above relation allows one to find the tensor $\epsilon_{\perp,ij}(\omega, \mathbf{k})$ by calculating the polarization **P**, induced in the crystal by the total transverse field $\mathbf{E}^\perp$, neglecting the local counterpart of the transverse field which is very small.[48] As unperturbed states in this case we have to use the Coulomb exciton states which are obtained taking full account of the Coulomb interaction between charges. Importantly in this case we can assume that charge transfer excitons are also taken into account. If the unperturbed states, obtained by taking into account the full Coulomb interaction, are known, considering the field $\mathbf{E}^\perp$ as a perturbation we can determine the polarization

$$P_i(\omega, \mathbf{k}) = \chi_{\perp,ij}(\omega, \mathbf{k}) E_j^\perp(\omega, \mathbf{k}), \tag{7.4}$$

and thus the transverse dielectric tensor $\epsilon_{\perp,ij}(\omega, \mathbf{k})$ (see Pekar (4), who, however, did not note that only a transverse dielectric tensor had been calculated in his paper). These methods are based on knowledge of the mechanical or Coulomb states of excitons. What do we have to do if dissipation has to be included in the dielectric tensor, for example for the calculation of the absorption of light waves? The usual procedure to consider the corresponding energy of a photon as a value with some imaginary part. This imaginary part depends on the frequency of the light and appears as a result of scattering of light in a medium by phonons or impurities. But this scattering is dependent on dispersion relations for polaritons which in these calculations of the dielectric tensor are unknown.[49] Below, another method of calculation of the dielectric tensor, is described. Its main feature is that these calculations for the exciton region of the spectrum are based on using the polariton state and not states of mechanical or Coulomb excitons. On the calculation of the dissipation rate in this case, see Section 7.4.

---

[48]It can be shown that taking into account this field leads, in the case of the exciton energy, to small corrections of order $(a/\lambda)^2$, where a is the lattice constant, and λ the wavelength.

[49]The same question arises in the theory of the dielectric tensor in the framework of semiphenomenological local field method described in Ch. 5.

7.2 The Pitaevsky–Dzyaloshinsky formula for the dielectric tensor

Let us consider a medium (crystal), where the density of external charges is zero, and the density of external electric currents $\mathbf{j}_{\text{ext}}(\mathbf{r},t) \neq 0$. In this case the equation for the field $\mathbf{E}(\omega, \mathbf{k})$ takes the form

$$\left[k^2 \delta_{ij} - k_i k_j - \frac{\omega^2}{c^2} \epsilon_{ij}(\omega, \mathbf{k})\right] E_j(\omega, \mathbf{k}) = i\frac{4\pi}{c^2} \omega j_{\text{ext},i}(\omega, \mathbf{k}), \tag{7.5}$$

so that

$$E_i(\omega, \mathbf{k}) = i\frac{4\pi}{c^2} \omega \Delta_{in}^{-1}(\omega, \mathbf{k}) j_{\text{ext},n}(\omega, \mathbf{k}), \tag{7.6}$$

with

$$\Delta_{ij}(\omega, \mathbf{k}) = k^2 \delta_{ij} - k_i k_j - \frac{\omega^2}{c^2} \epsilon_{ij}(\omega, \mathbf{k}). \tag{7.7}$$

By applying a gauge with vanishing scalar potential, then

$$\mathbf{E}^{\text{ext}}(\mathbf{r}, t) = -\frac{1}{c} \frac{\partial \mathbf{A}^{\text{ext}}(\mathbf{r}, t)}{\partial t}, \tag{7.8}$$

where $\mathbf{A}^{\text{ext}}(\mathbf{r}, t)$ is a vector potential then from (7.6) we obtain

$$A_i^{\text{ext}}(\omega, \mathbf{k}) = \frac{4\pi}{c} \Delta_{in}^{-1}(\omega, \mathbf{k}) j_{\text{ext},n}(\omega, \mathbf{k}). \tag{7.9}$$

The above relation gives the vector potential induced by external currents. On the other hand, the same potential is given as the mean value of the operator

$$\hat{\mathbf{A}}^{\text{ext}}(\mathbf{r}, t) = e^{\frac{i}{\hbar}(\hat{H}+\hat{H}_{\text{ext}})t} \hat{\mathbf{A}}(\mathbf{r}) e^{-\frac{i}{\hbar}(\hat{H}+\hat{H}_{\text{ext}})t}, \tag{7.10}$$

$\hat{H}$ being the total system Hamiltonian, including the radiation field when external charges and currents are absent, and

$$\hat{H}_{\text{ext}} = -\frac{1}{c} \int \mathbf{j}_{\text{ext}}(\mathbf{r}, t) \hat{\mathbf{A}}(\mathbf{r}, t) \mathrm{d}^3 r. \tag{7.11}$$

Assuming that $j_{\text{ext}}(t \to -\infty) \to 0$, we can put the operator

$$\exp\left\{-\frac{it}{\hbar}\left(\hat{H} + \hat{H}_{\text{ext}}\right)\right\}$$

into the form

$$\exp\left\{-\frac{it}{\hbar}\left(\hat{H} + \hat{H}_{\text{ext}}\right)\right\} = e^{-i\hat{H}t/\hbar} S_{\text{ext}}(t), \tag{7.12}$$

where (see (5), Ch. II)

$$S_{\text{ext}}(t) = T \exp\left\{-\frac{i}{\hbar} \int_{-\infty}^{t} \hat{H}_{\text{ext}}(t') \mathrm{d}t'\right\}. \tag{7.13}$$

Here T is the time-ordering operator. In consequence, the mean value of the vector potential in the case, when external currents are accounted for, takes the form [50]

$$\langle \hat{\mathbf{A}}^{\text{ext}}(\mathbf{r},t)\rangle = \langle S_{\text{ext}}^{-1}(t)\hat{\mathbf{A}}(\mathbf{r},t)S_{\text{ext}}(t)\rangle, \quad (7.14)$$

where $\hat{\mathbf{A}}(\mathbf{r},t)$ is the vector potential operator in a medium where external charges and currents are absent.

Expanding (7.14) into a series with respect to $\mathbf{j}_{\text{ext}}$ and retaining only the first order terms we get

$$\langle \hat{A}_n^{\text{ext}}(\mathbf{r},t)\rangle = -\frac{i}{\hbar c}\int_{-\infty}^{t}dt'\int d^3r'\, j_{\text{ext},k}(\mathbf{r}',t')$$
$$\times \langle \{\hat{A}_k(\mathbf{r}',t')\hat{A}_n(\mathbf{r},t) - \hat{A}_n(\mathbf{r},t)\hat{A}_k(\mathbf{r}',t')\}\rangle. \quad (7.15)$$

For spatially homogeneous media and under stationary conditions the function

$$-\frac{i}{\hbar}\langle\{\hat{A}_k(\mathbf{r}',t')\hat{A}_n(\mathbf{r},t) - \hat{A}_n(\mathbf{r},t)\hat{A}_k(\mathbf{r}',t')\}\rangle \equiv \tilde{\psi}_{nk}(\mathbf{r}-\mathbf{r}',t-t'), \quad (7.16)$$

depends only on the differences $\mathbf{r}-\mathbf{r}'$ and $t-t'$. Therefore, going over in (7.15) to Fourier transforms, we obtain

$$A_m^{\text{ext}} = -\frac{1}{c}\psi_{mn}(\omega,\mathbf{k})j_{\text{ext},n}(\omega,\mathbf{k}), \quad (7.17)$$

where

$$\psi_{mn}(\omega,\mathbf{k}) = \int_0^\infty d\tau \int d^3R\, e^{-i(\mathbf{kR}-\omega\tau)}\tilde{\psi}_{mn}(\mathbf{R},\tau)$$
$$\equiv \int_{-\infty}^\infty d\tau \int d^3R\, e^{-i(\mathbf{kR}-\omega\tau)}\tilde{\psi}_{mn}(\mathbf{R},\tau)\theta(\tau), \quad (7.18)$$

and $\theta(\tau)$ is the Heaviside function equal to zero for $\tau < 0$ and equal to 1 for $\tau > 0$. In consequence, the quantity $\psi_{mn}(\omega,\mathbf{k})$ is just the Fourier transform of the retarded Green's function $\psi_{mn}(\mathbf{r}-\mathbf{r}',t-t')$ given by the expression

$$\psi_{mn}(\mathbf{r}-\mathbf{r}',t-t') = -\frac{i}{\hbar}\langle\{\hat{A}_m(\mathbf{r},t)\hat{A}_n(\mathbf{r}',t') - \hat{A}_n(\mathbf{r}',t')\hat{A}_m(\mathbf{r},t)\}\rangle\theta(t-t'). \quad (7.19)$$

Comparing the relations (7.9) and (7.17) and keeping in mind that $\mathbf{j}_{\text{ext}}(\omega,\mathbf{k})$ is arbitrary, we obtain

[50]The symbol $\langle\hat{c}\rangle$ means the mean value in the Gibbs sense

$$\langle\hat{c}\rangle = \text{Tr}\left\{e^{\frac{F-\hat{H}}{kT}}\hat{c}\right\},$$

where F is the system free energy, and T its temperature.

$$\Delta_{mn}(\omega, \mathbf{k}) = -4\pi \psi_{mn}^{-1}(\omega, \mathbf{k}), \tag{7.20}$$

and hence, from (7.7)

$$\epsilon_{ij}(\omega, \mathbf{k}) = \frac{c^2 k^2}{\omega^2} \eta_{ij} + \frac{4\pi c^2}{\omega^2} \psi_{ij}^{-1}(\omega, \mathbf{k}), \tag{7.21}$$

where

$$\eta_{ij} = \delta_{ij} - \frac{k_j k_j}{k^2}. \tag{7.22}$$

The expression for $\psi_{mn}(\mathbf{r} - \mathbf{r}', t - t')$ (see eqn 7.19) was obtained by using a gauge with vanishing scalar potential. To obtain this function for an arbitrary gauge we first differentiate the function $\psi_{mn}(\mathbf{r} - \mathbf{r}', t - t')$ with respect to t, then with respect to t' and perform the Fourier transformation over the obtained expression. Since

$$\frac{\mathrm{d}\theta(t-t')}{\mathrm{d}t} = \delta(t-t'),$$

and the components of the vector potential taken at the same time commute, the first differentiation of (7.19) yields

$$-\frac{1}{c} \frac{\partial \psi_{mn}(\mathbf{r} - \mathbf{r}', t - t')}{\partial t}$$
$$= -\frac{i}{\hbar} \langle \{\hat{E}_m(\mathbf{r},t) \hat{A}_n(\mathbf{r}',t') - \hat{A}_n(\mathbf{r}',t') \hat{E}_m(\mathbf{r},t)\} \rangle \theta(t-t').$$

Now, differentiating the above expression with respect to t', we obtain

$$-\frac{1}{c^2} \frac{\partial^2 \psi_{mn}(\mathbf{r} - \mathbf{r}', t - t')}{\partial t \, \partial t'}$$
$$= -\frac{i}{c\hbar} \langle \{\hat{E}_m(\mathbf{r},t) \hat{A}_n(\mathbf{r}',t) - \hat{A}_n(\mathbf{r}',t) \hat{E}_m(\mathbf{r},t)\} \rangle \delta(t-t') \tag{7.23}$$
$$- \frac{i}{\hbar} \langle \{\hat{E}_m(\mathbf{r},t) \hat{E}_n(\mathbf{r}',t') - \hat{E}_n(\mathbf{r}',t') \hat{E}_m(\mathbf{r},t)\} \rangle \theta(t-t').$$

Since (cf. (6))

$$\frac{\partial \hat{A}_m(\mathbf{r},t)}{\partial t} \hat{A}_n(\mathbf{r}',t') - \hat{A}_n(\mathbf{r}',t') \frac{\partial \hat{A}_m(\mathbf{r},t)}{\partial t}$$
$$= -4\pi i \hbar c^2 \delta_{mn} \delta(\mathbf{r} - \mathbf{r}'), \tag{7.24}$$

then, performing in (7.23) the Fourier transformation, we obtain

$$\frac{\omega^2}{c^2} \psi_{mn}(\omega, \mathbf{k}) = D_{mn}(\omega, \mathbf{k}), \tag{7.25}$$

$D_{mn}(\omega, \mathbf{k})$ being the Fourier transform of the retarded Green's function $D_{mn}(\mathbf{r} - \mathbf{r}', t - t')$

$$D_{mn}(\mathbf{r} - \mathbf{r}', t - t')$$

$$= -\frac{i}{\hbar}\langle\left\{\hat{E}_m(\mathbf{r},t)\hat{E}_n(\mathbf{r}',t') - \hat{E}_n(\mathbf{r}',t')\hat{E}_m(\mathbf{r},t)\right\}\rangle\theta(t-t') \quad (7.26)$$
$$+ 4\pi\delta_{ij}\delta(\mathbf{r}-\mathbf{r}')\delta(t-t'),$$

which is already gauge invariant. Hence

$$\epsilon_{ij}(\omega,\mathbf{k}) = \frac{c^2 k^2}{\omega^2}\eta_{ij} + 4\pi D_{ij}^{-1}(\omega,\mathbf{k}) \quad (7.27)$$

so that the computation of $\epsilon_{ij}(\omega,\mathbf{k})$ is reduced to establishing the quantities $D_{ij}(\omega,\mathbf{k})$ (Dzyaloshinsky and Pitayevsky (1)).

Polariton states can be used as a complete set of eigenfunctions, which is required when computing the trace [51]

$$\mathrm{Tr}\left\{e^{\frac{F-\hat{H}}{k_B T}}\left[\hat{E}_i(\mathbf{r},t),\hat{E}_j(\mathbf{r}',t')\right]\right\}. \quad (7.28)$$

These states are the solution of the Schrödinger equation with the total Hamiltonian of the considered system. It reads $\hat{H} = \hat{H}_{\mathrm{cr}} + \hat{H}_\perp + \hat{H}_{\mathrm{int}}$, where the instantaneous interaction between charges is totally included into the Hamiltonian of the crystal $\hat{H}_{\mathrm{cr}}$, $\hat{H}_\perp$ is the Hamiltonian describing the free field of transverse photons, and $\hat{H}_{\mathrm{int}}$ describes the interaction of charges with transverse photons. Therefore in deriving $D_{ij}(\mathbf{r}-\mathbf{r}',t-t')$ it is more opportune to use the crystal eigenfunctions obtained when the total Coulomb interaction between charges and the retardation is taken into account, i.e. to use the polariton eigenfunctions of the operator $\hat{H}$, assuming that such functions are known. Note that for the Frenkel exciton spectral region these eigenfunctions and eigenvalues were found in Ch. 4. However, here we can obtain more general relations for the dielectric tensor assuming that states of charge transfer excitons as well as other types of excitations of the crystal (for instance, bound states of excitons and intramolecular phonons) are also taken into account in the calculation of the polariton states.

7.3 Polariton states in the calculation of the dielectric tensor in the region of Frenkel exciton resonances

The total crystal Hamiltonian $\hat{H}$ in the excitonic spectral region using polariton operators can be put in the form (see Section 4.2)

$$\hat{H} = \sum_{\rho,\mathbf{k}}\mathcal{E}_\rho(\mathbf{k})\xi_\rho^\dagger(\mathbf{k})\xi_\rho(\mathbf{k}). \quad (7.29)$$

Using the electric field operator $\hat{\mathbf{E}}$ (see eqn 4.47) in the Heisenberg representation, we obtain

$$\hat{\mathbf{E}}(\mathbf{r},t) = e^{\frac{i}{\hbar}\hat{H}t}\hat{\mathbf{E}}(\mathbf{r})e^{-\frac{i}{\hbar}\hat{H}t}$$

---

[51] The quantity $\left[\hat{A},\hat{B}\right]$ is, by definition, equal to the commutator of two operators $\hat{A}$ and $\hat{B}$, i.e. $\left[\hat{A},\hat{B}\right] \equiv \hat{A}\hat{B} - \hat{B}\hat{A}$.

$$= \sum_{\rho,\mathbf{k}} \mathbf{S}_\rho(\mathbf{k})\xi_\rho(\mathbf{k})e^{i[\mathbf{kr}-\omega_\rho(\mathbf{k})t]} + \text{h.c.}, \qquad (7.30)$$

where

$$\omega_\rho(\mathbf{k}) = \frac{\mathcal{E}_\rho(\mathbf{k})}{\hbar}.$$

In the derivation of the expression (7.29) in Ch. 4 the damping of normal waves was not taken into account. In this approximation the energies $\mathcal{E}_\rho(\mathbf{k})$ are real quantities.

As, however, the frequency of exciton–phonon or exciton–impurity collisions, giving rise to the scattering of waves and dissipation, is small compared to the frequencies $\omega_\rho(\mathbf{k})$, then the expression (7.29) can also be used in the presence of damping, but the energies $\mathcal{E}_\rho(\mathbf{k})$ must be considered as complex

$$\mathcal{E}_\rho(\mathbf{k}) = \mathcal{E}'_\rho(\mathbf{k}) + i\mathcal{E}''_\rho(\mathbf{k}), \qquad \text{where} \qquad |\mathcal{E}''_\rho(\mathbf{k})| \ll |\mathcal{E}_\rho(\mathbf{k})|.$$

The quantities $\mathcal{E}''_\rho(\mathbf{k})$ depend not only on the wavevector but also on the crystal temperature and on the concentration of defects, and so on. Here the complex quantities $\mathcal{E}_\rho(\mathbf{k})$ are supposed to be known.

Substituting (7.30) into the first term on the r.h.s. of (7.26), we obtain

$$-\frac{i}{\hbar}\langle \left[\hat{E}_m(\mathbf{r},t),\hat{E}_n(\mathbf{r}',t')\right]\rangle \theta(t-t')$$

$$= -\frac{i}{\hbar}\sum_{\rho,\mathbf{k}}\Big\{S_{\rho,m}(\mathbf{k})S^*_{\rho,n}(\mathbf{k})e^{i[\mathbf{kR}-\omega_\rho(\mathbf{k})\tau]} \qquad (7.31)$$

$$- S^*_{\rho,m}(\mathbf{k})S_{\rho,n}(\mathbf{k})e^{-i[\mathbf{kR}-\omega_\rho(\mathbf{k})\tau]}\Big\}\theta(\tau),$$

where

$$\mathbf{R} = \mathbf{r} - \mathbf{r}', \qquad \tau = t - t'.$$

Hence

$$D_{mn}(\omega,\mathbf{k}) = \int_{-\infty}^{\infty} d\tau \int d^3R\, e^{-i(\mathbf{KR}-\omega\tau)} D_{mn}(\mathbf{R},\tau) \qquad (7.32)$$

$$= -\frac{V}{\hbar}\sum_\rho \left[\frac{S_{\rho,m}(\mathbf{k})S^*_{\rho,n}(\mathbf{k})}{\omega - \omega_\rho(\mathbf{k}) + i\delta} - \frac{S^*_{\rho,m}(-\mathbf{k})S_{\rho,n}(-\mathbf{k})}{\omega + \omega_\rho(\mathbf{k}) + i\delta}\right] + 4\pi\delta_{mn},$$

where the relation

$$\int_0^\infty e^{i\omega\tau}d\tau = \lim_{\delta\to 0}\int_0^\infty e^{i\omega\tau - \delta\tau}d\tau = i\lim_{\delta\to +0}\frac{1}{\omega + i\delta},$$

has been used; the quantity δ indicates the direction in which we have to go around the pole.

The expression (7.32) gives $D_{mn}(\omega, \mathbf{k})$ in terms of the amplitudes of the electric field in the polariton $\mathbf{S}_\rho(\mathbf{k})$ and polariton energy $\mathcal{E}_\rho(\mathbf{k})$ which were derived in Ch. 4 and thus this expression permits us, by using (7.27), to calculate the tensor $\epsilon_{ij}(\omega, \mathbf{k})$.

For simplicity we consider below the case of an isotropic medium with inversion center, and neglect the absorption. In this case the quantities $\mathbf{S}_\rho(\mathbf{k})$, $\mathcal{E}_\rho(\mathbf{k})$ are real and do not depend on the sign of $\mathbf{k}$, so that

$$D_{mn}(\omega, \mathbf{k}) = \frac{V}{\hbar} \sum_\rho \frac{2 S_{\rho,m}(\mathbf{k}) S_{\rho,n}(\mathbf{k})}{\omega^2 - \omega_\rho^2(\mathbf{k})} \omega_\rho(\mathbf{k}) + 4\pi \delta_{mn}. \qquad (7.33)$$

The relations (7.27) and (7.33) express the dielectric tensor of the crystal in terms of polariton states. Below we apply it for the case of cubic crystals. In cubic crystals the normal waves ρ with small wavevectors are either transverse, or longitudinal. Therefore the product $S_{\rho,m}^\perp(\mathbf{k}) S_{\rho,n}^\parallel(\mathbf{k})$ in this medium always vanishes. In consequence (see also eqn 4.48)

$$D_{mn}(\omega, \mathbf{k}) = \frac{V}{\hbar} \left\{ \sum_{\rho = \rho_\parallel} \frac{2 S_{\rho,m}^\parallel(\mathbf{k}) S_{\rho,n}^\parallel(\mathbf{k})}{\omega^2 - \omega_\rho^2(\mathbf{k})} \omega_\rho(\mathbf{k}) \right.$$
$$\left. + \frac{2 S_{\rho,m}^\perp(\mathbf{k}) S_{\rho,n}^\perp(\mathbf{k})}{\omega^2 - \omega_\rho^2(\mathbf{k})} \omega_\rho(\mathbf{k}) \right\} + 4\pi \delta_{mn}. \qquad (7.34)$$

Since the vectors $\mathbf{S}_\rho^\parallel(\mathbf{k})$ and $\mathbf{k}$ are collinear, the following relation holds

$$S_{\rho,m}^\parallel(\mathbf{k}) S_{\rho,n}^\parallel(\mathbf{k}) = \frac{k_m k_n}{k^2} \left| \mathbf{S}_\rho^\parallel(\mathbf{k}) \right|^2. \qquad (7.35)$$

In contrast to the bands of longitudinal waves, which are nondegenerate, each band of transverse waves is two-fold degenerate, so that to the same frequency $\omega_{\rho_\perp}(\mathbf{k})$ there correspond two vectors $\mathbf{S}_\alpha(\alpha, \mathbf{k})$, $\alpha = 1, 2$. These vectors can be chosen to be perpendicular. In addition, they are of equal length. Thus

$$\sum_{\alpha = 1,2} S_{\rho,m}^\perp(\alpha, \mathbf{k}) S_{\rho,n}^\perp(\alpha, \mathbf{k}) = \left| \mathbf{S}_\rho^\perp(\mathbf{k}) \right|^2 \eta_{mn}, \qquad (7.36)$$

where the tensor η_{mn} is defined by (7.22). Inserting (7.35) and (7.36) into (7.34) and making use of the relation (see eqn 7.22)

$$\delta_{mn} = \frac{k_m k_n}{k^2} + \eta_{mn},$$

we get

$$D_{mn}(\omega, \mathbf{k}) = a(\omega, \mathbf{k}) \frac{k_m k_n}{k^2} + b(\omega, \mathbf{k}) \eta_{mn}, \qquad (7.37)$$

where

$$a(\omega, \mathbf{k}) = 4\pi + \frac{V}{\hbar} \sum_{\rho = \rho_\parallel} \frac{2 \left| \mathbf{S}_\rho^\parallel(\mathbf{k}) \right|^2 \omega_\rho(\mathbf{k})}{\omega^2 - \omega_\rho^2(\mathbf{k})},$$

$$b(\omega, \mathbf{k}) = 4\pi + \frac{V}{\hbar} \sum_{\rho = \rho_\perp} \frac{2 \left| \mathbf{S}_\rho^\perp(\mathbf{k}) \right|^2 \omega_\rho(\mathbf{k})}{\omega \rho^2 - \omega_\rho^2(\mathbf{k})}. \tag{7.38}$$

The values $a(\omega, \mathbf{k})$ and $b(\omega, \mathbf{k})$ have resonances at frequencies corresponding to longitudinal and transverse polaritons. If one takes into account the dissipation, the imaginary parts of the polariton energies would appear in the denominators of these expressions. As longitudinal excitons do not interact with the transverse electric field, the resonances of $a(\omega, \mathbf{k})$ coincide with the frequencies of longitudinal Coulomb excitons.

Since, owing to the orthogonality of the tensors $k_m k_n / k^2$ and η_{mn},

$$D_{mn}^{-1}(\omega, \mathbf{k}) = \frac{1}{a(\omega, \mathbf{k})} \frac{k_m k_n}{k^2} + \frac{1}{b(\omega, \mathbf{k})} \eta_{mn}, \tag{7.39}$$

the tensor $\epsilon_{mn}(\omega, \mathbf{k})$ (see eqn 7.27) is given in the form

$$\epsilon_{mn}(\omega, \mathbf{k}) = \left(\frac{c^2 k^2}{\omega^2} + \frac{4\pi}{b(\omega, \mathbf{k})} \right) \eta_{mn} + \frac{4\pi}{a(\omega, \mathbf{k})} \frac{k_m k_n}{k^2}. \tag{7.40}$$

It can be seen from this relation that, as could be expected, the dielectric tensors for transfer and longitudinal wave are different. The zeros of the value $b(\omega, \mathbf{k})$ determine the resonances of the dielectric tensor for transverse waves and the resonances of the value $a(\omega, \mathbf{k})$ determine the frequencies of longitudinal waves (i.e. zeros of dielectric tensor for longitudinal waves).

A further simplification of the above expression for $\epsilon_{mn}(\omega, \mathbf{k})$ can be obtained if we neglect the spatial dispersion, i.e. letting $\mathbf{k}$ in (7.40) go to zero at a fixed value of ω (in this case $c^2 k^2 / \omega^2 \to 0$ and in this transition one has to use the dependence on the wavevector of the polariton states). It can be shown that by this transition the relation

$$a(\omega, 0) = b(\omega, 0), \tag{7.41}$$

is satisfied, so that, as could be expected,

$$\epsilon_{mn} = \epsilon(\omega) \delta_{mn},$$

where

$$\frac{1}{\epsilon(\omega)} = 1 + \frac{1}{4\pi} \left(\frac{V}{\hbar} \right) \sum_{\rho = \rho_\parallel} \frac{2 \left| \mathbf{S}^\parallel(0) \right|^2 \omega_\rho(0)}{\omega^2 - \omega_\rho^2(0)}. \tag{7.42}$$

In this formula the quantity $1/\epsilon(\omega)$ is given as an expansion over poles corresponding to frequencies of the longitudinal waves and indeed for $\omega = \omega_{\rho_\parallel}(0)$ the quantity $\epsilon(\omega)$, as expected, is equal to zero, in agreement with the phenomenological theory (see eqn 4.67). It is clear also that in the considered approximation the dielectric constant $\epsilon(\omega)$ can be obtained as the inverse value of the function $1/\epsilon(\omega)$ with expansion over poles, corresponding to mechanical transverse excitons. The frequencies of these excitons obey the equation

$$1 + \frac{1}{4\pi}\left(\frac{V}{\hbar}\right)\sum_{\rho=\rho_{\|}} \frac{2\left|\mathbf{S}^{\|}(0)\right|^2 \omega_\rho(0)}{\omega^2 - \omega_\rho^2(0)} = 0. \tag{7.43}$$

This relation establishes the well-known relation from macroscopic electrodynamics between frequencies of longitudinal and transverse dipole-allowed excitations in cubic crystals if spatial dispersion is neglected. This relation, as well as expressions for $1/\epsilon(\omega)$ and $\epsilon(\omega)$, contain only the characteristics of longitudinal waves in crystals which can be considered without taking account of retardation and thus explicitly does not depend on the oscillator strengths of exciton transitions. However, we will show below that taking into account only longitudinal waves even in cubic crystals does not give the possibility of considering correctly the dependence of dissipation of transverse polaritons in the region of resonances. This process depends on the dispersion of transverse polaritons and depends strongly nonlinearly on the exciton transition oscillator strength.

Anisotropic crystals can be considered analogously, including also crystals with spatial dispersion. In these cases the general analysis of the structure of the dielectric tensor is more complicated and has to be done separately for different crystals, taking into account its symmetry.

7.4 The transverse dielectric tensor and dissipation of light waves

7.4.1 *The transverse dielectric tensor*

The consideration of the strong dependence of dissipation of polaritons near exciton resonances will be performed below with the use of transverse dielectric tensor $\epsilon_{\perp,ij}(\omega, \mathbf{k})$. This tensor will be calculated assuming that the excitonic states, with complete account of the Coulomb interaction, are known. Since the derivation of the expression for $\epsilon_{\perp,ij}(\omega, \mathbf{k})$ is given in the monograph (3), see also the articles (4), below we report only its basic points. The semiclassical procedure which will be used is well-known and consists of the determination of the current density, induced by an external electromagnetic field.

As unperturbed states we use the states obtained by accounting for the Coulomb interaction (Coulomb excitons), and the transverse part of the macroscopic field in the medium, i.e. $\mathbf{E}^\perp(\mathbf{r}, t) = -(1/c)\partial \mathbf{A}/\partial t$, where $\mathbf{A}(\mathbf{r}, t)$ is the vector potential of the transverse field, is considered as the perturbing field. Denote by Ψ_{n0} and E_n the wavefunctions and eigenenergies of the unperturbed states, and by $\Psi_n = \Psi_{n0} + \delta \Psi_n$ the corresponding wavefunctions when $\mathbf{E}^\perp$ is taken into account. The operator of interaction of charges with the external field, in the linear approximation, is given as

$$\hat{U} = -\sum_\alpha \frac{e_\alpha}{2m_\alpha c}\left\{\hat{\mathbf{J}}^\alpha \mathbf{A}\left(\mathbf{r}^\alpha, t\right) + \mathbf{A}\left(\mathbf{r}^\alpha, t\right)\hat{\mathbf{J}}^\alpha\right\}, \tag{7.44}$$

where $\mathbf{r}^\alpha$ is the radius-vector of particle α with charge e_α and mass m_α, the summation $\sum_\alpha$ runs over all crystal particles, and

$$\hat{\mathbf{j}}^\alpha = -i\hbar \frac{\partial}{\partial \mathbf{r}^\alpha}.$$

The operator $\hat{U}$ describes not the total perturbation, but only the time-dependent part. Taking

$$\mathbf{A}(\mathbf{r},t) = -\frac{ic}{2\omega}\left[\mathbf{E}_0 e^{i(\mathbf{kr}-\omega t)} - \mathbf{E}_0^* e^{-i(\mathbf{kr}-\omega t)}\right],$$

we find

$$\hat{U} = \hat{F}e^{-i\omega t} + \hat{G}e^{i\omega t}, \qquad \hat{F}(\mathbf{k}) = \hat{G}^\dagger(\mathbf{k}) = -\frac{ic}{2\omega}\hat{\mathbf{M}}(\mathbf{k})\mathbf{E}_0,$$

$$\hat{\mathbf{M}}(\mathbf{k}) = -\sum_\alpha \frac{e_\alpha}{2m_\alpha c}\left\{\hat{\mathbf{j}}^\alpha e^{i\mathbf{kr}^\alpha} + e^{i\mathbf{kr}^\alpha}\hat{\mathbf{j}}^\alpha\right\}. \qquad (7.45)$$

Applying perturbation theory, we obtain for the wavefunctions to first order (see (7), § 40)

$$\Psi_n = \Psi_{n0} - \sum_{m\neq n}\left\{\frac{F_{mn}e^{-i\omega t}}{\hbar(\omega_{mn}-\omega)} + \frac{F_{mn}^* e^{i\omega t}}{\hbar(\omega_{mn}+\omega)}\right\}\Psi_{m0} = \Psi_{n0} + \delta\Psi_n, \qquad (7.46)$$

where

$$F_{mn} = \langle\Psi_{m0}\hat{F}\Psi_{n0}\rangle, \qquad \hbar\omega_{mn} = E_m - E_n \equiv \hbar(\omega_m - \omega_n). \qquad (7.47)$$

The mean value of the current density in the state Ψ_n is given by

$$\mathbf{j}^{(n)}(\mathbf{r},t) = \int \Psi_n^* \hat{\mathbf{j}}(\mathbf{r},t)\Psi_n \, d\tau, \qquad (7.48)$$

where

$$\hat{\mathbf{j}}(\mathbf{r},t) = \sum_\alpha \frac{e_\alpha}{2m_\alpha}\left\{\left[-i\hbar\frac{\partial}{\partial \mathbf{r}^\alpha} - \frac{e_\alpha}{c}\mathbf{A}(\mathbf{r}^\alpha,t)\right]\delta(\mathbf{r}-\mathbf{r}^\alpha)\right.$$
$$\left. + \delta(\mathbf{r}-\mathbf{r}^\alpha)\left[-i\hbar\frac{\partial}{\partial \mathbf{r}^\alpha} - \frac{e_\alpha}{c}\mathbf{A}(\mathbf{r}^\alpha,t)\right]\right\}. \qquad (7.49)$$

The current density Fourier transform which is linear with respect to the field has the form

$$j_i^{(n)}(\omega,\mathbf{k}) = \delta j_i^{(n)}(\omega,\mathbf{k}) = \sigma_{\perp,ij}^{(n)}(\omega,\mathbf{k})E_j^\perp(\mathbf{k},\omega), \qquad (7.50)$$

$\sigma_{\perp,ij}^{(n)}(\omega,\mathbf{k})$ being the electroconductivity tensor. The tensor $\epsilon_{\perp,ij}(\omega,\mathbf{k})$ and the tensor $\sigma_{\perp,ij}(\omega,\mathbf{k})$ are connected by the relation

$$\epsilon^{(n)}_{\perp,ij}(\omega,\mathbf{k}) = \delta_{ij} + i\frac{4\pi}{\omega}\sigma^{(n)}_{\perp,ij}(\omega,\mathbf{k}). \tag{7.51}$$

Hence the expression for $\epsilon_{\perp,ij}(\omega,\mathbf{k})$ has the following form (see (3) for details)

$$\epsilon^{(n)}_{\perp,ij}(\omega,\mathbf{k}) = \left(1 - \frac{4\pi}{\omega^2 V}\sum_\alpha \frac{e_\alpha^2}{m_\alpha}\right)\delta_{ij} \tag{7.52}$$
$$-\frac{4\pi c^2}{\omega^2 V}\sum_\mu \left\{\frac{M^i_{0,\mathbf{k}\mu}(-\mathbf{k})M^j_{\mathbf{k}\mu,0}(\mathbf{k})}{\hbar\omega - E_\mu(-\mathbf{k})} - \frac{M^i_{\mathbf{k}\mu,0}(-\mathbf{k})M^j_{0,\mathbf{k}\mu}(\mathbf{k})}{\hbar\omega + E_\mu(\mathbf{k})}\right\}.$$

In eqn (7.52) $E_\mu(\mathbf{k}) \equiv \hbar\Omega_\mu(\mathbf{k})$ denotes the energy of a Colomb exciton with the wavevector $\mathbf{k}$ in the μth band,

$$M^i_{0,\mathbf{k}\mu}(\mathbf{k}) = \langle \Psi_0, M^i(\mathbf{k})\Psi_{\mathbf{k}\mu}\rangle,$$

where $\Psi_{\mathbf{k}\mu}$ is the wavefunction of the exciton $(\mathbf{k}\mu)$, and V is the crystal volume.

If the spatial dispersion is ignored, then we can put $\mathbf{k} = 0$ in (7.52). We keep in mind that in virtue of (7.45) [52]

$$M^j_{0;0\mu}(\mathbf{s}) = i\frac{W_\mu(\mathbf{s})}{\hbar c}\sqrt{N}\,P_j(0\mu;0), \tag{7.53}$$

where $\mathbf{P}(0\mu;0)$ is the matrix element of the dipole moment of the crystal elementary cell, N is the number of elementary cells per crystal unit volume, and $\mathbf{s} = \mathbf{k}/k$. The quantities $\mathbf{P}(0\mu;0)$ are different from zero only for those excitonic bands whose wavefunctions for $\mathbf{k} = 0$ transform like the components of a polar vector (dipole-allowed excitonic states). Consequently, only these bands contribute to the tensor $\epsilon_{\perp,ij}(\omega,0)$, which has the form

$$\epsilon_{\perp,ij}(\omega,0) = \left(1 - \frac{4\pi}{\omega^2 V}\sum_\alpha \frac{e_\alpha^2}{m_\alpha}\right)\delta_{ij} - \frac{8\pi}{v\hbar\omega^2}\sum_\mu \frac{\Omega^3_\mu(\mathbf{s})P_i(0;0\mu)P_j(0\mu;0)}{\omega^2 - \omega^2_\mu(\mathbf{s})},$$

v being the elementary cell volume. Now we apply the sum rule (4.28) and finally obtain

$$\epsilon_{\perp,ij}(\omega,\mathbf{s}) = \delta_{ij} - \frac{8\pi}{v\hbar}\sum_\mu \frac{\Omega_\mu(\mathbf{s})P_i(0;0\mu)P_j(0\mu;0)}{\omega^2 - \omega^2_\mu(\mathbf{s})}. \tag{7.54}$$

We now demonstrate that the tensor (7.54) yields the same dispersion relation for normal waves as obtained in Ch. 4 in the framework of the microscopic theory including retardation effects.

---

[52] In eqn (7.53) it is assumed (see Ch. 2) that the energy of a Coulomb exciton depends on $\mathbf{s} = \mathbf{k}/k$ when $k \to 0$. The same holds, in general, for the quantity $\mathbf{P}(0\mu;0)$.

Indeed, the refractive index $n(\omega, \mathbf{s})$ for normal electromagnetic waves satisfies the relation (7.54). We can verify, making use of (7.54), that the tensor components $\epsilon_{\perp,11}, \epsilon_{\perp,22}$, and $\epsilon_{\perp,12}$, which enter in (7.54), are given by the expression

$$\epsilon_{\perp,\alpha\beta}(\omega, \mathbf{s}) = \delta_{\alpha\beta} - \frac{8\pi}{v\hbar} \sum_\mu \frac{\Omega_\mu(\mathbf{s})\left(\mathbf{e}^\alpha \mathbf{P}(0; 0\mu)\right)\left(\mathbf{P}(0\mu; 0)\mathbf{e}^\beta\right)}{\omega^2 - \Omega_\mu^2(\mathbf{s})},$$

where $\mathbf{e}^\alpha (\alpha = 1, 2)$ are mutually orthogonal unit vectors, lying in the plane which is perpendicular to the vector $\mathbf{k}$. Hence, solving eqn (4.74) we can see that the roots, i.e. the quantities $n_{1,2}(\omega, \mathbf{s})$, are given by the relations (4.29), previously obtained by means of a microscopic theory. This coincidence of dispersion relations is not surprising. It only demonstrates that the use of the crystal Hamiltonian containing terms up to quadratic order with respect to Bose amplitudes on the one hand, and linear phenomenological Maxwell equations on the other, corresponds to the same type of approximation.

7.4.2 *The dissipation of polaritons in the vicinity of exciton resonances*

In the above discussion we have not considered dissipative processes. To include these processes we can, in the framework of the above applied perturbation theory, aside from the perturbation operator $\hat{U}$, include a time-independent operator which induces transitions between states Ψ_{n0}. In this case the tensor $\epsilon_{\perp,ij}$ is again defined by the expression (7.52), where in the resonant denominators the energy $\hbar\omega$ must be replaced by a complex quantity $\hbar\omega + i\hbar\gamma(\omega, \mathbf{k})$, $\gamma = \gamma' + i\gamma''$, with $|\gamma| \ll \omega$. Knowledge of the function $\gamma(\omega, \mathbf{k})$ is important, for example, in the analysis of a lineshape. Below we take into account the exciton–phonon interaction and, for simplicity, consider only the first order of perturbation theory.

Let us assume that the light frequency ω is in the vicinity of a well-isolated dipole-allowed exciton resonance. In this case the operator of the exciton–phonon interaction, linear with respect to the operator of displacements of molecules from their equilibrium positions, has the form (3.155). It can be shown that in the first order of perturbation theory the real part of $\gamma(\omega, \mathbf{k})$ is given by the following formula (we assume here that the crystal temperature $T = 0$):

$$\gamma'(\omega, \mathbf{k}) = \operatorname{Re} \gamma(\omega, \mathbf{k})$$
$$= \frac{\pi}{\hbar} \sum_{\mathbf{q}r} |F(\mathbf{k} + \mathbf{q}, \mathbf{k}, \mathbf{q}r)|^2 \, \delta\left[\hbar\omega - E_\mu(\mathbf{k} - \mathbf{q}) - \hbar\omega(\mathbf{q}, r)\right]. \quad (7.55)$$

The imaginary part of the quantity $\gamma(\omega, \mathbf{k})$ shifts the resonance position of the tensor $\epsilon_{\perp,ij}(\omega, \mathbf{k})$, whereas the real part causes a damping of light waves and for us only the ω dependence of the real part $\gamma'(\omega, \mathbf{k})$ will be important.

It is evident that an explicit form of this dependence can be given only if we know explicitly the expressions $F(\mathbf{k} + \mathbf{q}, \mathbf{k}, \mathbf{q}r)$ and the energies $E_\mu(\mathbf{k})$ of the Coulomb excitons. However, independently of the concrete form of those functions, it follows from (7.55) that $\gamma'(\omega, \mathbf{k})$ vanishes for $\omega < E_\mu^{\min}/\hbar$, where

$E_\mu^{\min}$ is the minimal value of the excitonic energy in the μth band. Hence, for example, if the minimum of the band is achieved for $\mathbf{k} = 0$, then $E_\mu^{\min} = E_\mu(0)$ and $\gamma'(\omega, \mathbf{k}) = 0$ for $\omega < E_\mu(0)/\hbar$. This conclusion remains valid if we consider high temperatures and also include higher order terms in the exciton–photon interaction operator. In reality, however, the frequency $\omega = E_\mu^{\min}/\hbar$ does not represent the low-frequency limit of excitonic absorption, because the exciton–photon interaction (the retardation) changes the dispersion of the normal wave transforming excitons into polaritons. As we know, the polariton states exists also for $\omega < E_\mu^{\min}/\hbar$ which makes possible the dissipation below the exciton minimal energy (see (9), (10), and (3), § 14.3). To describe more correctly the dissipation of polaritons we have, from the very beginning, to consider polaritons as the zeroth-order states, so that after introducing creation and annihilation operators for polaritons we deal with the crystal energy operator in the form

$$\hat{H} = \sum_{\mathbf{k},\rho} \mathcal{E}_\rho(\mathbf{k}) \xi_\rho^\dagger(\mathbf{k}) \xi_\rho(\mathbf{k}) + \hat{H}_{eL}, \qquad (7.56)$$

where

$$\hat{H}_{eL} = \sum_{\rho\rho'r\mathbf{kq}} \tilde{F}(\mathbf{k}\rho; \mathbf{k}-\mathbf{q}, \rho'; \mathbf{q}r) \xi_\rho^\dagger(\mathbf{k}) \xi_{\rho'}(\mathbf{k}-\mathbf{q}) \left(b_{\mathbf{q}r} + b_{-\mathbf{q}r}^\dagger\right) + \ldots \qquad (7.57)$$

In (7.57) $\mathcal{E}_\rho(\mathbf{k})$ are polariton energies, and $\tilde{F}$ are renormalized constants determining the polariton–phonon interaction. Making use of (7.57) we find by $(\omega, \mathbf{k}) \equiv (\rho, \mathbf{k})$ that

$$\gamma'(\omega, \mathbf{k}) = \frac{\pi}{\hbar} \sum_{\mathbf{q},r,\rho'} \left|\tilde{F}(\mathbf{k}\rho; \mathbf{k}-\mathbf{q}, \rho'; \mathbf{q}r)\right|^2 \delta[\hbar\omega - \mathcal{E}_{\rho'}(\mathbf{k}-\mathbf{q}) - \hbar\omega(\mathbf{q}, r)]. \qquad (7.58)$$

In (7.58) the exciton–photon interaction as well as scattering by phonons, is taken into account. The expression (7.58) is finite for an arbitrary ω. It corresponds to a result which would be obtained by summation of probabilities of photon absorption in all orders of perturbation theory, formulated with respect to the constant of the exciton–photon interaction. If in (7.58) we let the light velocity go to infinity, and thus if we neglect the retardation, the expression of (7.58) attains the form (7.55).

Calculation of the absorption at the long-wavelength edge $\left(\omega < E_\mu^{\min}/\hbar\right)$ was first performed for a simple model of molecular crystal with one molecule per unit cell (10). The authors have shown that the long-wavelength absorption, as could be expected, increases with increasing oscillator strength of the excitonic transition and in some cases such absorption occurs below the frequency $E_\mu(0)/\hbar$ in a frequency region of the order of a few hundreds of cm^{-1}. Further calculations of the long-wavelength absorption have been performed by Demidenko (11). In contrast with (10) he considered crystals of cubic symmetry. However both in (10) and (11) multiphoton processes were not taken into account, although they could be of importance in the theory of long-wavelength absorption, as has been pointed out by Knox (see (12), § 10 B).

Note also that the effect of long-wavelength absorption in the vicinity of exciton resonances is automatically taken into account in expressions for the dielectric tensor which were obtained in Sections 7.2 and 7.3. This approach is formulated in terms of polaritons and consequently the dissipation in the expression for the dielectric tensor has to be taken as the dissipation of polaritons and not as the dissipation of excitons.

The detailed theory of the frequency dependence of $\gamma(\omega)$ for the regions of exciton resonances has been developed by Toyozawa (14), (15). However, in this work the Coulomb excitons and transverse photons were used as zeroth-approximation states, so that the above discussed effect of the long-wavelength edge was not considered.

7.5 Macroscopic and microscopic theories of optical nonlinearity in the region of exciton resonances

The theory of nonlinear optical processes in crystals is based on the phenomenological Maxwell equations, supplemented by nonlinear material equations. The latter connect the electric induction vector $\mathbf{D}(\mathbf{r},t)$ with the electric field vector $\mathbf{E}(\mathbf{r},t)$. In general, the relations are both nonlocal and nonlinear. The property of nonlocality leads to the so-called spatial dispersion of the dielectric tensor. The presence of nonlinearity leads to the interaction between normal electromagnetic waves in crystals, i.e. makes conditions for the appearance of nonlinear optical effects.

The relation between the vectors $\mathbf{D}(\mathbf{r},t)$ and $\mathbf{E}(\mathbf{r},t)$ for a medium uniform in space and time and up to terms, for example, quadratic with respect to the field, can, in analogy to (4.54), be put into the form

$$D_i(\mathbf{r},t) = \int_{-\infty}^{t} dt' \int d^3 r' \hat{\epsilon}_{ij}\left(\mathbf{r}-\mathbf{r}', t-t'\right) E_j\left(\mathbf{r}',t'\right)$$
$$+ \int_{-\infty}^{t} dt' \int_{-\infty}^{t} dt'' \int d^3 r' \int d^3 r'' \hat{\epsilon}_{ij\ell}\left(\mathbf{r}-\mathbf{r}', \mathbf{r}-\mathbf{r}''; t-t', t-t''\right)$$
$$\times E_j\left(\mathbf{r}',t'\right) E_\ell\left(\mathbf{r}'',t''\right). \tag{7.59}$$

Here, as also in (4.54), the causality principle holds, owing to which the induction at time t is determined only by the present field and the field at previous times $t', t'' \leq t$. Applying now the Fourier transformation we obtain the following relation between vectors $\mathbf{D}(\mathbf{r},t)$ and $\mathbf{E}(\mathbf{r},t)$

$$D_i(\mathbf{k},\omega) = \epsilon_{ij}(\omega,\mathbf{k})E_j(\omega,\mathbf{k})$$
$$+ \int \epsilon_{ij\ell}\left(\mathbf{k}\omega,\mathbf{k}'\omega',\mathbf{k}''\omega''\right) E_j\left(\omega',\mathbf{k}'\right) E_\ell\left(\omega'',\mathbf{k}''\right) \delta\left(\omega - \omega' - \omega''\right)$$
$$\times \delta\left(\mathbf{k} - \mathbf{k}'' - \mathbf{k}''\right) d\omega'\, d\omega''\, d^3k'\, d^3k''. \quad (7.60)$$

Here $\epsilon_{ij}(\omega,\mathbf{k})$ is the dielectric tensor, when the spatial dispersion is taken into account (see Section 4.4). The tensor $(1/4\pi)\epsilon_{ij\ell}\left(\mathbf{k}\omega,\mathbf{k}'\omega',\mathbf{k}''\omega''\right)$ is termed the nonlinear polarizability tensor. It can easily be verified that this tensor is invariant upon transposition of its arguments $\mathbf{k}'\omega'$ and $\mathbf{k}''\omega''$.

Including in the expression for the vector $\mathbf{D}$ not only quadratic, but also cubic terms with respect to $\mathbf{E}$, we obtain in the expansion of $\mathbf{D}(\omega,\mathbf{k})$ besides the tensor $\epsilon_{ij\ell}\left(\mathbf{k}\omega,\mathbf{k}'\omega',\mathbf{k}''\omega''\right)$ also the fourth-rank tensor $\epsilon_{ij\ell m}$, etc.

Usually the dependence of the nonlinear polarizability tensors on the wavevectors $\mathbf{k},\mathbf{k}',\mathbf{k}''$, etc. is neglected. In this approximation in crystals with an inversion center the third-rank tensor $\epsilon_{ij\ell}\left(\omega,\omega',\omega''\right)$, which changes sign by inversion, vanishes and nonlinear processes involving three photons are absent. Explicit expressions for the nonlinear polarizability tensors, similar to the tensor $\epsilon_{ij}(\omega,\mathbf{k})$, can be obtained within a microscopic theory. Above we presented various methods for calculation of the tensor $\epsilon_{ij}(\omega,\mathbf{k})$. The important point was that to obtain this tensor it was necessary to establish the crystal polarizability in the linear approximation with respect to the electric field. The nonlinear polarizability tensors can be obtained in a similar way, but now the crystal polarizability must be determined by taking into account higher order terms with respect to the electric field (see (16)).

To calculate the nonlinear responses, for example, the nonlinear polarizabilities $\chi^{(2)},\chi^{(3)}$ of organic solids it is necessary to take into account the multilevel structure of organic molecules and to have information on the electric and in some cases also on the magnetic or quadrupole transition dipole moments between participating molecular states. These data for an isolated molecule give the possibility of calculating the molecular linear as well as nonlinear polarizabilities of isolated molecules which independently can be investigated by measurements of the optical properties of molecules in the gas phase or in transparent solutions. Such calculated molecular polarizabilities $\beta_{ijl}(\omega_1,\omega_2,\omega_3), \gamma_{ijlm}(\omega_1,\omega_2,\omega_3,\omega_4)$ have molecular resonances depending on the type of nonlinear processes. Of course, it is necessary to take into account effects similar to the gas–condensed matter phase shift of levels. However, if the tensors of molecular polarizability can be considered as known, calculation of the nonlinear polarizabilities $\chi^{(2)},\chi^{(3)}$ of organic solids can be performed in the framework of the local field approximation which we used in Ch. 5 and which, as we already mentioned, was introduced by Lorentz (17) in the calculation of the dielectric tensor of solids (18), (19). This approximation helps to account for a local electric field that is different from the macroscopic field which occurs in the macroscopic Maxwell equations. It can be applied to molecular crystals because the electrons in molecular crystals are localized in molecules and, in consequence, the molecules mainly conserve their individuality (here we neglect the contribution of intermolecular charge-transfer excitons). Note also that in this approach only the dipole–dipole intermolecular resonance interaction is taken into account. For example, in cubic crystals the expression for nonlinear polarizabilities obtains additionally the Lorentz local field correction $(\epsilon(\omega_i) + 2)/3$ for each frequency ω_i of a photon participating in nonlinear process, where $\epsilon(\omega)$ is the crystal dielectric constant. It is assumed in such calculations that the contribution of higher multipoles to the local field correction can be neglected. The calculation of the local field for anisotropic organic crystals is described in Ch. 5 and below we assume that the results of such calculations are known.

In a molecular crystal with $\sigma > 1$ molecules in unit cell the local field $\mathbf{E}^\alpha$ is different for different molecules $\alpha = 1, 2, \ldots \sigma$. The relation between the amplitudes of the local and macroscopic fields for a given frequency ω can be expressed by the relation

$$E_i^\alpha(\omega) = L_{ij}^\alpha(\omega) E_j(\omega),$$

where $L_{ij}^\alpha(\omega)$ is a local field tensor for molecule α. Using this definition the macroscopic tensor, for example, of third-order optical nonlinearity $\chi^{(3)}(\omega_1, \omega_2, \omega_3)$, can be given by the relation

$$\chi^3(\omega_1, \omega_2, \omega_3) = \frac{1}{v} \sum_{\alpha=1}^{\alpha=\sigma} \beta_{ij\ell}^\alpha(\omega_1, \omega_2, \omega_3),$$

where v is the volume of the unit cell and

$$\beta_{ij\ell}^\alpha(\omega_1, \omega_2, \omega_3) = L_{ip}^\alpha(\omega_1) L_{jr}^\alpha(\omega_2) L_{\ell s}^\alpha(\omega_3) \beta_{prs}(\omega_1, \omega_2, \omega_3).$$

In nonlinear processes when the frequency of the incident light or at least one of the frequencies arising due to optical nonlinearity, occurs in the vicinity of the molecular or exciton frequency, resonance enhancement of nonlinear processes can be expected. Of course, such enhancement can take place only if the dissipation of light waves is small enough. An excellent review of different resonance processes in organic material can be found in (20).

7.5.1 *On polariton anharmonicity in the nonlinear optical response*

Nonlinear optical effects in crystals can be investigated also microscopically without using the phenomenological Maxwell equations. In the framework of this approach one has to keep, in the Hamiltonian of the crystal (formed, for example, by multilevel molecules), not only quadratic but also terms of third, fourth, etc. order with respect to the Bose amplitudes of excitons and photons. The part of the Hamiltonian which is quadratic with respect to the Bose amplitudes (see Ch. 4), can be diagonalized by making use of new Bose operators $\xi_s(\mathbf{k})$ and $\xi_s^\dagger(\mathbf{k})$ (see eqn 4.16) so that

$$\hat{H}_0 = \sum_{s,\mathbf{k}} \mathcal{E}_s(\mathbf{k}) \xi_s^\dagger(\mathbf{k}) \xi_s(\mathbf{k}), \qquad (7.61)$$

where $\hat{H}_0$ is the crystal Hamiltonian (see eqn 4.2) obtained in the quadratic approximation, and $\xi_s(\mathbf{k})$ and $\xi_s^\dagger(\mathbf{k})$ are polariton Bose creation and annihilation operators. The nonlinear effects arise due to the presence in the total polariton Hamiltonian of terms of third, fourth, etc. order with respect to the operator $\xi_s, \xi_s^\dagger$. All these anharmonicity terms are strongly dependent on the spectra of molecules forming the crystal as well as on the intramolecular interaction and mixing of molecular configurations. These terms can easily be found for a crystal containing two-level molecules and this was first done by Ovander (21). However, the consideration of crystals formed by multilevel molecules is much more complicated but at the same time is more interesting. Such crystals can support nonlinear optical processes resonant not only with the pumping radiation or generated radiation, but with both simultaneously. It is not a surprise that due to such double resonance one usually observes a strong enhancement of optical nonlinearity. The general form of the polariton Hamiltonian with only terms cubic with respect to ξ and $\xi^\dagger$ is

$$\hat{H} = \hat{H}_0 + \hat{H}^{\mathrm{III}}, \tag{7.62}$$

where the operator $\hat{H}_0$ is given in (7.61), and the second term is given by the expression [53]

$$\hat{H}^{\mathrm{III}} = \sum_{s\,s'\,s''} \sum_{\mathbf{k},\mathbf{k}'} W\left(\mathbf{k}s; \mathbf{k}'s'; \mathbf{k}+\mathbf{k}', s''\right) \xi_s(\mathbf{k}) \xi_{s''}^\dagger \left(\mathbf{k}+\mathbf{k}'\right)$$
$$\times \left[\xi_{s'}\left(\mathbf{k}'\right) + \xi_{s'}^\dagger\left(-\mathbf{k}'\right)\right]. \tag{7.63}$$

In the relation (7.63) the coefficients W exhibit symmetry properties which follow from the hermiticity of the operator $\hat{H}^{\mathrm{III}}$. In particular

$$W\left(\mathbf{k}s; \mathbf{k}'s'; \mathbf{k}+\mathbf{k}', s''\right) = W^*\left(\mathbf{k}+\mathbf{k}', s''; -\mathbf{k}'s'; \mathbf{k}s\right). \tag{7.64}$$

The coefficients W determine the probabilities of third-order nonlinear optical processes in an unbounded crystal. An analogous expression can be derived for the coefficients determining the probabilities of fourth-order nonlinear optical processes. As already mentioned the derivation for multilevel molecules is rather complicated and has not yet been obtained. However, the simplicity of the final result, that is the simplicity of the nonlinear Hamiltonian, determines the simplicity of the calculations of nonlinear processes. Note also that a similar polariton approach can be applied for consideration of nonlinear processes in low-dimensional nanostructures (chains, quantum wells). For such structures just resonances of the pumping radiation with polaritons of low-dimensional structure and not with excitons will determine the resonances in the absorption of light as well as resonances in nonlinear processes.

---

[53]In the expression determining $\hat{H}^{\mathrm{III}}$ cubic terms of the form $\xi\xi\xi$ and $\xi^\dagger\xi^\dagger\xi^\dagger$ are omitted. These terms give only small contributions to the probability of third-order nonlinear processes as compared to (7.63).

8

DIELECTRIC TENSOR OF SUPERLATTICES

8.1 Long-period superlattices

Superlattices are of great interest for the study of phenomena which arise at interfaces of media. These artificial layered crystals are systems with "condensed" interfaces since the total area of the interfaces in them is proportional to the volume. In these conditions specific surface and quasi-two-dimensional effects must make an important contribution to the bulk crystal optics. Some of them are analyzed in the present chapter.

First we shall consider the simplest case of a superlattice formed of two types of alternating layers with thicknesses ℓ_1 and ℓ_2 whose electromagnetic properties are characterized by isotropic dielectric constants ϵ_1 and ϵ_2, respectively. This means that ℓ_1 and ℓ_2 are greater than the crystal lattice constants so that macroscopic electrodynamics can be applied to such layers. Such multilayer structures are named here long-period supperlattices. If the electromagnetic wave has wavelength $\lambda = 2\pi/k$ much greater than ℓ_1 and ℓ_2, then the electromagnetic properties of the superlattice can be described by some averaged dielectric tensor whose components can be expressed in terms of ϵ_1, ϵ_2 and ℓ_1, ℓ_2 (1). To this end, we use in each layer the relations

$$\mathbf{D}^{(1)} = \epsilon_1(\omega)\mathbf{E}^{(1)}, \qquad \mathbf{D}^{(2)} = \epsilon_2(\omega)\mathbf{E}^{(2)}, \tag{8.1}$$

and the boundary conditions which read that the tangential component $\mathbf{E}_t$ of $\mathbf{E}$ and the normal component $\mathbf{D}_n$ of $\mathbf{D}$ are continuous at the interfaces (2). Let us take tangential components of (8.1) and average them over a period of the superlattice,

$$\mathbf{D}_t \equiv \frac{1}{\ell_1+\ell_2}\left(\ell_1 \mathbf{D}_t^{(1)} + \ell_2 \mathbf{D}_t^{(2)}\right) = \frac{1}{\ell_1+\ell_2}\left(\ell_1\epsilon_1(\omega)\mathbf{E}_t^{(1)} + \ell_2\epsilon_2(\omega)\mathbf{E}_t^{(2)}\right). \tag{8.2}$$

Since $\mathbf{E}_t$ is continuous at the interfaces and the variation of $\mathbf{E}_t$ inside each layer at $\ell_1, \ell_2 \ll \lambda$ can be neglected, then $\mathbf{E}_t^{(1)} \cong \mathbf{E}_t^{(2)} \cong \mathbf{E}_t$, where $\mathbf{E}_t$ is the value of the electric field strength averaged over the superlattice. Hence we obtain

$$\mathbf{D}_t = \epsilon_\perp(\omega)\bar{\mathbf{E}}_t, \tag{8.3}$$

where

$$\epsilon_\perp = \frac{\ell_1\epsilon_1(\omega) + \ell_2\epsilon_2(\omega)}{\ell_1+\ell_2}. \tag{8.4}$$

In a similar way, averaging of normal components of eqn (8.1) gives

$$\mathbf{E}_n \equiv \frac{1}{\ell_1+\ell_2}\left(\ell_1 \mathbf{E}_n^{(1)} + \ell_2 \mathbf{E}_n^{(2)}\right)$$

$$= \frac{1}{\ell_1 + \ell_2} \left(\frac{\ell_1}{\epsilon_1(\omega)} \mathbf{D}_n^{(1)} + \frac{\ell_2}{\epsilon_2(\omega)} \mathbf{D}_n^{(2)} \right),$$

and since $\mathbf{D}_n$ is continuous at the interfaces and one can neglect its change inside the layers, we obtain

$$\mathbf{E}_n = \frac{1}{\epsilon_\parallel(\omega)} \mathbf{D}_n, \qquad (8.5)$$

where

$$\frac{1}{\epsilon_\parallel(\omega)} = \frac{1}{\ell_1 + \ell_2} \left(\frac{\ell_1}{\epsilon_1(\omega)} + \frac{\ell_2}{\epsilon_2(\omega)} \right). \qquad (8.6)$$

This means that the dielectric tensor of the superlattice consisting of layers with optically isotropic material has the form characteristic of a uniaxial crystal. Assuming that the layers are parallel to the (x, y) plane and normal to the z-axis, then the dielectric tensor of the superlattice has components

$$\epsilon_{ij} = \epsilon_i \delta_{ij}, \qquad (8.7)$$

where

$$\epsilon_x = \epsilon_y = \epsilon_\perp(\omega), \qquad \epsilon_z = \epsilon_\parallel(\omega). \qquad (8.8)$$

If the wavevector of light propagating in the superlattice $\mathbf{k}$ lies in the (x, z) plane, then

$$\mathbf{k} = (k \sin\theta, 0, k \cos\theta), \qquad (8.9)$$

where θ is the angle between $\mathbf{k}$ and the z-axis. Hence, for waves polarized perpendicular to the plane ($\mathbf{k}$, z-axis) (for ordinary electromagnetic waves) having electrical field $\mathbf{E} = (0, E_y, 0)$, the same as in the usual uniaxial crystals, the dispersion relation is

$$k^2 = \frac{\omega^2}{c^2} \epsilon_\perp. \qquad (8.10)$$

Analogously, for extraordinary waves with vector $\mathbf{E}$ lying in the same plane as the wavevector $\mathbf{k}$ and the optical axis of the superlattice, the dispersion relation is

$$\frac{\cos^2\theta}{\epsilon_\perp} + \frac{\sin^2\theta}{\epsilon_\parallel} = \frac{\omega^2}{c^2 k^2} = \frac{1}{n^2}, \qquad (8.11)$$

where n is the refractive index for the extraordinary wave.

8.2 Spatial dispersion in superlattices

8.2.1 *Spatial dispersion in the vicinity of an excitonic resonance* $(\ell_{1,2} > a_B)$

Below we assume that superlattice layers 1 or 2 can be of different natures and can be, for example, organic or inorganic. In the vicinity of excitonic resonances even at $\ell_{1,2} \gg a_B$ (a_B is the Bohr radius of an exciton, of the order of the lattice constant in the case of Frenkel excitons) the nonlocality of the dielectric permeability can be taken into account. Consider, for instance, the frequency region near the excitonic resonance in layer 1. We assume that in this frequency region the dielectric constant of layer 2 is local $\epsilon_2 = \epsilon_2(\omega)$, but we shall make allowance for the nonlocality of the optical response of layer 1.

When nonlocality is not taken into account, the value of the x-component of dielectric displacement vector averaged over the superlattice period is (see also Section 8.1)

$$\bar{D}_x = \frac{1}{\ell_1 + \ell_2} \left(\epsilon_1 E_x^{(1)} \ell_1 + \epsilon_2 E_x^{(2)} \ell_2 \right), \tag{8.12}$$

and the average value of E_z is

$$\bar{E}_z = \frac{1}{\ell_1 + \ell_2} \left(\frac{D_z^{(1)}}{\epsilon_1} \ell_1 + \frac{D_z^{(2)}}{\epsilon_2} \ell_2 \right). \tag{8.13}$$

In the nonlocal case, instead of (8.12) and (8.13) one should write, respectively:

$$\bar{D}_x = \frac{1}{\ell_1 + \ell_2} \left\{ (\epsilon_2 \ell_2 + \epsilon_\infty \ell_1) E_x + 4\pi \int_0^{\ell_1} P_x(z) dz \right\} \tag{8.14}$$

$$\bar{E}_z = \frac{1}{\ell_1 + \ell_2} \left\{ \left(\frac{\ell_2}{\epsilon_2} + \frac{\ell_1}{\epsilon_\infty} \right) D_z - \frac{4\pi}{\epsilon_\infty} \int_0^{\ell_1} P_z(z) dz \right\}, \tag{8.15}$$

where $\mathbf{P}(z)$ is the vector of the excitonic polarization in layer 1; ϵ_∞ is the dielectric constant of layer 1 far from the excitonic resonance.

To proceed we have to express the x-component of the polarization in layer 1 in terms of E_x and its z-component in terms of D_z. We can obtain these relations if we take into account that the displacement vector by definition is $\mathbf{D} = \epsilon_1(\omega, k)\mathbf{E} = \epsilon_\infty \mathbf{E} + 4\pi \mathbf{P}$. For the dielectric tensor in layer 1 we can take the expression

$$\epsilon_1(\omega, K) = \epsilon_\infty \frac{\omega_\parallel^2 - \omega^2 + \hbar\omega \frac{K^2}{m^*}}{\omega_\perp^2 - \omega^2 + \hbar\omega \frac{K^2}{m^*}} \tag{8.16}$$

where m^* is the exciton effective mass, $\omega_{\parallel(\perp)}$ are the frequencies of the longitudinal and transverse excitons, and $\mathbf{K}$ is a 3D wavevector. Thus, we use an effective mass approximation for the exciton and also neglect as unimportant here the difference of the effective masses of transverse and longitudinal excitons. Using all these relations we find the following relation which gives us the possibility to express the x-component of the polarization in layer 1 in terms of E_x:

$$\left(\omega_\perp^2 - \omega^2 + \hbar\omega \frac{K^2}{m^*} \right) \mathbf{P} = \frac{\epsilon_\infty}{4\pi} \left(\omega_\parallel^2 - \omega_\perp^2 \right) \mathbf{E}.$$

Analogously, using $\mathbf{E} = \mathbf{D}/\epsilon_1(\omega, \mathbf{K})$ we can find from the last relation that

$$\left(\omega_\parallel^2 - \omega^2 + \hbar\omega \frac{K^2}{m^*} \right) \mathbf{P} = \frac{1}{4\pi} \left(\omega_\parallel^2 - \omega_\perp^2 \right) \mathbf{D}.$$

Using these relations "the equation of motion" for the excitonic polarization $\mathbf{P}(z)\exp(ikx - i\omega t)$ may be written as (3):

$$\frac{\mathrm{d}^2 P_x}{\mathrm{d}z^2} - Q_\perp^2 P_x = -\frac{q_\perp^2}{4\pi}\epsilon_\infty E_x, \tag{8.17}$$

$$\frac{\mathrm{d}^2 P_z}{\mathrm{d}z^2} - Q_\parallel^2 P_z = -\frac{q_\perp^2}{4\pi} D_z, \tag{8.18}$$

where

$$Q_{\parallel(\perp)}^2 = \frac{m^*}{\hbar\omega_\perp}\left(\omega_{\parallel(\perp)}^2 - \omega^2\right) + k^2,$$

$$q_\perp^2 = \frac{m^*}{\hbar\omega_\perp}\left(\omega_\parallel^2 - \omega_\perp^2\right).$$

As has already been noted, for $\ell_1 \ll \lambda$ one may assume that E_x and D_z are independent of z inside layer 1. With regard to this assumption the solutions of eqns (8.17) and (8.18) have the form:

$$P_x = A_+ e^{Q_\perp z} + A_- e^{-Q_\perp z} + \frac{q_\perp^2 \epsilon_\infty}{4\pi Q_\perp^2} E_x, \tag{8.19}$$

$$P_z = B_+ e^{Q_\parallel z} + B_- e^{-Q_\parallel z} + \frac{q_\perp^2}{4\pi Q_\parallel^2} D_z. \tag{8.20}$$

The constants $A_\pm$ and $B_\pm$ must be found from the boundary conditions at $z=0$ and $z=\ell_1$. These so-called additional boundary conditions (ABC) are determined by the value of the surface energy connected with the excitonic polarization and are strongly dependent on the nature of the contacting layers. If we write this energy in the form

$$W_s = \alpha g \int P^2 \mathrm{d}S \tag{8.21}$$

and include also the term which leads to nonlocality of the equation of motion

$$W = g \int |\boldsymbol{\nabla} P|^2 \, \mathrm{d}V, \tag{8.22}$$

then the variation of (8.21) and (8.22) gives, in addition to the nonlocal terms in the equation of motion, the following boundary conditions:

$$\frac{\partial P}{\partial n} + \alpha P = 0. \tag{8.23}$$

Determining the constants $A_\pm$, $B_\pm$ with the use of (8.23) and substituting (8.20) and (8.20) into (8.14) and (8.15) we obtain

$$\epsilon_\perp(\omega, k) = \frac{1}{\ell_1 + \ell_2} \Bigg\{ \epsilon_2 \ell_2 + \epsilon_1(\omega, k) \ell_1 \\ - \frac{2\alpha q_\perp^2 \epsilon_\infty}{Q_\perp^3 \left(Q_\perp + \alpha \coth\left(\frac{1}{2} Q_\perp \ell_1\right)\right)} \Bigg\}, \qquad (8.24)$$

$$\epsilon_\parallel^{-1}(\omega, k) = \frac{1}{\ell_1 + \ell_2} \Bigg\{ \frac{\ell_2}{\epsilon_2} + \frac{\ell_1}{\epsilon_1(\omega, k)} \\ + \frac{2\alpha q_\perp^2}{\epsilon_\infty Q_\parallel^3 \left(Q_\parallel + \alpha \coth\left(\frac{1}{2} Q_\parallel \ell_1\right)\right)} \Bigg\}, \qquad (8.25)$$

where the nonlocal dielectric function in layer 1, $\epsilon_1(\omega, k)$, is determined by the expression (8.16).

First of all it should be noted that at $\alpha \neq 0$ the resonances in $\epsilon_\perp$ and $\epsilon_\parallel^{-1}$ which correspond to $Q_\perp = 0$ and $Q_\parallel = 0$ disappear since the pole terms $\sim Q_{\perp(\parallel)}^{-2}$ in (8.24) and (8.25) are cancelled. Thus, the nonlocal response of the layer 1 leads to a qualitative change in the behavior of $\epsilon_\perp(\omega)$ and $\epsilon_\parallel(\omega)$. In the case of local dielectric constants $\epsilon_1(\omega)$ and $\epsilon_2(\omega)$ the poles of $\epsilon_\perp(\omega)$ and $\epsilon_\parallel^{-1}(\omega)$ coincide with the poles of $\epsilon_{1,2}(\omega)$ and $\epsilon_{1,2}^{-1}(\omega)$. When the spatial dispersion in layer 1 is taken into account, the poles of $\epsilon_\perp(\omega)$ and $\epsilon_\parallel^{-1}(\omega)$ appear at frequencies which satisfy the equations

$$Q_{\perp(\parallel)}(\omega, k) + \alpha \coth\left(\frac{1}{2} Q_{\perp(\parallel)}(\omega, k) \ell_1\right) = 0 \qquad (8.26)$$

and not at the frequencies $\omega(k) = \omega_{\perp(\parallel)} + \hbar k^2 / 2m^*$ at which there is pole in ϵ_1 and ϵ_1^{-1}.

It may easily be shown that these frequencies correspond to even eigenmodes of polarization in the layer 1 for which $P(\ell_1/2 + z) = P(\ell_1/2 - z)$. Odd modes whose frequencies satisfy the equation

$$Q_{\perp(\parallel)}(\omega, k) + \alpha \tanh\left(\frac{1}{2} Q_{\perp(\parallel)}(\omega, k) \ell_1\right) = 0 \qquad (8.27)$$

are not excited by the field homogeneous in the z-direction easily since the average dipole moment for these modes equals zero. However, with regard to variation of the fields E_x and D_z along the z-axis odd modes are also excited (see (1)).

8.2.2 Spatial dispersion in the vicinity of an excitonic resonance ($\ell_1 < a_B, \ell_2 > a_B$)

Consider now the inverse limiting case $\ell_1 \ll a_B = \epsilon_\infty \hbar^2 / m^* e^2$ typical of inorganic quantum wells. In this case the excitons in layer 1 are two-dimensional and the excitonic polarization corresponding to the lowest level of the dimensional quantization has the form (4):

$$P_i = \ell_1 \chi_{ij}(\omega, k) \varphi(z) \int_0^{\ell_1} E_j^{(1)}(z) \varphi(z)\, dz, \qquad (8.28)$$

where the function $\varphi(z)$ describes the dimensionally quantized motion of electrons and holes (transverse with respect to the film):

$$\varphi(z) = \frac{2}{\ell_1} \sin^2\left(\frac{\pi z}{\ell_1}\right). \qquad (8.29)$$

The resonant part of the susceptibility χ_{ij} for optically isotropic layers includes only two different values $\chi_{xx} = \chi_{yy} = \chi_\perp$ and $\chi_{zz} = \chi_\parallel$ which are proportional to the value (4):

$$\frac{\epsilon_\infty f}{\omega^2(k) - \omega^2}; \qquad f = \left(\frac{e^2}{\hbar \ell_1 \epsilon_\infty}\right)^2 \qquad (8.30)$$

where $\hbar\omega(k)$ is the resonance energy of the 2D Wannier–Mott exciton. Since by (8.29)

$$\int_0^{\ell_1} \varphi(z) dz = 1 \qquad (8.31)$$

and E_x for $q\ell_1 \ll 1$ is weakly dependent on z, it follows from (8.28) and (8.31) that

$$\int_0^{\ell_1} P_x(z) dz = \ell_1 \chi_\perp E_x. \qquad (8.32)$$

Substitution of (8.32) into (8.14) gives

$$\epsilon_\perp = \frac{1}{\ell_1 + \ell_2} \left\{ \epsilon_2 \ell_2 + (\epsilon_\infty + 4\pi \chi_\perp(\omega, k)) \ell_1 \right\}. \qquad (8.33)$$

This expression formally coincides with (8.4), if instead of $\epsilon_1(\omega)$ we assume

$$\epsilon_\infty \left(1 + \frac{4\pi f_\perp}{\omega^2(k) - \omega^2}\right). \qquad (8.34)$$

However, for $\ell_1 \ll a_B$ the strength of the oscillator $f_\perp$ is determined by the formula (8.30) and increases due to squeezing of the exciton state with decreasing ℓ_1. Quite analogously we can calculate the quantity $\epsilon_\parallel$ at $\ell_1 \ll a_B$ (1).

8.2.3 Gyrotropy in superlattices

In concluding we shall consider briefly the calculation of the superlattice linear response with regard to gyrotropy. As before, we assume that the optical properties inside each of the superlattice layers are isotropic. Therefore, for each of the layers with regard to gyrotropy we have (3):

$$\mathbf{D}^{(1,2)} = \epsilon_{1,2}\mathbf{E}^{(1,2)} + \gamma^{(1,2)}\boldsymbol{\nabla} \times \mathbf{E}^{(1,2)}$$
$$+ \frac{1}{2}\left[\boldsymbol{\nabla}\gamma^{(1,2)} \times \mathbf{E}^{(1,2)}\right], \qquad (8.35)$$

where $\gamma^{(1,2)}$ is the gyrotropy constant. The terms with $\boldsymbol{\nabla}\gamma^{(1,2)}$ differ from zero only near the interfaces of the layers. The multiplier $1/2$ in these terms provides the fulfilment of the symmetry principle for kinetic coefficients.

Since $\boldsymbol{\nabla}\gamma^{(1,2)} \times \mathbf{E} \propto \mathbf{H}$ is continuous at the boundaries of the layers, and $\boldsymbol{\nabla}\gamma$ is directed along the z-axis and, therefore, E_z does not contribute to the last term in (8.35), one obtains

$$\mathbf{D} = \hat{\epsilon}\mathbf{E} + \gamma\boldsymbol{\nabla} \times \mathbf{E} + \frac{1}{2}\left[\boldsymbol{\nabla}\gamma \times \mathbf{E}\right],$$

where

$$\gamma = \frac{1}{\ell_1 + \ell_2}\left(\ell_1\gamma^{(1)} + \ell_2\gamma^{(2)}\right). \qquad (8.36)$$

The last term differs from zero only at the superlattice boundary.

In conclusion of this section one may say that the linear optics of superlattices bear rich information on the dynamics of interfaces. Such investigations may give an idea of the nature of the interaction of bulk excitations (excitons) with interfaces which manifests itself in additional boundary conditions (ABC) and determines the value of the constant α. Finally, the study of dispersion laws for polaritons in superlattices with semiconductor layer thicknesses small in comparison with the Bohr radius of exciton permits one to follow variations in the properties of excitons.

8.3 Dielectric tensor of superlattices with anisotropic layers

8.3.1 Dielectric tensor of a superlattice

Formulas (8.4) and (8.6) can be generalized to the case of an anisotropic dielectric constant in the layers (5).

Let us assume that the dielectric properties of the layers are determined by tensors $\epsilon_{ij}^\mu(\omega)$, μ is the layer number, $\mu = 1, 2, \ldots, \sigma$, σ being the number of the layers in a unit cell; the lattice period is $L = \sum_\mu \ell_\mu$ and ℓ_μ is the thickness of the μth layer. Again we assume that the layers are parallel to the (x,y) plane and perpendicular to the z axis. As before, at interfaces of the layers the field components E_1^μ, E_2^μ, D_3^μ are continuous, and therefore it is convenient to express the field components D_1^μ, D_2^μ, E_3^μ in terms of E_1^μ, E_2^μ, D_3^μ. To this end we write the anisotropic generalization of eqn (8.1):

$$D_1^\mu = \epsilon_{11}^\mu E_1^\mu + \epsilon_{12}^\mu E_2^\mu + \epsilon_{13}^\mu E_3^\mu,$$
$$D_2^\mu = \epsilon_{21}^\mu E_1^\mu + \epsilon_{22}^\mu E_2^\mu + \epsilon_{23}^\mu E_3^\mu, \qquad (8.37)$$

$$D_3^\mu = \epsilon_{31}^\mu E_1^\mu + \epsilon_{32}^\mu E_2^\mu + \epsilon_{33}^\mu E_3^\mu,$$

and from the last equation obtain

$$E_3^\mu = \frac{1}{\epsilon_{33}^\mu} D_3^\mu - \frac{\epsilon_{31}^\mu}{\epsilon_{33}^\mu} E_1^\mu - \frac{\epsilon_{32}^\mu}{\epsilon_{33}^\mu} E_2^\mu. \quad (8.38)$$

Then substitution of this equation into the first two equations (8.37) gives

$$D_1^\mu = \left(\epsilon_{11}^\mu - \frac{\epsilon_{13}^\mu \epsilon_{31}^\mu}{\epsilon_{33}^\mu}\right) E_1^\mu + \left(\epsilon_{12}^\mu - \frac{\epsilon_{13}^\mu \epsilon_{32}^\mu}{\epsilon_{33}^\mu}\right) E_2^\mu + \frac{\epsilon_{13}^\mu}{\epsilon_{33}^\mu} D_3^\mu,$$

$$D_2^\mu = \left(\epsilon_{21}^\mu - \frac{\epsilon_{23}^\mu \epsilon_{31}^\mu}{\epsilon_{33}^\mu}\right) E_1^\mu + \left(\epsilon_{22}^\mu - \frac{\epsilon_{23}^\mu \epsilon_{32}^\mu}{\epsilon_{33}^\mu}\right) E_2^\mu + \frac{\epsilon_{23}^\mu}{\epsilon_{33}^\mu} D_3^\mu. \quad (8.39)$$

Equations (8.38) and (8.39) provide a starting point for the averaging procedure similar to one used above. With an accuracy of order $\sim L/\lambda \ll 1$ the fields $\mathbf{E}^\mu$ and $\mathbf{D}^\mu$ inside the layers can be considered as constant and due to the continuity of E_1^μ, E_2^μ, D_3^μ we have for average fields $E_{1,2}$ and D_3

$$E_1 = E_1^\mu,\ E_2^\mu, \qquad D_3 = D_3^\mu, \qquad \mu = 1, 2, \ldots, \sigma. \quad (8.40)$$

Then averaging over the superlattice period according to the rules

$$E_i = \frac{1}{L} \sum_\mu E_i^\mu \ell_\mu, \qquad D_i = \frac{1}{L} \sum_\mu D_i^\mu \ell_\mu, \quad (8.41)$$

yields at once

$$\begin{aligned} D_1 &= a_{11} E_1 + a_{12} E_2 + a_{13} D_3, \\ D_2 &= a_{21} E_1 + a_{22} E_2 + a_{23} D_3, \\ E_3 &= b_{31} E_1 + b_{32} E_2 + b_{33} D_3, \end{aligned} \quad (8.42)$$

where

$$a_{11} = \frac{1}{L} \sum_\mu \left(\epsilon_{11}^\mu - \frac{\epsilon_{13}^\mu \epsilon_{31}^\mu}{\epsilon_{33}^\mu}\right) \ell_\mu \equiv \left\langle \epsilon_{11} - \frac{\epsilon_{13} \epsilon_{31}}{\epsilon_{33}} \right\rangle,$$

$$a_{12} = \left\langle \epsilon_{12} - \frac{\epsilon_{13} \epsilon_{32}}{\epsilon_{33}} \right\rangle, \qquad a_{13} = \left\langle \frac{\epsilon_{13}}{\epsilon_{33}} \right\rangle,$$

$$a_{21} = \left\langle \epsilon_{21} - \frac{\epsilon_{23} \epsilon_{31}}{\epsilon_{33}} \right\rangle, \quad (8.43)$$

$$a_{22} = \left\langle \epsilon_{22} - \frac{\epsilon_{23} \epsilon_{32}}{\epsilon_{33}} \right\rangle, \qquad a_{23} = \left\langle \frac{\epsilon_{23}}{\epsilon_{33}} \right\rangle,$$

$$b_{31} = -\left\langle \frac{\epsilon_{31}}{\epsilon_{33}} \right\rangle, \qquad b_{32} = -\left\langle \frac{\epsilon_{32}}{\epsilon_{33}} \right\rangle, \qquad b_{33} = \left\langle \frac{1}{\epsilon_{33}} \right\rangle.$$

The angle brackets denote an arithmetic average over the superlattice period. It is clear that this average is equivalent to averaging of the fields over the thickness of the superlattice $d \ll \lambda$, where $d \gg L$.

Thus we find that
$$D_i = \epsilon_{ij}^{\text{SL}} E_j, \tag{8.44}$$
where the superlattice dielectric tensor is determined as follows:

$$\epsilon_{11}^{\text{SL}} = \left\langle \epsilon_{11} - \frac{\epsilon_{13}\epsilon_{31}}{\epsilon_{33}} \right\rangle + \left\langle \frac{\epsilon_{13}}{\epsilon_{33}} \right\rangle \left\langle \frac{\epsilon_{31}}{\epsilon_{33}} \right\rangle \left\{ \left\langle \frac{1}{\epsilon_{33}} \right\rangle \right\}^{-1},$$

$$\epsilon_{12}^{\text{SL}} = \left\langle \epsilon_{12} - \frac{\epsilon_{13}\epsilon_{32}}{\epsilon_{33}} \right\rangle + \left\langle \frac{\epsilon_{23}}{\epsilon_{33}} \right\rangle \left\langle \frac{\epsilon_{31}}{\epsilon_{33}} \right\rangle \left\{ \left\langle \frac{1}{\epsilon_{33}} \right\rangle \right\}^{-1},$$

$$\epsilon_{13}^{\text{SL}} = \left\langle \frac{\epsilon_{13}}{\epsilon_{33}} \right\rangle \left\{ \left\langle \frac{1}{\epsilon_{33}} \right\rangle \right\}^{-1}, \tag{8.45}$$

$$\epsilon_{21}^{\text{SL}} = \left\langle \epsilon_{11} - \frac{\epsilon_{23}\epsilon_{31}}{\epsilon_{33}} \right\rangle + \left\langle \frac{\epsilon_{23}}{\epsilon_{33}} \right\rangle \left\langle \frac{\epsilon_{31}}{\epsilon_{33}} \right\rangle \left\{ \left\langle \frac{1}{\epsilon_{33}} \right\rangle \right\}^{-1},$$

$$\epsilon_{22}^{\text{SL}} = \left\langle \epsilon_{22} - \frac{\epsilon_{23}\epsilon_{32}}{\epsilon_{33}} \right\rangle + \left\langle \frac{\epsilon_{23}}{\epsilon_{33}} \right\rangle \left\langle \frac{\epsilon_{32}}{\epsilon_{33}} \right\rangle \left\{ \left\langle \frac{1}{\epsilon_{33}} \right\rangle \right\}^{-1},$$

$$\epsilon_{23}^{\text{SL}} = \left\langle \frac{\epsilon_{23}}{\epsilon_{33}} \right\rangle \left\{ \left\langle \frac{1}{\epsilon_{33}} \right\rangle \right\}^{-1},$$

$$\epsilon_{31}^{\text{SL}} = \left\langle \frac{\epsilon_{31}}{\epsilon_{33}} \right\rangle \left\{ \left\langle \frac{1}{\epsilon_{33}} \right\rangle \right\}^{-1}, \qquad \epsilon_{32}^{\text{SL}} = \left\langle \frac{\epsilon_{32}}{\epsilon_{33}} \right\rangle \left\{ \left\langle \frac{1}{\epsilon_{33}} \right\rangle \right\}^{-1},$$

$$\epsilon_{33}^{\text{SL}} = \left\{ \left\langle \frac{1}{\epsilon_{33}} \right\rangle \right\}^{-1}.$$

It follows from the above relations that the tensor $\epsilon_{ij}^{\text{SL}}$ is symmetric with respect to interchange of the indices i and j only if the tensors ϵ_{ij}^{μ} are symmetric (i.e., in the absence of an external magnetic field). Moreover, these relations allow one to investigate the decrease of symmetry of the tensor $\epsilon_{ij}^{\text{SL}}$ that can occur in some cases.

In the previously studied case with $\epsilon_{ij}^{\mu} = \epsilon^{\mu}\delta_{ij}$, the tensor $\epsilon_{ij}^{\text{SL}}$ acquires the symmetry of an uniaxial crystal (see relations 8.7, 8.8). However, no decrease in symmetry occurs if the layers have a uniaxial crystal symmetry with the optical axis perpendicular to the interfaces of the layers. If in at least one of the layers the optical axis is directed in a different way, the superlattice symmetry decreases. Thus if this axis in one of the layers is parallel to the interface of the layers, the superlattice has an orthorombic symmetry. Equation (8.45) can also be used for the treatment of other cases.

8.3.2 Magnetooptical effects in superlattices

Let us discuss the influence of a static magnetic field assuming that the dielectric constant in the layer is a scalar ϵ_0^{μ}. In the presence of a static magnetic field and up to terms linear in this field the dielectric constant in the layer is

$$\epsilon_{ij}^{\mu} = \epsilon_0^{\mu}\delta_{ij} + i\gamma_{ij\ell}^{\mu}H_\ell^0, \tag{8.46}$$

where $\mathbf{H}^0$ is the static magnetic field, and the pseudotensor $\gamma^\mu_{ij\ell}$ for isotropic material in layer can be written as

$$\gamma^\mu_{ij\ell} = \gamma^\mu e_{ij\ell}, \tag{8.47}$$

$e_{ij\ell}$ being the totally antisymmetric tensor of third rank. It is clear that the dielectric tensor of a superlattice in the approximation linear in the field $\mathbf{H}^0$ must have the form

$$\epsilon^{SL}_{ij} = \epsilon^{SL}_{ii}\delta_{ij} + ie_{ijm}\gamma^{SL}_{m\ell}H^0_\ell, \tag{8.48}$$

where the gyration tensor $\gamma^{SL}_{m\ell}$ is yet unknown. In a uniaxial crystal we have $\gamma_{m\ell} = \gamma_{mm}\delta_{m\ell}$, $\gamma_{11} = \gamma_{22} \equiv \gamma^\perp$, $\gamma_{33} \equiv \gamma^\parallel$, so that the problem is that of finding the quantities $\gamma^\perp$ and $\gamma^\parallel$. In order to find these quantities it is sufficient to consider two particular cases of orientation of the magnetic field $\mathbf{H}^0$:

(1) Assume that the magnetic field $\mathbf{H}^0$ is directed along the z-axis. Then the nondiagonal nonzero elements of the tensor ϵ^μ_{ij} are

$$\epsilon^\mu_{12} = -\epsilon^\mu_{21} = i\gamma^\mu H^0,$$

so that according to (8.45) we obtain

$$\epsilon^{SL}_{ij} = \epsilon^{SL}_{ii}\delta_{ij} + i\gamma^\parallel e_{ij3}H^0,$$

where

$$\epsilon^{SL}_{11} = \epsilon^{SL}_{22} \equiv \epsilon^\perp = \langle\epsilon\rangle, \qquad \epsilon^{SL}_{33} \equiv \epsilon^\parallel = \left\langle\frac{1}{\epsilon}\right\rangle^{-1}, \tag{8.49}$$

and

$$\gamma^\parallel = \frac{1}{L}\sum_\mu \ell_\mu \gamma^\mu \equiv \langle\gamma\rangle. \tag{8.50}$$

(2) If the magnetic field is parallel to the interfaces of layers and is directed, for example, along the y-axis, then the only nondiagonal nonzero components of the tensor are:

$$\epsilon^\mu_{31} = -\epsilon^\mu_{13} = i\gamma^\mu H^0.$$

In this case, according to (8.45), the tensor ϵ^{SL}_{ij} is determined by the relation

$$\epsilon^{SL}_{ij} = \epsilon^{SL}_{ii}\delta_{ij} + i\gamma^\perp e_{ij\ell}H^0, \tag{8.51}$$

where

$$\gamma^\perp = \left\langle\frac{\gamma}{\epsilon}\right\rangle\left\langle\frac{1}{\epsilon}\right\rangle^{-1}. \tag{8.52}$$

It is clear that for an arbitrary orientation

$$\epsilon^{SL}_{ij} = \epsilon^{SL}_{ii}\delta_{ij} + ie_{ij\ell}g_\ell, \tag{8.53}$$

where the gyration vector $\mathbf{g}$ is determined by the relation

$$g_\ell = \gamma_{\ell\ell}H^0_\ell. \tag{8.54}$$

8.3.3 Influence of a static electric field

Int the derivation of (8.45) we assumed that the medium was nonmagnetic. Only in this case can one consider the field $\mathbf{H}^0$ to be independent of μ and thus having the same value in all layers of the superlattice. In the case of an applied electric field such approximation cannot be justified. Therefore, in the presence of a static electric field instead of (8.46) we have to use the following relation

$$\epsilon_{ij}^\mu = \epsilon_0^\mu \delta_{ij} + \chi_{ij\ell}^\mu E_\ell^{0\mu}, \tag{8.55}$$

where $\chi_{ij\ell}$ is a third-rank tensor symmetric with respect to interchange of indices i and j. Let us assume that the superlattice is formed by layers with cubic crystal symmetry, so that the components of the tensor $\chi_{ij\ell}$ can be written in the form

$$\chi_{ij\ell}^\mu = \chi^\mu \left|e_{ij\ell}\right|, \tag{8.56}$$

where χ^μ is some constant and $|e_{ij\ell}|$ is the absolute value of the component of the totally antisymmetric tensor $e_{ij\ell}$. Since nondiagonal components of the tensor ϵ_{ij}^μ are linear in the field $\mathbf{E}^{0\mu}$, according to (8.45) all diagonal components of the tensor ϵ^{SL} become quadratic in the field E^0, corrections which in our approximation have to be neglected. For nondiagonal components of the tensor $\epsilon_{ij}^{\mathrm{SL}}$ linear in the field E^0 we find from (8.45)

$$\begin{aligned}
\epsilon_{12}^{\mathrm{SL}} &= \epsilon_{21}^{\mathrm{SL}} = \langle \chi E_3^0 \rangle, \\
\epsilon_{13}^{\mathrm{SL}} &= \epsilon_{31}^{\mathrm{SL}} = \left\langle \frac{\chi E_2^0}{\epsilon} \right\rangle \left\langle \frac{1}{\epsilon} \right\rangle^{-1}, \\
\epsilon_{23}^{\mathrm{SL}} &= \epsilon_{32}^{\mathrm{SL}} = \left\langle \frac{\chi E_1^0}{\epsilon} \right\rangle \left\langle \frac{1}{\epsilon} \right\rangle^{-1}.
\end{aligned} \tag{8.57}$$

Since for a static electric field we may take for any μ that $E_1^{0\mu} = E_1^0$, $E_2^{0\mu} = E_2^0$, $E_3^{0\mu} = (1/\epsilon^\mu)\epsilon_{33}^{\mathrm{SL}} E_3^0$, where $\mathbf{E}^0$ is the static field averaged over the superlattice period and $\epsilon^\mu \equiv \epsilon_0^\mu(\omega = 0)$, we find that the tensor $\epsilon_{ij}^{\mathrm{SL}}$ can be written as

$$\epsilon_{ij}^{\mathrm{SL}} = \epsilon_{ii}^{\mathrm{SL}} \delta_{ij} + \chi_{ij\ell}^{\mathrm{SL}} E_\ell^0, \tag{8.58}$$

where the tensor $\chi_{ij\ell}^{\mathrm{SL}}$ is determined as follows:

$$\begin{aligned}
\chi_{ij\ell}^{\mathrm{SL}} &= |e_{ij\ell}| \chi_{\ell\ell}^{\mathrm{SL}}, \\
\chi_{11}^{\mathrm{SL}} &= \chi_{22}^{\mathrm{SL}} = \left\langle \frac{\chi}{\epsilon} \right\rangle, \quad \chi_{33}^{\mathrm{SL}} = \left\langle \frac{\chi}{\epsilon} \right\rangle \left\langle \frac{1}{\epsilon} \right\rangle^{-1}.
\end{aligned} \tag{8.59}$$

The above relations can become useful in all situations when the influence of an external static or low-frequency electric field on the dielectric tensor has to be taken into account.

8.4 Optical nonlinearities in organic multilayers

8.4.1 $\chi^{(2)}$ optical nonlinearities in superlattices

Consider now one of the examples which shows how a tensor of nonlinear polarizability $\chi_{ij\ell}$ is formed in a superlattice. Suppose this tensor, the same as in cubic crystals of the group T_d, takes the form

$$\chi_{ij\ell} = \chi |e_{ij\ell}|,$$

where $e_{ij\ell}$ is the totally antisymmetric tensor of the third rank. Consequently, for example, in the presence of incident light with frequency ω,

$$D_x^{(1,2)}(2\omega) = \epsilon_{1,2}(2\omega)E_x^{(1,2)}(2\omega) + 4\pi\chi^{(1,2)}(2\omega;\omega,\omega)E_y^{(1,2)}(\omega)E_z^{(1,2)}(\omega), \quad (8.60)$$

$$D_z^{(1,2)}(2\omega) = \epsilon_{1,2}(2\omega)E_z^{(1,2)}(2\omega) + 4\pi\chi^{(1,2)}(2\omega;\omega,\omega)E_x^{(1,2)}(\omega)E_y^{(1,2)}(\omega), \quad (8.61)$$

so that the average (over the lattice period) value is

$$\bar{D}_x(2\omega) = \epsilon_\perp(2\omega)\bar{E}_x(2\omega) + 4\pi\chi_{123}(2\omega;\omega,\omega)\bar{E}_y(\omega)\bar{E}_z(\omega)$$

where

$$\chi_{123} = \left(\frac{\ell_1 \chi^{(1)}(2\omega;\omega,\omega)}{\epsilon_1(\omega)} + \frac{\ell_2 \chi^{(2)}(2\omega;\omega,\omega)}{\epsilon_2(\omega)}\right) \frac{\epsilon_\parallel(\omega)}{\ell_1+\ell_2}. \quad (8.62)$$

An analogous relation may easily be obtained also for $\bar{D}_y$ which leads to $\chi_{213} = \chi_{123}$. However, the expression for χ_{312} is quite different. One can easily show that

$$\bar{D}_z(2\omega) = \epsilon_\parallel(2\omega)\bar{E}_z(2\omega) + 4\pi\chi_{312}(2\omega;\omega,\omega)\bar{E}_x(\omega)\bar{E}_y(\omega)$$

where

$$\chi_{312} = \frac{1}{\ell_1+\ell_2}\left\{\ell_1\chi^{(1)} + \ell_2\chi^{(2)}\right\} \neq \chi_{123}. \quad (8.63)$$

The difference between χ_{312} and χ_{123} is especially noticeable near the frequencies where $\epsilon_\parallel^{-1}(\omega) \approx 0$, i.e. when

$$\frac{\ell_1}{\epsilon_1(\omega)} \approx -\frac{\ell_2}{\epsilon_2(\omega)}.$$

In this case χ_{123} has a pole whereas χ_{312} has no such peculiarity.

In the case of quadratic nonlinearity we shall consider such a simple example of the interaction of two waves with frequencies ω and 2ω, so that in each layer there are two material relations

$$D_\mu(2\omega) = \epsilon_\mu(2\omega)E_\mu(2\omega) + \chi_\mu^{(2)}\left(E_\mu(\omega)\right)^2,$$

$$D_\mu(\omega) = \epsilon_\mu(\omega)E_\mu(\omega) + \chi_\mu^{(2)} E_\mu(2\omega)E_\mu^*(\omega), \tag{8.64}$$

where all field variables have the same direction. For the case of tangential polarization we obtain

$$\chi_\|^{(2)}(2\omega;\omega,\omega) = \frac{1}{L}\sum_\mu \ell_\mu \chi_\mu^{(2)}(2\omega;\omega,\omega), \tag{8.65}$$

and

$$\chi_\|^{(2)}(\omega;2\omega,-\omega) = \frac{1}{L}\sum_\mu \ell_\mu \chi_\mu^{(2)}(\omega;2\omega,-\omega). \tag{8.66}$$

where all field variables have the same direction. For normal polarization of fields a calculation yields

$$\begin{aligned}D(2\omega) &= \epsilon_\|(2\omega)E(2\omega) + \chi_\|^{(2)}\left(E(\omega)\right)^2, \\ D(\omega) &= \epsilon_\|(\omega)E(\omega) + \chi_\|^{(2)} E(2\omega)E^*(\omega),\end{aligned} \tag{8.67}$$

where

$$\chi_\|^{(2)} = \frac{1}{L}\sum_\mu \ell_\mu \frac{\epsilon_\|(2\omega)}{\epsilon_\mu(2\omega)}\left(\frac{\epsilon_\|(\omega)}{\epsilon_\mu(\omega)}\right)^2 \chi_\mu^{(2)}. \tag{8.68}$$

This formula was obtained in (6) by a different method. In the mentioned paper the averaging of nonlinear equations with optical nonlinearity $\chi^{(3)}$ for superlattices was also performed. The predicted enhancement of the Kerr susceptibility was confirmed experimentally (7).

9

EXCITATIONS IN ORGANIC MULTILAYERS

9.1 Gas–condensed matter shift and the possibility of governing spectra of Frenkel excitons in thin layers

Now we turn to a discussion of the properties of excitons in layered molecular structures (1). First we consider the properties of excitations at the boundary of an anthracene crystal with the vacuum. Of course, this is a particular case of a boundary. However, this case has been investigated in many experiments and therefore can be considered as some kind of experimental confirmation of the approach we will use in our more general discussion. It can be considered now as well-established that the 2D exciton state – the lowest electronic excitation of the outermost monolayer of the anthracene crystal – is blue-shifted by 204 cm^{-1} with respect to the bottom of the exciton band in the bulk. Thus, the frequency of this electronic transition in the first monolayer lies between the bulk value of the exciton frequency and the frequency of excitation in an isolated molecule because the value of this molecular frequency in anthracene is blue-shifted by 2000 cm^{-1} with respect to the frequency of bulk excitation (see Fig. 9.1).

The excited electronic state of the outermost molecular monolayer in anthracene is clearly seen in emission at low temperature. The monolayer next to the surface is blue-shifted by 10 cm^{-1} and the following one by 2 cm^{-1}. The nature of these blue-shifts is now well understood and is related with the absence of neighbors for molecules located near the surface from the vacuum side. Therefore, for these molecules the gas–condensed matter (G–CM) shift of the electronic transition frequency is smaller than the G–CM shift in the bulk (see Fig. 9.1; we assume that the surface corresponds to the $(\mathbf{a}, \mathbf{b})$ plane of the anthracene crystal). For temperatures low compared with the blue-shift, the

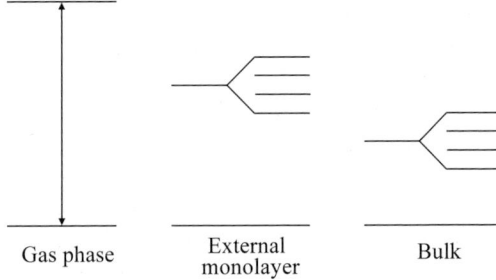

FIG. 9.1. The levels of a molecule near the boundary of a molecular crystal.

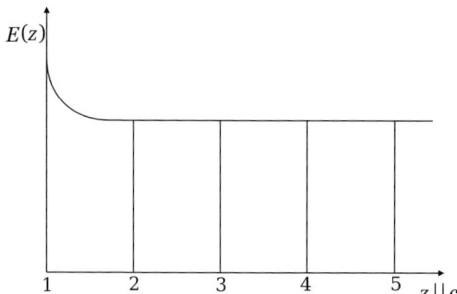

FIG. 9.2. The dependence $E(z)$ at the boundary with the vacuum. Reprinted with permission from Agranovich et al. (6). Copyright Elsevier (2003).

outermost monolayer acts as an isolated 2D structure and as a result is an ideal system for the investigation of the properties of two-dimensional excitons. Such excitons at weak dephasing should exhibit a superrradiant radiative decay of excitons (2). This ultrafast decay of the 2D exciton in anthracene films was indeed observed in picosecond measurements conducted by Aaviksoo et al. (3).

The picosecond time-scale observed in these experiments was the first example of superradiance of two-dimensional Frenkel excitons. Relative quantum yield measurements of the photoluminescence from bulk and the photoluminescence from the outermost monolayer indicate that the decay of excitons in the monolayer is purely radiative with a very small contribution from relaxation to the bulk. Later the same phenomenon for a 2D Wannier–Mott exciton in a semiconductor quantum well was observed by Deveaud et al. (4).

After these remarks let us return again to the first monolayer of the anthracene crystal. Note, first of all, that the width of the exciton band in this crystal for wavevectors directed along the C'-axis (i.e., along the normal to the $(\mathbf{a}, \mathbf{b})$ plane) is very small (~ 5 cm^{-1}) in comparison with bandwidth of excitons with wavevectors parallel to the $(\mathbf{a}, \mathbf{b})$ plane and it is very small in comparison with the blue-shift of the exciton level in the outermost monolayer. For this reason, first of all, the mobility of excitons located in monolayers in the direction towards the surface is rather small. On the other hand, as the energy of he exciton located at the outermost monolayer is larger than the energy of the exciton in the bulk the interaction of bulk excitons with the surface is repulsive. We have here a type of dead layer for the bulk exciton. Qualitatively this picture helps explain why the experimental observation of photoluminescence of excitons located at the outermost monolayer of the anthracene crystal was possible.

To go further we recollect how the gas–condensed matter shift can be calculated. It is known from the theory of molecular (Frenkel) excitons (5) that this shift appears due to the difference between the energies of interaction of the excited molecule (molecular state f) and the unexcited molecule (molecular ground state 0) with all other molecules of the same crystal in the ground state:

$$D^A = \sum_m D^A_{nm}, \qquad (9.1)$$

where

$$D^A_{nm} = \langle \phi_n^{fA} \phi_m^0 | V_{nm} | \phi_n^{fA} \phi_m^0 \rangle - \langle \phi_n^{0A} \phi_m^0 | V_{nm} | \phi_n^{0A} \phi_m^0 \rangle,$$

ϕ_n^r ($r = 0$, f) denotes the wavefunctions of a molecule in the ground (0) and in the excited (f) state, and V_{nm} is the operator of the Coulomb interaction between the molecules n and m determined by the coordinates of their electrons and nuclei $m \equiv \mathbf{m} \equiv (m_1, m_2, m_3)$; usually the values $D_{nm} < 0$ are negative. For molecules n lying on the boundary with the vacuum (to which corresponds, for example, empty space with $m_3 < 0$) the summation over molecules m with $m < 0$ is absent. Therefore, the respective value $D \equiv D_S^A$ (S denotes a surface, A a molecule) can be written as

$$D^A = \sum_{m_3 > 0} D^A_{nm}, \qquad (9.2)$$

where n determines the location of the molecule in outermost monolayer (here $n_3 = 1$, for simplicity, we consider crystals with one molecule per unit cell), $|D_S| < |D|$ and we obtain the case presented in Figs. 9.1 and 9.2. In this particular case the blue-shift value (the difference of energy of the molecule in the outermost layer and in the bulk) is equal to the positive value

$$\Delta D \equiv D_S^{0A} - D = -\sum_{m_3 \leq 0} D^{AA}_{nm} > 0, \qquad (9.3)$$

equal, for instance, for anthracene to the value 204 cm^{-1}.

Now let us consider what will occur if the crystal with some molecules A under consideration has a boundary with another molecular crystal B. Obviously, the value of the G–CM shift will change. The shift in this case can be written as

$$D_S \equiv D_S^{BA} = \sum_{m_3 > 0} D^{AA}_{nm} + \sum_{b_3 < 0} D^{AB}_{nm}, \qquad (9.4)$$

where

$$D^{AB}_{nm} = \langle \phi_n^{fA} \phi_m^{0B} | V_{nm} | \phi_n^{fA} \phi_m^{0B} \rangle - \langle \phi_n^{0A} \phi_m^{0B} | V_{nm} | \phi_n^{0A} \phi_m^{0B} \rangle.$$

Therefore, the shift of the level for molecules in the first monolayer in comparison to the bulk value of the crystal containing molecules A is equal to

$$\Delta D = D_S^{BA} - D^{AA} = \sum_{m_3 < 0} \left(D^{AB}_{nm} - D^{AA}_{nm} \right). \qquad (9.5)$$

Although each of the values D^{AA}_{nm} and D^{AB}_{nm} for the lowest electronic molecular excitation is negative, as a rule, we no longer have the possibility to make a definite statement with respect to the sign of the molecular level shift ΔD:

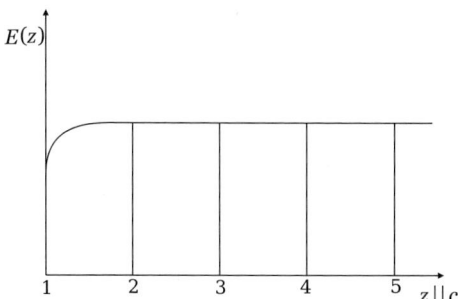

FIG. 9.3. The dependence $E(z)$ in the case $S_B > S_A$. Reprinted with permission from Agranovich et al. (6). Copyright Elsevier (2003).

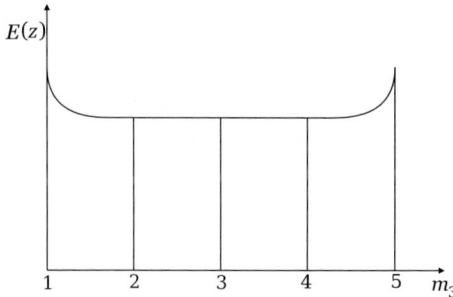

FIG. 9.4. The dependence $E(z)$ for the A-layer in the case $S_A > S_B$. Reprinted with permission from Agranovich et al. (6). Copyright Elsevier (2003).

this shift can be for different pairs of molecules A and B either positive or negative. Thus, if the molecules B possess, in the ground state, relatively small static multipoles (let us denote their value conditionally by S_B) the shift ΔD should be positive, as happens for anthracene at the boundary with the vacuum (by definition $S_B \gg S_0$) (Figs. 9.1, 9.2). If we have the opposite case and the respective multipoles of type B molecules are large enough ($S_B \gg S_A$) the shift ΔD can become negative. In this case instead of the situation expressed in Fig. 9.2 we obtain the attraction of excitons to the surface (Fig. 9.3). Let us recall now that we are interested in organic superlattices and assume that we have under discussion a superlattice of type B A B A B. Then we have the case shown in Fig. 9.4 for the exciton energy $E(z)$ at $S_A > S_B$ and for the case $S_A < S_B$ the dependence $E(z)$ is different (see Fig. 9.5; for definiteness Figs. 9.4 and 9.5 correspond to five lattice constants in the A-layer).

Let us stress one more circumstance now. The molecule and the crystal of anthracene which we have considered above as a well-investigated example, possess an inversion center. The dipole moment operator for such molecules can have

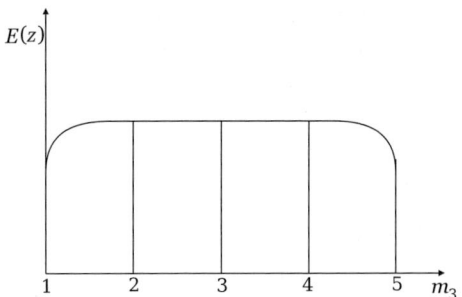

FIG. 9.5. The dependence $E(z)$ for the A-layer in the case $S_B > S_A$. Reprinted with permission from Agranovich et al. (6). Copyright Elsevier (2003).

only nondiagonal nonzero matrix elements so that the quantities D_{nm} depend only on the quadrupole and higher moments. Therefore, the quantities D_{nm} in this case (containing only diagonal matrix elements of the operators of the intermolecular interaction) decrease rapidly with increasing $n - m$. In this cases it proves sufficient to take into account the interaction between nearest neighbors in order to calculate the contribution of static multipoles to the exciton energy. That is why the shift of molecular levels in comparison to the bulk value (the change of the gas–condensed matter shift) for anthracene molecules is large only for the first outermost monolayer in the anthracene crystal. As we mentioned, for the second monolayer this shift is much smaller.

For the crystals formed by molecules without an inversion center the long–range dipole–dipole interaction of static dipoles becomes important. For such crystals the thickness of the surface layer, where the change of the gas–condensed matter shift can be important, increases and the special role of the outermost monolayer becomes weaker.

Let us consider now the conditions under which the dependencies of the type displayed in Figs. 9.4 and 9.5 can be realized. Note, first of all, that in layered crystals which we discuss now, the width of an exciton band Δ for wavevectors directed along the normal to the $(\mathbf{a}, \mathbf{b})$ plane is rather small and in this consideration can be neglected. For this reason Figs. 9.4 and 9.5 determine the minimum energy of an exciton in a monolayer of a crystal as a function of its distance to the surface.[54] However, the large homogeneous width γ of the lowest exciton transition can destroy this picture of spectra. The picture will survive only for such layered crystals like the anthracene crystal in which the inequalities

$$|\Delta D| \gg \gamma, \Delta,$$

are fulfilled. For such crystals the peculiarity of the exciton spectra presented in Figs. 9.4 and 9.5 can have an important influence on optical and electrooptical

[54]We assume here that the exciton band for states with wavevectors directed along the $(\mathbf{a}, \mathbf{b})$ plane has a positive effective mass as in all crystals of the anthracene family.

processes in organic multilayer structures. For instance, in crystals to which Fig. 9.4 corresponds the excitons created as a result of nonresonant absorption of the high-energy photons $\hbar\omega > E(z)$ after subsequent relaxation to the lowest energy states will be concentrated after such energy relaxation in the region of the minimum of the function $E(z)$, that is in one of the middle monolayers (the monolayer $m_3 = 3$ in Fig. 9.4). In the structures of another type to which Fig. 9.5 corresponds the excitons should be concentrated in the interface region of the outermost monolayer. It is clear that this effect, for instance, can be interesting for devices with charge separation at the interfaces.

Let us now consider organic multilayers in which molecules have no center of inversion and where, as a result, they possess static dipole moments in stationary states. If, in this case, the dipole–dipole interaction dominates, the relations for a G–CM shift including this shift at interfaces will change. Consider the case of an interface between crystals formed by molecules A and B. In this case the quantity D_{nm}^{AB} can be written in the following way:

$$D_{nm}^{AB} = \left(p_i^{fA} - p_i^{0A}\right) t_{ij}(n,m) \left(p_j^{0B}\right),$$

where p^{fA} and p^{0A} are static dipole moments of molecule A in the ground (0) and excited (f) state, p^{0B} the static dipole moment of molecule B in the ground state, and $t_{ij}(n,m)$ the tensor determining the interaction of dipoles situated at points n and m. Therefore, according to eqn (9.5) we obtain

$$\Delta D = \sum_{m(m_3<0)} \mu_i^{A(0,f)} t_{ij}(n,m) \mu_j^{B(0)}, \qquad (9.6)$$

where

$$\mu_i^{A(0,f)} = \left(p_i^{fA} - p_i^{0A}\right), \qquad \mu_j^{B(0)} = \left(p_j^{0B} - p_j^{0A}\right).$$

Thus, in this case the shift ΔD is equal to the energy of interaction of the difference dipole moment $\mu^{A(0,f)}$ at the site n with the set of dipoles $\mu^{B(0)}$ located on the lattice sites m with negative component m_3. Note again that in dependence on the directions of $\mu^{A(0,f)}$ and $\mu^{B(0)}$ the shift ΔD can be either positive or negative. Let $\mu^{A(0,f)} \simeq \mu^{B(0)} \simeq 5$ Debye, lattice constant $a \simeq 0.5$ nm. Then the equation for ΔD gives an order-of-magnitude estimate $\Delta D \simeq 1000$ cm^{-1}. This very crude estimation indicates nevertheless that for molecules without inverse symmetry the shifts ΔD will be determined by the dipole–dipole interaction and in some cases can be very large.

9.2 Fermi resonance interface modes in organic superlattices

9.2.1 *Fermi resonance in molecules*

Fermi resonance[55] is a phenomenon which can take place in vibrational or electronic spectra of molecules. We already discussed this phenomenon considering

---

[55] In this section we partly follow (6). The author is deeply grateful to Anatoly Kamchatnov for collaboration.

Fermi resonance of vibrational excitons with polaritons. In this section we discuss the same effect only for molecular vibrations in multilayers. Below we present a few preliminary remarks which can be useful in the following discussion.

Let a molecule have two vibrational modes with frequencies ω_a and ω_b. If the second-order resonance condition $2\omega_a \simeq \omega_b$ is fulfilled, then the $\hbar\omega_b$ transition in the infrared spectrum can split if the interaction is allowed by symmetry of molecule into two lines of comparable intensity. The second line cannot be explained as a result of the interaction of light with the vibrational a mode because the transition with excitation of two $\hbar\omega_a$ quanta is forbidden due to the well-known $n \to n \pm 1$ selection rule for a harmonic oscillator. Fermi explained (7) this experimental observation as a result of a nonlinear resonance interaction of two vibrational modes with each other. Since that time the notion of Fermi resonance has been generalized to processes with participation of different types of quanta (e.g. $\omega_1 + \omega_2 \simeq \omega_3$, $\omega_1 + \omega_2 \simeq \omega_3 - \omega_4$, and so on) and to electronic types of excitations as well. Further generalizations were suggested for Fermi resonance interactions of collective modes in molecular crystals and other macroscopic systems, so that the Fermi resonance phenomenon became part of not only molecular physics but solid-state physics also. In the case of multilayer crystalline organic structures their spectrum is created by the "overlapping" of the spectra of different crystalline compounds and new Fermi resonances arise due to the anharmonicity across the interface, which may be quite interesting from various points of view in discussion of its optical properties. First, we shall consider Fermi resonance theory in molecules from the classical mechanics point of view. Let q_a and q_b be two normal coordinates corresponding to the vibrational degrees of freedom of the molecule with eigenfrequencies ω_a and ω_b (we neglect all other degrees of freedom of the molecule). Then the Lagrangian of the vibrational motion can be written as

$$L = \frac{M_a \dot{q}_a^2}{2} + \frac{M_b \dot{q}_b^2}{2} - U(q_a, q_b), \tag{9.7}$$

where the overdot stands for the derivative with respect to time t, M_a and M_b are "mass" coefficients, and the potential energy $U(q_a, q_b)$ can be expanded into a series with respect to powers of the variables q_a and q_b,

$$U(q_a, q_b) = U_2(q_a, q_b) + U_3(q_a, q_b) + \ldots \tag{9.8}$$

Since q_a and q_b are normal coordinates, in the harmonic approximation the potential energy reads:

$$U_2(q_a, q_b) = \frac{M_a \omega_a^2}{2} q_a^2 + \frac{M_b \omega_b^2}{2} q_b^2. \tag{9.9}$$

The next nonlinear term in (9.8) corresponds to the third powers of the q's:

$$U_3(q_a, q_b) = \sum_{m+n=3} \alpha_{mn} q_a^m q_b^n \tag{9.10}$$

(the sum over cubic terms $\alpha_{03}q_b^3 + \alpha_{12}q_aq_b^2 + \ldots$). It is clear that not all terms in the sum (9.10) are equally important under the Fermi resonance condition

$$2\omega_a \simeq \omega_b. \tag{9.11}$$

Indeed, in the equations of motion

$$M_a\left(\ddot{q}_a + \omega_a^2 q_a\right) = -\frac{\partial U_3}{\partial q_a} = -3\alpha_{30}q_a^2 - 2\alpha_{21}q_aq_b - \alpha_{12}q_b^2,$$

$$M_b\left(\ddot{q}_b + \omega_b^2 q_b\right) = -\frac{\partial U_3}{\partial q_b} = -\alpha_{21}q_a^2 - 2\alpha_{12}q_aq_b - 3\alpha_{03}q_b^2, \tag{9.12}$$

the different terms on the right-hand side have a different physical sense. For example, the terms

$$-3\alpha_{30}q_a^2, \quad \text{and} \quad -3\alpha_{03}q_b^2$$

describe a weak nonlinearity of separate eigenmodes and are not responsible for their interaction at all; hence we can remove them from the sum (9.10). The other terms differ from each other by their time dependence. In the harmonic approximation we have

$$q_a = q_a^0 \cos\left(\omega_a t + \phi_a\right), \quad q_b = q_b^0 \cos\left(\omega_b t + \phi_b\right), \tag{9.13}$$

and the potential (9.10) is a small perturbation leading to a slow variation of q_a^0 and q_b^0. Substitution of (9.13) into the right-hand sides of (9.12) leads to resonant and nonresonant terms. For example, in the first equation (9.12) the left-hand side oscillates with approximate harmonic frequency ω_a whereas the term

$$q_b^2 \propto \cos^2\omega_b t = \frac{1 + \cos 2\omega_b t}{2}$$

does not contain such a harmonic and describes a nonresonant interaction of modes a_a and q_b. Hence we must hold in this equation only the term

$$q_a q_b \propto \cos\omega_a t \cos\omega_b t = \frac{\cos\left(\omega_b + \omega_a\right)t + \cos\left(\omega_b - \omega_a\right)t}{2},$$

which according to the condition (9.11), contains the resonant "force" oscillating with frequency $\omega_b - \omega_a \simeq \omega_a$. Analogously, in the second equation (9.12), we must hold only the term $\alpha_{21}q_a^2$ which contains the resonant "force" oscillating approximately with frequency $2\omega_a \simeq \omega_b$ of the mode q_b. Both these terms arise from the potential energy $\alpha_{21}q_a^2 q_b$, which includes in addition some nonresonant terms which must be removed in the so-called *rotating wave approximation* we use here. To this end, it is convenient to introduce into the classical equations of motion

$$M_a\left(\ddot{q}_a + \omega_a^2 q_a\right) + 2\alpha_{21}q_aq_b = 0,$$

$$M_b \left(\ddot{q}_b + \omega_b^2 q_b \right) + \alpha_{21} q_a^2 = 0, \tag{9.14}$$

where we took into account the relevant $\alpha_{21} q_a^2 q_b$ interaction term only, the complex variables

$$A = \sqrt{\frac{M_a}{2\omega_a}} \left(\omega_a q_a + i\dot{q}_a \right), \qquad B = \sqrt{\frac{M_b}{2\omega_b}} \left(\omega_b q_b + i\dot{q}_b \right), \tag{9.15}$$

and their complex conjugate A^* and B^*, so that

$$q_a = \frac{1}{\sqrt{2M_a\omega_a}} (A + A^*), \qquad q_b = \frac{1}{\sqrt{2M_b\omega_b}} (B + B^*), \tag{9.16}$$

$$\dot{q}_a = i\sqrt{\frac{\omega_a}{2M_a}} (A^* - A), \qquad \dot{q}_b = i\sqrt{\frac{\omega_b}{2M_b}} (B^* - B). \tag{9.17}$$

Simple transformation yields the following equations for complex amplitudes A and B:

$$\dot{A} = -i\omega_a A - 2 \frac{\alpha_{21} i}{2M_a \omega_a \sqrt{2M_b \omega_b}} (A + A^*)(B + B^*), \tag{9.18}$$

$$\dot{B} = -i\omega_b B - \frac{\alpha_{21} i}{2M_a \omega_a \sqrt{2M_b \omega_b}} (A + A^*)^2. \tag{9.19}$$

In the harmonic approximation the last term in these equations can be omitted, so in this case we get

$$A(t) = A(0) \exp(-i\omega_a t), \qquad B(t) = B(0) \exp(-i\omega_b t). \tag{9.20}$$

As we stressed above, the rotating wave approximation consists of taking into account only the terms varying with time at approximately the same frequency, that is, according to the resonance condition (9.11). Then we neglect in eqn (9.18) all interaction terms except $A^*B \propto \exp[-i(\omega_b - \omega_a)t]$ and in eqn (9.19) all interaction terms except $A^2 \propto \exp(-2i\omega_a t)$. As a result, we arrive at the following simple system of equations

$$i\dot{A} = \omega_a A + 2\Gamma A^* B, \qquad i\dot{B} = \omega_b B + \Gamma A^2, \tag{9.21}$$

where the notation for the interaction constant

$$\Gamma = \frac{\alpha_{21}}{2M_a \omega_a \sqrt{2M_b \omega_b}} \tag{9.22}$$

is introduced.

Equations (9.21) arise in various physical contexts. In particular, they describe the process of second harmonic generation in nonlinear optics (see, e.g. (8)). In fact, they can be applied to any process in which two classical oscillating modes (waves) transform into one and other under the resonance condition $2\omega_a \simeq \omega_b$.

It is important that eqn (9.21) can be derived from the Lagrangian

$$L = \frac{i}{2}\left(\dot{A}^*A - A^*\dot{A} + \dot{B}^*B - B^*\dot{B}\right) + \omega_a A^*A + \omega + bB^*B$$
$$+ \Gamma\left(A^{*2}B + A^2B^*\right), \tag{9.23}$$

where A, A^*, B, B^* are considered as independent variables. Then the Lagrange equations

$$\frac{d}{dt}\frac{\partial L}{\partial \dot{A}^*} - \frac{\partial L}{\partial A^*} = 0, \qquad \frac{d}{dt}\frac{\partial L}{\partial \dot{B}^*} - \frac{\partial L}{\partial B^*} = 0$$

reproduce eqns (9.21). The classical Hamiltonian has the form

$$H = \omega_a A^*A + \omega_b B^*B + \Gamma\left(A^{*2}B + A^2B^*\right), \tag{9.24}$$

where pairs of canonically conjugate variables are A, A^* and B, B^*, respectively. Then the Hamiltonian form of the equations of motion is as follows:

$$i\dot{A} = \frac{\partial H}{\partial A^*}, \qquad i\dot{A}^* = -\frac{\partial H}{\partial A},$$
$$i\dot{B} = \frac{\partial H}{\partial B^*}, \qquad i\dot{B}^* = -\frac{\partial H}{\partial B}. \tag{9.25}$$

Having the Hamiltonian treatment, it is easy to proceed to a quantum mechanical formulation of the Fermi resonance problem. The classical variables (9.15) and their complex conjugates correspond to the "annihilation" and "creation" operators of a quantum oscillator:

$$A \to \hbar^{1/2}\hat{a}, \qquad A^* \to \hbar^{1/2}\hat{a}^\dagger, \qquad B \to \hbar^{1/2}\hat{b}, \qquad B^* \to \hbar^{1/2}\hat{b}^\dagger, \tag{9.26}$$

where the "dagger" denotes hermitian conjugation and the operators $\hat{a}$, $\hat{a}^\dagger$, and $\hat{b}$, $\hat{b}^\dagger$ obey the usual commutation relations

$$\left[\hat{a},\hat{a}^\dagger\right] = 1, \qquad \left[\hat{b},\hat{b}^\dagger\right] = 1, \qquad \left[\hat{a},\hat{b}\right] = \left[\hat{a}^\dagger,\hat{b}^\dagger\right] = 0. \tag{9.27}$$

The classical Hamiltonian (9.24) converts into a quantum mechanical one in the following way

$$\hat{H} = \hbar\omega_a \hat{a}^\dagger\hat{a} + \hbar\omega_b \hat{b}^\dagger\hat{b} + \hbar^{3/2}\Gamma\left(\hat{a}^2\hat{b}^\dagger + \hat{a}^{\dagger 2}\hat{b}\right), \tag{9.28}$$

where the product of noncommuting variables is replaced by their mean value:

$$A^*A \to \frac{\hbar}{2}\left(\hat{a}^\dagger\hat{a} + \hat{a}\hat{a}^\dagger\right) = \hbar\left(\hat{a}^\dagger\hat{a} + \frac{1}{2}\right),$$

and zero oscillation terms are dropped.

Now we are ready to discuss the energy splitting of molecular Fermi resonance states. From the classical point of view, the stationary states correspond to a purely periodic dependence of the amplitudes A and B on time:

$$A = A_0 \exp\left(-\frac{i\Omega}{2}t\right), \qquad B = B_0 \exp\left(-i\Omega t\right). \tag{9.29}$$

So we seek the solution of eqn (9.21) in the form (9.29) which leads to the system

$$\left(\omega_a - \frac{\Omega}{2}\right) A_0 + 2\Gamma A_0^* B_0 = 0,$$
$$(\omega_b - \Omega) B_0 + \Gamma A_0^2 = 0. \tag{9.30}$$

We multiply the first equation by A_0, introduce the "intensity" of vibration

$$I = |A_0|^2, \tag{9.31}$$

analogous to the number of quanta, and eliminate A_0^2 and B_0 from this system, which gives the equation

$$\left(\omega_a - \frac{\Omega}{2}\right)(\omega_b - \Omega) = 2\Gamma^2 I \tag{9.32}$$

for calculation of the frequency Ω. This equation has two roots

$$\Omega_{1,2} = \omega_a + \frac{\omega_b}{2} \pm \sqrt{\left(\omega_a - \frac{\omega_b}{2}\right)^2 + 4\Gamma^2 I}. \tag{9.33}$$

We see that our classical system has two "eigenfrequencies" depending on the intensity I of the A-mode. In quantum terms, the intensity I is proportional to the number of a-quanta, $n_a = \hat{a}^\dagger \hat{a} = I/\hbar$, $n_b = \hat{b}^\dagger \hat{b} = I/(2\hbar)$, which in the classical limit are much greater than unity, $n_a = 2n_b \gg 1$. However, in the usual infrared experiments the lines observed correspond to the excitation of the state with $n_a = 2$, $n_b = 1$. Let us try to apply our classical formula (9.33) to this quantum region, i.e. we substitute $I = A^*A = 2\hbar$ (two a-quanta). As a result we find that the energies of these two states are equal to

$$E_{1,2} = \hbar\Omega_{1,2} = \hbar\omega_a + \frac{\hbar\omega_b}{2} \pm \sqrt{\left(\hbar\omega_a - \frac{\hbar\omega_b}{2}\right)^2 + 8\hbar^3\Gamma^2}. \tag{9.34}$$

It is interesting that this semiclassical formula almost reproduces the exact quantum result. Indeed, the interaction term in the quantum Hamiltonian (9.28) couples the states

$$|\psi_1\rangle = |2_a 0_b\rangle \quad \text{and} \quad |\psi_2\rangle = |0_a 1_b\rangle, \tag{9.35}$$

where $|n_a m_b\rangle = |n\rangle_a |m\rangle_b$ is the oscillators' state with n a-quanta and m b-quanta. Using the well-known relations

$$\hat{a}|n\rangle_a = \sqrt{n}|n-1\rangle_a, \qquad \hat{a}^\dagger |n\rangle_a = \sqrt{n+1}|n+1\rangle_a,$$

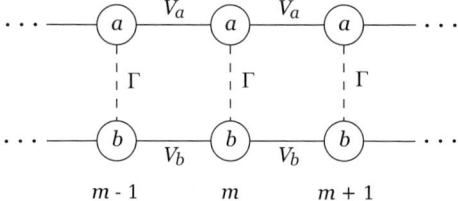

FIG. 9.6. Two-layer one-dimensional model of a crystal with one molecule in an elementary cell of each layer. Vibrational modes in molecules a and b satisfy the Fermi resonance condition $2\omega_a \simeq \omega_b$ and interact with excitations in neighboring cells with constants V_a and V_b, respectively. The constant Γ describes the nonlinear interaction between molecules in adjacent layers. Reprinted with permission from Agranovich et al. (6). Copyright Elsevier (2003).

$$\hat{b}|m\rangle_b = \sqrt{m}|m-1\rangle_b, \qquad \hat{b}^\dagger|m\rangle_b = \sqrt{m+1}|m+1\rangle_b, \qquad (9.36)$$

we obtain the matrix elements of the Hamiltonian in the basis (9.35):

$$H = \begin{pmatrix} 2\hbar\omega_a & \sqrt{2}\hbar^{3/2}\Gamma \\ \sqrt{2}\hbar^{3/2}\Gamma & \hbar\omega_a \end{pmatrix}. \qquad (9.37)$$

The "eigenstate" energies are determined by the secular equation

$$(2\hbar\omega_a - E)(\hbar\omega_b - E) = 2\hbar^3\Gamma^2$$

and differ by the factor 2 from the classical result (9.32) with $I = 2\hbar$. Thus, we see that such a molecule has two energy levels $E_{1,2}$ each connected with the ground state by a one-quantum transition due to the large b-component in both eigenfunctions. This leads to the observed splitting of the infrared and Raman spectral lines. Just such a splitting of the Raman spectra was explained by Fermi in his pioneering article (7).

9.2.2 Fermi-resonance wave in a two-layer system

The next natural step is the discussion of Fermi resonance effects in molecular crystals. Let molecules having Fermi resonance between intramolecular vibrations form a molecular crystal due to weak (van der Waals) forces. Then the individual molecular vibrational excitations discussed above become coupled to each other and form collective Fermi resonance bands. We shall consider here a simple two-layer 1D model with intermolecular interaction only between nearest neighbors (see Fig. 9.6).

The quantum-mechanical Hamiltonian can be written in the form

$$\hat{H} = \sum_m \left[\hbar\omega_a \hat{a}_m^\dagger \hat{a}_m + \hbar\omega_b \hat{b}_m^\dagger \hat{b}_m + \hbar^{3/2}\Gamma \left(\hat{a}_m^2 \hat{b}_m^\dagger + \hat{a}_m^{\dagger 2}\hat{b}_m \right) \right]$$

$$+ \sum_m \left[\hbar V_a \left(\hat{a}^\dagger_{m+1} \hat{a}_m + \hat{a}^\dagger_m \hat{a}_{m+1} \right) + \hbar V_b \left(\hat{b}^\dagger_{m+1} \hat{b}_m + \hat{b}^\dagger_m \hat{b}_{m+1} \right) \right],$$
(9.38)

where m determines the location of molecule in the lattice, the first sum represents molecular Hamiltonians in each site (see eqn 9.28), and the second sum corresponds to the intermolecular interaction of vibrations; the term $\hbar V_a \hat{a}^\dagger_{m+1} \hat{a}_m$ describes the transition of one a-quantum from the site m to the site $m+1$, with analogous interpretations for the other terms. In the classical approximation we obtain the Hamiltonian

$$H = \sum_m \left[\omega_a A^*_m A_m + \omega_b B^*_m B_m + \Gamma \left(A^2_m B^*_m + A^{*2}_m B_m \right) \right]$$
$$+ \sum_m \left[V_a \left(A^*_{m+1} A_m + A^*_m A_{m+1} \right) \right. \quad (9.39)$$
$$\left. + V_b \left(B^*_{m+1} B_m + B^*_m B_{m+1} \right) \right],$$

and the following equations of motion for complex amplitudes A_m and B_m:

$$i \frac{\partial A_m}{\partial t} = \frac{\partial H}{\partial A^*_m} = \omega_a A_m + V_a (A_{m-1} + A_{m+1}) + 2\Gamma A^*_m B_m,$$
$$i \frac{\partial B_m}{\partial t} = \frac{\partial H}{\partial B^*_m} = \omega_b B_m + V_b (B_{m-1} + B_{m+1}) + \Gamma A^2_m. \quad (9.40)$$

Let us look for the solution in the form of a plane wave

$$A_m = A \exp\left[-i \left(\Omega t - \frac{Km}{2} \right) \right], \quad B_m = B \exp\left[-i \left(\Omega t - \frac{Km}{2} \right) \right], \quad (9.41)$$

(we assume that the lattice constant is equal to unity). Then the infinite system reduces to a simple system of two algebraic equations

$$\left(\omega_a - \frac{\Omega}{2} + 2V_a \cos\left(\frac{K}{2} \right) \right) A + 2\Gamma A^* B = 0,$$
$$(\omega_b - \Omega + 2V_b \cos K) B + \Gamma A^2 = 0, \quad (9.42)$$

which actually coincides with the system (9.30). If there were no Fermi resonance interaction ($\Gamma = 0$), then we would have two linear modes with the well-known dispersion laws

$$\Omega_1(K) = 2\omega_a + 4V_a \cos\left(\frac{K}{2} \right), \quad \Omega_2(K) = \omega_b + 2V_b \cos K. \quad (9.43)$$

Fermi resonance coupling between molecular vibrations leads to the interaction of these linear modes with each other, which gives rise to mixed waves. To obtain their dispersion law, we again introduce the intensity

$$I = |A|^2$$

and reduce the system (9.42) to the equation

$$\left(\omega_a - \frac{\Omega}{2} + 2V_a \cos\left(\frac{K}{2}\right)\right)(\omega_b - \Omega + 2V_b \cos K) = 2\Gamma^2 I \qquad (9.44)$$

with solutions

$$\Omega_{1,2}(K) = \omega_a + \frac{\omega_b}{2} + 2V_a \cos\left(\frac{K}{2}\right) + V_b \cos K$$

$$\pm \left[\left(\omega_a - \frac{\omega_b}{2} + 2V_a \cos\left(\frac{K}{2}\right) - V_b \cos K\right)^2 + 8\Gamma^2 I\right]^{1/2}.$$

These expressions define the dispersion laws of normal modes arising from linear plane waves due to the nonlinear Fermi resonance interaction. It is important that the nonlinearity leads to the dependence of the dispersion laws on the intensity I of vibrations. Such a dependence gives rise to soliton solutions discussed in the following sections.

The relations (9.44) can be considered as a good enough approximation only in the limit of large intensity

$$I \gg \hbar. \qquad (9.45)$$

Nevertheless, as we saw in the preceding section, semiclassical formulas give exact enough results even in the quantum region $I \geq \hbar$. More exact relations can be easily derived by means of a quantum-mechanical treatment (9; 10).

9.2.3 *Fermi resonance interface waves*

Now we shall consider the case when one interface separates two simple cubic 2D crystals composed of different molecules of types a and b interacting across the interface. In two dimensions with a linear interface (the 3D generalization is straightforward) the equations of motion for the amplitudes A_{mn} and B_{mn} read (compare with eqn 9.40)

$$\vdots \qquad \vdots$$
$$n = 2: \quad i\dot{A}_{2,m} = \omega_a A_{2,m} + V_a(A_{2,m-1} + A_{2,m+1} + A_{1,m} + A_{3,m}),$$
$$n = 1: \quad i\dot{A}_{1,m} = \omega_a A_{1,m} + V_a(A_{1,m-1} + A_{1,m+1} + A_{2,m}) + 2\Gamma A_{1,m}^* B_{0,m},$$
$$n = 0: \quad i\dot{B}_{0,m} = \omega_b B_{0,m} + V_b(B_{0,m-1} + B_{0,m+1} + B_{-1,m}) + \Gamma A_{1,m}^2,$$
$$n = -1: \quad i\dot{B}_{-1,m} = \omega_b B_{-1,m} + V_b(B_{-1,m-1} + B_{-1,m+1} + B_{-2,m} + B_{0,m}),$$
$$\vdots \qquad \vdots \qquad (9.46)$$

where the equations with $n > 2$ and $n < -1$ have the same structure as at $n = 2$ and $n = -1$, respectively. They describe the propagation of excitations in bulk a and b crystals above and below the interface. New interface ("surface") modes, including Fermi resonance solitons, arise due to Fermi resonance interaction across the interface (11), (12). We look for the solution in the form

$$A_{n,m} = A \exp\left\{-\kappa_a(n-1) + \left[-i\left(\Omega t - \frac{Km}{2}\right)\right]\right\}, \quad \text{for} \quad n \geq 1,$$
$$B_{n,m} = B \exp\left\{\kappa_b n + [-i(\Omega t - Km)]\right\}, \quad \text{for} \quad n \leq 0, \quad (9.47)$$

where the amplitudes of vibrations exponentially decay as we go away from the interface. Then eqns (9.46) with $n \geq 2$ and $n \leq -1$ are solved by these functions provided

$$\frac{\Omega}{2} = \omega_a + 2V_a \cos\left(\frac{K}{2}\right) + 2V_a \cosh \kappa_a,$$
$$\Omega = \omega_b + 2V_b \cos K + 2V_b \cosh \kappa_b. \quad (9.48)$$

These equations give the values of κ_a and κ_b as functions of K and Ω. Then eqns (9.46) reduce to

$$V_a \exp(\kappa_a) A = 2\Gamma A^* B, \quad (9.49)$$
$$V_b \exp(\kappa_b) B = \Gamma A^2. \quad (9.50)$$

We multiply (9.49) by A and introduce the intensity of vibrations

$$I = |A|^2, \quad (9.51)$$

so that eqns (9.49), (9.50) give

$$V_a V_b \exp(\kappa_a + \kappa_b) = 2\Gamma^2 I. \quad (9.52)$$

Elimination of $V_a e^{\kappa_a}$ and $V_b e^{\kappa_b}$ from this equation can be done with the use of eqn (9.48), and yields the relation defining implicitly the dependence of Ω on K, i.e. the dispersion laws of Fermi resonance modes

$$\left\{\frac{\Omega}{2} - \omega_a - 2V_a \cos\left(\frac{K}{2}\right) + \mathrm{sgn}\left(\frac{\Omega}{2} - \omega_a - 2V_a \cos\left(\frac{K}{2}\right)\right)\right.$$
$$\left. \times \sqrt{\left(\frac{\Omega}{2} - \omega_a - 2V_a \cos\left(\frac{K}{2}\right)\right)^2 - 4V_a^2}\right\} \quad (9.53)$$
$$\times \left\{\Omega - \omega_b - 2V_b \cos K + \mathrm{sgn}(\Omega - \omega_b - 2V_b \cos K)\right.$$
$$\left. \times \sqrt{(\Omega - \omega_b - 2V_b \cos K)^2 - 4V_b^2}\right\} = 8\Gamma^2 I.$$

We see that they are modified compared to the case of Fermi resonance waves (9.44), (9.45) in an infinite 1D cubic crystal. At $I = 0$ (or $\Gamma = 0$), when the interaction across the interface disappears, eqn (9.53) reproduces two dispersion relations for surface waves in half-infinite crystals.

Note that the addition of a new dimension to a 1D cubic crystal with "point" interface leads to the replacements

$$\omega_a \to \omega_a + 2V_a \cos\left(\frac{K}{2}\right), \qquad \omega_b \to \omega_b + 2V_b \cos K, \tag{9.54}$$

which are a result of translational invariance along the interface following the interaction terms in the Hamiltonian (we omit the index n numbering sites along the axis perpendicular to the interface)

$$\sum_m \left[V_a \left(a_m^\dagger a_{m-1} + a_{m-1}^\dagger a_m \right) + V_b \left(b_m^\dagger b_{m-1} + b_{m-1}^\dagger b_m \right) \right].$$

If we add the third dimension (labelled by index ℓ) along the interface, we can take it into account by means of the following replacements in the formulas for the 1D case

$$\omega_a \to \omega_a + 2V_a \left(\cos\frac{K_m}{2} + \cos\frac{K_\ell}{2} \right),$$
$$\omega_b \to \omega_b + 2V_b \left(\cos K_m + \cos K_\ell \right), \tag{9.55}$$

K_m and K_ℓ being the wavevectors along m and ℓ axes, respectively. Thus, we conclude that in calculations of dispersion laws of waves propagating in superlattices with plane interfaces, it is sufficient to consider first the one-dimensional models with the axis directed perpendicular to the interfaces. Then the general formulas for dispersion laws can be obtained by means of replacements (9.54) or (9.55). So we reduce 2D or 3D problems to 1D problem with only one coordinate directed perpendicular to interfaces.

In the above we explored the nonlinear processes that occur when two molecular excitations on one side of the interface have a total energy in resonance with that of a single excitation on the other side of the interface. Coupling between these excitations across the interface can then give rise to Fermi resonance interface waves. In the paper (13) a variation of the Fermi resonance situation was explored: it arises when two excitations on one side of an interface have a difference in energy that matches the energy of a single excitation on the other side of the interface. Given suitably strong nonlinear coupling across the interface, this situation gives rise to what we call difference Fermi resonance interface waves. These have energies that lie outside the bands for the uncoupled excitations, and amplitudes that decay exponentially away from the interface. The model Hamiltonian for the system can be found in the paper (13) and also consideration of some of the simplest cases which should suffice to indicate the main new qualitative features for most practical purposes, given that interactions beyond nearest-neighbor molecules are negligible.

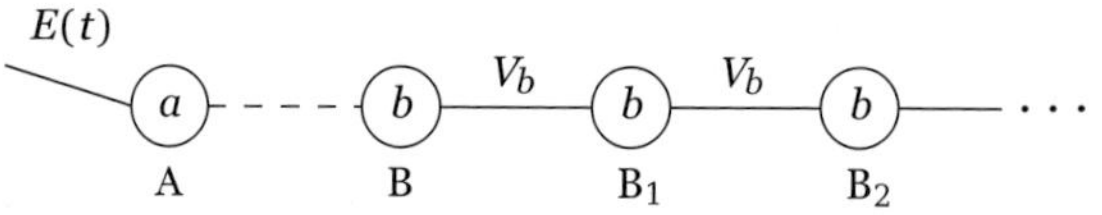

FIG. 9.7. Sketch of a 1D interface structure under the influence of an electromagnetic field. Reprinted with permission from Agranovich et al. (6). Copyright Elsevier (2003).

9.2.4 *Bistable energy transmission through the interface with Fermi resonance interaction*

Interesting phenomena can take place in systems with Fermi resonance under the influence of an external electromagnetic field. Here we consider one example of such behavior – bistable energy transmission through the interface with the Fermi resonance interaction (14). To make the calculations easier, we shall consider the following simplified model. Let a monomolecular layer of a molecules be deposited on the plane surface of a crystal made of b molecules. For the 1D case such a system is shown in Fig. 9.7.

Let a molecules interact with the electromagnetic field

$$\mathcal{E}(t) = E + E^*, \qquad E = E_0 \exp(-i\omega_L t), \tag{9.56}$$

in resonance with ω_a vibrations, $\omega_L \approx \omega_a \simeq \omega_b/2$, so that we can neglect the direct pumping of ω_b excitations. We take the dipole moment of a molecules to be linear in their coordinates q_a so that the electromagnetic interaction term in the Lagrangian is proportional to (see eqn 9.16)

$$q_a \mathcal{E}(t) = \frac{1}{\sqrt{2M_a\omega_a}}(A+A^*)(E+E^*) \simeq \frac{1}{\sqrt{2M_a\omega_a}}(AE^* + A^*E),$$

where we have omitted the rapidly oscillating terms $AE \propto \exp(-2i\omega_L t)$ and $A^*E^* \propto \exp(2i\omega_L t)$ according to our rotating wave approximation. Thus we have

$$L_{\text{int}} = \mu(AE^* + A^*E), \tag{9.57}$$

where μ, up to a constant factor, is the dipole moment of a molecules. Correspondingly, the equation of motion of molecule a in our 1D model reads

$$i\frac{\partial A}{\partial t} = \omega_a A + 2\Gamma A^* B + \mu E(t). \tag{9.58}$$

The molecules b do not interact with the electromagnetic field and their equations of motion have the usual form

$$i\frac{\partial B}{\partial t} = \omega_b B + \Gamma A^2 + V_b B_1,$$

$$i\frac{\partial B_1}{\partial t} = \omega_b B_1 + V_b\left(B + B_2\right), \ldots. \tag{9.59}$$

Now we have a driving force $E(t) = E_0 \exp(-i\omega_L t)$ so that the molecule a oscillates with this laser field frequency ω_L and due to the Fermi resonance interaction across the interface this leads to oscillations of molecules b with frequency $2\omega_L$. As a result, we obtain an algebraic system of equations for the amplitudes $A, B, B_1, \ldots$:

$$\begin{aligned}
(\omega_a - \omega_L) A + 2\Gamma A^* B + \mu E_0 &= 0, \\
(\omega_b - 2\omega_L) B + \Gamma A^2 + V_b B_1 &= 0, \\
(\omega_b - 2\omega_L) B_1 + V_b\left(B + B_2\right) &= 0, \ldots.
\end{aligned} \tag{9.60}$$

The equations for the amplitudes $B_1, B_2, \ldots$ are linear and solved by

$$B_n = \exp\left(ipn\right) B_{n-1}, \qquad B_1 = \exp(ip) B, \tag{9.61}$$

where the wavevector p is determined by the equation

$$2\omega_L = \omega_b + 2V_b \cos p. \tag{9.62}$$

Then we get from the first two equations of the system (9.60)

$$\begin{aligned}
B &= \frac{\exp(ip)\,\Gamma A^2}{V_b} \\
&= \frac{2\Gamma A^2}{2\omega_L - \omega_b + \operatorname{sgn}\left(2\omega_L - \omega_b\right) \sqrt{\left(2\omega_L - \omega_b\right)^2 - 4V_b^2}},
\end{aligned} \tag{9.63}$$

and

$$A\left[\omega_L - \omega_a - \frac{4\Gamma^2|A|^2}{2\omega_L - \omega_b + \operatorname{sgn}\left(2\omega_L - \omega_b\right)\sqrt{\left(2\omega_L - \omega_b\right)^2 - 4V_b^2}}\right]$$
$$= \mu E. \tag{9.64}$$

Now we introduce the "pumping" intensity

$$I_{\text{pump}} = \frac{|\mu E|^2}{\left(\omega_a - \omega_L\right)^2} \tag{9.65}$$

and arrive at the following equation

$$I(1 - DI)^2 = I_{\text{pump}} \tag{9.66}$$

which determines implicitly the intensity of vibrations

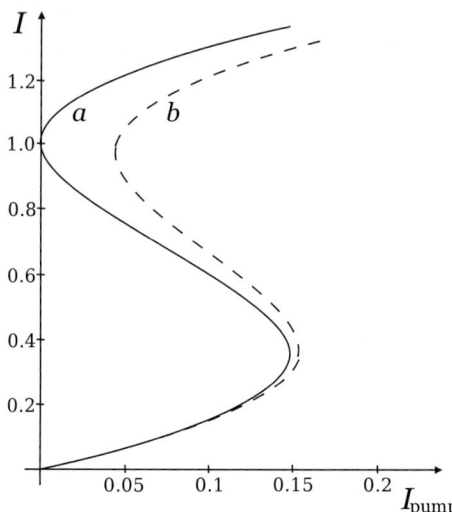

FIG. 9.8. Dependence of the intensity I of a vibrations on the pumping intensity: (a) without damping; (b) with damping. Reprinted with permission from Agranovich et al. (6). Copyright Elsevier (2003).

$$I = |A|^2 \qquad (9.67)$$

as a function of the pumping intensity I_{pump}, where

$$D = \frac{4\Gamma^2}{(\omega_L - \omega_a)\left[2\omega_L - \omega_b + \text{sgn}\,(2\omega_L - \omega_b)\sqrt{(2\omega_L - \omega_b)^2 - 4V_b^2}\right]}. \qquad (9.68)$$

The cubic equation (9.66) can have three roots which indicates bistability – two values of I correspond to one pumping intensity (the third root corresponds to an unstable state). The plot of the function $I(I_{\text{pump}})$ is shown in Fig. 9.8, where one can see the bistability region $0 \leq I_{\text{pump}} \leq 0.15\,I_c$ (with $I - c = 1/D = 1$).

More details can be found in (14). Here we note only that there exists a nonzero solution of (9.66) even at vanishing pumping

$$I = 1/D, \qquad (9.69)$$

i.e. when the oscillation frequency Ω of molecules b satisfies the following equation

$$\left(\frac{\Omega}{3} - \omega_a\right)\left[\Omega - \omega_b + \text{sgn}\,(\Omega - \omega_b)\sqrt{(\Omega - \omega_b)^2 - 4V_b^2}\right] = 4\Gamma^2 I. \qquad (9.70)$$

This vibrational state which exists without pumping is just the Fermi resonance interface state.

In conclusion we note that interfaces introduce many new phenomena in the physics of multilayer structures which can find numerous applications in science and technology.

10

CAVITY POLARITONS IN ORGANIC MICROCAVITIES

10.1 Giant Rabi splitting in organic microcavities

As already discussed in Ch. 4, polaritons in 3D cubic crystals have a gap in the dispersion curve in the vicinity of the exciton resonance equal to the value of the so-called longitudinal–transverse (LT) splitting of the exciton. The longitudinal–transverse splitting in the spectrum of excitons in cubic crystal results from the long-range instantaneous Coulomb interaction which splits the three-fold degenerate level of the exciton in a cubic crystal into a band of longitudinal excitons and a two-fold degenerate band of transverse excitons. The nature of the splitting in the polariton spectrum is different. This splitting arises due to the interaction of 3D transverse photons and 3D transverse excitons (or optical phonons) and both splittings can be observed only if the value of the splitting is large in comparison with the dissipative width of the exciton resonance. Using modern language accepted in the nanoscience literature we can say that in the case of polaritons we have a case of strong coupling of excitons and photons. The new states (polaritons) are coherent superpositions of states of excitons and photons. It is clear that analogous effect should exist also for interacting of 2D light and 2D excitations where in both cases the wavevector has only two components. The same holds for 1D photons and 1D excitations. Indeed, this splitting (the effect of strong coupling) was first found for surface polaritons, which are 2D excitations, interacting with dipole-active excitation in thin films on the surface supporting surface waves (1). The magnitude of the splitting is $\sim \sqrt{d/\lambda}$ where d is the film thickness and λ is the surface polariton wavelength. Then the splitting in the spectrum of surface waves was observed in the infrared, visual, and ultraviolet regions of the spectrum (for a review see (2)). More recently a strong coupling between surface plasmons and excitons in a layer of organic cyanine dye J-aggregates, deposited on a silver film, was demonstrated in Ref. (3). At room temperature the value of the splitting at resonance of a 2D Frenkel exciton and a surface plasmon was found equal to 180 meV. The work by Bellessa (3) is a second demonstration of the strong coupling of a Frenkel exciton and a surface plasmon after work by Pockrand *et al.* (4), (5) (the dye Langmuir–Blodgett bimonolayer on silver) where, however, the observed splitting was found to be much smaller. Later, in experiments of the same Bellessa group, the splitting 300 meV was observed (6) and, as could be expected, the $(d/\lambda)^{1/2}$ dependence of the splitting value on the thickness of the organic layer was also observed.

We mentioned here the strong coupling of a 2D electromagnetic wave (surface wave) and a 2D excitation because, as could be expected from the general

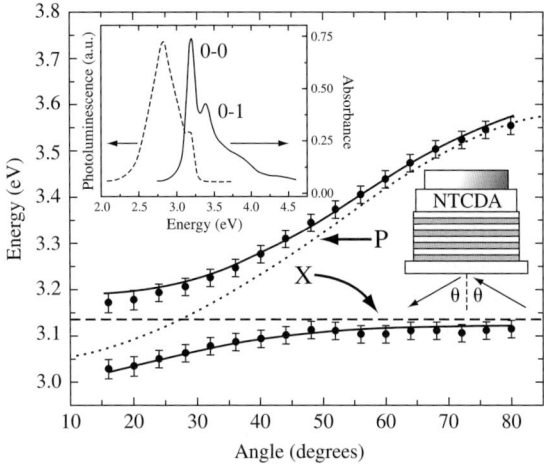

FIG. 10.1. Dispersion relation for an organic microcavity with a 20 nm thick NTCDA active layer. The broken curves are the uncoupled cavity photon (P) and Frenkel exciton (X) dispersion relations as determined by the fitting parameters. Inset: room temperature photoluminescence and absorbance spectra for a 50 nm thick film of NTCDA. Absorption peaks at 3.19 and 3.39 eV are designated as the 0-0 and 0-1 transitions. Reprinted with permission from Holmes et al. (12). Copyright 2004, American Physical Society.

point of view, the same physical phenomenon should take place also in a plane microcavity for microcavity photons which are 2D excitations and 2D excitons in a quantum well embedded in the microcavity. Indeed, this effect has also been observed in microcavities by Weisbuch et al. (7), where the experimental observation of polariton splitting between a quantum well exciton and a cavity mode was reported for the first time. Since then the regime of strong light–matter coupling in planar microcavities has been examined extensively (8).

The estimation made in (9) showed that in a microcavity with organics the Rabi splitting for an appropriate selection of organic material can be about two orders of magnitude larger than the Rabi splitting in a microcavity containing an inorganic semiconductor quantum well. This prediction was very soon confirmed in experiments performed by Bradley and his collaborators (10) even for microcavities with disordered organics where for the Rabi splitting values of the order of 100 meV have been observed. In all these experiments a strong coupling regime has been achieved, with using amorphous organics (porphyrin dyes, J-aggregated cyanine dyes and σ-conjugated silane polymers) in the microcavity. An excellent discussion of the similarities and differences between inorganic and organic semiconductor microcavities operating within the strong coupling regime and also the discussion of prospects for inorganic/organic hybrid materials for optoelectronic devices can be found in (11). Now this field of investigations is quickly growing (see, for example (12), (13), (14)). In these papers the authors examine the influence of singlet–triplet intersystem crossing and excited state molecular relaxation on strong exciton–photon coupling in an optical microcavity filled with small molecular weight organic materials. The dispersion of cavity polaritons in this case is presented in Fig. 10.1. As intersystem crossing and molecular relaxation quench strong coupling, the interplay between strong coupling and relaxation processes offers a unique opportunity to probe directly fundamental ultrafast excitonic phenomena and can be used for the estimation of their relative transition rates. We will explain below why, in such disordered microcavities in the region of exciton resonances, the coexistence of coherent and localized excitations in the spectrum of excitations can be expected (15). This effect creates a rather complicated structure of the cavity exciton–polariton spectrum and results a very unusual the cavity polariton dynamics. The first investigations of polariton dynamics have been performed in (16).

It is important that the relaxation processes in organic microcavities under nonresonant pumping can be much faster than in inorganic microcavities. Indeed, in organic microcavities all the characteristic energies (such as the Davydov splitting and Rabi splitting) are usually of the order of hundreds of meV. This means that the relaxation of cavity polaritons can be determined by the fast emission of intramolecular optical phonons with various energies. As the bands of these phonons are usually rather narrow ($\sim$ 1 meV), conservation of the polariton wavevector in the relaxation processes with participation of intramolecular vibrations can be disregarded (see also (9)) which will also speed up the polariton relaxation.

The spectra of cavity polaritons in crystalline organics and estimations of the rate of their relaxation processes have been been studied theoretically (17), (18) including the polariton nonlinear dynamics in such a microcavity (19), (20). Having in mind that the kinematic interaction between Frenkel excitons has a different nature in comparison with such an interaction between Wannier–Mott excitons in inorganic microcavities we can expect that collective properties of polaritons in organic microcavities will have many peculiarities in its dynamics and condensation. The first successful investigation of a microcavity with polycrystalline organics (tetracene) was performed by Kèna-Cohen and Forrest (21). The Rabi splitting in this microcavity has been observed for each component of the Davydov splitting as well as for transitions to several vibronic states. The orientation of polycrystals in the plane of the microcavity was random and both Davydov components have been seen for any polarization. Work is in progress and one can expect interesting results in the study of polariton spectra, bimolecular quenching, nonlinear optical processes, and condensation of cavity polaritons as a function of the intensity of the pumping.

In the following sections of this chapter we consider in more detail the properties of excitations in organic microcavities with crystalline as well as with amorphous materials.

10.2 Microcavities with crystalline organics

10.2.1 *Introduction*

Recently, interest in cavity polaritons has been revived due to the experimental observation of stimulated and self-stimulated polariton–polariton scattering (22)–(27). The accumulation of lower branch polaritons in the lowest energy state was observed under different experimental conditions in inorganic microcavities and the possibility of condensation of low-energy polaritons was extensively discussed. Evidently, the same types of ideas and problems can be formulated also for organic microcavities with large values of the Rabi splitting. Large values of the Rabi splitting are attractive, since they may provide stability of the polaritonic picture at higher (room) temperature and at higher pump densities. In widely used inorganic semiconductor quantum wells the Rabi splitting of cavity polaritons is typically of the order of 10 meV or even less. However, as mentioned, in recently fabricated organic microcavities it can be much larger.

The wide knowledge accumulated in the field of polaritons in inorganic materials is, of course, very important, but it cannot be directly transferred to the case of organic systems. Organic materials possess small-radius Frenkel excitons, which interact typically much more strongly with phonons and disorder. The spectrum of excitations in organic microcavities may show new significant properties and therefore some processes with the participation of exciton–polaritons in organic cavities may differ even qualitatively from the corresponding processes in inorganic semiconductors. As has been shown (15), in organic ordered or disordered materials with a broad excitonic line and rather weak intermolecular resonance Coulomb interaction, the coherent (polaritonic) states coexist with incoherent (localized) states, and the latter form the majority of the excited states in the microcavity. The excited states, which are localized wavepackets of large spatial size, can be described with the help of Maxwell's equations, while the majority of incoherent states (the states which are strongly spatially localized) require a microscopic treatment. As a consequence of this structure of the spectrum, the relaxation dynamics of the excitations differs strongly from that in inorganic microcavities. One may expect also that the exact statistics of the excitations (i.e. the fact that Frenkel excitons are not bosons, but paulions) will be especially important for these small-radius localized states.

In this section we examine the spectrum of exciton–polaritons in a microcavity, which utilizes as an optically active material an organic crystal with large oscillator strengths of the lowest energy Frenkel excitons and with rather narrow excitonic resonances. This picture is typical for well-investigated molecular crystals like anthracene, tetracene, and also for many others in the region of lowest energy exciton resonances. It allows one to avoid the complications related to the aforementioned complex structure of the excitation spectrum, since in the framework of this model incoherent localized states do not form. However, even in this simple case the difference between organic and inorganic materials is sufficiently large.

The main point is that organic crystals show strong optical anisotropy. Due to the anisotropy, s- and p-polarized waves no longer propagate independently in the optically active material. In isotropic materials the quantum well excitons can be divided into T-, L- and Z-modes (28), and an s-polarized cavity photon interacts with a T-exciton, while a p-polarized cavity polariton interacts with L- and Z-excitons only. In contrast, in organic (anisotropic) microcavities, with the exceptions of some special directions, both polarizations of the cavity photon participate in the formation of polaritonic states on an equal footing. It will be shown that the combined effect of the two modes results in an almost isotropic expression for the Rabi splitting energy. Another distinguishing feature of organic materials is that, instead of Wannier–Mott excitons, they possess Frenkel excitons. The number of excitonic bands in these crystals in the region of nondegenerate molecular transition is equal to the number of molecules in the unit cell of the crystal (29). These two features of organic crystals are important in the consideration of polariton spectra in organic microcavities.

Electronic excitations in crystals (either organic or inorganic), obtained with taking account of the total Coulomb interaction between the molecules forming the crystal (but neglecting the retardation), are known as Coulomb excitons (See Chs. 2 and 3). The polarizations of the excitonic resonances in anisotropic crystals have fixed directions determined by the symmetry of the crystal and by the transition dipole moments of the molecules forming the crystal. Their dispersion is anisotropic, and in the bulk material it shows as a nonanalytic behavior at small wavevectors (30). However, in thin organic films the energies of Coulomb excitons are analytical functions of small in-plane wavevectors and their dependence on the direction of the wavevector is much weaker than in the bulk. This effect can be easily understood taking into account that the energies of Coulomb excitons in a thin film are functions of the excitonic in-plane wavevector $\mathbf{q}$ when the z-component of the total wavevector is fixed at a finite value by the boundary conditions at the microcavity mirrors. As will be shown the anisotropy of Coulomb excitons in thin films is seen only at large $\mathbf{q}$ where, however, the exciton–cavity photon interaction is already not important.

In Subsection 10.2.2, the Maxwell equations in an anisotropic microcavity are discussed and some important facts concerning Coulomb and mechanical excitons are summarized. In Subsection 10.2.3 the dispersion equation of cavity polaritons is derived. Its solutions for the cases of crystals with one and two molecules in the unit cell are discussed in Subsections 10.2.4 and 10.2.5, respectively. The main results are summarized in the conclusions.

10.2.2 *Cavity photons and Coulomb excitons*

In this subsection we introduce some notations for the " "bare" excitations whose interaction leads to the formation of cavity polariton states. These "bare" excitations are cavity photons and Coulomb excitons.

We start with the notation for the cavity photon modes considering the solutions of Maxwell's equations for an isotropic microcavity. Let L_c be the microcavity thickness, and ϵ_c the dielectric constant of the microcavity nonresonant material. If the mirrors surrounding the microcavity are made from an ideal conductor, the boundary conditions at the microcavity boundaries are (31): $\mathbf{E}_\tau = 0, H_n = 0$ (the subscripts τ and n denote respectively the tangential and the normal components of the fields). Then the value of the wavevector in the growth direction (z-axis) for the lowest mode is $k_z = \pi/L_c$. Let $\mathbf{q}$ be the in-plane wavevector, $\boldsymbol{\kappa}_\mathbf{q}$ a unit vector in the direction of $\mathbf{q}$, and $\mathbf{n}_\mathbf{q}$ a unit vector in the direction perpendicular to $\mathbf{q}$: $\mathbf{n}_\mathbf{q} = [\mathbf{e}_z \times \boldsymbol{\kappa}_\mathbf{q}]$. Then the system of Maxwell's equations has two independent orthogonal solutions known as s- (or TE-) and p- (or TM-) cavity photons:

$$\mathbf{E}_s = \mathbf{n}_\mathbf{q}\, C_s \cos\frac{\pi z}{L_c}\, e^{i\mathbf{q}\mathbf{r}_\|},$$

$$\mathbf{E}_p = \left[\boldsymbol{\kappa}_\mathbf{q} \cos\frac{\pi z}{L_c} - \mathbf{e}_z\, \frac{iqL_c}{\pi}\sin\frac{\pi z}{L_c}\right] C_p\, e^{i\mathbf{q}\mathbf{r}_\|}, \qquad (10.1)$$

with the same dispersion law:

$$\omega_{\text{cav}}(q) = \omega_c\sqrt{1 + \left(\frac{qL_c}{\pi}\right)^2}, \qquad \omega_c \equiv \omega_{\text{cav}}(0) = \frac{c\pi}{L_c\sqrt{\epsilon_c}}. \qquad (10.2)$$

Now let us consider the fields in a microcavity made from an anisotropic organic crystal with excitonic resonances. We suppose that the microcavity is a slab of an organic crystal surrounded by ideal mirrors. The system of Maxwell's equations is:

$$\Delta \mathbf{E} + \frac{\omega^2}{c^2}\mathbf{D} - \text{grad div } \mathbf{E} = 0, \quad \text{(a)}$$

$$\text{div } \mathbf{D} = 0, \quad \text{(b)} \qquad (10.3)$$

$$D_i = \epsilon_{ik}(\omega) E_k. \quad \text{(c)}$$

To proceed, one has to know the dielectric tensor $\epsilon_{ik}(\omega)$. It can be found either from experiment, or with the help of a microscopic theory. Below we recall several useful facts concerning the relation existing between the excitonic resonances and the dielectric tensor in organic crystals (see also Chs. 2 and 3).

First of all, in an organic crystal the number of excitonic bands in the region of a nondegenerate molecular transition is equal to the number of molecules in the unit cell of the crystal. In particular, a unit cell of anthracene or tetracene type crystals contains two molecules with the same excitation energy but with two differently oriented transition dipole moments $\mathbf{p}_1$ and $\mathbf{p}_2$. Then due to the intermolecular interaction two excitonic bands appear, and the dipole moments of the components of the doublet are: $\mathbf{P}_1 = \mathbf{p}_1 + \mathbf{p}_2$, $\mathbf{P}_2 = \mathbf{p}_1 - \mathbf{p}_2$, with $\mathbf{P}_1 \perp \mathbf{P}_2$ (29). The dispersion relation of these modes is obtained taking into account the total electrostatic interaction between the molecules, which is the sum of the short-range electric field and the macroscopic (long-wavelength) electric field. The excited states of the crystal obtained in this way are the solutions of the

Coulomb problem, and thus are Coulomb excitons. The Coulomb excitons can be found as solutions of Maxwell's equations in the limit $c \to \infty$, i.e. neglecting the retardation.

Besides Coulomb excitons, it is convenient to use also the so-called mechanical excitons. These are the excited states of the crystal found in the approximation that the contribution of the macroscopic electric field to the intermolecular interaction is omitted, and only the contribution of the short-range intermolecular interaction is taken into account. The energies of the mechanical excitons $\hbar\tilde{\omega}_\gamma$ determine the poles of the dielectric tensor:

$$\epsilon_{ik}(\omega) = \epsilon_c \delta_{ik} - \frac{8\pi}{a^3 \hbar} \sum_\gamma \frac{P_{\gamma i} P_{\gamma k} \tilde{\omega}_\gamma}{\omega^2 - \tilde{\omega}_\gamma^2}, \qquad (10.4)$$

where a^3 is the volume of the unit cell of the crystal, ϵ_c is the background dielectric constant, $P_{\gamma i}$ is the ith projection of the dipole moment of the transition for the γ-th mechanical exciton, and the sum runs over all the mechanical excitons which are taken into account. When one knows the energies of the mechanical excitons, the corresponding transition dipole moments and the background dielectric constant (and, consequently, the tensor ϵ_{ik}), one can find the energies of the Coulomb excitons. Indeed, consider eqn (10.3a) in the limit $c \to \infty$. In this limit the electric field becomes purely longitudinal ($\mathbf{E} \parallel \mathbf{Q}$, where $\mathbf{Q}$ is the total wavevector). Then from eqn (10.3 b–c) the dispersion relation of Coulomb excitons follows (see also eqn 2.47):

$$\epsilon_{ik}(\omega) Q_i Q_k / Q^2 = 0. \qquad (10.5)$$

Let us return to the structure of the dielectric tensor of a crystal of anthracene, tetracene type (i.e. with two molecules in the unit cell). Based on the above arguments and supposing that we know the energies $\hbar\tilde{\omega}_1$ and $\hbar\tilde{\omega}_2$ of the mechanical excitons, which for simplicity we take to be dispersionless, we can write the tensor ϵ_{ik} in the form:

$$\epsilon_{ik}(\omega) = \begin{pmatrix} \epsilon_c(1 - \chi_1(\omega)) & 0 & 0 \\ 0 & \epsilon_c(1 - \chi_2(\omega)) & 0 \\ 0 & 0 & \epsilon_c \end{pmatrix}, \qquad (10.6)$$

where $\chi_\gamma(\omega) = W_\gamma^2/(\omega^2 - \tilde{\omega}_\gamma^2)$, $W_\gamma^2 = (8\pi P_\gamma^2 \tilde{\omega}_\gamma^2 / a^3 \hbar \epsilon_c)$. Here and below the x and y axes are chosen along $\mathbf{P}_1$ and $\mathbf{P}_2$, respectively. The transition to crystals with one molecule in the unit cell can be done by setting $\chi_2 \equiv 0$.

10.2.3 Cavity polaritons

We suppose that the microcavity was grown in such a way that the dipole moments $\mathbf{P}_1$ and $\mathbf{P}_2$ are parallel to the microcavity plane, and the z-axis denotes the growth direction. Let the microcavity of width L_c be surrounded by two perfect mirrors, so that the boundary conditions at the microcavity boundaries are: $\mathbf{E}_\tau = 0$, $H_n = 0$. We look for the normal modes in such a system solving the system of equations (10.3) with the dielectric tensor given by (10.6). For such modes the tangential components of the electric field can be found to be proportional to $\exp[i\mathbf{q}\mathbf{r}_\|] \cos(k_z z)$. From the boundary conditions we find that for the lowest mode $k_z = \pi/L_c$.

We shall describe the electric field by its in-plane longitudinal ($\mathbf{E}_l \| \mathbf{q}$), in-plane transversal ($\mathbf{E}_t \perp \mathbf{q}$), and z-components (E_z). Let φ denote the angle between the in-plane wavevector $\mathbf{q}$ and the x-axis. From eqn (10.3 a) we find that $E_z = (iq/\kappa_c^2)\partial E_l/\partial z$, where $\kappa_c = \sqrt{\omega^2 \epsilon_c/c^2 - q^2}$. The fields E_l and E_t are related to each other by the system of equations obtained from eqn (10.3 a):

$$E_l \cos\varphi \frac{\omega^2}{c^2 \kappa_c^2} \left[\epsilon_{xx} \kappa_c^2 - \frac{\pi^2}{L_c^2} \epsilon_c \right] - E_t \sin\varphi \left[\frac{\omega^2}{c^2} \epsilon_{xx} - \left(q^2 + \frac{\pi^2}{L_c^2} \right) \right] = 0,$$

$$E_l \sin\varphi \frac{\omega^2}{c^2 \kappa_c^2} \left[\epsilon_{yy} \kappa_c^2 - \frac{\pi^2}{L_c^2} \epsilon_c \right] + E_t \cos\varphi \left[\frac{\omega^2}{c^2} \epsilon_{yy} - \left(q^2 + \frac{\pi^2}{L_c^2} \right) \right] = 0. \tag{10.7}$$

The determinant of this system of equations yields the dispersion law of the cavity polaritons.

The dispersion equation (10.2) of bare cavity photons can be obtained from eqn (10.7) in the limit $\epsilon_{xx} = \epsilon_{yy} = \epsilon_c$. To find from eqn (10.7) the dispersion equation of cavity polaritons in an isotropic resonant medium, one has to replace ϵ_c by ϵ_{zz} and set:

$$\epsilon_{xx} = \epsilon_{yy} = \epsilon_{zz} \equiv \epsilon(\omega) = \epsilon_0 - \frac{W^2 \epsilon_0}{\omega^2 - \omega_\perp^2}. \tag{10.8}$$

Then the system of equations (10.7) yields, near the resonance:

$$(\omega - \omega_{\text{cav}}(q))(\omega - \omega_\perp) = (W^2/4) \equiv \Delta_0^2. \tag{10.9}$$

We stress that in the case of an isotropic material, the fields E_l and E_t are independent. In these materials an s-polarized cavity photon excites a T-exciton (28) in the quantum well material and interacts with it, while a p-polarized cavity photon excites L- and Z-excitons and interacts with them. In contrast, as seen from eqn (10.7), E_l and E_t become coupled when $\epsilon_{xx} \neq \epsilon_{yy}$. Except for waves propagating along the x and y axes, both E_l and E_t exist in the cavity material and interact with excitonic resonance on an equal footing.

We introduce $Q^2 = q^2 + k_z^2$ and write the dispersion equation of the cavity polaritons in the form:

$$\left(\frac{\omega^2}{c^2} \epsilon_c - Q^2 \right)^2 \left[1 - \frac{q^2}{Q^2} (\chi_1(\omega) \cos^2\varphi + \chi_2(\omega) \sin^2\varphi) \right]$$

$$-\left(\frac{\omega^2}{c^2}\epsilon_c - Q^2\right)\frac{\omega^2}{c^2}\epsilon_c\left\{\chi_1(\omega)\left(1 - \frac{q^2}{Q^2}\cos^2\varphi\right) + \chi_2(\omega)\left(1 - \frac{q^2}{Q^2}\sin^2\varphi\right)\right\}$$
$$+ \frac{\omega^2}{c^2}\epsilon_c\kappa_c^2\chi_1(\omega)\chi_2(\omega) = 0. \tag{10.10}$$

It is easy to see that the zeros of the square bracket in the first term of this equation are the frequencies of the Coulomb excitons in a thin organic film. Indeed, as was argued above, the Coulomb excitons are the solutions of Maxwell's equations when the retardation is neglected. When $c \to \infty$, eqn (10.10) reduces to:

$$0 = \left[1 - \frac{q^2}{Q^2}(\chi_1(\omega)\cos^2\varphi + \chi_2(\omega)\sin^2\varphi)\right]$$
$$\equiv \frac{(\omega^2 - \omega_{ex1}^2(\mathbf{q}))(\omega^2 - \omega_{ex2}^2(\mathbf{q}))}{(\omega^2 - \tilde{\omega}_1^2)(\omega^2 - \tilde{\omega}_2^2)}, \tag{10.11}$$

where $\omega_{ex1,2}(\mathbf{q})$ are the frequencies of the Coulomb excitons in the thin organic film. The same equation can also be obtained directly from eqn (10.5) for $\mathbf{Q} = (\mathbf{q}, k_z)$. The solutions of this equation are:

$$\omega_{ex1,2}^2(\mathbf{q}) = \frac{1}{2}\left\{\tilde{\omega}_1^2 + \tilde{\omega}_2^2 + W_1^2\frac{q^2}{Q^2}\cos^2\varphi + W_2^2\frac{q^2}{Q^2}\sin^2\varphi\right.$$

$$\tag{10.12}$$

$$\left.\pm\sqrt{\left[\tilde{\omega}_2^2 - \tilde{\omega}_1^2 + W_2^2\frac{q^2}{Q^2}\sin^2\varphi - W_1^2\frac{q^2}{Q^2}\cos^2\varphi\right]^2 + 4W_1^2W_2^2\frac{q^4}{Q^4}\sin^2\varphi\cos^2\varphi}\right\}.$$

10.2.4 One molecule in the unit cell

In this section we discuss the solutions of eqn (10.10) for the case of a crystal with one molecule in the unit cell. The results of this section will also be helpful when we consider crystals with two molecules in the unit cell. Let $\chi_2 \equiv 0$. Then one of the solutions of eqn (10.10) is: $\omega^2 = c^2Q^2/\epsilon_c \equiv \omega_{cav}^2(q)$. It is just the dispersion relation of the "bare" cavity photon (see eqn 10.2). From the first of eqns (10.7) it follows for this solution that the fields E_l and E_t are related by: $E_l = E_t \tan\varphi$. The total electric field $\mathbf{E}_c$ in this mode and in our coordinate system reads:

$$\mathbf{E}_c = \left\{0, \ (\sin\varphi E_l + \cos\varphi E_t)\cos\frac{\pi z}{L_c}, \ -\frac{iqL_c}{\pi}E_l\sin\frac{\pi z}{L_c}\right\}, \tag{10.13}$$

i.e. for any $\mathbf{q}$ it is perpendicular to the transition dipole moment $\mathbf{P}_1$, and therefore it does not couple to the excitonic resonance.

The second solution of eqn (10.10) for the region of the Rabi splitting where $\omega \approx \omega_{\text{cav}}(q) \approx \tilde{\omega}_1$ can be found from the equation

$$\left(\omega - \omega_{\text{cav}}(q)\right)\left(\omega - \omega_{ex1}(\mathbf{q})\right) = \frac{W_1^2}{4}\left(1 - \frac{q^2}{Q^2}\cos^2\varphi\right) \equiv \Delta^2(\mathbf{q}), \qquad (10.14)$$

where

$$\omega_{ex1}(\mathbf{q}) = \sqrt{\tilde{\omega}_1^2 + W_1^2 \frac{q^2}{Q^2}\cos^2\varphi} \approx \tilde{\omega}_1 + \frac{W_1^2}{2\tilde{\omega}_1}\frac{q^2}{Q^2}\cos^2\varphi. \qquad (10.15)$$

It follows from eqn (10.15) that the energy of the Coulomb exciton in the quasi-two-dimensional film is an analytical function of its in-plane wavevector, and the anisotropy of the exciton energy enters only through the term quadratic in q (in other words, the Coulomb exciton has an anisotropic effective mass $m^*(\varphi) = (\hbar k_z^2 \tilde{\omega}_1 / W_1^2 \cos^2 \varphi)$). This means that the anisotropy in the dispersion of the Coulomb exciton is important in the region of large enough wavevectors only. However, even at very large q, the anisotropy is only a small correction to the energy of the mechanical exciton $\tilde{\omega}_1$ (the correction is of the order of $(W_1^2/2\tilde{\omega}_1^2)$). In turn, the angular dependence in the expression for the Rabi splitting energy in (10.14) for small q is also negligibly small (it is of the order of $q^2/2Q^2$). We conclude that in the region of wavevectors small in comparison with k_z, which corresponds to the wavevector interval usually examined in microcavities using inorganic quantum wells, the spectrum of cavity polaritons in an anisotropic organic microcavity with high accuracy is isotropic. The solutions of eqn (10.14) are

$$\omega_{U,L}(\mathbf{q}) = \frac{1}{2}\left[\omega_{\text{cav}}(q) + \omega_{ex1}(\mathbf{q}) \pm \sqrt{(\omega_{\text{cav}}(q) - \omega_{ex1}(\mathbf{q}))^2 + 4\Delta^2(\mathbf{q})}\right]$$
$$\simeq \frac{1}{2}\left[\omega_{\text{cav}}(q) + \tilde{\omega}_1 \pm \sqrt{(\omega_{\text{cav}}(q) - \tilde{\omega}_1)^2 + W_1^2}\right]. \qquad (10.16)$$

Thus the cavity polariton dispersion has a simple interpretation. Let us calculate the electric fields $\mathbf{E}_{U,L}$ from eqn (10.7) for the modes (10.16). Neglecting small terms of the order of q^2/κ_c^2, the fields E_l and E_t are related in these modes by $E_l = -E_t \cot \varphi$. Then the y-component of the fields $\mathbf{E}_{U,L}$ is equal to zero. In other words, with accuracy up to small terms (of the order of q^2/κ_c^2) the total in-plane electric field in the polaritonic modes is parallel to the dipole moment $\mathbf{P}_1$ for any direction of the wavevector $\mathbf{q}$, and the value of the Rabi splitting energy thus does not depend on the wavevector direction.

10.2.5 *Two molecules in the unit cell*

When the unit cell of an organic crystal contains two or more molecules, the spectrum of the cavity polaritons strongly depends on the relation between (i) the detuning $u \equiv \omega_c - \tilde{\omega}_1$, (ii) the energy of the Rabi splittings W_1 and W_2, and

(iii) the energy of the Davydov splitting $\Delta_D = \tilde{\omega}_{ex2} - \tilde{\omega}_{ex1}$. Let us consider the solutions of eqn (10.10) for crystals with two molecules in the unit cell. In this discussion, having in mind small values of the in-plane wavevector, we neglect terms of the order of $(q/Q)^4$. In this approximation, which we have checked by numerical comparison with the exact results, eqn (10.10) can be rewritten in the form:

$$\left[(\omega^2 - \omega_{\text{cav}}^2(q))(\omega^2 - \omega_{\text{ex1}}^2(\mathbf{q})) - \omega^2 W_1^2 \left(1 - \frac{q^2}{Q^2}\cos^2\varphi\right)\right]$$

$$\times \left[(\omega^2 - \omega_{\text{cav}}^2(q))(\omega^2 - \omega_{\text{ex2}}^2(\mathbf{q})) - \omega^2 W_2^2 \left(1 - \frac{q^2}{Q^2}\sin^2\varphi\right)\right] = 0, \quad (10.17)$$

where, with the same accuracy, $\omega_{\text{ex1}}(\mathbf{q}) = \tilde{\omega}_1 + (W_1^2 q^2 / 2\tilde{\omega}_1 Q^2)\cos^2\varphi$, $\omega_{\text{ex2}}(\mathbf{q}) = \tilde{\omega}_2 + (W_2^2 q^2 / 2\tilde{\omega}_2 Q^2)\sin^2\varphi$ (see eqn 10.12).

Applying the results of the previous subsection to each of the two brackets on the left-hand side of this equation, we find that for $q \ll Q$ the solutions of eqn (10.10) are two pairs of independent polaritonic branches with the total in-plane electric fields parallel to the transition dipole moments $\mathbf{P}_1$ and $\mathbf{P}_2$:

$$\omega_{U1,L1}(\mathbf{q}) = \frac{1}{2}\left[\omega_{\text{cav}}(q) + \omega_{ex1}(\mathbf{q}) \pm \sqrt{(\omega_{\text{cav}}(q) - \omega_{ex1}(\mathbf{q}))^2 + 4\Delta_1^2(\mathbf{q})}\right]$$

$$\approx \omega_{U1,L1}^{(0)}(q) = \frac{1}{2}\left[\omega_{\text{cav}}(q) + \tilde{\omega}_1 \pm \sqrt{(\omega_{\text{cav}}(q) - \tilde{\omega}_1)^2 + W_1^2}\right],$$

(10.18)

$$\omega_{U2,L2}(\mathbf{q}) = \frac{1}{2}\left[\omega_{\text{cav}}(q) + \omega_{ex2}(\mathbf{q}) \pm \sqrt{(\omega_{\text{cav}}(q) - \omega_{ex2}(\mathbf{q}))^2 + 4\Delta_2^2(\mathbf{q})}\right]$$

$$\approx \omega_{U2,L2}^{(0)}(q) = \frac{1}{2}\left[\omega_{\text{cav}}(q) + \tilde{\omega}_2 \pm \sqrt{(\omega_{\text{cav}}(q) - \tilde{\omega}_2)^2 + W_2^2}\right],$$

where

$$\Delta_1^2(\mathbf{q}) = \frac{W_1^2}{4}\left(1 - \frac{q^2}{Q^2}\cos^2\varphi\right), \quad \Delta_2^2(\mathbf{q}) = \frac{W_2^2}{4}\left(1 - \frac{q^2}{Q^2}\sin^2\varphi\right). \quad (10.19)$$

Typical examples of the dispersion curves are shown in Fig. 10.2 for $\varphi = \pi/4$, small positive detuning $u = \omega_c - \tilde{\omega}_1 = 35$ meV, and different relations between W_1, W_2 and $\Delta_D = \tilde{\omega}_2 - \tilde{\omega}_1$. It is clear that when one of the coupling parameters W is small, the spectrum consists of a doublet of polariton branches and of two branches which are close to the "bare" cavity photon and the "bare" exciton (Fig. 10.2a). When the Davydov splitting is small and $W_1 \sim W_2$, then the pairs of the dispersion curves almost overlap (Fig. 10.2b). The electric fields in the overlapping curves are (with accuracy up to terms of the order of q^2/Q^2) perpendicular to each other (see the discussion in Subsection 10.2.4). When all the parameters are of the same order, the spectrum consists of four well-pronounced polaritonic branches (Fig. 10.2c).

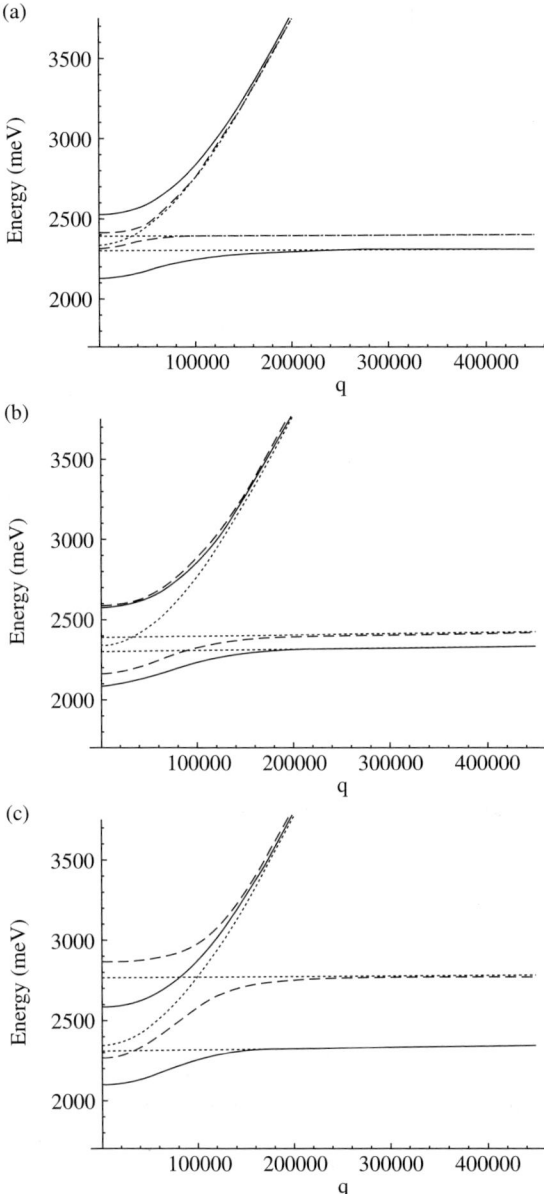

FIG. 10.2. (a) dispersion of polaritons for small coupling parameter of higher Davydov component; (b) Disperssion curves when the Davydov splitting is small but coupling parameters of both Davydov components are the same order; (c) the dispersion of polaritons when all parameters including the Davydov splitting are of the same order. Reprinted with permission from Litinskaia *et al.* (18). Copyright 2004, Wiley VCH Verlag.

10.2.6 *Conclusions*

We have found that, in spite of the strong anisotropy present in the problem, the dispersion equations of cavity polaritons for small in-plane wavevectors ($q \ll \pi/L_c$) with high accuracy are not dependent on the direction of in-plane wavevector. The reason is two-fold. First, the two polarizations of the "bare" cavity photon take part in the formation of polaritonic states and their combined effect is such that all the effects of anisotropy for these q-values are actually washed out in the expression for the Rabi splitting. This happens because the electric field in the microcavity is such a combination of s- and p-polarized cavity photons, that (with accuracy up to small terms) the total electric field is parallel to the excitonic transition dipole moment for each given **q**. Second, the z-component of the total wavevector of the interacting coherent cavity photons and the crystalline thin film excitons are fixed by the boundary conditions at the microcavity mirrors, and this destroys the nonanalytical dependence of the exciton energy on its wavevector existing in bulk anisotropic crystals. As a result the anisotropy in the dispersion relation of the quasi-two-dimensional "bare" (Coulomb) excitons in a crystalline thin film enters only through small terms (proportional to powers of q^2/Q^2), and is too small to exhibit itself in the dispersion of the cavity polaritons. An observable anisotropy ($\sim W^2/\tilde{\omega}$) in the dispersion of the lower polariton branch can be found for large wavevectors only, where, however, the lower polariton is already exciton-like.

The predicted in-plane isotropization of the polariton dispersion in an anisotro -pic crystalline organic microcavity can be observed in the spectra of reflection, transmission and photoluminescence.

10.3 Microcavities with disordered organics

10.3.1 *Qualitative picture*

Many organic materials (disordered and crystalline as well) have rather broad and dispersionless electronic resonances. The excitations in these materials are localized. They propagate by jumps and not by wavepackets as coherent excitations. Thus for these excitations the wavevector is not a "good" quantum number and the number of excited states in organics (states of excitons) in approximations of two-level system is equal to the number of organic molecules. We analyze in this section the optical properties of exciton–polaritons in a microcavity, which utilizes such organic materials as the optically active semiconductor. We begin our consideration of different processes in an organic microcavity with a very qualitative discussion.

Recall that the cavity photons are coherent excitations with a continuous spectrum and with the energy dependent on the wavevector. The interaction of these coherent excitations with localized excitons determines the spectrum of new coherent excitations (cavity polaritons) appearing under the influence of the exciton–photon interaction. However, if this interaction is strong enough it produce a splitting (Rabi splitting) in the vicinity of the resonance where the energy of the cavity photon is close to the energy of the exciton. The number of cavity photon states strongly interacting with exciton states is small. Nevertheless, as a result of the light–matter interaction in the vicinity of the resonant frequency two branches of coherent states (cavity polaritons) appear, which are analogous to the cavity polariton branches observed in inorganic semiconductor microcavities. In contrast to the case of polaritons in inorganic semiconductors, the lower polariton branch (lower branch of coherent states) appears only in a certain restricted interval of wavevectors. The number of such new states is much smaller than the total number of bar molecular excited states in the microcavity. As a result the majority of the molecular electronic excited states do not strongly couple to the cavity photon, and these states, the same as the states in the organic material outside of the cavity, can be regarded as essentially incoherent.

The numerical simulations performed in (32), (33) for one-dimensional microcavity containing organic material with diagonal disorder qualitatively confirm the conclusions made above on the structure of the polaritons spectrum in an organic microcavity. Future simulations for realistic 2D organic microcavities would be very topical.

The origin of the difference of this picture of the spectrum in a disordered organic microcavity with the spectrum of polaritons in a microcavity containing a crystalline semiconductor is rather simple. Indeed, the number of cavity photon states strongly interacting in the semiconductor microcavity with Wannier–Mott excitons is the same as in an organic microcavity and is small in comparison with the total number of exciton states. However, for the exciton states in a semiconductor microcavity the wavevector is a good quantum number and these states in the semiconductor form the band of coherent states. The light–matter interaction creates Rabi splitting and hybrid exciton–photon states (cavity exciton–polaritons) for some range of wavevectors and at the same time does not destroy the coherence of exciton states in the band. In such a microcavity all the states are coherent (of course, we neglect here different types of disorder, and so on).

The above physical picture of the coexistence of coherent and incoherent (localized) states in a microcavity with disordered organics can also be expected to occur outside of the microcavity in bulk or at surfaces for many different disordered structures strongly interacting with light wave mode including, for instance, the surface plasmon–polariton modes. As an example we can mention some recent experiments (34). This paper investigated in the fluorescence of a structure containing periodically conjugated J-aggregated cyanine dye (TDBC) layer deposited on a surface of a silver film. It demonstrated the coexistence of emission from coherent states – polaritons – and incoherent localized cavity excitations. Namely, it was found that in the case of emission from coherent states the energy depends on the angle of observation. In the case of incoherent states the maximum of the luminescence is located at the same energy as the uncoupled TDBC excitons.

As another example, we mention the hybrid plasmon–polariton modes in chains of noncontacting noble metal nanoparticles where the interaction of a photon and nanoparticles leads toe delocalized plasmon–polaritons (see (35) and

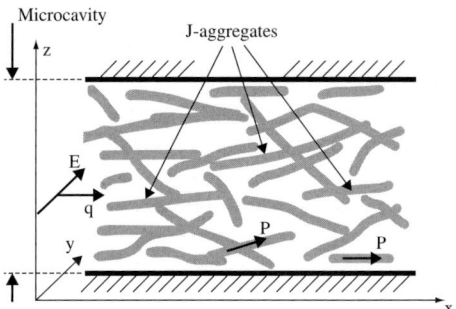

FIG. 10.3. A microcavity with *J*-aggregates formed from a cyanine dye suspended in a transparent polymer matrix. The dipole moments of *J*-aggregates are indicated for two particular chains. Reprinted with permission from Agranovich *et al.* (15). Copyright 2003, American Physical Society.

references herein). One can expect for this structure also the coexistence of coherent (delocalized) and incoherent (localized) excitations. This conclusion follows directly from our numerical simulation on states in a 1D microcavity.

10.4 Macroscopic classical theory

In this section we examine the solutions of Maxwell's equations for a system with a broad and dispersionless electronic resonance. We show that these conditions result in the appearance of the end-points of the lower and upper polariton branches. These end-points restrict the intervals in which the polariton states have well-defined wavevectors. This consideration is applicable, in particular, to the disordered system of *J*-aggregates since each *J*-aggregate chain possesses rather narrow electronic transitions instead of broad dispersion (Fig. 10.3). The disorder present in the system does not influence the following arguments, since for small-cavity photon wavevectors, the system can be treated as effectively homogeneous.

10.4.1 *General expressions*

For simplicity we treat the microcavity with an optically active material as a slab of width L_c bounded by two perfect mirrors. Let the z-axis be directed normal to the cavity plane, with $z = 0$ corresponding to the middle of the microcavity. The eigenmodes of the empty microcavity are characterized by a two-dimensional (in-plane) wavevector $\mathbf{q}$, and the values of the wavevector in the z-direction are quantized, with $k_z = \pi/L_c$ for the lowest energy eigenmode. The mode which we will consider is polarized in the xy plane, and is normal to the photon wavevector $\mathbf{q}$. Its dispersion is given by:

$$E_{\text{cav}}(q) = \frac{\hbar c}{\sqrt{\epsilon_b}} \sqrt{k_z^2 + q^2} = \frac{\hbar c}{\sqrt{\epsilon_b}} \sqrt{\frac{\pi^2}{L_c^2} + q^2}, \qquad (10.20)$$

where ϵ_b is the background dielectric constant. Let the microcavity be filled with a medium with a single dispersionless resonance at the energy E_0 (the generalization for the case of a few resonances is straightforward). Then the Maxwell equations for the transverse wave read:

$$\frac{\hbar^2(q^2+k_z^2)c^2}{E^2} = \epsilon_b + \frac{A}{E_0^2 - E^2 - 2iE\gamma_0}, \tag{10.21}$$

where γ_0 is the total (homogeneous plus inhomogeneous) broadening of the excitonic resonance and A is proportional to the oscillator strength of the transition. Then for $E \approx E_{\text{cav}}(q) \approx E_0$:

$$[E - E_{\text{cav}}(q)]\,[E - E_0 + i\gamma_0] = \Delta^2, \tag{10.22}$$

where $\Delta^2 = (A/4\epsilon_b)$. Equation (10.22) is just a well-known dispersion relation for microcavity polaritons which accounts for the broadening of the excitonic (dispersionless, in our case) resonance. Its well-known solutions are the upper and the lower polariton branches:

$$E_{U,L}(q) = \frac{E_{\text{cav}}(q) + E_0 - i\gamma_0}{2} \pm \sqrt{\Delta^2 + \frac{(E_{\text{cav}}(q) - E_0 + i\gamma_0)^2}{4}}. \tag{10.23}$$

If $\gamma_0 \ll \Delta$ (the case of strong light–matter coupling), eqn (10.23) reduces to

$$E_{U,L}(q) = E_{U,L}^{(0)}(q) - i\,\delta E_{U,L}(q); \qquad \delta E_{U,L}(q) = \gamma_0 \left|c_{\text{ex}}^{(U,L)}(q)\right|^2, \tag{10.24}$$

where $\delta E_{U,L}(q)$ is the polariton energy broadening, and

$$E_{U,L}^{(0)}(q) = \frac{E_{\text{cav}}(q) + E_0}{2} \pm \sqrt{\Delta^2 + \frac{(E_{\text{cav}}(q) - E_0)^2}{4}} \tag{10.25}$$

are the polariton dispersion curves in the absence of exciton broadening (i.e. with $\gamma_0 = 0$ in eqn (10.23)), and

$$\left|c_{\text{ex}}^{(U,L)}(q)\right|^2 = \frac{\Delta^2}{\Delta^2 + \left(E_0 - E_{U,L}^{(0)}(q)\right)^2} \tag{10.26}$$

are the weights of the exciton state in the polaritonic states shown in the inset of Fig. 10.5.

Before analyzing the above equations, we would like to make several short remarks. For a given system the Maxwell equations yield a dispersion equation in the form of a relation between ω and q. If from this dispersion equation it follows that for some interval of the real frequencies the wavevector is also real, one can say that the relation between ω and q describes the dispersion of so-called classical normal modes. As a result of further quantum description one comes to the quasiparticles picture and in this case the dispersion relation describes the dependence of the quasiparticle energy $\hbar\omega$ on the wavevector q (*coherent quasiparticles*).

However, in a realistic situation some dissipation processes are involved and the dielectric tensor becomes complex. Then, generally speaking, both the energy and the wavevector have to be considered as complex variables: $\omega = \omega' + i\omega''$, $q = q' + iq''$. Nevertheless, the picture of quasiparticles is still applicable if $\omega'' \ll \omega'$ and $q'' \ll q'$, i.e. when, first, the uncertainty of the energy $\delta E = \hbar\omega''$ is small in comparison with the quasiparticle energy and second, the uncertainty of the wavevector $\delta q = q''$ is small in comparison with the wavevector.

In the optical region of the spectrum the uncertainty of the energy (the energy broadening) of a quasiparticle is usually much smaller than the quasiparticle energy. But the uncertainty of the wavevector (wavevector broadening) in some cases can be very large. In terms of the uncertainty relation for the momentum (for the wavevector) and the coordinate of a quasiparticle, one can say that a wavevector uncertainty corresponds to some degree of excited state localization. The spatial localization of a quasiparticle can be described in terms of wavepackets, and the character of the wavepacket propagation strongly depends on its size. As is known, if the size of the wavepacket is much larger than the wavelength of the quasiparticle (i.e. when $q'' \ll q'$) and the dissipation is weak, the wavepacket propagates with group velocity $v_g = d\omega/dq$. In the following, we refer to such quasiparticles when speaking about coherent states with well-defined wavevector. In contrast, in the limit of strong localization of excitations typical, for instance, for disordered media, the excitations can propagate only by means of random hops. The energy of such excitations cannot be characterized in terms of wavevectors at all, and the excited states of this type will be called incoherent states in the following. We recall these well-known arguments, since in the materials which we consider here the incoherent and the coherent excited states can coexist.

Considering the energy and the wavevector broadenings, one has to take into account that these broadenings are mutually dependent quantities, since both E and q obey the same dispersion equation. For our case of microcavity polaritons, the relation between them is illustrated in Fig. 10.4, where solid lines are the polariton dispersion curves in the absence of dissipation. Equation (10.24) shows that, as long as only the broadening of the excitonic resonance is taken into account, the energy broadening of the polaritonic state is the excitonic broadening times the excitonic weight in the collective exciton–photon state. This broadening δE is plotted for each real q (dotted lines). Thus, each polaritonic branch is surrounded by a stripe. The width of the stripe for a given real q tends to γ_0 when $|c_{ex}^{(U,L)}|^2 \to 1$, and it vanishes when $|c_{ex}^{(U,L)}|^2 \to 0$. This means that the energy broadening of the lower polariton state grows moderately with the increase of the wavevector, while the upper polariton broadening decreases with the increase of the wavevector. As $E_{U,L}^{(0)}$ is of the order of 2 eV, and γ_0 is of the order of 20 meV, we see that the energy is well-defined, since the uncertainty of

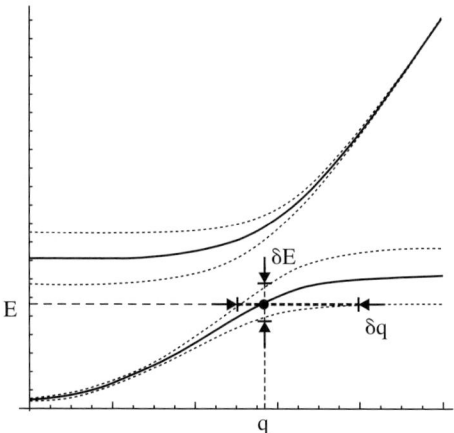

FIG. 10.4. An illustration of the wavevector and energy broadenings of polariton branches. Reprinted with permission from Agranovich *et al.* (15). Copyright 2003, American Physical Society.

the energy is $\delta E_{U,L}(q) \ll E_{U,L}^{(0)}(q)$.

In turn, from Fig. 10.4 it is seen that for each given real energy E there exist a set of states with different values of the wavevectors. The uncertainty in the energy δE and the uncertainty in the wavevector δq are marked in Fig. 10.4 with arrows for some given pair E and q. The strong increase of the lower polariton wavevector broadening for large wavevectors seen in Fig. 10.4 suggests that for large wavevectors the coherent lower branch polaritonic states do not form anymore. In the following we estimate quantitatively the borders of the intervals where polaritonic states have well-defined wavevectors.

For the regions where the wavevector is a "good" quantum number, the wavevector broadening of the polaritonic states can be estimated (36) as $\delta q = (\delta E/\hbar v_g)$, where v_g is the group velocity of the polaritonic state:

$$v_g^{(U,L)} = \frac{\hbar c^2 q}{\epsilon_b E_{\mathrm{cav}}(q)} \frac{\left(E_0 - E_{U,L}^{(0)}(q)\right)^2}{\Delta^2 + \left(E_0 - E_{U,L}^{(0)}(q)\right)^2}. \tag{10.27}$$

After the expressions for $v_g^{(U,L)}$ and $\delta E^{(U,L)}$ are substituted into the expression for δq, we find (eqn 10.22 with $\gamma_0 = 0$ was used for the calculation of a group velocity):

$$\delta q_{(U,L)} = \frac{\gamma_0 \left|c_{\mathrm{ex}}^{(U,L)}(q)\right|^2}{\hbar v_g^{(U,L)}} = \frac{\gamma_0 \epsilon_b E_{\mathrm{cav}}(q)}{c^2 \hbar^2 q} \frac{\left(E_{\mathrm{cav}}(q) - E_{U,L}^{(0)}(q)\right)^2}{\Delta^2}. \tag{10.28}$$

The polaritonic state has a well-defined wavevector as long as $\delta q \ll q$. However, δq diverges (i) for $q \to 0$ for both branches, and (ii) for large q for lower polariton, since in both these cases $v_g \to 0$. Below we consider both these cases in detail.

10.4.2 The case of vanishing q

Let us consider first the vicinity of $q = 0$. Since $\delta q_{(U,L)}$ diverges for $q \to 0$, it is clear that at some value of the wavevector the wavevector broadening becomes larger than the wavevector itself. We can estimate the characteristic values $q_{\min}^{(U,L)}$ of the lower edge of the spectrum of coherent excitations with well-defined wavevector from the condition $\delta q \sim q$. For the J-aggregate-based microcavities investigated in Ref. (10), assuming the detuning $u = E_{\text{cav}}(0) - E_0$ is -130 meV, and $\gamma_0 = 20$ meV, we find: $q_{\min}^{(U)} = 3.5 \cdot 10^4$ cm^{-1} and $q_{\min}^{(L)} = 4.2 \cdot 10^3$ cm^{-1} for $2\Delta = 80$ meV; $q_{\min}^{(U)} = 2.2 \cdot 10^4$ cm^{-1} and $q_{\min}^{(L)} = 10^4$ cm^{-1} for $2\Delta = 300$ meV. We note that $q_{\min}^{(U)} > q_{\min}^{(L)}$, since the upper branch is exciton-like at small q. As a result, the group velocity $v_g^{(U)}(q)$ grows more slowly than $v_g^{(L)}(q)$ with the increase of q, and also $|c_{\text{ex}}^{(U)}(q)|^2 \approx 1$ for small q (see eqn 10.28).

From this estimation it follows that for states with small $q < q_{\min}$ the wavevector is not a "good" quantum number for both polaritonic branches. The excited states in this part of the spectrum are also the wavepackets formed by the hybridized cavity photon mode and electronic excitations of the optically active medium, but the expression for the group velocity (10.27) used in our estimations is already inapplicable. We discuss the nature of these lowest energy states at the end of this chapter. The properties of low-energy excited states should be manifested in the angular dependencies of reflectivity and photoluminescence measurements, e.g. when analyzing the radiation which is nearly normal to the plane of the microcavity. Since $q_{\min}^{(U)} > q_{\min}^{(L)}$, this effect should be more pronounced for the upper branch. It is also clear that analogous arguments can be applied for the usual inorganic microcavities. However, the value of $q_{\min}$ may be strongly reduced for inorganic structures where the width γ_0 at low temperature is much smaller. We address this point below in more detail.

Now let us consider also states with energies below $E_L(q_{\min}^{(L)})$, where the imaginary part of the wavevector is no longer small in comparison with its real part. These states are of interest, since, first, they determine the distribution of photoluminescence for the directions near to the normal to the surface of the microcavity, and second, because just these low-energy states can be important in the discussion of condensation of cavity polaritons.

The structure of the lowest energy polaritonic state in the presence of dissipation can be examined directly from the dispersion relation (10.22). In the absence of dissipation, for the lower branch this state is characterized by the energy $E = E_L^{(0)}(0)$ and $q = 0$. In this approximation the photoluminescence from this state is directed strictly normal to the microcavity surface. If the dissipation is taken into account, for the same value of energy $E = E_L^{(0)}(0)$ the wavevector becomes complex, $q = q' + iq''$. For small wavevectors, $E_{\text{cav}}(q) = E_c + (\hbar^2 q^2/2\mu)$, where $\mu = = (\pi\hbar\sqrt{\epsilon_b}/cL_c)$ and $E_c = E_{\text{cav}}(0)$. Then from eqn (10.22) it follows that

$$q^2 = i\frac{2\mu\gamma_0}{\hbar^2}\frac{\sqrt{\Delta^2 + u^2/4} + u/2}{\left[\sqrt{\Delta^2 + u^2/4} - u/2\right] - i\gamma_0}, \tag{10.29}$$

where $u = E_c - E_0$ is the detuning. For zero or negative detuning, the term $i\gamma_0$ can be neglected in the denominator, and we find that the real and imaginary parts of the wavevector are equal.

For zero detuning, i.e. when $E_c = E_0$ (which is often the case for many experiments with inorganic quantum wells inserted into a microcavity), we have:

$$q' \approx q'' \approx \sqrt{\frac{\mu\,\gamma_0}{\hbar^2}}. \tag{10.30}$$

Taking the parameters typical for usual inorganic structures ($\gamma_0 = 1$ meV, $\mu = = 0.5 \cdot 10^{-5} m_e$, where m_e is the electron mass), we find: $q' \sim q'' \sim 3 \cdot 10^3$ cm^{-1}. This means that the photoluminescence from the sample excited at the energy $E = E_L^{(0)}(0)$ comes out within the angle $\theta \simeq q'/Q$, where $Q = (E_L^{(0)}(0)/\hbar c)$ is the wavevector of the light outside the microcavity. For inorganic structures, the estimate gives: $\theta \sim 2°$.

In turn, the parameters of the organic microcavities investigated experimentally in Ref. (10) are: the Rabi splitting $\Delta = 40$ meV, the detuning $u = -130$ meV, $\gamma_0 = 20$ meV, $\mu = 0.7 \cdot 10^{-5}$ m_e. For this structure with large negative detuning, eqn (10.29) gives:

$$q' \approx q'' \approx \sqrt{\frac{\mu\,\gamma_0}{\hbar^2}\frac{\Delta^2}{u^2}}, \tag{10.31}$$

so that $q' \sim q'' \sim 4500$ cm^{-1}. The angular broadening near the normal direction then is about 3°, i.e. it is again rather narrow.

In the same way the angular broadening of photoluminescence from the upper branch can be considered. Assume that $E = E_U^{(0)}(0)$ in eqn (10.22). For zero detuning, the angular broadening for the upper branch is the same as for the lower branch. For the organic material of Ref. (10), we obtain: $q' \sim 48000$ cm^{-1}, $q'' \sim 13600$ cm^{-1}. Thus, $\theta \sim 30°$, i.e. the angular distribution of photoluminescence from the upper branch is well noticeable.

The broadenings δE and δq can be measured directly in different types of optical measurements. First, for example, for the samples with a plane structure (like for a plane microcavity or a plane surface) at pumping one can fix the tangential component of the wavevector and collect the radiation (PL) at different energies. As a result the energy broadening of the excited state in the microcavity with a given tangential component of the wavevector can be measured. Second, one may fix the energy of the excitation and collect the radiation for different angles. The radiation collected in a certain angle around some value of the central wavevector will then characterize the wavevector broadening of the state which is excited by pumping.

10.4.3 The case of large q

Now let us consider the case of large q. We start with the lower polariton. The right-hand side of eqn (10.28) behaves as q^2 for $q \to \infty$, and thus at some point the broadening in the wavevector again becomes large in comparison with the wavevector itself. Note that this is not the case for usual (inorganic) cavity polaritons, since there the excitonic resonance has a finite mass m^* as a result of the resonance intermolecular interaction, so that for large q the lower polariton group velocity tends to $q\hbar/m^*$, and not to zero. Again, for the samples of Ref. (10) we can roughly estimate the upper edge $q_{\max}^{(L)}$ of the spectrum of coherent states with well-defined wavevector from the condition $\delta q \sim q$. For $2\Delta = 80$ meV and $2\Delta = 300$ meV we find, respectively: $q_{\max}^{(L)} = 9.3 \cdot 10^4 \text{cm}^{-1}$ and $q_{\max}^{(L)} = 2.6 \cdot 10^5 \text{cm}^{-1}$. We note that from eqn (10.28) it is seen that the larger is Δ, the smaller is δq, and thus the wider is the coherent part of the spectrum.

Now let us consider the wavevector broadening of the upper polariton states for large q. At large wavevectors the upper polariton dispersion curve tends to that of the cavity photon, and $\delta q \simeq \gamma_0(\Delta^2 \epsilon_b{}^{3/2}/q^2\hbar^3 c^3) \ll q$. Thus, independently of the value of the Rabi splitting, for large q the upper cavity polariton branch contains the coherent states only.

From this macroscopic consideration it is seen that the states for which the wavevector is not a "good" quantum number do not form in a certain vicinity of $q = 0$ for both branches, and for $q > q_{\max}^{(L)}$ for the lower branch. In other words, the states with the well-defined wavevector exist in the intermediate region of the wavevectors only: $q_{\min}^{(L)} < q < q_{\max}^{(L)}$ for the lower branch, and $q > q_{\min}^{(U)}$ for the upper branch. However, in contrast to the case of vanishing q, one can say that for $q \gg q_{\max}^{(L)}$ the coherent polaritonic states do not form at all. The excited states from this part of the spectrum are not resonant with the cavity photon, and as a result no hybridization happens. Instead, these excited states are similar to the localized excited states in a non-cavity material, i.e. they are to be treated just as incoherent excited states.

The spectrum of the excitations is shown in Fig. 10.5 for $2\Delta = 80$ meV. The dashed lines show the uncoupled molecular excitons and photons, and the solid lines show the coherent part of the spectrum with well-defined wavevector. The crosses show the end-points of the spectrum of excitations for which q is a "good" quantum number. The spectrum of incoherent (weakly coupled to light) states is shown by a broadened line centered at the energy E_0. It follows from the expression for the dielectric tensor that this spectrum is the same as the spectrum of out-of-cavity organics. The spectrum of absorption as well as the dielectric tensor depend on temperature. This means that in the calculation of the temperature dependence of the polariton spectrum we have to use the temperature dependence of the resonance frequency E_0 as well as the temperature dependence of γ determining the width of the absorption maximum. However, the spectrum of emission of local states which pump polariton states can be different from the spectrum of absorption. The Stokes shift in many cases

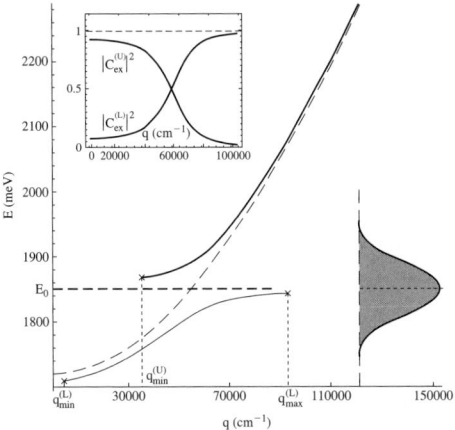

FIG. 10.5. The dispersion curves of the coherent polaritonic states (solid lines) and of uncoupled cavity photons and the molecular excitation (dashed lines). The crosses show the end-points of the part of the spectrum with well-defined wavevector. On the right, the broadened line of the molecular resonance is shown. The inset shows the excitonic weights (10.26) for upper ($|c_{ex}^{(U)}|^2$) and lower ($|c_{ex}^{(L)}|^2$) polaritonic branches. Reprinted with permission from Agranovich et al. (15). Copyright 2003, American Physical Society.

can be significant for organic materials and one has to take this into account in the discussion of the pumping of polaritons and the dynamics of the polariton relaxation in cavity emission.

Finally, let us estimate the relative number of lower polariton states (the absence of such states with "good" wavevector in the vicinity of $q \simeq 0$ in this estimation can be neglected). If we treat the molecules as two-level systems, then the total number of excited states is just equal to the number N of the molecules. In turn, it is clear that the number of new coherent states is of the order of $(\bar{R}q_{\max})^2 N \ll N$ ($\bar{R}$ is the mean distance between the molecules). For the samples of Ref. (10) the order of magnitude of $\bar{R}$ can be obtained from the concentration η of J-aggregates: $\bar{R} \simeq \eta^{-1/3}$. We estimate the concentration from (i) the data on the concentration of the monomers used for the preparation of the film, and (ii) from the estimation of the number n of the monomers in a single J-aggregate. This last figure was obtained from the comparison between the inhomogeneous broadening of a monomer γ_{inh} and the inhomogeneous broadening of a J-aggregate $\Gamma_{\text{inh}} \sim \gamma_{\text{inh}}/\sqrt{n}$ (37). The rough estimation gives: $n \sim 100$. Then $\eta = 1.3 \cdot 10^{17}$ cm^{-3}, and $\bar{R} = 200$ Å. We find that for $2\Delta = 80$ meV and $2\Delta = 300$ meV the fractions of the coherent states are 0.035 and 0.27, respectively. Thus, the incoherent states form the majority of the electronic excited states of the system. As we shall show below, for this reason they play a very important role in the excitations dynamics in a microcavity.

10.4.4 Microscopic quantum theory

In this section we develop a microscopic quantum model which accounts both for positional and orientational disorder in such a system. Using the bosonic Hamiltonian for the system, we find the structure of the eigenstates (i.e. the weights of the electronic excitations on different molecules of the disordered medium) in the intervals where the wavevector of the cavity polaritons is a good quantum number. These weights will be used in Ch. 13 in consideration of the upper polariton nonradiative decay and also for estimations of the rate transition from incoherent states to the lowest energy polariton states.

Again, we treat the microcavity as a slab of thickness L_c and dielectric constant ϵ_b, bounded by two perfect mirrors. However, the slab now contains J-aggregate chains having both position and orientation distributed chaotically throughout the microcavity volume (see Fig. 10.3). Again, we suppose that each J-aggregate has a single intense and dispersionless optical transition at an energy E_0. We are interested in the eigenstates in the system, which appear as a result of the interaction between the cavity photons and the excitations supported by the J-aggregates. As the length of each J-aggregate chain is assumed to be much shorter than the typical photon wavelength, we do not account for the finite extension of the J-aggregates. Instead we model each J-aggregate as a single "molecule", which possesses a point dipole moment, again distributed homogeneously and randomly throughout the cavity.

Let $a^\dagger(\mathbf{q})$ and $a(\mathbf{q})$ be the creation and annihilation operators of cavity photons, and $B_i^\dagger$ and B_i be the creation and annihilation operators of an excitation on the ith "molecule". We do not account for the dipole–dipole interactions between different "molecules", as we consider the coupling by light to be the principal coupling mechanism. Thus the Hamiltonians of noninteracting cavity photons and molecular excitations are

$$H_c = \sum_{\mathbf{q}} E_{\text{cav}}(q) a^\dagger(\mathbf{q}) a(\mathbf{q}); \qquad H_J = \sum_i E_0 B_i^\dagger B_i. \qquad (10.32)$$

Following the usual procedure (see Ch. 4), we write the Hamiltonian of the exciton–photon interaction near the anticrossing region in the form:

$$H_{\text{int}} = \sum_i \sum_{\mathbf{q}} \left[T_i(\mathbf{q}) a(\mathbf{q}) B_i^\dagger + T_i^*(\mathbf{q}) a^\dagger(\mathbf{q}) B_i \right], \qquad (10.33)$$

$$T_i(\mathbf{q}) = i \sqrt{\frac{4\pi}{S L_c \epsilon_b E_{\text{cav}}(q)}} p_{0f} E_0 \, (\mathbf{n_q s}_i) \cos\left(\frac{\pi z_i}{L_c}\right) e^{i \mathbf{q} \mathbf{r}_\parallel^{(i)}}, \qquad (10.34)$$

where $\langle 0|\mathbf{P}_i|f\rangle \equiv \mathbf{s}_i p_{0f}$ ($\mathbf{s}_i$ is a unit vector directed along ith "molecule", and $\mathbf{P}_i$ is the operator of the dipole moment of the "molecule"), $\mathbf{n_q} = [\mathbf{e}_z \times \mathbf{q}]/q$ ($\mathbf{e}_z$ is the unit vector in the direction of the z-axis), and the coordinates $(\mathbf{r}_\parallel^{(i)}, z_i)$ determine the position of the ith "molecule".

To find the eigenstates of the full Hamiltonian, which is the sum of the Hamiltonians (10.32) and (10.33), we introduce new Bose operators ξ and $\xi^\dagger$ for the mixed exciton–photon states as follows:

$$\xi^\dagger = \sum_\mathbf{q} \alpha(\mathbf{q}) a^\dagger(\mathbf{q}) + \sum_i \beta_i B_i^\dagger; \quad \xi = \sum_\mathbf{q} \alpha^*(\mathbf{q}) a(\mathbf{q}) + \sum_i \beta_i^* B_i. \quad (10.35)$$

The coefficients $\alpha(\mathbf{q})$ and β_i determine the fraction of light (with wavevector $\mathbf{q}$) and the fraction of the excitation on the ith "molecule" in the polaritonic states respectively. From the Hamiltonian eigenvalue problem we find that the coefficients $\alpha(\mathbf{q})$ and β_i obey the following system of equations:

$$\alpha(\mathbf{q})\left[E - E_{\text{cav}}(q)\right] = \sum_i T_i^*(\mathbf{q}) \beta_i;$$
$$\beta_i \left[E - E_0\right] = \sum_\mathbf{q} T_i(\mathbf{q}) \alpha(\mathbf{q}). \quad (10.36)$$

Substituting β_i from the second of eqn (10.36) into the first equation, we find that the relation which determines the energies and amplitudes of cavity polaritons is:

$$\alpha(\mathbf{q})[E - E_{\text{cav}}(q)] = \sum_i T_i^*(\mathbf{q}) \sum_{\mathbf{q}_1} T_i(\mathbf{q}_1) \alpha(\mathbf{q}_1) / [E - E_0]. \quad (10.37)$$

The amplitude of cavity photon $\alpha(\mathbf{q})$ enters in the right-hand side of eqn (10.37) with all possible values of its argument $\mathbf{q}_1$ due to the fact that the medium does not possess true translation symmetry. However, those values of the argument, which differ from $\mathbf{q}$ appearing in the left-hand side of eqn (10.37), just describe the scattering of light by the inhomogeneities in the cavity medium. If this scattering is neglected, then only the term with $\mathbf{q}_1 = \mathbf{q}$ survives in the sum over $\mathbf{q}_1$, and the zero-order relation which determines the dispersion equation for the cavity polaritons is:

$$[E - E_{\text{cav}}(q)] [E - E_0] = \sum_i |T_i(\mathbf{q})|^2. \quad (10.38)$$

In this approximation the in-plane positions of the "molecules" $\mathbf{r}_\parallel^{(i)}$ do not enter into the solution. The final form for the dispersion equation of cavity polaritons repeats eqn (10.22) obtained in the classical formalism with $\gamma_0 = 0$ and

$$\Delta^2 = \sum_i |T_i(\mathbf{q})|^2 \simeq \frac{4\pi p_{0f}^2 E_0^2}{SL_c \epsilon_b E_c} \sum_i (\mathbf{n_q s}_i)^2 \cos^2\left(\frac{\pi z_i}{L_c}\right), \quad (10.39)$$

where $E_c = E_{\text{cav}}(0)$. In the particular case when the dipole moments of all the "molecules" are completely randomly distributed with the centers of the "molecules" being distributed homogeneously in the z-direction, the sum over i in eqn (10.39) can be calculated using

$$\sum_i \rightarrow \frac{N}{2\pi L_c} \int_0^{2\pi} d\phi \int_{-L_c/2}^{L_c/2} dz, \qquad (10.40)$$

where N is the total number of "molecules", and ϕ is the angle between $\mathbf{q}$ and $\mathbf{s}_i$. Then

$$\Delta^2 = \frac{\pi p_{0f}^2 E_0^2}{\epsilon_b E_c} \eta, \qquad (10.41)$$

where $\eta = (N/SL_c)$ is the concentration of the "molecules". The dispersion relations of the cavity polaritons are again given by eqn (10.25), and the expressions for the weight coefficients $\alpha(\mathbf{q})$ and β_i can be found for each branch from eqn (10.36) together with the normalization condition. If we neglect the scattering of the polaritons by the inhomogeneities of the medium, then $\beta_i^{(U,L)} = \beta_i^{(U,L)}(\mathbf{q})$, and the normalization condition for each polariton mode reads:

$$|\alpha(\mathbf{q})|^2 + \sum_i |\beta_i(\mathbf{q})|^2 = 1. \qquad (10.42)$$

Then, from eqns (10.36) and (10.42) we find:

$$|\alpha^{(U,L)}(\mathbf{q})|^2 = \frac{(E_{U,L} - E_0)^2}{(E_{U,L} - E_0)^2 + \Delta^2};$$
$$|\beta_i^{(U,L)}(\mathbf{q})|^2 = \frac{\Delta^2 (4/N)(\mathbf{n_q s}_i)^2 \cos^2\left(\frac{\pi z_i}{L_c}\right)}{(E_{U,L} - E_0)^2 + \Delta^2}. \qquad (10.43)$$

In this section we completely ignored the damping of the molecular and the cavity photon states. In other words, the cavity polariton wavevector was treated as a "good" quantum number. Therefore, based on the results of the previous Section, we conclude that the relations we have obtained are only applicable for the wavevectors $q_{\min}^{(L)} < q < q_{\max}^{(L)}$ for the lower branch, and $q > q_{\min}^{(U)}$ for the upper branch.

10.5 The localized end-point polariton states

10.5.1 *Dynamics of a low-energy wavepacket in a perfect microcavity*

One of the unsolved problems arising in the framework of the macroscopic approach is the nature of the lowest energy local states in the microcavity with disordered organics. The nature of states which formally appear in the region of molecular resonance are clear: they do not interact with the cavity photon and for this reason are similar to states in non-cavity organic material. However, the nature of cavity polaritons near the lowest end-points deserves careful analysis (see also (38)). It is interesting to clarify their properties because, for instance, just the lowest energy states have to be populated in the processes of condensation of cavity polaritons. These states, first of all, strongly depend on detuning. Most interesting is the case of small detuning because at small detuning these states, which are the same as states of coherent polaritons, are hybrid light–matter states. To simplify our consideration of the polariton wavefunction let us begin with a known situation with the wavepacket of normal waves:

$$U(\mathbf{r},t) = \int A(\mathbf{k})\exp[i(\omega(\mathbf{k})t - \mathbf{kr})]d^3k,$$

where the dependence of the function $\omega(\mathbf{k})$ on wavevector determines the dispersion of normal waves in the medium. If in the wavepacket the main contribution into the considered integral arises due to the contribution of wavevectors from the vicinity of the wavevector $\mathbf{k}_0$ we can use, in the calculation of the integral, the expansion:

$$\omega(\mathbf{k}) = \omega(k_0) + (k - k_0)\frac{d\omega}{dk}$$

and as a result we obtain:

$$U(\mathbf{r},t) = \exp\left[i\left(\omega(k_0)t - k_0 r\right)\right]\int d^3\kappa A(\kappa)\exp\left[i\kappa\left(\frac{d\omega}{dk_0}t - \mathbf{r}\right)\right]$$
$$= \exp\left[i\left(\omega(k_0)t - k_0 r\right)\right]\psi\left(\frac{d\omega}{dk_0}t - \mathbf{r}\right).$$

Thus, it follows from this expression that the amplitude of the wavepacket depends only on $\left(\frac{d\omega}{dk_0}t - \mathbf{r}\right)$, moves with group velocity $\frac{d\omega}{dk_0}$, and is a correct and known result only if the group velocity does not equal zero. We will briefly continue to discuss the time evolution of wavepackets in perfect microcavities where all polaritons are coherent states with well characterized wavevectors. Not only will this establish a comparative benchmark but it is useful in itself as such dynamics reflect features of the polariton spectrum, and hence of the exciton–photon hybridization. To consider eigenstates near the end-points we have to take into account that for lowest energy end-points $\mathbf{k}_0 = 0$ and that the group velocity at $k_0 = 0$ is also equal zero. Taking into account the next term quadratic in $\mathbf{k}$ in the expansion of $\omega(\mathbf{k})$ we obtain

$$\omega(\mathbf{k}) = \omega(k_0) + \alpha k^2. \qquad (10.44)$$

Just such a situation takes place for microcavity dispersion at the bottom of the lower and upper polariton branches in a microcavity with $\alpha = \hbar/2M$ where M is the effective mass of the cavity polariton. Of course, specific features of the low-energy wavepackets stem from the fact that the polariton dispersion near the bottom of the LP branch ($\mathbf{k} \simeq 0$) is manifestly parabolic. Consider a wavepacket formed with the states close to the branch bottom:

$$U(\mathbf{r},t) = e^{i\omega_0 t} \int d^3k A(\mathbf{k})\, e^{i(\mathbf{k}\cdot\mathbf{r}-\alpha k^2 t)}. \tag{10.45}$$

It is convenient to choose the weight amplitude function $A(\mathbf{k})$ Gaussian: $A(\mathbf{k}) = (\beta/2\pi^3)^{1/2} \exp\left[-\beta(\mathbf{k}-\mathbf{k}_0)^2\right]$, centered at wavevector $\mathbf{k}_0$. With this amplitude function, eqn (10.45) yields

$$|U(\mathbf{r},t)|^2 = C(t)\exp\left[-\frac{\beta(\mathbf{r}-2\alpha\mathbf{k}_0 t)^2}{2(\beta^2+\alpha^2 t^2)}\right] \tag{10.46}$$

for the time evolution of the spatial "intensity" of the wavepacket,

$$C(t) = \frac{\beta}{2\pi(\beta^2+\alpha^2 t^2)}.$$

Equation (10.46) describes a Gaussian-shaped wavepacket in 2D whose center

$$\mathbf{r}_c(t) = \mathbf{v}_g t, \quad \mathbf{v}_g = 2\alpha\mathbf{k}_0,$$

moves with the group velocity consistent with the dispersion (10.44), and whose linear width increases with time in accordance with the 1D variance

$$s(t) = \left(\beta + \alpha^2 t^2/\beta\right)^{1/2}. \tag{10.47}$$

The total energy in the wavepacket is conserved: with our choice of the amplitude function,

$$\int d^3 r\, |U(\mathbf{r},t)|^2 = 1.$$

Of course, eqn (10.46) can be derived as a product of two independent 1D normalized evolutions such as

$$|U(x,t)|^2 = \sqrt{C(t)}\exp\left[-\frac{\beta(x-2\alpha k_0 t)^2}{2(\beta^2+\alpha^2 t^2)}\right], \tag{10.48}$$

which will be important in our discussion of 1D microcavities, in this case the 1D packet amplitude function

$$A(k) = (\beta/2\pi^3)^{1/4}\exp\left[-\beta(k-k_0)^2\right]. \tag{10.49}$$

The spatial broadening (10.47) features an initial value of $\beta^{1/2}$ and a characteristic time $t_b = \beta/\alpha$ such that, at times $t \gg t_b$, the variance grows linearly with velocity $v_b = \alpha\beta^{-1/2}$. To appreciate the scales, some rough estimates can be made. So for the effective polariton mass M on the order of $10^{-5}m_0$ (m_0 being the free electron mass), the parameter $\alpha = \hbar/2M \simeq 5\cdot 10^4\ \text{cm}^2/\text{s}$. Estimates in

Ref. (15) made with the Rabi splitting and detuning ~ 100 meV yielded, for microcavities with disordered organics, $k_{\min} \sim 10^4$ cm^{-1}. Then for the wavepackets satisfying $1 \leq \beta^{1/2} k_{\min} \leq 10$, the characteristic time would be $0.2 \leq t_b \leq 20$ ps and the corresponding velocity $5 \cdot 10^8 \geq v_b \geq 5 \cdot 10^7$ cm/s. In our 1D numerical example below we will use the value of parameter β within the segment just discussed.

We note that by changing the physical parameters of the microcavity and organic material, as well as conditions for polariton excitation, one can influence the dynamics described above. One should also be aware that the evolution times are limited by the actual lifetimes τ of small wavevector cavity polaritons. Long lifetimes τ on the order of 10 ps can be achieved only in microcavities with high quality factors $Q = \omega \tau$.

10.5.2 *Time evolution of the lowest wavepacket*

Finding polariton states in disordered planar microcavities microscopically is a difficult task which do not attempt here. As a first excursion into the study of disorder effects on polariton dynamics, here we will follow (32) to explore the dynamics in a simpler microscopic model of a 1D microcavity. Such microcavities are interesting in themselves and can have experimental realizations; from the results known in the theory of disordered systems (39) one can also anticipate that certain qualitative features may be common for 1D and 2D systems (38).

The microscopic model we study is set up in the following Hamiltonian:

$$H = \sum_n (\varepsilon + \varepsilon_n) a_n^\dagger a_n + \sum_k \varepsilon_k b_k^\dagger b_k$$
$$+ \gamma \sum_{nk} \sqrt{\frac{\varepsilon}{N \epsilon_k}} \left(e^{ikna} a_n^\dagger b_k + e^{-ikna} b_k^\dagger a_n \right). \quad (10.50)$$

It consists of a lattice of N "molecular sites" spaced by distance a and comprises the exciton part (a_n is the exciton annihilation operator on the site n), photon part (b_k is the photon annihilation operator with the wavevector k and a given polarization) as well as the ordinary exciton–photon interaction. The cavity photon energy ε_k is defined by eqn (10.44), ε represents the average exciton energy, while ε_n are the on-site exciton energy fluctuations.

We will use uncorrelated normally distributed ε_n with zero mean and variance σ:

$$\langle \varepsilon_n \varepsilon_m \rangle = \sigma^2 \delta_{nm}. \quad (10.51)$$

The exciton–photon interaction is written in such a form that 2γ yields the Rabi splitting energy in the perfect system. We chose to use the same number N of photon modes, and the wavevectors k are discrete with $2\pi/Na$ increments. Our approach is to straightforwardly find the normalized polariton eigenstates $|\Psi_i\rangle$ (i is the state index) of the Hamiltonian (10.50) and then use them in the site-coordinate representation:

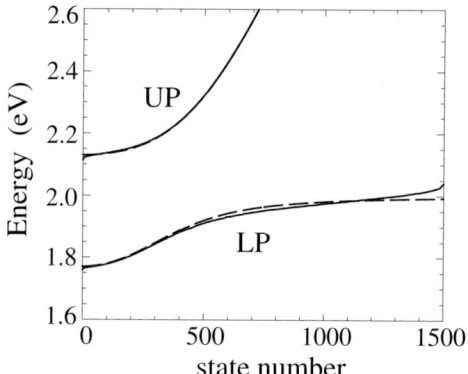

FIG. 10.6. The energy spectrum of the polaritonic eigenstates in a 1D model microcavity described by the Hamiltonian (10.50) with $N = 1500$ molecular sites and $N = 1500$ photon modes. Parameters of the systems are as follows: average exciton energy $\varepsilon = 2$ eV, cavity photon cutoff energy $\Delta = 1.9$ eV, dielectric constant $\epsilon = 3$, the exciton–photon interaction energy $\gamma = 0.15$ eV. The dashed lines show the energy eigenvalues in the system without exciton disorder, $\sigma = 0$, the solid lines correspond to the spectrum in the system with disorder, $\sigma = 0.03$ eV. Here the energies are shown as a function of state "number" sorted in increasing energy order, separately for the lower (LP) and upper (UP) polariton branches. In the system without disorder, the state numbers would be immediately convertible to the corresponding wavevectors. On this scale, the UP branches of perfect and disordered systems are hardly distinguishable. Reprinted with permission from Agranovich *et al.* (22). Copyright 2007, American Physical Society.

$$\Psi(n) = (\Psi_p(n), \Psi_e(n)), \quad \sum_n |\Psi(n)|^2 = 1, \qquad (10.52)$$

where Ψ_p and Ψ_e, respectively, describe the photon and exciton parts of the polariton wavefunction and n denotes the nth site.

In (38) various numerical parameters have been tried in the model Hamiltonian with the results being qualitatively consistent; the parameters exploited in these calculations have been chosen, on one hand, to be reasonably comparable with the experimental data in the output and, on the other hand, to better illustrate our point within a practical computational effort. It should be kept in mind though that we consider a model system and the numerical values of results may differ, likely within an order of magnitude, for various systems.

The numerical parameters are indicated in the caption to Fig. 10.6 and have been used to calculate the eigenstates of the Hamiltonian (10.50) with $N = 1500$ for a cavity of physical length $L = Na = 0.150$ mm and a small negative detuning $(\Delta - \varepsilon) = -0.1$ eV. Figure 10.6 compares the energy spectra in the perfect microcavity and in the cavity with one realization of the excitonic disorder, eqn (10.51), $\sigma = 0.03$ eV. It is apparent that the effect of this amount of disorder on the polariton energy spectrum *per se* is relatively small, except in the higher-energy region of the LP branch where eigenstates, as we discussed, are practically of a pure exciton nature.

The lower-energy part of the LP branch, however, corresponds to the polariton states Ψ (10.52) in which the exciton and photon are strongly coupled ($\gamma = 0.15 > |\Delta - \varepsilon| = 0.1$ eV) with comparable weight contributions in Ψ_p and Ψ_e. A dramatic effect of the disorder is in the strongly localized character of the polaritonic eigenstates near the bottom of the LP branch, as illustrated in Fig. 10.7(a) (needless to say the same behavior is observed for the states near

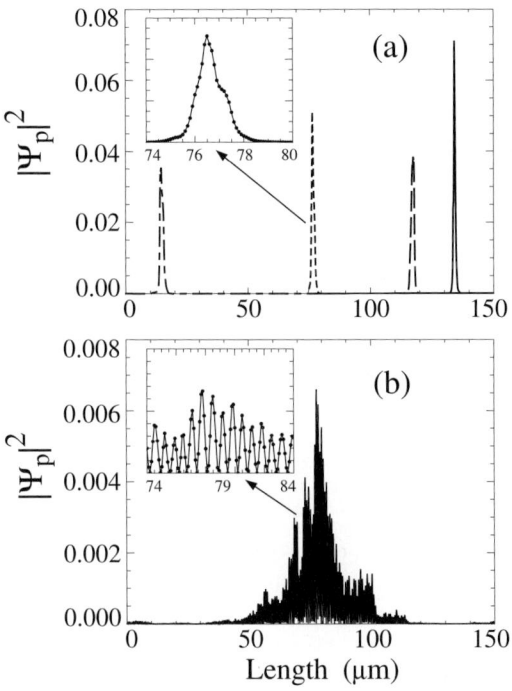

FIG. 10.7. Examples of the spatial structure of the the photon part $|\Psi_\mathrm{p}|^2$ of the polariton eigenstates in a 1D microcavity with disorder. (a) Four states, shown by different lines, from the very bottom of the LP polariton branch with energies within the range of 1.76–1.77 eV (see the spectrum in Fig. 10.6). The inset shows one of these states in more detail. Dots in the inset correspond to the sites of the underlying lattice. (b) One state with a higher energy close to 1.82 eV. The inset shows part of the spatial structure of this state in more detail. Dots correspond to the lattice sites and spatial oscillations of the wavefunction are clearly seen. Reprinted with permission from Agranovich *et al.* (22). Copyright 2007, American Physical Society.

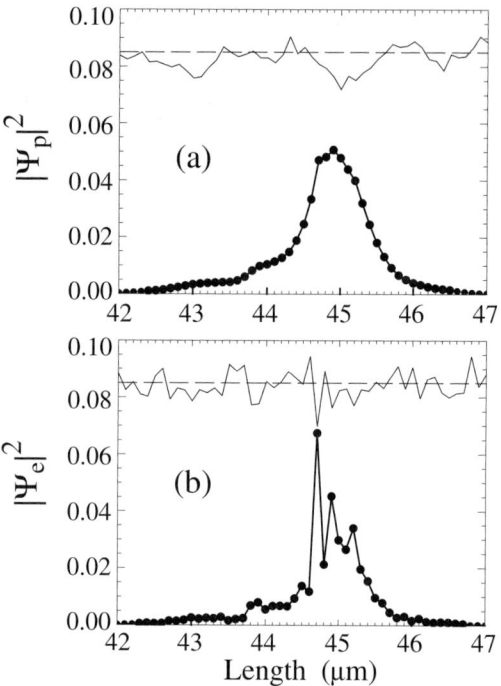

FIG. 10.8. Comparison of the spatial structure of the (a) photon $|\Psi_p|^2$ and (b) exciton $|\Psi_e|^2$ components of one of the localized polariton eigenstates near the bottom of the LP branch. Wavefunctions are shown by the thick lines with dots corresponding to the lattice sites. Thin lines in the top parts of both panels display *arbitrarily scaled* diagonal disorder energies along with their baseline shown with dashes. The diagonal disorder shown in panel (b) is the original uncorrelated site disorder of eqn (10.51), and the exciton component $|\Psi_e|^2$ is clearly seen to reflect these individual site energy fluctuations. The photon component $|\Psi_p|^2$, however, "responds" to spatially smoothed energetic disorder, which is illustrated in panel (a) with the diagonal disorder that is a result of box-averaging of the original site disorder with the box size of seven lattice sites. Reprinted with permission from Agranovich *et al.* (22). Copyright 2007, American Physical Society.

the bottom of the UP branch which we are not concerned with here). Of course, both the photon Ψ_p and exciton Ψ_e parts of the localized polariton state are localized on the same spatial scale, however the exciton part of the spatial wavefunction is more "wiggly" reflecting more the individual site energy fluctuations, Figure 10.8. For better clarity, in both Figs. 10.7 and 10.9 we show only the smoother behaving photon parts Ψ_p.

Panel (a) of Fig. 10.7 displays examples of $|\Psi_p|^2$ for four states from the very bottom of the LP branch in a realization of the disordered system that are localized at different locations of the 0.150 mm sample. The inset to this panel shows the spatial structure of one of these states in more detail; it demonstrates both the spatial scale ℓ of localization in this energy range ($\ell \sim 1 \times 10^{-6}$ m with the assumed parameters) as well as the "macroscopic" size of the localized state in comparison with the lattice spacing: $\ell \gg a = 100$ nm. A more detailed comparison of the photon and exciton components within a localized polariton state is illustrated in Fig. 10.8. for one of the states near the bottom of the LP branch. The wavefunctions in this figure are shown along with the actual diagonal disorder realization used to calculate this state. The difference in the displayed disorder patterns is that in panel (b) of Fig. 10.8 we show the original site energy fluctuations ε_n, while in panel (a) the disorder is smoothed by spatial box-averaging fluctuations ε_n with the box size of seven sites. It is evident from Fig. 10.8(b) that the exciton component indeed "responds" to the individual site energy fluctuations. On the contrary, the photon component, as suggested by Fig. 10.8(a), is more reflective of a spatially smoothed behavior of the disorder, responding, so to say, to whole groups of neighboring sites. It is interesting that the resulting spatial smoothing of the photon component occurs in a "self-consistent" way as determined by the interplay of the exciton–photon coupling and exciton–disorder interactions.

The states at the bottom of the LP branch can be characterized as strongly localized in the sense of $k\ell \leq 1$ where k is a typical wavevector of the parent polariton states in the perfect system. This feature may be contrasted to the behavior at somewhat higher energies and at higher $k > k_{\min}$ of parent states, where the disorder-induced indeterminacy of the k-vector becomes small, satisfying the equation

$$\delta k \ll k, \tag{10.53}$$

so that k would appear as a good quantum number. As is known, (39) however, multiple scattering should still lead to spatial localization of the eigenstates, now on the spatial scale l such that $k\ell \gg 1$. Panel (b) of Fig. 10.7 illustrates the spatial structure of such a state with much larger ℓ than in panel (a). The inset to panel (b) shows that the wavefunction in this case, within the localization length, exhibits multiple oscillations with a period on the order of $1/k$, which produces the black appearance on the scale of the whole panel (b).

Having all the eigenstates of the system calculated, we are now in a position to study the time evolution of an initial polariton excitation, which we choose in the form of a wavepacket $|\Psi^0\rangle$ built out of the low-energy polariton states $|\Psi_i^0\rangle$ of the perfect system:

$$|\Psi^0\rangle = \sum_i A_i |\Psi_i^0\rangle = \sum_i B_i |\Psi_i\rangle. \tag{10.54}$$

Polaritons in the perfect system are ordinary plane waves and we used a discretized analog of eqn (10.49) for the amplitude function A_i. The result is a Gaussian-shaped wavepacket as illustrated in Fig. 10.9 by the long-dashed lines

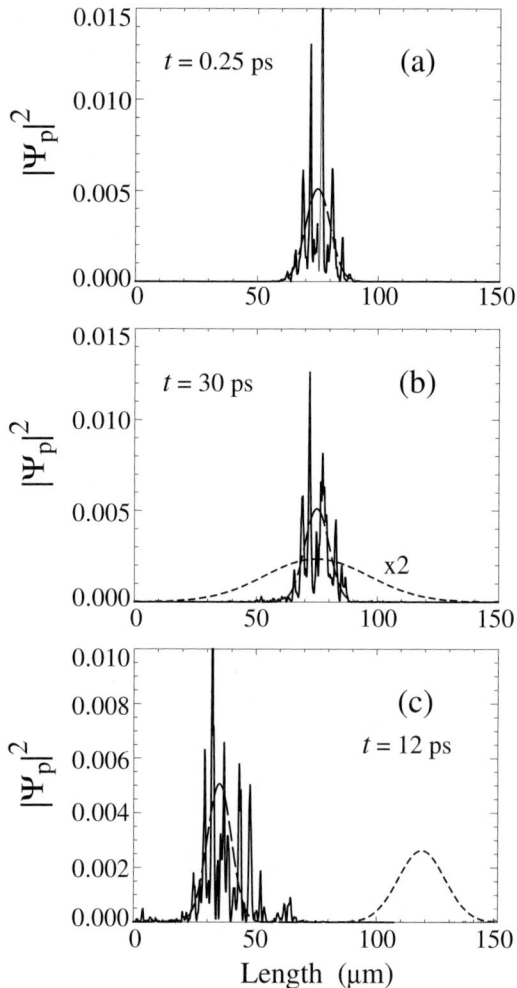

FIG. 10.9. Examples of the time evolution of spatially identical wavepackets built out of the polariton eigenstates of a perfect 1D microcavity as in eqn (10.49) with the parameter $\beta^{1/2} = 5 \times 10^{-6}$ m. For panels (a) and (b), the initial packet has zero total momentum, $k_0 = 0$; for panel (c) the initial packet has a finite momentum determined by $k_0 = 10^4$ cm^{-1}. Only the photon part $|\Psi_p|^2$ of the polariton wavefunction is displayed. The initial packets are shown by long-dashed lines, results of the evolution after indicated times t are shown by solid lines for the disordered system and by short-dashed lines for the perfect microcavity (except in panel (a), where the latter practically coincides with the initial packet.) Reprinted with permission from Agranovich et al. (22). Copyright 2007, American Physical Society.

for the photon part of the polariton wavefunction. Amplitudes B_i in eqn (10.54) are, on the other hand, expansion coefficients of the same initial excitation over the eigenstates $|\Psi_i\rangle$ of the system with disorder. The time evolution of the initial excitation in the perfect system is then given by

$$|\Psi^0(t)\rangle = \sum_i A_i\, e^{\mathrm{i}E_i^0 t/\hbar}\, |\Psi_i^0\rangle, \qquad (10.55)$$

while the evolution in the disordered system is given by

$$|\Psi(t)\rangle = \sum_i B_i\, e^{\mathrm{i}E_i t/\hbar}\, |\Psi_i\rangle, \qquad (10.56)$$

where E_i^0 and E_i are the respective eigenstate energies.

Of course, the evolution of the low-energy wavepacket (10.55) in the perfect microcavity takes place in accordance with our continuum generic description in eqn (10.48) (barring small differences that may be caused by deviations from the purely parabolic spectrum). This is clearly seen in panels (b) and (c) of Fig. 10.9 where the photon part of $|\Psi^0(t)\rangle$ at indicated times t is displayed by the short-dash lines: mere broadening of the wavepacket with no momentum ($k_0 = 0$) in panel (b), and both broadening and translational displacement ($k_0 \neq 0$) in panel (c). On the time-scale of panel (a), the $|\Psi^0(t)\rangle$ state has not practically evolved yet from $|\Psi^0\rangle$ and is not shown on that panel.

The time evolution of exactly the same initial polariton packets is drastically different in the disordered system; the corresponding spatial patterns of the photon part of $|\Psi(t)\rangle$ are shown in Fig. 10.9 with solid lines. First of all, the initial packet is quickly (faster than a fraction of ps) transformed into a lumpy structure reflecting the multitude of the localized polariton states within the spatial region of the initial excitation. Note that in our illustration here we intentionally chose the initial amplitude function (10.49) with the parameter $\beta^{1/2} = 5 \times 10^{-6}$ m large enough for the spatial size of the initial excitation to be much larger than the size of the individual localized polaritons at these energies (compare to Fig. 10.7a). Importantly, however, note that, while displaying some internal dynamics (likely resulting from the overlap of various localized states), this lumpy structure does not propagate well beyond the initial excitation region over longer times. This is especially evident in comparison, when the broadening and motion of the packets in the perfect system is apparent (panels (b) and (c) of Fig. 10.9). We have run simulations over extended periods of time (~ 100 ps) with the result that $|\Psi(t)\rangle$ remains essentially localized within the same spatial region. Of course, some details of $|\Psi(t)\rangle$ depend on the initial excitation, see, e.g. a somewhat broader localization region in Fig. 10.9(c) for the initial excitation with an initial momentum corresponding to $k_0 = 10^4$ cm^{-1} but the long-term localization in the disordered system appears robust in all our simulations. It would be interesting to extend the dynamical studies with participation of higher energy states such as in Fig. 10.7(b).

10.5.3 Concluding remarks

The nature and dynamics of low-energy cavity polariton states are important for various physical processes in microcavities, particularly for the problem of condensation of polaritons into the lowest energy state (s). As was demonstrated in Ref. (15), low-energy polaritons in organic microcavities should be especially susceptible to effects of scattering/disorder in the exciton subsystem. The problem of disorder effects on polaritons in organic microcavities appears quite interesting as organic materials would typically feature both strong exciton–photon coupling and substantial static and/or dynamic exciton scattering. In these calculations we have continued a line of study in Ref. (32) to look in some more detail at disorder effects on polaritons in a 1D model microcavity. Our numerical analysis has brought further evidence that low-energy polariton states in organic microcavities can be strongly localized in the sense of $\ell \leq \lambda$, where ℓ is the spatial size of localized states and λ the wavelength of parent polariton waves. (We have also found indications of weaker localization at higher polariton energies in the sense of $\ell \gg \lambda$.) Our illustrations have included demonstrations of localization not only via the spatial appearance of polariton eigenstates but also via the time evolution of different low-energy wavepackets.

On physical grounds (see Section 10.3 and (15)) one should expect that low-energy polaritons in 2D organic microcavities would also be rendered strongly localized by disorder, as it would also follow from the general ideas of the theory of localization (39). Further work on microscopic models of two-dimensional polariton systems is required to quantify their localization regimes.

The strongly localized nature of low-energy polariton states should affect many processes such as light scattering and nonlinear phenomena as well as temperature-induced diffusion of polaritons. Manifestations of the localized polariton statistics (Frenkel excitons are paulions exhibiting properties intermediate between Fermi and Bose particles) in the problem of condensation also appear interesting and important.

We note that one can exercise experimental control over the degree of exciton–photon hybridization and disorder by modifying the size of the microcavity for various organic materials making such systems fertile ground for detailed experimental and theoretical research into their physics.

11

CHARGE TRANSFER EXCITONS

11.1 Introduction

As we already mentioned, excitons are distinguished in two main groups: small-radius Frenkel excitons, which basically are delocalized molecular excitations, and large-radius Wannier-Mott excitons, in which the electron and hole have a hydrogenlike relative motion with a radius much larger than the lattice constant. The charge-transfer excitons (CTE) occupy an intermediate place in this classification (1)–(3). The lowest energy CTE usually extends over two nearest neighbor molecules and creates a socalled "donor–acceptor (DA) complex". This is currently considered as an important intermediate state in the creation of free carriers in the photoconductivity of organic crystals, a process in which the first step is the photogeneration of a Frenkel exciton. In a CTE, the electron is localized on the acceptor and the hole on the donor. In organic crystals, such CTE localization over nearest-neighbor molecules is usually stable, because the electron-hole attraction energy is large compared to the widths of the conduction and valence bands. The localization is further stabilized by the strong tendency of the CTE to undergo self-trapping (4). Nevertheless, such an ionic pair as a whole can be mobile; the corresponding band theory of CTEs has been discussed in many papers (see, e.g. (5)). Due to the electron-hole separation in a CTE, the static dipole moment created by the positive and negative ions can assume values as large as 10–25 D. This is responsible for some of the most characteristic properties of the CTE. For example, due to the CTEs large dipole moment, the CTEs can strongly contribute to a large secondorder nonlinear polarizability $\chi^{(2)}$ (6).

To qualitatively explain in the simplest way the structure and excitation energy of CTEs it suffices to consider a donor–acceptor pair of molecules that are neutral in their ground states. When exciting a CTE, an electron leaves the donor and occupies one of the empty energy levels of the acceptor. The energy of this excited state can be estimated from the relation $E = I - A + C + P$, where I, A, C, and P are the donor's ionization energy, the acceptor's electron affinity, the electrostatic Coulomb attraction between the electron and hole, and the polarization energy of the crystal for an infinitely separated positive and negative ion pair. An excellent review of the optical and photoconductive properties of organic solids composed of different donor and acceptor molecules arranged in an alternating way in quasione-dimensional arrays, has been published by Haarer and Philpott (7). They discussed molecular crystals with neutral ground states and with rather small ionization potentials ($I < 7$ eV) and large electron affinities ($A > 2$ eV). In such crystals the CTEs are the lowest energy electronic excita-

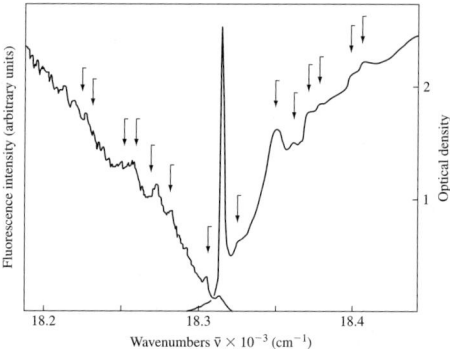

FIG. 11.1. Absorption (right side) and emission spectrum (left side) of a 170 micron thick crystal at 2 K. The arrows at the right-hand side indicate phonon sidebands in the absorption spectrum. The pattern is symmetrically repeated at the low energy side of the zero-phonon line. Reprinted with permission from Rose *et al.* (23). Copyright Elsevier (1984).

tions. As a rule these excitations are self-trapped, have very broad absorption bands, and their optical properties can be understood in the framework of a local picture, ignoring the dispersion of states in a CT exciton band. Typical examples of such crystals are anthracene–PMDA, which has the structure of segregated stacks of different molecules (D–A–D–A–), and the crystal of TTF–TCNQ, which has the structure of segregated stacks of identical molecules, A–A–A–A– and D–D–D–. It was shown by Haarer (8) (Fig. 11.1) that high resolution spectra of the 1:1 CT–crystal anthracene–PMDA show at low temperatures sharp structures in absorption and in emission. These structures are attributed in (8) to a zero-phonon transition which is accompanied by vibronic satellites. The spectra yield the coupling strength of the CT-transition to the phonons and to the vibrational modes of both the donor and acceptor molecule. The situation is different for molecular crystals composed of identical molecules. An excellent review of the state of the art in this field has recently been published by Petelenz (9). Usually, the energy of CTE states in such crystals is larger than the energy of the lowest Frenkel exciton state. However, for some quasione-dimensional crystals with a small distance between the nearest neighbor molecules, like the crystal of 3,4,9,10-perylenetetracarboxylic dianhydride (PTCDA), the energy separation between the lowest energy of Frenkel exciton and CTE states can be very small. For such crystals the nature of the lowest energy electronic excitations is determined by the mixing of Frenkel exciton and CTE states and strongly depends on the orbital overlap between the nearest molecules, see Hoffmann (5). If the energy separation between Frenkel and CTE states is small and orbital overlap is substantial the mixing (hybridization) of states occurs due to the possible virtual transformation of a Frenkel exciton into a CTE and vice versa. In the processes of such transformation the Frenkel exciton localized at molecule n can dissociate due to electron transfer to molecule m to give the CT state n^+m^- or by hole transfer to give n^-m^+. On the other hand, the CT state can recombine by subsequent hole or electron transfer to give a Frenkel exciton located on molecule n or m.

In the modern literature on collective and nonlinear properties of excitons (see, for instance, (6), (10), (11)) one may find mainly the investigations devoted to Wannier-Mott or Frenkel excitons (1), (12). The studies of collective properties of charge transfer excitons are only at the very beginning. On the other hand, this type of exciton also plays an important role in the understanding of the optical and photoelectric properties of many organic materials, including nanostructures (see (7) and also (3), (13)). For this reason in this chapter we describe some unusual collective properties of CTEs at high pumping which can arise due to the CTE dipole–dipole interaction and their possible large dipole moments. Among these phenomena we discuss below the dielectric-conductor phase transition at the donor-acceptor interface, the photovoltaic effect in asymmetrical donor–acceptor superlattices and the resonant as well as offresonant $\chi^{(3)}$ optical nonlinearity. In all considered cases of donor–acceptor multilayers it will be assumed that CTEs at D–A interfaces between alternating layers of donor and acceptors are the lowest energy electronic excited states. We assume that these states can be strongly populated as the result of fast lattice relaxation from higher-energy Frenkel-type electronic or vibronic states.

11.2 Stark effect and electroabsorption of CTEs

An externally applied static electric field **F** shifts the exciton levels and these shifts are reflected in the absorption spectra.[56] In the case of CTEs the shift is linear in the field:

$$E_-^{\mathrm{CT}} = I - A + C + P - \boldsymbol{\mu}_p \cdot \mathbf{F}, \qquad (11.1)$$

where $\boldsymbol{\mu}_p$ is the static dipole moment of the CTE. If the crystal has inversion symmetry, a CTE state with opposite direction of $\boldsymbol{\mu}_p$ exists as well and this exciton acquires the energy

$$E_+^{\mathrm{CT}} = I - A + C + P + \boldsymbol{\mu}_p \cdot \mathbf{F}. \qquad (11.2)$$

In these expressions we neglected the corrections quadratic in the field **F**. For the Frenkel exciton, which has no permanent dipole moment, such quadratic contributions are the lowest-order ones and the energy reads

$$E(\mathbf{F}) = E(\mathbf{0}) + \frac{1}{2}\sum_{i,j} a_{ij} F_i F_j, \qquad (11.3)$$

where $a_{ij} = a_{ij}^{\mathrm{exc}} - a_{ij}^{\mathrm{g}}$ is the excess polarizability tensor, i.e. the difference of the molecular excited state and ground state static polarizability tensors.

[56]In this subsection we use (14).

The dielectric constant in the frequency region of the CT transition can be expressed by the formula

$$\epsilon(\omega, \mathbf{F}) = \epsilon_{0,0} + \frac{A}{E_+^{CT} - \hbar\omega - i\hbar\gamma} + \frac{A}{E_-^{CT} - \hbar\omega - i\hbar\gamma}, \quad (11.4)$$

where A is a constant proportional to the oscillator strength and γ is the dissipative width. It follows from this relation that up to the second order in the field

$$\epsilon(\omega, \mathbf{F}) = \epsilon(\omega, \mathbf{0}) + \delta\epsilon(\omega, \mathbf{F}), \quad (11.5)$$

with

$$\delta\epsilon(\omega, \mathbf{F}) = \frac{1}{2}\left[\boldsymbol{\mu}_p \cdot \mathbf{F}\right]^2 \frac{d^2\epsilon(\omega, \mathbf{0})}{d\omega^2}. \quad (11.6)$$

Thus, for CTEs the leading effect of a static electric field on $\epsilon(\omega)$ is seen to be quadratic in the field and proportional to the second derivative of $\epsilon(\omega, \mathbf{0})$.

In the frequency region of the Frenkel exciton, we have

$$\epsilon(\omega, \mathbf{F}) = \epsilon_{0,0} + \frac{A}{E(\mathbf{F}) - \hbar\omega - i\gamma} = \epsilon(\omega, \mathbf{0}) + \delta\epsilon(\omega, \mathbf{F}), \quad (11.7)$$

with

$$\delta\epsilon(\omega, \mathbf{F}) = -\frac{1}{2}\sum_{i,j} a_{ij} F_i F_j \frac{d\epsilon(\omega, \mathbf{0})}{d\omega}. \quad (11.8)$$

Thus, $\delta\epsilon(\omega, \mathbf{F})$ is proportional to $\sum_{i,j} a_{ij} F_i F_j$ and to the first derivative of $\epsilon(\omega, \mathbf{0})$.

For the interpretation of experiments it is important to know the corrections to the real and imaginary parts of the refractive index, n and κ, respectively, where κ is measured as a function of frequency and gives the absorption spectrum. These corrections can be found easily from the relations $\epsilon = (n + i\kappa)^2$. Thus, the corrections to n and κ, which we denote as δn and $\delta\kappa$, read

$$\delta\kappa = \frac{n_0 (\delta\epsilon)'' - \kappa_0 (\delta\epsilon)'}{2\left(n_0^2 + \kappa_0^2\right)}, \quad (11.9)$$

and

$$\delta n = \frac{\kappa_0 (\delta\epsilon)'' + n_0 (\delta\epsilon)'}{2\left(n_0^2 + \kappa_0^2\right)}, \quad (11.10)$$

where $(\delta\epsilon)'$ and $(\delta\epsilon)''$ are the real and imaginary parts of $\delta\epsilon$, respectively ($\delta\epsilon = (\delta\epsilon)' + i(\delta\epsilon)''$, while $n_0(\omega)$ and $\kappa_0(\omega)$ are the real and imaginary parts, respectively, of the refractive index in the absence of the static electric field.

If in the spectral region under consideration the absorption is not very strong, so that $n_0(\omega) \gg \kappa_0(\omega)$, the expressions for $\delta\kappa$ and δn are reduced to

$$\delta\kappa = \frac{(\delta\epsilon)''}{2n_0}, \quad \delta n = \frac{(\delta\epsilon)'}{2n_0}. \quad (11.11)$$

The formula for $\delta\kappa$ forms the basis of electroabsorption spectroscopy, which is an experimental technique in which one measures the change of the absorption spectrum induced by a slowly varying external electric field. As we demonstrated, for Frenkel excitons the change of the absorption spectrum is proportional to the first-derivative signal, which is an antisymmetric function of ω relative to the resonance frequency. On the other hand, a second-derivative signal, which is symmetric around the resonance frequency, is commonly associated with CTEs (see, for example, (15)–(17)). This may be made explicit by considering the simple generic form for $\epsilon''(\omega, \mathbf{0})$ close to the resonance frequency ω_0:

$$\epsilon''(\omega, \mathbf{0}) = \frac{g}{(\omega - \omega_0)^2 + \gamma^2}. \tag{11.12}$$

Using this expression and eqn (11.8) and (11.11), we find for a Frenkel exciton resonance

$$\delta\epsilon(\omega, \mathbf{F}) = \frac{1}{2} \sum_{i,j} a_{ij} F_i F_j \frac{2g(\omega - \omega_0)}{\left[(\omega - \omega_0)^2 + \gamma^2\right]^2}, \tag{11.13}$$

which indeed is seen to be an antisymmetric function relative to the resonance frequency. On the other hand, if ω_0 is a CTE resonance, we find from eqn (11.6) and (11.11)

$$\delta\kappa(\omega, \mathbf{F}) = \frac{1}{2} [\boldsymbol{\mu}_p \cdot \mathbf{F}]^2 \left[-\frac{2g}{\left[(\omega - \omega_0)^2 + \gamma^2\right]^2} + \frac{4g(\omega - \omega_0)^2}{\left[(\omega - \omega_0)^2 + \gamma^2\right]^4} \right], \tag{11.14}$$

which is symmetric relative to the resonance frequency.

It should be noted that the above classification of the electroabsorption spectrum is valid only approximately, because first of all eqn (11.11) is correct only in the case of weak absorption and, second, the Frenkel and CT exciton states usually mix. We finally mention that the change of the refractive index δn is of the same order as $\delta\kappa$; new experimental techniques are required to measure this change, however. Good candidates for such methods have been proposed by Warman and coworkers (18). The success of such measurements could be the basis of electrorefraction spectroscopy, complementary to the existing electroabsorption spectroscopy.

11.3 Phase transition from the dielectric to the conducting state (cold photoconductivity)

In this section, following the papers by Agranovich et al. (19) and Kiselev et al. (20), we consider the stability of interacting CTERs at a (donor–acceptor) D–A interface and the possibility of a transition to a conducting state.

The realistic possibility considering such organic crystalline structures has only recently appeared due to progress in the development of organic molecular beam deposition and related techniques (13). Such progress has led to monolayer control over the growth of organic thin films and superlattices with extremely high chemical purity and structural precision. This opens a wide range of possibilities for creating new types of ordered organic multilayer structures including ordered interfaces. It is well known that the necessity for lattice matching places strong restrictions on the materials which can be employed to produce high quality interfaces using inorganic semiconductor materials. This occurs since inorganic semiconductor materials are bonded by weak van der Waals forces. This fact relaxes the above restrictions and broadens the choice of materials that can be used to prepare organic crystalline layered structures with the required properties (for more details and many examples see (13), (21), (22)). Note that the D–A interfaces can be created also in Langmuir–Blodgett films (23), (24). We can mention the paper (25) where in order to study the nonlinear optical properties of the multilayer organic superlattices have been grown with a structure of the type $\ldots AAA|DDD|AAA|DDD\ldots$, the vertical dashes indicating the donor–acceptor interfaces. In this paper, the molecule C_{60} was used as acceptor (A) and molecules of perylene, coronene, and others, as donors of electrons (D).

11.3.1 Analytical approach

Consider the CTEs on a single D–A interface with a highly ordered structure. To explain the main collective effects in the physics of CTEs at a D–A interface, we assume that the static dipoles of the CTEs are aligned approximately normal to the interface plane, resulting in mutual repulsion. For example, if the static CTE dipole moment is equal to 20 D and the distance between them is 0.5 nm (the lattice constant at the interface) the repulsion energy is near 1 eV. If the distance between CTEs increases to 1 nm the repulsion energy decreases to about 0.1 eV. It is important to note that for crystals in which the lowest energy electronic excitations are CTEs, the repulsion energies are of the order of the energy difference B from the CTE level to the lowest conduction band ($B < 0.5$ eV, see (7)). Thus, at high CTE concentrations we can expect that due to the repulsion energy the higher energy states are populated with free carriers, thus producing photoconductivity even at very low temperature (cold photoconductivity). Of course, for example, multiphoton ionization or other nonlinear processes can produce photoconductivity at low temperature. However, such processes are universal, they take place in condensed matter of any nature, and they are not relevant to CTEs and their interaction, which we discuss below.

We assume that the lowest conductivity band, which is responsible for the conductivity along the D–A interface, has a lower energy than the Frenkel excitations in the donor and acceptor materials. In this particular case, the interface at low temperature provides the lowest energy site for the CTE and interface free carriers. In (19) cold photoconductivity at the D–A interface was considered under the assumption that the time required for a phase transition to the conducting state is smaller than the CTE lifetime and a phase transition was obtained by minimazing the total energy of bond (CTE) and dissociated excitations (free carriers). Following (19) let us calculate, first, the energy of a 2D array of self-trapped CTEs at $T = 0$.

TRANSITION FROM DIELECTRIC TO CONDUCTING STATE

The energy of CTEs (of concentration n_1) and the energy of dissociated e–h pairs (of concentration n_2), can be calculated by assuming that the total number of excitations determined by optical pumping is constant:

$$n_1 + n_2 = n. \tag{11.15}$$

The energy of the CTE array is therefore:

$$E_1 = n_1 \Delta + E_{\text{int}}, \tag{11.16}$$

where Δ is the energy of a single CTE and E_{int} is the total repulsion energy of their interaction. This energy can be estimated from the average distance ρ between CTEs and their dipole p as:

$$E_{\text{int}} = A \frac{p^2}{\rho^3} \frac{n_1}{2}, \tag{11.17}$$

where A is a geometric constant depending on the CTE distribution in the interface plane. For example, for a square lattice $A \approx 10$. Since the CTE concentration by definition is $n_1 = 1/\rho^2$ the total electrostatic energy of the interaction between the dipole moments is (see also the result of numerical simulation shown in Fig. 11.5 below):

$$E_{\text{int}} = \frac{A p^2 n_1^{5/2}}{2}. \tag{11.18}$$

We can approximate the energy of the dissociated pairs at $E_2 = (\Delta + B) n_2$, where the kinetic energy of the free carriers has been neglected (due to the self-trapping and narrow electronic bands). Assuming that we consider the region near the threshold where the concentration $n_2 \ll n_1$, we can also neglect the interaction of the free carriers with the CTEs. The total energy of the system is then written as

$$E = E_1(n_1) + E_2(n_2) = n\Delta + \frac{A p^2 (n - n_2)^{5/2}}{2} + B n_2. \tag{11.19}$$

Minimizing the above expression with respect to n_2, gives:

$$n_2 = n - \left(\frac{4B}{5 A p^2} \right)^{2/3}. \tag{11.20}$$

It is clear from eqn (11.20) that n_2 is positive at $n > n_{\text{cr}} = [4B/(5Ap^2)]^{2/3}$ (see Fig. 11.2). The appearance of free carriers at $n > n_{\text{cr}}$ is considered to be a phase transition from the dielectric to conducting state. This transition corresponds to photoconductivity at low temperatures (i.e. to cold photoconductivity) and is due to long-range dipole–dipole interactions between CTEs. In this simplified picture, we neglect the randomness in the CTE distribution and the transient to

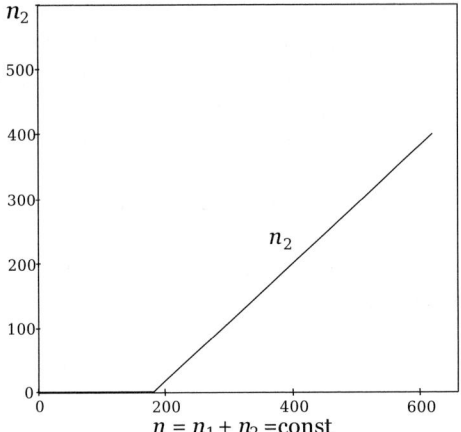

FIG. 11.2. The number of the dissociated pairs (n_2) as a function of the total number of excitations ($n_1 + n_2$) at the donor–acceptor interface (according to the simplified model of eqn (11.20) and in units consistent with the numerical simulations discussed below. Reprinted with permission from Agranovich et al. (14). Copyright Elsevier (1993).

establish an equilibrium state. The establishment of a steady state depends on the pump intensity and the CTE lifetime which above was considered as infinite.

The phase transition in the system of interacting CTEs at finite temperatures was considered in the paper (19) for 1D donor–acceptor interface. Such 1D interfaces can be found, for example, among CT crystals containing segregated stacks. In such crystals, the 1D interfaces separate the columns of donors and acceptors. However, all the segregated stacks organic solids have an ionic ground electronic state and the excited CT state in this case corresponds to a transition of two neighboring ions (positive donor and negative acceptor) to the neutral state (7).

For 1D structures, the interaction between CTEs can be taken into account only for nearest neighbors. Besides, as follows from (7), due to strong self-trapping of CTEs the bandwidth of CTEs can be neglected. Due to these assumptions an exactly solvable model was obtained in (19).

As could be expected, it follows from the results of this model that the phase transition region broadens with increasing temperature and disappears at high temperatures. The same picture was obtained for 2D structures in the paper (26) in the Weiss molecular field approximation.

The transition from a dielectric to a conducting state can be also investigated using computer simulations. Following (20) we describe below such numerical results for a more realistic model where the random distribution in space and also the finite lifetime of CTEs are explicitly taken into account.

11.3.2 *Numerical simulations*

To numerically simulate the time evolution of CTEs distributed over a two-dimensional donor–acceptor interface the D–A sites were arranged in a square lattice. It was assumed that the D–A interface is uniformly irradiated with a time independent source of intensity I. Only one CTE can be generated at any site, so every D–A site can be either occupied or not. The CT exciton generated at a given lattice site will stay there and it cannot move to another D–A site because of self-trapping.

Once generated, there are two mechanisms for the CTE to disappear. First, recombination occurs because of the finite lifetime τ of the CTE. The second mechanism is via dissociation. The CT exciton dissociates when, due to the dipole–dipole interaction, the energy of the particular exciton exceeds some threshold. If there are n_1 CTEs occupying the D–A interface, the electrostatic energy of the ith exciton in the electric field of the other excitons surrounding this site is:

$$V_i = \sum_{j=1}^{n_1} \frac{p^2}{r_{ij}^3} \quad (j \neq i). \tag{11.21}$$

The ith CT exciton dissociates when the repulsion energy, V_i, is larger than the energy B. This condition is satisfied and dissociation occurs when more CTEs appear on sites adjacent to that occupied by the ith exciton. The electrostatic potential energy of the exciton strongly increases when a few CTEs occupy the nearest lattice sites. If this occurs, one or more CTEs will dissociate, thereby reducing the total system energy. Such a mechanism should result in correlations between exciton positions, and an ordering of the system of nonmobile CT excitons can be expected. Such a spatial ordering suggests the existence of a critical pump light intensity above which there is an onset of cold photoconductivity.

In contrast to the thermodynamical theory, in simulations the process of recombination of free carriers which can result in the creation of CTEs was neglected. Near the threshold where the concentration of free carriers is small the contribution of this process to the number of CTEs will indeed be negligible. However, even at higher concentration, the effect of free carrier recombination can be reduced by applying an electric field along the interface. This field will separate electrons and holes and thus will create the photocurrent which has to be measured (see also subsection 11.3.4 below).

Computer simulations were performed for a two-dimensional square lattice containing 600×600 sites. Under continuous pumping of the sample with a constant intensity, the CTEs are generated in the process described above. In order to avoid the influence of boundary conditions, we simulate the evolution of only the central part of the lattice. This square, central sublattice consists of 200×200 D–A sites, $N_{\text{sites}} = 40\,000$. Next, we replicate the central sublattice by adding 8 more square sublattices surrounding the central one. That is, the exciton positions calculated for the central 200×200 sites square lattice is reflected via a mirror symmetry operation to the other surrounding eight squares.

To simulate the time-evolution, we run the system through equally spaced time steps separated by the interval Δt. The value of Δt is chosen to be much shorter than the CTE lifetime, i.e. ($\Delta t \ll \tau$). Here, we choose $\Delta t = \tau/50$.

We start the simulations when there are no CTEs at the interface. Under the influence of the pumping the excitons begin to appear. After the time $\approx \tau$, the number of CT excitons occupying the lattice reaches the steady state value. In our current work we take a time interval of $5\,\tau$ to ensure that the steady state is reached. From this time on the necessary statistical information is collected.

The time-evolution of the system is simulated as follows. At every time step a few CTEs (depending on the pumping intensity, I) are created at randomly chosen positions at the central sublattice with $N_{\text{sites}} = 40\,000$ sites. Then we go over the central sublattice sites and check every D–A molecule. With some probability the exciton at this site can recombine, as explained above. It also can dissociate if its electrostatic energy is high enough. The rules for these events to happen at one particular D–A site are:

1. If the site is empty the charge–transfer exciton can be created with probability $P_{\text{create}} = I\Delta t/N_{\text{sites}}$.
2. If a charge-transfer exciton already occupies this site it can recombine with probability $P_{\text{rec}} = \Delta t/\tau$.

Next, during the same time step, we calculate the energy of every CTE in the electrostatic field produced by the dipole moments of all other excitons. The energy of the ith CT exciton can be found using eqn (11.21). If this energy is greater than the dissociation threshold, B, the CT exciton dissociates. Finally, we recalculate the energies of all CT excitons that remain at the D–A interface.

11.3.3 *Results of numerical simulations*

All results reported below are collected after the steady state is achieved. Figure 11.3 shows the dependence of the number of CTEs (n_1) on the value S which is the product of generation intensity of the CTEs I and the CTEs lifetime τ : $S = I\tau$. The steady state number of dissociated pairs is determined by its own lifetime but we do not estimate here the concentration of carriers and conductivity. Nevertheless, Fig. 11.3 shows the value S_2 which is equal to the number of dissociations which take place at given S in steady state during the time τ. We find qualitative agreement with the analytical theory that the CTEs populate the D–A interface only up to some saturation concentration. Further increase of the pumping S results mainly in the dissociation of CTEs into electron–hole pairs. When the number of the CTEs (n_1) reaches the saturation density, the number of dissociations during interval τ into electron–hole pairs S_2 increases linearly with the pumping S. It is interesting to compare the critical concentration of CTEs from the analytical model (see above) in which the CTEs are assumed to be in thermal equilibrium and having a spatially ordered structure (square lattice) with the results of the numerical simulation. In the simulation with a random CTE distribution, dissociated pairs appear even at low pumping. Nevertheless, following qualitatively the results of the analytical model (Fig. 11.2), as a critical

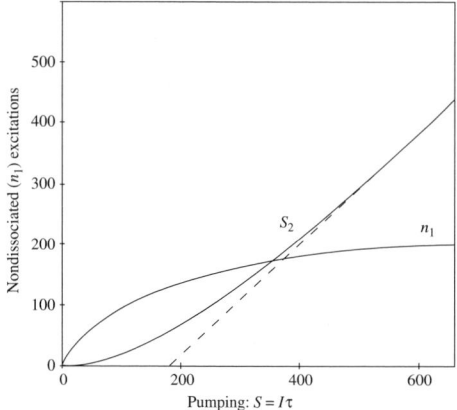

FIG. 11.3. The steady state number of CTEs (n_1) occupying the donor–acceptor interface and the number of dissociated pairs (S_2) as a function of the pumping intensity. The pumping intensity S is equal to the number of charge-transfer excitons produced at the interface during a CTE lifetime in the absence of dissociation processes. The results are from the numerical simulations of the CTE system described in the text. Reprinted with permission from Kiselev et al. (20). Copyright Elsevier (1998).

concentration of CTEs we can take the concentration corresponding to the saturation of CTEs at the interface or, what is nearly the same, the concentration of CTEs which corresponds to intersection of the linear S_2 asymptote with the horizontal axis. In Fig. 11.3 the value of n_1 is approximately 200 and thus the corresponding critical dimensionless concentration $C_{cr} = 200/40\,000 = 0.5\%$. The curve in Fig. 11.2 corresponds to the value $M = Ba^3/p^2 = 0.01$. From the analytical theory it follows that for the same value M the critical dimensionless concentration $C_{cr} = (4Ba^3/5Ap^2)^{2/3} = (4M/5A)^{2/3} = 0.85\,\%$. Thus, a random CTE distribution decreases the critical concentration for the transition to the conducting state. This effect could be expected, because, for a random distribution, in contrast to the analytical model of ordered CTEs, the occurrence of small distances between CTEs is allowed even at low CTE concentration. In both approaches, the critical concentration strongly depends on the values of B, P and a and below we compare the critical concentrations for different values of these parameters. For example, for $B = 0.2$ eV, $p = 20$ D and $a = 0.5$ nm, corresponding to $M = 1$, the analytical model gives $C_{cr} = 4\%$, computer simulations give $C_{cr} = 2.5\%$. For $M = 0.05$, the analytical approach gives $C_{cr} = 2.5\%$, and the computer simulation gives $C_{cr} = 1.5\,\%$ and so on. Thus, for a random CTE distribution at the D–A interface, the critical concentration is almost twice as small as that predicted by the analytical model of ordered CTEs with infinite lifetime. As we already mentioned above the dissociation of nearest CTEs results in a change of their spatial distribution. This affects the repulsion energy distri-

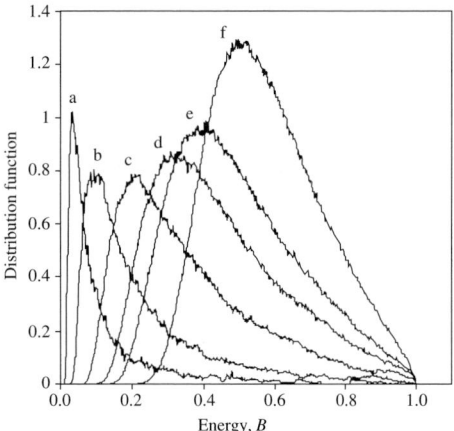

FIG. 11.4. Steady state distribution of the CTE electrostatic energies for different pumping intensities. The pumping intensity S is the number of CTEs produced at the interface during the CTE lifetime in the absence of dissociation processes. The values of S and the corresponding ones of n_1 are given by: (a) $S = 50$, $n_1 = 45$; (b) $S = 100$, $n_1 = 82$; (c) $S = 200$, $n_1 = 133$; (d): $S = 350$, $n_1 = 177$; (e) $S = 500$, $n_1 = 202$; (f) $S = 1500$, $n_1 = 247$. The repulsion energy of the CTEs is given in units of the dissociation energy B. Reprinted with permission from Kiselev et al. (20). Copyright Elsevier (1998).

bution of the CTEs. If the CTE dissociation is absent their energy distribution would have a peak (associated with the average distance between the excitons) and a tail extending to high energies (such a tail is associated with the CTEs occupying nearby lattice sites). The dissociation prevents the creation of clusters of the closely placed CT excitons and, especially, it prohibits CTEs from occupying adjacent sites at the D–A interface and it cuts off that high energy tail. This is shown in Fig. 11.4 which presents the distribution of the CTE repulsion energies for different pumping intensities S. In this figure the dissociation threshold is represented by the energy B and the origin corresponds to the situation when the CTEs occupy infinitely remote sites. It is seen from this figure that the peak of the energy distribution increases with the number of CTEs and so does the width of the distribution. It is interesting to note also that the position of the peak of the repulsion energy distribution (corresponding to the energy of highest probability) varies with the steady state number of CTEs approximately in the way which the theoretical model of ordered CTEs lattice predicts. As follows from eqn (11.19) the CTE energy as a function of the number of CTEs should vary as $n_1^{3/2}$. Figure 11.5 demonstrates that in computer simulations such a dependence takes place with high accuracy. Figure 11.6 shows the distribution of the nearest-neighbor (n-n) distances between the CTEs in the steady state when the pumping intensity S varies. We took $B = 0.02$ eV in Fig. 11.6, with

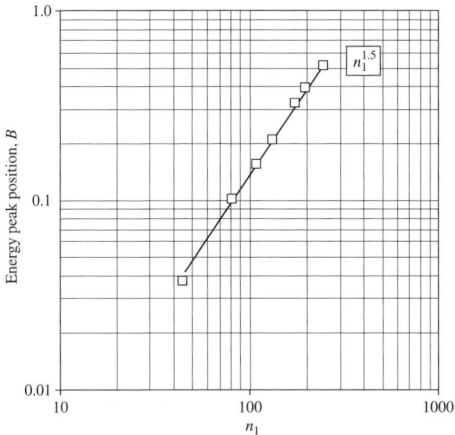

FIG. 11.5. Position of the energy distribution peak as a function of the number of CT excitons, n_1. The position of the energy peak obtained from the numerical simulation appears to be proportional to $n_1^{1.5}$. Reprinted with permission from Kiselev et al. (20). Copyright Elsevier (1998).

$C_{\text{cr}} = 0.5\%$ as discussed above; the closest approach of any two CTEs is then about five lattice spacings. At low intensity (curve (a)) the CTE distribution is broad and there are isolated CTEs at sites up to $20a$ from any other CTE. At higher S the distribution sharpens. At high S the CTE coverage of the D–A interface is roughly uniform (n–n distances vary from $5a$ to $16a$). For the parameters used, CTEs closer than $5a$ dissociate and there is always a lower limit on separations.

11.3.4 Concluding remarks

In this section, we have considered the transition to a conducting state at $T = 0$ due to the CT exciton–exciton repulsion. Using computer simulations in which the randomness of the CTE distribution and their finite lifetime were taken into account, we have found how the repulsive interactions between CTEs populates higher energy states of dissociated e–h pairs and thus creates free carriers. It is clear that at a finite temperature, the repulsion can also be important because it decreases the activation energy. This decrease of activation energy depends on the concentration of CTEs. The computer simulation demonstrates also that the critical concentration of CTEs should depend on their mobility. If this mobility is small and the CTE distribution at a D–A interface is random, the critical concentration that leads via a phase transition to the conducting state will be smaller than in the case of mobile CTEs. In the case in which their lifetime is long and their repulsion is strong an ordered state is realized. An interesting problem is how to observe the predicted photoconductivity. It is evident that for crystalline D–A multilayers the conductivity along the interfaces can be measured. Alternatively, the optical properties of the interface near the conductivity

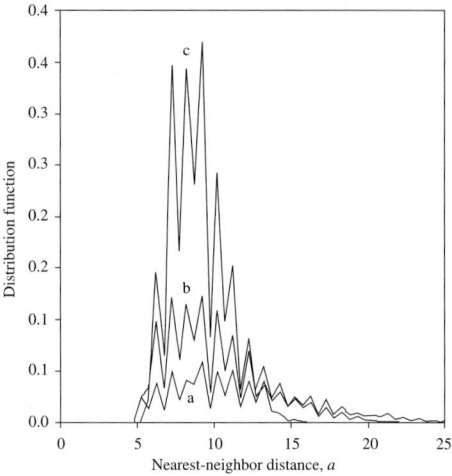

FIG. 11.6. Distribution of the nearest-neighbor distances, in units of the lattice constant a, between the charge-transfer excitons for three different pumping intensities. The pumping intensity S and the corresponding number n_1 are: (a) $S = 100$, $n_1 = 82$:, (b) $S = 200$, $n_1 = 133$; (c) $S = 1500$, $n_1 = 247$. Reprinted with permission from Kiselev et al. (20). Copyright Elsevier (1998).

transition can be observed. For such experiments, methods for observing photoconductivity parallel to the plane of dipoles developed in the investigations of Langmuir–Blodgett films can be used. However, for such measurements we need to have nearly perfectly ordered crystalline D–A multilayers with a large interface area.

The states and mobility of free carriers at D–A interfaces are also important for the observation of photoconductivity. In the discussion of these problems, it is tempting to use the analogy with Tamm states of electrons or holes at the crystal interface. In the case of Tamm states, the wavevector is a good quantum number. The states are coherent and they form the conduction and valence bands of surface states. This is typical for surface states of inorganic semiconductors or metals, since the electron–phonon coupling is too weak to destroy the bands. On the other hand, for D–A organic structures as well as for organic crystals in general, the width of energy bands at room temperature is small (order 0.1 eV), and electron–phonon coupling can be rather strong, giving rise to self-trapping of free carriers and hopping conductivity. We expect that the same is true for free carriers at an organic interface. In the discussion of carrier mobility at the D–A interface we have to take into account not only self-trapping, but also Coulomb interactions between free carriers, and with CTEs and recombination processes. In semiconductors the recombination of electron–hole pairs is usually slow due to spatial delocalization. To suppress the recombination of free carriers at a D–A interface, an electric field can be applied parallel to the interface. This electric field can also be used for observation of photoconductivity along the D–A interface. Thus to observe the phenomena considered here it is necessary to have a high quality D–A structure, and to resolve many difficult technical problems usually connected to this type of experiments.

11.4 Cumulative photovoltage in asymmetrical donor–acceptor organic superlattices

11.4.1 *Introduction*

Photovoltaic energy conversion is an important component of a future network of renewable energy sources that could provide a sustainable energy supply without greenhouse gas emission (27). That is the reason behind the intensive investigations in the field of photovoltaics which also demonstrate a very peculiar competition between the use of inorganic and organic materials for the realization of solar cells with increasing efficiency. No inorganic material matches the absorption coefficients of organic dyes, which are in the range of 10^5 cm^{-1}, and give rise to the hope of producing organic based very thin solar cells with low energy and material consumption.

From the point of view of macroscopic electrodynamics the appearance of a constant current under the influence of light (photovoltaic effect) can be understood in the framework of the theory of perturbation (for a small light intensity) from the nonlinear relation (see also (28)):

$$J_i(0) = \sigma^{(2)}_{ij\ell}(0;\omega,-\omega)E_j(\omega)E_\ell(-\omega) \tag{11.22}$$
$$+\sigma^{(4)}_{ij\ell mn}(0;\omega_1,-\omega_1,\omega_2,-\omega_2) E_j(\omega_1) E_\ell(-\omega_1) E_m(\omega_2) E_n(-\omega_2) + \ldots,$$

where **J** is a constant electrical current, $\mathbf{E}(\omega)$ is the amplitude of the electrical field of light with frequency ω and $\sigma^{(2),(4)}$ are the tensors of nonlinear conductivity. These tensors are constant in space for infinite homogeneous media and are different from zero only for noncentrosymmetric structures. The calculation of these tensors is a problem of microscopic theory. Such a theory needs the analysis of the main mechanisms which can be responsible for the appearance of the photovoltaic effect. In the case of asymmetric organic donor–acceptor superlattices with periods of the order of 10 nm considered below it is necessary to carry out the analysis of possible mechanisms of photovoltaics in organic nanostructured materials. It is also worth mentioning also that in some cases the electronic excitations which appear under the influence of light absorption may be responsible for the structural and chemical rearrangement of organic materials. In this case, of course, perturbation theory is not applicable and as a result the tensors of nonlinear conductivity depend, themselves, on the light intensity.

In this section, following (29), we discuss the electrooptical properties of an asymmetric stack of organic donor–acceptor (D–A) interfaces. As we mentioned, the technological progress in molecular organic beam deposition is very fast and there is little doubt that a variety of such systems will be synthesized in the near future. With this in mind, we discuss the properties of a superlattice of the type ... $DDD|AAA|NNN|DDD|AAA|NNN$ ..., where N stands for a material which is neither a good donor nor a good acceptor, and all molecules in the ground state are neutral. In such a noncentrosymmetric structure, when CTEs are generated by pumping light, all the interface CTE dipoles point in the same direction and the potential differences due to the dipole layers at each D–A interface add up to a macroscopic voltage across the superlattice creating a macroscopic potential drop. The corresponding electric field will drive the free electrons and holes produced by the absorption of light and, thus, will provide a photovoltaic current. It is interesting to mention that in the case of D–A interfaces with negative charge-transfer exciton energy or, in other words, in the case of donor and acceptor for which the charge transfer exists already in ground state, a macroscopic potential drop also has to appear in the asymmetrical structure which we consider here but in this case it will be independent on the intensity of the incident light. For both cases the appearing electric field will be directed from the donor side to the acceptor side as we will also show below by direct calculations.

Before discussing the origin of electrooptical effects in organic asymmetrical multilayer structures, it is worth mentioning that in 1996 Rolf Landauer in the discussion of author's talk in Sweden pointed out that the photovoltaic mechanism which we proposed in (29) for organic D–A asymmetrical superlattices is similar to those considered to generate cumulative photovoltages in some inorganic crystals. He had in mind the papers by Cheroff and Keller (30), Pensak (31), and Goldstein (32) which independently discovered the large photovoltages in ZnS and CdTe crystals with periodic intrinsic inhomogeneities, and also the paper by Swanson (33). In the Introduction in the paper (33) it was mentioned that "the large photovoltages observed in (30) in some insulators are due to numerous internal electrostatic barriers which under illumination act as $p - n$ junctions connected in series. The crystal structure of these materials does not exhibit inversion symmetry, and it is conceivable that the directionality in the crystal induces directionality of the internal barriers. Such an intrinsic directionality is necessary, since structure of randomly alternating conductivity type could provide no basis for a preferred direction in which the voltage accumulates". It follows from these remarks that, indeed, the structures which have been investigated in (30), (31) and (32) have many similar features with the asymmetrical D–A organic structure which is under discussion in this section, particularly in the case where charge transfer takes place already in the ground state of a donor–acceptor pair. However, in the paper (29) it was assumed that the charge transfer takes place only in the excited state of the D–A pair and that the main role in the appearance of photovoltage is not played by an intrinsic ground state asymmetry of the asymmetrical D–A multilayer structure, but rather by the asymmetry of the sheets of dipoles at all donor–acceptor interfaces which arise only due to the generation of CTEs via light absorption. The density of CTEs on the sheets is dependent on the light intensity and this should transform the photovoltaic effect in asymmetrical donor–acceptor multilayers into a strongly nonlinear function of the light intensity.

It is known that a photocurrent may be measured from the response to external voltages under illumination. In materials having a large enough photovoltage, this voltage itself can be used without an external bias to measure a photocurrent (short–circuit photocurrent). One is not only interested in measuring the short-circuit photocurrent, but it is of interest to measure independently the open-circuit photovoltage which is the other main characteristic of photovoltaics (see, for example, (30), (31) and (32)). Both these characteristics of photo–voltaics, generally speaking, are dependent on the concentration and mobility of charge carriers. We will demonstrate below that in asymmetrical stack of organic donor–acceptor (D–A) interfaces the open-circuit photovoltage may arise even in the absence of free carriers.

11.4.2 On the mechanisms of the photovoltaic effects in organics

To discuss in more detail the possible peculiarities of photovoltaics in an asymmetrical D–A multilayer structure, we will make a few remarks concerning possible mechanisms of photo–voltaic effect in multilayered structures containing organic dyes (see (34) and also a recent issue of *Chemical Reviews* devoted to organic electronics and optoelectronics (35)).

One model (see, for example, (36)) is based on the assumption that the contacting organic layers have dark n- or p-type conductivity and that the band structure of the organic D–A interface is similar to that of a p–n junction between inorganic semiconductors. In this case in the region near the interface between these layers a depletion region is formed, where the internal electric field results in the dissociation of an excited electron–hole pair into free carriers. In an alternative model (37)–(41), the photovoltaic effect occurs exclusively at the interface of an organic dye with a second dye, the charge separation takes place at the interface, and an extended region with an electric field is not required and may even be absent (see also (34)). Natural photosynthesis may be considered as an important example of this type of carrier generation. Indeed, excitons are formed by light absorption in antenna pigments and diffuse to the reaction center where they dissociate with almost unit quantum efficiency by injecting an electron and hole in opposite spatial directions down a chain of acceptors and donors, respectively (see (42), (43)). It is important to mention that in this mechanism of photovoltaics the electrons and holes after the process of interface charge separation diffuse away from the D–A interface in opposite directions as if driven by an effective electric field directed from the acceptor A to the donor D, or in other words along the same direction of the CTE dipole. This direction, as will be shown below, is opposite to the direction of the electric field created by a sheet of CTE dipoles mentioned above.

In our discussion below, we assume that the binding energy of a CTE is large, and the CTEs are rather stable and do not participate in the photogeneration of free carriers. We will show that the asymmetrical stack of D–A interfaces under pumping of CTEs provides a macroscopic potential drop that can be considered as a cumulative photovoltage. However, the open problem is how to use this cumulative photovoltage. One of the solutions could be the electron or hole doping of all layers. The discussion of this possibility is important, however it is outside of the scope of this book.

11.4.3 Cumulative photovoltage in an asymmetrical stack of D–A interfaces

To estimate the potential profile determined by the interface CTEs, we consider first a single D–A interface with a two-dimensional (2D) density of CTs n of order 10^{12} cm^{-2}, each one having an electric dipole moment μ of about 20 Debye. Such large dipole moments are not unusual for CTEs and the 2D density above, taking a superlattice period of a few tens of monolayers, would correspond to a bulk concentration of excited molecules of order 10^{-4}, which is not problematic in relation to the photochemical stability of the organic materials. In a first approximation, this CTE configuration corresponds to a uniform static dipole moment per unit area μn perpendicular to the D–A interface. As the CTEs repel each other through the dipole–dipole interaction $\Delta H = \mu^2/\rho^3$ (ρ being the exciton–exciton distance) and as their mobility is not negligible, they tend indeed to be uniformly spaced along the interface; for instance, a similar system of dipoles moving classically along a plane and oriented perpendicular to it have been shown (44) to order in a 2D lattice at low temperatures and to form a homogeneous liquid at high temperatures. The CTE repulsive interaction considered here is, of course, the same as that at a higher concentration and would give rise to the dielectric–conductor transition as discussed in the previous section.

If we assume that the dipole layer is uniform and that we can neglect its thickness the corresponding polarization per unit volume is $\mathbf{P}(z) = \mathbf{p}_0 \delta(z)$, where $\mathbf{p}_0$ is the polarization per unit area and the z-axis is directed along the normal to the interface planes from the donor to the acceptor side. In our case, the vector $\mathbf{p}_0 \equiv (0, 0, -p_0)$, $p_0 = \mu n$. Using the equation div $\mathbf{D}(\mathbf{r}) = 0$ where $\mathbf{D}(\mathbf{r})$ is the Maxwell displacement vector, $\mathbf{D} = \mathbf{E} + 4\pi \mathbf{P}$, and taking into account that all these values depend only on z we have:

$$\frac{dD_z(z)}{dz} = 0, \quad E_z(z) - 4\pi p_0 \delta(z) = 0, \quad E_z = -\frac{dV}{dz}, \tag{11.23}$$

where $V(z)$ is the potential. It follows from these equations that

$$V(z) = V(-\infty) - \int_{-\infty}^{z} 4\pi p_0 \delta(z) dz. \tag{11.24}$$

Thus, in the approximation of a uniform dipole layer, the resulting electrostatic potential has the form of a sharp step of hight $\Delta V = 4\pi \mu n$. This effect is well-known and is used in the surface double layer model for the work function of metals (see (45), (46)). In our case the height of the step $\Delta V = 4\pi \mu n \simeq 0.1$ V at the interface. Thus, the electric field in the region of a step with an effective thickness a is equal $E_z = 4\pi \mu n/a$, opposite to the direction of the CTE dipole

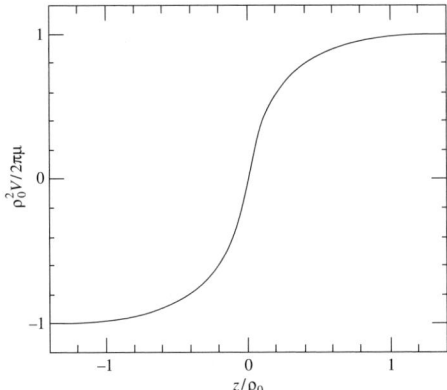

FIG. 11.7. Normalized potential profile across a D–A interface in the presene of a CTE density $n \simeq 1/\rho_0^2$.

moment. Probably, it is impossible to create a density of CTEs much larger than that used in our estimation above because the insulator-to-metal transition (19) leading to the ionization of the CTEs, sets a limit on n, as discussed above. For a typical intermolecular spacing a of about 0.5 nm and the exciton densities considered here, the average distance ρ_0 between two excitons along the interface plane $(n \simeq 1/\rho_0^2)$ is much larger than the interface thickness; as a consequence, the potential does not exhibit an abrupt jump and the electric field corresponding to ΔV is not restricted to the interface, but extends on either side over a layer of width comparable to ρ_0, as shown in Fig. 11.7. Taking, for sake of simplicity the CTEs located at sites of a square lattice, this figure shows $\phi(z) = V(z)/(2\pi\mu n)$ where $V(z)$ is the electrostatic potential along a direction perpendicular to the D–A interface and passing through the center of a square unit cell. Even in the case of a homogeneous disordered 2D distribution of CTEs, the average electric field profile is not expected to be much different than that for a 2D square lattice of equal density. When considering an asymmetrical superlattice in which all the interface voltage drops add up, the qualitative shape of the total electrostatic potential profile will be determined by the ratio between the distance between successive D–A interfaces, i.e. the superlattice period L and ρ_0: if L is larger than ρ_0 (i.e., at high density n) the potential will resemble a staircase as shown in Fig. 11.8, otherwise (i.e. at low density n) it will have a rather uniform slope. In either case, the average electric field in the direction of growth will be given by $E_0 = \Delta V/L$ and can be comparable to the electric field in the depletion layer of a typical semiconductor p–n junction (for the values of μ estimated above, $L \simeq 30$ nm and $n \simeq 10^{12}$ cm^{-2}, for instance, E_0 is about $3 \cdot 10^4$ V/cm). Of course, the uniformity along the superlattice planes will never be perfect in a real structure and, in general, a rather complicated spatial pattern of electric field force lines (and therefore current filaments) can be expected.

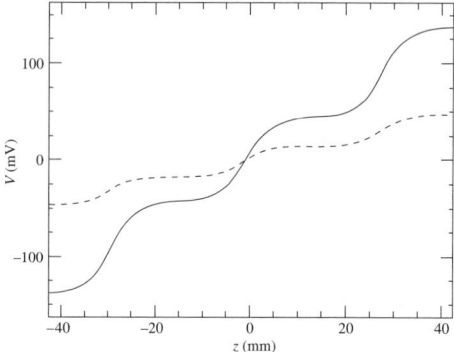

FIG. 11.8. Electrostatic potential along a D–A–N superlattice in the presence of 2D CTEs at the D–A interfaces with different densities: $n \simeq 10^{12}$ cm^{-2} (solid line) and $n \simeq 3.5 \times 10^{11}$ cm^{-2} (dashed line).

Under steady illumination, a condition of dynamical equilibrium will be reached with a constant 2D density of interface CTEs. Free carriers will also be present, either photogenerated directly or, for instance, as a result of the thermal ionization of the CTEs (see, for example, (37)–(39)) at room temperature. The electric field above will effectively separate the electrons and holes and drive a current along the superlattice crossing the D–A interfaces from the donor to the acceptor side. We are here neglecting the effects of the periodic changes of the band edges due to the superlattice compositional changes (such discontinuities could be minimized by a proper choice of materials) with respect to the additive effect of the potential variations at the D–A interfaces. Of course, these effects would be important to estimate the electron and hole mobilities along the growth axis. In the structures considered here, electrons could be effectively injected into the valence band on the acceptor side through a contact with a large work function metal and extracted from the conduction band on the donor side through a contact with a small work function metal (as done in the reverse direction in LED devices); such a current would deliver power to the external circuit load at expense of the light absorbed.

The expected efficiency of photovoltaic conversion is strongly dependent on the dark conductivity, on the processes of carrier photogeneration, on the kinetics of excitons and charge carriers, and can be estimated only in the framework of a complete theory which properly takes into account also the structure of the heterojunction.

However, the qualitative features of its dependence on the light intensity can be established on the grounds of simpler considerations. At low intensity of light the steady state CTE concentration, which is dependent on the pumping intensity I of CTEs and their lifetime, will be small and the macroscopic potential drop will be negligible. The macroscopic potential drop $V \sim I$ will be important with increasing CTE concentration. For a constant density of free carriers, the photovoltaic power will be proportional to I. At still higher pumping intensities, when the presence of CTE leads via their "cold" ionization to an increase in the density of free carriers as described above, the photovoltage will no longer increase, but the photovoltaic power dependence on I will still be approximately linear due to the increase in free carrier density.

11.5 Nonlinear optical response of charge-transfer excitons at donor–acceptor interface

In this section following the paper (40), we discuss the resonant and off–resonant optical nonlinearities of a system of charge-transfer excitons (CTEs) at a donor–acceptor (D–A) interface. We continue to consider the interface between organic materials (neutral in the ground state) for which the lowest electronic excitations are CTEs corresponding to the displacement of an electron from the donor to the acceptor side. We assume that the photogenerated CTEs have large static electric dipole moments perpendicular to the D–A interface plane. We show that this system may exhibit strong resonant optical nonlinearities induced by the dipole–dipole repulsion among the CTEs. We show that on account of this long-range interaction, the excitation intensity dependence of the Kerr nonlinearity is nonanalytic: it depends on the two-dimensional density n of CTEs as $n^{3/2}$. Therefore, the dependence of the CW nonlinear polarization on the laser electric field is beyond the usual power expansion. We also point out that the static electric field produced by the CTEs modifies the hyperpolarizabilities of nearby molecules. Therefore, in media with CTEs the intensity dependence of the nonlinear optical response can be stronger than usually expected, and this theoretical prediction could be easily experimentally tested. First we consider the effects of the dipole–dipole repulsion among CTEs on the nonlinear optical response and, then, the influence on the nonlinear hyperpolarizabilities of nearby molecules exerted by the large static field associated with a CTE.

11.5.1 *Resonant optical nonlinearity of CTEs: the role of the exciton–exciton repulsion*

11.5.1.1 *Exciton–exciton interaction at a D–A interface*

We have already discussed some peculiarities of the exciton–exciton interaction at a D–A interface above. Nevertheless, for the convenience of the readers we repeat here some results of this discussion to also make this section self-contained.

The dipole–dipole interaction energy between two CTEs having a dipole moment μ (which is typically 20 Debye) at a distance ρ along the D–A interface plane is $U = \mu^2/\rho^3$ and is rather large for small distances and decreases with a long range as ρ^{-3}; for instance, for $\rho \simeq 0.5$ nm, $U \simeq 1$ eV and for $\rho \simeq 1$ nm, $U \simeq 0.3$ eV. As the number of excitons at a distance ρ along the interface also scales as ρ the interaction energy of a given exciton with all other excitons which are at distance ρ decreases more slowly and scales as ρ^{-2}, rather than as ρ^{-3}.

The average exciton–exciton distance ρ_0 is related to the two-dimensional (2D) density of CTEs n by $n \simeq 1/\rho_0^2$. The ensuing repulsion increases the energy of the CTEs and the corresponding energy shift $\Delta\varepsilon$ is given by the average interaction of one exciton with all the others. We can expect that CTEs having a nonnegligible mobility will tend to order in such a way as to minimize the energy of repulsion. As discussed above, we assume a simple 2D square lattice structure for the CTE spatial distribution. Then

$$\Delta\varepsilon \simeq 10\,\mu^2/\rho_0^3 = 10\,V_0(n/N)^{3/2}, \quad (11.25)$$

with $V_0 = \mu^2/a^3$ and N the total 2D density of molecules given by $N = 1/a^2$ where a is the molecular crystal lattice constant along the interface (which is typically 0.5 nm). It is clear that such an additional repulsion energy is required to create a CTE at an empty site (i.e. at a vacancy in the CTE lattice).

To estimate the energy shift $\Delta\varepsilon$ we can use the expression

$$\Delta\varepsilon = 10\,V_0(a/\rho)^3. \quad (11.26)$$

Assuming that the mean distance between CTEs $\rho \simeq 10\,a$, e.g. that the concentration of CTEs is of the order of 10^{-2} we obtain that $\Delta\varepsilon \simeq 150$ cm^{-1}, which is significant on the scale of the homogeneous width of a CTE transition.

As discussed above, the scaling dependence $\Delta\varepsilon \simeq n^{3/2}$ follows also from the results of numerical simulations and, thus, the estimate made is expected to be valid even for a disordered homogeneous distribution of CTEs of comparable 2D density n. We wish to stress here the nonanalytic dependence of $\Delta\varepsilon$ on n: a perturbation theory expansion in terms of n, n^2, n^3, etc. would be inadequate as the leading correction scales like $n^{3/2}$.

11.5.1.2 Nonlinear D–A interface polarizability

The D–A interface polarizability due to the CTE contribution can be written as $\chi(\omega) \simeq A/((\varepsilon_0 + \Delta\varepsilon)^2 - \omega^2)$, where ε_0 is the CTE energy for $n = 0$ and A is a constant proportional to the CTE oscillator strength. Expanding $\chi(\omega)$ in series of $\Delta\varepsilon/\varepsilon_0$ we find that $\chi(\omega) = \chi_0^{(1)}(\omega)(1 - 2\varepsilon_0\Delta\varepsilon(n))/(\varepsilon_0^2 - \omega^2) = \chi_0^{(1)} + \Delta\chi(n)$, where $\chi_0^{(1)}$ is the polarizability for $n = 0$. Thus, for the nonlinear correction to the polarizability usually corresponding to the Kerr nonlinearity, we have $\Delta\chi/\chi_0^{(1)} =$ $= -2\varepsilon_0\Delta\varepsilon/(\varepsilon_0^2 - \omega^2) \simeq -\Delta\varepsilon/(\varepsilon_0 - \omega)$, where for resonant pumping $|\varepsilon_0 - \omega| \simeq$ $\simeq \delta$, δ being the exciton linewidth. Assuming the concentration $n \simeq N/100$ and using the previous estimates, we have $\Delta\varepsilon \simeq 150$ cm^{-1} which, even for $\delta \simeq 500$ cm^{-1}, gives for the resonant nonlinearity $\Delta\chi/\chi_0^{(1)} \simeq 0.3$. Such a large change in polarizability is not due to the two-level system-like anharmonicity which is the main nonlinear mechanism for Wannier–Mott excitons in semiconductors (phase space filling (10), (47)–(50)), but it is caused by the exciton–exciton interaction which in the present case is particularly large; in fact, we have that $\Delta\chi/\chi_0^{(1)}$ is one order of magnitude larger than n/n_S where the CTE saturation density n_S is given by N itself. A 2D concentration along each interface of

0.01 can hardly be dangerous for organic crystals as, taking into account the thickness of the layer between successive interfaces, it may correspond to a 3D concentration of order 10^{-3}. We note again that the continuous wave (CW) optical nonlinearity of 2D CTEs considered here is also interesting as it goes beyond the usual perturbation theory (51)–(53). As a matter of fact, because in steady state equilibrium conditions $n \propto I \propto |E(\omega)|^2$ (I being the pump light intensity), the shift of the CTE energy $\Delta\varepsilon(n)$ and the nonlinear correction to the polarizability turn out to be proportional to $|E|^3$. Thus, the nonlinear part of the polarization is $\Delta \mathbf{P}(\omega) \propto |E(\omega)|^3 \mathbf{E}(\omega)$ and such a term cannot be found in the usual expansion of the polarization $\mathbf{P}$ in powers of the components of the electric field $\mathbf{E}$. This peculiarity stems from the long-range exciton–exciton interaction that shifts the CTE energy in a nonanalytic way with respect to n. The unusual dependence of $\Delta\mathbf{P}$ on $\mathbf{E}$ considered here should be easily experimentally observed for CW resonant pumping.

11.5.2 Photogenerated static electric field: influence on the nonresonant optical response

In the above discussion, we have only considered the effects due to the CTE–CTE repulsion, which contribute to the resonant nonlinear absorption (as well as to other resonant nonlinearities) by the CTE themselves. Here, however, we want to mention a more general mechanism by which the nonlinear optical properties of media containing CTEs in the excited state can be enhanced. This influence is due to the strong static electric field arising in the vicinity of an excited CTE, If, for example, the CTE (or CT complex) static electric dipole moment is 20 Debye, at a distance of 0.5 nm it creates a field E^{CTE} of order 10^7 V/cm. Such strong electric fields have to be taken into account in the calculation of the nonlinear susceptibilities, because they change the hyperpolarizabilities α, β, γ, etc. of all molecules close to the CTE. For instance, in the presence of these CTE induced static fields, the microscopic molecular hyperpolarizabilities are modified as follows

$$\alpha_{ij} = \alpha_{ij}^{(0)} + \alpha_{ij\ell}^{(1)} E_\ell^{\mathrm{CTE}} + \alpha_{ij\ell k}^{(2)} E_\ell^{\mathrm{CTE}} E_k^{\mathrm{CTE}} + \cdots,$$
$$\beta_{ij\ell} = \beta_{ij\ell}^{(0)} + \beta_{ij\ell k}^{(1)} E_k^{\mathrm{CTE}} + \cdots,$$

etc. The changes in the molecular hyperpolarizabilities are reflected in all the macroscopic nonlinear optical constants of a medium. As the CTE induced electric fields increase with the concentration of CTEs, a stronger dependence of the nonlinear optical response on the intensity of light is obtained. For example, $\chi^{(3)}$ will include a contribution from $\chi^{(4)}$ and the corresponding polarization will be given by

$$\Delta P_i \simeq \chi_{ij\ell m}^{(3)} E_j E_\ell E_m + \chi_{ij\ell m q}^{(4)} E_j E_\ell E_m E_q^{\mathrm{CTE}},$$

where E is the pump light electric field; therefore, if the second term is not negligible, $|\Delta P|^2$ will depend on the pump intensity $I \propto |E(\omega)|^2$ more strongly like I^3, through the dependence of E^{CTE} on I. Of course, the macroscopic susceptibilities will depend on an average over the positions and orientations of the molecular species involved with respect to the static electric dipoles of the CTE photogenerated in the medium. For the 2D distribution of CTEs at the D–A interface considered above, the presence of a static macroscopic electric field extending on either side of the interface from the donor to the acceptor side may affect the molecular hyperpolarizabilities enough to reduce the symmetry of the optical response tensors; for instance, centrosymmetric molecules may acquire a second-order polarizability $\beta_{ij\ell} = 0 + \beta^{(1)}_{ij\ell k} E_k^{\text{CTE}} + \cdots$. The influence of static electric fields on the molecular hyperpolarizabilities has long been known, but in our case the static field effects are controlled by the pumping light as they are associated to the presence of the CTEs; this can lead to a novel class of *all-optical* nonlinearities. The calculation of second harmonic generation induced by charge-transfer excitations in a centrosymmetric medium can be found in (54). These calculations can be easily adapted to the case of D–A interfaces.

In conclusion, we have studied in this subsection the nonlinear optical response of 2D CTEs at a D–A interface: such a structure belongs to a class of novel systems of current interst to material scientists (13). The long-range exciton–exciton interaction leads to a large resonant nonlinear polarization exhibiting an unusual dependence on the light electric field which goes beyond the standard perturbation expansion. The static electric fields induced by the photogenerated CTEs affect the off-resonant hyperpolarizabilities of nearby molecules giving rise to new and stronger all-optical nonlinearities.

12

SURFACE EXCITONS

12.1 Introduction

12.1.1 *Surface excitons and polaritons*

The elementary surface excited states of electrons in crystals are called surface excitons. Their existence is due solely to the presence of crystal boundaries. Surface excitons, in this sense, are quite analogous to Rayleigh surface waves in elasticity theory and to Tamm states of electrons in a bounded crystal. Increasing interest in surface excitons is provided by the new methods for the experimental investigation of excited states of the surfaces of metals, semiconductors and dielectrics, of thin films on substrates and other laminated media, and by the extensive potentialities of surface physics in scientific instrument making and technology.

In the experimental study of surface excitons various optical methods have been used successfully, including the methods of linear and nonlinear spectroscopy of surface polaritons. A particularly large body of information has been obtained by the method of attenuated total reflection of light (ATR), introduced by Otto (1; 2) (Fig. 12.1) to study surface plasmons in metals. Later the useful modification of ATR method also was introduced by Kretschmann (3) (the so-called Kretschmann configuration, see Fig. 12.2). The different modification of ATR method has opened the way to an important development in the optical studies of surface waves and later was used by numerous authors for investigations of various surface excitations.

Experiments have also been started that use the inelastic light scattering and include the methods of coherent anti-Stokes Raman spectroscopy (CARS), as well as electron energy loss spectroscopy (EELS). The methods (see, for instance, (4)(5)) are based on the application of various physical processes, as can be seen from their names. Accordingly, they complement one another and enable us to study the elementary excitations of a surface over a wide range of energies and wavevectors.

In this section we mainly discuss the theoretical aspects of surface excitons. In this connection we should emphasize, first of all, that the determination of the wavefunctions and energies of surface excitons for different crystal models in general requires the application of microtheoretical methods. An exception is the case when the thickness of the subsurface layer L in which the surface excitons are localized substantially exceeds the lattice constant of the crystal a. Such surface states can be investigated within the framework of macroscopic phenomenological electrodynamics, which uses one or another model of the crys-

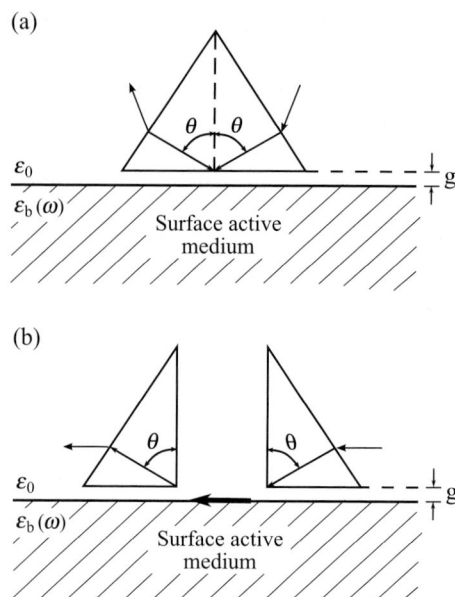

FIG. 12.1. Experimental Otto configuration for (a) single prism ATR measurements and (b) double prism measurement of propagation length L. Prism are spaced a variable distance g above surface (from (6)).

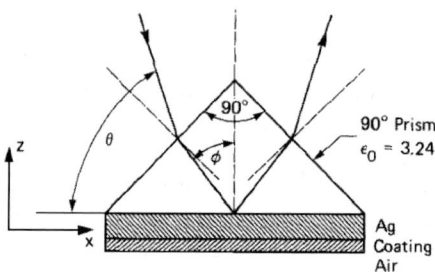

FIG. 12.2. Kretschmann configuration for investigation of surface wave at interface of metal (silver, in this case) and coating.

tal dielectric tensor and, in general, takes spatial dispersion into account (4). But phenomenological electrodynamics enables us to consider only long-wavelength excitations and those for which the wavelength $\lambda \gg a$. With regard to waves having a two-dimensional wavevector k, this means that within the framework of phenomenological electrodynamics only those states on the dispersion curve of surface excitons can be investigated for which the inequality $L \equiv L(k) \gg a$ mentioned above, as well as the inequality $k \ll 2\pi/a$, are valid. Yet, in the region of small values of k the phase velocity of the surface exciton $v = \omega/k$ can be of the order of light velocity c. This fact means that for long-wavelength surface excitons as well as for long-wavelength bulk excitons we should take into account the retarded interaction between charges. As is known (4), this part of the interaction can be taken into account in the most natural way within the framework of macroscopic phenomenological electrodynamics. The retarded interaction and transition from excitons to polaritons can be taken into account using Maxwell's equations. However, if the limit for light velocity $c \to \infty$ is carried out in Maxwell's equations, i.e. if we turn our attention to a more approximated consideration, then only the instantaneous (nonretarded) Coulomb interaction will be taken into account. In this approximation we go from polaritons to bulk and surface exciton states.

12.1.2 Coulomb surface excitons

Following (4), we shall call such surface states Coulomb surface excitons. The surface excitons obtained with retardation taken into account, in the region of small values of $k \sim \omega/c$ differ essentially from Coulomb surface excitons, and, consequently, such surface excitons are known as surface polaritons. For large values of k ($k \gg \omega/c$) the contribution of the retarded part of the interaction is negligible, the surface polaritons transform into Coulomb excitons with the growth of the value of k. To construct a theory of Coulomb excitons for any value of k the microtheory approach is required. The so-called cyclic boundary conditions can be applied for two coordinates of particles (electrons and nuclei) that are parallel to a crystal surface or an interface between two media. However, the boundary conditions for the third coordinate, corresponding to the displacement of particles along the direction normal to the crystal surface, are to be stated in such a way as to satisfy realistic physical conditions. It is quite evident that these conditions may differ significantly for surfaces with different structures as well as for excitons of various kinds. Accordingly, in the following discussion we shall consider separately the specific properties of surface excitons of small radius (Frenkel excitons) and charge transfer surface excitons. Under such conditions we take into account the fact that a semi-infinite crystal bounded by a plane, coinciding with one of its crystallographic planes, is invariant against discrete translations through vectors that are parallel to the surface of the crystal. These translations comprise a group, so that the states of surface excitons on such an ideal surface might be classified by irreducible representations of this group. The preceding applies of course not only to surface excitons but also to surface quasiparticles of any nature.

Since the irreducible representations of the group of translations are one-dimensional and are determined by the given wavevector, among the set of quantum numbers characterizing a surface exciton there is always a quasicontinuous quantum number – the wavevector. This vector differs from the corresponding one in the case of "bulk" excitons: it may be directed only along the crystal surface and assumes values only within a two-dimensional Brillouin zone. In connection with what has just been said, we note an interesting, but not yet investigated, theoretical problem, viz. that of finding the states of surface excitons at the interface of two crystals separated, for example, by a thin layer of vacuum and having different commensurate or incommensurate surface structures.[57]

One of the main theoretical problems is to determine the dependence of the energy of a surface exciton on its wavevector or, in other words, to obtain the dispersion law for surface excitons. Then the next problems arise in consideration of the interactions of these excitons with light, phonons and with surface defects.

12.1.3 *The exciton-phonon interaction and the role of surface defects*

We discussed in Ch. 3 the role of exciton–phonon interaction in the bulk. Below we stress some effects important also for surface excitons.

It should be born in mind that the idea of coherent surface excitons is valid only in those cases when the exciton–phonon interaction and the scattering of surface excitons by surface defects are enough weak. In this case the wavevector proves to be a "good" quantum number for an exciton, and its transfer of the energy is realized by wavepackets with an uncertainty of the wavevector $\Delta k \ll k \leq 1/a$. Consequently, the wavepacket covers an area of the surface with the dimensions $\Delta x \sim \Delta y \sim 1/\Delta k$, which are many times larger than the lattice constant. Specifically, under conditions of weak scattering there might be displayed some typical coherent properties of surface exciton states, and in this sense there is a complete analogy with the conditions for the existence of bulk coherent excitons. However, the kinetics of surface excitons may differ significantly from those of bulk excitons. A priori this difference is rather natural if one takes into account the two-dimensional nature of surface excitons, some features of their dispersion, an increased tendency to localization, and the peculiarities of the radiative width (see Subsection 12.3.2).

In cases where the exciton–phonon interaction is strong enough, another limiting situation may be realized. The physical substance of this situation has been discussed already by Frenkel (9), (10) in an investigation of the properties of bulk excitons. As applied to surface excitons of small radius, this situation arises in crystals where the time, τ_1, for the transfer of excitation from one molecule to a neighboring one ($\tau_1 \approx \hbar/\Delta E_{\text{exc}}$, where ΔE_{exc} is the width of an excitonic band) is large in comparison with the time τ_2, during which nuclei are displaced into new equilibrium states ($\tau_2 \approx \hbar/E_b$, where E_b is the energy of local deformation, and equals the "polaron" shift of the energy of an exciton). The displacement of the nuclei in this case might be due to the change of the interaction forces in the medium that takes place on molecular excitation. The local deformation of the lattice should travel together with an exciton. In this case, as Frenkel figuratively said, implying bulk excitons, the surface exciton while moving along the lattice "might also drag with itself the entire load of atomic displacements".

---

[57]In the discussion of the problem arising here the articles by Suslov (7; 8) may be useful. He considered electron localization in the field of two incommensurate potentials.

The wavevector is not always a "good" quantum number for such surface excitons (i.e. for $E_b > E_{\text{exc}}$). The band picture for such excitons is no longer valid, especially if the exciton mean free time T decreases to such a degree that the corresponding uncertainty of the energy $\hbar/T$ becomes larger than the width of the surface exciton band as well as width of bulk exciton band. In this case the exciton becomes noncoherent ("localized" in the terminology of Frenkel), and it walks randomly over the lattice sites on the surface of the crystal. In the considered situation the exciton will continue to be locate at surface only if the gas–condensed matter shift, discussed in Ch. 9, is much smaller at the surface than in the bulk. In opposite case the exciton with large probability will propagate from the surface to the bulk of crystal. The influence of the exciton–phonon interaction on the properties of surface excitons might be observed in investigations of the electronic excitation energy transfer along crystal surfaces.

12.2 Phenomenological theory of surface Coulomb excitons and polaritons

12.2.1 Surface polaritons at the sharp interface between media

If the thickness of the surface layer in which a surface exciton–polariton is localized considerably exceeds the lattice constant of a crystal, the electric and magnetic field strength vectors, i.e. vectors $\mathbf{E}$ and $\mathbf{H}$ of a wave with energy $\hbar\omega$ in both media (in vacuum and in the crystal; the crystal is assumed to be nonmagnetic so that the magnetic induction vector $\mathbf{B} = \mathbf{H}$), satisfy Maxwell's equations

$$\begin{aligned}
\boldsymbol{\nabla} \times \mathbf{E} &= \frac{i\omega}{c}\mathbf{H}, \\
\boldsymbol{\nabla} \times \mathbf{H} &= -\frac{i\omega}{c}\mathbf{D}, \\
\boldsymbol{\nabla}\mathbf{D} &= 0, \\
\boldsymbol{\nabla}\mathbf{H} &= 0,
\end{aligned} \qquad (12.1)$$

where $\mathbf{D}$ is the electric displacement vector (electric induction), whose Fourier components are connected with those of the vector $\mathbf{E}$ as in unbounded crystal, i.e.

$$D_i = \epsilon_{ij}(\omega, \mathbf{k})E_j. \qquad (12.2)$$

Let us assume that the crystal surface coincides with the plane $z = 0$, that the half-space $z > 0$ is vacuum ($\epsilon_{ij} = \delta_{ij}$), and that the half-space $z < 0$ is a cubic crystal with a dielectric function $\epsilon(\omega)$ (spatial dispersion is neglected so far). Hence, we are interested in the solutions of Maxwell's equations that vanish as $|z| \to \infty$; we shall look for them in the following form:

$$\mathbf{E}^{(1)} = \mathbf{E}_0^{(1)} \exp\{i(k_1 x + k_2 y + i\kappa_1 z)\} \qquad (\operatorname{Re} \kappa_1 > 0, \; z > 0),$$

$$\mathbf{E}^{(2)} = \mathbf{E}_0^{(2)} \exp\{i(k_1 x + k_2 y - i\kappa_1 z)\} \quad (\text{Re } \kappa_2 > 0,\ z < 0), \quad (12.3)$$

and similarly for $\mathbf{H}^{(1)}$ and $\mathbf{H}^{(2)}$. Let us first consider surface waves ignoring retardation, i.e. let us consider Coulomb surface excitons. In this case we must neglect the terms proportional to $1/c$ in eqn (12.1), so that the latter assume the form

$$\nabla \times \mathbf{E} = 0, \quad \nabla \mathbf{D} = 0, \quad \nabla \times \mathbf{H} = 0, \quad \nabla \mathbf{H} = 0. \quad (12.4)$$

Expressions (12.3) satisfy eqn (12.4) in both media, if $\mathbf{H} \equiv 0$ and moreover,

$$\mathbf{K}^{(1)} \times \mathbf{E}_0^{(1)} = 0, \quad \mathbf{K}^{(2)} \times \mathbf{E}_0^{(2)} = 0,$$
$$\mathbf{K}^{(1)} \cdot \mathbf{E}_0^{(1)} = 0, \quad \mathbf{K}^{(2)} \cdot \mathbf{D}_0^{(2)} = 0, \quad (12.5)$$

where

$$\mathbf{K}^{(1)} = k_1 \mathbf{e}_1 + k_2 \mathbf{e}_2 + i\kappa_1 \mathbf{e}_3,$$
$$\mathbf{K}^{(2)} = k_1 \mathbf{e}_1 + k_2 \mathbf{e}_2 - i\kappa_1 \mathbf{e}_3. \quad (12.6)$$

In relations (12.6) $\mathbf{e}_1$, $\mathbf{e}_2$ and $\mathbf{e}_3$ are real unit vectors directed along the x-, y- and z-coordinate axes, respectively. From the first two equations (12.5) it follows that the vector $\mathbf{E}_0^{(1)}$ is collinear with the vector $\mathbf{K}^{(1)}$ and the vector $\mathbf{E}_0^{(2)}$ with the vector $\mathbf{K}^{(2)}$, i.e.

$$\mathbf{E}_0^{(1)} = C_1 \mathbf{K}^{(1)}, \quad \mathbf{E}_0^{(2)} = C_2 \mathbf{K}^{(2)}, \quad (12.7)$$

where C_1 and C_2 are certain scalars. Substituting these relations into the remaining equations of the system (12.6) and using (12.2) we obtain the following relations

$$C_1 \mathbf{K}_i^{(1)} \mathbf{K}_i^{(1)} = 0, \quad C_2 \epsilon(\omega) \mathbf{K}_i^{(2)} \mathbf{K}_i^{(2)} = 0. \quad (12.8)$$

The electric field vectors at a sharp interface between media must satisfy the boundary conditions

$$\mathbf{E}_t^{(1)} = \mathbf{E}_t^{(2)}, \quad E_n^{(1)} = D_n^{(2)}, \quad (12.9)$$

where the indices t and n denote the tangential and normal components of a vector, respectively. Therefore, as a consequence of relations (12.6) and (12.7) and the boundary condition for the tangential components we find that

$$C_1 = C_2, \quad (12.10)$$

so that if for the Coulomb exciton the electric field strength vector is nonzero, this is the case in both media at the same time (i.e. in both the crystal and vacuum). Assuming that $C_1 = C_2 \neq 0$ and $\epsilon(\omega) \neq 0$, from eqn (12.8) we obtain

$$\mathbf{K}_i^{(1)} \mathbf{K}_i^{(1)} = 0, \quad \mathbf{K}_i^{(2)} \mathbf{K}_i^{(2)} = 0,$$

so that in view of (12.6)

$$\kappa_1 = \kappa_2 = \left(k_1^2 + k_2^2\right)^{1/2}. \tag{12.11}$$

Applying now the boundary conditions (12.9) for the normal components, and also taken relations (12.7), (12.10), (12.6) and (12.11) into account, we obtain the equation

$$\epsilon(\omega) = -1, \tag{12.12}$$

that determines the frequency of the Coulomb surface exciton. This frequency is beyond the region of the resonance (where $\epsilon(\omega) = \infty$) and can be obtained if an expression for $\epsilon(\omega)$ is assumed. In particular, for a cubic crystal in the vicinity of a dipole resonance at $\omega = \omega_\perp$, where

$$\epsilon(\omega) = \epsilon_\infty - \frac{A}{\omega^2 - \omega_\perp^2}, \quad A > 0, \tag{12.13}$$

we have for the frequency of the surface exciton, $\omega = \omega_s$,

$$\omega_s = \left[\omega_\perp^2 + A\left(\epsilon_\infty + 1\right)^{-1}\right]^{1/2}. \tag{12.14}$$

Since the condition $\epsilon(\omega) = 0$ is the condition for the existence of a longitudinal bulk exciton, whose frequency is $\omega_\parallel = \left(\omega_\perp^2 + A/\epsilon_\infty\right)^{1/2}$, it follows from (12.14) that the frequency of the Coulomb surface exciton is between the frequencies of the transverse $\omega_\perp$ and longitudinal $\omega_\parallel$ bulk Coulomb excitons, i.e. in the region where there is the total reflection of light from a crystal, if absorption and spatial dispersion are ignored. It is important to point out that the frequency ω_s in the approximation used does not depend on the wavevector, and the group velocity vanishes: $\partial \omega_s/\partial k_1 = \partial \omega_s/\partial k_2 = 0$. The situation changes when the retarded interaction is taken into account. In this case the frequency $\omega_s = \omega_s(\mathbf{k})$; $\mathbf{k} \equiv (k_1, k_2)$ is obtained from the equation (5)

$$\frac{\kappa_1}{\epsilon_1} + \frac{\kappa_2}{\epsilon_2} = 0, \tag{12.15}$$

where

$$\kappa_1 = \left(k^2 - \frac{\omega^2}{c^2}\epsilon_i(\omega)\right)^{1/2}, \quad i = 1, 2, \tag{12.16}$$

and $\epsilon_i(\omega)$ are the dielectric functions of the media in contact. As values $\kappa_{1,2}$ are positive the one of dielectric constants ϵ_1 or ϵ_2 has be negative. If for instance the dielectric constant $\epsilon_1 = 1$ (air) the function ϵ_2 has to be negative in the frequency interval where surface polaritons can be formed. For surface of metal this condition determines region where frequency ω is less than plasma frequency. For dielectrics the surface polaritons for the same reason can be formed only in the region of transverse–longitudinal splitting of exciton or optical phonon. It is clear also that incident from vacuum light with frequency equal to frequency of surface polariton will be completely reflected assuming that we neglected dissipation and possible roughness of surface.

Relations (12.15) and (12.16) can also be written in the form

$$k^2 = \frac{\omega^2}{c^2} \frac{\epsilon_1 \epsilon_2}{\epsilon_1 + \epsilon_2}. \tag{12.17}$$

They transform into (12.11) and (12.12) only for large values of $k \gg (\omega_s/c)\,\epsilon_i\,(\omega_s)$.

As a matter of fact, as can be seen from eqn (12.17), the limit $k \to \infty$ is consistent with the sum $\epsilon_1(\omega) + \epsilon_2(\omega)$ approaching zero, which in the particular case of the boundary with vacuum ($\epsilon_1 = 1, \epsilon_2 \equiv \epsilon$) agrees with eqn (12.12). The result of this limiting transition confirms once more the remark made above, viz. that surface polaritons for large values of k transfrom into Coulomb surface excitons. The dispersion law for Coulomb surface excitons at a sharp boundary and without taking spatial dispersion into account has the form

$$\omega_s(\mathbf{k}) = \text{const.}$$

In order to derive the first few terms in the expansion of the function $\omega_s(\mathbf{k})$ in powers of $k = 2\pi/\lambda$ for $ka \ll 1$ a more detailed consideration is required.

12.2.2 *Observation of exciton surface polaritons at room temperature*

It is interesting to note here (see also (11)) that there is a number of organic crystals that are insulators but which look like metals, that is, they have faces that reflect visible light well enough to give the crystal a metallic lustre. This phenomenon was first investigated by Anex and Simpson (11), in 1960, and since then the number of organic solids known to exhibit this phenomenon of "metallic reflection" has grown steadily. High reflectivity implies a frequency range within which the components of the dielectric tensor responsible for the effect is negative. These materials are therefore prime candidates in a search for insulators that will support at optical frequencies and at room temperature surface electromagnetic waves that are the counterparts of plasmon surface polaritons of real metals. Since it is a transition to a molecular exciton state that gives rise to the wide reflection bands, these surface electromagnetic waves are called exciton surface polaritons.

The first experimental observation of exciton surface polaritons at room temperature was presented in the paper by Pockrand et al. (12). The measurements have been made on the (110) face of crystals of the cationic organic dye CTIP, full name 3'-cyclopropyl- bis (1,3,3-trimethyl-indolenine-2-yl) pentamethinium-tetrafluoroborate. This material is ionic with a very intense electronic transition with oscillator strength f= 1.89 at 1.95 eV in methanol solution. The CTIP crystal was therefore chosen in (12) because the (110) face has a polariton stop–band extending from 1.7 eV to 2.9 eV. It is the great width of this stop-band, spanning most of the visible wavelengths of light, that gives the (110) face a striking silver reflectivity. For the purpose of comparison we note that anthracene, the only other organic shown to support exciton polaritons, is a crystal of neutral molecules with a first singlet exciton transition at 3.1 eV which is much less

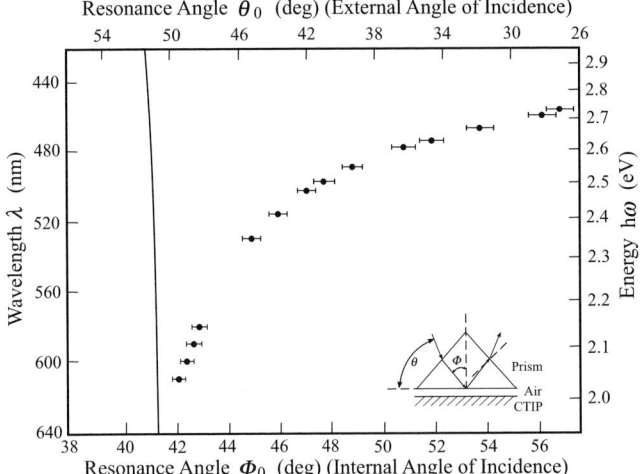

FIG. 12.3. Experimental dispersion curve of wavelength in nm and energy in eV as a function of internal and external angles of incidence. The almost vertical line is the critical angle, i.e., the light line. A lower right inset shows the experimental configuration.

intense with an oscillator strength off ≈ 0.1. In Fig. 12.3 the result of this experimental study is demonstrated (for details see (12)). Analogous results were demonstrated also in (13). However, in this paper the first observation of exciton surface polaritons at 295 K was presented on crystals of poly-2,4-hexadiyne-1,6-diol bis (toluene sulfonate). The very useful calculations of the attenuated reflectivity as a function of angle of incidence at wavelengths within the surface polariton spectrum of orthorhombic crystals, CTIP, TCNQ and single crystal polymer PTS can be found in (14).

12.2.3 Surface excitons in the presence of a transition layer

In the preceding derivation of the frequencies of surface polaritons and surface excitons the boundary conditions were applied at a sharp boundary without surface currents and charges. In this simplest version of the theory the so-called transition subsurface layer has been ignored; however, this layer is always present at the interface between two media, and its dielectric properties differ from the dielectric properties of the bulk. Transition layers may be of various origins, even created artificially, e.g. by means of particular treatment of surfaces or by deposition of thin films of thickness $d \ll \lambda$. Nevertheless, within the framework of the phenomenological theory it is rather easy to take account of their effects on surface wave spectra in an approximation linear in k (15).

It follows from eqn (12.4) for **E** and **D**, when the transition layer is taken into account, that the tangential component $\mathbf{E}_t$ and the normal component D_n are discontinuous. Within the framework of the linear theory, at an interface with vacuum and for fields of the form $\mathbf{D} = \mathbf{D}(z)\exp\{ik_1 x + ik_2 y\}$, it is found that

$$D_n^{(2)} - D_n^{(1)} = i\gamma \mathbf{k}_t \cdot \mathbf{E}_t^{(1)}, \quad \mathbf{E}_t^{(2)} - \mathbf{E}_t^{(1)} = -i\mu E_n^{(1)} \mathbf{k}_t, \qquad (12.18)$$

where γ and μ are certain characteristics of the transition layer; the details of the derivation of this result are given in (15) and (16).

Now substituting (12.7) into (12.18) and taking into account that for Coulomb surface excitons $\kappa_1 = \kappa_2 = k$ (see eqn 12.11), we obtain a system of two equations for the quantities C_1 and C_2:

$$\epsilon C_2 + C_1 = -\gamma k C_1, \quad C_2 - C_1 = \mu k C_1. \qquad (12.19)$$

The condition that the determinant of this system be equal to zero has the form

$$\epsilon(\omega) + 1 = -(\gamma - \mu)k, \qquad (12.20)$$

to the first order in k, and thus extends relation (12.12). In this approximation

$$\omega_s(\mathbf{k}) = \omega_s(\mathbf{0}) + \alpha k + 0\left(k^2\right), \qquad (12.21)$$

where the frequency $\omega_s = \omega_s(\mathbf{0})$ is determined from (12.12) while the quantity α is defined by

$$\alpha = (\mu - \gamma)\left(\frac{d\epsilon}{d\omega}\right)^{-1}_{\omega_s(\mathbf{0})}.$$

The quantity α for real μ, γ and $\epsilon(\omega)$ represents the group velocity of a surface exciton. Since it follows from (12.13) that

$$\left(\frac{d\epsilon}{d\omega}\right)^{-1}_{\omega_s(\mathbf{0})} = \frac{\epsilon_\infty}{(\epsilon_\infty + 1)^2}\frac{\left(\omega_\parallel^2 - \omega_\perp^2\right)}{2\omega_s(\mathbf{0})} \approx \frac{\epsilon_\infty}{(\epsilon_\infty + 1)^2}\delta,$$

where $\delta = \omega_\parallel - \omega_\perp$ is the longitudinal–transverse splitting, we find that

$$\alpha \approx \frac{\epsilon_\infty \delta}{(\epsilon_\infty + 1)^2}(\mu - \gamma).$$

In general, the quantities μ and γ are functions of the frequency ω and may have resonances. If the resonances lie far from the frequency $\omega_s(\mathbf{0})$, the dependence of the quantities μ and γ on ω may be ignored. For a rough evaluation of α in this case we assume that $\mu - \gamma \sim d$, where d is the thickness of the transition layer. Specifically, for a microscopic transition layer $d = 10^{-7}$ cm, and when $\delta = 10^{13} \text{s}^{-1}$, $\epsilon_\infty \approx 5$, we obtain $\alpha = 10^5$ cm/s. In the case when the transition layer is macroscopic ($d \gg a$) and, for example, homogeneous (with a dielectric permittivity $\tilde{\epsilon}(\omega)$), $\gamma = d\tilde{\epsilon}(\omega)$ and $\mu = d/\tilde{\epsilon}(\omega)$.

The possibility of resonance is of particular interest, because it may lead to nonanalytic dependence of ω_s on $\mathbf{k}$. Assume, for instance, that in the spectral region being considered, i.e. for $\omega \approx \omega_s = \omega_s(\mathbf{0})$, there is a resonance of the quantity $\mu(\omega)$, i.e. that $\mu(\omega) = \tilde{\mu}\omega_0/(\omega - \omega_0)$, $\tilde{\mu} > 0$ and the frequencies $\omega_s = \omega_s(\mathbf{0})$ and ω_0 are quite close. In this case the frequency of $\omega_s = \omega_s(\mathbf{k})$ is obtained from

$$\left(\frac{d\epsilon}{d\omega}\right)^{-1}_{\omega_s(\mathbf{0})} [\omega - \omega_s(\mathbf{0})] = \tilde{\mu}\frac{\omega_0}{\omega - \omega_0}k,$$

so that $\omega - \omega_s(0) = \pm bk^{1/2}$, $b = \left[\tilde{\mu}\omega_0 \left(d\epsilon/d\omega\right)_{\omega_s(0)}\right]^{1/2}$ for $\omega_0 = \omega_s(0)$. This relation qualitatively explains the splitting of the dispersion curve for surface plasmons on a random rough surface: for a macroscopic transition layer which replaces the surface roughness, $\mu = d/\tilde{\epsilon}$, where $\tilde{\epsilon}$ is equal to the half–sum of the dielectric functions of the medium and vacuum: $\tilde{\epsilon} = \frac{1}{2}[\epsilon(\omega) + 1]$; for this case $\omega_0 = \omega_s(0)$ and $\tilde{\mu} = 2d/\left(d\epsilon/d\omega\right)_{\omega_s(0)} \cdot \omega_0$ (for more details see Maradudin (17)). In a similar manner, together with the frequency dependence of the quantities μ and γ we may also take into account their dependence on the wavevector $\mathbf{k}$. It should be emphasized that the linear dispersion law (12.21) always occurs for Coulomb surface excitons for sufficiently small values of k.

12.2.4 Nonlinear surface polaritons

During 1980–1981 the possibility of the existence of nonlinear surface polaritons of various types was predicted in the literature (18), (19)–(20). In particular, Tomlinson (19) and Maradudin (21) derived s-polarized surface polaritons at a plane interface between two dielectrics, one of which has an isotropic and linear $\left(\epsilon^I\right)$ dielectric constant whereas the dielectric constant of the other is that of a nonlinear uniaxial medium

$$\epsilon_{11}\left(\omega, |\mathbf{E}|^2\right) = \epsilon_{22}\left(\omega, |\mathbf{E}|^2\right) \equiv \epsilon_0 + a(\omega)\left(|E_1|^2 + |E_2|^2\right),$$

$$\epsilon_{33} = \epsilon(\omega) + b(\omega)|E_3|^2.$$

Nonlinear p-polarized polaritons for $b = 0$ were considered under the same conditions (20). The papers cited revealed the dispersion laws of nonlinear waves of s and p types. Their most significant features are the non–monotonic dependence of the field amplitudes on z and the dependence of the frequencies of these waves on the field amplitude at the boundary (i.e. at $z = 0$). We do not go into details of the calculations here. We emphasize only that the transition layer, which can also be optically nonlinear, was ignored in these papers. Taking into account the nonlinearity in the transition layer results in the dependence of surface wave characteristics on the field at the interface, and this special case has been dealt with (22), (23) in connection with the problem of self-focusing of surface polaritons.

As stressed earlier, the transition from surface polaritons to Coulomb surface excitons corresponds to the limiting transition $c \to \infty$. For p-polarized waves it yields the dispersion relation

$$(\epsilon^I)^2 = \epsilon \left[\epsilon_0 + \frac{1}{2} a |E_1(0)|^2 \right],$$

which for $a \to \infty$ transforms into the analogous relation for linear Coulomb excitons at the boundary of uniaxial dielectric (4). In the same simple fashion we can carry out the limiting transition from s–polarized to Coulomb waves which, in contrast to p–polarized waves, do not exist within the framework of the linear theory.

The s-polarized nonlinear surface polaritons (21; 19) as far as we know have not yet been observed experimentally. As to p-polarized nonlinear surface polaritons, the situation here appears to have changed recently. The first observations of this type of slightly nonlinear surface polaritons have been performed by Chen and Carter (243). Namely, the dispersion of p-polarized surface waves on the flat surface of semiconductors (Si and GaAs were chosen) was investigated and a dependence of the dispersion on the electric field intensity was observed. Let us pay attention to one more type of nonlinear surface electromagnetic wave, viz. to the surface waves of self-induced transparency (25). In this work the self-induced transparency of surface polaritons under conditions of resonance with oscillations in the transition layer was considered. This layer was described in (25) within the framework of a gas model of two-level systems. A solution in the form of a surface soliton (i.e. a 2π surface pulse) was found, and the dependence of its duration on the dielectric constant of the media in contact and on the characteristics of the layer was investigated. Let us first give the expression for the duration of a 2π pulse, τ_p. According to (25) the value of τ_p and the value of the electric field strength in a 2π pulse, E_p, are determined by the relationships

$$\tau_p^{-2} = \frac{2\pi n_0 p^2/\hbar}{v_g/u - 1} \left(\frac{\partial f}{\partial \omega} \right)^{-1}_{\substack{\omega=\Omega \\ k=Q}}, \qquad E_p \approx \frac{2\pi \hbar}{p \tau_p},$$

where $u < v_g$ and

$$f(\omega, k) = \frac{\epsilon_1}{\kappa_1} + \frac{\epsilon_2}{\kappa_2}, \qquad \kappa_i = \left(k^2 - \frac{\omega^2}{c^2} \epsilon_i \right)^{1/2};$$

ϵ_i ($i = 1, 2$) are the dielectric constants of the media in contact; Ω is the resonance frequency in the transition layer; the value of Q is determined from the relation $f(\Omega, Q) = 0$; p is the transition dipole moment in a two-level molecule; n_0 is the molecular surface density in the transition layer; u is the velocity of pulse motion; and v_g is the group velocity of a surface polariton at the frequency Ω.

In these relations, for a boundary with vacuum, we make the transition $c \to \infty$, then $f \to (1/k)(\epsilon + 1)$ and $v_g = d\omega/dk = \alpha$ (see eqn 12.21). Thus, in this limit we have

$$\tau_p^{-2} = \frac{2\pi n_0 p^2 Q/\hbar}{(\alpha/u) - 1} \left(\frac{d\epsilon}{d\omega} \right)^{-1}_{\omega=\Omega}, \qquad \alpha > u.$$

It is expected within the framework of the theory of self–induced transparency that the pulse duration is either small or of the order of the relaxation time T_2 in

12.3 Site shift surface excitons in molecular crystals

12.3.1 *Site shift surface excitons (SSSE) in anthracene*

In this subsection we go back to more detail than in Section 9.1 discussion of excitations in first monolayers of organic crystals of anthracene family (naphthalene, tetracene and so on). As we explained the spectrum of this states depends on the distance to surface of crystal. For this reason these excitons are called site shift surface excitons (SSSE) what is reflected in title of this subsection.

In anthracene crystal as it follows from data collected in Table 3.1 the distance between **a, b** plane is large in comparison with distance between nearest molecules in this plane. As the result, as we already mentioned in Section 9.1, the interactions between molecules in different planes is smaller than interaction between molecules inside the same plane. This means that anthracene crystal the same as other crystals of the its family have layered structure which we explicitly take into account in microscopical theory of surface states. We will use the same Hamiltonian (2.2) as we used in consideration of bulk states in simplest Heitler–London approximation. However, now we have to take into account that translational symmetry exists only along the surface of crystal which we assume parallel to **a, b** plane. In an infinite crystal the diagonalization of Hamiltonian leads to two exciton bands $E_{1,2}(k)$, so that the general pattern of levels is the one shown schematically in Fig. 12.4b.

In the given case and in Heitler–London approximation the diagonalization of Hamiltonian (2.2) can be carried out by means of (2.8) with functions $u_{n\alpha}$ taken in the form:

$$u_{n\alpha} = C_\alpha(n_3) \exp\{i(k_1 n_1 a + k_2 n_2 b)\}, \quad (12.22)$$

where

$$n_1, n_2 = 0, \pm 1, \pm 2, \ldots, \quad n_3 = 1, 2, \ldots$$

are the integers determining the value of the vector **n** in a semi-infinite crystal and (k_1, k_2) are the components of the in-plane wavevector. Using this relation and recalling that in anthracene we deal with a nondegenerate molecular transition, we obtain the following system of equations:

$$[E - \Delta\varepsilon - D(n_3)] C_\alpha(n_3) = \sum_{\beta m_3} T_{\alpha\beta}(k_1, k_2; |n_3 - m_3|) C_\beta(m_3), \quad (12.23)$$

giving the quantities $C_\alpha(n_3)$ with dependence on n_3 of value $D(n_3)$ taken into account, and where

$$T_{\alpha\beta}(k_1, k_2; |n_3 - m_3|)$$

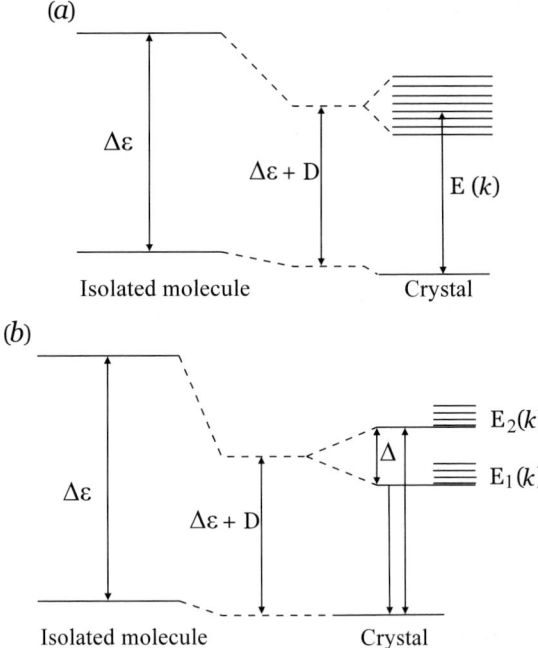

FIG. 12.4. The molecular and exciton levels in a crystal with (a) one molecule and (b) two molecules in a crystal unit cell.

$$= \sum_{m_1 m_2} M_{\mathbf{n}\alpha,\mathbf{m}\beta} \exp\{ik_1 a(m_1 - n_1) + ik_2 b(m_2 - n_2)\}. \quad (12.24)$$

To simplify the analysis here we confine ourselves to discussing the simplest case, when $k_1 = k_2 = 0$. In this case the quantity (12.24) is equal to the energy of interaction of the dipole p^α in the plane $n_3 = $ const with the lattice dipoles p^β located in the plane $m_3 = $ const. It was shown long ago (26; 27) that the electric field created by such a lattice of dipoles decreases exponentially at distances of the order of the lattice constant. Thus in (12.23) when summing over m_3, one can confine oneself to the nearest-neighbor approximation. In this approximation the system of equations (12.23, 12.24) can be written as follows

$$[E - \Delta\varepsilon - D(n_3)] C_\alpha(n_3) = \sum_\beta \{T_{\alpha\beta}(0,0;0) C_\beta(n_3)$$
$$+ T_{\alpha\beta}(0,0;1) [C_\beta(n_3+1) + C_\beta(n_3-1)]\}. \quad (12.25)$$

The matrix elements $T_{\alpha\beta}$ which occur in (12.25) were evaluated for anthracene by Philpott (28). According to his results

$$T_{11}(0,0;0) = -211 \text{ cm}^{-1}, \quad T_{12}(0,0;0) = 114.4 \text{ cm}^{-1},$$

$$T_{11}(0,0;1) = -0.44 \text{ cm}^{-1}, \qquad T_{12}(0,0;1) = 0.8 \text{ cm}^{-1}.$$

If follows that with a reasonable accuracy we may ignore the terms containing the factors $C_\beta(n_3 \pm 1)$ in (12.25). In this approximation the exciton states, both bulk and surface, are formed as a result of the interaction of molecules localized in the same plane $n_3 = \text{const}$. The latter result shows that even bulk excitons in anthracene are actually quasi-two-dimensional in the spectral region treated.

According to (12.25) far from the crystal boundary the coefficients C_α $\alpha = 1,2$, satisfied the system of equations

$$[E - \Delta\varepsilon - D - T_{11}(0,0;0)]\,C_1(n_3) = T_{12}(0,0;0)C_2(n_3),$$
$$[E - \Delta\varepsilon - D - T_{22}(0,0;0)]\,C_2(n_3) = T_{21}(0,0;0)C_1(n_3). \qquad (12.26)$$

Since $T_{11} = T_{22}$ and $T_{12} = T_{21}$, we find that the energy of a bulk exciton is given by

$$E_{1,2} = \Delta\varepsilon + D + T_{11}(0,0;0) \pm T_{12}(0,0;0), \qquad (12.27)$$

and the value of the Davydov splitting is

$$\Delta = 2T_{12}(0,0;0) \approx 230 \text{ cm}^{-1}, \qquad (12.28)$$

which is rather close to the value estimated experimentally. However, as we explained in Section 9.1, the quantity D may depend on n_3. This value depends only on the short-range interaction and in estimation of D we may use the nearest-neighbor approximation. Far from the boundary this quantity may be represented as the sum

$$D = D^{(0)} + 2D^{(1)} + 2D^{(2)} + \ldots \approx D^{(0)} + 2D^{(1)} + 2D^{(2)},$$

where the contribution of intermolecular interaction in the same plane is taken into account in $D^{(0)}$, the interaction with the molecules of the neighboring planes is taken into account in $D^{(1)}$, and so on.

For the excitons localized in the plane $n_3 = 1$

$$D = D^{(0)} + D^{(1)} + D^{(2)}.$$

At the same time for the excitons localized in the plane $n_3 = 2$

$$D = D^{(0)} + 2D^{(1)} + D^{(2)},$$

and so on.

It follows from the preceding discussion that the energies of SSSE are given by the relation (12.27), their Davydov splitting by relation (12.28). This splitting and the polarization in considered approximation are independent of the value of n_3 and, consequently, must be the same as the corresponding quantities for bulk excitons. If we take into account that all the quantities $D^{(i)}$, as well as their sum, are negative, then we have to conclude that the spectrum of excitons

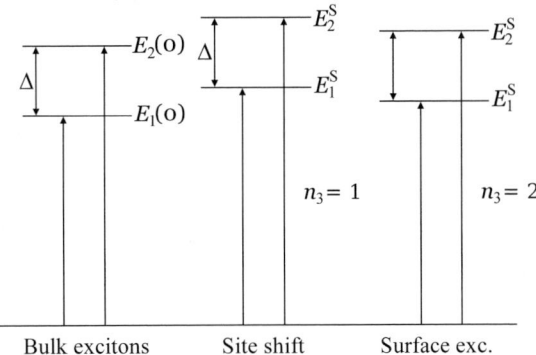

FIG. 12.5. The bulk exciton and the site shift surface exciton levels in anthracene-type crystals with two molecules in a crystal unit cell.

localized, e.g. in the plane of $n_3 = 1$, must be shifted as a whole with respect to the spectrum of bulk excitons into the region of higher energies by the amount $|D^{(1)} + D^{(2)}|$ (see Fig. 12.5).

Similarly, the spectrum of excitons localized in the plane $n_3 = 2$ must be shifted by the quantity $|D^{(2)}|$, and so on.

These simple and rather precise predictions of the theory, developed mainly by Sugakov (29)–(31) and Philpott (28) (the last reference includes a review of experimental investigations), are in good agreement with the results of experimental studies of the spectra of absorption and reflection of light and fluorescence (32). We are not able to discuss in detail here the problem how SSSE influence the optical effects mentioned above. For such a discussion see the works by Sugakov (29)–(31), Turlet and Philpott (33) and Orrit et al. (34).

Note here only the experimental and theoretical results reported in (35). In this paper the polarized reflection spectrum of the first singlet transition of crystalline anthracene at 2K have been investigated for light at near normal angles of incidence on the (001) face (Fig. 12.6). The optical properties of the bulk crystal have been calculated by a Kramers–Kronig transformation of the reflectivity data. The complex refractive index so obtained was used to calculate the effect of surface transitions on the bulk reflection spectrum using equations given by a microscopic theory of surface and bulk exciton states. Such approach gave an excellent agreement between the calculated and experimental spectra (see Fig. 12.7).

Since SEEE are localized in one or two subsurface monolayers, the discussion is actually about the effects of electronic excitations in the transition layer on optical properties of a crystal. Note also that even in pure crystals without impurities could be localized adsorbed molecules on the surface that makes the value of $D(n_3)$ a random function. The process of attaching foreign molecules to the surface lead to an analogous effect. As SSSE can be experimentally observed,

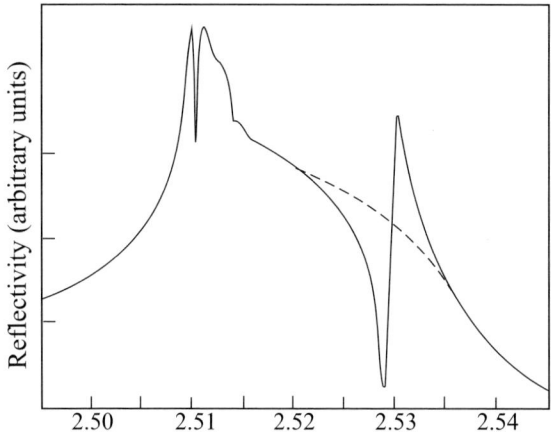

FIG. 12.6. b-polarized reflection spectrum of anthracene in the region of 0-0 transition. Broken line shows the reflectivity assumed for the bulk crystal with the absence of surface transition (adopted from (35)).

the consideration of the influence of the random perturbation on these surface states may be of not only theoretical interest. The spectra of surface excitons in disordered media are considered briefly in Section 12.6.

12.3.2 On the radiative width of site shift surface excitons

The macroscopical surface excitons obtained when retardation is taken into account, i.e. surface polaritons, cannot spontaneously transform into bulk emitted photons. Therefore, surface polaritons are sometimes said to have zero radiation width (it goes without saying that a plane boundary without defects it implies). At the same time the Coulomb surface excitons and polaritons in two-dimensional crystals possess, as was shown in Ch. 4, the radiation width $\Gamma \approx \Gamma_0 (\lambda/2\pi a)^2$, where λ is the radiation wavelength, a is the lattice constant, and Γ_0 the radiative width in an isolated molecule. For example, for λ=500 nm and $a = 0.5$ nm the factor $(\lambda/2\pi a)^2 \approx 2 \times 10^4$, which leads to enormous increase of the radiative width. For dipole allowed transitions $\Gamma_0 \approx 5 \times 10^{-4} \text{cm}^{-1}$, so that the value of $\Gamma \approx 10 \text{ cm}^{-1}$ corresponds to picosecond lifetimes $\tau = 2\pi\hbar/\Gamma \approx 10^{-12}$s.

Usually the states of surface excitations show up primarily in optical absorption and reflection spectra as well as in luminescence spectra. This fact has been used by Sugakov (29)–(31), by Turlet and Philpott (33) and by Philpott and Turlet (35) to interpret the results of experimental investigations of the optical properties of anthracene in the region of its lowest electronic excitation, see the reviews by Orrit et al. (34). and by Turlet et al. (36). In the reflection of light from the (**a, b**) surface of this crystal, and also in the fluorescence spectra, specific features are observed at energies higher than the lowest bulk exciton by 206

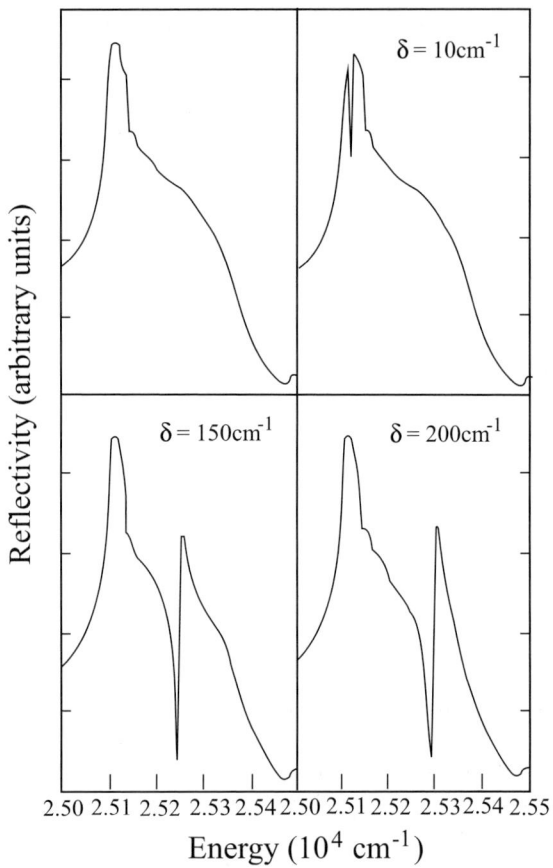

FIG. 12.7. The effect of a single SSSE transition on the bulk crystal reflection spectrum (top left) calculated for surface barrier hight $\delta = 10$, 150, and 200 cm^{-1}. For outermost monolayer $\delta = |D^{(1)} + D^{(2)}|$. These calculated spectra are qualitatively identically to the measured **b**-polarized reflection spectra (Fig. 12.6) (adopted from (35)).

cm^{-1}, 8.7 cm^{-1} and 0.7 cm^{-1}. These features were connected with the SSSE discussed in Subsection 12.3.1 that are localized on the first, second and third subsurface (**a, b**) layers. As these states possess an enormous radiation width and, consequently, a short lifetime, one can understand why they are visible in fluorescence spectra. As the energy of these states exceeds the energy of bulk excitons of the lowest band, SSSE decay accompanied by formation of bulk excitations is possible. However, in this case the decay has no time to occur and therefore plays no significant role, since the matrix element of the resonant energy transfer between the nearest (**a, b**) planes is equal to ≈ 1 cm$^{-1} \ll \Gamma$. In the following employing a simple semiclassical approach we consider the radiative width of SSSE within the framework of classical macroscopic electrodynamics assuming, however, that the surface currents created by microscopic 2D Coulomb excitons are derived on the basis of the quantum theory. Such a method allows one to take account of the dependence of the radiation width on the dielectric constants of the media in contact and thus to generalize the results described in Ch. 4.

For the sake of definiteness we assume that a state of SSSE is localized in the single crystallographic plane $z = 0$. Let this plane separate media with positive dielectric constants $\epsilon_1(\omega)$ ($z > 0$) and $\epsilon_2(\omega)$ ($z < 0$), respectively. We ignore an anisotropy in the plane $z = 0$ and assume that the dipole moment of the transition of the Coulomb 2D exciton is located in the plane $z = 0$.

To consider this state when retardation is taken into account, the solutions of Maxwell's equations (12.1) should be obtained in the form (12.3) taking into account the presence of a surface polarization current $\mathbf{j} = \dot{\mathbf{P}}$. Assuming that the x-axis is directed along the direction of the surface exciton wavevector (in this case $k_2 = 0$, $k_1 \equiv k$), we come to the conclusion that in this case the amplitudes of the fields E_x and H_y at $z = 0$ satisfy the conditions:

$$H_y^{II} - H_y^{I} = \frac{4\pi}{c}\dot{P}_x,$$

$$E_x^{II} - E_x^{I} = 0.$$

As for the waves of the form (12.3), $H_y^I = -(i\omega\epsilon_1/c\kappa_1)E_x^I$, $H_y^{II} = (i\omega\epsilon_2/c\kappa_2)E_x^{II}$, where $\kappa_i = [k^2 - (\omega^2/c^2)\epsilon_i]^{1/2}$, $i = 1, 2$ and, moreover, $\dot{P}_x = -i\frac{\omega}{c}\chi(\omega)E_x^I$; $\chi(\omega)$ is the layer polarizability; from the given system of equations for the amplitudes of E_x and H_y we derive the required dispersion relation

$$\frac{\epsilon_1}{\kappa_1} + \frac{\epsilon_2}{\kappa_2} = -4\pi\chi(\omega). \tag{12.29}$$

The polarizability of the two-dimensional crystal along the x-axis for the frequency region $\omega \approx \omega_0$, where $\omega_0 \equiv E(0)/\hbar$ and $E(0)$ is the energy of the two-dimensional exciton with $\mathbf{k} = 0$, is defined by

$$\chi(\omega) = \frac{2|p_x^{01}|^2 \omega_0 (\hbar a^2)^{-1}}{\omega_0^2 - \omega^2 - 2i\gamma\omega}, \tag{12.30}$$

where p_x^{01} is the matrix element of the molecular dipole moment, γ is the damping constant due to the exciton–phonon interaction, and a is the distance between molecules in the crystallographic plane $z = 0$.

The assumption that $\epsilon_1 = \epsilon_2 = 1$, $\kappa_1 = \kappa_2 = \kappa = [k^2 - (\omega^2/c^2)]^{1/2}$ leads to the situation of a two-dimensional crystal in the vacuum (see (37)). In this case the dispersion equation (12.29) assumes the form

$$\frac{A}{2}\left(k^2 - \frac{\omega^2}{c^2}\right)^{1/2} = \omega - \omega_0, \tag{12.31}$$

for $|\omega_0 - \omega| \ll \omega_0$, where $A = 4\pi|p_x^{01}|^2/a^2\hbar$. This relation implies that the branch of nonradiative states (corresponding to real value of wavevector k) of $\omega(k)$ $(k > (\omega/c))$ starts at $k_0 = \omega_0/c$, and for large $k \gg k_0$ has a linear dispersion law, viz. $\omega = \omega_0 + (A/2)k$.

We now determine the radiation width of the exciton state assuming that in (12.29) the wavevector $k = 0$. For normal incidence of light just the states with $k = 0$ determine the features of the absorption and reflection spectra and, in the presence of radiative decay, these states give rise to fluorescent photons.

When $k = 0$, (12.29) implies that for positive $\epsilon_1(\omega)$ and $\epsilon_2(\omega)$

$$\omega = \omega_0 - i(\gamma + \Gamma), \tag{12.32}$$

where $\Gamma = 4\pi\omega_0|p_x^{01}|^2/ca^2(n_1 + n_2)$, and n_1 and n_2 are the refractive indices of light at the frequency ω_0 in the media in contact (in (12.29) at $k = 0$ for outgoing waves (see 12.3) $\kappa_1 \to i(\omega/c)\epsilon_1^{1/2}$, $\kappa_2 \to -i(\omega/c)\epsilon_2^{1/2}$).

The radiation width of the isolated molecule is

$$\Gamma_0 = 4\pi \left|p_x^{01}\right|^2 \left(\frac{\omega}{c}\right)^3.$$

Thus

$$\Gamma = \Gamma_0 \left(\frac{\lambda}{2\pi a}\right)^2 (n_1 + n_2)^{-1}. \tag{12.33}$$

Note that for $\epsilon_1 > 0$ and $\epsilon_2 > 0$ the fluorescent photons can propagate in both half-spaces. If one of the values of ϵ_i, e.g. $\epsilon_2(\omega_0)$, is negative, then the fluorescent photons can propagate in the half-space $z > 0$ only. In this case alongside the enormous width of the two-dimensional exciton with $k = 0$ a shift of its energy occurs as well. In fact, for this case (12.29) yields

$$\omega = \omega_0 - i(\gamma + \Gamma) + \delta,$$

where

$$\Gamma = \Gamma_0 \left(\frac{\lambda}{2\pi a}\right)^2 \frac{\sqrt{\epsilon_1}}{\epsilon_1 + |\epsilon_2|},$$

$$\delta = \frac{\Gamma\sqrt{|\epsilon_2|}}{\sqrt{\epsilon_1}}.$$

These relations generalized results obtained in (37) (see also Ch. 3) for the case of two-dimensional excitons in the vacuum and demonstrate the dependence of the quantity Γ on the refractive indices of the contacting media. They show, for example, that when the frequency ω_0 tends to the frequency of the bulk dipole resonance (in this case $|\epsilon_2(\omega_0)|$ can be large) the radiation width Γ should decrease; within the scope of the microscopic theory this problem was discussed by Tovstenko (38). Unlike the radiation width Γ, which under the indicated conditions is much larger than Γ_0, the shift δ is usually of the order of, or smaller than the shift of the exciton energy due to the exciton–phonon interaction. For this reason the radiative shift generally plays a minor role.

12.4 Edge exciton states in molecular chains

12.4.1 *Introduction. Tamm states*

In most well-studied molecular crystals, such as anthracene, tetracene, and other crystals of the aromatic series, the intermolecular distances are quite large in all directions and, hence, the overlap of the wavefunctions of neighboring molecules is very small. In such crystals, the deviations from electroneutrality of the molecules in the lowest energy excitations in the majority of cases can be neglected. The Frenkel excitons and weak exciton–phonon interaction are typical for the lowest energy states of this crystals.

On the other hand, there are crystals in which the distances between molecules in one direction is much smaller than in the others; i.e. crystals which are quasi-one-dimensional along a stacking axis. Examples are the perylene derivatives 3,4:9,10–perylenetetracarboxylic dianhydride (PTCDA) and the N-N′ - dimethylperylene–3,4:9,10–dicaboximide (MePTCDI). During the last few years, the nature of their lowest energy electronic states was intensively investigated (39) –(41). Analogous examples can be found among molecular aggregates and polymers.

In the analysis of the lowest electronic excitations in quasi-one-dimensional crystals, it is natural to take into account not only Frenkel excitons, but also one-dimensional charge-transfer (CT) excitons. We will show below that the spectrum of excited states in the molecular chain is strongly sensitive to the mixing of Frenkel and CT states.

It is well known that for ideal linear molecular chains of finite length without mixing of Frenkel with CT states, only the usual space quantization of exciton states inside the band appears. Nearest neighbor approximation (this model is often used for analysis of spectra of *J*-aggregates, see, e.g. (42)) yields the energy of Frenkel exciton states:

$$E_j = E_0 + 2M \cos\left(\frac{\pi j}{N+1}\right), \qquad (12.34)$$

where N is the number of molecules in the chain, M the matrix element of transferring the excitation to the neighboring site, $j = 1, \ldots, N$ numbers the states, and E_0 is the energy of the molecular excitation. The width of the exciton band (12.34) is equal to $4M$. For $M < 0$ the $j = 1$ state lies at the bottom of the band (direct band edge), whereas for $M > 0$ (the *H*-aggregates) this state is located at the top of the band.

However, even in perfect 1D structures, the mixing of Frenkel and CT excitons destroys this simple picture. Below, following (43), we show that this mixing is responsible for the appearance of new excitonic states, which are localized at the ends of a one-dimensional crystal chain and which are analogous to Tamm surface states of electrons. Their energy can be blue- or red-shifted in comparison with the "bulk" states. In the case of red-shift, these states can determine the fluorescence spectrum of a molecular chain. They can also play an important role in quantum confinement of the states in the molecular chain. For the description of the states in finite molecular chains, we use the same approach which was developed in (41), discussing the mixing of Frenkel and CT excitons of infinite one-dimensional crystals.

12.4.2 Mixing of Frenkel and charge-transfer excitons in a finite molecular chain

For the discussion of the excitonic spectrum in a one-dimensional molecular crystal (with one molecule per unit cell) we use the following Hamiltonian:

$$\hat{H} = \hat{H}^{\mathrm{F}} + \hat{H}^{\mathrm{C}} + \hat{H}^{\mathrm{FC}},$$

$$\hat{H}^{\mathrm{F}} = \sum_n E_{\mathrm{F}} B_n^\dagger B_n + \sum_{nn'} M_{nn'} B_n^\dagger B_{n'},$$

$$\hat{H}^{\mathrm{C}} = \sum_{n\sigma} E_{\mathrm{CT}} C_{n\sigma}^\dagger C_{n\sigma}, \qquad (12.35)$$

$$\hat{H}^{\mathrm{FC}} = \sum_n \left\{ \varepsilon_e \left(B_n^\dagger C_{n,+1} + B_n^\dagger C_{n,-1} \right) \right.$$
$$\left. + \varepsilon_h \left(B_n^\dagger C_{n+1,-1} + B_n^\dagger C_{n-1,+1} \right) \right\} + \mathrm{h.c.}$$

In these relations the operator $B_n^\dagger(B_n)$ describes the creation (annihilation) of a molecular excitation at lattice site n. We assume below that $n = 1, 2, \ldots, N$, where N is the number of molecules in the chain and we consider one electronically excited molecular state. Then E_{F} is the on-site energy of a Frenkel exciton and $M_{nn'}$ is the hopping integral for molecular excitation transfer from molecule n to molecule n'. In the summation in $\hat{H}^{\mathrm{F}}$ the terms with $n = n'$ are omitted. The Hamiltonian $\hat{H}^{\mathrm{F}}$ describes the Frenkel excitons in the Heitler–London approximation.

Besides the Frenkel excitons, we additionally included nearest-neighbor charge-transfer excitons in our consideration. A localized CT exciton with the hole at lattice site n and the electron at lattice site $n + \sigma$ ($\sigma = -1, +1$) is created (annihilated) by the operator $C_{n\sigma}^\dagger (C_{n\sigma})$. This approximation is justified if the nearest-neighbor CT exciton is energetically well separated from higher CT excitons. Indeed this is the case, for example, in PTCDA and MePTCDI crystals where the next CT exciton lies already ≈ 0.5 eV higher (44). This energetic separation is larger than the charge-transfer integrals which in these crystals are in the order of 0.1 eV (see below subsection 12.4.3). Therefore, the necessary conditions for an exciton model which includes only the lowest Frenkel and CT states are fulfilled for such crystals. Similar to Frenkel excitons, only the vibrational ground state is considered for the CT excitons with E_{CT} as their on-site energy. Hopping of CT states will not be considered. However, as will be shown, the Frenkel–CT exciton mixing creates a finite width of exciton bands even if the Frenkel hopping matrix element is neglected. The mixing of Frenkel and CT excitons arises due to the last part $\hat{H}^{\mathrm{FC}}$ of the Hamiltonian. Here, the nonlocal transformation of a CT state into a Frenkel state at the lattice site of either hole or electron is allowed. The relevant transfer integrals $\varepsilon_e(\varepsilon_h)$ can be visualized as a transfer of an electron (hole) from the excited molecule n to its nearest neighbor.

In the same way as in the theory of Frenkel exciton mixing (Ch. 3), we have to diagonalize the total Hamiltonian $\hat{H}$ to consider the mixing of Frenkel and CT states and to find the new mixed states. We use the linear transformation to new operators ζ and $\zeta^\dagger$, where

$$\zeta = \sum_{n=1}^{N} u_n B_n + \sum_{n=1}^{N-1} v_{n,+1} C_{n,+1} + \sum_{n=2}^{N} v_{n,-1} C_{n,-1}. \tag{12.36}$$

In this relation, the terms with $v_{1,-1}$ and $v_{N,+1}$ are absent, because the corresponding CT states are missing in a finite chain.

The coefficients $u_n, V_{n,+1}$ and $v_{n,-1}$ have to obey the system of equations:

$$(E_\mathrm{F} - E) u_n + M (u_{n+1} + u_{n-1}) + \varepsilon_e (v_{n,+1} + v_{n,-1})$$
$$+ \varepsilon_h (v_{n-1,+1} + v_{n+1,-1}) = 0,$$
$$(E_\mathrm{CT} - E) v_{n,+1} + \varepsilon_e u_n + \varepsilon_h u_{n+1} = 0, \tag{12.37}$$
$$(E_\mathrm{CT} - E) v_{n,-1} + \varepsilon_e u_n + \varepsilon_h u_{n-1} = 0,$$

for $1 < n < N$. The equations for $n = 1$ and $n = N$ describe the boundary conditions:

$$(E_\mathrm{F} - E) u_1 + M u_2 + \varepsilon_e v_{1,+1} + \varepsilon_h v_{2,-1} = 0,$$
$$(E_\mathrm{F} - E) u_N + M u_{N-1} + \varepsilon_e v_{N,-1} + \varepsilon_h v_{N-1,+1} = 0,$$
$$(E_\mathrm{CT} - E) v_{1,+1} + \varepsilon_e u_1 + \varepsilon_h u_2 = 0, \tag{12.38}$$
$$(E_\mathrm{CT} - E) v_{N,-1} + \varepsilon_e u_N + \varepsilon_h u_{N-1} = 0.$$

The wavefunction of the crystal in the state with one mixed F–CT exciton is

$$\Phi = \zeta^\dagger |0\rangle, \tag{12.39}$$

where $|0\rangle$ is the wavefunction of the crystal in the ground state. The wavefunction Φ is normalized if

$$\sum_{n=1}^{N} |u_n|^2 + \sum_{n=1}^{N-1} |v_{n,+1}|^2 + \sum_{n=2}^{N} |v_{n,-1}|^2 = 1. \tag{12.40}$$

Eliminating the coefficients v_n from eqns (12.37) and (12.38), we obtain the equations for u_n:

$$(E_\mathrm{F} - E) u_n + \left[M - \frac{2\varepsilon_e \varepsilon_h}{E_\mathrm{CT} - E} \right] (u_{n+1} + u_{n-1})$$
$$- 2 \frac{\varepsilon_e^2 + \varepsilon_h^2}{E_\mathrm{CT} - E} u_n = 0, \quad (1 < n < N),$$

$$(E_F - E)u_1 + Mu_2 - \frac{\varepsilon_e^2 + \varepsilon_h^2}{E_{CT} - E}u_1 - \frac{2\varepsilon_e\varepsilon_h}{E_{CT} - E}u_2 = 0, \tag{12.41}$$

$$(E_F - E)u_N + Mu_{N-1} - \frac{\varepsilon_e^2 + \varepsilon_h^2}{E_{CT} - E}u_N - \frac{2\varepsilon_e\varepsilon_h}{E_{CT} - E}u_{N-1} = 0.$$

Substituting the relation $u_n = Ae^{\kappa n}$ into eqn (12.41.1), we obtain:

$$(E_F - E) + 2\left[M - \frac{2\varepsilon_e\varepsilon_h}{E_{CT} - E}\right]\cosh\kappa$$
$$- 2\frac{\varepsilon_e^2 + \varepsilon_h^2}{E_{CT} - E} = 0. \tag{12.42}$$

This equation gives the dependence of the energy E on κ. Since eqn (12.42) does not depend on the sign of κ, the general solution for u_n is

$$u_n = Ae^{\kappa n} + Be^{-\kappa n},$$

where the coefficients A and B have to be determined from the boundary conditions (12.41.2), (12.41.3) and the normalization condition (12.40).

For further treatment, we assign the labels n of the molecules in the chain as follows:

For the case that the total number N of molecules is odd, we use $n = 0, \pm 1, \pm 2, \ldots, \pm m$ with $N = (2m+1)$, and, for the even case $n = \pm\frac{1}{2}, \pm\frac{3}{2}, \ldots, \pm(m-\frac{1}{2})$ with $N = 2m$. We can then simplify our considerations if we take into account that the total Hamiltonian $\hat{H}$ has inversion symmetry; the solutions for u_n and $v_{n,\pm 1}$ have to be either symmetrical or antisymmetrical:

$$u_n = \pm u_{-n}, \quad v_{n,+1} = \pm v_{-n,-1}. \tag{12.43}$$

Thus, we can conclude that $B = \pm A$ and that

$$u_n = A\cosh\kappa n, \quad \text{for symmetrical solutions} \tag{12.44}$$

and

$$u_n = A\sinh\kappa n, \quad \text{for antisymmetrical solutions.} \tag{12.45}$$

We assume now that the length of the chain is small in comparison with the optical wavelength corresponding to the energy of electronic excitation. In this case we can neglect the effects of retardation (polaritonic effects). Additionally, the change of the optical electric field along the chain can be neglected in the calculation of the exciton transition dipole moment. In this approximation, the transition dipole moment from the ground to the excited crystal state μ is:

$$\mathbf{P}^{0,\mu} = \mathbf{p}^F \sum_{n=1}^{N} u_n + \mathbf{p}^{+CT}\sum_{n=1}^{N-1} v_{n,+1} + \mathbf{p}^{-CT}\sum_{n=2}^{N} v_{n,-1}, \tag{12.46}$$

where $\mathbf{p}^F$ and $\mathbf{p}^{+CT}, \mathbf{p}^{-CT}$ are the molecular transition dipole moment and transition dipole moments for $\pm$CT states, respectively. It follows from this relation that the transition dipole moment $\mathbf{P}^{0,\mu}$ is nonzero only for symmetrical solutions.

By substituting the symmetrical and antisymmetrical solutions for u_n in eqn (12.41), we obtain two equations which yield the energy E and κ of the excited states.

The first equation, which corresponds to the "bulk"-equation (12.41.1), can be directly obtained from (12.42)

$$\Delta_1 + 2\cosh\kappa - \frac{2(\alpha + \beta\cosh\kappa)}{\Delta_2} = 0, \qquad (12.47)$$

with the new parameters

$$\Delta_1 = \frac{E_F - E}{M}, \qquad \Delta_2 = \frac{E_{CT} - E}{M},$$

$$\alpha = \frac{\varepsilon_e^2 + \varepsilon_h^2}{M^2}, \qquad \beta = 2\frac{\varepsilon_e \varepsilon_h}{M^2}.$$

The second equation, into which the boundary equations (12.41.2), (12.41.3) will be transformed, depends on the symmetry of the states which we are interested in, and on the number of molecules N within the chain.

For symmetrical solutions, eqns (12.41.2) and (12.41.3) for an odd-N-chain become (with u_n relabeled as mentioned above $u_1 \to u_{-m}, u_N \to u_m$):

$$\Delta_1 \cosh m\kappa + \cosh(m-1)\kappa - \frac{\alpha\cosh m\kappa + \beta\cosh(m-1)\kappa}{\Delta_2} = 0. \qquad (12.48)$$

For an even-N-chain (with u_n relabeled $u_1 \to u_{-(m-\frac{1}{2})}, u_N \to u_{m-\frac{1}{2}}$) we get instead:

$$\Delta_1 \cosh\left(m - \frac{1}{2}\right)\kappa + \cosh\left(m - \frac{3}{2}\right)\kappa$$
$$- \frac{\alpha\cosh\left(m - \frac{1}{2}\right)\kappa + \beta\cosh\left(m - \frac{3}{2}\right)\kappa}{\Delta_2} = 0. \qquad (12.49)$$

Analogously the boundary condition for the antisymmetrical solutions can be expressed as follows:

The odd-N-case yields

$$\Delta_1 \sinh m\kappa + \sinh(m-1)\kappa$$
$$- \frac{\alpha\sinh m\kappa + \beta\sinh(m-1)\kappa}{\Delta_2} = 0, \qquad (12.50)$$

whereas in the even-N-case we get

$$\Delta_1 \sinh\left(m - \frac{1}{2}\right)\kappa + \sinh\left(m - \frac{3}{2}\right)\kappa$$
$$- \frac{\alpha\sinh\left(m - \frac{1}{2}\right)\kappa + \beta\sinh\left(m - \frac{3}{2}\right)\kappa}{\Delta_2} = 0. \qquad (12.51)$$

As can be exemplarily seen from eqns (12.48), (12.49), and (12.50), (12.51) respectively, all equations which describe even-N-chains can be easily obtained from the corresponding equations of odd-N-chains when substituting m by $m-\frac{1}{2}$. Therefore we restrict our further demonstrations to chains with an odd number of molecules.

We can assume that ε_e and ε_h are real. That means that $\alpha > 0$ and $\alpha \geq \beta$.

In the case that the interaction between Frenkel and CT excitons is absent and therefore $\alpha = \beta = 0$, we obtain from eqn (12.47) the solutions

$$\Delta_1 = -2\cosh\kappa, \qquad (12.52)$$

$$\Delta_2 = 0. \qquad (12.53)$$

Substituting into eqn (12.48), these energies yield

$$\begin{aligned}0 &= -2\cosh\kappa\cosh m\kappa + \cosh(m-1)\kappa \\ &= -\cosh(m+1)\kappa - \cosh(m-1)\kappa \\ &\quad + \cosh(m-1)\kappa\end{aligned} \qquad (12.54)$$

and therefore the condition:

$$\cosh(m+1)\kappa = 0 \qquad (12.55)$$

for the symmetrical case. If we put the energies (12.52) and (12.53) into eqn (12.50), we obtain the condition

$$\sinh(m+1)\kappa = 0 \qquad (12.56)$$

for the antisymmetrical case. The product of the two equations (12.55), (12.56) gives a general condition for κ:

$$2\sinh(m+1)\kappa\cosh(m+1)\kappa = \sinh 2(m+1)\kappa = 0. \qquad (12.57)$$

Hence the solution κ of eqn (12.57) has the form $\kappa = ik$ (k real) with

$$k = \frac{\pi j}{N+1}, \quad j = 1, \ldots, N, \qquad (12.58)$$

and the energy Δ_1 transforms into

$$\Delta_1 = -2\cosh\kappa = -2\cos k. \qquad (12.59)$$

With $E_F \equiv E_0$ and $E_j \equiv E = E(k)$ this expression corresponds to the relation (12.34), which describes the energies of Frenkel states.

On the other hand, if a coupling between Frenkel and CT excitons ($\alpha \neq 0$, $\beta \neq 0$) exists and the hopping integral M for Frenkel excitons is very small, the width of exciton bands appears only due to the nonlocality in the interaction between Frenkel and CT excitons.

It follows from eqns (12.47), (12.48) for symmetrical and (12.47), (12.50) for antisymmetrical solutions that the energy of new excited states and the corresponding value of κ can be found from the following equations:

(a) for symmetrical solutions:

$$\Delta_2 = \beta + \alpha \frac{\cosh m\kappa}{\cosh(m+1)\kappa}, \tag{12.60}$$

$$\Delta + 2\cosh\kappa + \frac{\alpha \cosh m\kappa}{\cosh(m+1)\kappa} = 2\cosh(m+1)\kappa$$
$$\times \frac{\alpha + \beta \cosh \kappa}{\alpha \cosh m\kappa + \beta \cosh(m+1)\kappa}. \tag{12.61}$$

(b) for antisymmetrical solutions:

$$\Delta_2 = \beta + \alpha \frac{\sinh m\kappa}{\sinh(m+1)\kappa}, \tag{12.62}$$

$$\Delta + 2\cosh\kappa + \frac{\alpha \sinh m\kappa}{\sinh(m+1)\kappa} = 2\sinh(m+1)\kappa$$
$$\times \frac{\alpha + \beta \cosh \kappa}{\alpha \sinh m\kappa + \beta \sinh(m+1)\kappa}, \tag{12.63}$$

where

$$\Delta = \Delta_1 - \Delta_2 + \beta = \frac{E_\mathrm{F} - E_\mathrm{CT} + M\beta}{M}.$$

For purely imaginary solutions of κ ($\mathrm{Re}\,\kappa = 0$, $\kappa = ik$, k–real), the function u_n becomes proportional to $\cos(kn)$ for symmetrical solutions, and proportional to $\sin(kn)$ for antisymmetrical solutions, respectively. In these cases we obtain "bulk" symmetrical and antisymmetrical solutions which are similar to the solutions in chains with Frenkel excitons only.

However, if $\mathrm{Re}\,\kappa \neq 0$, we obtain a new type of solutions. We will see below that for these solutions the function u_n is localized near the ends of the chain. Below, such type of states will be called edge states. They are similar to Tamm states of electrons in 3D crystals. It is important to note that as with Tamm states, the edge states arise even in ideal chains which have neither diagonal nor nondiagonal disorder.

12.4.3 Edge and bulk states in a finite molecular chain with mixing of Frenkel and charge-transfer excitons

Let us assume that eqn (12.48) has solutions κ with $\mathrm{Re}\,\kappa > 0$.[58] Hence, $|e^{-\kappa}|$ must be smaller than 1. For long chains ($m \gg 1$) the term $|e^{-2m\kappa}| \ll 1$ can

---

[58]It is easy to check that in case $\mathrm{Re}\,\kappa < 0$ the same results can be obtained.

be neglected and the further equations become independent of the difference between odd-N-chains and even-N-chains.

With $Z = e^{-\kappa} < 1$ and $2\cosh\kappa = Z + 1/Z$, eqns (12.47) and (12.48) reduce to:

$$\Delta_2 = \beta + \alpha Z, \tag{12.64}$$

$$Z^2 + Z\frac{\beta + \Delta}{1+\alpha} - (\alpha - \Delta\beta) = 0. \tag{12.65}$$

It follows from eqn (12.64) that the energy of an edge state E_s is given by

$$E_s = E_{\text{CT}} - M(\beta + \alpha Z). \tag{12.66}$$

The solutions of eqn (12.65) for Z are:

$$Z_{1,2} = -\frac{\beta + \Delta}{2(1+\alpha)}$$
$$\pm \frac{1}{2(1+\alpha)}\sqrt{(\beta+\Delta)^2 + 4(\alpha - \Delta\beta)(1+\alpha)}. \tag{12.67}$$

It is obvious that the condition $|Z_{1,2}| < 1$ occurs only for some combinations of parameters α, β and Δ.

For illustration, we consider below a few typical examples of combinations of these parameters.

Let us first assume that $|\beta| \ll \alpha$ (which is realized if $\varepsilon_e \approx 0$ or $\varepsilon_h \approx 0$). In this case we can put in our estimations $\beta = 0$. Additionally, we assume that the energies E_F and E_{CT} coincide. Then, Δ is equal to zero and eqn (12.65) has the solutions

$$Z_{1,2} = \pm\sqrt{\alpha}, \qquad \kappa_1 = \frac{1}{2}\ln(1+\alpha)$$
$$\text{and} \qquad \kappa_2 = i\pi + \frac{1}{2}\ln(1+\alpha). \tag{12.68}$$

Both solutions fulfill the condition $|Z_{1,2}| < 1$. As a consequence near the ends of the chain where $n = \pm m, \pm(m-1), \pm(m-2)\ldots$ for odd-N-chains or $n = \pm(m-1/2), \pm(m-3/2), \pm(m-5/2)\ldots$ for even-N-chains, respectively, the function u_n decreases exponentially with decreasing n. For the first solution this decay is monotonic:

$$u_n \sim \cosh(n\kappa_1) \approx \frac{1}{2}e^{\kappa_1|n|}; \tag{12.69}$$

the second solution has an oscillating decay:

$$u_n \sim \cosh(n\kappa_2) \approx \frac{(-1)^{|n|}}{2}e^{\kappa_1|n|}. \tag{12.70}$$

Using eqn (12.66) we can calculate the energies of edge states. Substituting $Z_{1,2}$ in eqn (12.55), we obtain for the lowest edge state energy:

$$E_s^{\text{low}} = E_{\text{CT}} - \frac{\alpha|M|}{\sqrt{1+\alpha}}. \qquad (12.71)$$

It is interesting to compare these values with the energies of the lowest energy bulk states with the high oscillator strength. It is seen from the expression for the transition dipole moment $\mathbf{P}^{0,\mu}$ that such states, similar to analogous states in molecular chains with Frenkel excitons, should have the smallest permitted wavevector k. The smallest permitted wavevector k, as can be concluded from eqn (12.60), is very close to zero for a long chain ($N \gg 1$).

Using eqn (12.47) and remembering that here we consider the structure with $\beta = 0$ and $\Delta_1 = \Delta_2$, we obtain that the lowest bulk state energy E_b with $k \approx 0$ is equal

$$E_b = E_{\text{CT}} + M \left[1 - \sqrt{1+2\alpha}\right], \qquad \text{for} \qquad M > 0,$$

and

$$E_b = E_{\text{CT}} - |M| \left[1 - \sqrt{1+2\alpha}\right], \qquad \text{for} \qquad M < 0. \qquad (12.72)$$

Comparing these energies with the energy of the lowest edge state, we can conclude that for both positive and negative M, the lowest energy surface state is blue-shifted in comparison with the lowest energy bulk state, independent of α. In the case of $M > 0$, this shift becomes negligibly small for $\alpha \ll 1$ (see Fig. 12.8a). As a second situation, we mention here the resonance of molecular and CT states ($E_F = E_{\text{CT}}$) with $\beta = \pm\alpha$. Analogous calculations give the following results:

$$Z = \frac{1-\alpha}{1+\alpha}, \qquad E_s = E_{\text{CT}} - 2M\frac{\alpha^2}{1+\alpha} \qquad \text{for} \qquad \beta = \alpha$$

and

$$Z = \frac{\alpha-1}{1+\alpha}, \qquad E_s = E_{\text{CT}} + 2M\frac{\alpha^2}{1+\alpha} \qquad \text{for} \qquad \beta = -\alpha. \qquad (12.73)$$

The lowest bulk state energy is

$$E_b = E_{\text{CT}} + M \left[1 - \sqrt{1+4\alpha}\right], \qquad \text{for} \qquad \beta = \alpha, \qquad (12.74)$$

and

$$E_b = E_{\text{CT}}, \qquad \text{for} \qquad \beta = -\alpha.$$

For $M > 0$, the edge state is red-shifted for $\beta = \alpha$ and blue-shifted for $\beta = -\alpha$, compared to the lowest bulk state (see Fig. 12.8b).

The edge levels for negative M and $E_F \neq E_{\text{CT}}$, respectively, can be considered in an analogous way to the cases we have treated above.

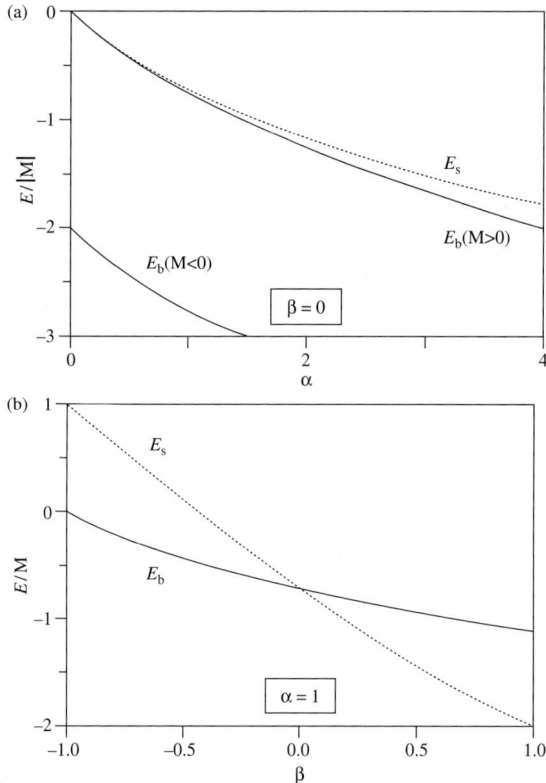

FIG. 12.8. Lowest bulk and edge state energies for resonance ($E_F = E_{CT} = 0$). (a) Energies E_s (dotted) and E_b (solid) as a function of α ($\beta = 0$). For $M < 0$ or $M > 0$, the edge state is blue-shifted with respect to the bulk state ($E_s > E_b$); for small α and $M > 0$ this shift becomes negligibly small. (b) Energies E_s (dotted) and E_b (solid) as function of β with fixed α ($\alpha = 1$, $-\alpha \leq \beta \leq \alpha$). For $\beta > 0$ the edge state is red-shifted with respect to the bulk state; for $\beta < 0$ the edge state is blue-shifted in comparison with the bulk state.

12.5 Application to PTCDA and MePTCDI crystals

The strong dependence of the energies of edge states on the parameters α, β, Δ and M clearly demonstrates the possibility of edge exciton spectroscopy for the study of Frenkel and CT exciton mixing in polymer chains. Of course, the significant contribution of edge states can be expected only in experimental investigations of absorption spectra of rather short chains. Fortunately, the application of organic molecular beam deposition (39) opens interesting perspectives for such studies.

APPLICATION TO PTCDA AND MEPTCDI CRYSTALS

In this subsection, we apply the relations obtained above for calculations of the energies of edge excitons in MePTCDI and PTCDA crystals. As we already mentioned, these crystals are quasi-one-dimensional. The distance between the molecules within the one-dimensional stacks is 0.34 nm for MePTCDI (45) and 0.337 nm for PTCDA (46)–(48).

This distance is small in comparison with other lattice constants and also small in comparison with the size of molecules. This causes strong interactions of the π-electron systems within the stacks and a very weak interaction in the other directions. Due to these strong interactions in the stack, we can expect that in such crystals the qualitative difference between Frenkel and CT excitations becomes smaller, their energies approach each other and their strong mixing determines the nature of the lowest energy states (40), (41).

To compare the energies of edge and lowest energy bulk states in MePTCDI and PTCDA, we use for these crystals the fitting parameters obtained in (41) to explain the frequency dependence of the polarization ratio in absorption spectra of MePTCDI crystals. These fitting parameters are collected in Table 12.1.

Using these data (without any new fitting parameter) we obtain the parameters α, β, Δ and M, which we need for the calculation of energies of edge and lowest energy bulk states in PTCDA and MePTCDI crystals.

Substituting these parameters in eqn (12.67), we obtain

$$Z_1 = -0.495 \quad \text{and} \quad Z_2 = -1.236 \quad \text{for MePTCDI,}$$
$$Z_1 = 0.178 \quad \text{and} \quad Z_2 = -1.074 \quad \text{for PTCDA.}$$

For each crystal only one solution fulfills $|Z| < 1$ and therefore we have only one edge state. Using relation (12.65), we can now obtain the energy of edge states. These energies and the energy of the lowest bulk state E_b are collected in Table 12.2.

Table 12.1 *Fitting parameters which were obtaind in (41). Differing from (41), the values ε_+, ε_- and M listed here already contain the vibronic overlap factor.*

	E_F (eV)	E_{CT} (eV)	ε_+ (eV)	ε_-	M (eV)
MePTCDI	2.23	2.15	0.064	$0.29|\varepsilon_+|$	0.045
PTCDA	2.34	2.27	0.062	$0.79|\varepsilon_+|$	0.042

Of course, these results of calculations of edge state energies in crystals of PTCDA and MePTCDI are preliminary, because the accuracy of the fitting parameters is very limited (see (41)). These results can be considered as a demonstration of the possible existence of small-radius edge states in quasi-one-dimensional crystals. For MePTCDI crystals, the calculated lowest energy edge state is red-shifted in comparison with the bulk state, with an estimated shift of about 240 cm^{-1}. For PTCDA crystals, the calculated lowest energy edge state is blue-shifted in comparison with the bulk state and this shift following our estimations is ≈ 80 cm^{-1}. Due to limited accuracy of the fitting parameters taken from (41), our qualitative agreement might be fortuitous. Further experimental investigation of edge states can reveal additional and independent information about the origin of this blue-shift. The edge states which we discussed in this section can play an important role in quantum confinement of exciton states in short chains or stacks. The theory of quantum confinement should be developed beyond the approximation of long chains which was used above. We can expect that the characteristic quantum length in this case is equal to the effective size of the edge state and for small-length molecular chains (or stacks) the edge states from both ends of the chain will overlap. The comparison of the theoretical predictions with available experimental data for quantum confinement in quasi-one-dimensional crystals can give additional information on the Frenkel and CT exciton mixing. Now data are available only from the experiments done by So et. al. (49) and Leonhardt et. al. (50), where very thin films of PTCDA with thickness of the order of a few lattice constants were investigated. It was found that the lowest absorption maximum was shifted ≈ 200 cm^{-1} to higher energies when the layer thickness was reduced from more than 60 to 3 lattice constants. Just these experiments stimulated the attempts to explain the nature of the observed blue-shift (see (50), (51)), each starting from completely different points of view. However, new experimental investigations are needed to clarify the origin of this phenomenon.

Table 12.2 *Comparison of the energies of the lowest bulk and surface states. In MePTCDI the lowest surface state is red-shifted in comparison with the bulk; in PTCDA the surface state is blue-shifted.*

	E_s (eV)	E_b (eV)
MePTCDI	2.14	2.11
PTCDA	2.24	2.23

12.6 Frenkel surface excitons in disordered media

12.6.1 *Macroscopic surface excitons and polaritons in isotopically mixed crystalline solutions*

We will start the discussion of the problem of surface excitons in disordered media considering macroscopic states. It is necessary to know the dielectric constant tensor to determine these states. We will use here the mean polarizability method. which we discussed in Chapter5 and which we apply for a cubic molecular crystal whose molecules differ only in the isotopic composition of the nuclei.

In the mean polarizability approximation indicated above there exists a Lorenz–Lorentz relation (Lorenz–Lorentz formula)

$$\frac{\epsilon - 1}{\epsilon + 1} = \frac{4\pi}{3v} a, \qquad (12.75)$$

where v is the volume per molecule, and a is the mean polarizability defined by

$$a(\omega, c) = \sum_i c_i a_i(\omega), \qquad (12.76)$$

where $a_i(\omega)$ is the polarizability of a molecule of type i, and c_i is the concentration of these molecules, $\sum_i c_i = 1$. From relationship (12.75) we find that

$$\epsilon(\omega, c) = 1 + \frac{4\pi}{3v} a(\omega, c) \left(1 - \frac{4\pi}{3v} a(\omega, c)\right)^{-1}. \qquad (12.77)$$

Using this relation and also eqn (12.17) and assuming a boundary with vacuum, we obtain the dispersion relation for surface polaritons:

$$k^2 = \frac{\omega^2}{2c^2} \frac{1 + 8\pi a/3v}{1 + 2\pi a/3v}. \qquad (12.78)$$

If the polarizabilities a_i are known, eqn (12.78) determines the surface polariton dispersion as a function of the isotopic composition of the mixture.

Equation (12.78) implies that the frequency of a Coulomb surface exciton satisfies the equation

$$1 + \frac{2\pi a(\omega, c)}{3v} = 0. \qquad (12.79)$$

In the vicinity of one of the resonances of the i-component of the isotopic mixture the corresponding polarizability has the form

$$a_i(\omega) = \frac{F_i}{\omega_i^2 - \omega^2} + a_0, \qquad (12.80)$$

where

$$F_i = \frac{2\omega_i \left|\mathbf{p}^{01}\right|^2}{\hbar},$$

and a_0 is the contribution to the molecular polarizability from more distant resonances.

For electronic transitions in a dielectric the frequencies of Coulomb surface excitons are in the region of the longitudinal–transverse splitting, which is always small in comparison with the frequencies ω_i. Taking this fact into account, eqn (12.80) can be changed into

$$a_i(\omega) = a_0 + \frac{\left|\mathbf{p}^{01}\right|^2}{\hbar (\omega_i - \omega)}.$$

Substituting this relation into eqn (12.79), we find that the frequency sought obeys the equation

$$A + \sum_i \frac{c_i}{\omega_i - \omega} = 0,$$

where

$$A = \frac{(1 + 2\pi a_0/3v)\hbar}{\left(2\pi \left|\mathbf{p}^{01}\right|^2 /3v\right)}.$$

For the case of a binary mixture ($i = 1, 2$), $c_1 = c$ and $c_2 = 1 - c$, and the above equation reduces to a quadratic equation for ω. When it is solved for a given composition of the mixture, we obtain two values of the frequency $\omega_\alpha(c)$, $\alpha = 1, 2$. We confine ourselves to the preceding simple relations and proceed to consider some features of the microscopic surface states.

We emphasize first of all that the disorder in the subsurface region, i.e. in the region of localization of the surface exciton, may result from its own internal disorder or may be caused by other, external reasons (e.g. by absorbed molecules). The microscopic surface states under consideration are strongly affected by both types of disorder. This circumstance should be borne in mind even in the cases when SSSE are treated in isotopically disordered crystalline solutions. In such states, which interact weakly with internal crystal monolayers, the effect of "internal" and "external" disorder can result in equally serious consequences. We will now make some qualitative remarks on the Anderson localization of surface excitons. As before let us assume a molecular crystal, ignore the exciton–phonon interaction, but take into account, for instance, the diagonal disorder (i.e. the random energy distribution of molecular excitations).

The Anderson localization (AL) of excitons on the surface as well as AL in the bulk may be studied by measurement of the exciton energy transfer. If at some time t' the molecule $\mathbf{n}$ is excited on a crystal surface, then, due to diffusion of excitons, there is a nonzero probability $W_{\mathbf{nm}}(t - t')$ to find the excitation on the molecule $\mathbf{m}$ at time t, $t > t'$. If the molecule $\mathbf{m}$ is also on the surface, then the value of $W_{\mathbf{nm}}(t - t')$ will just characterize the rate of transfer along the surface. However, in some cases which we discuss below the surface exciton may move away from the surface and continue the diffusion as a bulk exciton. Indeed, assume that on the given surface of the crystal there is a band of surface excitons, e.g. of SSSE type, and that the separation of this band from the band of bulk excitons is equal to Δ. Consider how the energy transfer along the surface should change with an increase of the diagonal disorder δ. If $\delta \ll \Delta \leq M$, where M is the bandwidth of the bulk excitons, then the increasing of disorder may lead to the strong or weak Anderson localization of the states only in the surface band. The presence of disorder satisfying the inequality $\delta \ll \Delta$ will not result in mixing the surface and bulk states. However, when $\delta \sim \Delta$, part of the surface excitation may pass into the bulk, and the transfer along the surface can be realized also due to diffusion of bulk excitons. Under these conditions the effective decrease of the transfer along the surface for a certain interval $\delta \sim \Delta$ can either cease, or become less strong, or even be changed into an increase. For subsequent growth of the disorder δ, when its value becomes of the order or larger than M, AL of the bulk excitations starts. The energy transfer along the surface as well as the transfer in the bulk for this region of δ ought to decrease down to values that are determined by only thermally activated jumps. Further analysis of the stability of Anderson localization of surface excitons and its diffusion is of particular interest. (For two-dimensional electrons such an analysis has been performed, see (52).)

12.7 Conclusion

Due to the general progress in experimental techniques, a great variety of experimental methods for investigating surface spectra has been developed. Many of them use the techniques and processes discovered in different fields of physics, and hence these methods complement each other. At present along with the optical and photoelectric methods one frequently studies the spectra of the inelastic scattering of slow electrons and neutrons, the inelastic scattering of molecular beams, etc. It is hoped that the experimental investigations will soon demonstrate the role that can be played by surface excitons in the physics and chemistry of surfaces. It goes without saying that the same investigations can produce a series of new theoretical and experimental problems, extending considerably the area of research available. However, it is probably not worth trying to foresee all the consequences to which the development of experiments will lead. Below we mention those theoretical problems which are just now raised by the logic of development of the investigations and which are currently practicable.

Apparently, in the near future there will be developed: (a) a detailed theory of surface excitons not only at the crystal boundary with vacuum but also at the interfaces of various condensed media, particularly of different symmetry; (b) a theory of surface excitons including the exciton–phonon interaction and, in particular, the theory of self-trapping of surface excitons; (c) the features of surface excitons for quasi-one-dimensional and quasi-two-dimensional crystals; (d) the theory of kinetic parameters and, particularly, the theory of diffusion of surface excitons; (e) the theory of surface excitons in disordered media; (f) the features of Anderson localization on a surface; (g) the theory of the interaction of surface excitons of various types with charged and neutral particles; (h) the evaluation of the role of surface excitons in the process of photoelectron emission; (i) the electronic and structural phase transitions on the surface with participation of surface excitons. We mention here also the theory of exciton–exciton interactions at the surface, the surface biexcitons, and the role of defects (see, as example, (53)). The above list of problems reflects mainly the interests of the author and thus is far from complete. Referring in one or another way to surface excitons we enter into a large, interesting, and yet insufficiently studied field of solid-state physics.

13

EXCITONS IN ORGANIC-BASED NANOSTRUCTURES

13.1 Introduction

The majority of optoelectronic devices (such as LEDs, solar cells, and nonlinear-optical devices) are built on the basis of traditional inorganic semiconductors. Over the last couple of decades, however, a lot of progress has been made in producing devices based on organic electronic materials, which, for many applications, may become less expensive alternatives to inorganic counterparts. Particularly, as a result of the fast development of organic material science the tremendous advances in organic-based electronic materials applications have been reported recently. The progress in the field is impressive due to the use of innovative growth techniques and the realization of systems in low-dimensional confined geometries. Various organic compounds (such as small molecules, conjugated polymers, carbon nanotubes) have been shown to be of high interest and utility for electronic applications. Advances have been reported on a variety of device types: from OLEDs to organic lasers and devices of nonlinear optics, molecular-scale transistors, etc.

Nevertheless, the need for systems having even better optoelectronic properties to be used in applications has driven researchers in materials science to develop novel compounds and novel structures. As the results of such activity the advances in organic material science have generated a vital and growing interest in organic materials research which could potentially revolutionize future electronic applications. The current development prospects of organic materials are, however, mostly limited in their scope to relatively low-performance areas. One of the reasons for this is, for instance, the low mobility of charge carriers.

In the following sections we discuss a qualitatively distinct idea to develop and use *resonant* hybrid organic–inorganic nanostructures containing semiconductor and organic components with nearly equal energies of excitons. We will demonstrate that such structures with properly selected materials of organic and inorganic components can occupy a suitable place in the development of organic-based material science. They are interesting from the point of view of basic science as well as for applications in optoelectronics because in one and the same hybrid structure we can combine ingeniously the high conductivity of semiconductor components, (important for electric pumping) with the strong light–matter interaction of organic components. In this way, the desirable properties of both the organic and inorganic material unite and in many cases can overcome the basic limitation of each. We will describe the most important qualitative features of different types resonant hybrid nanostructures and also some

INTRODUCTION

recent experimental observations of predicted effects.

It is important to have in mind that in consideration of the resonance interaction of systems the dissipation of excited states plays a very important role. At the resonance of two states the splitting of these states in the energy spectrum can be formed but this takes place only if the dissipative width of resonating states is small in comparison with the splitting value. This limiting case (a strong coupling regime) has been considered in the beginning of our studies. However, for organic materials, the comparatively large dissipative width of excited electronic states is typical. Therefore the weak coupling can be expected for many hybrid organic–inorganic resonant structures. The consideration of both limiting cases is interesting because the physics behind them and also the possibilities of applications are very different. The hybrid structures in the limit of strong coupling between Frenkel and Wannier–Mott excitons (W–M) have been investigated in different geometrical configurations such as quantum wells (QW) (1), (2), quantum wires (QWW) (3), and quantum dots (QD) (4). The resonant interaction between Frenkel excitons (FE) in the organic layer and Wannier–Mott excitons in semiconductor quantum wells was considered also in a microcavity configuration (5), where organic and inorganic materials are separated. The hybrid structures in weak coupling regime have been investigated in (6)–(9). In these papers for different configurations the electronic excitation energy transfer from an inorganic quantum well to an organic overlayer have been considered and it was shown that this process in many cases is very fast, much faster than radiative or nonradiative processes in such inorganic quantum wells. Having in mind that an inorganic quantum well can be easily pumped electrically, this observation gave the possibility for the formulation of a new concept of light-emitting devices (10). Before considering these results (more information can be found in (11)-(13)) we recall and compare below some important peculiarities of excitons in organic and inorganic materials.

The internal structure of W–M excitons is represented by hydrogen-like wavefunctions (14). The mean electron–hole distance for this type of exciton is typically large (in comparison with the lattice constant). On the other hand, the Frenkel exciton is represented as an electronic state of a crystal in which electrons and holes are placed at the same molecule. We can say that Frenkel excitons in organic crystals have radii a_F, comparable to the lattice constant $a_F \sim a \sim 5$ Å. In contrast, weakly bound W–M excitons in semiconductor QWs have large Bohr radii ($a_B \sim 100$ Å in III–V materials and $a_B \sim 30$ Å in II–VI ones, in both cases $a_B \gg a$). The oscillator strength of a Frenkel exciton is close to a molecular oscillator strength F and due to strong overlapping of electron and hole wavefunctions in the same molecule can sometimes be very large (the order of unity), whereas the oscillator strength f of a W–M exciton is usually much weaker: in a quantum well $f \sim a^3 a_B^{-2} L^{-1} F$ where L is the QW width ($a_B > L > a$). In high-quality semiconductors as well as in organic crystalline materials, the optical properties near and below the band gap are dominated by the exciton transitions and the same situation takes place also for organic and inorganic

QWs (or wires or dots). The excitonic optical nonlinearities in semiconductor QWs can be large because the ideal bosonic approximation for W–M excitons breaks down as soon as their 2D density n becomes comparable to the saturation density $n_S \sim 1/(\pi a_B^2)$ (as the Bohr radius is large the saturation density is small; n_S is, typically, 10^{12} cm^{-2}). Then, due to phase space filling (PSF), exchange and collisional broadening, the exciton resonance is bleached. The generic figure of merit for all-optical nonlinearities scales like $I_P^{-1}(\Delta\chi/\chi)$ where $\Delta\chi$ is the nonlinear change in the susceptibility in the presence of the pump of intensity I_P. As $\Delta\chi/\chi \sim n/n_S \sim n\,a_B^2$, but also $n \propto f\,I_P \propto a_B^{-2} I_P$, such a figure of merit in semiconductors is nearly independent of the exciton Bohr radius (15). This conclusion, made nearly 20 years ago, did not leave any hope that the transition to organic materials could be useful for the creation of devices with increased all-optical nonlinearity. We will show below that in hybrid structures the described situation can be changed.

13.2 Hybrid 2D Frenkel–Wannier–Mott excitons at the interface of organic and inorganic quantum wells. Strong coupling regime

13.2.1 Configuration of heterostructure and general relations

We will study here the effects of resonant interaction between an organic quantum well (OQW) and an inorganic (IQW) one and demonstrate how new hybrid states arise (1). The configuration we consider is the following. A plane semiconductor IQW of thickness L_w occupies the region $|z| < L_w/2$, the z-axis being chosen along the growth direction. All the space with $z > L_w/2$ is filled by the barrier material and that with $z < -L_w/2$ by the organic material in which the OQW is placed (Fig. 13.1a). For simplicity, we treat the interaction of IQW excitations with organic molecules in the dipole approximation, neglecting the contribution of higher multipoles to the interaction, and we consider the OQW as a single monolayer, i.e. as a 2D lattice of molecules at discrete sites $\mathbf{n}$, placed at $z = -z_0 < -L_w/2$ (the generalization to the case of several monolayers is easy). All the semiconductor well–barrier structure ($z > -L_w/2$) is assumed to have the same background dielectric constant ϵ, while the organic half-space ($z < -L_w/2$) has the dielectric constant $\tilde{\epsilon}$ (corresponding to the organic substrate). For example, the role of the OQW can be played by the outermost monolayer of the organic crystal. Due to the gas–condensed matter shift the exciton transition in such a monolayer can be blue-shifted with respect to the bulk excitonic transition (as, for instance, this takes place for anthracene crystals), thus giving rise to two-dimensional Frenkel excitonic states (16). In this case the difference $L_w/2 - z_0$ is of the order of the organic crystal lattice constant.

Due to the different electronic structure of the two QWs under consideration and the rather large organic crystal lattice constant, the OQW and the IQW states are assumed to have zero wavefunction overlap. It is known that this is a rather good approximation for organic crystals in the bulk for the ground and for the lowest energy excited states. Thus, we assume that the same takes place also at the interface between organic and inorganic QWs. Assuming perfect

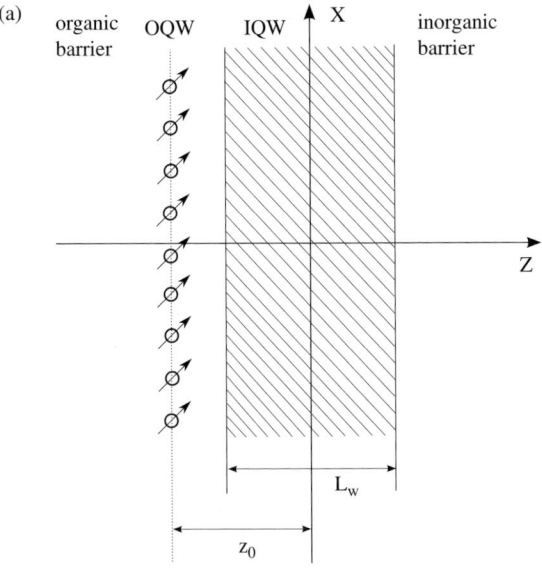

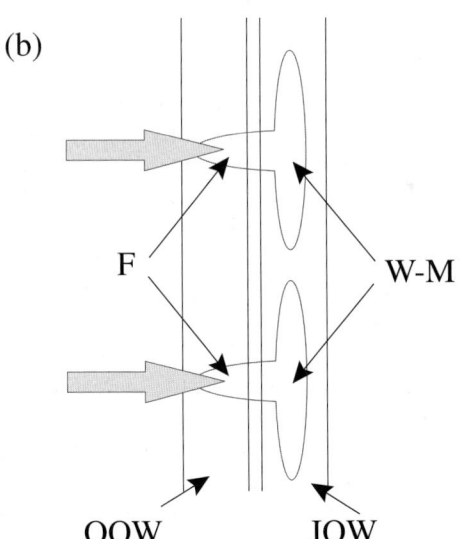

FIG. 13.1. (a) The physical configuration under study; (b) the Frenkel and Wannier–Mott excitons in nanostructure (from (13)).

2D translational invariance of the system, we classify the excitons by their in-plane wavevector **k**. Supposing that for some bands of Frenkel excitons in the OQW and Wannier–Mott excitons in the IQW the energy separation is much less than the distance to other exciton bands we take into account only the hybridization between these two bands. We choose as a basis set the "pure" Frenkel and Wannier states, i.e. the state (denoted by $|F, \mathbf{k}\rangle$) when the OQW is excited, while the IQW is in its ground state, and vice versa (denoted by $|W, \mathbf{k}\rangle$), their energies being $E_F(\mathbf{k})$ and $E_W(\mathbf{k})$. We seek the new hybrid states in the form

$$|\alpha, \mathbf{k}\rangle = A_\alpha(\mathbf{k})|F, \mathbf{k}\rangle + B_\alpha(\mathbf{k})|W, \mathbf{k}\rangle, \qquad (13.1)$$

where $\alpha = $ "u", "ℓ" labels the two resulting (upper and lower) branches. The Schrödinger equation for the coefficients A, B is then written as:

$$\begin{aligned}(E_F(\mathbf{k}) - E)\, A(\mathbf{k}) + \langle F, \mathbf{k}|\hat{H}_{\text{int}}|W, \mathbf{k}\rangle B(\mathbf{k}) &= 0, \\ \langle W, \mathbf{k}|\hat{H}_{\text{int}}|F, \mathbf{k}\rangle A(\mathbf{k}) + (E_W(\mathbf{k}) - E)\, B(\mathbf{k}) &= 0, \end{aligned} \qquad (13.2)$$

where $\hat{H}_{\text{int}}$ is the Hamiltonian of the interaction between the QWs. Solution of (13.2) gives the energies of the upper and lower branches and the splitting $\Delta(\mathbf{k})$:

$$E_{u,\ell}(\mathbf{k}) = \frac{E_F(\mathbf{k}) + E_W(\mathbf{k}) \pm \Delta(\mathbf{k})}{2},$$

$$\Delta(\mathbf{k}) = \sqrt{(E_F(\mathbf{k}) - E_W(\mathbf{k}))^2 + 4\Gamma^2(\mathbf{k})}, \qquad (13.3)$$

where we use the notation $\Gamma(\mathbf{k}) \equiv |\langle W, \mathbf{k}|\hat{H}_{\text{int}}|F, \mathbf{k}\rangle|$ for the coupling matrix element. If, for some $\mathbf{k} = \mathbf{k}_r$ energies, $E_W(\mathbf{k}_r) = E_F(\mathbf{k}_r)$, the splitting $\Delta(\mathbf{k}_r) = 2\Gamma(\mathbf{k}_r)$. For the orthonormalized new states the weighting coefficients are given by

$$|A_u(\mathbf{k})|^2 = |B_\ell(\mathbf{k})|^2 = \frac{1}{2}\left(1 + \frac{E_F(\mathbf{k}) - E_W(\mathbf{k})}{\Delta(\mathbf{k})}\right), \qquad (13.4)$$

$$|A_\ell(\mathbf{k})|^2 = |B_u(\mathbf{k})|^2 = \frac{1}{2}\left(1 - \frac{E_F(\mathbf{k}) - E_W(\mathbf{k})}{\Delta(\mathbf{k})}\right). \qquad (13.5)$$

It follows from these relations that in the vicinity of resonance the exciton in a hybrid structure, roughly speaking, "lives" half-time as a Frenkel exciton and half-time as a Wannier–Mott exciton. Thus, in a hybrid structure we can expect a strong exciton–light interaction typical for Frenkel excitons as well as strong resonance optical nonlinearity typical for Wannier–Mott excitons.

13.2.2 The coupling matrix element

To evaluate the matrix element $\Gamma(\mathbf{k})$ determining the resonance interaction between Frenkel and Wannier–Mott excitons we write down the interaction Hamiltonian as

$$\hat{H}_{\text{int}} = -\sum_{\mathbf{n}} \hat{\mathbf{p}}^F(\mathbf{n}) \cdot \hat{\boldsymbol{\mathcal{E}}}(\mathbf{n}), \qquad (13.6)$$

where $\hat{\mathbf{p}}^F(\mathbf{n})$ is the operator of the dipole moment of the organic molecule situated at the lattice site $\mathbf{n}$, and $\hat{\boldsymbol{\mathcal{E}}}(\mathbf{n})$ is the operator of the electric field at the point n, produced by the IQW exciton. If we introduce the operator of the IQW polarization $\hat{\mathbf{P}}^W(\mathbf{r})$, then the operators $\hat{\mathcal{E}}(\mathbf{n})$ and $\hat{\mathbf{P}}^W(\mathbf{r})$ are related to each other in exactly the same way as the corresponding classical quantities in electrostatics:

$$\hat{\mathcal{E}}_i(\mathbf{r}) = \int d^3r' \mathcal{D}_{ij}\left(\mathbf{r}_\| - \mathbf{r}'_\|, z, z'\right) \hat{P}_j^W(\mathbf{r}'), \qquad (13.7)$$

where $i, j = x, y, z$, $\mathbf{r}_\| \equiv (x, y)$ and $\mathcal{D}_{ij}(\mathbf{r}, \mathbf{r}')$ is the Green's function appearing in the analogous problem of classical electrostatics. It is equal to the ith Cartesian component of the classical static electric field at the point $\mathbf{r}$, produced by the jth component of the classical point dipole, situated at the point $\mathbf{r}'$ and is connected to the Green's function G of the Poisson equation in an inhomogeneous medium with the dielectric constant $\epsilon_{ij}(\mathbf{r})$:

$$\mathcal{D}_{ij}(\mathbf{r}, \mathbf{r}') = -\frac{\partial}{\partial x_i} \frac{\partial}{\partial x'_j} G(\mathbf{r}, \mathbf{r}'), \qquad (13.8)$$

$$\frac{\partial}{\partial x_i} \epsilon_{ij}(\mathbf{r}) \frac{\partial}{\partial x_j} G(\mathbf{r}, \mathbf{r}') = -4\pi \delta(\mathbf{r} - \mathbf{r}'). \qquad (13.9)$$

Since our system is translationally invariant in two dimensions, it is convenient to consider the Fourier transform:

$$\mathcal{D}_{ij}\left(\mathbf{r}_\| - \mathbf{r}'_\|, z, z'\right) = \int \frac{d^2k}{(2\pi)^2} \mathcal{D}_{ij}(\mathbf{k}, z, z') e^{i\mathbf{k}(\mathbf{r}_\| - \mathbf{r}'_\|)}, \qquad (13.10)$$

and analogously for $G\left(\mathbf{r}_\| - \mathbf{r}'_\|, z, z'\right)$. Then $G(\mathbf{k}, z, z') \exp(i\mathbf{k}\mathbf{r}_\|)$ is the potential, produced by a charge density wave $\rho(\mathbf{r}) = \delta(z - z') \exp(i\mathbf{k}\mathbf{r}_\|)$. In our case the dielectric constant is a simple step function

$$\epsilon_{ij}(\mathbf{r}) = \begin{cases} \tilde{\epsilon}\delta_{ij}, & z < -L_w/2, \\ \epsilon\delta_{ij}, & z > -L_w/2, \end{cases} \qquad (13.11)$$

and the potential may be readily found from Poisson's equation

$$\left(\frac{d^2}{dz^2} - k^2\right) G(\mathbf{k}, z, z') = -\frac{4\pi\delta(z - z')}{\epsilon(z)}, \qquad (13.12)$$

with the usual electrostatic boundary conditions at the interface $z = -L_w/2$ (continuity of the tangential component of the electric field $-i\mathbf{k}G$ and the normal component of the electric displacement $-\epsilon \partial G/\partial z$). The Green's function $\mathcal{D}_{ij}$ for $z < -L_w/2$, $z' > -L_w/2$ is then given by:

$$\mathcal{D}_{ij}(\mathbf{k}, z, z') = \frac{4\pi}{\epsilon + \tilde{\epsilon}} k e^{k(z-z')} \left(\frac{ik_i}{k} + \delta_{i,z}\right)\left(\frac{ik_j}{k} + \delta_{j,z}\right). \qquad (13.13)$$

Thus, the matrix element of $\hat{H}_{\text{int}}$ we are interested in can be written as

$$\langle F, \mathbf{k}|\hat{H}_{\rm int}|W, \mathbf{k}\rangle$$
$$= -\sum_{\mathbf{n}} \int d^3r \langle F, \mathbf{k}|\hat{p}_i(\mathbf{n})|0\rangle \mathcal{D}_{ij}(\mathbf{n}-\mathbf{r}_\|, -z_0, z)\langle 0|\hat{P}_j^W(\mathbf{r})|W, \mathbf{k}\rangle. \quad (13.14)$$

The matrix element of the IQW polarization between the ground state $|0\rangle$ and $|W, \mathbf{k}\rangle$ for a 1s-exciton with Bohr radius a_B is equal to (18), (19)

$$\langle 0|\hat{\mathbf{P}}^W(\mathbf{r})|W, \mathbf{k}\rangle = \sqrt{\frac{2}{\pi}}\frac{\mathbf{d}^{vc}}{a_B}\frac{e^{i\mathbf{k}\mathbf{r}_\|}}{\sqrt{S}}\chi^e(z)\chi^h(z), \quad (13.15)$$

where $\sqrt{2/(\pi a_B^2)}$ is the value of the 1s-wavefunction of the relative motion of the electron and hole, taken at $r_\| = 0$; $\chi^e(z)$, $\chi^h(z)$ are the envelope functions for the electron and hole in the IQW confinement potential (we assume the IQW to be thin, so that the transverse and the relative in-plane motion of the electron and hole are decoupled) and S is the in-plane normalization area. Finally,

$$\mathbf{d}^{vc} = \int_{u.c.} u_v^*(\mathbf{r})(-e\mathbf{r})u_c(\mathbf{r})d^3r \quad (13.16)$$

is the matrix element of the electric dipole moment between the conduction and valence bands ($\mathbf{d}^{vc}$ is taken to be independent of $\mathbf{k}$, $u_{c(v)}$ are the Bloch functions for the conduction (valence) band extremum and the integration in (13.16) is performed over the unit cell). Its Cartesian components $d_i^{vc}(i = x, y, z)$ may be expressed in terms of Kane's energy (18):

$$|d_i^{vc}|^2 = \frac{e^2\hbar^2 E_0 c_i^2}{2m_0 E_g^2}, \quad (13.17)$$

where m_0 is the free electron mass, E_g is the energy gap between the conduction and valence bands and c_i is the appropriate symmetry coefficient. In semiconductors of zinc-blende structure $c_x^{hh} = c_y^{hh} = 1/\sqrt{2}$, $c_z^{hh} = 0$ (heavy holes) and $c_x^{lh} = c_y^{lh} = 1/\sqrt{6}$, $c_z^{lh} = \sqrt{2/3}$ (light holes). We see that only light holes can contribute to the z-component of the IQW polarization. For the Frenkel exciton the dipole matrix element, contributing to the matrix element (13.14), is given by

$$\langle F, \mathbf{k}|\hat{p}_i(\mathbf{n})|0\rangle = \frac{e^{-i\mathbf{k}\mathbf{n}}}{\sqrt{N}}d_i^{F*} = \frac{e^{-i\mathbf{k}\mathbf{n}}}{\sqrt{S}}a_F d_i^{F*}, \quad (13.18)$$

where $\mathbf{d}^F$ is the transition dipole moment for a single organic molecule (analogous to $\mathbf{d}^{vc}$ in the semiconductor), N is the total number of sites in the lattice, a_F is the lattice constant, which may be considered as the radius of the Frenkel exciton.

Now we can write the final expression for the coupling matrix element:

$$\langle F, \mathbf{k}|\hat{H}_{\rm int}|W, \mathbf{k}\rangle = -\sqrt{\frac{2}{\pi}}\frac{d_i^{F*} d_j^{vc}}{a_F\, a_B} \int \mathcal{D}_{ij}(\mathbf{k}, -z_0, z)\chi^e(z)\chi^h(z) dz. \quad (13.19)$$

From eqns (13.13) and (13.19) we see that the only contributing polarizations for the semiconductor are those along $\mathbf{k}$ (L-modes) and along the growth direction

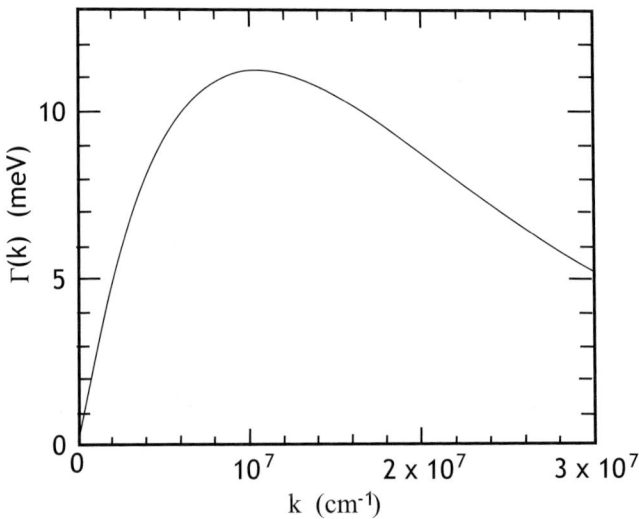

FIG. 13.2. The interaction parameter $\Gamma(k)$ for $d^{vc} = 12$ Debye, $d^F = 5$ Debye, $a_B = 2.5$ nm, $a_F = 0.5$ nm, $L_w = 1$ nm, $z_0 = 1$ nm, $\epsilon_\infty = 6$, $\tilde{\epsilon}_\infty = 4$.

z (Z-modes, only for the light holes, according to eqn (13.17)). For simplicity, we take the electron and hole confinement wavefunctions for the lowest subbands in the approximation of an infinitely deep IQW:

$$\chi^e(z)\chi^h(z) = \frac{2}{L_w}\cos^2\left(\frac{\pi z}{L_w}\right), \qquad (13.20)$$

and assume the transition dipole moment in the organics $\mathbf{d}^F$ to be real (which is always possible with an appropriate choice of molecular wavefunctions). Without loss of generality we may take the vector $\mathbf{k}$ along the x-axis. Evaluating the integral in (13.19), we obtain the interaction parameter $\Gamma_{L,Z}$ for the L- and Z-modes:

$$\Gamma_{L(Z)} = \frac{8\sqrt{2\pi}}{\epsilon+\tilde{\epsilon}}\frac{e^{-kz_0}\sinh(kL_w/2)}{1+\left(\frac{kL_w}{2\pi}\right)^2}\frac{\left|d^{vc}_{x(z)}\right|\sqrt{(d^F_x)^2+(d^F_z)^2}}{a_F a_B L_w}. \qquad (13.21)$$

It is seen that $\Gamma(k)$ has a maximum $\Gamma_{\max}$ at $k = k_{\max}$. The value of $k_{\max}$ for arbitrary z_0 and L_w may be found numerically, for $z_0 - L_w/2 > 0.1\, L_w$ it is well described by the formula (see also Fig. 13.2)

$$k_{\max} \simeq \frac{1}{L_w}\ln\left(\frac{2z_0+L_w}{2z_0-L_w}\right), \qquad (13.22)$$

while in the limit $z_0 \simeq L_w/2$ we have $k_{\max} \simeq 2.4\, L_w^{-1}$.

13.2.3 Dispersion relations of hybrid states

To calculate the dispersion relation of the hybrid excitons we approximate the WE energy by a parabola with the in-plane effective mass $m_W = m_e + m_h$, $m_{e(h)}$ being the electron (hole) mass, and neglect the FE dispersion since the typical masses are (5–100) m_0:

$$E_W(\mathbf{k}) = E_W(0) + \frac{\hbar^2 k^2}{2m_W}, \quad E_F(\mathbf{k}) = E_F(0), \quad E_F(0) - E_W(0) = \delta. \quad (13.23)$$

We will measure all energies with respect to $E_W(0)$. The dispersion of the hybrid states (13.3) can be written as

$$E_{u,l}(\mathbf{k}) - E_W(0) = \frac{\delta}{2} + \frac{\hbar^2 k^2}{4m_W} \pm \sqrt{\left(\frac{\delta}{2} - \frac{\hbar^2 k^2}{4m_W}\right)^2 + \Gamma^2(\mathbf{k})}. \quad (13.24)$$

To perform numerical estimates we choose the following values of the parameters. For the IQW those representative of II–VI semiconductor (e.g. ZnSe/ZnCdSe) quantum wells are taken (20): $\epsilon = \epsilon_\infty = 6$, $d^{vc}/a_B \approx 0.1e$ (which corresponds to $d^{vc} \simeq 12$ Debye and a Bohr radius of 2.5 nm), the exciton mass $m_W = 0.7\, m_0$ and the well width $L_w = 1$ nm. For the organic part of the structure, we take parameters typical for such media (e.g. see (17), (21), (22)): $\tilde{\epsilon} = \tilde{\epsilon}_\infty = 4$, the transition dipole for the molecules in the monolayer $d^F = 5$ Debye, $a_F = 0.5$ nm and $z_0 = 1$ nm. We plot $\Gamma(k)$ for these values of parameters in Fig. 13.2.

We see that $\Gamma_{\max} \simeq 11$ meV. The dispersion curves $E_{u,\ell}(k)$ along with the FE weight in the lower branch $|A_\ell(k)|^2$ for three different detunings $\delta = 10$ meV, $\delta = = 0$ and $\delta = -10$ meV are plotted on Figs. 13.3 –13.5. For $\delta > 0$ the properties of the excited states are changed drastically. In this case the zeroth approximation dispersion curves for FE and Wannier excitons (WE) cross at the point $k = k_0 = = \sqrt{2m_W \delta/\hbar^2}$. At $k = 0$ the upper states are purely FE-like (Frenkel exciton-like) and the lower states WE-like (Wannier exciton-like), at $k \sim k_0$ they are strongly mixed and a large splitting of their dispersion curves is present, $\Delta(k_0) \sim 2\Gamma(k_0)$, and for large k ($k \gg k_0$) they "interchange": the upper branch becomes W-like with the quadratic dispersion and excitations of the lower branch tend to FE. If $\delta < 0$ then $E_W(k) > E_F(k)$ for small k and no crossing occurs, $E_u(k)$ closely follows the WE dispersion and $|A_u(k)|^2 \ll 1$, the lower state is FE-like. A nontrivial feature of the lower branch dispersion is a minimum away from $k = 0$, which is always present for $\delta < 0$ as well as for some positive values of δ, $0 < \delta < \delta_{\rm cr}$ and is the deepest for $\delta = 0$. The critical value of δ may be found if one looks at the values of the derivatives of $E_\ell(k)$ at $k = 0$. It turns out that

$$\frac{dE_\ell(0)}{dk} = 0 \quad (\delta \neq 0), \quad \frac{d^2 E_\ell(0)}{dk^2} < 0 \quad (\delta < \delta_{\rm cr}),$$
$$\frac{d^2 E_\ell(0)}{dk^2} > 0 \quad (\delta > \delta_{\rm cr}), \quad (13.25)$$

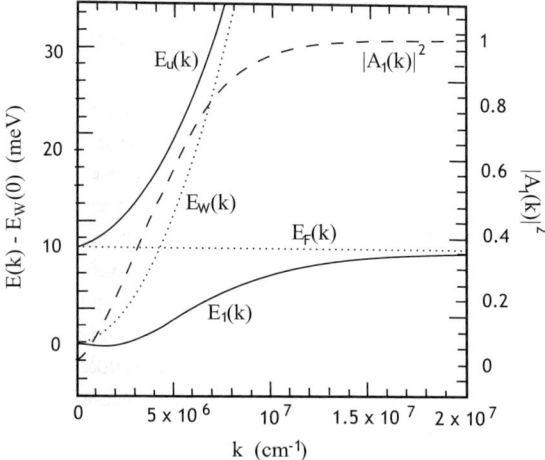

FIG. 13.3. The dispersion $E_{u,\ell}(k)$ of the upper and lower hybrid exciton branches (solid lines) and that of the unperturbed Frenkel and Wannier excitons (dotted lines). The "weight" of the FE component in the lower branch $|A_\ell(k)|^2$ is shown by the dashed line. The parameters are the same as in Fig. 13.2 ($m_W = 0.7\ m_0$), the detuning $\delta = 10$ meV (from (13)).

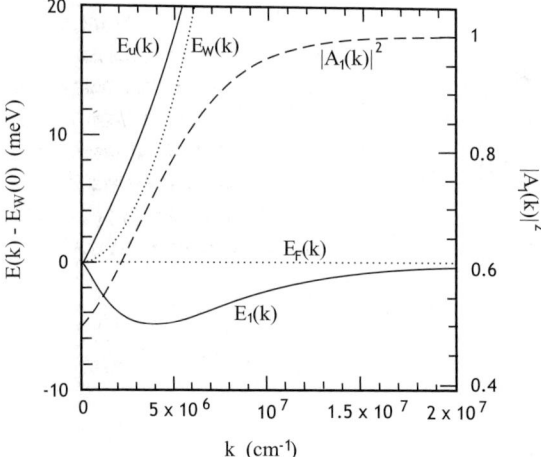

FIG. 13.4. The same as in Fig. 13.3, but $\delta = 0$, (from (13)).

and $\delta_{\rm cr}$ when the minimum "splits" off $k = 0$ is given by

$$\delta_{\rm cr} = \left(\frac{d\Gamma(0)}{dk}\right)^2 \frac{2m_W}{\hbar^2}. \tag{13.26}$$

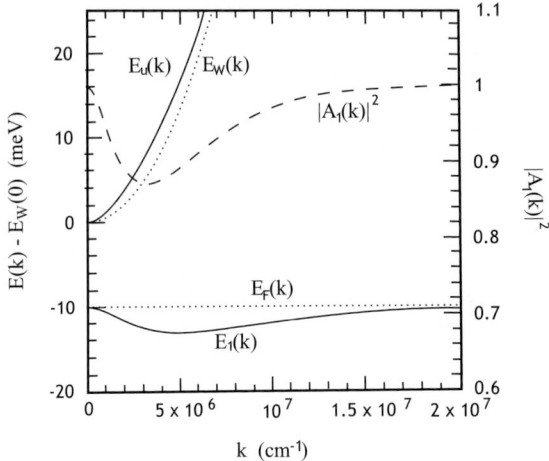

FIG. 13.5. The same as in Fig. 13.3, but $\delta = -10$ meV (from (13)).

For our parameters $\delta_{cr} \simeq 16$ meV. For large negative values of $\delta \ll -\Gamma_{max}$ the lower branch dispersion at $k \ll \sqrt{2m_W \delta/\hbar^2}$ may be approximated by

$$E_\ell(k) - E_W(0) \simeq -|\delta| - \frac{\Gamma^2(k)}{|\delta|}. \qquad (13.27)$$

So, the depth of the minimum for large δ is $\Gamma^2_{max}/|\delta|$ while for small δ it is of the order of Γ_{max} and we see that the effective range of δ, when the minimum is the most pronounced, is $-\Gamma_{max} \leq \delta < \delta_{cr}$. As a consequence, at low temperatures and under optical pumping at frequencies above the excitonic resonance excitons will accumulate in this minimum, which can be detected, for example, by pump–probe experiments. The fluorescence from these states should increase with temperature since states with small k become populated.

13.2.4 Linear optical response of hybrid states

If an incident electromagnetic wave with the electric field $\boldsymbol{\mathcal{E}}(\mathbf{r}) = \boldsymbol{\mathcal{E}}_0 e^{i\mathbf{Q}\mathbf{r}}$ is present, then the interaction with the hybrid structure is described by the Hamiltonian (neglecting local field corrections)

$$\hat{H}_{em} = -\boldsymbol{\mathcal{E}}_0 \cdot \left(\sum_{\mathbf{n}} \hat{\mathbf{p}}^F(\mathbf{n}) e^{i\mathbf{Q}_\| \mathbf{n}} + \int dz \int d^2 r_\| \hat{\mathbf{P}}^W(\mathbf{r}) e^{i\mathbf{Q}_\| \mathbf{r}} \right), \qquad (13.28)$$

where we have neglected the z-dependence of the incident field since the thickness of our structure is much less than the light wavelength. The corresponding matrix element is different from zero only if $\mathbf{k} = \mathbf{Q}_\|$ and in this case equal to

$$\langle \alpha, \mathbf{k} | \hat{H}_{em} | 0 \rangle \equiv -\boldsymbol{\mathcal{E}}_0 \mathbf{M}_\mathbf{k}^\alpha = -\boldsymbol{\mathcal{E}}_0 \cdot \left(A_\alpha^*(\mathbf{k}) \mathbf{M}^F + B_\alpha^*(\mathbf{k}) \mathbf{M}_\mathbf{k}^W \right), \qquad (13.29)$$

where

$$\mathbf{M}^F = \sqrt{N}\mathbf{d}^{F*} = \frac{\sqrt{S}}{a_F}\mathbf{d}^{F*}, \qquad (13.30)$$

$$\mathbf{M}_{\mathbf{k}}^{W} = \sqrt{\frac{2}{\pi}}\frac{\sqrt{S}}{a_B}\mathbf{d}^{vc*}\int \chi^{e*}(z)\chi^{h*}(z)\mathrm{d}z \qquad (13.31)$$

are the optical matrix elements for the isolated OQW and IQW, respectively, which are independent of $\mathbf{k}$. Usually we have $M^F \gg M^W$ since $a_F \ll a_B$, and in the region of strong mixing the oscillator strengths f^α of a hybrid state is determined by its FE component:

$$f^\alpha(\mathbf{k}) \simeq |A_\alpha(\mathbf{k})|^2 f^F. \qquad (13.32)$$

At the crossing point $k = k_0$ (for $\delta > 0$) we have $|A_\alpha(k_0)|^2 = 1/2$ and the FE oscillator strength is equally distributed between the two hybrid states. For the hybrid exciton radii the opposite relation holds. Calculating the expectation value of the exciton radius squared $\hat{r}^2$ in the state $|\alpha, \mathbf{k}\rangle$ we obtain

$$\langle \alpha, \mathbf{k}|\hat{r}^2|\alpha, \mathbf{k}\rangle = |A_\alpha(\mathbf{k})|^2 \langle F, \mathbf{k}|\hat{r}^2|F, \mathbf{k}\rangle + |B_\alpha(\mathbf{k})|^2 \langle W, \mathbf{k}|\hat{r}^2|W, \mathbf{k}\rangle$$
$$\simeq |B_\alpha(\mathbf{k})|^2 a_B^2, \qquad (13.33)$$

since $a_B \gg a_F$ and we neglect the latter. Cross-terms do not appear since we neglect the single-particle wavefunction mixing between the two QWs. We see that the new states can possess both large oscillator strengths and exciton radii. This effect is especially pronounced if the crossing of the FE and WE dispersion curves occurs for a value of the wavevector close to that of the maximum of the coupling strength: $k_0 \simeq k_{\max}$. Since k_0 is determined by the detuning δ, and $k_{\max}$, in turn, depends on L_w and z_0 (eqn 13.22), a special choice of these parameters should be made for maximizing the effect. Also, in order to take advantage of the hybrid states in optics, the wavevector of light in the medium $q = n_\infty \omega/c$ (n_∞ being the background refractive index) should not be far from k_0. Usually near excitonic resonances, $q < k_0$ and special care should be taken to overcome this difficulty (e.g. using a coupled diffraction grating with period $2\pi/k_0$ (23) or a prism). We mention, however, that even in the region of small wavevectors in which the 2D excitons are radiative, the hybridization may be realized not due to the instantaneous dipole–dipole interaction, but due to the retarded interaction stemming from the exchange of photons. Such a situation has been analyzed (even in the nonlinear regime) with an appropriate transfer matrix approach, which is equivalent to the solution of the full Maxwell equations (2).

Concerning the choice of materials for the implementation of the system considered here, examples of molecular substances having small-radius (≤ 0.5 nm) excitons with energies of a few eV, among those already successfully grown (17) as crystalline layers on a variety of inorganic (including semiconductor) crystals, are the acenes, such as tetracene (2 eV) or pentacene (1.5 eV), the metal phthalocyanines, such as VOPc (1.6 eV) or CuPc (1.8 eV), and the tetracarboxilic compounds, such as NTDCA or PTCDA (2.2 eV). Semiconductors having large-radius excitons with matching energies are, for instance, the III–V and II–VI ternary solid solutions such as GaAlAs, ZnCdSe, and ZnSe (24); besides a judicious choice of alloy composition and well thickness, a fine tuning of the resonance condition could be achieved applying an external static field along the growth direction (quantum confined Stark effect (25); for hybdrid excitons it has been considered in Ref. (26)). A major experimental problem is the control of the interface quality: the inhomogeneous broadening should remain small and the in-plane wavevector **k** a (sufficiently) good quantum number; organic superlattices with high-quality interfaces have been demonstrated (17).

A necessary condition for the hybrid states to be observable is that the exciton linewidths must be smaller than the splitting $\Delta(k)$. This is the case in the present calculations, where for $k_0 = k_{max}$ we have $\Delta(k_0) = 2\Gamma_{max} \simeq 20$ meV, while in inorganic QWs the homogeneous linewidth at low temperatures is ~ 1 meV (27), (28). The nonradiative linewidth of a 2D Frenkel exciton in a OQW can also be small: in the case of a 2D-exciton in the outermost monolayer of anthracene this linewidth at low temperatures is ~ 2 meV (16). In principle, apart from the resonance condition and the large difference in excitonic radii, the present model demands no specific requisite and the rapid progress in the growth of organic crystalline multilayers justifies some optimism about its concrete realization.

We also mention here the work (29), where the effects of the exciton–phonon interaction in hybrid systems were studied. In this work resonant Raman spectroscopy is also suggested as a tool for studying hybrid organic–inorganic QWs.

13.3 Hybrid excitons in parallel organic and inorganic semiconductor quantum wires

In this section (see also (3)) we study hybrid exciton states not in a two-dimensional (2D) but in a one-dimensional (1D) system of parallel organic and inorganic semiconductor quantum wires. In particular, we will show that the interwire hybridization strength is nonzero even for small exciton wavevectors along the wires and decays rather slowly with increasing of the spacing between the wires. This is in contrast with 2D quantum wells considered in the preceding section, where the dipole–dipole coupling decays rapidly with increasing interwell distance and is nonzero only for nonzero exciton wavevectors.

Although there has already been considerable progress in the preparation of organic–inorganic semiconductor structures, the achievements are mostly connected with the fabrication of planar heterostructures. As to laterally modulated composite organic–semiconductor heterostructures, we realize that the fabrication of such systems is a more formidable task. However, the expected unique physical properties of such systems may justify the efforts. A motivation for studying laterally modulated hybrid systems, such as a system of organic–inorganic wires, is the following. As has been shown in the preceding section for a 2D organic molecular layer in contact with a neighboring semiconductor 2D quantum well, the electrostatic coupling of Frenkel and Wannier–Mott vanishes in the range of small two-dimensional exciton wavevectors (see eqn 13.21). This is a consequence of the well-known fact that the electric field of a uniformly polarized layer vanishes outside the layer. Thus, in 2D systems, the conditions for the manifestation of hybridization effects are not favorable just for the most interesting case of small wavevector excitons which interact actively with light. To lift this restriction, here we consider another organic semiconductor system where the hybridization of Frenkel and Wannier–Mott excitons is especially effective just for excitons with small wavevectors. Namely, we consider a system of parallel organic and semiconductor quantum wires. The Frenkel $|F, k, \ell\rangle$ and Wannier–Mott $|W, k, \ell\rangle$ exciton states in the wire are characterized by a one-dimensional (1D) wavevector k along the wires and by the label ℓ counting quantized states of the transverse motion of excitons within the wires. To simplify the consideration, below we restrict our analysis to the lowest transverse state of excitons and omit the label ℓ.

Our task is to calculate the hybridization parameter

$$\Gamma(k) = \langle F, k | \hat{H}_{\text{int}} | W, k \rangle, \tag{13.34}$$

that determines the resonance coupling of the Frenkel and Wannier–Mott excitons, see eqns (13.1)–(13.5). The Hamiltonian H_{int} of the dipole–dipole interaction between the two systems is given by eqn (13.6) which may be represented also in the form

$$H_{\text{int}} = -\int P_i^F(\mathbf{r}) \mathcal{D}_{ij}(\mathbf{r}, \mathbf{r}') P_j^W(\mathbf{r}') \mathrm{d}^3 r \mathrm{d}^3 r', \tag{13.35}$$

where the integration goes over the organic and semiconductor wires, $\mathcal{D}_{ij}(\mathbf{r}, \mathbf{r}')$ is the Green's function determined by eqn (13.8) and (13.9); indices i, j denote vector components in Cartesian coordinates with the z-axis chosen along the wires, and the x- and y-axes perpendicular to the wires; the y-axis is perpendicular to the plane determined by the wires. The polarization operator $\mathbf{P}^F$ for Frenkel excitons is given by

$$P_i^F(\mathbf{r}) = \sum_{\mathbf{n}} \mu_i^F \left(A_{\mathbf{n}}^\dagger + A_{\mathbf{n}} \right) \delta \left(\mathbf{r} - \mathbf{r}_{\mathbf{n}} \right), \tag{13.36}$$

where the summation runs over sites $\mathbf{n}$ (with the radius-vector $\mathbf{r}_{\mathbf{n}}$) of the molecular lattice, $A_{\mathbf{n}}^\dagger$ is the creation operator of the Frenkel exciton at the site $\mathbf{n}$, and $\boldsymbol{\mu}^F$ is the transition dipole moment for Frenkel excitons. The transition polarization operator $\mathbf{P}^W$ for Wannier–Mott excitons is given by

$$\mathbf{P}^W = \boldsymbol{\mu}^W \Psi_e(\mathbf{r}) \Psi_h(\mathbf{r}) + \text{h.c.}, \tag{13.37}$$

where $\Psi_{e(h)}(\mathbf{r})$ is the electron (hole) annihilation operator and $\boldsymbol{\mu}^W$ is the (intracell) optical transition dipole moment.

The state corresponding to the Frenkel exciton is represented as

$$|F,k\rangle = \frac{1}{\sqrt{N_F}} \sum_{\mathbf{n}} \exp\left(\mathrm{i}k z_{\mathbf{n}}\right) A_{\mathbf{n}}^{\dagger} |0\rangle, \qquad (13.38)$$

where N_F is the total number of molecular sites and $|0\rangle$ is the exciton vacuum state. For the state of the Wannier–Mott exciton we take the following representation:

$$|W,k\rangle \qquad (13.39)$$
$$= \frac{1}{\sqrt{L}} \int \mathrm{d}z_e \mathrm{d}z_h \exp\left(\mathrm{i}k \frac{m_e z_e + m_h z_h}{m_e + m_h}\right) \Phi_0\left(z_e - z_h\right) \Psi_{e0}^{\dagger}\left(z_e\right) \Psi_{h0}^{\dagger}\left(z_h\right) |0\rangle.$$

Here L is the length of the wires, m_e and m_h are the effective electron and hole masses, respectively; the function $\Phi_0\left(z_e - z_h\right)$ describes the relative 1D motion of the bound electron and the hole. The operators $\Psi_{e(h)0}^{\dagger}(z)$ in eqn (13.39) correspond to the lowest $(\ell = 0)$ state of the transverse motion in the operator expansion over transverse modes ϕ_ℓ:

$$\Psi_{e(h)}^{\dagger}(\mathbf{r}) = \sum_{\ell} \Psi_{e(h)\ell}^{\dagger}(z) \phi_{e(h)\ell}^{*}(\boldsymbol{\rho}), \qquad (13.40)$$

where $\mathbf{r} = (\boldsymbol{\rho}, z)$.

In calculating the hybridization parameter Γ (see eqn 13.34) we meet the following matrix elements of the polarization operators between the ground ($|0\rangle$) and the corresponding excited states $|F,k\rangle$ or $|W,k\rangle$. With the use of eqn (13.36) and (13.38) we obtain:

$$\langle F,k|P_i^F(\mathbf{r})|0\rangle = \frac{\mu_i^F}{\sqrt{N_F}} \sum_{\mathbf{n}} \exp\left(-\mathrm{i}k z_{\mathbf{n}}\right) \delta\left(\mathbf{r} - \mathbf{r}_{\mathbf{n}}\right)$$
$$\approx \frac{\mu_i^F}{\sqrt{N_F}} \exp(-\mathrm{i}k z), \qquad (13.41)$$

where we have used the long-wavelength approximation ($k a_F \ll 1$, a_F is the organic lattice constant) and substituted the summation over the lattice sites $\mathbf{n}$ by integration over $\mathrm{d}^3 r_{\mathbf{n}}/v_F$, where v_F is the volume of the organic lattice elementary cell. Similarly, with the use of eqn (13.37), (13.39), and (13.40) we find:

$$\langle 0|P_j^W(\mathbf{r}')|W,k\rangle = \frac{\mu_j^W}{\sqrt{L}} \phi_{e0}(\boldsymbol{\rho}') \Phi_0(0) \exp(\mathrm{i}k z'). \qquad (13.42)$$

Combining eqns (13.34), (13.41), and (13.42) we arrive at the following expression for the hybridization parameter Γ from eqn (13.34):

$$\Gamma(k) = -\mu_i^F \mu_j^W \Phi_0(0) \sqrt{\frac{N_F}{L}} \int \frac{\mathrm{d}^2\rho \, \mathrm{d}^2\rho'}{S_F} \mathcal{D}_{ij}\left(\boldsymbol{\rho}, \boldsymbol{\rho}'; k\right) \phi_{e0}(\boldsymbol{\rho}') \phi_{ho}(\boldsymbol{\rho}'), \qquad (13.43)$$

where S_F is the organic wire cross-sectional area, $\mathcal{D}_{ij}(\boldsymbol{\rho},\boldsymbol{\rho}';k)$ is the Fourier transform of $\mathcal{D}_{ij}(\boldsymbol{\rho},z;\boldsymbol{\rho}',z')$ with respect to the difference $z-z'$ (the surrounding medium is assumed to be homogeneous in the direction parallel to the wires). Equation (13.43) determines the hybridization parameter of interest for an arbitrary system geometry and dielectric tensor of the surrounding medium. Below we consider in more detail the particular case of thin wires ($d_w \ll R$, $d_w \ll 1/k$, where d_w is the thickness of the wires, R is the distance between the wires) embedded into a medium with an isotropic dielectric tensor $\epsilon_{ij} = \epsilon \delta_{ij}$. In this case the Green's function $\mathcal{D}_{ij}(\mathbf{r},\mathbf{r}') = \epsilon^{-1} \partial_i \partial_j G_0(\mathbf{r}-\mathbf{r}')$, where $G_0(\mathbf{r}-\mathbf{r}') = 1/|\mathbf{r}-\mathbf{r}'|$. The Fourier transform $G_0(\boldsymbol{\rho}-\boldsymbol{\rho}';k)$ of $G_0(\mathbf{r}-\mathbf{r}')$ with respect to $z-z'$ is given by

$$G_0(\boldsymbol{\rho}-\boldsymbol{\rho}';k) = 2K_0(k|\boldsymbol{\rho}-\boldsymbol{\rho}'|), \tag{13.44}$$

where K_0 is the modified Bessel function of zeroth order (30). For thin wires, we may neglect the variation of the transverse coordinates $\boldsymbol{\rho}$ and $\boldsymbol{\rho}'$ in the argument of the function $\mathcal{D}_{ij}(\boldsymbol{\rho}-\boldsymbol{\rho}';k)$, substituting $\boldsymbol{\rho}-\boldsymbol{\rho}'$ by the vector $(R,0)$ (of course, this can be done only after taking spatial derivatives of $G_0(\boldsymbol{\rho}-\boldsymbol{\rho}';k)$).

As a result, eqn (13.43) for the hybridization parameter $\Gamma(k)$ takes the form:

$$\Gamma(k) = f_{eh}\Phi_0(0)\sqrt{\frac{N_F}{L}\frac{\mu_i^F \mu_j^W}{\epsilon}} C_{ij}, \tag{13.45}$$

where

$$f_{eh} = \int \phi_{e0}(\boldsymbol{\rho})\phi_{h0}(\boldsymbol{\rho})\mathrm{d}^2\rho, \tag{13.46}$$

$$C_{ij} = -2\left[\left(\nabla_i^\perp - ik\delta_{iz}\right)\left(\nabla_j^\perp - ik\delta_{jz}\right)K_0(k\rho)\right]_{(\rho_x,\rho_y)=(R,0)}. \tag{13.47}$$

Here $\nabla_i^\perp$ denotes the derivative with respect to the transverse variables ($i = x, y$). The function K_0 possesses the following limiting behavior (30):

$$K_0(x) = \begin{cases} \sqrt{\frac{\pi}{2x}}\exp(-x), & x \gg 1, \\ -\ln(x/2), & x \ll 1. \end{cases} \tag{13.48}$$

As follows from eqn (13.48), the interwire coupling is suppressed exponentially for excitons with wavevectors $k \geq 1/R$, i.e. for a major part of the Brillouin zone. In contrast, coupling of excitons with relatively small wavevectors $k \leq 1/R$ is quite efficient. This is different from the case of a 2D system of quantum wells where the coupling at small wavevectors is suppressed because the electric field outside of a uniformly polarized layer vanishes.

The range of small wavevectors $k \sim 1/\lambda \ll 1/R$ is of special interest as excitons with such wavevectors may be created straightforwardly by light of wavelength λ. To leading order $kR \ll 1$, the hybridization parameter $\Gamma(k)$ of eqn (13.45) has the following form

$$\Gamma(k) = \frac{f_{eh}}{\epsilon R^2}\sqrt{\frac{2S_F}{a_{1B}v_F}}\left(\mu_y^F \mu_y^W - \mu_x^F \mu_x^W\right), \tag{13.49}$$

where S_F and v_F are the cross-section and the volume of an elementary lattice cell for the molecular wire; the 1D exciton ground state wavefunction $\Phi_0(0) =$

$= 1/\sqrt{2a_{1B}}$ in the strong-confinement limit has been expressed in terms of the 1D Bohr radius $a_{1B} = (a_0/2)\sqrt{E_0/E_1}$, with a_0 and E_0 being the Bohr radius and the ground state energy of the bulk exciton, respectively; E_1 is the ground state energy of the 1D exciton, see (31), (32).

Note that the excitonic polarization component along the wires does not contribute to Γ to leading order in kR. This is due to the obvious fact that a uniform longitudinal polarization is not accompanied by the appearance of an electric charge. To estimate the value of Γ we use the following parameter values: $a_{1B} = 3$ nm, $\mu^F = 5$ Debye, $\mu^W = 10$ Debye, $S_F = 25$ nm^2, $v_F = 10^{-1}$nm^3, $R = 5$ nm, $f_{eh} = 1$, $\epsilon = 3$, and we obtain for value of splitting equal $2\Gamma \approx 10.8$ meV. Similar to the 2D case, the resonance coupling of 1D Frenkel and Wannier–Mott excitons results in the appearance of hybrid states described by eqns (13.1)–(13.5).

If the energies of Frenkel and Wannier–Mott excitons are in resonance the size of the hybrid state is comparable with that for Wannier–Mott excitons, i.e. it is much larger than the radius of Frenkel excitons. In this case we can expect that the saturation concentration of excitons in hybrid structure will be of the same order as in a semiconductor quantum wire. Outside the resonance range, the coupling is governed by the parameter $\Gamma^2/(E_F - E_W)$ and is rather small. The condition of resonance is rather strict for the considered range of parameters and requires a careful choice of materials for both wires. And, naturally, the exciton linewidths should be small compared to 2Γ. For these parameters these linewidths have to be smaller than the resonant splitting $2\Gamma \approx 11$ meV of the hybrid excitations.

To summarize, we have demonstrated the possibility of strong resonance hybridization of 1D Frenkel and Wannier–Mott excitons in parallel organic and semiconductor wires. Like the 2D case, the new states possess the properties of both types of excitons. They have a relatively large size (along the wires) like Wannier–Mott excitons, but they also have a large transition dipole moment which is typical for Frenkel excitons. Thus, one may expect the same as for 2D structures (see Fig. 13.1b) the strong nonlinear optical effects in such structures.

13.4 On the hybridization of "zero-dimensional" Frenkel and Wannier–Mott excitons

In the previous sections we have considered the hybridization of Frenkel and Wannier–Mott excitons in two-dimensional (quantum wells) and one-dimensional (quantum wires) geometries. For the sake of completeness, in this subsection we shall briefly and qualitatively discuss the zero-dimensional (0D) case that corresponds to a quantum dot geometry. We have in mind a configuration where a semiconductor QD is located near a small size organic cluster or is just covered by a thin shell of an organic material.

There are the following qualitative differences in the hybridization scenario in this geometry as compared to those in the 2D and 1D cases. The presence of a translational geometry along planes or wires has resulted in a selection rule restricting the hybridization to states $|F, k\rangle$ and $|W, k\rangle$ with the same wavevectors. The absence of translational symmetry in the 0D case leads, generally speaking, to the coupling of all exciton states. This circumstance is not favorable for efficient hybridization: as was discussed earlier, the condition for strong hybridization is an energy resonance between the mixed exciton states. Therefore, it is desirable to deal with a situation where there are only two resonant (Frenkel and Wannier–Mott) exciton states strongly coupled to each other and weakly connected with other states. This may be achieved in small-size QDs and clusters where the exciton motion is quantized. As soon as the resonance between two (say, the lowest) exciton states $|F\rangle$ and $|W\rangle$ has been achieved, the further description in terms of an effective two-level model is similar to what has been done in previous sections. The hybridization parameter $\langle F|H_{\text{int}}|W\rangle$ is determined by properties of the resonant states $|F\rangle$ and $|W\rangle$ for a concrete system configuration, i.e. on the shape and symmetry of the semiconductor QD and organic cluster, on the orientation of organic molecules, etc. Just as one example of a great variety of possible geometries, we mention here the case of a spherical semiconductor QD covered with a thin shell of organic molecules (see (4) for more details).

Interesting results of the studies of the strong coupling regime of Wannier–Mott excitons in a quantum dot lattice embedded in organic medium and in dendrites and also unusual nonlinear properties of such structures can be found in the articles by Birman and coworkers (33)–(37).

13.5 Nonlinear optics of 2D Hybrid Frenkel–Wannier–Mott excitons

13.5.1 *The resonant $\chi^{(3)}$ nonlinearity*

From the results of Section 13.2 we may expect that the exciton hybridization should strongly modify the nonlinear optical properties of the structure under consideration. Indeed, hybrid excitons can combine both a large oscillator strength, which makes it easy to produce large populations, and a large radius, which, in turn, leads to low saturation densities (Fig. 13.1b). In this section we analyze the situation quantitatively (2), (38), calculating the response of the interband polarization $\mathbf{P} = \mathbf{P}^W + \mathbf{P}^F$ on the external driving electric field (corresponding to a cw experiment)

$$\mathcal{E}(\mathbf{r}, t) = \mathcal{E}_0 e^{i\mathbf{Q}\mathbf{r} - i\omega t} + \text{c.c} \qquad (13.50)$$

in the presence of a large density of excitations using the standard technique of semiconductor Bloch equations (25), (39). Since we are considering a cw experiment, the populations are stationary and may be treated as parameters in the equation for the time-dependent interband polarization.

First, we express the operator of the electron–hole interband polarization $\hat{\mathbf{P}}^W$ in terms of the electron and hole creation and annihilation operators in the envelope function approximation, following the standard procedure (18), (39):

$$\hat{\mathbf{P}}^W(\mathbf{r}) = \frac{\mathbf{d}^{vc}}{S}\chi^e(z)\chi^h(z)\sum_{\mathbf{k},\mathbf{q}} e^{i\mathbf{k}\mathbf{r}_\parallel}\hat{h}_{-\mathbf{q}}\hat{c}_{\mathbf{k}+\mathbf{q}} + \text{h.c.} \quad (13.51)$$

Here $\chi^e(z)$, $\chi^h(z)$ are electron and hole wavefunctions in the given IQW subbands (resonant with the FE), $\hat{c}_{\mathbf{k}}$ and $\hat{h}_{\mathbf{k}}$ are annihilation operators for an electron and hole with the in-plane wavevector $\mathbf{k}$ in the subbands under consideration, S is the in-plane normalization area and $\mathbf{d}^{vc}$ is the matrix element (28). We do not take into account the spin degeneracy, considering thus the polarization produced by electrons and holes with a given spin (thus, the final expression for the susceptibility should be multiplied by two). An analogous expression for the OQW polarization is

$$\hat{\mathbf{P}}^F(\mathbf{r}) = \frac{\mathbf{d}^F}{a_F\sqrt{S}}\delta(z+z_0)\sum_{\mathbf{k}} e^{i\mathbf{k}\mathbf{r}_\parallel}\hat{B}_{\mathbf{k}} + \text{h.c.}, \quad (13.52)$$

where $\hat{B}_{\mathbf{k}}$ is the annihilation operator for the Frenkel exciton, which is assumed to be tightly bound. Besides the term of the Hamiltonian describing free Frenkel excitons and free electron–hole pairs (with the single-particle energies $E_F(\mathbf{k}), \varepsilon_e(\mathbf{k})$ and $\varepsilon_h(\mathbf{k})$, respectively) the Hamiltonian we consider here includes the following.
(i) The Coulomb interaction between electrons and holes

$$\hat{H}_{\text{coul}} = \frac{1}{2S}\sum_{\mathbf{q}\neq 0} v(\mathbf{q}) \quad (13.53)$$

$$\times \sum_{\mathbf{k},\mathbf{k}'}\left(\hat{c}^\dagger_{\mathbf{k}+\mathbf{q}}\hat{c}^\dagger_{\mathbf{k}'-\mathbf{q}}\hat{c}_{\mathbf{k}'}\hat{c}_{\mathbf{k}} + \hat{h}^\dagger_{\mathbf{k}+\mathbf{q}}\hat{h}^\dagger_{\mathbf{k}'-\mathbf{q}}\hat{h}_{\mathbf{k}'}\hat{h}_{\mathbf{k}} - 2\hat{c}^\dagger_{\mathbf{k}+\mathbf{q}}\hat{h}^\dagger_{\mathbf{k}'-\mathbf{q}}\hat{h}_{\mathbf{k}'}\hat{c}_{\mathbf{k}}\right),$$

$$v(\mathbf{q}) = \frac{2\pi e^2}{\epsilon_0 q}, \quad (13.54)$$

ϵ_0 being the static dielectric constant of the IQW.
(ii) The dipole–dipole interaction between the QWs, as follows from eqns (13.6) and (13.7)

$$\hat{H}_{\text{hyb}} = \sum_{\mathbf{k}} V_{\text{hyb}}(\mathbf{k})\sum_{\mathbf{q}} \hat{h}_{-\mathbf{q}}\hat{c}_{\mathbf{k}+\mathbf{q}} + \text{h.c.}, \quad (13.55)$$

$$V_{\text{hyb}}(\mathbf{k}) = -\frac{d_i^{F*}d_j^{vc}}{a_F\sqrt{S}}\int \mathcal{D}_{ij}(\mathbf{k},-z_0,z)\chi^e(z)\chi^h(z)dz, \quad (13.56)$$

which corresponds to (13.19) with $\sqrt{2/(\pi a_B^2)}$ replaced by $1/\sqrt{S}$ since we use plane waves as the basis for the semiconductor states. Of course, this interaction is also of Coulomb nature, but since we treat the OQW and the IQW as completely different systems and neglect all effects of electronic exchange between them, these pieces of the Hamiltonian come separately.
(iii) The interaction with the driving electric field (13.50)

$$\hat{H}_{\text{dr}} = -\left(\mathcal{E}_0\cdot\mathbf{M}^F\right)e^{-i\omega t}\hat{B}^\dagger_{\mathbf{Q}_\parallel}$$

$$-\left(\boldsymbol{\mathcal{E}}_0 \cdot \mathbf{M}^{eh}\right) e^{-i\omega t} \sum_{\mathbf{q}} \hat{c}^{\dagger}_{\mathbf{Q}_{\parallel}+\mathbf{q}} \hat{h}_{-\mathbf{q}} + \text{h.c.}, \tag{13.57}$$

$$\mathbf{M}^{eh} = \mathbf{d}^{vc*} \int \chi^{e*}(z) \chi^{h}(z) dz, \qquad \mathbf{M}^F = \frac{\sqrt{S}}{a_F} \mathbf{d}^{F*}, \tag{13.58}$$

and we again neglect the z-dependence of the field and the wavevector dependence on $\mathbf{M}^{eh}$.

Given the Hamiltonian, we can write the equations of motion for the Heisenberg operators. The polarization is obtained by averaging the expressions (13.51), (13.52) over the equilibirium density matrix. The result is expressed in terms of the polarization functions

$$\langle \hat{h}_{-\mathbf{q}}(t) \hat{c}_{\mathbf{k}+\mathbf{q}}(t) \rangle = \mathcal{P}^W_{\mathbf{k}}(\mathbf{q}), \qquad \langle \hat{B}_{\mathbf{k}}(t) \rangle = \mathcal{P}^F_{\mathbf{k}}. \tag{13.59}$$

Average values of the four-operator terms are factorized in the Hartree–Fock approximation and are expressed in terms of the polarization functions and the populations defined by

$$\langle \hat{c}^{\dagger}_{\mathbf{q}}(t) \hat{c}_{\mathbf{q}'}(t) \rangle = \delta_{\mathbf{q}\mathbf{q}'} n^e_{\mathbf{q}}, \qquad \langle \hat{h}^{\dagger}_{\mathbf{q}}(t) \hat{h}_{\mathbf{q}'}(t) \rangle = \delta_{\mathbf{q}\mathbf{q}'} n^h_{\mathbf{q}}. \tag{13.60}$$

Here the averages with different wavevectors correspond to the intraband polarization, which is far off resonance and may be neglected. Since the electric field excites only states with the given total in-plane wavevector $\mathbf{Q}_{\parallel}$, from now on we set $\mathbf{k} = \mathbf{Q}_{\parallel}$. As a result, we obtain the equations for the polarization functions:

$$i\hbar \frac{d\mathcal{P}^F_{\mathbf{k}}}{dt} = E_F(\mathbf{k}) \mathcal{P}^F_{\mathbf{k}} + V_{\text{hyb}}(\mathbf{k}) \sum_{\mathbf{q}} \mathcal{P}^W_{\mathbf{k}}(\mathbf{q}) - \left(\boldsymbol{\mathcal{E}} \cdot \mathbf{M}^F\right) e^{-i\omega t}, \tag{13.61}$$

$$i\hbar \frac{d\mathcal{P}^F_{\mathbf{k}}}{dt} = \hat{\mathcal{H}}_0 \mathcal{P}^F_{\mathbf{k}}(\mathbf{q}) + \hat{\mathcal{H}}_1 \mathcal{P}^W_{\mathbf{k}}(\mathbf{q})$$
$$+ \left(1 - n^e_{\mathbf{k}+\mathbf{q}} - n^h_{-\mathbf{q}}\right) \left[V^*_{\text{hyb}}(\mathbf{k}) \mathcal{P}^F_{\mathbf{k}} - \left(\boldsymbol{\mathcal{E}} \cdot \mathbf{M}^{eh}\right) e^{-i\omega t}\right],$$

$$\hat{\mathcal{H}}_0 \mathcal{P}^F_{\mathbf{k}}(\mathbf{q}) \equiv \left[\varepsilon_e(\mathbf{k}+\mathbf{q}) + \varepsilon_h(-\mathbf{q})\right] \mathcal{P}^F_{\mathbf{k}}(\mathbf{q}) - \sum_{\mathbf{q}'} \frac{v(\mathbf{q}-\mathbf{q}')}{S} \mathcal{P}^F_{\mathbf{k}}(\mathbf{q}'),$$

$$\hat{\mathcal{H}}_1 \mathcal{P}^W_{\mathbf{k}}(\mathbf{q}) \equiv -\left[\sum_{\mathbf{q}'} \frac{v(\mathbf{q}-\mathbf{q}')}{S} \left(n^e_{\mathbf{k}+\mathbf{q}'} + n^h_{-\mathbf{q}'}\right)\right] \mathcal{P}^W_{\mathbf{k}}(\mathbf{q}) \tag{13.62}$$
$$+ \left(n^e_{\mathbf{k}+\mathbf{q}} + n^h_{-\mathbf{q}}\right) \sum_{\mathbf{q}'} \frac{v(\mathbf{q}-\mathbf{q}')}{S} \mathcal{P}^W_{\mathbf{k}}(\mathbf{q}').$$

Here the "Hamiltonian" $\hat{\mathcal{H}}_0$ describes the evolution of the polarization in an isolated IQW in the absence of electron–hole populations and corresponds to the Wannier equation (39). The resonant Wannier exciton wavefunction in momentum space $\Phi_{\mathbf{k}}(\mathbf{q})$ is its eigenfunction with the eigenvalue $E_W(\mathbf{k})$. The "Hamiltonian" $\hat{\mathcal{H}}_1$ describes the nonlinear many-particle corrections. It is proportional to the populations n^e, n^h and we treat it perturbatively, keeping only the first-order corrections to the eigenfunction $\delta\Phi_{\mathbf{k}}(\mathbf{q})$ and to the eigenvalue $\delta E_W(\mathbf{k})$. Since populations are proportional to the intensity of the applied field $|\mathcal{E}_0|^2$, our calculation describes a third-order nonlinearity.

We seek the solutions depending on time as $\exp(-i\omega t)$. The solution for $\mathcal{P}_{\mathbf{k}}^W(\mathbf{q})$ may be expressed in terms of the orthonormal basis of eigenfunctions of $\hat{\mathcal{H}}_0 + \hat{\mathcal{H}}_1$. Picking up only the resonant term, we may write

$$\mathcal{P}_{\mathbf{k}}^W(\mathbf{q}) = u_{\mathbf{k}}^W \left[\Phi_{\mathbf{k}}(\mathbf{q}) + \delta\Phi_{\mathbf{k}}(\mathbf{q})\right] e^{-i\omega t}, \qquad (13.63)$$

$$u_{\mathbf{k}}^W e^{-i\omega t} = \sum_{\mathbf{q}} \left[\Phi_{\mathbf{k}}^*(\mathbf{q}) + \delta\Phi_{\mathbf{k}}^*(\mathbf{q})\right] \mathcal{P}_{\mathbf{k}}^W(\mathbf{q}), \qquad (13.64)$$

$$\mathcal{P}_{\mathbf{k}}^F = u_{\mathbf{k}}^F e^{-i\omega t}. \qquad (13.65)$$

Then eqns (13.61) and (13.63) are reduced to

$$\begin{aligned}(\hbar\omega - E_F(\mathbf{k}))\, u_{\mathbf{k}}^F &= F_{FW}(\mathbf{k}) u_{\mathbf{k}}^W - J_F, \\ (\hbar\omega - E_W(\mathbf{k}) - \delta E_W(\mathbf{k}))\, u_{\mathbf{k}}^W &= \beta_{\mathbf{k}} V_{FW}^*(\mathbf{k}) u_{\mathbf{k}}^F - \beta_{\mathbf{k}} J_W, \end{aligned} \qquad (13.66)$$

where we have introduced the coupling matrix element

$$V_{FW}(\mathbf{k}) + \delta V_{FW}(\mathbf{k}) = V_{\text{hyb}}(\mathbf{k}) \sum_{\mathbf{q}} \left[\Phi_{\mathbf{k}}(\mathbf{q}) + \delta\Phi_{\mathbf{k}}(\mathbf{q})\right], \qquad (13.67)$$

the effective driving forces

$$J_F = \boldsymbol{\mathcal{E}}_0 \cdot \mathbf{M}^F, \qquad J_W + \delta J_W = \boldsymbol{\mathcal{E}}_0 \cdot \mathbf{M}^{eh} \sum_{\mathbf{q}} \left[\Phi_{\mathbf{k}}^*(\mathbf{q}) + \delta\Phi_{\mathbf{k}}^*(\mathbf{q})\right], \qquad (13.68)$$

and the Pauli blocking factor which is given by

$$\beta_{\mathbf{k}} = \frac{\sum_{\mathbf{q}} \left(1 - n_{\mathbf{k}+\mathbf{q}}^e - n_{-\mathbf{q}}^h\right) \Phi_{\mathbf{k}}^*(\mathbf{q})}{\sum_{\mathbf{q}} \Phi_{\mathbf{k}}^*(\mathbf{q})}. \qquad (13.69)$$

In the low-density limit and with $\mathcal{E}_0 = 0$ these equations correspond to the eigenvalue equation (13.2) with the coupling matrix element given by (13.19) since for 1s-exciton

$$\Phi_{\mathbf{k}}(\mathbf{q}) = \sqrt{\frac{8\pi a_B^2}{S}} \frac{1}{(q^2 a_B^2 + 1)^{3/2}}, \qquad \sum_{\mathbf{q}} \Phi_{\mathbf{k}}(\mathbf{q}) = \sqrt{\frac{2S}{\pi a_B^2}}. \qquad (13.70)$$

Solving the system (13.66), we obtain for the polarization of the structure under consideration (per unit area):

$$P_i^s(\mathbf{r}_\parallel) \equiv \langle \hat{P}_i^F(\mathbf{r}) + \hat{P}_i^W(\mathbf{r}) \rangle dz \simeq \frac{u_{\mathbf{k}}^F M_i^{F*}}{S} e^{-it\omega + i\mathbf{k}\mathbf{r}_\parallel} + \text{c.c.}$$

$$= \chi_{ij}(\omega)\mathcal{E}_{0j}e^{-it\omega+i\mathbf{k}\mathbf{r}_\parallel} + \text{c.c.}, \tag{13.71}$$

where we retained only the term proportional to $|\mathbf{M}^F|^2$ since $|J_W| \ll |J_F|$. Finally, we obtain for the susceptibility (not forgetting the factor 2 originating from spin degeneracy as mentioned in the beginning of this section):

$$\chi_{ij}(\omega, \mathbf{k}) = 2\frac{d_i^{F*}d_j^F}{a_F^2} \tag{13.72}$$

$$\times \frac{E_W(\mathbf{k}) + \delta E_W(\mathbf{k}) - \hbar\omega}{[E_W(\mathbf{k}) + \delta E_W(\mathbf{k}) - \hbar\omega][E_F(\mathbf{k}) - \hbar\omega] - \beta_\mathbf{k}|V_{FW}(\mathbf{k}) + \delta V_{FW}(\mathbf{k})|^2}.$$

In eqn (13.72) the nonlinearities appear through the blue-shift δE_W, the blocking factor β and the modification of the hybridization δV_{FW} due to the correction $\delta \Phi$; all these effects are typical of Wannier excitons (25), but here they belong to the hybrid excitons which also have a large oscillator strength characteristic of Frenkel excitons. When only excitons are present (i.e. under resonant excitation at low temperature), the nonlinear correction can be calculated to first order in the n's with

$$n_{\mathbf{k}+\mathbf{q}}^e = n_{-\mathbf{q}}^h \simeq \frac{n_T S}{4}|\Phi_\mathbf{k}(\mathbf{q})|^2, \tag{13.73}$$

where n_T is the total density of electron–hole pairs and the factor $1/4$ takes into account electron (and hole) spin degeneracy of two and an equal population of resonant FE and WE. In terms of Section 13.2, this corresponds to the situation when $k \simeq k_0$, thus $E_F(\mathbf{k}) \simeq E_W(\mathbf{k})$, $|A_\alpha|^2 \simeq |B_\alpha|^2 \simeq 1/2$, and $|u_\mathbf{k}^F|^2 \simeq |u_\mathbf{k}^W|^2 \simeq (n_T S)/4$. The blue-shift δE_W is given by the expectation value of $\mathcal{H}_1$ on $\Phi_\mathbf{k}(\mathbf{q})$ and reduces to

$$\delta E_W \simeq 0.48 E_b \pi a_B^2 n_T, \tag{13.74}$$

where E_b is the binding energy of a 2D Wannier exciton. The blocking factor is calculated from eqn (13.69) and results as

$$\beta_\mathbf{k} \simeq 1 - 0.57 \pi a_B^2 n_T. \tag{13.75}$$

The effect of δV_{FW} can be estimated (25) writing $\delta \Phi_\mathbf{k}(\mathbf{q})$ as a sum over all continuous and discrete excitonic states which are then approximated by plane waves in the expression $|V_{FW}(\mathbf{k}) + \delta V_{FW}(\mathbf{k})|^2$ obtaining

$$|V_{FW}(\mathbf{k}) + \delta V_{FW}(\mathbf{k})|^2 \simeq \left(1 - 0.48 \pi a_B^2 n_T\right)|V_{FW}|^2. \tag{13.76}$$

Close to resonance (denoting the detuning $\hbar\omega - E_W(\mathbf{k})$ by ΔE) eqn (13.72) can be approximated by

$$\chi_{ij}(\omega, \mathbf{k}) = -2\frac{d_i^{F*}d_j^F}{a_F^2}\frac{\Delta E}{\Delta E^2 - |V_{FW}|^2} \tag{13.77}$$

$$\times \left[1 - \pi a_B^2 n_T \left(\frac{1.05|V_{FW}|^2 - 0.48 E_b \Delta E}{\Delta E^2 - |V_{FW}|^2} + \frac{0.48 E_b}{\Delta E}\right)\right]$$

$$= \chi_{ij}(\omega, \mathbf{k})\left(1 - \frac{n_T}{n_S}\right),$$

$\chi_{ij}(\omega, \mathbf{k})$ being the susceptibility of the hybrid structure at $n_T = 0$ (the linear susceptibility) and n_S is the saturation density. The characteristic feature of the expression (13.77) is the presence of the factor $(d^F/a_F)^2$ in $\chi^{(1)}$ instead of $(d^{vc}/a_B)^2$ in the analogous expression for an isolated IQW. This leads to the enhancement of absorption, determined by $\mathrm{Im}\,\chi^{(1)}$. Thus, while the saturation density is comparable to that of Wannier excitons ($n_S \sim 1/a_B^2$), the density of photogenerated electron–hole pairs, for a given light intensity, can be two orders of magnitude larger (by a factor$\sim (a_B/a_F)^2$); for the same reason, also the linear susceptibility $\chi^{(1)}$ can be two orders of magnitude larger. Therefore, the present theory substantiates the intuitive expectation of very pronounced nonlinear optical properties of the hybrid excitons.

While the range of validity of eqn (13.77) (with respect to variations of ΔE and n_T) is rather limited, the expression for χ given by eqn (13.72) holds true as long as the basic approximations of the present approach are tenable. These are, in addition to first-order perturbation theory with respect to the excitation density n_T, the usual Hartree–Fock decoupling in the equations of motion adopted in eqns (13.61) and (13.63), the subsistence of well-defined individual excitons (valid only for $n_T \leq n_S$) and the neglect of screening due to the reduced screening efficiency of a two-dimensional exciton gas (25), (39). Numerical examples of the predictions of eqn (13.72) have been obtained (see also (2)) using the values of semiconductor parameters representative of III–V semiconductor (e.g. GaAs/AlGaAs) quantum wells, since the necessary information on homogeneous linewidths of excitons in II–VI QWs is not presently available. Namely, we set $\epsilon_\infty = 11$, $d^{vc} = 20$ Debye, the Bohr radius $a_B = 6$ nm and the binding energy is taken to be $E_b \simeq 20$ meV, all the rest are the same as in the previous section. This gives $|V_{FW}| \simeq 4$ meV at $k = 10^7$ cm^{-1}. Assuming a phenomenological linewidth $\hbar\gamma_W = 2$ meV, Fig. 13.6 shows the effect of a density-dependent broadening of the Wannier exciton: $\hbar\gamma_W = 1$ meV at low excitation densities and $\hbar\gamma_W = 3$ meV at high excitation density (40) ($\hbar\gamma_F$ being fixed at 2 meV). From numerical estimates such as those shown in Fig. 13.6, we obtain for the relative nonlinear change in the absorption coefficient close to resonance $|\Delta\alpha|/\alpha \sim 10^{-11}$ cm$^2 n_T$, which is analogous to the case of a semiconductor multiple quantum well. However, for a given pump intensity the 2D density of photogenerated excitons n_T in our case of hybrid excitons is about two orders of magnitude larger because the oscillator strength of hybrid excitons is comparable to that of Frenkel excitons rather than that of Wannier excitons. A similar theoretical approach can be used to calculate the dynamical Stark effect for hybrid excitons which shows qualitative and quantitative differences with respect to the case of the usual inorganic semiconductor QWs (41).

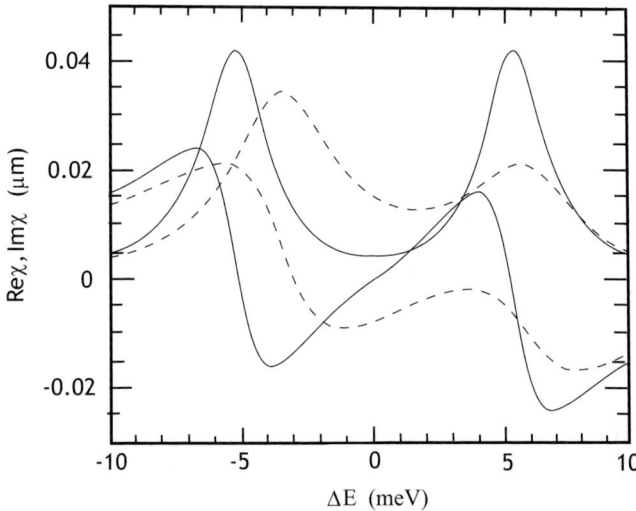

FIG. 13.6. Real and imaginary parts of χ in the linear regime (solid lines), and high excitation ($n = 10^{11}\text{cm}^{-2}$, long dashed lines), in the first case $\hbar\gamma_W = 1$ meV, in the second case $\hbar\gamma_W = 3$ meV.

13.5.2 Second-order susceptibility $\chi^{(2)}$

As was already mentioned, the calculations, performed here, correspond to the third-order nonlinearity. But the hybrid system considered here also has a nonzero second-order susceptibility $\chi^{(2)}$. For such a structure $\chi^{(2)} \neq 0$ even if the original OQW and IQW are centrosymmteric and the second-order processes are forbidden by parity conservation. Such a phenomenon can take place because the resonant dipole–dipole coupling breaks the symmetry along the growth direction. Of course, any interaction between the OQW and the IQW can be responsible for symmetry breaking. However, the resonant dipole–dipole coupling considered here is probably the strongest among others. In a geometrical sense, this system is analogous to an asymmetric semiconductor QW. The calculation of $\chi^{(2)}$ for such a system can be found in Ref. (42) and the calculation of $\chi^{(2)}$ for the hybrid system may be performed following the lines of the latter work and here we restrict ourselves only to some qualitative remarks.

The general microscopic expression for the nth-order susceptibility contains $n+1$ dipole moment matrix elements, involving n intermediate states. For the linear susceptibility there is only one intermediate state, and if the latter is a hybrid one, the corresponding dipole matrix elements are determined mainly by the Frenkel component of the hybrid state. Thus, the linear susceptibility of the hybrid structure contains the factor $(d_F/a_F)^2$, as is seen from eqn (13.77). For the second-order nonlinear susceptibility $\chi^{(2)}$ one must have two intermediate states or three virtual transitions. One of them may be a hybrid one, and as long as the materials under consideration have no static dipole moment, the other intermediate state has to be an excited state of the IQW, which is not resonant with the Frenkel exciton. Hence, the result will be proportional to d_F/a_F, the other two virtual transitions will give a factor, coinciding with that for an isolated IQW. One may apply analogous arguments to the case of the third-order nonlinearity: of the three required intermediate states one may be the hybrid one, the second may be the ground state, and the third one – again the hybrid state (such a scheme corresponds to the Kerr nonlinearity). Thus, one should obtain a factor $(d^F/a_F)^4$. Indeed, in eqn (13.77) we have $(d^F/a_F)^2$ in $\chi^{(1)}$ and another $(d^F/a_F)^2$ comes from n_T when the latter is expressed in terms of the incident electric field. There exists another mechanism for second-harmonic generation. It does not require parity breaking, since the optical quadratic nonlinearity appears due to the contribution of spatial derivatives of the electric field to the nonlinear response (43), (44). It works also in the case of an isolated symmetric QW and corresponds to the higher multipole contribution rather than the dipole one, which is usually considered. The hybrid system will again have an advantage here because of the increase in the oscillator strength due to the Frenkel exciton component.

An interesting first example of hybridization of exciton states with different symmetries was demonstrated in the recent paper by Roslyak and Birman (45). They predicted substantial enhancement of the second-harmonic generation (SHG) generic to the cuprous oxide crystals by resonant hybridization with appropriate organic material. The quadrupole origin of the inorganic part of the quadrupole–dipole hybrid provides inversion symmetry breaking and the organic part contributes to the oscillator strength of the hybrid. They have shown that the enhancement of the SHG, compared to the bulk cuprous oxide crystal, is proportional to the ratio of the transition dipole moment in the organic component and the effective dipole moment of the quadrupole transitions in the cuprous oxide. It is also inversely proportional to the linewidth of the hybrid and bulk excitons. The enhancement may be regulated by proper organic blend (mutual concentration of both components) and pumping conditions (varying the angle of incidence in the case of optical pumping or populating the minimum of the lower branch of the hybrid in the case of electric pumping).

13.6 Weak coupling regime in hybrid nanostructures

13.6.1 The Förster energy transfer

It is useful in discussion of weak coupling between nanostructures to remember the nonradiative mechanism of Förster resonant energy transfer from an excited molecule (a donor) to some other molecule (an acceptor) which can be in the ground or in an excited state. The probability of such a transfer is determined by the Coulomb nonretarded (instantaneous) dipole–dipole interaction between molecules and is proportional to R_F^6/R^6 where R_F is the Förster radius and R is the distance between molecules. For organic materials the Förster radius is usually about several nanometers and strongly depends on the overlapping of the fluorescence spectrum of the donor with the absorption spectrum of the acceptor. In this formula any footprints of dissipation are absent because actually this formula is correct only in the limit of very strong dissipation in the acceptor. Thus, in using the Förster mechanism for describing the energy transfer between molecules we assume that dissipation in the acceptor is much faster than the time of energy transfer in the opposite direction, from acceptor to donor (for more details see (46)).

Just such a physical picture can be applied for hybrid nanostructures in the weak coupling regime. In this case the dissipation of the excited state in the acceptor (organic or inorganic) component of the nanostructure is very strong but the energy of the resonance Coulomb coupling between the semiconductor nanostructure and the organic material is not strong enough to produce a stable coherent superposition of excited states of the organic and the inorganic components of the nanostructure. The dependence of the probability of energy transfer on the distance between components of the nanostructure strongly depends on the geometry of nanostructures and is different for quantum well or quantum dots, for example. Below we describe the results of calculations for the energy transfer only in planar geometry from a semiconductor quantum well to an organic overlayer.

13.6.2 Förster energy transfer in a planar geometry

Calculations of the energy transfer rate in this case can be simplified, because the nonradiative (Förster-like) energy transfer from a quantum well to organics is nothing but the Joule losses in the organic material. These are produced by the penetration into the organics of the electric field generated by the semiconductor exciton polarization which has to be calculated in the framework of microscopic exciton theory (see Appendix B). Thus, to consider the energy transfer from a semiconductor quantum well to an overlayer we can use a semiclassical approach and characterize the overlayer (it can be organic or some other material) as a medium with a dielectric constant $\epsilon_{ij}(\omega)$. Then as the result of quantum-mechanical calculations it is necessary to find a density of the transition dipole moment of the exciton transition in a semiconductor microcavity $\mathbf{P}(\omega, \mathbf{r}, t)$. In the case considered we are unable to use the approximation of a point transition dipole because the distance between semiconductor quantum well and the overlayer is of order of the quantum well thickness. Having in mind that $\rho = -\mathrm{div}\mathbf{P}$ is the coupled charge density in the quantum well and using the Poisson equation with appropriate boundary conditions we can determine the distribution of the potential and electric field $\mathbf{E}(x, y, z)$ in the overlayer induced by exciton polarization. Knowledge of this field and also a knowledge of the dielectric function of the overlayer material gives us the possibility of calculating the Joule losses of the electromagnetic field created by exciton polarization in the overlayer using the known formula from electrodynamics of continuous media (47). For the rate of energy dissipation of a monochromatic electromagnetic field we have:

$$G = \frac{1}{2\pi\hbar}\int \mathrm{Im}\epsilon_{ij}(\mathbf{r},\omega_{\mathrm{exc}})E_i E_j \mathrm{d}^3 r$$

and the corresponding time of energy transfer of the exciton $\hbar\omega_{\mathrm{exc}}$ from the semiconductor quantum well to the overlayer can now be found from the relation:

$$\tau = \hbar\omega_{\mathrm{exc}}/G.$$

Although in such an estimation we assumed an infinitely thick layer of organic molecules the obtained results are correct with rather high accuracy even for layers with thickness of order 10 nm because the electric field created by exciton polarization penetrates into the organic layer only on lengths of the order of the semiconductor quantum well thickness and the barrier between the quantum well and the organic layer.

Such a procedure for the calculation was used in (6), as well as in (1) and up to now was used in many other papers. For the case considered here, weak coupling, it exactly corresponds to results of consecutive microscopic quantum-mechanical calculations in the framework of Fermi's golden rule (see Appendix C) and can also be applied for the derivation of energy transfer in the weak coupling regime for the case of other types of nanostructures, quantum wires, quantum dots, and so on. As the organic material is described here only by its dielectric function, the same theory can be used if, instead of organic material as the acceptor, another material is used exhibiting fast relaxation and a broad spectrum in the region of the Wannier–Mott exciton transition. It was used recently, for instance, for the study of regimes of energy transfer from an epitaxial quantum well to a proximal monolayer of semiconductor nanocrystals (49).

The numerical estimations in (6)–(9) (see Appendix D) have been performed (for the structure shown in Fig. 13.7) for electron–hole excitations in a semiconductor quantum well including free and localized excitons and unbound e–h pairs in typical quantum wells of II–VI and III–V semiconductor materials and typical organic materials. These calculations (an excellent review was published by Basko (48)) demonstrate that nonradiative energy transfer from a semiconductor quantum well to an organic overlayer in combination with strong dissipation of excited states in organic materials occurs on time-scales of several tens of picoseconds for II–VI semiconductors and several hundreds of picoseconds for III–V semiconductors, which in both cases is significantly less than the semiconductor excitation lifetime in the absence of such a transfer. The authors of similar calculations concerning the energy transfer from an epitaxial quantum well to a proximal monolayer of semiconductor nanocrystals (49) indicate that with a careful design of the system (geometrical and electric parameters) the Förster transfer can be used as an efficient "noncontact" pumping mechanism of nanocrystal QD-based light-emitting devices.

For studies of the dynamics of energy transfer from a semiconductor quantum well to organics it is very important to know the origin of exciton luminescence of the semiconductor quantum well under electric or optical nonresonant pumping

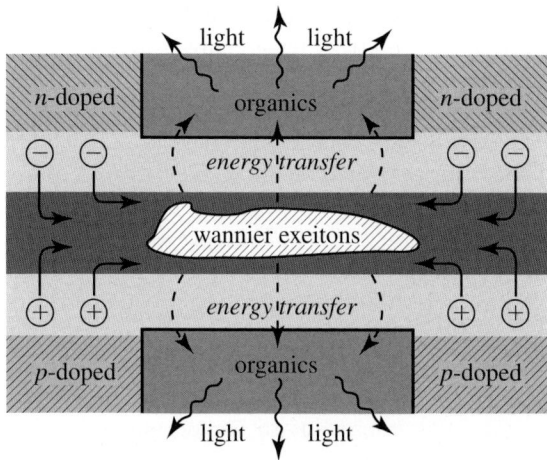

FIG. 13.7. The sketch of the planar structure under discussion (from (13)).

and to know the characteristic time of build-up of the exciton line as a function of excitation density and temperature. More generally it is important to know the dynamics which determines the concentrations of free carriers and also free and localized Wannier–Mott excitons. An interesting discussion of these problems can be found in the paper by Deveaud *et al.* (50) where experiments are reported on the time-resolved photoluminescence and time dependence of exciton versus free carrier luminescence in a very high-quality InGaAs quantum well sample.

13.6.3 Noncontact pumping of light emitters via nonradiative energy transfer: new concept for light-emitting devices

Over the past few years, considerable progress has been made in the development of organic light-emitting devices (LEDs) in the visible range using polymers as well as small molecular compounds. In all such organics-based devices the fundamental role is played by electroluminescence, the generation of light by electric excitations. Already, in early studies of Pope *et al.* (51), it was established that the process responsible for the electroluminescence requires injection of electrons from one electrode and holes from the other, transport of one or both charges, capture of oppositely charged carriers on the same molecule (or recombination center), and radiative decay of the resulting excited electron–hole state. For inorganic semiconductors, where the recombining electron–hole state may be a Wannier–Mott exciton, all the above mentioned processes are investigated and well documented (52). In organics, in contrast to inorganic semiconductors, scientists have met many problems in the use of electroluminescence for the creation of devices conceptually similar to what was done with the use of semiconductor materials. The main reasons are the small mobility of carriers and strong chemical interaction between organics and metals which prevents the injection of charges into organics. Due to the small mobility of charge carriers in organic materials the current is bulk-limited, principally through the build-up of a space charge.

To avoid these problems for light-emitting devices it was proposed in (6) (see also (10)) to use hybrid organic–inorganic structures to combine the comparatively good transport properties of semiconductors (e.g. pumped electrically) and good light-emitting properties of organic substances. If an organic material having a broad absorption band in the optical range and overlapping with the semiconductor exciton resonance is placed near the semiconductor, the resonant dipole–dipole interaction between the two substances in combination with the fast dephasing, common for many organic substances (e.g. due to the scattering by phonons) will result as we discussed above in the efficient nonradiative, noncontact transfer of the semiconductor excitation energy to the organics. Of course, in this case it is important to use organic materials with the high-luminescence quantum yield. Due to the energy transfer from a semiconductor quantum well to the organics, the nonradiative processes with characteristic times larger than that of the energy transfer will be suppressed. As a result, we can expect that the quantum efficiency of the considered layer structure will be increased in comparison with the quantum efficiency of the semiconductor quantum well in the absence of the energy transfer to the organics, if the luminescence quantum yield of emissive organic layer is high enough.

13.6.4 First experiments

Although in the papers (6), (10) organic materials were used as a model acceptor, the first experimental demonstration of the new concept of LEDs was performed by Victor Klimov's (53) group by investigating the energy transfer from a quantum well (donor) to a monolayer of semiconductor nanocrystals (acceptor) having, however, a broad absorption spectrum (as most organic materials do). The structure studied in these experiments is depicted in Fig. 13.8a.

It consists of an InGaN quantum well, on top of which a close-packed monolayer of highly monodisperse CdSe/ZnS core/shell nanocrystals is assembled, using the Langmuir–Blodgett technique. The nanocrystals consist of a CdSe core (radius 1.9 nm) overcoated with a shell of ZnS (0.6 nm thickness), followed by a final layer of the organic molecules trioctylphosphine (TOP) and trioctylphosphine oxide (TOPO). These nanocrystals show efficient emission centered near 575 nm and a structured absorption spectrum with the lowest 1S absorption maximum at 560 nm (Fig. 13.8b). Quantum-well samples were grown on sapphire substrates by metal–organic chemical-vapor deposition. They consisted of a 20 nm GaN nucleation layer, a 3 mm GaN bottom barrier, and a 3 nm InGaN quantum well. The concentration of In in the quantum wells was $5-10\%$, which corresponds to an emission wavelength of ~ 400 nm (Fig. 13.8b). This wavelength is in the range of strong nanocrystal absorption, which provides strong coupling of quantum-well excitations to the absorption dipole of nanocrystals, and should allow efficient energy transfer. The emission color of semiconduc-

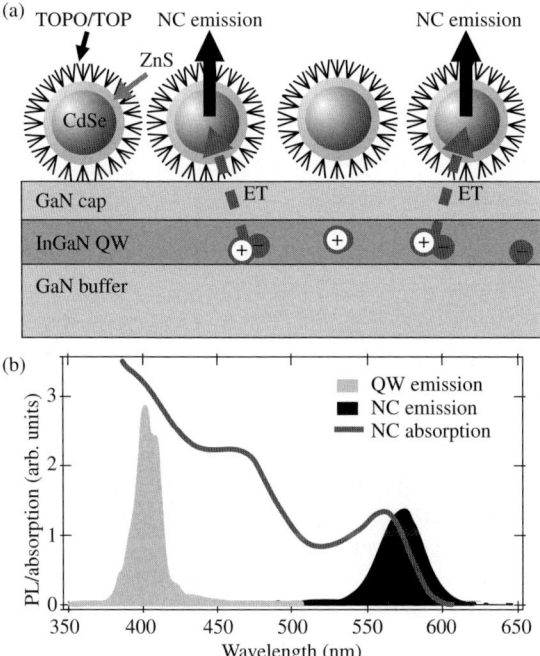

FIG. 13.8. Schematic and optical properties of the hybrid quantum-well/nanocrystal structure. (**a**). The structure consists of an InGaN/GaN quantum-well heterostructure with a monolayer of TOPO/TOP–capped CdSe/ZnS core/shell nanocrystals on top of it. Electron–hole pairs in the quantum well can experience nonradiative resonant transfer into nanocrystals. The nanocrystals excited by energy transfer produce emission with a wavelength determined by the nanocrystal size. (**b**). The emission of the quantum well (blue) spectrally overlaps with the absorption of the nanocrystals (green). For CdSe nanocrystals with 1.9 nm radius, the emission wavelength is around 575 nm (red) (from (13)).

tor nanocrystals as a result of quantum-confinement effects, can be modified by changing their size. Such spectral tunability, together with large photoluminescence quantum yields and high photostability, makes nanocrystals attractive for use in a variety of light-emitting technologies. It is difficult to achieve electric pumping in the case of nanocrystals largely due to the presence of an insulating organic capping layer on its surface.

An approach with indirect injection of electron–hole excitations into nanocrystals by the above described noncontact nonradiative Förster-like energy transfer from a proximal quantum well that can in principle be pumped either electrically or optically, can solve the problem of pumping of nanocrystals. The result obtained by the Klimov group indicate that this energy transfer is fast enough to compete with electron–hole recombination in the quantum well, and results in greater than 50% energy-transfer efficiencies in the tested structures. The measured energy-transfer rates as indicated in (53) are sufficiently large to provide pumping even in the stimulated emission regime, indicating the feasibility of nanocrystal-based optical amplifiers and lasers based on this approach (see also (54)).

The monolayer of the nanocrystals has a rather small thickness. The electric field created by quantum well exciton polarization spreads out on a length-scale much larger than the thickness of the monolayer of the nanocrystals and can pump a larger volume of emissive material. In the case of an organic overlayer this volume can be easily increased. The use of organic materials in the creation of new types of LED will enrich this field of investigation. There is no doubt that the new type of organic LEDs will be brighter, cheaper and will have a wider color range.

Reassuringly for our original ideas, the effect of nonradiative energy transfer has been recently experimentally demonstrated in hybrid structure with an actual organic overlayer by Bradley and his collaborators (Heliotis et al. (56)). In this paper the fabrication and study of a hybrid inorganic–organic semiconductor structure has been reported, employing an organic polyfluorene thin film with the opportunity to place it in sufficiently close proximity to the underlying InGaN QW. As we mentioned, this architecture has the potential to take advantage of the complementary properties of the two classes of semiconductors that it contains and thus, might lead to high-performance devices with desirable electric and optical characteristics. A schematic of the fabricated structures is shown in Fig. 13.9. The UV light-emitting InGaN QW (2 nm thick) is spaced from the blue-light-emitting poly(9,9- dioctylfluorene-co-9,9-di(4-methoxy)phenylfluorene) film (5 nm thick) by GaN cap layers of variable thickness. Energetic alignment (as illustrated in the energy-level diagram of Fig. 13.9) is thus needed in order to maximize the resonant coupling between the inorganic and organic excitations. Radiative decay of the latter should then result in light emission with the characteristic spectrum of the organic layer. Figure 13.9 shows the corresponding InGaN QW photoluminescence (PL) emission and the absorption and PL spectra of the organic film. The QW has a narrow PL band that peaks at 385 nm, coincident with the organic layer absorption maximum. The organic layer, in turn, emits blue light with a broad structured PL band (vibronic features at 425, 450, and 480 nm). For these experiments, three hybrid heterostructures were fabricated with different GaN cap thicknesses (namely, 15, 4, and 2.5 nm) between the QW and the organics. The variation in the GaN cap-layer thickness allowed some tuning of the strength of the dipole–dipole interaction between the QW and organics and, hence, to look for the expected improvement in organics emission efficiency (relative to the radiative transfer) under nonradiative Förster-like energy transfer from the QW to the organics at small distances from the QW and the organics. In concluding remarks the authors, Heliotis et al. (56), stress that by adjusting the separation between the inorganic and organic layers they found an intensity enhancement

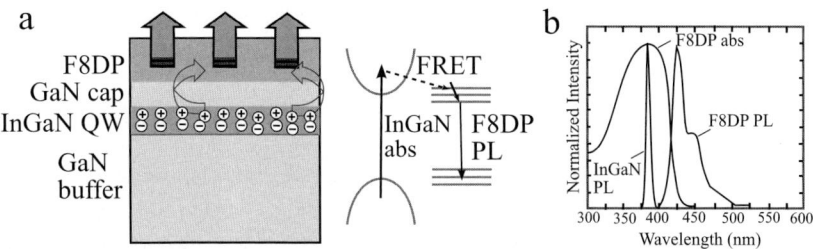

FIG. 13.9. (a) Schematic of the hybrid inorganic–organic semiconductor heterostructures and a simplified energy-level diagram illustrating the proposed energy-transfer scheme: the excitations initially generated by absorption (ABS) in the QW layer (InGaN) can resonantly Förster transfer (FRET) their energy to the singlet excited states of the poly(9,9-dioctylfluoreneco- 9,9-di(4-methoxy)phenylfluorene) (F8DP) polymer layer. Following internal conversion (vibrational relaxation), emission should then occur via radiative decay to the polymer ground state (photoluminescence, F8DP PL). (b) Absorption spectrum (labeled F8DP ABS) of F8DP and photoluminescence (labelled PL) emission spectra of the InGaN QW and the F8DP thin film. Reprinted with permission from Heliotis *et al.* (56). Copyright 2006, American Physical Society.

of organic layer radiation by 20 times, at least in the case of purely radiative transfer from the QW to organics. Such structures, as stressed by Heliotis *et al.*, are able to take advantage of the complementary properties of organic and inorganic semiconductors, which may lead to devices with highly efficient emission across the entire visible range of the electromagnetic spectrum.

In a more recent publication of the F. Henneberger group (57) the next important step in the studies of hybrid structures in the weak coupling regime was made. Electronic coupling between Wannier and Frenkel excitons in an inorganic–organic semiconductor hybrid structure is experimentally observed with the use of time-resolved photoluminescence and excitation spectroscopy. These techniques directly demonstrated that the electronic excitation energy can be transferred with an efficiency of up to 50% from an inorganic ZnO quantum well to an organic (2,2-*p*-phenylenebis-(5-phenyloxazol) (POPOP), or α-sexithiophene) (6T) overlayer (recently an efficiency value of 80% has been obtained (private communication)). Energy transfer is observed up to about 100 K. The coupling as in previous papers has been mediated via dipole–dipole-interaction analog to the Förster transfer in donor–acceptor systems. The investigated structure is displayed in Fig. 13.10. More systematic studies of the coupling mechanism are necessary in order to optimize the length dimensions in the hybrid structure and to increase the coupling strength. Further experimental work in this direction would be extremely valuable. Needless to say, new direct experimental evidence is also required of the hybrid heterostructures with organic layers and with electric pumping of inorganic quantum wells. The first practical

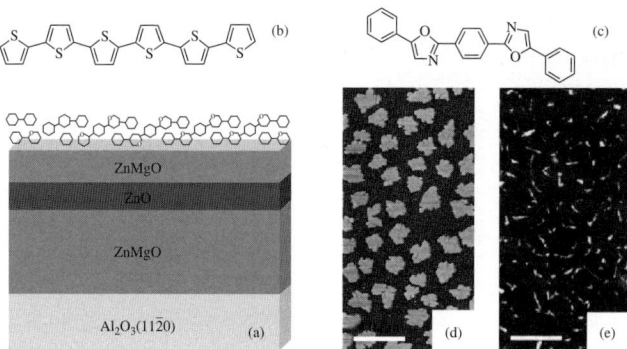

FIG. 13.10. (a) Sketch of the hybrid nanostructure. The 4 nm wide ZnO QW is situated on top of a 600 nm thick $Zn_{0.9}Mg_{0.10}O$ barrier layer grown on a Al_2O_3 (1120) substrate. The growth direction is along the wurtzite **c** axis. (b), (c) Structures of 6T and POPOP. (d), (e) AFM images of 6T and POPOP deposited on ZnO(0001)(4×4) at a substrate temperature of 100^0 (6T) and 25^0 (POPOP), respectively, and a deposition rate of 1 Å/min. Reprinted with permission from Blumstengel et al. (57). Copyright 2006, American Physical Society.

implementation of high-efficient nonradiative energy transfer in an electrically pumped hybrid LED has already been performed by the Klimov group (55) in the study of the energy transfer from a quantum well (donor) to a monolayer of semiconductor nanocrystals (acceptor).

13.6.5 Förster energy transfer from quantum dots to organics

One type of semiconductor nanostructure which has been actively studied in the last few years is quantum dots. In the present subsection we analyze Förster energy transfer from a quantum dot to organic material surrounding the dot. For simplicity, we consider a spherical quantum dot and we employ the effective-mass approximation to describe the Wannier–Mott exciton. For the organic subsystem, similar to the case of a plane microcavity, we use a macroscopic description using some value of the permittivity. We perform the analysis for a nanostructure where a spherical quantum dot with radius R_1 and permittivity ϵ is surrounded by a concentric semiconductor barrier with outer radius R_2 and the same permittivity, while the material outside the barrier is surrounded by organic material with permittivity $\tilde{\epsilon}$. The permittivity $\tilde{\epsilon}$ includes only the contribution from higher resonances with respect to the excitonic resonance under study, and we assume that $\tilde{\epsilon}$ is real. The permittivity $\tilde{\epsilon}$, however, is the total permittivity of the organic material and is substantially complex. The general scheme for calculation the rate of Förster energy transfer in a hybrid nanostructure is the same as used for plane microcavities. In this case this transfer rate can be found by calculating the Joule losses of the electric field that is produced by the excitonic polarization of a quantum dot in an organic medium. Let an exciton with energy $\hbar\omega$ be described by the envelope of the wavefunction $\psi(\mathbf{r}_e, \mathbf{r}_h)$, and let $\mathbf{d}^{vc}$ be the matrix element of the electric dipole moment between Bloch functions of the valence band and the conduction band. The matrix element of excitonic polarization in the quantum dot is

$$\mathbf{P}(\mathbf{r}) = \mathbf{d}^{vc}\psi(\mathbf{r},\mathbf{r}). \tag{13.78}$$

The electric field $\mathbf{E}(\mathbf{r})$ corresponding to this polarization can be found from the Poisson equation for the potential $\phi(\mathbf{r})$:

$$\epsilon(r)\boldsymbol{\nabla}^2\phi(\mathbf{r}) = 4\pi\boldsymbol{\nabla}\cdot\mathbf{P}(\mathbf{r}), \tag{13.79}$$

where $\epsilon(r) = \epsilon$ for $r < R_2$ and $\epsilon(r) = \tilde{\epsilon}$ for $r > R_2$, with appropriate boundary conditions at $r = R_2$. Knowing the electric field, we can calculate the power dissipated in the organic material

$$W = \frac{\omega \operatorname{Im}\tilde{\epsilon}}{2\pi}\int_{r>R_2}|\mathbf{E}(\mathbf{r})|^2 d^3r. \tag{13.80}$$

Hence we find that the rate of energy transfer (the reciprocal of the transfer time) is given by the expression

$$\frac{1}{\tau} = \frac{W}{\hbar\omega} = \frac{\operatorname{Im}\tilde{\epsilon}}{2\pi\hbar}\int_{r>R_2}|\mathbf{E}(\mathbf{r})|^2 d^3r. \tag{13.81}$$

In our spherical geometry the states of an electron–hole pair can be classified according to the values of the total angular momentum of the electron and hole. We shall assume that the electron–hole pair occupies the lowest excited state. In this state the total angular momentum is zero, and therefore the function $\psi(\mathbf{r},\mathbf{r})$ is spherically symmetric and $\mathbf{P}(\mathbf{r}) = \mathbf{P}(r)$. Orienting the z-axis parallel to the vector $\mathbf{d}^{vc}$, we obtain for the charge density

$$\rho(\mathbf{r}) = -\boldsymbol{\nabla}\cdot\mathbf{P}(\mathbf{r}) = -\frac{\partial P(r)}{\partial z} = \rho^r(r) Y_{10}(\theta). \tag{13.82}$$

Here $Y_{10}(\theta) \propto \cos\theta$ is the spherical harmonic with $\ell = 1$ and $m = 0$ that is obtained when a spherically symmetric function is differentiated with respect to z. Expanding the potential $\phi(\mathbf{r})$ in the organic material, where there are no charges, in terms of multipoles, we see that only the dipole term, which is proportional to $Y_{10}(\theta)$, will be different from zero. Therefore the potential $\phi(\mathbf{r})$ in the organic medium is formally identical to the potential of an effective point dipole with dipole moment $\mathbf{p}_{\text{eff}}$:

$$\phi(\mathbf{r}) = (\mathbf{p}_{\text{eff}}\cdot\mathbf{r})/\tilde{\epsilon}r^3. \tag{13.83}$$

Calculations similar to those in Ref. (46), Ch. 1, show that

$$p_{\text{eff}} = \frac{3\tilde{\epsilon}}{\epsilon + 2\tilde{\epsilon}}\int_0^{R_2} P(r) 4\pi r^2 dr. \tag{13.84}$$

Using this expression and calculating the field created by the dipole we can find for the energy transfer time τ the expression

$$\frac{1}{\tau} = \frac{W}{\hbar\omega} = \frac{\operatorname{Im}\tilde{\epsilon}}{2\pi\hbar} \int_{r>R_2} |\mathbf{E}(\mathbf{r})|^2 d^3r = \frac{4\operatorname{Im}\tilde{\epsilon}}{3\hbar\tilde{\epsilon}^2 R_2^3} |p^{\text{eff}}|^2. \tag{13.85}$$

To calculate the effective transition dipole moment we need to know the wavefunction of the exciton. In a quantum dot it depends on two interactions: (i) the electron and hole confinement potential, which we shall assume to be infinite for $r > R_1$ and zero for $r < R_1$ and (ii) the electron and hole Coulomb attraction. For these interactions we have to consider the following characteristic lengths: R_1 – the radius of the quantum dot, and a_B – the Bohr radius of an exciton in a macroscopic three-dimensional semiconductor. The problem of solving the two-particle Schrödinger equation for an arbitrary ratio of these lengths is quite difficult but the situation simplifies substantially in two important limiting cases.

If $R_1 \gg a_B$ (weak confinement), the splitting between the levels in a spherical well is much smaller than the binding energy of a three-dimensional Wannier–Mott exciton. In this case the latter can be assumed to be a rigid particle moving in a spherical well. The variables corresponding to the relative motion of an electron and a hole and the motion of the center of mass effectively separate, and the wavefunction factorizes. Then

$$\psi(\mathbf{r},\mathbf{r}) = \frac{1}{\sqrt{\pi a_B^3}} \chi_0(\mathbf{r}), \tag{13.86}$$

where the function χ_0 has the shape

$$\chi_0(\mathbf{r}) = \frac{1}{\sqrt{2\pi R_1}} \frac{\sin(\pi r/R_1)}{r}. \tag{13.87}$$

The corresponding transition dipole moment is

$$p_{\text{eff}} = \frac{1}{\pi} \left(\frac{2R_1}{a_B}\right)^{3/2} \frac{3\tilde{\epsilon} \, d^{vc}}{\epsilon + 2\tilde{\epsilon}}. \tag{13.88}$$

If, however, $R_1 \ll a_B$ (strong confinement), then the Coulomb interaction can be completely neglected, and the wavefunction of the exciton will be simply the product of two (identical) single-particle electron and hole functions:

$$\psi(\mathbf{r}_e, \mathbf{r}_h) = \chi_0(\mathbf{r}_e) \chi_0(\mathbf{r}_h), \tag{13.89}$$

and the corresponding transition dipole moment will be

$$p_{\text{eff}} = \frac{3\tilde{\epsilon} d^{vc}}{\epsilon + 2\tilde{\epsilon}}. \tag{13.90}$$

The energy transfer rates in these limiting cases are

$$\frac{1}{\tau_1} = \frac{1}{\hbar} \frac{96 \operatorname{Im} \tilde{\epsilon}}{\pi^2 |\epsilon + 2\tilde{\epsilon}|^2} \frac{|d^{vc}|^2}{a_B^3} \left(\frac{R_1}{R_2}\right)^3, \qquad R_1 \gg a_B, \qquad (13.91)$$

and

$$\frac{1}{\tau_1} = \frac{1}{\hbar} \frac{12 \operatorname{Im} \tilde{\epsilon}}{|\epsilon + 2\tilde{\epsilon}|^2} \frac{|d^{vc}|^2}{R_2^3}, \qquad R_1 \ll a_B. \qquad (13.92)$$

These formulas demonstrate a different and strong dependence of the energy transfer time on the radius of a quantum dot. To obtain a numerical estimate we employ the typical parameter values for the elements of II–VI compounds (for example, ZnSe) and organic substances used in similar experiments: $a_B = 5$ nm, $d^{vc} = = 12$ D, $\epsilon = 6$, and $\tilde{\epsilon} = 4 + 3i$. For $R_2 = 1.5\, R_1$ these parameters give a transfer time $\tau_1 \simeq 20$ ps, and the time τ_2 is even shorter. This estimate shows that the energy transfer time is much shorter than the exciton lifetime (hundreds of picoseconds in II–VI semiconductors) determined by all the other processes in a quantum dot. Thus, the energy transfer mechanism considered above, just as in the case of quantum wells, is quite rapid and makes possible the transfer of a substantial part of the energy from a semiconductor quantum dot to the organic molecules surrounding the dot.

13.6.6 *Exciton energy transfer from a quantum dot to its surface states*

The theory of energy transfer considered in this subsection was used to interpret the experiments with PbSe quantum dots (58) on the size-dependent energy relaxation in a quantum dot. In this paper it was shown that smaller dots have faster relaxation. In the theoretical paper by Hong *et al.* (59) it was assumed that the above energy transfer from a quantum dot exciton to surface states of the dot is a dominant channel of the electronic energy relaxation. Hong *et al.* considered in their calculations a spherical quantum dot of radius R and the transfer rate was obtained from the calculation of the power dissipation W on the surface of the quantum dot by the relation

$$W = \frac{\omega \operatorname{Im} \tilde{\epsilon}}{2\pi} \int_{R < r < R+\Delta} |\mathbf{E}(\mathbf{r})|^2 \mathrm{d}^3 r, \qquad (13.93)$$

where Δ is the thickness of the surface wall (boundary) and the dielectric constant of the surface of the $\tilde{\epsilon}$ was employed to introduce absorption in the frequency range of interest. Having the power W the transfer time τ is obtained by the relation $\tau = \hbar \omega / W$.

In two extreme cases, when the wall thickness Δ of the quantum dot is much smaller than the radius R of the quantum dot, the dependence of the relaxation rate on the dot radius is approximated to be $w = 1/\tau \propto \Delta/R^4$, whereas when the wall thickness $\Delta \sim R$, the proportionality of the relaxation rate to the dot radius is $w \propto 1/R^3$, thus the actual exponent c in the dependence of the relaxation rate on the dot radius $w \propto 1/R^c$ will be a number between 3 and 4, in good agreement with experiments by Okuno *et al.* (58). The excited surface states can emit a luminescence light with spectrum reflecting the structure of the spectrum of surface excitations in quantum dots.

13.6.7 Exciton energy transfer from organics to semiconductor nanocrystals and carrier multiplication

The possibility of energy transfer from organics to a quantum well was noted in consideration of the interaction of quantum well excitations with a resonant localized excitation in organics (60). Such a process is opposite to energy transfer from a quantum well (or quantum dot) to organic material which we considered above in this chapter and which can be important for the creation of a new type of light-emitting devices. The process of electronic energy transfer from organic material to a semiconductor nanostructure which we consider in this subsection can be interesting for a new type of solar cell. It results in the creation of free electron–hole pairs in the inorganic nanostructure which, in contrast to organic materials, typically have a high mobility of carriers. Thus, in this case, the most complicated process in organic solar cells, namely the process of transformation of a Frenkel exciton into a charge-transfer exciton with subsequent charge separation, does not appear.

It is known that the main limitation of solar power right now is the cost, because the crystalline silicon used to make most solar photovoltaic cells is expensive. One approach to overcome this cost factor is to concentrate light from the sun using mirrors or lenses, thereby reducing the total area of silicon needed to produce a given amount of electricity. But traditional light concentrators are bulky and unattractive – less than ideal for use on suburban rooftops. Here we note the possibility to create a new type of concentration of sunlight using the process of nonradiative energy transfer from organics to semiconductor nanostructures. As a model nanostructure we consider the results of a recently published paper (64). In this paper the electronic energy transfer from organics to semiconductor nanocrystals (NCs) has been investigated using the multilayer structure presented in Fig. 13.11.

The structure was obtained by a layer-by-layer (LBL) assembly approach (64). Figure 13.11 schematically shows the typical structure of the hybrid organic–inorganic (J-aggregate/QD) LBL films that was synthesized for this study, along with the chemical structures of the constituents. In the hybrid film a single monolayer J-aggregate of cyanine dye (TDBC) was sandwiched between two monolayers of CdSe–ZnS core–shell structured NCs (quantum dots), with polyelectrolyte (PDDA) acting as the ultrathin "molecular glue". Two types of hybrid films were considered in (64), one with NC emission centered at 548 nm (referred to as film I) and the other at 653 nm (referred to as film II), thus providing contrasting cases of excitation coupling with respect to the fixed J-aggregate emission at 594 nm. Representative atomic force microscopy (AFM) images of hybrid film II and two additional NCs and J-aggregate reference films are shown in Fig. 13.11 (b)-(d). It was found for films II that resonance coupling of electronic excitations in J-aggregates with electronic excitations in the two monolayers of NCs can reach

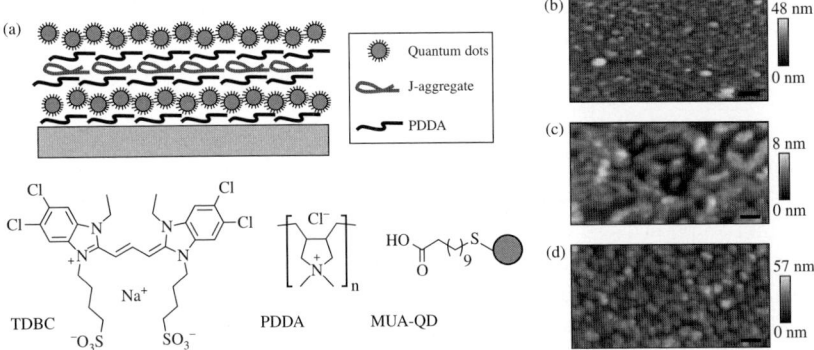

FIG. 13.11. Hybrid organic–inorganic (J-aggregate/QD) multilayer film deposited by LBL assembly. (a), Schematic of the hybrid film layer structure: a monolayer J-aggregate of TDBC is sandwiched between two monolayers of CdSe-ZnS QDs, joined by monolayers of poly(diallyldimethylammonium chloride) (PDDA). Molecular structures of TDBC, PDDA and MUA are illustrated in the lower panel. (b)-(d), AFM images of the LBL-grown films, consisting of a monolayer of QDs (λ_{em} = 653 nm) (PDDA/QD) (b), a monolayer of J-aggregate (PDDA/TDBC/PDDA) (c), and the hybrid film II (d). The scale bars in (b)-(d) are 100 nm.

efficiencies of energy transfer from J-aggregates to NCs of up to 98% at room temperature. This result can be especially interesting for the properties of solar cells if, in a structure similar to that presented in Fig. 13.11, instead of layers of CdSe-ZnS core–shell structured NCs, to use NCs with small energy gap (PbSe or similar), that are usually used in the studies of so-called carrier multiplication, the effect that is the direct photogeneration of multiexcitons in NCs by single photons.

NCs contain approximately 100–10,000 atoms. Because of the strong spatial confinement of electronic wavefunctions and reduced electronic screening, the effects of carrier–carrier Coulomb interactions are greatly enhanced in NCs compared with those in bulk materials. These interactions open a highly efficient decay channel via Auger recombination and just such a strong carrier–carrier interaction in NCs is responsible for carrier multiplication (61)–(63).

It is clear that the Förster resonant energy transfer (FRET) of electronic energy excitation with energy 2–4 eV from organic material to NCs with a small energy gap of order of 0.5 eV in structures similar to that presented in Fig. 13.11 can also give carrier multiplication. However, this process can have a few interesting peculiarities in comparison with carrier multiplication under the influence of photons of the same energy.

First of all a high efficiency of energy transfer from organic material to NCs can drastically increase the number of carriers in NCs, because the absorbtion of light by rather dense organic material can be much larger than the absorption of light by the system of separated NCs (in the case of the structure created in (64) the absorption in organics is larger than the absorption of NCs by more than ten times, see Fig. 2 in (64)). One can expect that absorption of the NCs based on an equivalent bulk volume is very close to absorption of the bulk semiconductor. However, the NCs have to be separated and the relatively small absorption of light by NCs is mainly the result of the small density of NCs which cannot be larger than some critical value.

The second important feature of the carrier multiplication under the influence of this energy transfer is the structure of the electric field in the volume of the NC. In the simplest approximation this field is created by a transition dipole moment of the organic molecules. This field, even in spherical NCs, is very inhomogeneous and nonspherical. It is clear that this asymmetry of the electric field could be responsible for the change of the selection rules, determining the population of electron higher energy states in NCs. This effect needs a careful analysis and is interesting because it can decrease the minimal energy of excitation which produces carrier multiplication (65). In the case, for example, of pumping of NCs (PbSe and PbS) by light this minimal energy is equal to the threefold energy gap (63). One can expect that for such NCs and with pumping of them by Förster energy transfer this threshold will be smaller.

13.7 Hybridization of Frenkel and Wannier–Mott excitons in a 2D microcavity in the regime of strong coupling

13.7.1 *Microcavity embedded resonant organic and inorganic quantum wells*

If in a microcavity an organic and an inorganic quantum well separated in space are present (Fig. 13.12; see also Fig. 13.13 for coupled or separated microcavities) we can expect that the resonant interaction between exciton states may appear through the exchange of virtual cavity photons. As for the case of a single quantum well in a microcavity, the interaction between organic and inorganic quantum wells can be strong. This takes place if the cut-off frequency of the cavity photon is close enough to the exciton resonances. In this case the hybrid states are substantial superpositions of three states: two exciton states and cavity photon states. In this case new hybrid Frenkel–Wannier–Mott exciton + cavity photon states can be tailored to engineer the fluorescence efficiency and relaxation processes in the microcavity (66). Another design compared to that of Fig. 13.12 is depicted in Fig. 13.13: here two coupled microcavities are used.

The strong coupling of semiconductor Wannier–Mott excitons with cavity photons is a mechanism working only in a small fraction of k space close to the light line, with ks much smaller than the typical value of thermalized excitons. Relaxation of excitons to these small k is a slow and inefficient process (~ 1 ns) mediated by the weaker acoustic phonons (67). This bottleneck precludes the use of strong coupling to speed up spontaneous emission whose dynamics remains dominated by uncoupled excitons. Relaxation processes in a hybrid organic–inorganic microcavity capitalize on the strong interaction with organic's phonons.

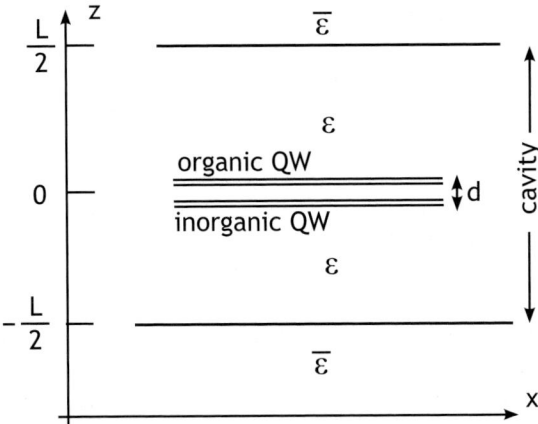

FIG. 13.12. Schematics of the microcavity structure. Reprinted with permission from Agranovich et al. (66). Copyright Elsevier (1997).

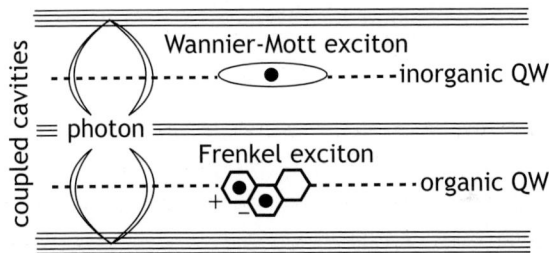

FIG. 13.13. Scheme of hybrid cavity with a semiconductor quantum well in one cavity and, in the coupled cavity, an organic quantum well. Excitons interact through the cavity photon (from (13)).

Both for intrabranch and interbranch relaxation and for relaxation processes the time constant can be estimated to be ~ 100 ps in order of magnitude. The strong optical coupling of two different Frenkel excitons from two separated organic layers in the microcavity has been observed by Lidzey et al. (68)–(71). As could be expected, in these papers three cavity polariton branches have been found with the middle branch containing a significant component of the cavity photon and both of the two excitons states. Following the paper (66) we will discuss in the next subsection the peculiarities of polariton states arising in a hybrid organic–inorganic microcavity due to resonance of a Frenkel exciton in the organic layer and a Wannier–Mott exciton in a semiconductor inorganic quantum well.

13.7.2 Dispersion and relaxation of a polariton in a hybrid microcavity containing a crystalline organic layer and a resonant inorganic quantum well

To attain the strong coupling regime with organic materials, we need compounds combining large oscillator strength with an absorption width smaller than the cavity-mode splitting. Good candidates are thin film crystals of aromatic molecules like tetracene, terrylene, etc. For example, five monolayers of terrylene ($d \sim 5$ nm) exhibit an oscillator strength per unit area as large as 10^{15}cm^{-2}, a few hundred times that of a GaAs quantum well exciton. Using a classical approach we derive below the spectrum and widths (radiative as well as nonradiative) of the coupled QW exciton–MC (microcavity) photon modes (cavity-polaritons), and also present a complementary QED approach which gives the structure of hybrid states used later to find the operators of the cavity polariton–phonon interaction.

13.7.3 Classical formalism

We will examine a model MC with two quantum wells (QWs) with polarized in-plane exciton resonances and located near $z = 0$ within a distance $d < \lambda$ (Fig. 13.12). We can consider both QWs as one very thin layer with polarization

$$P_y(\omega, k) = [\chi_1(\omega, k) + \chi_2(\omega, k)] E_y(\omega, k)$$

embedded in the microcavity. We assume that the dielectric constants ϵ and $\bar{\epsilon}$ indicated in Fig. 13.12 obey the inequality $\bar{\epsilon} \gg \epsilon$ which create mirrors for cavity photons. We assume that in this region of the spectrum we have to take into account only transverse electromagnetic waves. To consider eigenmodes in the microcavity (cavity polaritons) we assume that only outgoing waves are present for $z > L/2$ and $z < -L/2$ and that in the calculation of fields we can assume that both quantum wells are placed at $z = 0$. In this case for a given value of the in-plane wavevector k the dependence of fields on z can be presented by the relations:

$$E_y = \begin{cases} Ae^{iq(z-L/2)}, & z > L/2, \\ A_1 e^{i\beta z} + A_2 e^{-i\beta z}, & L/2 > z > 0, \\ A_3 e^{i\beta z} + A_4 e^{-i\beta z}, & 0 > z > -L/2, \\ A_5 e^{-iq(z+L/2)}, & z < -L/2, \end{cases} \quad (13.94)$$

where

$$q = \sqrt{\bar{\epsilon}\frac{\omega^2}{c^2} - k^2}, \qquad \beta = \sqrt{\epsilon\frac{\omega^2}{c^2} - k^2},$$

with $k = k_x$.

At $z = \pm L/2$ we take the usual Maxwell boundary conditions

$$E_y^< - E_y^> = 0, \qquad H_x^> - H_x^< = 0 \quad (13.95)$$

but at $z = 0$, following (72), we take into account the QWs polarization $P_y(\omega, k)$. As a result we obtain:

$$E_y^> - E_y^< = 0,$$

$$\left.\frac{\partial E_y}{\partial z}\right|^>_{z=0} - \left.\frac{\partial E_y}{\partial z}\right|^<_{z=0} = -4\pi\frac{\omega^2}{c^2}P_y(\omega, k),$$

$$P_y(\omega, k) = [\chi_1(\omega, k) + \chi_2(\omega, k)] E_y(\omega, k),$$

$$E_y(\omega, k) = A_1 + A_2, \quad (13.96)$$

$$\chi_1(\omega, k) = \frac{f^W}{\omega^W(k) - \omega - i\gamma^W(\omega)},$$

$$\chi_2(\omega, k) = \frac{f^F}{\omega^F(k) - \omega - i\gamma^F(\omega)},$$

χ_1 being the contribution of the semiconductor QW, χ_2 that of the organic QW.

Using the boundary conditions (13.95) at $z = L/2$ and $H_x = (ic/\omega)(\partial E_y/\partial z)$ we find for the amplitude of the outgoing wave in the region $z > L/2$

$$A = A_1 e^{i\beta(L/2)} + A_2 e^{-i\beta(L/2)}, \qquad A_1 = -rA_2, \quad (13.97)$$

where the factor

$$r = \frac{q + \beta}{q - \beta} e^{-i\beta L}, \quad (13.98)$$

is the inverse of the usual Fresnel reflection coefficient r_{12}. Substituting the fields (13.94) for $|z| < L/2$ into eqn (13.96) we find at $z = 0$

$$F(\omega, k)E_y(\omega, k) = -4i\pi P_y(\omega, k)$$
$$= -4i\pi (\chi_1 + \chi_2) E_y(\omega, k), \quad (13.99)$$

hence for $E_y \neq 0$

$$F(\omega, k) = -4i\pi (\chi_1 + \chi_2), \quad \text{where} \quad F(\omega, k) = \frac{2\beta(1+r)c^2}{(1-r)\omega^2}. \quad (13.100)$$

Equation (13.100) gives the dispersion of polaritons for a cavity with two QWs. The function $F(\omega, k)$ represents the modification of the electric field by the microcavity and may easily be generalized for example if d is of the order of λ. In an empty cavity $F(\omega, k)$ yields the cavity photon dispersion $\omega = \Omega = \Omega(k)$. If then $\Omega(k)$, ω^W and ω^F are close we expand $F(\omega, k)$ for $\omega \approx \Omega$

$$F(\omega, k) \approx F(\Omega, k) + \left.\frac{\partial F}{\partial \omega}\right|_\Omega (\omega - \Omega). \quad (13.101)$$

As, by definition, $F(\Omega, k) = 0$, we have for cavity polaritons

$$\left.\frac{\partial F}{\partial \omega}\right|_\Omega (\omega - \Omega) = -4i\pi (\chi_1 + \chi_2). \quad (13.102)$$

For $F(\omega, k) = 0$, $r = -1$ (see eqn 13.100)

$$\left.\frac{\partial F}{\partial \omega}\right|_\Omega = \frac{c^2}{\omega^2}\left.\frac{\partial r}{\partial \omega}\right|_\Omega. \quad (13.103)$$

Assume $\bar{\epsilon} \gg \epsilon$. In this case $q \gg \beta$ and $r = e^{-i\beta L}$. For cavity modes $r = -1$ and $\beta L = \pi, 3\pi, \ldots$. As $\partial r/\partial \omega = (iL/\beta)(\omega\epsilon/c^2)$, we have from eqn (13.103) $\partial F/\partial \omega|_\Omega = iL\epsilon/\Omega$. This ω-dependence justifies our development of F around $\omega = \Omega$ because all splittings are small in comparison with ω. Equation (13.96) then takes the form (neglecting damping)

$$\frac{iL\epsilon}{\Omega}(\omega - \Omega) = -4i\pi\left[\frac{f^W}{\omega^W - \omega} + \frac{f^F}{\omega^F - \omega}\right]$$

or

$$[\omega - \Omega(k)] + \frac{|\Gamma_{13}|^2}{\omega^W(k) - \omega} + \frac{|\Gamma_{23}|^2}{\omega^F(k) - \omega} = 0, \quad (13.104)$$

where

$$|\Gamma_{13}|^2 = \frac{4\pi}{L\epsilon}\Omega f^W, \qquad |\Gamma_{23}|^2 = \frac{4\pi}{L\epsilon}\Omega f^F,$$

$2|\Gamma_{13}|$ is the polariton splitting if only the semiconductor QW is present, $2|\Gamma_{23}|$ if only the organic QW is present. Equation (13.104) gives the dispersion of the cavity polaritons. We assumed above that $\bar{\epsilon} \gg \epsilon$ to neglect the radiative width of states. However, eqn (13.100) also contains information on the radiative widths. For example, taking for the factor r (eqn 13.98), the more exact expression $r = e^{-i\beta L}(1 + 2\beta/q)$, we obtain for $\Omega(k) = \Omega'(k) - i\Omega''(k)$ for the lowest cavity modes

$$\Omega'(k) = \omega_{\text{cav}}(0)\sqrt{1 + \bar{k}^2},$$
$$\Omega''(k) = \frac{\omega_{\text{cav}}(0)}{2\pi}\frac{1 - R}{\sqrt{1 + \bar{k}^2}} \quad (13.105)$$

where R is the reflection coefficient at normal incidence

$$R = 1/|r|^2 \approx 1 - 4\sqrt{\epsilon/\bar{\epsilon}}, \quad \omega_{\text{cav}}(0) = \pi c/L\sqrt{\epsilon}, \quad \text{and} \quad \bar{k} = kL/\pi.$$

To discuss phonon relaxation processes, it is better to develop a quantum-mechanical treatment still with $\bar{\epsilon} \gg \epsilon$.

13.7.4 QED approach

We introduce for an MC containing inorganic and organic QWs the Hamiltonian

$$\hat{H} = \sum_k \left\{\hbar\omega^W(k)A_k^\dagger A_k + \hbar\omega^F(k)B_k^\dagger B_k + \hbar\Omega(k)a_k^\dagger a_k\right\}$$
$$- \sum_k \left\{\hbar\Gamma_{13}\left(A_k^\dagger a_k + a_k^\dagger A_k\right) + \hbar\Gamma_{23}\left(B_k^\dagger a_k + a_k^\dagger B_k\right)\right\}, \quad (13.106)$$

where $A_k(A_k^\dagger)$, $B_k(B_k^\dagger)$, $a_k(a_k^\dagger)$ are the usual boson operators, $\Gamma_{13} = (1/\hbar)P_W^{01}E_0$, $\Gamma_{23} = (1/\hbar)P_F^{01}E_0$, $P_{W,F}^{01}$ is the dipole matrix element for a Wannier and Frenkel exciton from the ground state, E_0 the amplitude of the vacuum electric field at $z = 0$. In the dipole approximation used in eqn (13.106) the interaction is determined by the local amplitude of the vacuum electric field, a key factor to ensure interaction for possibly distant QWs without any direct Frenkel–Wannier–Mott interaction. To diagonalize this Hamiltonian we introduce the new Bose operators $\xi_\rho(k)$, $\xi_\rho^\dagger(k)$ by using the transformation

$$A_k = \sum_\rho N_\rho^W(k)\xi_\rho(k), \qquad B_k = \sum_\rho N_\rho^F(k)\xi_\rho(k),$$
$$a_k = \sum_\rho N_\rho^C(k)\xi_\rho(k). \qquad (13.107)$$

From the Hamiltonian eigenvalue problem, N_ρ^W, N_ρ^F, and $N_\rho^C(k)$ obey the equations

$$\begin{aligned}
\left(\omega^W - \omega\right) N^W & & + & \Gamma_{13} N^C = 0, \\
& \left(\omega^F - \omega\right) N^F & + & \Gamma_{23} N^C = 0, \qquad (13.108) \\
N^W \Gamma_{13}^* + & \Gamma_{23}^* N^F & + & (\Omega - \omega) N^C = 0.
\end{aligned}$$

The eigenfrequencies $\omega_\rho(k)$ are determined from the vanishing determinant

$$\begin{vmatrix} \omega^W - \omega & 0 & \Gamma_{13} \\ 0 & \omega^F - \omega & \Gamma_{23} \\ \Gamma_{13}^* & \Gamma_{23}^* & \Omega - \omega \end{vmatrix} = 0, \qquad (13.109)$$

which yields the cubic equation

$$\left(\omega^W - \omega\right)\left(\omega^F - \omega\right)(\Omega - \omega) \\ - |\Gamma_{13}|^2 \left(\omega^F - \omega\right) - |\Gamma_{23}|^2 \left(\omega^W - \omega\right) = 0 \qquad (13.110)$$

with three solutions $\omega = \omega_\rho(k)$, $\rho = 1, 2, 3$. New boson operators are given by

$$\xi_\rho(k) = N_\rho^W A_k + N_\rho^F B_k + N_\rho^C a_k. \qquad (13.111)$$

The quantity $|N_\rho^W|^2$ gives us the Wannier–Mott exciton component, $|N_\rho^F|^2$ the part of the Frenkel exciton, and $|N_\rho^C|^2$ the part of the cavity photon mode in the (ρ, k) cavity polariton state with energy $\hbar\omega_\rho(k)$. The diagonal Hamiltonian is

$$\hat{H} = \sum_{\substack{\rho=1,2,3 \\ k}} E_\rho(k) \xi_\rho^\dagger(k) \xi_\rho(k), \qquad E_\rho = \hbar\omega_\rho(k).$$

Equations (13.104) and (13.110) coincide. However, the quantum treatment is useful in discussing polariton–phonon scattering rates. This scattering is essential to draw the benefits of the properties of W–M excitons which may be electrically injected and Frenkel excitons which have high oscillator strength.

13.7.5 Exciton–phonon scattering in a microcavity

In this subsection we take into account the exciton–phonon interaction and show that the exciton–cavity photon interaction also gives rise to a nonelastic Wannier–Frenkel interaction with phonons. Let us begin with the exciton–phonon interaction Hamiltonian

$$\hat{H}_{\rm in} = \sum_{k,q} \left[T^W_{k,q} A^\dagger_k A_{k-q} \left(b_q + b^\dagger_{-q} \right) \right.$$
$$\left. + T^F_{k,q} B^\dagger_k B_{k-q} \left(b_q + b^\dagger_{-q} \right) \right], \qquad (13.112)$$

where $b_q (b^\dagger_q)$ are phonon operators. Using the new states (13.111), we have

$$\hat{H}_{\rm in} = \sum_{k,q,\rho,\rho'} T^{\rho\rho'}_{k,q} \xi^\dagger_\rho(k) \xi_{\rho'}(k-q) \left(b_q + b^\dagger_{-q} \right), \qquad (13.113)$$

where

$$T^{\rho\rho'}_{k,q} = N^{*W}_\rho(k) N^W_{\rho'}(k-q) T^W_{k,q} + N^{*F}_\rho(k) N^F_{\rho'}(k-q) T^F_{k,q}. \qquad (13.114)$$

Operator (13.113) has terms with $\rho \neq \rho'$ which give us the transitions from one branch ρ to another ρ'. It is seen from eqn (13.114) that in the region of strong renormalization of cavity modes, the polariton–phonon interaction will be determined by the structure of the ρ, ρ' modes. For example, to have a large contribution of the exciton–phonon interaction from organic QWs to the transition $\rho \to \rho'$ with emission of a phonon we need large enough values of $|N^F_\rho(k)|$ and $|N^F_{\rho'}(k-q)|^2$.

13.7.6 Estimation of transfer rates

In this subsection we give some illustrative calculations. We use data from feasible experimental systems because the QED approach can account for any real MC with proper constants in eqn (13.106). We assume that $\omega^F(k=0) = \Omega(k=0)$ and $\omega^W(k=0) = \omega^F(0)(1+\delta)$, i.e. a Frenkel exciton tuned to the cavity photon mode and a Wannier exciton detuned by δ. Again, in units of $k_{\rm cav} = \pi/L$, $\bar{k} = k/k_{\rm cav} = k_x$, we have $\omega_{\rm cav}(k)/\omega_{\rm cav}(0) - \sqrt{1+\bar{k}^2}$, $\omega^W(k)/\omega_{\rm cav}(0) = 1+\delta+a\bar{k}^2$, where $a = \hbar k^2_{\rm cav}/2M\omega_{\rm cav}(0)$ and $\omega^F(k)/\omega_{\rm cav}(0) = 1$ (we neglect the dispersion of Frenkel exciton). For resonance at $\hbar\omega_{\rm cav} = 1.5$ eV and $\epsilon \approx 10$, we have $k_{\rm cav} = \pi/L = 2.4 \times 10^5$ cm^{-1}, and a typical exciton mass leads to $a = 10^{-5}$.

Assuming that $\Gamma_{13} = \omega_{\rm cav} \times 10^{-3} \delta = 10^{-2}$ and, for example, a ratio $|P^{01}_F|/|P^{01}_W| = 8$, we have for $\Gamma_{23} = \omega_{\rm cav}(0) \times 8 \times 10^{-3} \approx 12$ meV and $\Delta_2 = = 24$ meV. To show that this ratio of 8 is by no means unusual, let us recall that the corresponding ratio of oscillator strength $|P^{01}_F|^2/|P^{01}_W|^2 = 64$ and even larger ones are easily attained with many organic systems. For example, we know from the LT splittings (whose ratio is $|P^{01}_F|^2/|P^{01}_W|^2 \epsilon^F_\infty$), 0.08 meV for GaAs ($\epsilon^W_\infty = 12$)

and ~ 50 meV for a singlet exciton in tetracene ($\epsilon_\infty^F \approx 9$) (73), that the ratio $|P_F^{01}|^2/|P_W^{01}|^2$ can exceed 500. These large splittings give reasonable hope for reaching the strong coupling regime since the absorption linewidth may be as low as a few tens of meV in selected organic systems. We assume such a situation and neglect dissipation for both bare excitonic states. Dispersion of cavity-polaritons and of mixing coefficient are shown in Fig. 13.14(a)-(d). From Fig. 13.14(c) it is seen that the branch $\rho = 1$ (at $k \to \infty$ a pure Wannier–Mott exciton) contains a large part of the Frenkel state ($|N_1^F|^2$) for $\bar{k} \leq 0.1$. As is seen from Fig. 13.14(d) the branch $\rho = 2$ (at $k \to \infty$ a pure Frenkel exciton) for $\bar{k} \leq 0.25$ also retains a large part of the Frenkel state ($|N_1^F|^2$) while exhibiting a large cavity photon component on the same wide wavevector range.

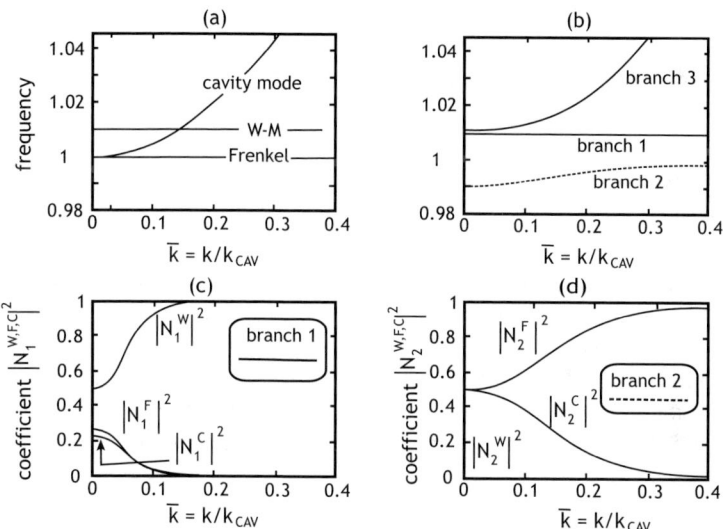

FIG. 13.14. (a) Bare reduced dispersion curves of cavity photon, Wannier–Mott exciton (W–M) and Frenkel exciton as a function of reduced wavevector k (see units in the text). A cavity photon is chosen to match the Frenkel exciton frequency at $k = 0$. Wannier–Mott has positive detuning. (b) Cavity polariton dispersion curves: branch 1 and branch 2 respectively transform for large wavevectors to Wannier–Mott and Frenkel excitons. Branch 3 transforms to the cavity photon mode. (c) Mixing coefficients for branch 1 in the small wavevector region. (d) Mixing coefficients for branch (2) in the small wavevector region. The better mixing of branch 2 with the cavity photon means faster radiative decay in a larger phase space. Reprinted with permission from Agranovich et al. (66). Copyright Elsevier (1997).

The Frenkel component is crucial in assisting nonelastic relaxation whereas the cavity component obviously shows the large radiative width of this branch. Let us recall that the order of magnitude of the photon lifetime is for $1 - R =$

$= 10^{-3}$, $\tau_c \approx 1$ ps. But such short lifetimes are indeed effective in a very narrow region of phase space ($k < 0.05\, k_{\text{cav}}$) for those splittings typical of inorganic QWs, as seen from the mixing coefficients of hybrid molecules in Fig. 13.15. This narrow region can be reached only with a lifetime of about 100 ps due to slowed-down relaxation (74) in the flat part of the dispersion curve, poorly coupled to the cavity mode.

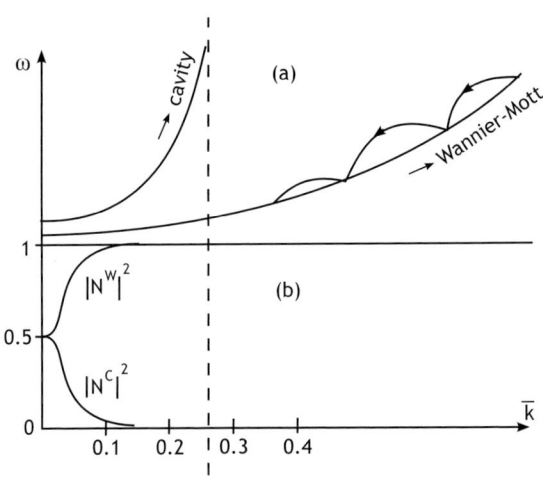

FIG. 13.15. (a) Dispersion curve for cavity polariton and energy relaxation of Wannier–Mott excitons; (b) coupling coefficient of the lower branch to cavity photon and to exciton.

To calculate the nonelastic transition rate $\mathbf{k}, \rho = 1 \to \mathbf{k}', \rho = 2$ (Fig. 13.16), we use Fermi's golden rule

$$P_{12}(\mathbf{k}) = \frac{2\pi}{\hbar} \sum_{\mathbf{q}_\parallel, q_z} \left| T^{1,2}_{\mathbf{k}, \mathbf{q}_\parallel, q_z} \right|^2$$
$$\times \delta \left[\hbar\omega_1(\mathbf{k}) - \hbar\omega_1(\mathbf{k} - \mathbf{q}_\parallel) - \hbar\Omega_{\text{ph}}(\mathbf{q}) \right], \qquad (13.115)$$

where $\hbar\Omega_{\text{ph}}(\mathbf{q})$ is the energy of an active optical phonon and its wavevector is three-dimensional: $\mathbf{q} = (\mathbf{q}_\parallel, q_z)$.

If the exciton–phonon coupling in organics is dominant we can neglect in eqn (13.113) the contribution of exciton–phonon coupling in semiconductor QWs. Then to populate the states of branch $\rho = 2$ with large radiative width (i.e. states with small $k < 0.2\, k_{\text{cav}}$, see Fig. 13.14c,d) we assume a dominant, resonant, intramolecular phonon to play the main role so that

$$P_{12}(k) = \left| N^F_1(k) \right|^2 \left| N^F_2(k) \right|^2 \frac{1}{\tau_{\text{ph}}(k)}, \qquad (13.116)$$

where

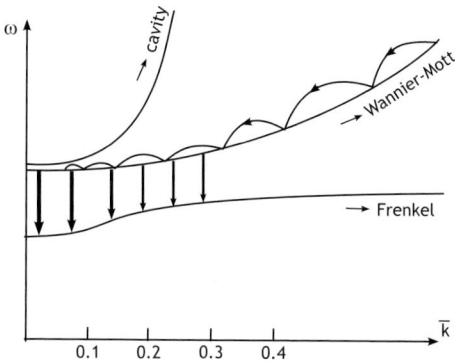

FIG. 13.16. Energy relaxation of Wannier–Mott excitons when coupled by the cavity to Frenkel excitons in an organic quantum well. The vertical arrows represent inelastic scattering of Wannier–Mott excitons to Frenkel excitons. For simplicity, possible changes in k are omitted. Reprinted with permission from Agranovich et al. (66). Copyright Elsevier (1997).

$$\frac{1}{\tau_{\rm ph}(k)} = \frac{2\pi}{\hbar} \sum_{\mathbf{q}_\|,q_z} \left|T^F_{\mathbf{k},\mathbf{q}_\|,q_z}\right|^2$$
$$\times \delta\left(\hbar\omega_1(\mathbf{k}) - \hbar\omega_1\left(\mathbf{k} - \mathbf{q}_\|\right) - \hbar\Omega_{\rm ph}(\mathbf{q})\right). \quad (13.117)$$

We can make some estimations of this time using, e.g. data from the vibrational energy relaxation rate of electronic states of organic molecules (see (75) and references therein). In the majority of cases however the question of the competition between vibrational (our case) and electronic relaxation unfortunately arises.

More appropriate for this discussion are investigations of exciton energy relaxation in J-aggregates (76), where the width of exciton band is of the order of 1000 cm^{-1}. In (76) an investigation of accumulated photon echoes as a function of wavelength in the J-band of organic TC aggregates (TC = thiacarbocyanine, a dye) is reported. Typical decay times-range from 2 ps high excitation energy to 15 ps at low energies (lower optical phonon energy).

In organics the Debye temperature is usually near 100 K and the transition may involve a few phonons. For the case of resonance with two optical phonons (77) it was reported that in GaAs QWs, the intraband transition with energy $\varepsilon_i \simeq 100$ meV $\geq 2\hbar\omega_0$, $\hbar\omega_0 \approx 40$ meV in GaAs relaxes, with short time $\tau \approx 0.4$ ps. Usually in organics the electron–phonon coupling is stronger and for the same resonance situation we can expect even shorter times. We thus assume for our estimation that $\tau_{\rm phon} \leq 1$ ps.

Let us see how the process $\rho = 1 \rightarrow \rho = 2$ competes with that of W–M exciton scattering by acoustic phonons along the $\rho = 1$ branch.

For the case of the strong-coupling regime in a semiconductor, studies show (74) that the slowest relaxation of cavity W–M polaritons (with a time of the order of 100 ps), appears in the region of $k < 0.2\ k_0$, $k_0 = \omega_{\text{W–M}}/c \approx 0.8 \times 10^5 \text{cm}^{-1}$. A novel situation is realized when using an MC with organic and inorganic QWs (Figs. 13.12 and 13.14). For k states of the W–M branch where mixing with the Frenkel exciton is strong enough, a new channel for exciton relaxation arises which corresponds to a W–M–F transformation accompanied by emission of an energy-conserving phonon. Such processes can be rather fast, as estimated in the following. For the case we considered in Fig. 13.14, $\pi/L = \omega_c(0)\sqrt{\epsilon}/c \approx 2.4 \times 10^5 \text{cm}^{-1}$ and for $k < 0.2\ k_0$, we have $\bar{k} \equiv k_x/k_{\text{cav}} \leq 0.07$. It follows from Fig. 13.14(c),(d) that for small $\bar{k} < 0.07$, we have $|N_1^F|^2 \sim \frac{1}{5}$; $|N_2^F|^2 \approx \frac{1}{2}$. Thus, for such wavevectors, using eqn (13.116), we obtain an estimated $P_{12} \approx 2 \times 10^{11} \text{s}^{-1}$ and for the corresponding time we have $\tau \leq 10$ ps. The inclusion of phonon-assisted transitions with $k' \neq k$ should only increase the interbranch transfer rate.

These estimates demonstrate the new possibilities which may appear in microcavities containing resonating organic and inorganic QWs. We can expect in such a structure a drastic shortening of the relaxation time of excitons to states with large radiative width and short fluorescence decay times. We can also expect that the combination of electric pumping of excitons in inorganic QWs with the fast relaxation and fluorescence of excitons in organic QWs opens a new scenario of excitonic processes in microcavities for which the choice of materials and technology will be decisive.

The first demonstration of strong exciton–photon coupling with photon-mediated hybridization of Frenkel and Wannier–Mott excitons in an organic–inorganic microcavity has been demonstrated by the groups of Forrest (78) and Lidzey (79). In the paper (78) the hybridization occurs between the Frenkel excitons of the small molecular weight organic tetraphenyl–porphirin and the Wannier–Mott excitons of InGaP quantum wells. In the microcavity used in (79) the optical hybridization was observed for the Frenkel exciton in an organic–dye zinc tetraphenyl–porphirin and Wannier–Mott excitations in a thin film of a self-assembled perovskite. In the future we can expect the continuation of these studies and demonstration of unusual processes of fluorescence kinetics, nonlinear optical properties and condensation of polaritons in hybrid microcavities.

13.7.7 *Conclusion*

We have discussed in this chapter the linear and nonlinear optical properties of novel hybrid excitons in organic–inorganic heterostructures. With respect to those of the usual semiconductor quantum wells, quantum wires, and quantum dots, a very strong enhancement of both the linear and nonlinear parts of the susceptibility is predicted for structure where hybrid coherent Frenkel–Wannier–Mott excitations can be created (the strong-coupling regime). These results are rationalized in terms of the large oscillator strength of Frenkel excitons and the low saturation density of Wannier–Mott excitons. If successfully synthesized, structures of the type considered here would exhibit pronounced nonlinearities of potential technological interest. We also discussed the strong-coupling regime in microcavities containing organic and inorganic quantum wells. It was shown that in such a microcavity we can expect fast polariton relaxation. A drastic reduction of the relaxation time of cavity polaritons to produce states having a large radiative width and short fluorescence decay time can be of interest for application in optoelectronics. However, to achieve the strong-coupling regime the dissipative width of both exciton states should be small. For this reason the case of strongly broadened exciton resonances in the organic material was also considered. The coherent superposition of Frenkel–Wannier–Mott exciton states in a hybrid structure does not form. However, it was shown that in this case the resonant dipole–dipole interaction of excitons in organic and inorganic nanostructures leads to very fast and efficient Förster energy transfer from excitation in an inorganic quantum well to the organic medium. The combination of electric pumping of electron–hole pairs in inorganic quantum wells with the fast relaxation and fluorescence of excitons in organics opens, due to fast energy transfer, a new scenario of excitonic processes, of interest for both basic science and device applications. For instance, new concept for light-emitting devices, discussed in this chapter, has been formulated. In this chapter we discussed also the first successful experiments which actually confirm the possibility of modern nanotechnology to create and investigate the proposed structures. One can hope that, as noted in (80), "In the future, the combination of both materials systems will allow the rules that apply to matter and light to be further stretched, potentially seeding a new paradigm in optoelectronic devices".

14

MOBILITY OF FRENKEL EXCITONS

14.1 Diffusion of Frenkel excitons

As the mobility of excitons causes transport of the electronic excitation energy through the system, this mobility has been in the focus of interest for many years. This interest concerns excitation energy transport in bulk crystals, but also in smaller molecular aggregates, such as the chlorophyll aggregates that occur in the photosynthetic systems of bacteria and higher plants (1). If a molecular system is excited in one of its exciton eigenstates and no interactions occur with other degrees of freedom, the only evolution in the system will be a periodic phase change of the exciton wavefunction as a whole, which is not associated with spatial motion. In a realistic situation one excites a wavepacket of excitons, which will propagate through the crystal with a group velocity. The wavepacket (with some mean value of the wavevector) will scatter on static disorder (e.g. on the gas–condensed medium shifts D_n), on lattice vibrations, and on other excitons. At low excitation densities, we may neglect the latter scattering events. The scattering leads to a finite exciton mean free path. If the scattering is weak, the mean free path may be very large compared to the lattice constant and a description in terms of weakly perturbed wavepackets is appropriate. This situation is referred to as the case of coherent excitons and may be described using the Boltzmann equation. In the other extreme case, the scattering is strong and basically the exciton already loses its phase information when propagating from one molecule to the next. This situation is referred to as the incoherent case (Förster energy transfer (2)) and is mostly described using a set of coupled rate equations for the excitation probabilities of the individual molecules.

Various methods have been developed that interpolate between the coherent and incoherent regimes (for reviews see, e.g. (3)–(5)). Well-known approaches use the stochastic Liouville equation, of which the Haken–Strobl–Reineker (3) model is an example, and the generalized master equation (4). A powerful technique, which in principle deals with all aspects of the problem, uses the reduced density matrix of the exciton subsystem, which is obtained by projecting out all degrees of freedom (the bath) from the total statistical operator (6). This reduced density operator obeys a closed non-Markovian (integrodifferential) equation with a memory kernel that includes the effects of (multiple) interactions between the excitons and the bath. In practice, one is often forced to truncate this kernel at the level of two interactions. In the Markov approximation, the resulting description is known as Redfield theory (7).

It should be realized that, independent of whether the system is in the coherent, incoherent, or intermediate regime, the motion at distances larger than the mean free path is always diffusive. Alternatively stated, on time-scales large compared to the typical scattering time, the exciton motion is described by a diffusion equation. In this chapter, we will restrict ourselves to this diffusive regime, which presumes that the exciton lifetime is long compared to the scattering time, so that enough scattering events may occur before the exciton decays through spontaneous emission, internal conversion, trapping by an impurity, or any other decay channel.

In the diffusive regime, the quantity of interest is the exciton concentration $c(\mathbf{r}, t)$ as a function of position $\mathbf{r}$ and time t. It obeys the diffusion equation

$$\frac{\partial c}{\partial t} = D\nabla^2 c - \frac{1}{\tau_0} c + I_0(t) \kappa e^{-\kappa z}, \tag{14.1}$$

where D is the exciton diffusion coefficient (considered isotropic here), τ_0 denotes the exciton lifetime, $I_0(t)$ is the intensity of the pumping light incident on the sample, and κ is the absorption coefficient in the crystal. Thus, the last term in eqn (14.1) gives the number of excitons generated by the external radiation per unit volume and per unit time. We assume here that the sample is a parallel-sided slab with boundaries at $z = 0$ and $z = d$ and that the incident radiation propagates from the region $z < 0$ along the normal to the plane $z = 0$. The slab thickness d should be large compared to the exciton mean free path in order for the diffusion equation to be of any use. To solve the diffusion equation we must define boundary conditions. In the steady state, the number of excitons that arrive at the boundary surface per second and per unit area is given by $D[dc/dz]_{z=0}$. In the steady state, this should equal the rate of surface annihilation, which we may write as $v_a c(z = 0)$, v_a being a characteristic surface annihilation velocity. The boundary condition at $z = 0$ is then written in the form

$$D \left[\frac{dc}{dz}\right]_{z=0} = v_a c(0), \tag{14.2}$$

which may also be formulated as $[dc/dz]_{z=0} = c(0)/\ell_0$, with $\ell_0 = v_a/D$. It should be kept in mind that this boundary condition only determines the asymptotic behavior of the exciton concentration, i.e., for z large compared to the exciton mean free path. The parameters D, τ_0, κ, and v may be taken from experiments or calculated in the framework of some microscopic theory (see, for example, (5), (8)).

A useful measure for exciton migration is the diffusion length $L = (D\tau_0)^{1/2}$. Experimental data show that for Frenkel excitons in molecular crystals at room temperature the diffusion coefficient $D \approx 10^{-3}$ cm^2/s and the lifetime of singlet excitons $\tau_0 \approx 10^{-8}$s. This gives a typical diffusion length $L \approx 10^{-6}$cm (for anthracene crystals $L \approx 5 \cdot 10^{-6}$cm).

In the remainder of this chapter, we will focus on various ways to calculate the diffusion constant from microscopic principles. We will start by considering the formal definition for quantum particles, first given by Kubo (Section 14.2). We will then consider the actual calculation in more detail for the case of coherent excitons (Section 14.3) and incoherent ones (Section 14.4). Finally, we briefly address the measurement of exciton transport properties and the effect of exciton–polariton formation on transport in molecular crystals (Section 14.5).

Note that in the right-hand side of eqn (14.1) we did not take into account the so-called bimolecular quenching of excitons (the term $-\alpha c^2$, where α is a constant) which can be important at high intensity of pumping (8). One of the channels of bimolecular quenching which can be interesting also at low intensity of pumping is the process of charge generation. In this process two excitons interact to form a pair of charge carriers and unexcited molecules. The intensity of such a mechanism of photoconductivity was calculated for the first time by Choi and Rice (9). It was shown for crystalline anthracene, in agreement with experiments, that the rate of generation of charge carriers is of the order of 10^8 cm^{-3} s^{-1} when the exciton concentration is of the order of 10^{10} cm^{-3}.

14.2 The diffusion tensor

According to Kubo (10), the general quantum-mechanical expression for the diffusion tensor is

$$D_{ij} = \frac{1}{\beta} \int_0^\infty dt e^{-\eta t} \int_0^\beta d\lambda \langle \hat{v}_i(-i\hbar\lambda) \hat{v}_j \rangle, \qquad (14.3)$$

where $\beta = 1/k_B T$, η denotes an infinitesimal positive constant, $\hat{v}_i(t)$ is the ith component of the velocity operator of the migrating particle in the Heisenberg representation, and the brackets $\langle \ldots \rangle$ denote taking the statistical equilibrium average. For classical particles ($\hbar = 0$) in an isotropic medium and using a relaxation time approximation, $v_i(t) = v_i(0)\exp(-t/\tau)$, eqn (14.3) leads to the well-known expression $D_{ij} = (1/3)\langle v^2 \tau \rangle \delta_{ij}$.

Let us consider the Kubo expression in somewhat more detail for the case of excitons in a molecular crystal. We will restrict ourselves to the presence of one exciton. The position of its "center of gravity" then reads

$$\mathbf{R} = \sum_n \mathbf{R}_n B_n^\dagger B_n, \qquad (14.4)$$

where n labels the molecules in a crystal, n is short for $(\mathbf{n}, s)$, the position of the unit cell and the index of the molecule within the unit cell. Furthermore, $\mathbf{R}_n$ denotes the position of the nth molecule. The velocity operator of the exciton may now be defined through

$$\hat{\mathbf{v}} = \frac{i}{\hbar} \left[\hat{H}, \mathbf{R} \right], \qquad (14.5)$$

where $\hat{H}$ is the system's total Hamiltonian, including the interactions with phonons and disorder. As we deal with one exciton only, we may neglect exciton–exciton interactions, and we have

WEAK EXCITON–PHONON COUPLING: COHERENT EXCITONS

$$\hat{H} = \sum_n \hat{H}_n B_n^\dagger B_n + \sum_{n,m} J_{nm} B_n^\dagger B_m + \sum_q \omega_q b_q^\dagger b_q. \quad (14.6)$$

Here, $b_q^\dagger$ and b_q are the creation and annihilation operators, respectively, for phonons in mode $q = (\mathbf{q}, r)$, where $\mathbf{q}$ denotes the vector of the phonon and r is the branch label. The energy of these phonon modes is given by ω_q. Furthermore, the single-molecule Hamiltonian as well as the intermolecular transfer interaction are still considered to be operators in phonon space.

From eqns (14.4)–(14.6), the velocity operator is found to be

$$\hat{\mathbf{v}} = \frac{i}{\hbar} \sum_{n,m} \mathbf{R}_{nm} \hat{J}_{nm} B_n^\dagger B_m, \quad (14.7)$$

where $\mathbf{R}_{nm} = \mathbf{R}_n - \mathbf{R}_m$. In eqn (14.7) we shall ignore the dependence of $\hat{J}_{nm}$ on the phonon operators. If the exciton–phonon interaction is weak, it is sufficient to include the dependence on the phonon operators only in the propagation of the coherences $B_n^\dagger B_m$ through the phonon dependencies in the operator $\hat{H}_n$. Accounting for the phonon dependence of the $\hat{J}_{nm}$ in eqn (14.7) only yields small corrections to the expression for D_{ij}. Thus, in eqn (14.7) we shall replace the operator $\hat{J}_{nm}$ by the scalar J_{nm}. Then, integration of eqn (14.3) over t and λ yields (11)

$$D_{ij} = \frac{1}{2\hbar^2} \sum_{n,m,n',m'} (\mathbf{R}_{nm})_i J_{nm} (\mathbf{R}_{n'm'})_j J_{n'm'}$$
$$\times \int_{-\infty}^{\infty} dt \langle B_n^\dagger(t) B_m(t) B_{n'}^\dagger(0) B_{m'}(0) \rangle. \quad (14.8)$$

As can be seen from eqn (14.8), the calculation of the tensor D_{ij} reduces to the calculation of two-particle correlation functions. The lack of sufficiently detailed data on the exciton band structure and the exciton–phonon coupling constants considerably complicates the accurate calculation of the two-particle correlation functions and the exciton diffusion coefficients. However, the temperature dependence of this coefficient differs significantly for coherent and incoherent excitons (see below). Therefore, studying the temperature dependence of diffusion has always been an important tool to analyze the character of the energy transfer in molecular crystals. In the remainder of this chapter, we will focus on the main characteristics of the diffusion constant and its temperature dependence

14.3 Weak exciton–phonon coupling: coherent excitons

14.3.1 General expressions

If the exciton–phonon coupling is sufficiently weak, the solution of the equation for the correlation function $\langle B_n^\dagger(t) B_m(t) B_{n'}^\dagger(0) B_{m'}(0) \rangle$ is equivalent to the solution of the Boltzmann equation (11). In this coherent limit, the exciton states of the ideal lattice serve as good zeroth-order states. In other words, the wavevector still proves to be a good quantum number and the notion of excitons propagating in between scattering events as wavepackets with a well-defined group velocity is useful. The kinetics of such coherent excitons under the influence of weak exciton–phonon coupling is described by the Boltzmann equation. One may show that this equation reduces to the diffusion equation (14.1), if the exciton concentration changes little over lengths of the order of magnitude of the exciton mean free path.

In this case the exciton–phonon interaction can be taken into account perturbatively. The influence of this interaction on the shape and position of the exciton band(s) is insignificant and can usually be ignored. Then, the only remaining effect of the interaction is the scattering of the excitons. This scattering results in a change of the wavevector and the energy of the excitons. Therefore, if $\delta E_{\mathbf{k}}$ is the width of the energy level of the exciton with the wavevector $\mathbf{k}$ determined by the exciton–phonon interaction, the related uncertainty $\delta \mathbf{k}$ of the wavevector is given by

$$\delta E_{\mathbf{k}} = \hbar \mathbf{v}(\mathbf{k}) \cdot \delta \mathbf{k} \tag{14.9}$$

where $\mathbf{v} = (1/\hbar)(dE/d\mathbf{k})$ is the exciton group velocity. The uncertainty $\delta \mathbf{k}$ indicates that the exciton state in the crystal is realized as a wavepacket rather than a plane wave. The dimensions of the wavepacket, δx, δy, and δz, can be estimated from the uncertainty relations $\delta x \, \delta k_x \simeq 1$, etc.

The motion of the exciton wavepacket causes the transport of energy. In order to find the appropriate energy diffusion coefficient we must estimate the mean free path and the mean free time of the wavepackets. This situation is quite similar to that of phonon heat conductivity (see, for example, (12)).

In analogy to the mobility of electrons and holes in crystals, the diffusion coefficient for coherent excitons is determined by the relaxation time τ. According to Frölich (13), we have

$$\frac{1}{\tau} = -\sum_{\mathbf{q},r} \frac{\Delta k_z(\mathbf{q}r)}{k_z} \left[W_a^{\mathbf{k}}(\mathbf{q}r) + W_e^{\mathbf{k}}(\mathbf{q}r) \right], \tag{14.10}$$

where $\mathbf{k}$ is the exciton wavevector before the collision, $\Delta k_z(\mathbf{q}r)$ is the change of the exciton wavevector in the z-direction due to the collision with the phonon $\mathbf{q}r$, and $W_a^{\mathbf{k}}(\mathbf{q}r)$ and $W_e^{\mathbf{k}}(\mathbf{q}r)$ are the probability per unit time of absorption and emission, respectively, of the phonon $\mathbf{q}r$ by the exciton system. After absorption or emission, the new exciton wavevectors, accurate to an integral reciprocal lattice vector, are $\mathbf{k}' = \mathbf{k} \pm \mathbf{q}$. According to eqn (14.10), we can write

$$\frac{1}{\tau} = \frac{1}{\tau^{\mathrm{ac}}} + \frac{1}{\tau^{\mathrm{op}}}, \tag{14.11}$$

where $1/\tau^{\mathrm{ac}}$ and $1/\tau^{\mathrm{op}}$ are the scattering rates of the exciton on acoustic and optical phonons, respectively. As is the case for electrons in semiconductors, τ is approximately equal to the mean free time of the excitons with respect to collisions with phonons. Typically, a few collisions are sufficient to reach a thermodynamic equilibrium between phonons and band excitons. Thus, if the exciton lifetime τ_0 is considerably larger than the relaxation time τ, as is typically the case at elevated temperatures, we can assume that the excitons are in thermodynamic equilibrium with the lattice prior to their decay. The exciton diffusion coefficient is then related to the relaxation time by

$$D = \frac{1}{3}\langle \tau v^2 \rangle \approx \langle \tau \rangle \langle v^2 \rangle, \qquad (14.12)$$

where $\langle v^2 \rangle$ is the (equilibrium) mean squared group velocity of the excitons, and $\langle \tau \rangle$ is the mean relaxation time. Strictly speaking, explicit expressions for wavepackets should be used when calculating the quantities in eqn (14.10). However, usually the absorption and emission rates vary only slightly with the exciton wavevector over the $\mathbf{k}$ interval spanned by the wavepacket. Therefore, the transition probabilities can be calculated by using exciton wavefunctions in the form of plane waves both before and after the scattering process. We will not discuss the details of such calculations, but rather address some typical and frequently used results.

14.3.2 Isotropic exciton effective mass and scattering by acoustic phonons

Using the Fermi's golden rule, the probabilities of absorption and emission of a phonon by the exciton system read

$$W_e^{\mathbf{k}}(\mathbf{q}r) = \frac{2\pi}{\hbar} |F(\mathbf{k}-\mathbf{q};\mathbf{k};\mathbf{q}r)|^2$$
$$\times (n_{\mathbf{q}r}+1)\, \delta\left[E(\mathbf{k}) - E(\mathbf{k}-\mathbf{q}) - \omega_r(\mathbf{q})\right] \qquad (14.13)$$

$$W_a^{\mathbf{k}}(\mathbf{q}r) = \frac{2\pi}{\hbar} |F(\mathbf{k}+\mathbf{q};\mathbf{k};\mathbf{q}r)|^2\, n_{\mathbf{q}r}\, \delta\left[E(\mathbf{k}) - E(\mathbf{k}+\mathbf{q}) + \omega_r(\mathbf{q})\right], \qquad (14.14)$$

where $n_{\mathbf{q}r} = [\exp\frac{\hbar\omega_r(\mathbf{q})}{k_B T} - 1]^{-1}$ is the thermal occupation of the phonon mode $\mathbf{q}r$ and $F(\mathbf{k}\pm\mathbf{q};\mathbf{k};\mathbf{q}r)$ is the exciton–phonon coupling constant.

At sufficiently low temperatures, when the thermal energy $k_B T$ is much smaller than the exciton bandwidth, most excitons at thermodynamic equilibrium are concentrated in the vicinity of the exciton band minimum in wavevector space. If this minimum corresponds to $\mathbf{k} = \mathbf{0}$ and we assume for simplicity that the excitons have an isotropic effective mass m, we have

$$E_{\mathbf{k}} = E_0 + \frac{\hbar^2 \mathbf{k}^2}{2m}. \qquad (14.15)$$

Moreover, under these conditions the relation $|\mathbf{k}|a \ll 1$ (a is the lattice constant) is satisfied for the overhelming majority of excitons.

The conservation of energy for the absorption and emission processes can now be written as
$$\frac{\hbar^2|\mathbf{k}|^2}{2m} \pm \hbar\omega_r(\mathbf{q}) = \frac{\hbar^2(\mathbf{k} \pm \mathbf{q})^2}{2m}. \tag{14.16}$$
Since we are dealing with relatively low temperatures, let us focus on the case of acoustic phonons. We then have $\hbar\omega_r(\mathbf{q}) = v_0|\mathbf{q}|$ ($r = 1, 2, 3$), where v_0 is the sound velocity. For the sake of simplicity we ignore the dependence of v_0 on the polarization r and the direction $\mathbf{q}$. Using eqn (14.16), we obtain
$$q = \mp 2k\cos\theta \pm 2mv_0/\hbar, \tag{14.17}$$
where θ is the angle between the vectors $\mathbf{k}$ and $\mathbf{q}$. From the Boltzmann statistics $\hbar\langle|\mathbf{k}|^2\rangle/2m = (3/2)k_B T$, we obtain for the typical value of $|\mathbf{k}|$ at a given temperature T:
$$\langle|\mathbf{k}|^2\rangle^{1/2} = \frac{1}{\hbar}[3mk_B T]^{1/2}. \tag{14.18}$$
Hence, we can ignore the second term on the right-hand side in eqn (14.17) relative to the first if
$$k_B T \gg \frac{1}{3}mv_0^2. \tag{14.19}$$
This criterion defines the temperature region where the scattering of excitons by phonons is almost elastic. We may make this estimate more quantitative by using $v_0 \approx 10^5$ cm/s as the typical velocity of sound for solids, leading to
$$T \gg \frac{0.025\,m}{m_0}\mathrm{K}, \tag{14.20}$$
m_0 being the electron mass in vacuum.

Below we shall assume that the criterion in eqn (14.20) is satisfied and, therefore, we shall neglect the phonon energy in the argument of the delta functions in eqn (14.13) and (14.14), and, likewise, we shall neglect the second term on the right-hand side of eqn (14.17).

Now we observe from eqn (14.17) that in our model for a given value of $|\mathbf{k}|$ the exciton can interact to the first approximation with phonons with wavevectors in the range $0 \leq |\mathbf{q}| \leq 2|\mathbf{k}|$, implying that, like $|\mathbf{k}|$, the phonon wavvector is small: $|\mathbf{q}|a \ll 1$. Using the smallness of $|\mathbf{k}|$ and $|\mathbf{q}|$, we can expand $|F(\mathbf{k}-\mathbf{q};\mathbf{k};\mathbf{q}r)|^2$ in powers of these wavevectors. Keeping only the lowest-order nonzero contribution, $|F(\mathbf{k}-\mathbf{q};\mathbf{k};\mathbf{q}r)|^2$ is then a linear function of $|\mathbf{q}|$ for acoustic phonons (14). Moreover, if we ignore the generally weak dependence of this quantity on the directions of the vectors $\mathbf{q}$ and $\mathbf{k}$ and neglect the dependence of the phonon frequency on the direction of $\mathbf{q}$ and the phonon polarization, we obtain
$$|F^{\mathrm{ac}}(\mathbf{k}+\mathbf{q};\mathbf{k};\mathbf{q}r)|^2 \approx |F_0^{\mathrm{ac}}|^2 a|\mathbf{q}|, \tag{14.21}$$
where F_0^{ac} is a constant. Using eqn (14.21), we find that (in a three-dimensional medium) the relaxation time for the exciton with wavevector $\mathbf{k}$ due to scattering by acoustic phonons is given by

$$\frac{1}{\tau^{\rm ac}} = \frac{3|F_0^{\rm ac}|^2 a^4 m}{4\pi\hbar^3|\mathbf{k}|^3}\left(\frac{k_BT}{\hbar v_0}\right)^5 \int_0^\xi x^4 \frac{e^{-x}-e^x}{(e^x-1)(e^{-x}-1)}\mathrm{d}x, \tag{14.22}$$

with $\xi = 2|\mathbf{k}|\hbar v_0/k_BT$.

14.3.3 Temperature dependence of the diffusion constant

If $\xi < 1$, the integrand in eqn (14.22) can be replaced with its value for small x. Using eqn (14.18) for the typical values for $|\mathbf{k}|$, this condition translates into

$$T > T_0 \equiv 12 m v_0^2/k_B. \tag{14.23}$$

If we now replace $|\mathbf{k}|$ with its mean value $\langle|\mathbf{k}|^2\rangle^{1/2}$ and perform the integration in eqn (14.22) using the small-x expansion, we obtain

$$\frac{1}{\tau^{\rm ac}} = \frac{|F_0^{\rm ac}|^2 a^4 m}{\pi\hbar^3}\left(\frac{k_BT}{\hbar v_0}\right)\langle|\mathbf{k}|^2\rangle^{1/2}. \tag{14.24}$$

Thus, for $T > T_0$, we find $\tau^{\rm ac} \sim 1/T^{3/2}$ which using eqn (14.12) for the diffusion constant leads to

$$D \sim T^{1/2}. \tag{14.25}$$

This relationship for Frenkel excitons was derived in (14); it can be seen from its derivation that it is independent of the model and, therefore, is valid also for ground state large-radius excitons as well as for electrons and holes in semiconductors.

In the region of very low temperatures when $\xi \gg 1$, that is, when the condition opposite to eqn (14.23) is satisfied but the inequality (14.19) still holds, the main contribution to the integral in eqn (14.22) comes from large x. Replacing the integrand with its asymptotic large-x value and performing the integration then yields

$$\frac{1}{\tau^{\rm ac}} = \frac{24|F_0^{\rm ac}|^2 a^4 m}{5\pi\hbar^3}\left(\frac{k_BT}{\hbar v_0}\right)\langle|\mathbf{k}|^2\rangle^{1/2}, \tag{14.26}$$

which implies (eqn 14.12) that the diffusion coefficient due to scattering by acoustic phonons ceases to be temperature dependent in the region of low temperatures. Thus, with decreasing temperature the relationship $D \sim 1/\sqrt{T}$ for the diffusion coefficient reduces to $D = $ constant. The exciton scattering by optical phonons becomes important only at sufficiently high temperatures. A discussion of the influence of such processes on the diffusion constant can be found in (8) and (5).

To end this section, we make a few remarks. First it should be noted that the applicability of the above $D(T)$ relationships is limited by the condition of applicability of the Boltzmann equation. This condition reads $\ell \gg \lambda$ where $\ell = \langle v\rangle\tau$ and λ is the thermal de Broglie wavelength of the excitons. Therefore, we should expect that this condition may be satisfied only for sufficiently low temperatures. Of course, the size of the corresponding temperature range depends on the exciton–phonon interaction constants and can be found only from experimental data.

The second remark concerns the role of crystal anisotropy. The majority of the well-studied molecular crystals, such as anthracene, naphthalene, pyrene, etc., are not cubic, but very anisotropic. It is therefore natural to consider to what extent the above qualitative results for the temperature dependence of the exciton diffusion coefficient in cubic crystals are valid for anisotropic crystals. It is clear, of course, that in anisotropic crystals the exciton diffusion coefficient can exhibit anisotropy owing, for instance, to anisotropy of the exciton's effective mass. However, as long as the effective masses in various directions are of the same order of magnitude (which seems to be just the case for naphthalene crystals), the temperature dependence of the diffusion constant maintains the same $D \sim 1/\sqrt{T}$ character (with the exclusion of the region of low temperatures). More interesting are the class of anisotropic crystals in which the lowest exciton band corresponds to high oscillator strengths. Then the exciton energy $E_\mathbf{k}$ is known to be a nonanalytic function of $\mathbf{k}$ for small $|\mathbf{k}|$ values and this has to be taken into account when calculating the diffusion tensor. Katalnikov (15) calculated the exciton diffusion tensor in uniaxial crystals taking into account the nonanalytic energy term and found, as could be expected, that in those crystals different components of the diffusion tensor have different temperature dependence. Strong anisotropy of D_{ij} in molecular crystals can lead to one- or two-dimensional exciton motion (16).

14.4 Strong exciton–phonon coupling: incoherent excitons

If the exciton–phonon coupling is strong, it results in the localization of the exciton at a lattice site, so that the exciton behaves as a classical particle that can hop from cell to cell. The hopping particle is in fact a self-trapped exciton, which, due to interactions with the phonon bath, loses its phase memory on a time-scale short compared to the time it takes to hop between unit cells. The resulting hopping process is known as incoherent energy transfer and, if the relevant excitation transfer interaction J_{nm} is of dipole–dipole type, it is often referred to as Förster energy transfer. To lowest order in the excitation transfer interaction, the exciton motion is then a series of uncorrelated hops (forming a Markov process), described by a random walk over the lattice sites. If $P(\mathbf{m}, t)$ is the probability that an exciton is at the lattice site $\mathbf{m}$ at the moment t, then the random-walk equation for $P(\mathbf{m}, t)$ has the form

$$\frac{\mathrm{d}P(\mathbf{m}, t)}{\mathrm{d}t} = \sum_\mathbf{n} [W(\mathbf{n}, \mathbf{m})P(\mathbf{n}) - W(\mathbf{m}, \mathbf{n})P(\mathbf{m})], \qquad (14.27)$$

where $W(\mathbf{n}, \mathbf{m})$ is the probability of hopping from site $\mathbf{n}$ to site $\mathbf{m}$ per unit time. In the next order of approximation the hopping process is not Markovian. We then have the following integrodifferential equation for $P(\mathbf{m}, t)$

$$\frac{dP(\mathbf{m},t)}{dt} = \int_0^t \Big[\tilde{W}(\mathbf{n},\mathbf{m},\tau)P(\mathbf{n},t-\tau)$$
$$-\tilde{W}(\mathbf{m},\mathbf{n},\tau)P(\mathbf{m},t-\tau)\Big]d\tau, \tag{14.28}$$

i.e. the hopping process acquires a "memory". For molecular crystals, this problem has been discussed by Kenkre and Knox (4), (17). The excellent review by Silbey (5) deals with the problem in a more general way, including exciton scattering by impurities, dispersive transport, and the Haken–Strobl–Reineker model of exciton–phonon scattering (3), (18). We also point out the recent book on transport by May and Kühn, where the problem of memory kernels is addressed (6).

If we assume that the variation of the function $P(\mathbf{m},t)$ over distances of the order of the lattice constant is small, then eqn (14.27) reduces to the diffusion equation. Indeed, assuming that the vector $\mathbf{m}$ varies continuously, we obtain

$$P(\mathbf{n}) = P(\mathbf{m}) + \sum_i (\mathbf{n}-\mathbf{m})_i \frac{\partial P(\mathbf{m})}{\partial m_i}$$
$$+ \frac{1}{2}\sum_{i,j}(\mathbf{n}-\mathbf{m})_i(\mathbf{n}-\mathbf{m})_j \frac{\partial^2 P(\mathbf{m})}{\partial m_i \partial m_j}. \tag{14.29}$$

If $W(\mathbf{n},\mathbf{m}) = W(\mathbf{m},\mathbf{n})$ (which holds if there is no static disorder in the crystal), substitution of the above expression into eqn (14.27) yields

$$\frac{\partial P(\mathbf{r},t)}{\partial t} = D_{ij}\frac{\partial^2 P(\mathbf{r},t)}{\partial x_i \partial x_j}, \tag{14.30}$$

where the diffusion coefficient tensor is

$$D_{ij} = \frac{1}{2}\sum_{\mathbf{m}}(\mathbf{n}-\mathbf{m})_i(\mathbf{n}-\mathbf{m})_j W(\mathbf{n},\mathbf{m}). \tag{14.31}$$

For crystals with several molecules (labeled by s) per unit cell, which have symmetry operations exchanging molecules with different s values, as holds for anthracene crystals, a similar procedure yields

$$D_{ij} = \frac{1}{2}\sum_{\mathbf{m},s'}(\mathbf{r}_{\mathbf{n}s}-\mathbf{r}_{\mathbf{m}s'})_i(\mathbf{r}_{\mathbf{n}s}-\mathbf{r}_{\mathbf{m}s'})_j W(\mathbf{n}s,\mathbf{m}s'). \tag{14.32}$$

Thus, under the given conditions, the calculation of the diffusion tensor reduces to the determination of the hopping rates $W(\mathbf{n}s,\mathbf{m}s')$.

For singlet excitons the probability $W(\mathbf{n}s,\mathbf{m}s')$ can be estimated using the results of strong exciton–phonon coupling theory, as has been done by Trlifaj (19). The intermolecular dipole–dipole interaction leads to $W(\mathbf{n}s,\mathbf{m}s') \sim 1/|\mathbf{r}_{\mathbf{n}s}-\mathbf{r}_{\mathbf{m}s'}|^6$, as is characteristic for Förster energy transfer (2), and in the summation equation (14.31) it suffices to take into account only nearest neighbors. The temperature dependence of this probability may be presented by the relation

$$D_{ij}(T) \approx D_{ij}^0 \exp\left(-U_a/k_B\overline{T}\right), \qquad (14.33)$$

where U_a is the activation energy for hopping, $\overline{T}$ is a constant for $T \ll T_D$ (T_D is the Debye temperature), and $\overline{T} = T$ for $T \gg T_D$. It follows from these qualitative considerations that in the strong-coupling regime the exciton diffusion constant, in contrast to the case of weak exciton–phonon coupling, increases with growing temperature.

In some molecular crystals a crossover from coherent excitons (exciton mean free path $\ell \gg \lambda$) to incoherent ones ($\ell \approx \lambda$, Ioffe–Regel criterion) takes place with increasing temperature. We then expect that upon increasing the temperature from very low values, at some threshold temperature the decreasing behavior of the diffusion constant for coherent excitons goes over into an increasing behavior.

A similar crossover phenomenon may be observed in heavily doped isotopically mixed crystals at low temperatures. In such crystals the impurity molecules are responsible for the scattering of excitons and at low temperature this scattering is almost elastic. When increasing the impurity concentration, a crossover occurs between so-called weak (at $\ell \gg \lambda$) and strong ($\ell \approx \lambda$) Anderson localization. This crossover is analogous to the change of the electron mobility in metals upon increasing the impurity concentration and should be expected to have a strong influence on the exciton transport. A different situation arises when the electronic excitation energy of the isotopic impurity is lower than the energy of the exciton in the crystalline host. The isotopic impurities enter the lattice substitutionally and they are randomly distributed in the matrix. The electronic excitation energy transfer from one impurity to another when decreasing the impurity concentration may be used to investigate the transition from impurity band to impurity hopping transfer. Such a transition is similar to the Anderson conductor–insulator transition in semiconductors.

An extensive discussion of experiments on exciton transport in isotopically disordered crystals and numerical simulations of this phenomenon in the framework of a percolation model may be found in the review paper by Kopelmann (20). A more recent review of this field, including the discussion of the Anderson model, may be found in the book by Pope and Swenberg (21).

14.5 Transport measurements and diffusion of polaritons

Many papers have been devoted to the experimental determination of the exciton diffusion constant D. In most of the studies, D was determined by observing how the diffusion of excitons results in their capture by impurities (sensitized fluorescence) or in bimolecular quenching of excitons (reviews of these experiments may be found in (8),(21) . The interpretation of such experiments requires that not only the diffusion of the excitons to the acceptor is taken into account, but also the character of the exciton interaction with the acceptor (i.e. with the impurity molecule or with another exciton). An alternative experimental technique that does not suffer from these problems is the picosecond transient grating (TG) method. This third-order nonlinear optical technique has been used abundantly for the study of various kinetic parameters of condensed media (liquids, semiconductors, etc.). Fayer and collaborators were the first to propose the use of TG experiments for the study of exciton transport in molecular crystals (22). They applied the method to anthracene thin films (23), (24).

In order to determine the exciton diffusion constant, one studies the decay kinetics of excitonic gratings, i.e. a spatially periodic variation of the exciton density, formed in a molecular crystal as a result of the interference of two coherent picosecond laser pulses. The periodic spatial distribution of the excitons, as well as its evolution, can be investigated by observing the diffraction of a short probe pulse sent into the crystal with some delay time t after creating the grating. As a result of the finite exciton lifetime and exciton diffusion, the grating amplitude decreases in time, so that the intensity $S(t)$ of the diffracted signal decreases with growing t. Thus, measuring $S(t)$ allows one to obtain information on the diffusion coefficient D and the exciton lifetime τ_0. In fact, it is easy to show that

$$S(t) = S(0)e^{-Kt}, \qquad (14.34)$$

with

$$K = 2\left(\frac{1}{\tau_0} + D\Delta^2\right). \qquad (14.35)$$

Here, $\Delta = 2\pi/L$, where L is the fringe spacing of the grating, which is given by $L = \lambda_e/[2n\sin(\theta/2)]$ (n is the crystal's refractive index, λ_e the wavelength of the two excitation pulses, and θ the angle between them). By measuring the diffracted-signal decay for various values of θ and plotting the observed value of K versus θ^2, the diffusion constant D can be obtained from the slope, while the $\Delta = 0$ intercept is $2/\tau_0$. As the lifetime τ_0 can also be found from other experiments (for example, from photoluminescence measurements) this provides a rigorous test of the assumption of diffusive propagation of the excitons as well.

In the above-mentioned TG experiments by the Fayer group (23), (24) it was found that the diffusion constant D in anthracene films at low temperature ($T = 1.8, 10, 20$ K) can reach values of the order of 1–10 cm^2/s (Fig. 14.1). Such very large values of the diffusion constant contradicted too strongly the typical room temperature value of $D \sim 10^{-3}$ cm^2/s and for this reason the experiments by Fayer attracted the attention of many investigators who tried to explain the large value of diffusion constant in terms of diffusion of excitons. However, in the interpretation of the experiments with anthracene films it is necessary to take into account that the lowest electronic transition in an anthracene crystal has a rather large oscillator strength, leading to a strong exciton–polariton formation. As a consequence, polaritons rather than excitons are the lowest-energy elementary excitations at low temperatures and the theoretical analysis of the decay time of the excitonic gratings should be associated with the diffusion of polaritons rather than excitons (25), (26).

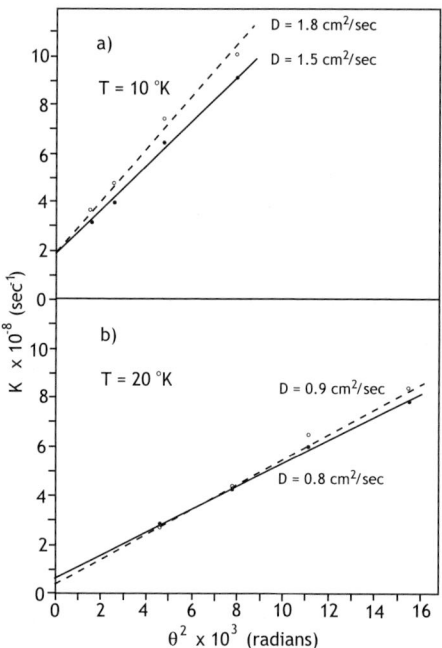

FIG. 14.1. The decay rate of the transient grating signal versus θ^2 (θ is the angle between the pump pulses) for anthracene crystals at 10 and 20 K (23). The magnitude of the slope is proportional to the diffusion constant of the excitations in the crystal. With increasing temperature, the diffusion constant decreases. The average diffusion constant obtained from these data is about 10 times larger than the value expected for incoherent exciton motion (25).

Indeed, at low temperature the polaritons are concentrated in the "bottle neck" region. Their diffusion constant along the surface of the film can be estimated as $D \approx v_p \ell_p$, where v_p is the group velocity of the polaritons in this region and ℓ_p is their mean free path. Since $\ell_p = c/(\omega\kappa)$, where κ is the imaginary part of the refractive index $((2\omega/c)\kappa \approx 10^4 - 10^5 \text{ cm}^{-1})$, and $v_p \approx 10^5$ cm/s, one obtains for the diffusion constant $D \geq 1$ cm^2/s, which agrees within an order of magnitude with the value obtained in the TG experiments.

The discussion of the influence of bimolecular quenching and reabsorption of exciton fluorescence on the decay of the exciton gratings in organic crystals may be found in (26). The microscopic formulation of TG and four-wave-mixing experiments in molecular crystals in terms of polaritons may be found in (27).

15

STATISTICS AND COLLECTIVE PROPERTIES OF FRENKEL EXCITONS

15.1 Approximate second quantization and kinematic interaction

In Ch. 3 we have applied the second quantization for investigation of exciton states. The first step was to express the crystal Hamiltonian in terms of creation and annihilation Pauli operators $P_s^{\dagger f}$ and P_s^f of single-molecule excited states, where the index s indicates the lattice points where the molecule is placed, and f labels the molecular excited states. When taking into account only one fth excited molecular state, the operators $P_s^{\dagger f}$ and P_s^f satisfy the following commutation rules (see eqn 3.28)

$$P_s^f P_s^{\dagger f} - P_s^{\dagger f} P_s^f = 1 - 2 P_s^{\dagger f} P_s^f, \qquad (15.1)$$

$$\left(P_s^f\right)^2 = \left(P_s^{\dagger f}\right)^2 = 0, \qquad (15.2)$$

$$\left.\begin{array}{l} P_s^f P_{s'}^f - P_{s'}^f P_s^f = 0, \\ P_s^f P_{s'}^{\dagger f} - P_{s'}^{\dagger f} P_s^f = 0. \end{array}\right\} s \neq s'. \qquad (15.3)$$

Equations (15.1)–(15.3) are combinations of commutation rules of Fermi type (when $s = s'$) and of Bose type (when $s \neq s'$). The appearance of commutation relations for Fermi operators when $s = s'$ means that the number of molecular excitations, i.e. the eigenvalue of the operator $P_s^{\dagger f} P_s^f$, takes either the value 0 (the molecule in the ground state) or 1 (the molecule is excited). In turn, the Bose commutation rules for $s \neq s'$ are related to the fact that the operators with different indices s are acting on different arguments of the crystal wavefunction.

As shown in Ch. 3, the crystal Hamiltonian, expressed in terms of the operators $P_s^{\dagger f}$ and P_s^f (in the following the index f will be omitted) when the lattice vibrations are not taken into account, has the following form (see also eqn 3.39)

$$\hat{H} = \hat{H}_0 + \hat{H}_{\text{int}}, \qquad (15.4)$$

where the operator

$$\hat{H}_0 = \sum_s \Delta P_s^\dagger P_s + \frac{1}{2} \sum_{s \neq s'} M_{ss'}^{\text{I}} P_s^\dagger P_{s'}$$

$$+ \frac{1}{2} \sum_{s \neq s'} M_{ss'}^{\text{II}} \left(P_s^\dagger P_{s'}^\dagger + P_s P_{s'}\right) \qquad (15.5)$$

is quadratic with respect to the operators $P_s^{\dagger f}$ and P_s^f and the operator $\hat{H}_{\text{int}}$ is a sum of terms of third and fourth order. For weakly excited crystal states the following inequality holds

$$c = \langle P_s^\dagger P_s \rangle \ll 1, \tag{15.6}$$

thus, for this reason, usually the operator $P_s^\dagger P_s$ on the r.h.s. of (15.1) can be neglected. The operators $P_s^{\dagger f}$ and P_s^f are in this case Bose operators ($P_s^\dagger = B_s^\dagger, P_s = B_s$). The above property is the basis for the method of approximate second quantization, developed by Bloch (1), (2) and in the following by Bogoljubov and Tyablikov (3), (4). Making use of this approach, the operator (15.5) has in Ch. 3 been put into the form

$$\hat{H}_0 = \sum_{\mu \mathbf{k}} E_\mu(\mathbf{k}) B_\mu^\dagger(\mathbf{k}) B_\mu(\mathbf{k}), \tag{15.7}$$

$E_\mu(\mathbf{k})$ being the energy of an elementary excitation–exciton $\mu, \mathbf{k}$, and $B_\mu^\dagger(\mathbf{k})$, $B_\mu(\mathbf{k})$ being the Bose creation- and annihilation operators of these elementary excitations, respectively. The operator (15.7) corresponds to a system of non-interacting excitons. Hence for consideration of processes of exciton–exciton interactions it is necessary to include in the Hamiltonian terms of third, fourth and higher orders with respect to the Bose operators. Some of these terms are contained in the operator $\hat{H}_{\text{int}}$ in (15.4). We shall consider below precisely the procedure for correctly separating these terms, since replacement of the Pauli operator by Bose operators ((5) and Ch. 3) gives rise, as it were, to an additional interaction between elementary excitations, which we shall designate, just as in magnetism theory, as kinematic. The kinematic exciton–exciton interaction determine the collective properties of ideal gas of paulions at low temperatures. We will find these terms using an exact transformation from paulions to bosons (Section 3.11).

When speaking of kinematic interaction, it should be noted that the problem of its separation in connection with the transition from Pauli operators to Bose operators is far from new. This problem arises, in particular, for the Heisenberg Hamiltonian, which corresponds, for example, to an isotropic ferromagnet with spin $\sigma = 1/2$ when spin waves whose creation and annihilation operators obey Bose commutation relations are introduced. This problem was dealt with by many people, including Dyson (6), who obtained the low-temperature expansion for the magnetization. However, even before Dyson's paper, Van Kranendonk (7) proposed to take into account of the kinetic interaction by starting from a picture where one spin wave produces an obstacle for the passage of another spin wave, since two flipped spins cannot be located at the same site (for Frenkel excitons this means that two excitons cannot be localized simultaneously on one and the same molecule).

In mathematical language, such an approach means adding to the initial Hamiltonian, in which the Pauli operators are replaced by Bose operators, a term that corresponds to the limiting strong repulsion of two bosons in one site.

The picture postulated by Van Kranendonk (7) leads to a molecular cross-section, corresponding to the "hard spheres" approximation.

The application of this approach to spin waves was called by Dyson "naive" and criticized as incorrect and leading to results different from those obtained by him (see (6), the end of § 3). We shall show in what follows, however, on the basis of an exact representation of Pauli operators in terms of Bose operators, that the picture described above does take place for Frenkel excitons. This takes place only because the excitation energy Δ for Frenkel excitons is large compared with the width of the exciton band. As for the spin waves, where the inequality indicated above is not satisfied, the cross-section for the scattering of long-wavelength spin waves by each other can indeed, in agreement with Dyson, differ substantially from a value that follows from the "hard sphere" approximation (7).

15.2 Collective properties of an ideal gas of paulions

In this section, using the results of Section 3.11, we consider collective properties of an ideal gas of paulions, i.e. of a system described by the Hamiltonian (15.5) corresponding to the case of a vanishing operator of the dynamic interaction between elementary excitations, i.e. with $\hat{H}_{\text{int}} = 0$.[59]

Substituting (3.192) and (3.193) in (15.5) and going over from Pauli operators to Bose operators we obtain, besides the Hamiltonian in the zeroth approximation (15.7), two types of contributions to the operator of the kinematic exciton–exciton interaction. The terms of the first type are those resulting from the fact, as seen from (3.193), that the operator $P_s^\dagger P_s \neq \hat{N}_s$. These terms are proportional to the excitation energy Δ and have the form

$$\hat{H}' = \Delta \sum_{\nu=1}^{\infty} \frac{(-2)^\nu}{(1+\nu)!} \sum_s B_s^{\dagger(\nu+1)} B_s^{\nu+1}. \tag{15.8}$$

Let us consider in detail the term with $\nu = 1$ in (15.8), corresponding to scattering of two excitons by each other. In accordance with (15.8) this term has the form

$$\hat{H}'(\nu = 1) = -\Delta \sum_{ss'} \delta_{ss'} B_s^\dagger B_{s'}^\dagger B_{s'} B_s, \tag{15.9}$$

i.e. corresponds to scattering of excitons by each other with a delta-like interaction potential

$$V_{ss'} = -2\Delta \delta_{ss'}.$$

If we go over to a coordinate frame connected with the mass center of the system consisting of the two excitons, then, in accord with (15.8) the problem of determining the cross-section for the scattering of the excitons by each other

---

[59] In crystals such as benzene, naphthalene, etc. having an inversion center, the operator $\hat{H}_{\text{int}}$ vanishes identically, if only dipole–dipole interactions between molecules is taken into account. For further details on the operator $\hat{H}_{\text{int}}$ see Section 15.3.

reduces the exciton–exciton scattering problem to the problem of the scattering of a quasiparticle by the potential

$$V_{ss'} = -2\Delta \delta_{s0}\delta_{s'0}. \tag{15.10}$$

A potential of this type cannot be considered as a weak perturbation since it may lead, in particular, to the appearance of local states. The appearance of local states means that the two excitons can form a bound state, i.e. can form a biexciton. For this reason we discuss the problem of local states in more detail.[60]

First we note that, inasmuch as the quantity 2Δ in molecular crystals is larger by more than one order of magnitude than the width of the exciton band, a local level at large depth, approximately equal to 2Δ, always appears under the influence of the potential (15.10).

No shallow local levels are produced under the influence of the perturbation (15.10) below the lowest exciton band (see Section 5.3), which is the very band that is essential for the study of Bose–Einstein condensation of excitons.

However, the process of free excitons binding in a deep local state need not be taken into consideration if no account is taken of the processes of nonradiative decay of single excitons, whereby the energy Δ goes over into the phonon energy. In crystals, where the quantum yield of exciton luminescence is close to unity (for instance, in anthracene crystals), the nonradiative decay of excitons cannot be realized within the exciton lifetime (otherwise we cannot regard the number of excitons in the crystal in consideration of collective processes as specified).

Clearly, in such crystals a binding of two excitons at a deep local state is even less probable, since it requires a transformation in phonons of double energy. Therefore, in spite of the fact that states in which two excitons are situated in the same site are formally possible, their formation out of individual excitons is in practice forbidden at time $t < \tau$.[61]

Hence, within the considered interval $t < \tau$, the potential (15.10) leads only to the scattering of excitons by excitons. The corresponding scattering length cannot be calculated within the Born approximation.

Accurate calculations of the cross-section of exciton scattering by an impurity molecule, which have been done by Konobeev and Dubovsky (8), can be used here. The results of this paper concluded that in the case at hand ($2\Delta \gg$ excitonic bandwidth) the amplitude of exciton–exciton scattering for long-wavelength excitons equals $-a/2$ so that for the cross-section we obtain

$$\sigma = \pi a^2 \equiv 4\pi \left(\frac{a}{2}\right)^2, \tag{15.11}$$

---

[60]The situation discussed in this section is similar to that of local states appearing in excitonic spectra in the presence of an impurity molecule having excitation energy which largely differs from the excitation energy of host molecules (see Section 5.3).

[61]The states of bound bosons are not unphysical, because by transitions to these states the number of paulions, which is not equal to the number of bosons at the lattice point (see eqn 3.193), is not larger than one. But in these processes the number of paulions decreases.

where a is the lattice constant. This result becomes more evident if we use the results of computation of scattering length of a slow particle on a rectangular potential well of depth 2Δ and width $a/2$ (see (9), Problem 1 in Section 130), under conditions that the inequality $2\Delta \gg 4\hbar^2/m^*a^2$ is satisfied, and furthermore the quantity $(a/2)(4\Delta m^*)^{1/2}$ is not very close to an odd multiple of $\pi/2$ (i.e. there are no shallow levels inside the well). In this case the scattering length is equal to the well width, taken with the opposite sign, so that the relation (15.11) holds. The same result will be obtained when the well is replaced by a potential "hump" of height 2Δ. In both cases the scattering length is negative, which means that at small distances a repulsion occurs. The repulsion results from the fact that electronic excitations in a molecular crystals are not bosons, but paulions, so that the appearance of repulsion at small distances (smaller than the lattice constant) compensate the error, owing to the transition from paulions to bosons.

The repulsion at small distances does not allow one to deduce the nature of the state of an excitonic system at low temperatures. Indeed, the existence of a sufficiently strong attraction of excitons at larger distances can lead generally speaking to formation of biexcitons; this makes the consideration of the exciton system at low temperatures more complicated, and requires a different approach.[62]

In connection with the aforesaid we take into account the next terms of the kinematic exciton–exciton interaction and consider the part not included in $\hat{H}'$ (see eqn 15.8) which, as $\hat{H}'$ does, results from (15.5) by transition to Bose operators and which is determined by matrix elements $M_{ss'}^{\mathrm{I,II}}$.

Substituting (3.197) into (15.5) we obtain that the contribution to the kinematic exciton–exciton interaction operator, not accounted for in (15.8), has the form[63]

$$\hat{H}_{\mathrm{int}}^c = -\frac{1}{2}\sum_{s \neq s'} M_{ss'}^{\mathrm{I}} \left(B_s^\dagger B_s^\dagger B_s B_{s'} + B_s^\dagger B_{s'}^\dagger B_s' B_{s'} \right). \tag{15.12}$$

Let us investigate in more detail the properties of the above operator. Acting on the exciton, placed at the point s, it moves it to the point s'. But the result of applying the operator on the corresponding eigenfunction differs from zero only if, besides the exciton at the point s, there exists also an additional exciton at point s' or at point s. Hence the matrix element of the operator (15.12) differs from zero only for such a pair of states, for which both excitons "sit" on one site either in the initial state or in the final one. Now we make use of the above-mentioned property of exciton repulsion at a small distance. It can be shown

---

[62]In this case we must consider the collective properties of biexctons, and not of excitons. The possibility of formation of biexcitons (excitonic molecules) has been indicated in (10). They have been found experimentally by Haynes (11).

[63]For the reason, which will be explained in the following section, in (15.12) we ignore terms which do not conserve the number of excitons.

that the modulus of the wavefunction corresponding to the case of small relative distance between excitons is of the order M/Δ, where M is comparable with the exciton bandwidth. If so, although the quantities $\left|M_{ss'}^{I}\right|$ in (15.12) are of the order of the exciton bandwidth, the corrections to the exciton interaction energy, appearing in virtue of (15.12) at various orders of perturbation, are proportional to powers of a small parameter M^{I}/Δ and are small compared to the exciton bandwidth even at distances of the order of the lattice constant. Since the matrix elements $M_{ss'}^{I}$ decrease with increasing $|\mathbf{n}-\mathbf{n}'|$ at least as $1/|\mathbf{n}-\mathbf{n}'|^3$ (we recall that $s \equiv (\mathbf{n}\alpha)$), the exciton–exciton interaction determined by the operator (15.12), with regard to the above discussion, for arbitrary distances between excitons satisfies the inequality

$$|\mathcal{V}_{ss'}^{c}| \ll \frac{\hbar^2}{m^*|\mathbf{n}-\mathbf{n}'|^2}, \qquad (15.13)$$

m^* being the exciton effective mass.

With regard to (9, Sections 45 and 125), the fulfilling of the inequality (15.13) means that the interaction $\mathcal{V}_{ss'}^{c}$, though corresponding to exciton–exciton attraction, does not lead to the appearance of bound states, and its contribution to the scattering amplitude may be calculated in the first-order of the Born approximation.

Since the interaction energy $\mathcal{V}_{ss'}^{c}$ is small also compared to the exciton bandwidth, taking account of this interaction, not leading to the appearance of bound states, can only give a small correction to the exciton–exciton scattering length, connected to the energy (15.10). Thus the exciton–exciton scattering length remains negative, which means that in the absence of the dynamic exciton–exciton interaction, the Bose–Einstein condensation of excitons is possible.

Making use of the scattering length above obtained and recalling the results of (12), (13), we find that if $\mathbf{k} = 0$ corresponds to the minimum of the excitonic band, the spectrum of an ideal paulion gas is given by

$$\epsilon(\mathbf{k}) = \sqrt{\left(\frac{\hbar^2 k^2}{2m^*}\right)^2 + \frac{4\pi N_0 \hbar^2 a}{m^*}\left(\frac{\hbar^2 k^2}{2m^*}\right)}, \qquad (15.14)$$

where the concentration $N_0 \ll a^{-3}$, i.e. coincides with the spectrum of excitations of the condensate in a weakly nonideal Bose gas with repulsion between molecules. The above performed transition from Pauli to Bose operators has allowed us to extract the kinetic exciton–exciton interaction and to establish the scattering length in (15.14); we can also use the results of the theory of a weakly nonideal Bose gas.

As follows from (15.14), the obtained quasiparticles have acoustic dispersion for $|\mathbf{k}| \ll \sqrt{4\pi a N_0}$ and for $|\mathbf{k}| \gg \sqrt{4\pi a N_0}$ become free particles with

$$\epsilon(\mathbf{k}) = \frac{\hbar^2 k^2}{2m^*} + \frac{2\pi \hbar^2 a N_0}{m^*}.$$

For concentrations $N_0 \leq 10^{18}$ 1/cm^3 and $a \simeq 5 \cdot 10^{-8}$ cm the wavevector $\sqrt{4\pi a N_0} \leq 10^5$ cm^{-1}, i.e. is of the order of the light wavevector. The total energy shift of states with $|\mathbf{k}| \ll \sqrt{4\pi a N_0}$ appearing because excitons are paulions and not bosons for the same parameters equals

$$\frac{2\pi\hbar^2 a N_0}{m^*} \leq 2 \cdot 10^{-4} \text{eV}.$$

The conclusion on the possibility of condensation of elementary excitations in momentum space with Hamiltonian (15.5) coincides with that made in the article by Bocchieri and Seneci (14) where the problem of condensation in a crystal lattice of an ideal paulion gas has been considered without the transformation from paulions to bosons. For this reason the spectrum of the elementary excitations of the condensate has not been obtained.

To conclude this section we note that higher terms of the kinetic interaction with $\nu > 1$, not accounted for in (15.8), are not relevant because of the assumed smallness of the exciton concentration.

The generalization of the representation of paulion operators in terms of bosonic ones for the case of truncated oscillators of higher ranks is derived in the paper (15). The authors of this paper used this generalization to introduce a new constraint–free bosonic description of truncated oscillator systems. This result can be important in consideration of collective properties of Frenkel excitons in organic crystals with account of multilevel molecular structures and mixing of molecular configurations.

15.3 Collective properties of Frenkel excitons in the presence of a dynamic interaction

The operator $\hat{H}_{\text{int}}$ in (15.4) contains, in general, terms of third and of fourth order with respect to the operators P_s and $P_s^\dagger$. Terms of third order with respect to the operators P_s and $P_s^\dagger$ always lead to a weak exciton–exciton interaction. Since in such crystals the exciton bandwidth is much smaller than the energy required for the exciton formation, the third-order terms, which do not preserve the number of excitons, contribute to the exciton–exciton interation energy only in even orders of perturbation theory.

If $M_{ss'}^{\text{III}}$ is the matrix element present in the cubic terms then, for example, the correction to the interaction energy $\sim \left|M_{ss'}^{\text{III}}\right|^2/\Delta$, i.e. is negligibly small compared to the exciton bandwidth even if $\left|M_{ss'}^{\text{III}}\right|$ is of the order of this bandwidth. Since the quantity $\left|M_{ss'}^{\text{III}}\right|$ decreases with increasing $|\mathbf{n}-\mathbf{n}'|$ faster than $|\mathbf{n}-\mathbf{n}'|^{-3}$, the inequality of type (15.13) for this exciton–exciton interaction energy can be assumed to be satisfied.

Making use of formulas (3.16) and (3.25) we find that the operator of the exciton–exciton interaction has the form

$$\hat{H}_{\text{int}}^{\text{IV}} = \frac{1}{2} \sum_{s \neq s'} M_{ss'}^{\text{IV}} P_s^\dagger P_{s'}^\dagger P_s P_{s'}, \qquad (15.15)$$

where, using the notation of Ch. 3,

$$M_{ss'}^{IV} = \langle ff | \hat{V}_{ss'} | ff \rangle + \langle 00 | \hat{V}_{ss'} | 00 \rangle - 2\langle 0f | \hat{V}_{ss'} | 0f \rangle. \qquad (15.16)$$

The first term in (15.16) represents the interaction energy of molecules s and s' being at the fth excited states; the second term gives the interaction energy of these molecules when both are in the ground state. The third term gives the interaction energy when only one molecule is in the excited state f. The terms in (15.16) can be found when the wavefunctions of an isolated molecule for the ground state and for the fth excited state are known.

The quantity $M_{ss'}^{IV}$ in crystals having an inversion center decreases with increasing $|\mathbf{n} - \mathbf{n}'|$ as $|\mathbf{n} - \mathbf{n}'|^{-5}$ or faster. In consequence, for large $|\mathbf{n} - \mathbf{n}'|$ the inequality (15.13) for $M_{ss'}^{IV}$ is always fulfilled. So the most significant for the formation of biexciton is the magnitude and the sign of the interaction $M_{ss'}^{IV}$ in the case when the molecules s and s' are nearest neighbors.[64] Since the Bose–Einstein condensation of excitons is accompanied by the appearance of the spectrum (15.14), satisfying the Landau superfluidity conditions, this condensation may be established by observing the contribution of the superfluid component to the energy transfer from the host material to the exciton detector by experiments, similar to that of Simpson (19). The energy transfer by overcondensate excitons can be estimated by the diffusion equation. Note that the motion of the condensate can be induced by a gradient of the exciton concentration. Its concentration near the detector can be small because the surface of the detector captures the excitons.[65]

In the opposite case of narrow exciton bands, when the interaction (15.15) leads to "sticking" together of excitons into pairs, triplets, and more numerous excitonic "droplets", a more appropriate description is that in terms of excitons–paulions. The process of exciton coagulation can be described with the well-known methods of colloid statistics (21). We notice here that the size distribution of exciton droplets will depend on the exciton lifetime, bimolecular quenching

---

[64]The quantity $M_{ss'}^{IV}$, in the case of biexciton formation, can be estimated from the energy shift between the biexciton energy and double exciton energy. Unfortunately, corresponding experimental data for molecular crystals are, so far, not available. We note also that apart from the exciton–exciton interaction, determined by the operator (15.15), another interaction appears which must be, in general, accounted for. It is the exciton–exciton interaction owing to exchange of virtual phonons. As is known, for electrons in superconducting metals such interaction leads in many cases to the formation of Cooper pairs. In the case of excitons such interaction will also be of the attraction type and can lead, in general, to the formation of biexcitons. For more details we refer to (16)–(18). However, the Bose–Einstein condensation of biexcitons will be possible if the amplitude of biexciton–biexciton scattering is negative when both kinetic and dynamic interaction are taken into account.

[65]If v is the condensate velocity, then during the excitonic lifetime τ the exciton moving with the condensate can cover a distance of $\ell \simeq v\tau$. If $v \sim 10^4$–10^5cm/s, then for $\tau \simeq 10^{-8}$s the displacement $\ell \sim 1$–10×10^{-6}m. At the same time the magnitude of the diffusional displacement of the overcondensate excitons $\lambda = \sqrt{D\tau}$ for singlet excitons is smaller by an order of magnitude. We also note that the distribution of the exciton concentration when considering of their superfluidity has been discussed by Gergel, Kazarinov, and Suris (20).

and mobility and must be considered separately in any specific case. If, however, the formation of exciton droplets takes place, the crystal also becomes optically nonhomogeneous since the polarizability of excited molecules differs from that in the ground state. This property may lead to additional light scattering and other similar effects. As concerns the energy transfer from the host material to impurities or to the exciton detector, this transfer will rapidly go down due to the small mobility of the droplets compared to the mobility of single excitons.

Above, we have not considered the possibility of exciton decay due to bimolecular quenching. This process is especially essential in the case when the excitonic system tends to the formation of droplets. However, if the exciton–exciton repulsion prevails then the repulsion will suppress the bimolecular quenching at small distances between excitons. Let N be the exciton concentration, so that the number of exciton decays per unit time equals γN^2, γ being the corresponding kinetic parameter. Simultaneously, the number of exciton decays with photon emission is given by N/τ where τ is the exciton lifetime. Thus the decays of singlet excitons owing to bimolecular quenching can be ignored if $N \leq 1/\gamma\tau$.

For lowest energy singlet excitons in anthracene (22) $\tau \simeq 10^{-9}$s, $\gamma \simeq 5\times$ $\times 10^{-9}$cm^3/s, so that only for $N \leq 10^{16}$cm^{-3} can the bimolecular quenching be not very significant.

So far, we have considered the Bose–Einstein condensation of Coulomb excitons and thus we neglected the influence of the retardation. Such a consideration is correct only for excitons with small (or zero) oscillator strengths. Thus the above-described consideration on the Bose–Einstein condensation can be applied, for example, to the case of triplet excitons, or to singlet excitons for which the exciton–photon interaction energy is smaller than the exciton level width, caused by, for example, scattering by phonons.

However, if a singlet exciton has a large oscillator strength, Bose–Einstein condensation is nevertheless possible for excitons but only if the minimum energy in the lowest excitonic band does not correspond to the center of the Brillouin zone. In the opposite case we have to take into account that the retardation drastically modifies the spectrum of excitons, transforming excitons into polaritons. In such 3D crystals the condensation of excitons is impossible because the spectrum of polaritons in the region of small $\mathbf{k}$ (see Ch. 4) coincides with the spectrum of photons with energy $E(\mathbf{k}) \to 0$ at $\mathbf{k} \to \mathbf{0}$. However, for polaritons in a microcavity Bose–Einstein condensation is possible because in a microcavity there exists a so-called cut-of frequency (see Ch. 10)

In concluding this section we consider a specific consequence of the exciton Bose–Einstein condensation. As shown by Gergel et al. (20), the formation of a condensate in the excitonic spectral region can be accompanied by the appearance of negative light absorption.

By absorption of a photon with frequency ω and wavevector $\mathbf{Q}$, an exciton is created having the energy

$$E(0) + \varepsilon(\mathbf{Q}) = \hbar\omega,$$

where $E(0)$ is the energy of the exciton in the condensate, and $\varepsilon(\mathbf{Q})$ is the energy of elementary excitation of the condensate with wavevector $\mathbf{Q}$. The probability of this process is proportional to $N_\mathbf{Q} + 1$ where $N_\mathbf{Q}$ is the exciton wavevector distribution (note that due to the exciton–exciton interaction also for $T = 0$ a "push out" of excitons from the state $\mathbf{Q} = 0$ occurs so that $N_\mathbf{Q} \neq 0$ also for $\mathbf{Q} \neq 0$).

A reverse process is also possible. It corresponds to light emission from condensed excitons with creation of an elementary condensate excitation having wavevector $-\mathbf{Q}$ and the energy $\varepsilon(-\mathbf{Q})$; the probability of this process is proportional to $N_\mathbf{Q}$, and the frequency of photon emission equals

$$\omega = \frac{1}{\hbar}\left[E(0) - \varepsilon(-\mathbf{Q})\right].$$

It follows from the above discussion that the emission and the absorption lines differ in energy by the value $2\varepsilon(\mathbf{Q})$. The observation of such an emission from the condensate may be difficult since simultaneously we require $2\varepsilon(\mathbf{Q})$ to be not very small compared to the excitonic absorption and emission linewidths. Nonetheless, searching for appropriate conditions and objects for observation of these effects is of interest. Note, in this connection, the recent claims on the observation of polariton condensation in microcavities. The effective mass of the polariton is very small in the microcavity and the energies of elementary excitations in the condensate can drastically increase.

15.3.1 On biexcitons in organic crystals

If molecules of a crystal have no inversion center, the static dipole moments of a molecule in the ground and in the excited states are usually not equal to zero. In this case and in the dipole–dipole approximation for $\hat{V}_{s,s'}$ the expression (15.16) can be reduced to the interaction energy of dipoles $\mathbf{P}_s$ and $\mathbf{P}_{s'}$

$$\mathbf{P}_s = \mathbf{P}_s^{ff} - \mathbf{P}_s^{00}$$

and so on, equal to the change of the dipole moment of a molecule in its excitation. In this approximation

$$M_{ss'}^{IV} = \frac{\mathbf{P}_s \mathbf{P}_{s'} - 3(\mathbf{P}_s \boldsymbol{\ell}_{s,s'})(\mathbf{P}_s \boldsymbol{\ell}_{s,s'})}{|\mathbf{s} - \mathbf{s}'|^3},$$

where $\boldsymbol{\ell}$ is the unit vector

$$\boldsymbol{\ell}_{s,s'} = \frac{\mathbf{s} - \mathbf{s}'}{|\mathbf{s} - \mathbf{s}'|}.$$

For dipoles of the order of 5 Debye and nearest distance of molecules of the order of 5 Å the energy $M_{s,s'}^{IV}$ is of the order of 1000 cm^{-1} which can be large or of the order of the width of the exciton bandwidth. In such crystals we can expect the existence of biexciton states. The theory of biexcitons in layered organic anthracene type crystals has been developed by Spano et al. (23), (24).

The first observation of biexciton states in a molecular structure has been performed by Chakrabarti et al. (25) for the 1D H-aggregate of an epitaxially grown trivalent–halogen–phthalocyanine with a unique packing arrangement of the monomers with nearest distance ≈ 4 Å. The distinct photoinduced absorption to a continuum of two-exciton states and to a bound biexciton state was observed. The value of the exciton–exciton interaction for this structures was estimated to be equal to ≈ 0.3 eV.

15.4 Kinematic interaction of exciton-polaritons in crystalline organic microcavities

Organic microcavities, as with inorganic microcavities, have attracted great interest for their ability to control the light–matter coupling. The discussion of the nonlinear effects in such a microcavity can also be very interesting for science as well as for applications. The most important and interesting thing is the comparison with inorganic microcavities where excitons are Wannier–Mott type and not Frenkel type. For organic microcavities this direction of research has taken only the first steps. Without going in details of recently published papers we mention here only some of the main predictions.

The exciton–exciton and polariton–polariton kinematic interactions in a crystalline organic monolayer and in an organic microcavity have been considered in (26). The kinematic interactions in this paper are derived using for Frenkel excitons the transformation from paulions to bosons (see Ch. 3).

For an organic crystalline monolayer it was found that the exciton–exciton kinematic interaction can be described as scattering, not by "hard spheres" as in 3D crystals, but as scattering by hard disks. It was shown also that, as in the case of a two-dimensional ultracold trapped atom boson gas, the excitons in a confined monolayer may behave as a dilute degenerate boson gas at low temperature. Then for a microcavity with an organic crystalline monolayer such as a resonant material the polariton–polariton kinematic interaction steming from the polariton excitonic part was derived.

The difficulty in the calculation of the polariton–polariton scattering amplitude is caused by a nonparabolic dispersion of cavity polaritons. Fortunately, the scattering of quasiparticles with an arbitrary dispersion law in the presence of degenerate perturbations (the delta-like potential appearing in the formulation of the problem of kinematic interaction belongs to this class) has been examined by Lifshitz in a series of works (27). It has been shown there that only specific (determined by the topology of the equipotential surfaces) wavevectors appear in the scattered wave, and that the scattering amplitude f of a quasiparticle is inversely proportional to the group velocity of the quasiparticle at these wavevectors.

The general Lifshitz theory was applied for two-dimensional excitons and polaritons in an organic microcavity by Litinskaia (28). Unfortunately, we are not able to go in details of these calculations here. Note only that also stud-

15.5 Fermionic character of Frenkel excitons in one-dimensional molecular crystals

In previous sections we have considered collective properties of Frenkel excitons in three-dimensional molecular crystals only. In one- and two-dimensional molecular crystals the kinematic interaction appearing in the transition from Pauli to Bose operators can be, generally speaking, quite large, so that the description of Frenkel excitons in terms of Bose excitations can be very rough. Below we consider this problem in more detail.

First we must keep in mind that if we are not interested in collective properties of excitons and, in particular, we not consider exciton–exciton scattering, then in a good approximation we can take into account only those crystal states where the number of excitons is 0 or 1. It is evident that in these states, independent of the crystal dimension, the kinematic interaction identically vanishes so that the Frenkel exciton can be considered both as a boson as well as a fermion (i.e. it can be described by Bose or Fermi operators) and both descriptions will give identical results.

Quite another situation occurs when we consider collective properties of excitons. In this case, as shown above, the kinematic interaction can play an important role. However, the mutual scattering of elementary excitations, which causes this interaction, is quite different in three-, two- and one-dimensional crystals. In three-dimensional crystals at not very large concentration of excitations this scattering at least does not change the statistics.

Another situation can be observed in one-dimensional crystals. Chesnut and Suna (30) have shown that if in a one-dimensional paulion gas only interactions with nearest-neighbors are taken into account, the energy operator for this case can be represented as the energy operator of an ideal gas of fermions.

Indeed, in this case the energy operator, in terms of Pauli operators, has the form

$$\hat{H} = \sum_{s=1}^{N} \Delta\, P_s^\dagger P_s + \sum_{s=1}^{N-1} M \left(P_s^\dagger P_{s+1} + P_{s+1}^\dagger P_s \right), \quad (15.17)$$

Δ being the excitation energy of an isolated molecule, M is the matrix element of the excitation transfer from molecule s to molecule $s+1$, and $N+1$ is the number of molecules in a linear chain.

If we pass from operators $P_s, P_s^\dagger$ to operators $\alpha_s, \alpha_s^\dagger$,

$$\alpha_s = P_s \hat{\epsilon}_s,$$
$$\alpha_s^\dagger = P_s^\dagger \hat{\epsilon}_s, \quad (15.18)$$

where the operator $\hat{\epsilon}_s$ is given by the relation

$$\hat{\epsilon}_s |n_1, \ldots, n_N\rangle = \left\{ \prod_{i=1}^{i=s} (-1)^{n_i} \right\} |n_1, \ldots, n_N\rangle, \tag{15.19}$$

then

$$\hat{H} = \sum_{s=1}^{N} \Delta \, \alpha_s^\dagger \alpha_s + \sum_{s=1}^{N} M \left(\alpha_s^\dagger \alpha_{s+1} + \alpha_{s+1}^\dagger \alpha_s \right), \tag{15.20}$$

where α_s, $\alpha_s^\dagger$ are just Fermi operators. Introducing new Fermi operators

$$a_k = \left(\frac{2}{N} + 1 \right)^{1/2} \sum_{s=1}^{N} \alpha_s \sin ks,$$

$$a_k^\dagger = \left(\frac{2}{N} + 1 \right)^{1/2} \sum_{s=1}^{N} \alpha_s^\dagger \sin ks, \tag{15.21}$$

we obtain

$$\hat{H} = \sum_k E_k \alpha_k^\dagger \alpha_k, \tag{15.22}$$

with

$$E_k = \Delta + 2M \cos k, \quad k = \frac{\pi j}{N+1}, \quad j = 1, 2, \ldots, N. \tag{15.23}$$

Hence in this approximation the Hamiltonian of the system can be presented as the Hamiltonian of a gas of noninteracting fermions so that the condensation in momentum space is not possible. The wide spread and successful application of the transformation from paulions to fermions described above (the Jordan–Wigner transformation) can be found in the theory of linear and nonlinear optical properties of 1D J-aggregates in the review papers (31)–(33). Recently has been published a review paper by Egorov and Alfimov (34). This review concerns the current status of the theory of formation of the J-band, an abnormally narrow, high-intensity, red-shifted optical absorption band arising from the aggregation of polymethine dyes. Two opposite approaches to explaining the physical nature of the J-band are given special attention. In the first of these, the old one based on Frenkel's statistical exciton model, and the J-band is explained by assuming that the quickly moving Frenkel exciton acts to average out the quasistatic disorder in electronic transition energies of molecules in the linear J-aggregate (see Knapp (35)). In the second approach, on the contrary, the specific structure of the dye (the existence of a quasilinear polymethine chain) is supposed to be very important. This approach is based on the theory of charge transfer.

If we take into account not only the nearest, but also the farther neighbors, the additional so-called kinematic interaction between Fermi quasiparticles appears. For molecular chains the contribution of such additional resonance intermolecular interactions to one exciton energy is relatively small (36) even if a transition from the ground to an excited state is dipole allowed. However, its influence on the statistics of elementary excitations in these molecular structures has never been investigated. Only if the kinematic interaction does not destroy the Fermi statistics of excitations can the above statement on the impossibility of excitonic Bose–Einstein type condensation in 1D molecular chains be proved. However, this problem for a 1D structure where the noninteracting excitations are paulions and not bosons needs further analysis. It is useful to note here that in 1D structures where noninteracting quasiparticles are bosons and not paulions or fermions then under the influence of the interaction between quasiparticles, new regimes of statistics as a function of boson concentration have been discovered (see (37) and references therein).

APPENDIX A

DIAGONALIZATION OF A HAMILTONIAN QUADRATIC IN THE BOSE-AMPLITUDES

Let the Hamiltonian $\hat{H}$ be a quadratic form with respect to the Bose amplitudes b_α and $b_\alpha^\dagger$, $\alpha = 1, \ldots, t$:

$$\hat{H} = E_0 + \frac{1}{2} \sum_{\alpha,\beta} \left[A_{\alpha\beta} b_\alpha^\dagger b_\beta^\dagger + A_{\alpha\beta}^* b_\alpha b_\beta \right] + \sum_{\alpha,\beta} B_{\alpha\beta} b_\beta^\dagger, \qquad (A.1)$$

$B_{\alpha\beta}$ being a hermitian matrix, and $A_{\alpha\beta}$ a symmetric matrix. According to the Tyablikov–Bogoljubov method[66] the diagonalization of the quadratic form (A.1) can be performed by introducing new operators

$$\xi_\mu = \sum_\alpha \left(b_\alpha u_{\alpha\mu}^* - b_\alpha^\dagger v_{\alpha\mu}^* \right); \quad \xi^\dagger = \sum_\alpha \left(b_\alpha^\dagger u_{\alpha\mu} - b_\alpha v_{\alpha\mu} \right), \qquad (A.2)$$

where the coefficients $u_{\alpha\mu}, v_{\alpha\mu}$ result from the following system of linear equations

$$Eu_\alpha = \sum_\beta A_{\alpha\beta} v_\beta + \sum_\beta B_{\alpha\beta} u_\beta,$$
$$-Ev_\alpha = \sum_\beta A_{\alpha\beta}^* u_\beta + \sum_\beta B_{\alpha\beta}^* v_\beta, \qquad (A.3)$$

subject to the normalization condition

$$\sum_\alpha (u_\alpha^* u_\alpha - v_\alpha^* v_\alpha) = 1. \qquad (A.4)$$

It can be shown, starting from the properties of the matrices $A_{\alpha\beta}$ and $B_{\alpha\beta}$, that the values of E for which the system (A.3) has a nontrivial solution satisfying (A.4), are real and positive. If so, they can be labeled by an index μ, $E = E_\mu$, $\mu = = 1, 2, \ldots, t$, and the corresponding values u_α, v_α become $U_{\alpha\mu}, v_{\alpha\mu}$. Having those coefficients, we obtain the operators ξ_μ and $\xi^\dagger$ by eqn (A.2). The coefficients $u_{\alpha\mu}, v_{\alpha\mu}$ satisfy the orthogonality conditions. For obtaining those coefficients, let us consider eqns (A.3) substituting $E = E_\mu$:

---

[66] See Bogoljubov, N. N. (1949). *Lectures on Quantum Statistics.* Kiev. § 11, and Tyablikov, S. V. (1975). *Methods in the Quantum Theory of Magnetism.* 2nd edition, Plenum, New York.

$$E_\mu u_\alpha = \sum_\beta A_{\alpha\beta} v_\beta + \sum_\beta B_{\alpha\beta} u_\beta,$$
$$-E_\mu v_\alpha = \sum_\beta A^*_{\alpha\beta} u_\beta + \sum_\beta B^*_{\alpha\beta} v_\beta, \tag{A.5}$$

It follows from the above equations that

$$E_\mu \sum_\alpha \left(u_{\alpha\mu} u^*_{\alpha\mu_1} - v_{\alpha\mu} v^*_{\alpha\mu_1} \right) \tag{A.6}$$
$$= \sum_{\alpha\beta} \left\{ A_{\alpha\beta} v_{\beta\mu} u^*_{\alpha\mu_1} + B_{\alpha\beta} u_{\beta\mu} u^*_{\alpha\mu_1} + A^*_{\beta\alpha} u_{\beta\mu} v^*_{\alpha\mu_1} + B^*_{\alpha\beta} v_{\beta\mu} v^*_{\alpha\mu_1} \right\}.$$

Now we take the complex conjugate of the above equation and change the indices $\mu \rightleftharpoons \mu_1$:

$$E_{\mu_1} \sum_\alpha \left(u_{\alpha\mu} u^*_{\alpha\mu_1} - v_{\alpha\mu} v^*_{\alpha\mu_1} \right) = \tag{A.7}$$
$$= \sum_{\alpha\beta} \left\{ A^*_{\alpha\beta} v^*_{\beta\mu_1} u_{\alpha\mu} + B^*_{\alpha\beta} u^*_{\beta\mu_1} u_{\alpha\mu} + A_{\beta\alpha} u^*_{\beta\mu_1} v_{\alpha\mu} + B_{\alpha\beta} v^*_{\beta\mu_1} v_{\alpha\mu} \right\}.$$

APPENDIX B

CALCULATIONS OF POLARIZATION IN INORGANIC QUANTUM WELLS AND IN ORGANICS

B.1 General relations

Consider[67] two spatially separated systems of charges a and b with charge densities $\rho^a(\mathbf{r})$ and $\rho^b(\mathbf{r})$, respectively. Then the interaction energy is given by

$$V = \int \rho^a(\mathbf{r}) G(\mathbf{r}, \mathbf{r}') \rho^b(\mathbf{r}') \mathrm{d}^3 r \mathrm{d}^3 r', \tag{B.1}$$

where $G(\mathbf{r}, \mathbf{r}')$ is the Green's function of the Poisson equation. In vacuum it is the usual $1/|\mathbf{r} - \mathbf{r}'|$, in an inhomogeneous anisotropic medium with the dielectric tensor $\epsilon_{ij}(\mathbf{r})$ it satisfies the equation

$$\frac{\partial}{\partial x_i} \epsilon_{ij}(\mathbf{r}) \frac{\partial}{\partial x_j} G(\mathbf{r}, \mathbf{r}') \rho^b(\mathbf{r}') = -4\pi \delta(\mathbf{r} - \mathbf{r}'). \tag{B.2}$$

If the total charges of the systems a and b are zero, one may introduce the polarizations, vanishing outside the regions of space, occupied by the systems (1), such that

$$\rho^a(\mathbf{r}) = -\mathrm{div}\, \mathbf{P}^a(\mathbf{r}), \qquad \rho^b(\mathbf{r}) = -\mathrm{div}\, \mathbf{P}^b(\mathbf{r}). \tag{B.3}$$

Then the energy may be rewritten as

$$V = \int P_i^a(\mathbf{r}) P_j^b(\mathbf{r}') \frac{\partial^2 G(\mathbf{r}, \mathbf{r}')}{\partial x_i \partial x_j'} \mathrm{d}^3 r \mathrm{d}^3 r' = -\int \mathbf{P}^a(\mathbf{r}) \mathcal{E}(\mathbf{r}) \mathrm{d}^3 r, \tag{B.4}$$

where $\mathcal{E}(\mathbf{r})$ is the electric field produced by the system b:

$$\mathcal{E}_i^b(\mathbf{r}) = \int \bar{\chi}(\mathbf{r}, \mathbf{r}') P_j^B(\mathbf{r}') \mathrm{d}^3 r', \qquad \bar{\chi}(\mathbf{r}, \mathbf{r}') \equiv -\frac{\partial^2 G(\mathbf{r}, \mathbf{r}')}{\partial x_i \partial x_j'}. \tag{B.5}$$

The Coulomb response function $\bar{\chi}(\mathbf{r}, \mathbf{r}')$ gives the ith component of the electric field at the point $\mathbf{r}$, produced by the jth component of a point dipole, situated at the point $\mathbf{r}'$; $\bar{\chi}(\mathbf{r}, \mathbf{r}')$ is nothing else but the limit of $\chi_{ij}(\mathbf{r}, \mathbf{r}', \omega)$ at $\omega \to 0$

---

[67] See also Agranovich, V. M., Basko, D. M., La Rocca, G. C., and Bassani, F. (1998). *J. Phys.: Condens. Matter* **10**, 9369, and Basko, D. M. (2003). Electronic energy transfer in a planar microcavity. In: *Thin Films and Nanostructures. Electronic Excitations in Organic Based Nanostructures* **31**, edited by V. M. Agranovich and G. F. Bassani. Elsevier, Amsterdam, pp. 403–446.

(electrostatics) or, equivalently, $c \to \infty$ (no retardation). Working in this limit is justified if the characteristic distances in the problem are much shorter than the resonant light wavelength, which is equivalent to $k \gg \omega/c$.

One should keep in mind that speaking of two different electron densities for the systems a and b already completely disregards the exchange effects, which is justified if the wavefunction overlap between the systems a and b is negligible.

B.2 Polarization in a semiconductor

For the problems under consideration the most interesting are the zinc-blende semiconductors formed by elements of the groups III and V or groups II and VI of the periodic table (like GaAs or ZnSe). In these materials the lowest conduction band is doubly degenerate, and the highest valence band is four-fold degenerate at $\mathbf{k} = 0$ due to the symmetry of the crystal, both bands having extremes at $\mathbf{k} = 0$ (the Γ point). At wavevector $\mathbf{k} \neq 0$ the four-dimensional manifold splits into two bands (each of them being doubly degenerate), having different effective masses. They are called the light hole and heavy hole bands (2)–(4). For our purposes it will be sufficient to consider the first one a nondegenerate conduction band and one a nondegenerate valence band, assuming the contribution of different bands at the end, when needed.

A system of many identical particles is most conveniently described by the $\hat{\psi}$ field operators (5). The generic form of the electronic charge density operator in a semiconductor is

$$\hat{\rho}^{\text{sem}}(\mathbf{r}) = -e\hat{\psi}^\dagger(\mathbf{r})\hat{\psi}(\mathbf{r}). \tag{B.6}$$

Taking into account only two nondegenerate bands, one can split the electronic ψ-operator into two parts, corresponding to the conduction and valence bands:

$$\hat{\psi}(\mathbf{r}) = \hat{\psi}_c(\mathbf{r}) + \hat{\psi}_v(\mathbf{r}) \equiv \hat{\psi}_e(\mathbf{r}) + \hat{\psi}_v^\dagger(\mathbf{r}), \tag{B.7}$$

where we introduced a hole creation operator as a valence electron destruction operator. In the operator of the electron density

$$\hat{\psi}^\dagger(\mathbf{r})\hat{\psi}(\mathbf{r}) = \hat{\psi}_v^\dagger(\mathbf{r})\hat{\psi}_c(\mathbf{r}) + \hat{\psi}_c^\dagger(\mathbf{r})\hat{\psi}_v(\mathbf{r}) + \hat{\psi}_c^\dagger(\mathbf{r})\hat{\psi}_c(\mathbf{r}) \\ + \hat{\psi}_v^\dagger(\mathbf{r})\hat{\psi}_v(\mathbf{r}), \tag{B.8}$$

the first two terms correspond to the interband transitions, the last two, to the intraband transitions.

Consider the first term (the second one being just its hermitian conjugate). The matrix element of the operator $\hat{\psi}_v^\dagger(\mathbf{r}')\hat{\psi}_c(\mathbf{r})$ between the ground state $|0\rangle$ and some eigenstate $|s\rangle$ of a single electron–hole pair is

$$\langle 0|\hat{\psi}_v^\dagger(\mathbf{r}')\hat{\psi}_c(\mathbf{r})|s\rangle = v_0 \int \frac{d^3k}{(2\pi)^3} \frac{d^3k'}{(2\pi)^3} \Psi_s(\mathbf{k},\mathbf{k}')u_{c,\mathbf{k}}(\mathbf{r})u^*_{v,-\mathbf{k}'}(\mathbf{r}')e^{i\mathbf{k}\mathbf{r}+i\mathbf{k}'\mathbf{r}'}$$

$$\approx v_0 u_c(\mathbf{r})u_v^*(\mathbf{r}')\Psi_s(\mathbf{r},\mathbf{r}'). \tag{B.9}$$

Here v_0 is the unit cell volume, $u_c(\mathbf{r})$ and $u_v(\mathbf{r})$ are the conduction and valence band extremes Bloch functions,[68] and $\Psi_s(\mathbf{r}_e,\mathbf{r}_h)$ is the envelope wavefunction of the electron–hole pair state $|s\rangle$, normalized to the unit integral

$$\int |\Psi_s(\mathbf{r}_e,\mathbf{r}_h)|^2\,\mathrm{d}^3 r_e \mathrm{d}^3 r_h = 1.$$

Now consider the interband matrix element of the charge density, corresponding to the first term in the expression (B.8). One is interested in the long-wavelength Fourier components of the density, given by

$$\rho_{0s}(\mathbf{q}) \equiv \langle 0|\hat\rho^{\mathrm{sem}}(\mathbf{q})|s\rangle$$
$$= -e\int \frac{\mathrm{d}^3 k}{(2\pi)^3}\Psi_s(\mathbf{k},\mathbf{q}-\mathbf{k})\int_{u.c.} u_{c,\mathbf{k}}(\mathbf{r})u_{v,\mathbf{k}-\mathbf{q}}^*(\mathbf{r})\mathrm{d}^3 r, \tag{B.10}$$

which is obtained from the first line of (B.9) using the periodicity of the Bloch functions. More specifically, due to the periodicity, the product $u_{c,\mathbf{k}}(\mathbf{r})u_{v,\mathbf{k}-\mathbf{q}}^*(\mathbf{r})$ contains plane waves with wavevectors equal to either zero or a reciprocal lattice vector. Being interested only in the long-wavelength part of ρ_{0s}, one should pick up only the zero wavevector contribution, which corresponds to integration over the unit cell, as done in (B.10).

If one simply approximates the Bloch functions in (B.10) by those for the band extremes, the result will be zero due to the orthogonality

$$\int_{u.c.} u_c(\mathbf{r})u_v^*(\mathbf{r})\mathrm{d}^3 r = 0.$$

Hence one can expand the Bloch functions using the $\mathbf{k}\cdot\mathbf{p}$ perturbation theory (2)–(4), to find the admixture to the functions $u_{c,\mathbf{k}}(\mathbf{r})$ of the Bloch functions $u_b(\mathbf{r})$ of all other bands b:

$$u_{c,\mathbf{k}}(\mathbf{r}) \approx u_c(\mathbf{r}) - ik_j\frac{\hbar^2}{m_0}\sum_b \frac{\langle b|\partial/\partial x_j|c\rangle}{E_c(0)-E_b(0)}u_b(\mathbf{r})$$
$$= u_c(\mathbf{r}) - ik_j\sum_b \langle b|x_j|c\rangle u_b(\mathbf{r}), \tag{B.11}$$

where the symbol $\langle b|\mathcal{O}|c\rangle$ denotes

$$\int_{u.c.} u_b^*(\mathbf{r})\mathcal{O}u_c(\mathbf{r})\mathrm{d}^3 r.$$

---

[68] We adopt the term "Bloch functions" for the cell-periodic part of the full electron wavefunction in the crystal.

In the transition from the first to the second line of (B.11) the quantum-mechanical relation $\dot{\mathbf{r}} = -\mathrm{i}(\hbar/m_0)\boldsymbol{\nabla}$ for the bare electron in the crystal was used, $\dot{\mathbf{r}}$ being related to the commutator of $\mathbf{r}$ with the crystal Hamiltonian. Expanding $u^*_{v,\mathbf{k-q}}(\mathbf{r})$ analogously, and substituting them into (B.10), one obtains

$$\rho_{0s}(\mathbf{q}) = -\mathrm{i}\mathbf{q}\mathbf{d}^{vc}\int\frac{\mathrm{d}^3 k}{(2\pi)^3}\Psi_s(\mathbf{k},\mathbf{q-k}),$$
$$\rho_{0s}(\mathbf{r}) = -\mathrm{div}\left(\mathbf{d}^{vc}\Psi_s(\mathbf{r},\mathbf{r})\right), \qquad (\mathrm{B.12})$$

where $\mathbf{d}^{vc} = \langle v|(-e\mathbf{r})|c\rangle$ is the transition dipole moment of the unit cell. The contributions of all the bands different from the conduction and valence bands vanish due to the orthogonality. According to (B.3), the obtained charge density corresponds to the interband polarization

$$\langle 0|\hat{\mathbf{P}}(\mathbf{r})|s\rangle = \mathbf{d}^{vc}\Psi_s(\mathbf{r},\mathbf{r}). \qquad (\mathrm{B.13})$$

The expression (B.13) is the basic one to be used below. The Cartesian components of the dipole moments d^{vc}_i ($i = x, y, z$) may be expressed in terms of Kane's energy E_0 (3) as

$$|d^{vc}_i|^2 = \frac{e^2\hbar^2 E_0 c_i^2}{2m_0 E_g^2} = c_i^2\,(ea_0)^2\,\frac{E_0 \mathrm{Ry}_0}{E_g^2} \equiv c_i^2\,|d^{vc}|^2, \qquad (\mathrm{B.14})$$

where a_0 and Ry_0 are the hydrogen atom Bohr radius and Rydberg constant, and c_i are the appropriate symmetry coefficients. In semiconductors with the zincblende structure $c^{hh}_x = c^{hh}_y = 1$, $c^{hh}_z = 0$ (heavy holes), and $c^{lh}_x = c^{lh}_y = 1/\sqrt{3}$, $c^{lh}_z = \sqrt{4/3}$ (light holes).

For an intraband transition between the states $|s\rangle$ and $|s'\rangle$ of the electron–hole pair one may simple average $\hat{\psi}^\dagger_c(\mathbf{r})\hat{\psi}_c(\mathbf{r})$ and $\hat{\psi}^\dagger_v(\mathbf{r})\hat{\psi}_v(\mathbf{r})$ over the unit cell using the Bloch functions at the band extremes, since the principal term does not already vanish. As a result, the corresponding matrix element of the charge density is given by the sum of the electron and hole contributions:

$$\langle s|\hat{\rho}^{\mathrm{sem}}(\mathbf{r})|s'\rangle = -e\int \Psi^*_s(\mathbf{r},\mathbf{r}_h)\,\Psi_{s'}(\mathbf{r},\mathbf{r}_h)\,\mathrm{d}^3 r_h$$
$$+ e\int \Psi^*_s(\mathbf{r}_e,\mathbf{r})\,\Psi_{s'}(\mathbf{r}_e,\mathbf{r})\,\mathrm{d}^3 r_e. \qquad (\mathrm{B.15})$$

B.3 Polarization in organics

We assume that the excitations in the organic medium are localized, corresponding to the excited states of a molecule or a group of strongly coupled molecules. Thus, the organic subsystem may be described by the ground state $|g_A\rangle$, and the excited states $|\mathbf{r},\nu\rangle$, where $\mathbf{r}$ is the continuous position of the excited state and ν is a continuous quantum number, labeling the excited states at the point $\mathbf{r}$. As we restrict ourselves to the linear regime, only "one-particle" excited states are considered, which means that two excitations $|\mathbf{r},\nu\rangle$ and $|\mathbf{r}',\nu'\rangle$ are not allowed to exist simultaneously. The particular dissipation mechanism, determining the structure of these states need not to be specified here. We use the following normalization of the states

$$\langle g_A|g_A\rangle = 1, \tag{B.16}$$

$$\langle g_A|\mathbf{r},\nu\rangle = 0, \tag{B.17}$$

$$\langle \mathbf{r},\nu|\mathbf{r}',\nu'\rangle = \delta(\nu-\nu')\,\delta(\mathbf{r}-\mathbf{r}'), \tag{B.18}$$

$$\hat{I}_A = |g_a\rangle\langle g_A| + \int d^3r \int d\nu\, |\mathbf{r},\nu\rangle\langle \mathbf{r},\nu|, \tag{B.19}$$

where $\hat{I}_A$ is the unit operator for the organic subsystem. The Hamiltonian and the polarization of the organic medium are written as

$$\hat{H}_A = |g_A\rangle 0\langle g_A| + \int d^3r \int d\nu\, |\mathbf{r},\nu\rangle E_\nu(\mathbf{r}) \langle \mathbf{r},\nu|, \tag{B.20}$$

$$\hat{\mathbf{P}}^A(\mathbf{r}) = \int d\nu\, |\mathbf{r},\nu\rangle\, \mathbf{d}^\nu(\mathbf{r})\langle \mathbf{r},\nu| + \text{h.c.}, \tag{B.21}$$

where $E_\nu(\mathbf{r})$ is the energy of the corresponding state and $\mathbf{d}^\nu(\mathbf{r})$ is the matrix element of the dipole moment between the excited and the ground state:

$$\mathbf{d}^\nu(\mathbf{r}) = \langle \mathbf{r},\nu| \left(\sum_i e_i \hat{\mathbf{r}}_i\right) |g_A\rangle, \tag{B.22}$$

with e_i, $\hat{\mathbf{r}}_i$ being the charge and the position operator of the ith charge in the medium and the sum is taken over all charges constituting the medium. Both $E_\nu(\mathbf{r})$ and $\mathbf{d}^\nu(\mathbf{r})$ are assumed to be slowly varying in space.

APPENDIX C

MICROSCOPIC QUANTUM-MECHANICAL CALCULATIONS OF THE ENERGY TRANSFER RATE

Typically the broadening of the Wannier–Mott exciton resonance (several meV) is negligible compared to the width of the absorption spectrum of organics (several hundred meV). Thus, one may consider the excited state in a semiconductor to be discrete with energy $\hbar\omega_D$. The excited state $|\Psi\rangle$ of the semiconductor is an electron–hole pair described by the two-particle envelope function $\Psi(\mathbf{r}_e, \mathbf{r}_h)$, assumed to be normalized to unity:

$$\int |\Psi_s(\mathbf{r}_e, \mathbf{r}_h)|^2 \, d^3r_e d^3r_h = 1. \tag{C.1}$$

The effective Hamiltonian and the polarization for the semiconductor are simply given by

$$\hat{H}_D = |\Psi\rangle\hbar\omega_D\langle\Psi| + |g_D\rangle 0 \langle g_D|, \tag{C.2}$$

$$\hat{\mathbf{P}}^D(\mathbf{r}) = \mathbf{d}^{vc}\Psi(\mathbf{r}, \mathbf{r})|g_D\rangle\langle\Psi| + \text{h.c.}, \tag{C.3}$$

where $|g_D\rangle$ is the semiconductor ground state, and $\mathbf{d}^{vc}$ was defined in Appendix B.2.

The Hamiltonian of the Coulomb interaction between the semiconductor and the organics is given by eqns (B.4), (B.5):

$$\hat{H}_{\text{int}} = -\int \hat{\mathbf{P}}^A(\mathbf{r})\hat{E}^D(\mathbf{r}) \, d^3r,$$

$$\hat{\mathcal{E}}_i^D(\mathbf{r}) = \int \bar{\chi}_{ij}(\mathbf{r}, \mathbf{r}') \hat{P}_j^D(\mathbf{r}') d^3r', \tag{C.4}$$

where $\hat{E}^D(\mathbf{r})$ is the operator of the electric field, produced by the semiconductor polarization $\hat{\mathbf{P}}^D(\mathbf{r})$ in the organics. The Förster transfer rate Γ (determining the partial electron–hole pair lifetime τ due to the transfer) is given by Fermi's golden rule:

$$\Gamma \equiv \frac{1}{\tau} = \frac{2\pi}{\hbar} \int d^3r \int d\nu \, |\mathbf{d}^\nu(\mathbf{r}) \cdot E^{vc}(\mathbf{r})|^2 \, \delta\left(E_\nu(\mathbf{r}) - \hbar\omega_D\right), \tag{C.5}$$

where $E^{vc}(\mathbf{r}) = \langle g_D|\hat{E}^D(\mathbf{r})|\Psi\rangle$. Considering the general expression for the dielectric function (6), which in our normalization of states may be written as

$$\epsilon_{ij}(\mathbf{r},\omega) = \delta_{ij} - 8\pi \int d\nu \frac{E_\nu(\mathbf{r})d_i^\nu(\mathbf{r})\left(d_j^\nu(\mathbf{r})\right)^*}{(\hbar\omega)^2 - (E_\nu(\mathbf{r}))^2 + i\eta\omega}, \tag{C.6}$$

where the infinitesimal $\eta \to +0$ prescribes the position of the poles in the ω complex plane, the expression (C.5) may be identically rewritten as

$$\Gamma = \frac{1}{2\pi\hbar} \int \operatorname{Im} \epsilon_{ij}(\mathbf{r},\omega_D) \, \mathcal{E}_i^{vc}(\mathbf{r}) \left(\mathcal{E}_i^{vc}(\mathbf{r})\right)^* d^3r, \tag{C.7}$$

where $\epsilon_{ij}(\mathbf{r},\omega_D)$ is the dielectric tensor of the organic medium at the frequency of the Wannier exciton.

APPENDIX D

ENERGY TRANSFER IN THE PLANAR GEOMETRY

In this appendix[69] some details and results of calculations of the Förster transfer in the planar geometry is presented, the donor being a semiconductor quantum well (QW), while the surrounding organic material plays the role of acceptor (7), (8). Three different possibilities are considered, corresponding to the initial excited state in the semiconductor being a free Wannier–Mott exciton, a localized Wannier–Mott exciton, or a dissociated electron–hole pair.

The geometry of the problem is the following. We consider a symmetric structure, consisting of a semiconductor QW of thickness L_w between two barriers of thickness L_b each, the whole semiconductor structure being surrounded by thick slabs of an organic material (actually, we assume each slab to be semi-infinite). We assume that in the frequency region here considered the semiconductor background dielectric constant ϵ is real (including only the contribution of higher resonances with respect to the exciton resonance under consideration) and the same for the well and the barrier, while that of the organic material $\tilde{\epsilon}$ is complex. For simplicity we assume the organic material to be isotropic (generalization to the anisotropic case is straightforward). So, the dielectric constant to be used in eqn (C.7) as well as in the Poisson equation below, is

$$\epsilon_{ij}(\mathbf{r}) = \begin{cases} \epsilon \delta_{ij}, & |z| < L_w/2 + L_b, \\ \tilde{\epsilon}, & |z| > L_w/2 + L_b, \end{cases} \tag{D.1}$$

where the z-axis is chosen to be along the growth direction, $z=0$ corresponding to the center of the QW.

D.1 Free excitons

We adopt a simplified microscopic quantum-mechanical model of a 2D Wannier–Mott exciton, in which the polarization (eqn C.3) can be taken to vanish for $|z| > L_w/2$ and inside the well to be given by the product of the 1s-wavefunction of the relative motion of the electron and hole at the origin, with the lowest subband envelope functions for the electron and hole in the approximation of

---

[69]See also Agranovich, V. M., Basko, D. M., La Rocca, G. C., and Bassani, F. 1998). *J. Phys.: Condens. Matter* **10**, 9369 , and Basko, D. M. (2003). *Energy transfer from a semiconductor nanostructure to an organic material and a new concept for light-emitting devices.* In: *Thin Films and Nanostructures. Electronic Excitations in Organic Based Nanostructures* **31**, edited by V. M. Agranovich and G. F. Bassani. Elsevier, Amsterdam. pp. 447–486.

the infinitely deep well and finally with the wavefunction of the center-of-mass motion, all of them normalized according to eqn (C.1). Thus, we have

$$\mathbf{P}(\mathbf{r}) = \mathbf{d}^{vc}\sqrt{\frac{2}{\pi a_{B2}^2}}\frac{2}{L_w}\cos^2\left(\frac{\pi z}{L_w}\right)\frac{e^{i\mathbf{k}\mathbf{r}_\|}}{\sqrt{S}}, \tag{D.2}$$

where S is the in-plane normalization area, $\mathbf{k}$ is the in-plane wavevector of the center-of-mass motion, $\mathbf{r}_\| \equiv (x,y)$ is the in-plane component of $\mathbf{r}$ and a_{B2} is the 2D 1s-exciton Bohr radius, which is half the bulk Bohr radius (9). We choose the x-axis as the direction of the in-plane component of the exciton dipole moment $\mathbf{d}^{vc}$, preferring to consider the polarization not with respect to the wavevector, but to some fixed frame. This slight complication is justified since next to the free exciton we intend to study the case of localized excitons, i.e. a system with broken 2D translational symmetry. Evidently, we need to consider two cases: $\mathbf{d}^{vc}$ being parallel and perpendicular to the QW plane. We will refer to them as X and Z polarizations, respectively. When dealing only with free excitons in a single well, three modes of different symmetry would be identified: longitudinal (L), transverse (T), and perpendicular (Z). The L and Z modes correspond to the X and Z polarizations above (their energies are split by the depolarization shift, but this is immaterial for the following). For the T mode the dipole–dipole interaction considered here vanishes (10).

The corresponding electric field $\mathbf{E}(\mathbf{r}) \equiv -\boldsymbol{\nabla}\phi(\mathbf{r})$ can be obtained from the solution of the Poisson equation for the potential $\phi(\mathbf{r})$ (the charge density being $\rho(\mathbf{r}) \equiv -\boldsymbol{\nabla}\mathbf{P}(\mathbf{r})$)

$$\epsilon(z)\boldsymbol{\nabla}^2\phi(\mathbf{r}) = 4\pi\boldsymbol{\nabla}\cdot\mathbf{P}(\mathbf{r}), \tag{D.3}$$

with the appropriate boundary conditions at $z = \pm L_w/2$ and at $z = \pm(L_w/2+ +L_b)$, i.e. continuity of the tangential component of the electric field $\mathbf{E}(\mathbf{r})$ and of the normal component of the electric displacement $\mathbf{D}(\mathbf{r}) = \epsilon(z)\mathbf{E}(\mathbf{r})$. Writing $\phi(\mathbf{r} = \phi(z)e^{i\mathbf{k}\mathbf{r}_\|}$, we have the equation for $\phi(z)$:

$$\left[\frac{d^2}{dz^2}-k^2\right]\phi(z) = \begin{cases}4\pi\rho(z)/\epsilon, & |z|<L_w/2, \\ \tilde{\epsilon}, & |z|>L_w/2,\end{cases} \tag{D.4}$$

where

$$\rho^{(X)}(z) = ik_x L_w \rho_0(1+\cos qz), \tag{D.5}$$

$$\rho^{(Z)}(z) = -qL_w\rho_0\sin qz, \tag{D.6}$$

$$\rho_0 = \sqrt{\frac{2}{\pi a_{B2}^2}}\frac{d^{vc}}{\sqrt{S}L_w^2}, \quad q \equiv 2\pi/L_w \tag{D.7}$$

with the boundary conditions that $\phi(z)$ and $\epsilon(z)d\phi(z)/dz$ should be continuous at the four interfaces. The corresponding solution in the organic material (for $z > (L_w/2+L_b)$) is given by

$$\phi(z) = \rho_0 C_\mathbf{k} e^{-k(z-L_b-L_w/2)}, \tag{D.8}$$

$$C_{\mathbf{k}}^{(X)} = -\frac{ik_x}{k}\frac{8\pi^2 q}{k(k^2+q^2)}$$
$$\times \frac{\sinh(kL_w/2)}{\epsilon \sinh(kL_b + kL_w/2) + \tilde{\epsilon}\cosh(kL_b + kL_w/2)}, \quad (D.9)$$
$$C_{\mathbf{k}}^{(Z)} = \frac{8\pi^2 q}{k(k^2+q^2)}\frac{\sinh(kL_w/2)}{\epsilon \sinh(kL_b + kL_w/2) + \tilde{\epsilon}\cosh(kL_b + kL_w/2)}. \quad (D.10)$$

Thus, the electric field penetrating the organic material is given by

$$\mathbf{E}(\mathbf{r}) = [-i\mathbf{k} + k\mathbf{e}_z]\,\phi(z)e^{i\mathbf{k}\mathbf{r}_\parallel}. \quad (D.11)$$

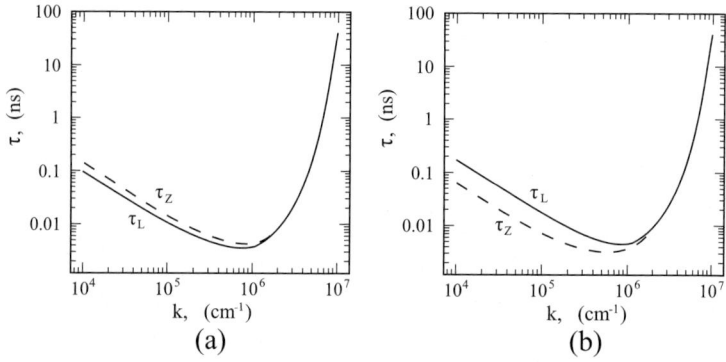

FIG. D.1. Free L-exciton (solid line) and Z-exciton (dashed line) lifetime τ (ns) versus the in-plane wavevector k (cm^{-1}), $d^{vc} = 0.1 e a_{B2}$, $L_w = 60$ Å, $L_b = 40$ Å, $\epsilon_b = 6$, $\tilde{\epsilon} = 4 + 3i$ (a); the same, but $\epsilon_b = 4$, $\tilde{\epsilon} = 6 + 3i$ (b).

Now we simply substitute this electric field into (C.7) and get the decay rate:

$$\Gamma \equiv \frac{1}{\tau} = \frac{S}{2\pi\hbar}\mathrm{Im}\,\tilde{\epsilon}\int_{L_b+L_w/2}^{+\infty} 2k^2|\phi(z)|^2 dz$$
$$= \frac{\mathrm{Im}\,\tilde{\epsilon}}{\pi^2\hbar}\frac{|d^{vc}|^2}{a_{B2}^2}\frac{k|C_{\mathbf{k}}|^2}{L_w^4}, \quad (D.12)$$

where we have considered the absorption only at $z > L_w/2 + L_b$ (considering also the organic material in $z < -L_w/2 - L_b$, τ would be twice as quick). For the numerical estimations we use the parameters, typical for II–VI semiconductor (e.g. ZnSe/ZnCdSe) quantum wells (11): $\epsilon \approx 6$, $d^{vc} \approx 0.1\ ea_{B2}$ (about 12 Debye, the Bohr radius is taken to be 25 Å). For the organic part one needs to know only the dielectric constant, taken to be $\tilde{\epsilon} = 4 + 3i$. This value is not even the most optimistic one: for PTCDA one has $\tilde{\epsilon} = 3.6 + 4.5i$ (12).

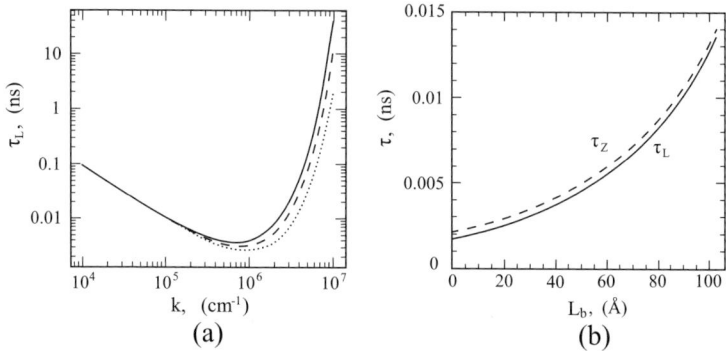

FIG. D.2. (a) Free L-exciton lifetime τ (ns) versus the in-plane wavevector k (cm^{-1}) for three well widths: $L_w = 20$ Å (dotted line), $L_w = 40$ Å (dashed line), $L_w = 60$ Å (solid line), other parameters being $L_b = 40$ Å, $\epsilon_b = 6$, $\tilde{\epsilon} = 4+3i$. (b) Free L-exciton (solid line) and Z-exciton (dashed line) lifetime τ (ns) versus the barrier width L_b (Å), $k = 10^6$ cm^{-1}, $L_w = 60$ Å, $\epsilon_b = 6$, $\tilde{\epsilon} = 4+3i$.

Two cases should be considered: $\mathbf{d}^{vc}$ lying in the QW plane, $\mathbf{k} \| \mathbf{d}^{vc}$ (L-exciton) and $\mathbf{d}^{vc}$ perpendicular to the QW plane (Z-exciton). Taking $L_w = 6$ nm, $L_b = 4$ nm, we plot τ_L and τ_Z as functions of k for $\epsilon = 6$, $\tilde{\epsilon} = 4+3i$ and $\epsilon = 4$, $\tilde{\epsilon} = 6+3i$ on Figs. D.1(a),(b). It is seen from the plot that the lifetime does not depend drastically on the polarization and the real parts of the dielectric constants. Figure D.2(a) shows that the dependence on L_w is also weak (though at large k, $kL_w \geq 2\pi$, $\tau \propto L_w^6$ from (D.10). The most interesting dependence is that on k. We see that τ exhibits a minimum at $k_{\min} \sim 1/L_b$. This dependence may be easily understood if one recalls that the dipole–dipole interaction between two planes behaves like

$$V(\mathbf{k}, z) \propto k e^{-kz}, \qquad (D.13)$$

which, when substituted into Fermi's golden rule, gives the correct asymptotics $\tau \sim 1/k$ at $k \to 0$ and exponential growth at $k \to \infty$.

Considering, first, a quasithermalized exciton distribution; typical values of k at a temperature ~ 100 K are $\sim 3 \cdot 10^6$ cm^{-1}. The corresponding energy transfer lifetime (tens of picoseconds) is much less than the exciton recombination lifetime which is about 200 ps in II–VI semiconductor QWs, as reported by different authors ((11) and references therein; (13)). We remark that for the case of free excitons in a quantum well, the effective radiative lifetime (which, assuming a thermal distribution, increases linearly with temperature) is determined by the population transfer from nonradiative excitons with large k to small k excitons undergoing fast radiative decay (14), (15). Thus, the dipole–dipole energy transfer mechanism considered here proves to be efficient enough to quench a large fraction of the semiconductor excitons, thereby activating the organic medium luminescence. Moreover, the intraband relaxation of excitons due to acoustic phonon scattering occurs at time-scales of the order of 20–30 ps at 10 K (13), which is larger than the minimal transfer time, obtained here (less than 10 ps for $k_{\min} \sim 10^6 \text{cm}^{-1}$). This makes it possible to excite QW in a way to produce the initial nonequilibrium distribution of excitons with $k = k_{\min}$, tuning the frequency of the excitation pulse to exceed the energy $\hbar\omega_{\text{exc}}(k_{\min})$ of the exciton with $k = k_{\min}$ by one LO-phonon frequency Ω_{LO} (since in II–VI semiconductors the free-carrier-to-exciton relaxation is governed mainly by LO-phonon scattering and happens in times of about 1 ps (13)–(18), or an integer multiple of Ω_{LO}, if the exciton binding energy is larger than $\hbar\Omega_{\text{LO}}$. A numerical estimate for ZnSe gives $\hbar\omega_{\text{exc}}(k_{\min}) - \hbar\omega_{\text{exc}}(k = 0) \sim 1$ meV, while $\hbar\Omega_{\text{LO}} \approx 31$ meV (13), so that the following kinetics of excitons at $k \sim k_{\min}$ is governed mainly by the acoustic phonons. Finally, another possibility would be to pump resonantly excitons with the appropriate k by using a coupling grating configuration (19).

Analogous calculations may be performed for the case of III–V semiconductor materials. We take $\epsilon \approx 11$, $d^{vc} \approx 0.05\ ea_{B2}$ and plot the L-exciton lifetime versus the wavevector k for several values of L_w (Fig. D.3, analogous to Fig. D.2(a) for II–VI materials). All other parameters are the same as in Fig. D.2(a). We see that the lifetime is longer compared to that in Fig. D.2(a) by about an order of magnitude, which is due to the larger values of a_{B2} and ϵ. However, the energy transfer discussed here is still efficient enough because the effective exciton recombination time in III–V materials is also larger (about 1 ns (20)).

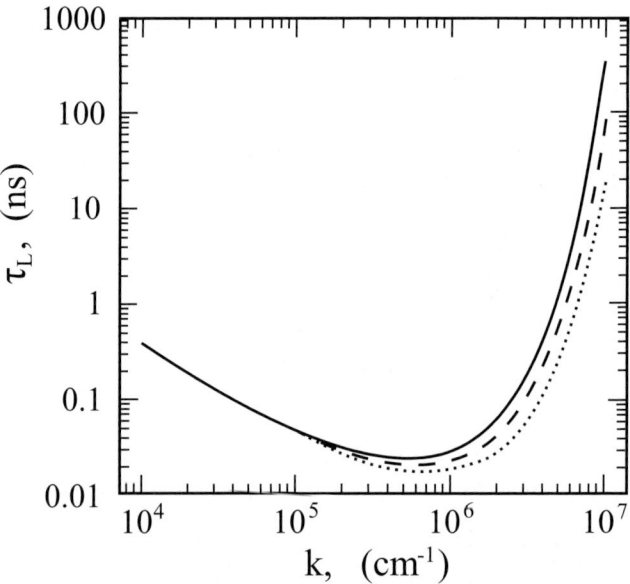

FIG. D.3. The same as in Fig. D.2, but for the III–V semiconductor compounds ($\epsilon_b = 11$, $d^{vc} = 0.05\ ea_{B2}$), all other parameters being the same as in Fig. D.2(a).

D.2 Localized excitons

Now we turn to the situation when the QW width fluctuations, alloy disorder or impurities localize the 2D exciton (such a situation is more frequent for II–VI semiconductor quantum wells than for III–V ones). Then, the wavefunction of the center-of-mass exciton motion $\Phi(\mathbf{r}_\parallel)$ is no longer just a plane wave, and the corresponding polarization is given by

$$\mathbf{P}(\mathbf{r}) = \mathbf{d}^{vc} \sqrt{\frac{2}{\pi a_B^2} \frac{2}{L_w}} \cos^2\left(\frac{\pi z}{L_w}\right) \Phi(\mathbf{r}_\parallel), \tag{D.14}$$

which implies that $\Phi(\mathbf{r}_\parallel)$ is normalized according to

$$\int d^2 r_\parallel \left|\Phi(\mathbf{r}_\parallel)\right|^2 = 1. \tag{D.15}$$

The solution of the Schrödinger equation for a particle in the random potential, caused by the QW width fluctuations and the alloy disorder is beyond the scope of the present work (much work has been done in this field, see, e.g. (21) and references therein). We can mention only some general properties that $\Phi(\mathbf{r}_\parallel)$ should have: (i) it should be localized within some distance $L \geq L_w$; (ii) it should be smooth and without nodes. As a consequence, its spatial Fourier expansion should contain mainly the components with wavevectors $k \leq 1/L$.

It is convenient to expand the wavefunction $\Phi(\mathbf{r}_\parallel)$ into plane waves:

$$\Phi(\mathbf{r}_\parallel) = \int \frac{d^2 k}{(2\pi)^2} \Phi_\mathbf{k} e^{i\mathbf{k}\mathbf{r}_\parallel}, \tag{D.16}$$

and analogously the charge density $\rho(\mathbf{r})$ and the potential $\phi(\mathbf{r})$. Then one again obtains eqn (D.4), but the charge density is now given by

$$\rho_\mathbf{k}^{(X)}(z) = ik_x L_w \tilde{\rho}_0 L \Phi_\mathbf{k}(1 + \cos qz), \tag{D.17}$$

$$\rho_\mathbf{k}^{(Z)}(z) = -q L_w \tilde{\rho}_0 L \Phi_\mathbf{k} \sin qz, \tag{D.18}$$

$$\tilde{\rho}_0 = \sqrt{\frac{2}{\pi a_B^2}} \frac{d^{vc}}{L L_w^2}. \tag{D.19}$$

The solution is

$$\phi_\mathbf{k}(z) = \tilde{\rho}_0 L \Phi_\mathbf{k} C_\mathbf{k} e^{-k(z - L_b - L_w/2)}, \tag{D.20}$$

with the same $C_\mathbf{k}$, given by (D.9), (D.10). For the decay rate we obtain:

$$\Gamma = \frac{1}{\tau} = \frac{\text{Im}\,\tilde{\epsilon}}{2\pi\hbar} \int_{L_b + L_w/2}^{\infty} dz \int \frac{d^2 k}{(2\pi)^2} 2k^2 |\phi_\mathbf{k}(z)|^2 \tag{D.21}$$

$$= \frac{\text{Im}\,\tilde{\epsilon}}{\pi^2 \hbar} \frac{|d^{vc}|^2}{a_{B2}^2} \frac{1}{L_w^4} \int \frac{d^2 k}{(2\pi)^2} k |\Phi_\mathbf{k}|^2 |C_\mathbf{k}|^2. \tag{D.22}$$

It is possible to get some information about the decay rate (D.22) based only on general properties of the wavefunction, mentioned above. We have three length-scales in our problem: L_w, L_b and L. First, we have the condition $L_w \leq L$. Since wavevectors with $kL \geq 1$, being cut off by $|\Phi_{\mathbf{k}}|^2$, do not contribute to the integral, we may set $kL_w \to 0$. The subsequent analysis depends on the relation between L_b and L.

If $L_b \ll L$, we may put $kL_b \to 0$ as well. Then we have

$$\frac{C_{\mathbf{k}}^{(X)}}{L_w^2} \to -\frac{2\pi i}{\tilde{\epsilon}} \frac{k_x}{k}, \qquad \frac{C_{\mathbf{k}}^{(Z)}}{L_w^2} \to \frac{2\pi}{\tilde{\epsilon}} \tag{D.23}$$

and the integral may be estimated as

$$\frac{1}{\tau_X} = A \frac{2 \operatorname{Im} \tilde{\epsilon}}{\hbar} \frac{|d^{vc}|^2}{|\tilde{\epsilon}|^2} \frac{1}{a_{B2}^2} \frac{1}{L}, \qquad \frac{1}{\tau_Z} = A \frac{4 \operatorname{Im} \tilde{\epsilon}}{\hbar} \frac{|d^{vc}|^2}{\epsilon^2} \frac{1}{a_{B2}^2} \frac{1}{L} \tag{D.24}$$

up to a numerical factor $A \sim 1$, determined by the detailed shape of $\Phi_{\mathbf{k}}$. We have set the average value of k over the wavefunction $\Phi_{\mathbf{k}}$ to be A/L and if for the X polarization we assume $\Phi_{\mathbf{k}}$ to be cylindrically symmetric (which may be considered as the average over the realizations of disorder), then the numerical factor A is the same for both cases.

In the opposite limit, $L_b \gg L$ (which also implies $L_b \gg L_w$), we may set $\Phi_{\mathbf{k}} = \Phi_{\mathbf{k}=0}$ since the values of $\mathbf{k}$, contributing to the integral, are determined by $C_{\mathbf{k}}$ (namely, $k \leq 1/L_b$) which in this limit takes the form

$$\frac{C_{\mathbf{k}}^{(X)}}{L_w^2} \to -\frac{ik_x}{k} \frac{4\pi}{(\tilde{\epsilon}+\epsilon) e^{kL_b} + (\tilde{\epsilon}-\epsilon) e^{-kL_b}}, \tag{D.25}$$

$$\frac{C_{\mathbf{k}}^{(Z)}}{L_w^2} \to \frac{4\pi}{(\tilde{\epsilon}+\epsilon) e^{kL_b} + (\tilde{\epsilon}-\epsilon) e^{-kL_b}}. \tag{D.26}$$

Estimating the integral, we have

$$\frac{1}{\tau_X} = A \frac{1}{\pi\hbar} \frac{\operatorname{Im} \tilde{\epsilon}}{|\tilde{\epsilon}+\epsilon|^2} \frac{|d^{vc}|^2}{a_{B2}^2} \frac{|\Phi_{\mathbf{k}=0}|^2}{L_b^3}, \tag{D.27}$$

where $|\Phi_{\mathbf{k}=0}|^2 \sim L^2$, which follows from the normalization condition. The expression for $1/\tau_Z$ differs from this by an additional factor of 2 and the factor A may be different in the two cases. It is determined by the values of $\tilde{\epsilon}$, ϵ:

$$A = \int_0^\infty \frac{4\xi^2 d\xi}{|\frac{\tilde{\epsilon}+\epsilon}{|\tilde{\epsilon}+\epsilon|} e^\xi \pm \frac{\tilde{\epsilon}-\epsilon}{|\tilde{\epsilon}+\epsilon|} e^{-\xi}|^2}, \tag{D.28}$$

and is bounded by

$$\frac{\pi^2}{12} = \int_0^\infty \frac{\xi^2 d\xi}{\cosh^2 \xi} < A < \int_0^\infty \frac{\xi^2 d\xi}{\sinh^2 \xi} = \frac{\pi^2}{6}. \tag{D.29}$$

So we see that at $L \ll L_b$ the decay rate is proportional to L^2, at $L \gg L_b$ – to L^{-1}, therefore it has a maximum at some $L \sim L_b$. This is in agreement

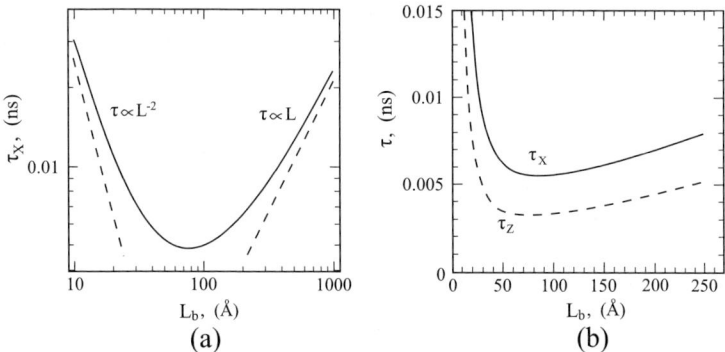

FIG. D.4. (a) Localized X-exciton lifetime τ versus the localization length L(Å) (solid line) along with the limiting cases $L \ll L_b$ and $L \gg L_b$ (dashed lines), $L_w \ll L$, $L_b = 40$ Å, $\epsilon_b = 6$, $\tilde{\epsilon} = 4 + 3$i. (b) Localized X-exciton (solid line) and Z-exciton (dashed line) lifetime τ (ns) versus the localization length L(Å), $L_w = 10$ Å, $L_b = 40$ Å, $\tilde{\epsilon} = 4 + 3$i.

with the results of the previous appendix, since the plane waves, giving the largest contribution to the wavefunction $\Phi(\mathbf{r}_\parallel)$ and thus determining the decay rate, have the wavevector values of the order $k \sim 1/L$ and we have seen that wavevectors corresponding to the shortest lifetimes were $k_{\min} \sim 1/L_b$.

To illustrate these considerations, we choose a specific example of the localized wavefunction – that of the ground state in the isotropic parabolic potential:

$$\Phi_\mathbf{k} = \sqrt{4\pi} L e^{-k^2 L^2/2}, \tag{D.30}$$

which obviously has all necessary features mentioned in the beginning of this appendix. For this wavefunction the integral in (D.22) may be evaluated numerically for arbitrary parameters L, L_w, L_b (we recall that only $L \geq L_w$ are physically relevant). The results of the calculation (τ versus L) are plotted in Fig. D.4 along with the asymptotic dependencies for $L_b = 40$ Å, $\epsilon = 6$, $\tilde{\epsilon} = 4+3$i (we have set $L_w \to 0$ for the plots in Fig. D.4(a), but a more specific value $L_w = 10$ Å was chosen for Fig. D.4(b). In the limit $L \gg L_b$ the coefficients $A = \sqrt{\pi}/2$, the coefficient for $L \ll L_b$, given by (D.28) was calculated numerically.

We also plot the dependence, analogous to that in Fig. D.4 (b), for parameters typical of III–V semiconductors: $d^{vc}/ea_{B2} = 0.05$ and $\epsilon = 11$ (Fig. D.5); analogously to the previous appendix, we obtain larger lifetimes than those for II–VI semiconductors.

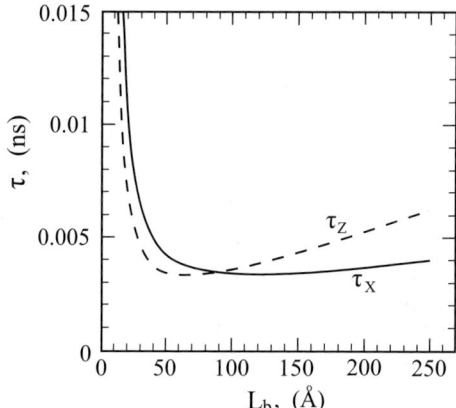

FIG. D.5. The same as in Fig. D.4 (b), but for the III–V semiconductor compounds, other parameters being the same as in Fig. D.4(b).

REFERENCES

Chapter 1
1. Frenkel, Ya. J. (1931). *Phys. Rev.* **37**, 17.
2. Frenkel, Ya. J. (1931). *Phys. Rev.* **37**, 1276.
3. Peierls, R. E. (1932). *Ann. Physik* **13**, 5, 905.
4. Frenkel, Ya. J. (1936). *Sov. Phys.* **9**, 158.
5. Wannier, G. H. (1937). *Phys. Rev.* **52**, 191.
6. Lyubarsky, G. Ya. (1958). *Group Theory and its Application in Physics*. Fizmatgiz, Moscow.
7. Gross, E. F. (1957). *Usp. Fiz. Nauk* **63**, 575; English translation: (1957). *Sov. Phys. Usp. Nauk* **63**, 576.
8. Knox, R. S. (1963). *Theory of Excitons*. Academic Press, New York.
9. Thomas, D. G. and Hopfield, J. J. (1960). *Phys. Rev. Lett.* **5**, 505; (1961) *Phys. Rev.* **124**, 657.
10. Agranovich, V. M. and Ginzburg, V. L. (1984). *Crystal Optics with Spatial Dispersion, and Excitons*. Springer, Berlin, Heidelberg (1962); *Usp. Fiz. Nauk* **76**, 643 and **77**, 663; (1963) *Fortschr. Phys.* **11**, 163; English translation: *Sov. Phys. Usp.* **5**, 323 and 675 (1962/63).
11. Agranovich, V. M. (1966). *Fiz. Tverd. Tela* **8**, 2801.
12. Broude, V. L., Rashba, E. I., and Sheka, E. F. (1961). *Dokl. Akad. Nauk SSSR* **139**, 1085. Rashba, E. I. (1957). *Optics and Spectroscopy* **2**, 576; (1962) *Fiz. Tverd. Tela* **4**, 330.
13. Agranovich, V. M. and Konobeev, Yu. V. (1959). *Optika i spektroskopija* **6**, 242.
14. Wolf, H. C. and Gallus, G. (1966). *Phys. Stat. Sol.* **16**, 277.
15. Davydov, A. S. (1951). *Theory of Light Absorption in Molecular Crystals*. UKSSR Akad. Nauk, Kiev;
(1964). *Usp. Fiz. Nauk* **82**, 393.
16. Ginzburg, V. L. (1967). *The Propagation of Electromagnetic Waves in Plasma*. Nauka, Moscow; English translation: (1970). Pergamon, Oxford.
17. Stepanov, B. I. (1957). *Dokl. Akad. Nauk SSSR* **112**, 839.
18. Ketskemety, I., Dombi, I. and Horvai, R. (1961). *Ann. Physik* **8**, 342.
19. Haken, H. (1958). *Fortschr. Phys.* **6**, 271; (1959) **68**, 565.
23. Philpott, M. R. (1973). *Advances in Chemical Physics* **23**, 227-342.
24. Turlet, J. M., Kottis, Ph., and Philpott, M. R. (1976). *Advances in Chemical Physics* **54**, 303-468.

Chapter 2

1. Obreimov, I. V. (1927). *J. Russia Soc. Phys. Chem.* **59**, 548.

2. Obreimov, I. V. and de Haas, W. J. (1928). *Proc. Akad. Sci. Amsterdam* **31**, 353.
3. Obreimov, I. V. and de Haas, W. J. (1929). *Proc. Akad. Sci. Amsterdam* **32**, 1324.
4. Obreimov, I. V. and Prikhotko, A. F. (1932). *Sov. Phys.* **1**, 203; (1936) **9**, 34 ; (1936) **9**, 48.
5. McClure, D. S. (1959). *Solid State Phys.* **8**, 1.
6. Wolf, H. C. (1959). *Solid State Phys.* **9**, 1.
7. Broude, V. L., Klimusceva, G. V. , Liberman, A. L., Onoprinenko, M. I., Prikhotko, A. F. and Shatenshtein, A. I. (1965). *Absorption spectra of molecular crystals (benzene and its homologs)*. Kiev.
8. Broude, V. L., Prikhotko, A. F. and Rashba, E. I. (1959). *Usp. Fiz. Nauk* **67**, 99.
9. Davydov, A. S. (1951). *Theory of Light Absorption in Molecular Crystals*. UKSSR Akad. Nauk, Kiev.
10. Davydov, A. S. (1968). *Theory of molecular excitons*. Nauka, Moscow; English translation: (1971). Plenum, New York.
11. Davydov, A. S. (1948). *Zh. Èksper. Theoret. Fiz.* **18**, 210.
12. Agranovich, V. M. and Ginzburg, V. L. (1966). (1984). *Crystal Optics with Spatial Dispersion, and Excitons*. Springer, Berlin, Heidelberg.
13. Landau, L. D. and Lifshitz, E. M. (1962). *Field Theory*. Gostekhizdat, Moscow.
14. Bethe, H. (1929). *Ann. Phys.* **3**, 133.
15. Ewald, P. P. (1912). *Dissertation*. München; (1917), *Ann. Phys.* **54**, 519, 557. (1924). *Ann. Phys.* **64**, 253.
16. Born, M. and Huang, K. (1954). *Dynamical Theory of Crystal Lattices*. Clarendon Press, Oxford.
17. Knox, R. S. (1963). *Theory of Excitons*. Academic Press, New York.
18. Winston, H. (1951). *J. Chem. Phys.* **19**, 156.
19. Winston, H., Halford, R. S. (1949). *J.Chem. Phys.* **17**, 607.
20. Lax, M. (1974). *Symmetry principles in solid state physics and molecular physics*. John Wiley, New York.
21. Lyubarsky, G. Ya. (1958). *Group Theory and its Application in Physics*. Fizmatgiz, Moscow.
22. Yanagawa, S. (1953). *Progr. Theor. Phys.* **10**, 83.
23. Herring, C. (1937). *Phys. Rev.* **52**, 361.
24. Koster, G. F. (1957). *Solid State Phys.* **5**, 211.
25. Prikhotko, A. F. and Soskin, M. S.(1962). *Optics and Cpectroscopy* **13**, 522.
26. Sternlicht, H. and McConnell, H. M. (1962). *J. Chem. Phys.* **35**, 1793.
27. Lynden-Bell R. M. and McConnell, H. M. (1962). *J. Chem. Phys.* **37**, 794.
28. Thomas, D. G., Keller, H. and McConnell, H. M. (1963). *J. Chem. Phys.* **39**, 2321.
29. Thomas, D. G., Merkl A. W., Heldebrandt A. F. and McConnell, H. M. (1964). *J. Chem. Phys.* **40**, 2588.

30. McConnell, H. M. and Lynden-Bell R. M. (1962). *J. Chem. Phys.* **36**, 2588.
31. Merrifield, R. E. (1955). *J. Chem. Phys.* **23**, 402.
32. Lyons, L. E. (1957). *J. Chem. Soc.* **541**, 5001.
33. Landau, L. D. and Lifshitz, E. M. (1963). *Quantum Mechanics*. Pergamon Press, Oxford.
34. Hutchison, C. A. and Mangum B. W. (1958). *J. Chem. Phys.* **29**, 952. (1961). **34**, 908.
35. Van der Waals, J. H. and de Groot, M. S. (1960). *Molecular Physics* **3**, 190; Kottis, P. and Lefebver, R. (1963). *J. Chem. Phys.* **39**, 393.
36. Deigen, M. F. and Pekar, S. I. (1958). *Zh. Èksper. Theoret. Fiz.* **34**, 684.
37. Grescishkin, W. S and Ajbinder, H. E. (1967). *Usp. Fiz. Nauk* **91**, 645.
38. Schwoerer, M. and Wolf, H. C. (1967). ESR- Untersuchungen an Naphtalin-d_8: Naphtalin-h_8-Mischkristallen in deren Triplettzustand. *Molecular Crystals* **3**, 177.
39. Jortner, J., Rice, S. A., Katz, I. L. and Choi, S-I. (1965). *J. Chem. Phys.* **42**, 309.
40. Choi, S-I., Jortner, J., Rice, S. A. and Silbey, R. (1964). *J. Chem. Phys.* **41**, 3294.
41. Khokhlov, Yu. K. (1968). *Trudy IF AN SSSR* **53**, 1.
42. Berk, N. F., Bizzaro, W., Rosenthal, J., and Yarmus, L. (1981). *Phys. Rev. B* **23**, 5661,5673.
43. Reineker, P. (1975). *Phys. Status Sol. B* **70**, 189; 471.
44. Reineker, P. (1982). *Exciton Dynamics in Molecular Crystals and Aggregates. Stochastic Liouville Equation Approach: Coupled Coherent and Incoherent Motion, Optical Lineshapes, Magnetic Resonance Phenomena*. Springer Tracts in Modern Physics, Vol. **94**, Springer, Berlin, Heidelberg.
45. Blume, M. (1968). *Phys. Rev.* **174**, 351.
46. Swenberg, C. E. and Giacintov, N. E. (1973). In *Organic Molecular Physics*. Edited by Birks, J. B. Wiley, New York, vol.**1**.
47. Hochstrasser, R. M. and Whiteman, J .D. (1972). *J. Chem. Phys.* **56**, 5945.
48. Hochstrasser, R. M. (1976). *International review of science: Physical Chemistry* vol. **3**, pp. 1–36.
49. Agranovich, V. M., Basko, D. M., Schmidt, K., La Rocca, G. C., Bassani, F., Forrest, S. R., Leo, K., and Lidzey, D. G. (2001). *Chem. Phys.* **272**, 159.

Chapter 3

1. Agranovich, V. M. (1959). *Zh. Èksper. Theoret. Fiz.* **37**, 430.
2. Agranovich, V. M. (1961). *Fiz. Tverd. Tela* **3**, 811; English translation: (1961). *Sov. Phys. Solid State* **3**, 592].
3. Demidenko, A. A. (1961). *Fiz. Tverd. Tela* **3**, 1164.
4. Hoffmann, R. (1963). *Radiation Research* **20**, 140.
5. Fano, U. (1956). *Phys. Rev.* **103**, 1202.

6. Hopfield, J. J. (1958). *Phys. Rev.* **112**, 1555.
7. Agranovich, V. M. and Basko, D. M. (2000). *J. Chem. Phys.* **112**, 8156.
8. Bakalis, L. D. and Knoester, J. (1997). *J. Chem. Phys.* **106**, 6964.
9. Chesnut, D. B. and Suna, A. (1963). *J. Chem. Phys.* **39**, 146.
10. Mahan, G. D. (1964). *J. Chem. Phys.* **41**, 2930.
11. Mahan, G. D. (1975). *NATO Advanced Study Institutes Series. Series B, Physics* **B9**, 79.
12. Craig, D. P. (1955). *J. Chem. Soc.* **539**, 2302.
13. Craig, D. P. and Hobbins, P. O. (1955). *J. Chem. Soc.* **539**, 2309.
14. Landau, L. D. and Lifshitz, E. M. (1963). *Quantum Mechanics*. Pergamon Press, Oxford.
15. Mavroyannis, C. (1965). *J. Chem. Phys.* **42**, 1772.
16. Toyozawa, Y. (1958). *Progr. Theor. Phys.* **20**, 53.
17. Agranovich, V. M. and Konobeev, Yu. V. (1963). *Fiz. Tverd. Tela* **5**, 2544; English translation: (1963). *Sov. Phys. Solid State* **5**, 1858.
18. Agranovich, V. M. and Konobeev, Yu. V. (1964). *Fiz. Tverd. Tela* **6**, 831; English translation: (1964). *Sov. Phys. Solid State* **6**, 644.
19. Davydov, A. S. (1951). *Theory of Light Absorption in Molecular Crystals*. UKSSR Akad. Nauk, Kiev.
20. Pekar, S. I. (1951). *Studies on Electronic Theory of Crystals*. Gostechizdat, 1951.
21. Klinger, M. I. (1961). *Izv. AN SSSR, Ser. Phys.* No. **11**, 1342.
22. Firsov, Yu. A. and Kudinov, E. K. (1964). *Zh. Èksper. Theoret. Fiz.* **47**, 601; (1965). *Zh. Èksper. Theoret. Fiz.* **49**, 867; (1965). *Fiz. Tverd. Tela* **7**, 546, (1965). *Fiz. Tverd. Tela* **7**, 2634.
23. Rashba, E. I. (1957). *Optics and spectroscopy* **2**, 658.
24. Weller, W. (1961). *Z. Naturforsch.* **16a**, 401.
25. Trlifai, M. (1956). *Czech. J. Phys.* **6**, 6.
26. Rashba, E. I. (1957). *Optics and spectroscopy* **2**, 75; 88; (1957). **3**, 568.
27. Ioselevich, A. S. and Rashba, E. I. (1992). In: Kagan, Yu. and Legget, A. J. (Eds.). *Quantum Tunneling in Condensed Media*. North-Holland, Amsterdam, p. 347.
28. Mizuno, K. and Matsui, A. (1987). *J. Lumin.* **38**, 323;
Matsui, A., Mizuno, K., Tamai, N. and Yamazaki, I. (1988). *J. Lumin.* **40-41**, 455;
Matsui, A. and Mizuno, K. (1988). *Proc. SPIE* **910**, 131; (1989). *Proc SPIE* **1054**, 224;
Furukawa, M., Mizuno, K., Matsui, A., Tamai, N., and Yamazaki, I. (1989). *Chem. Phys.* **138**, 423;
Matsui, A. (1990). *J. Opt. Soc. Am. B* **7**, 1615.
29. Kagan, Yu. and Prokof'ev, N. V. (1992). In: Kagan, Yu. and Legget, A. J. (Eds.). *Quantum Tunneling in Condensed Media*. North-Holland, Amsterdam, p. 37.

30. Birks, J. B. (1975). *Excimers. Rep. Prog. Phys.* **38**, pp. 903–974.
31. Deigen, M. F. and Pekar, S. I. (1951). *Zh. Èksper. Theoret. Fiz.* **21**, 803.
32. Rashba, E. I. (1982). In: Rashba, E. I. and Sturge, M. D. (Eds.) *Excitons.* North Holland, Amsterdam, p. 543.
33. Landau, L. D. (1933). *Phys. Z. Soviet.* **3**, 664.
34. Pekar, S. I. (1954). *Untersuchungen über die Elektronentheorie der Kristallen.* Akademie–Verlag, Berlin.
35. Peierls, R. E. (1932). *Ann. Physik* **13**, 905.
36. Frenkel, Ya. J. (1936). *Sov. Phys.* **9**, 158.
37. Davydov, A. S. (1971). *Theory of Molecular Excitons.* Plenum Press, New York.
38. Schwentner. N. Koch, E. E., and Jortner, J. (1985). *Electronic Excitations in Condensed Rare Gases.* Springer Tracts in Modern Physics, Vol. **107**. Springer, Berlin, Heidelberg.
39. Ueta, M., Kanzaki, H., Kabayashi, K., Toyozawa, Y., and Hanamura, E. (1985). *Excitonic Processes in Solids.* Springer Series in Solid State Sciences, Vol. **60**. Springer, Berlin, Heidelberg.
40. Zimmerer, G. (1987). In: Grassano, U. and Terzi, N. (Eds.) *Proceedings of the XCVI Course of the E. Fermi School on Excited State Spectroscopy in Solids.* North Holland, Amsterdam, p. 37.
41. Haarer, D. and Philpott, M. R. (1983). In: *Spectroscopy and Excitation Dynamics of Condensed Molecular Systems*, edited by V. M. Agranovich and R. M. Hochstrasser. North–Holland, Amsterdam, pp. 27–82.
42. Hoffmann, M. (2003). *Mixing of Frenkel and Charge-Transfer Excitons and their Quantum Confinement in thin Films.* In: Agranovich, V. M. and Bassani, G. F. (Eds.) *Electronic Excitations in Organic Based Nanostructures. Thin Films and Nanostructures* **31**. Elsevier Academic Press, Amsterdam, pp. 221–292.
43. Agranovich, V. M. and Zakhidov, A. A. (1977). *Chem. Phys. Lett.* **50**, 278.
44. Agranovich, V. M., Antonyuk, B. P., and Mal'shukov, A. G. (1976). *Pis'ma Zh. Èksper. Theoret. Fiz.* **23**, 492; English translation: (1976). *JETP Lett.* **23**, 448;
Agranovich, V. M., Antonyuk, B. P., Ivanova, E. P., and Mal'shukov, A. G. (1977). *Zh. Èksper. Theoret. Fiz.* **72**, 614; English translation: (1977). *Sov. Phys. JETP* **45**, 322.
45. Agranovich, V. M. and Kamchatnov, A. M. (1999). *Chem. Phys.* **245**, 175.
46. Knoester, J. (2002). *Optical properties of molecular aggregates.* In: *Organic Nanostructures: Science and Applications.* Proceedings of the International School of Physics Enrico Fermi, Course CXLIX. Edited by V. M. Agranovich and G. C. La Rocca, Bologna, pp. 149–86.
1 47. Knoester, J. and Agranovich, V. M. (2003). *Frenkel and charge-transfer excitons in organic solids.* In: Agranovich, V. M. and Bassani, G. F. (Eds.), *Electronic Excitations in Organic Based Nanostructures. Thin Films and Nanostructures* **31**. Elsevier Academic Press, Amsterdam, pp. 1–96.

48. Balagurov, D. B., La Rocca, G. C., and Agranovich, V. M. (2003). *Phys. Rev. B* **68**, 045418.
49. Embriaco, D., Balagurov, D. B., La Rocca, G. C., and Agranovich, V. M. (2004). *Phys. Stat. Sol. C* **1**, 1429.
50. Simpson, W. T. and Peterson, D. L. (1957). *J. Chem. Phys.* **26**, 588.
51. McClure, D. S. (1958). *Can. J. Chem.* **36**, 59.
52. Merrifield, R. E. (1964). *J. Chem. Phys.* **40**, 445.
53. Suna, A. (1964). *Phys. Rev. A* **135**, 111.
54. Rashba, E. I. (1966). *Zh. Èksper. Theoret. Fiz.* **50**, 1064; (1968). **54**, 542.
55. Philpott, M. R. (1967). *J. Chem. Phys.* **47**, 2534.
56. Merrifield, R. E. (1968). *J. Chem. Phys.* **48**, 3693.
57. Broude, V. L., Rashba, E. I., and Sheka, E. F. (1967). *Phys. Stat. Sol.* **19**, 395.
58. Hoffman, M. and Soos, Z. G. (2002). *Phys. Rev. B* **66**, 024305.
59. Holstein, T. (1959). *Ann. Phys. (N. Y.)* **8**, 325, 343.
60. Schlosser, D. W. and Philpott, M. R. (1980). *Chemical Physics* **49**, 181.
61. Schlosser, D. W. and Philpott, M. R. (1982). *J. Chem. Phys.* **77** (4), 1969.
62. Born, M. and Huang, K. (1954). *Dynamical Theory of Crystal Lattices*. Clarendon Press, Oxford.
63. Clark, L. B. and Philpott, M. R. (1970). *J. Chem. Phys.* **53**, 3790.
64. Chen, H. H. and Clark, L. B. (1969). *J. Chem. Phys.* **51**, 1862.
65. Cruickshank, D. W. J. (1957). *Acta Cryst.* **10**, 504.
66. Cruickshank, D. W. J. (1956). *Acta Cryst.* **9**, 915.
67. Campbell, R. B., Robertson, M. J., and Trotter, J. (1962). *Acta Cryst.* **15**, 289.
68. George, G. A. and Morris, G. C. (1968). *J. Mol. Spectry* **26**, 67.
69. Angus, J. G., Christ, B. J., and Morris, G. C. (1968). *Aust. J. Chem.* **21**, 2153.
70. Philpott, M. R. (1969). *J. Chem. Phys.* **50**, 5157.
71. Lyons, L. E. and Morris, G. C. (1965). *J. Chem. Soc.* 2764.
72. Bree, A. and Lyons, L. E. (1960). *J. Chem. Soc.* 5206.
73. Hellner, C., Lindquist, L., and Roberge, P. C. (1972). *J. Chem. Soc. Faraday II* **68**, 1928.
74. Silbey, R., Jortner, J., and Rice, S. A. (1965). *J. Chem. Phys.* **42**, 1515.
75. Fox, D. and Yatsiv, Sh. (1957). *Phys. Rev.* **108**, 938.
76. Davydov, A. S. and Sheka, E. F. (1965). *Phys. Stat. Sol.* **11**, 877.
77. Claxton, T. A., Craig, D. P., and Thirunamachadran, T. (1961). *J. Chem. Phys.* **35**, 1525.
78. Agranovich, V. M. and Ginzburg, V. L. (1984). *Crystal Optics with Spatial Dispersion, and Excitons*. Springer, Berlin, Heidelberg.
79. Brodin, M. S. and Marisova, S. V. (1961). *Optics and Spectroscopy* **10**, 473.
80. Wolf, H. C. (1959). *Solid State Phys.* **9**, 1; (1958). *Z. Naturforsch.* **13a**, 414.
81. Philpott, M. R. (1971). *J. Chem. Phys.* **54**, 111.
82. Lyons, L. E. (1957). *J. Chem. Soc.* **541**, 5001.

83. Tanaka, J. (1959). *Progr. Theor. Phys. Supplement* **12**, 183.
84. Kitajgorodskij, A. I. (1955). *Organical Crystallography*. Izd. AN SSSR, Moscow.
85. Merrifield, R. E., (1961). *J. Chem. Phys.* **34**, 1835.
86. Choi, S-I., Jortner, J., Rice, S. A. and Silbey, R. (1964). *J. Chem. Phys.* **41**, 3294.
87. Goldhirsch, I., Levich, E., and Yakhot, V. (1979). *Phys. Rev. B* **19**, 4780.
88. Goldhirsch, I. and Yakhot, V. (1980). *Phys. Rev. B* **21**, 2833.
89. Tyablikov, S. V. (1975). *Methods in the Quantum Theory of Magnetism*. 2nd edition, Plenum, New York.
90. Agranovich, V. M. and Toshich, W. S. (1968). *Sov. Phys. JETP* **26**, 104.
91. Holstein, T. and Primakoff, H. (1940). *Phys. Rev.* **58**, 1098.
92. Ovander, L. N. (1965). *Usp. Fiz. Nauk* **86**, 3.
93. Toshich, W. S. (1967). *Fiz. Tverd. Tela* **9**, 1713.
94. Keldysh, L. W. and Kozlov, A. N. (1967). *Zh. Èksper. Theoret. Fiz.*, Letters to the Redaction **5**, 538; (1968). *Zh. Èksper. Theoret. Fiz.* **54**, 978.
95. Gergel, W. A., Kazarinov, R. F., and Suris, R.A. (1968). *Zh. Èksper. Theoret. Fiz.* **54**, 978.
96. Moskalenko, S. A. (1962). *Fiz. Tverd. Tela* **4**, 276; Blatt. J. M., Boer, K. W., and Brandt, W. (1962). *Phys. Rev.* **126**, 1691.
97. Agranovich, V. M. and Galanin, M. D. (1982). *Electronic excitation transfer in condensed matter*. North–Holand, Amsterdam.
98. Agranovich, V. M. (1968). *Theory of Excitons*. Izd. Nauka, Moscow.
99. Agranovich, V. M., Dubovsky, O. A. and Kamchatnov, A. M. (2001). *Synth. Metal.* **116**, 293.
100. Agranovich, V. M., Dubovsky, O. A., Basko, D. M., La Rocca, G. C., and Bassani, F. (2000). *J. Lumin.* **85**, 221.

Chapter 4

1. Heitler, W. (1954). *The Quantum Theory of Radiation*. Clarendon Press, Oxford.
2. Born, M. and Huang, K. (1954). *Dynamical Theory of Crystal Lattices*. Clarendon Press, Oxford.
3. Tolpygo, K. B. (1950). *Zh. Èksper. Theoret. Fiz.* **32**, 520.
4. Huang, K. (1951). *Proc. Roy. Soc.* **A 208**, 352.
5. Agranovich, V. M. and Rukhadze, A. A. (1958). *Zh. Èksper. Theoret. Fiz.* **35**, 982; English translation: (1959). *Sov. Phys. JETP* **35**, 685.
6. Hopfield, J. J. and Thomas D. G. (1960). *J. Phys. Chem. Solids* **12**, 276.
7. Fano, U. (1956). *Phys. Rev.* **103**, 1202.
8. Neamtan, S. M. (1953). *Phys. Rev.* **92**, 1362; (1954). *Phys. Rev.* **94**, 327.
9. Agranovich, V. M. (1959). *Zh. Èksper. Theoret. Fiz.* **37**, 430;(1960). *Sov. Phys. JETP* **10**, 307.

10. Fano, U. (1960). *Phys. Rev.* **118**, 451.
11. Nicols, G. and Rice, S. A. (1967). *J. Phys. Chem.* **46**, 4445.
12. Hopfield, J. J. (1958). *Phys. Rev.* **112**, 1555.
13. Agranovich, V. M. (1959). *Role of defects in exciton luminescence of molecular crystals.* X All–Union Conf. on Luminescence. Leningrad, published 1960; Agranovich, V. M. (1960). *Uspekhi Fiz. Nauk* **71**, 141; (1960). *Sov. Phys.-Uspekhi* **3** (3), 427.
14. Ball, M. A. and McLachlan, A. D. (1964). *Proc. Roy. Soc.* **282**, 433.
15. Mavroyannis, C. (1967). *J. Math. Phys.* **8**, 1515, 1522.
16. Bassani, F. and Quattropani, A. (1985). *Helvetica Phys. Acta* **58**, 244.
17. Landau L. D. and Lifshitz E. M. (1960). *Electrodynamics of Continuous Media.* Pergamon, Oxford.
18. Pekar, S. I. (1957). *Zh. Èksper. Theoret. Fiz.* **33**, 1022.
19. Agranovich, V. M. and Ginzburg, V. L. (1984). *Crystal Optics with Spatial Dispersion, and Excitons.* Springer, Berlin, Heidelberg.
20. Toyozawa, Y.(1959). *Suppl. of Prog. Theor. Phys.***12**, 111
21. Agranovich, V. M. and Dubovsky, O. A. (1966). *Pis'ma Zh. Èksper. Theoret. Fiz.* **3**, 345; English translation: (1966). *JETP Lett.* **3**, 223.
22. Agranovich, V. M. and Gartstein, Yu. N. (2007). *Phys. Rev. B* **75**, 075302 (7 pages).
23. Orrit, M., and Kottis, Ph. (1986). *Phys. Rev. B* **34**, 680.
24. Citrin, D. S. (1992). *Phys. Rev. Lett.* **69**, 3393.
25. Andreani, L. C., Tassone, F., and Bassani, F. (1991). *Solid State Commun.* **77**, 641.
26. Weisbuch, C. Benisty, H., and Houdré, R., (2000). *J. Lumin.* **85**, 271.
27. Hanamura, E. (1973). *Solid State Commun.* **12**, 951.
28. Aaviksoo, Ya., Lippmaa, Ya., and Reinot, T. (1987). *Opt. Spectrosc. USSR* **62**, 419.
29. Deveaud, B., Clérot, F., Roy, N., Satzke, K., Sermage, B., and Katzer, D. S. (1991). *Phys. Rev. Lett.* **62**, 2355.
30. Bassani, F. (2003). *Polaritons.* In: *Electronic Excitations in Organic Based Nanostructures.* Edited by Agranovich, V. M. and Bassani, F. Elsevier, Amsterdam, pp. 129–183.
31. Perebeinos, V., Tersoff, J., and Avouros, P. (2005). *Nano Letters* **5**, 2495.
32. Spataru, C., Ismail-Beigi, S., Capaz, R. B., and Louie, S. G. (2005). *Phys. Rev. Lett.* **95**, 247402.
33. Schott,. M. (2003). *Synthetic Metals.* **139**. 739.
34. Lee, P. A. and Ramakrishnan, T. V. (1985). *Rev. Mod. Phys.* **57**, 287.
35. Gershoni, D., Katz, M., Wegscheider, W., Pfeiffer, L. N., Logan, R. A., and West, K. (1994). *Phys. Rev. B* **50**, 8930.
36. Lécuiller, R., Berrhar, J., Ganière, D., Lapersonne–Meyer, C., Lavallard, P., and Shott, M. (2002). *Phys. Rev. B* **66**, 125205.

Chapter 5

1. Agranovich, V. M. and Ginzburg, V. L. (1984). *Crystal Optics with Spatial Dispersion, and Excitons.* Springer, Berlin, Heidelberg.
2. Agranovich, V. M. (1968). *Theory of Excitons.* Izd. Nauka, Moscow.
3. Agranovich, V. M. (1974). *Usp. Fiz. Nauk* **112**, 143.
4. Born, M. and Huang, K. (1954). *Dynamical Theory of Crystal Lattices.* Clarendon Press, Oxford.
5. Dunmur, D. A. (1972). *Mol. Phys.* **23**, 109.
6. Cummins, P. G., Dunmur, D. A. and Munn, R. W. (1973). *Chem. Phys. Lett.* **22**, 519.
7. Wünsche, A. (1975). *Ann. Phys.* **32**, 7 Folge, Heft **2**, 122.
8. Wünsche, A. (1975). *Experimentale Technik der Physik* **XXIII**, Heft 3, 323.
9. Fulton, R. L. (1974). *J. Chem. Phys.* **61**, 4141.
10. Khokhlov, Yu. K. (1972). *Trudy FIAN SSSR* **59**, 221.
11. Philpott, M. R. and Mahan, G. D. (1973). *J. Chem. Phys.* **59**, 445.
12. Rashba, E. I. (1972). In: *Physics of impurity centres in crystals.* Proc. Intern. Seminar, Tallin, p. 427.
13. Agranovich, V. M. and Talanina, I. B. (1987). *Chem. Phys. Lett.* **134**, 525.
14. Bakhshiev, N. (1972). *Spectroscopy of intermolecular interactions.* Nauka, Leningrad (in Russian).
15. Vol'kenstein, M. W. (1951). *Molecular Optics.* Gostechizdat, Moscow.
16. Onodera, Y. and Toyozawa, Y. (1968). *J. Phys. Soc. Japan* **24**, 341.
17. Dubovsky, O. A. and Konobeev, Yu. V. (1970). *Fiz. Tverd. Tela* **12**, 406.
18. Hoschen, J. and Jortner, J. (1970). *Chem. Phys. Lett.* **5**, 351.
19. Hong, H. K. and Robinson, G. W. (1970). *J. Chem. Phys.* **52**, 825.
20. Obreimov, I. V. (1945). *Applications of Fresnel diffraction for physical and technical measurements.* AN SSSR, Moscow.
21. Broude, V. L. and Rashba, E. I. (1961). *Fiz. Tverd. Tela* **3**, 1941.
22. Agranovich, V. M., Doronina, V. I. and Konobeev, Yu. V. (1969). *Fiz. Tverd. Tela* **11**, 2607.
23. Kamenogradskiy, N. E. and Konobeev, Yu. V. (1970). *Phys. Status Solidi* **37**, 29.
24. Mahan, G. D. (1967). *Phys. Rev.* **153**, 983.
25. Mahan, G. D. and Mazo, R. M. (1968). *Phys. Rev.* **175**, 1191.
26. Nienhuis, G. and Deutch, J. M. (1972). *J. Chem. Phys.* **56**, 235, 1819, 5511.
27. Smith, D. Y. and Dexter, D. L. (1972). *Progr. Opt.***X**, 165.
28. Mahan, G. D. (1975). In: *Electronic Structure of Polymers and Molecular Crystals*, edited by Jean–Marie Andre and Janos Ladik. Plenum Press, New York, London, p. 79.
29. Landau L. D. and Lifshitz E. M. (1965). *Quantum mechanics.* Addison–Wesley, Reading, Mass.
30. Landau, L. D. and Lifshitz, E. M. (1962). *Field Theory.* Gostekhizdat, Moscow (in Russian).

31. Dunmur, D. A. and Munn, R. W. (1975). *Chem. Phys.* **11**, 297.
32. Hochstrasser, R. M. and Noe, L. J. (1969). *J. Chem. Phys.* **50**, 1684.

Chapter 6

1. Einstein, A. (1906). *Ann. der Phys.* **22**, 1800.
2. Einstein, A. (1911). *Ann. der Phys.* **35**, 679.
3. Debye, P. (1912). *Ann. der Phys.* **39**, 789.
4. Bethe, H. (1931). *Z. Phys.* **71**, 205.
5. Wortis, M. (1963). *Phys. Rev.* **132**, 85.
6. Hanus, J. (1962). *Phys. Rev. Lett.* **11**, 336.
7. Van Kranendonk, J. (1959). *J. Phys.* **25**, 1080.
8. Van Kranendonk, J. and Karl, G. (1968). *Rev. Mod. Phys.* **40**, 531.
9. Gush, H. P., Hare, W. F. I., Allin, E. I., and Welsh, H. L. (1960). *Can. J. Phys.* **38**, 176.
10. Bogani, F. (1978). *J. Phys. C.: Solid State Phys.* **11**, 1283, 1297.
11. Sheka, E. F. (1971). *Usp. Fiz. Nauk* **104**, 593.
12. Califano, S., Schettino, V. and Netto, N. (1981). *Lattice Dynamics of Molecular Crystals*. Springer, Berlin, Heidelberg.
13. Gardini, G. (1988). *Chemical Physics* **119**, 241.
14. Agranovich, V. M. (1970). *Fiz. Tverd. Tela* **12**, 562. English translation: (1970). *Sov. Phys. Solid State* **12**, 430.
15. Agranovich, V. M. (1973). *Supplement to the Russian edition*, Mir Publishers, Moscow, of H. Poulet and J. P. Mathieu, 1970, *Spectres de vibration et symetrie des cristaux*. Gordon and Breach, Paris.
15. Ruvalds, I. and Zawadowski, A. (1970). *Phys. Rev. B* **2**, 1172.
16. Ruvalds, I. and Zawadowski, A. (1970). *Phys. Rev. Lett.* **24**, 1111.
17. Maradudin, A. A. (1971). In: *Phonons*, edited by M. A. Nusimovici. Flammarion Sciences, Paris, pp. 427–443.
18. Agranovich, V. M. and Lalov, I. I. (1985). *Sov. Phys. Uspekhi* **28** (6), 484.
19. Agranovich, V. M. and Lalov, I. I. (1971). *Fiz. Tverd. Tela* **13**, 1032; English translation: (1971). *Sov. Phys. Solid State* **13**, 859.
20. Agranovich, V. M. and Lalov, I. I. (1971). *Zh. Èksper. Theoret. Fiz.* **61**, 656; English translation: (1972). *Sov. Phys. JETP* **34**, 350.
21. Agranovich, V. M. and Lalov, I. I. (1976). *Solid State Commun.* **19**, 503.
22. Agranovich, V. M., Ivanova E. P., and Lalov, I. I. (1979). *Fiz. Tverd. Tela* **21**, 1629.
23. Agranovich, V. M., Efremov, N. A., and Kaminskaya, E. P. (1971). *Opt. Commun.* **3**, 387.
24. Efremowv, N. A. and Kaminskaya, E. P. (1972). *Fiz. Tverd. Tela* **14**, 1185.
25. Efremowv, N. A. and Kaminskaya, E. P. (1972). *Fiz. Tverd. Tela* **14**, 2105.
26. Agranovich, V. M., Lozovik, Yu. E., and Mekhtiev, M. A. (1970). *Zh. Èksper. Theoret. Fiz.* **59**, 246.

27. Agranovich, V. M., Dubovsky, O. A., and Stoichev, K. Ts. (1979). *Fiz. Tverd. Tela* **21**, 3012.
28. Agranovich, V. M. and Dubovsky, O. A. (1980). *Mol. Cryst. Liq. Cryst.* **57**, 175.
29. Krauzman, M., Pick, R., Poulet, H., Hamel, G., and Prevot, B. (1974). *Phys. Rev. Lett.* **33**, 528.
30. Prevot, B., Hennion, B., and Dorner, B. (1977). *J. Phys. C.* **10**, 3999.
31. Agranovich, V. M. and Lalov, I. I. (1976). *Fiz. Tverd. Tela* **18**, 1971.
32. Krishnan, R. S. (1946). *Proc. Ind. Acad. Sci.* **24**, 25.
33. Tubino, R. and Birman, J. L. (1975). *Phys. Rev. Lett.* **35**, 670.
34. Ron, A. and Hornig, D. F. (1963). *J. Chem. Phys.* **39**, 1129.
35. Leibfried, G. and Ludwig, W. (1961). *Theory of Anharmonic Effects in Crystals.* Academic Press, New York.
36. Reisland, J. A. (1973). *The Physics of Phonons.* Wiley, New York.
37. Poulet, H. and Mathieu, J. P. (1970). *Spectres de vibration et symetrie des cristaux.* Gordon and Breach, Paris.
38. Pitaevsky, L. P. (1976). *Zh. Èksper. Theoret. Fiz.* **70**, 738.
39. Proville, L. (2005). *Phys. Rev. B* **71**, 104306 ; **72**, 184301.
40. Lalov, I. I. (1974). *Fiz. Tverd. Tela* **16**, 2476.
41. Agranovich, V. M., Efremov, N. A., and Kobozev, I. K. (1976). *Fiz. Tverd. Tela* **18**, 3421.
42. Lifshitz, E. M. (1956). *Nuovo Cimento Suppl.* **3** (X), 716.
43. Herzberg, G. (1945). *Infrared and Raman Spectra of Polyatomic Molecules.* Van Nostrand, New York.
44. Agranovich, V. M. and Ginzburg, V. L. (1984). *Crystal Optics with Spatial Dispersion, and Excitons.* Springer, Berlin, Heidelberg.
45. Wright, G. B. (editor). (1969). *Light Scattering Spectra of Solids.* Plenum Press, New York.
46. Agranovich, V. M. (1959). *Zh. Èksper. Theoret. Fiz.* **37**, 430.
47. Agranovich, V. M. and Mekhtiev, M. A. (1971). *Fiz. Tverd. Tela* **13**, 2424.
48. Mooradian, A., and McWhorter, A. L. (1969). *Phys. Rev.* **177**, 1231.
49. Bloembergen, N. (1965). *Nonlinear Optics.* Benjamin, New York.
50. Fain, V. M. and Khanin, Ya. I. *Quantum Radiophysics.* Moscow, 1965.
51. Hanamura, E. (1973). *Solid State Commun.* **12**, 951.
52. Flytzanis, C. (1978). In: *Treatise in Quantum electronics*, edited by N. Rabin and C. L. Tang, vol. **1 A**, p. 111. Plenum Press, New York.
53. Grun, J. B., Vu Duy Phach, Bivas, A., and Hönerlage, B. (1978). *Phys. Stat. Sol. (b)* **86**, 159.
54. Dows, D. A. and Schettino, V. (1973). *J. Chem. Phys.* **58**, 5009.
55. Schettino, V. and Salvi, P. R. (1975). *Spectrochim. Acta* bf 31 A, 399.
56. Mavrin, B. N. and Sterin, Kh. E. (1972). *Zh. Èksper. Theoret. Fiz. Pis'ma* **16**, 265.
57. Winter, F. X. and Claus, R. (1972). *Opt. Commun.* **6**, 22.

58. Gorelik, V. S., Mitin, G. G., and Sushinskii, M. M. (1975). *Zh. Èksper. Theoret. Fiz.* **69**, 1823.
59. Polivanov, Yu. N. (1979). *Zh. Èksper. Theoret. Fiz. Pis'ma* **30**, 415.
60. Polivanov, Yu. N. (1979). *Fiz. Tverd. Tela* **21**, 1884.
61. Gorelik, V. S., Maximov, O. P., Mitin, G. G., and Sushinskii, M. M. (1977). *Solid State Commun.* **21**, 615.
62. Polivanov, Yu. N. (1978). *Usp. Fiz. Nauk* **126**, 185.
63. Aktsipetrov, O. A., Kitaeva G. Kh., and Penin, A. N. (1977). *Fiz. Tverd. Tela* **19**, 1001.
64. Aktsipetrov, O. A., Kitaeva G. Kh., and Penin, A. N. (1977). *Fiz. Tverd. Tela* **20**, 402.
65. Gorelik, V. S., Mitin, G. G., and Polivanov, Yu. N. (1978). *Kristallogr.* **23**, 561.
66. Krauzman, M., Le Postollec, M., and Mathieu, J.-P. (1973). *Phys. Stat. Sol. (b)* **60**, 761.
67. Belousov, M. V. and Pogarev, D. E. (1978). *Zh. Èksper. Theoret. Fiz. Pis'ma* **28**, 692.
68. Gale, G. M., Vallee, F., and Flytzanis, C. (1986). *Phys. Rev. Lett.* **57**, 1867.
69. Vallee, F., Gale, G. M., and Flytzanis, C. (1988). *Phys. Rev. Lett.* **61**, 2102.
70. Vallee, F., Gale, G. M., and Flytzanis, C. (1988). *Chem. Phys. Lett.* **149**, 572.
71. Mitin, G. G., Gorelik, V. S., Kulevskii, L. A., Polivanov, Yu. N., and Sushinskii, M. M. (1975). *Sov. Phys. JETP* **41**, 882.
72. Lifshitz, E. M. (1956). *Il Nuovo Cimento Suppl.* **3**, 716.
73. Dubovsky, O. A. (1985). *Solid State Commun.* **54**, 261.
74. Agranovich, V. M. and Dubovsky, O. A. (1988). *Phonon multimode spectra: biphonons and triphonons in crystals with defects.* In: R. J. Elliott and I. P. Ipatova (Eds.), *Optical Properties of Mixed Crystals with defects.* North-Holland, Amsterdam, pp. 297–398.
75. Rashba, E. I. (1988). In: R. J. Elliott and I. P. Ipatova (Eds.), *Optical Properties of Mixed Crystals with defects.* North-Holland, Amsterdam, pp. 215–295.
Price, W. C., Sherman, W. F., and Wilkinson, G. R. (1960). *Proc. Roy. Soc.* **255**, 5.
77. Belousov, M. V., Wolf, B. E., Ivanova, E. A. and Pogarev, D. E. (1982). *Zh. Èksper. Theorct. Fiz. Lett.* **35**, 457.
78. Lalov, I. I. (1975). *Phys. Stat. Sol. (b)* **68**, 319, 681.
79. Krivenko, T. A., Sheka, E. F., and Rashba, E. I. (1978). *Mol. Cryst. Liq. Cryst.* **47**, 119.
80. Gotschev, I. and Lalov, I. I. (1979). *Bulg. J. Phys.* **6**, No. 4.
81. Green, D. E. (1973). *Science* **181**, 583.
82. Davydov, A. S. and Kislukha, N. I. (1973). *Phys. Stat. Sol. (b)* **59**, 465.
83. Davydov, A. S. and Kislukha, N. I. (1976). *Zh. Èksper. Theoret. Fiz.* **71**, 293.

REFERENCES

84. Elliot, R. S., Krumahnsl, J. A., and Leath, P. L. (1974). *Rev. Mod. Phys.* **46**, 465.
85. Agranovich, V. M. and Dubovsky, O. A. (1981). *Fiz. Tverd. Tela* **23**, 2197.
86. Agranovich, V. M. and Lalov, I. I. (1976). *Optics Commun.* **16**, 239.

Chapter 7

1. Dzyaloshinsky, I. E. and Pitayevsky, L. P. (1959). *Zh. Èksper. Theoret. Fiz.* **36**, 1797; English translation: (1959). *Sov. Phys. JETP* **9**, 1282.
2. Born, M. and Huang, K. (1954). *Dynamical Theory of Crystal Lattices.* Clarendon Press, Oxford.
3. Agranovich, V. M. and Ginzburg, V. L. (1984). *Crystal Optics with Spatial Dispersion, and Excitons.* Springer, Berlin, Heidelberg.
4. Pekar, S. I. (1957). *Zh. Èksper. Theoret. Fiz.* **33**, 1022; English translation: (1958). *Sov. Phys. JETP* **6**, 785; (1959). *Zh. Èksper. Theoret. Fiz.* **36**, 451; English translation: (1959). *Sov. Phys. JETP* **9**, 314.
5. Abrikosov, A. A., Gorkov, L. P., and Dzyaloshinsky, I. E. (1962). *Methods of Quantum Field Theory in Statistical Physics.* Fizmatgiz, Moscow (in Russian); English translation: (1975). *Methods of Field Theory in Statistical Physics.* Dover.
6. Schiff, L. I. (1968). *Quantum Mechanics.* Third Ed., McGraw Hill, New York.
7. Landau, L. D. and Lifshitz, E. M. (1963). *Quantum Mechanics.* Pergamon Press, Oxford.
8. Tolpygo, K. B. (1950). *Zh. Èksper. Theoret. Fiz.* **32**, 520.
9. Agranovich, V. M. (1960). *Usp. Fiz. Nauk* **71**, 141.
10. Agranovich, V. M. and Konobeev, Yu. V. (1961). *Fiz. Tverd. Tela* **3**, 360.
11. Demidenko, A. A. (1963). *Fiz. Tverd. Tela* **5**, 489.
12. Knox, R. S. (1963). *Theory of Excitons.* Academic Press, New York.
13. Agranovich, V. M. and Konobeev, Yu. V. (1963). *Fiz. Tverd. Tela* **5**, 2544; English translation: (1963). *Sov. Phys. Solid State* **5**, 1858.
14. Toyozawa, Y. (1958). *Progr. Theor. Phys.* **20**, 53.
15. Toyozawa, Y. (1962). *Progr. Theor. Phys.* **22**, 455.
16. Ilinskii, Yu. A. and Keldysh, L. V. (1994). *Electromagnetic Response of Material Media.* Plenum, New York.
17. Lorentz, H. A. (1916). *The Theory of Electrons.* Teubner, Leipzig, reprinted by Dover, New York (1951).
18. Bloembergen, N. (1965). *Nonlinear Optics.* Benjamin, New York.
19. Chemla, D. S. and Zyss, J. (1987). (Eds.) *Nonlinear Optical Properties of Organic Molecules and Crystals.* Academic Press, INC. Harcourt, Brace, and Jovanovich (Publishers).
20. Dick, B., Hochstrasser, R. M., and Trommsdorff, H. P. (1987). In: Chemla, D. S. and Zyss, J. (Eds.), *Nonlinear Optical Properties of Organic Molecules and Crystals.* Academic Press, INC. Harcourt, Brace, and Jovanovich (Publishers), vol. **2**, p. 159.
21. Ovander, L. N. (1965). *Usp. Fiz. Nauk* **86**, 3.

Chapter 8

1. Agranovich, V. M. and Kravtsov, V. E. (1985). *Solid State Commun.* **55**, 85; Born, M. and Wolf, E. (1968). *Principles of Optics.* Pergamon, Oxford.
2. Landau, L. D. and Lifshitz, E. M. (1960). *Electrodynamics of Continuous Media.* Pergamon, Oxford.
3. Agranovich, V. M. and Ginzburg, V. L. (1984). *Crystal Optics with Spatial Dispersion, and Excitons.* Springer, Berlin, Heidelberg.
4. Keldysh, L. V. (1979). *Zh. Èksper. Theoret. Fiz. Lett.* **29**, 716; **30**, 244; English translation: (1979). *Sov. Phys. JETP Lett.* **29**, 658; **30**, 224.
5. Agranovich, V. M. (1991). *Solid State Commun.* **78**, 747.
6. Boyd, R. W. and Sipe, J. E. (1994). *J. Opt. Soc. Am. B* **11**, 297.
7. Fisher, G. L., Boyd, R. W., Gehr, R. J. et al. (1975). *Phys. Rev. Lett.* **74**, 1871.

Chapter 9

1. Agranovich, V. M. (1993). *Mol. Cryst. Liq. Cryst.* **230**, 13.
2. Agranovich, V. M. and Dubovsky, O. A. (1966). *Pis'ma v Zh. Èksper. Theoret. Fiz.* **3**, 345; English translation: (1966). *Sov. Phys. JETP Lett.* **3**, 223.
3. Aaviksoo, Ya., Lipmaa, Ya., and Reinot, D. (1987). *Opt. Spectrosc. USSR* **62**, 419.
4. Deveaud, B., Clérot, F., Roy, N., Satzke, K., Sermage, B., and Katzer, D. S. (1991). *Phys. Rev. Lett.* **67**, 2355.
5. Davydov, A. S. (1971). *Theory of Molecular Excitons.* Plenum Press, New York.
6. Agranovich, V. M. and Kamchatnov, A. M. (2003). *Optics and Nonlinearities of Excitons in Organic Multilayered Nanostructures and Superlattices.* In: Agranovich, V. M. and Bassani, G. F. (Eds.), *Electronic Excitations in Organic Based Nanostructures. Thin Films and Nanostructures* **31**. Elsevier Academic Press, Amsterdam, pp. 185–220.
7. Fermi, E. (1931). *Z. Phys.* **71**, 250.
8. Newell, A. C. and Moloney, J. V. (1992). *Nonlinear Optics.* Addison–Wesley, Redwood City.
9. Agranovich, V. M. (1983). *Biphonons and Fermi resonance in vibrational spectra of crystals.* In: *Spectroscopy and Excitation Dynamics of Condensed Molecular Systems.* Edited by V. M. Agranovich and R. M. Hochstrasser. North-Holland, Amsterdam, pp. 83–138.
10. Agranovich, V. M. and Dubovsky, O. A. (1988). *Phonon multimode spectra: biphonons and triphonons in crystals with defects.* In: R. J. Elliott and I. P. Ipatova (Eds.), *Optical Properties of Mixed Crystals with defects.* North-Holland, Amsterdam.

11. Agranovich, V. M. and Kamchatnov, A. M. (1994). *JETP Lett.* **59**, 425.
12. Agranovich, V. M. and Dubovsky, O. A. (1993). *Chem. Phys. Lett.* **210**, 458.
13. Agranovich, V. M., Kamchatnov, A. M., and Munn, R. W. (2002). *Chem. Phys.* **282**, 399.
14. Agranovich, V. M. and Page, J. B. (1993). *Phys. Lett. A* **183**, 395.

Chapter 10

1. Agranovich, V. M. and Malshukov, A. G. (1974). *Opt. Commun.* **11**, 169.
2. Agranovich, V. M. and Leskova, T. A. (1988). *Progress in Surface Science* **29**, 169 ;
Vinogradov, E. A, and Leskova, T. A. (1990). *Physics Reports* **194**, 273.
3. Bellessa, J., Bonnand, C., Plenet, J. C., and Mugnier, J. (2004). *Phys. Rev. Lett.* **93**, 036402.
4. Pockrand, I., Swalen, J. D., Brillante, A., and Philpott, M. R. (1978). *J. Chem. Phys.* **69**, 4001.
5. Pockrand, I., Brillante, A., and Möbius, D. (1982). *J. Chem. Phys.* **77**, 6289.
6. Bonnand, C., Bellessa, J., and Plenet, J. C. (2006). *Phys. Rev. B* **73**, 245330.
7. Weisbuch, C., Nishioka, M., Ishikawa, A. and Arakawa, Y. (1992). *Phys. Rev. Lett.*, **69**, 3314.
8. Burstein, E. and Weisbuch, C. (1995). (Eds.), *Confined Electrons and Photons: New Physics and Applications*. Plenum Press, New York;
Weisbuch, C. and Rarity, J. G. (Eds.) (1996). *Microcavities and Photonic Bandgaps: Physics and Applications*. Kluwer, Dordrecht;
Skolnick, M. S., Fisher, T. A., and Whittaker, D. M. (1998). *Semicond. Sci. Technol.* **13**, 645;
Khitrova, G., Gibbs, H. M., Jahnke, F., Kira, M., and Koch, S. W. (1999). *Rev. Mod. Phys.* **71**, 1591;
Kavokin, A. and Malpuech, G. (2003). *Cavity Polaritons*. Elsevier, Amsterdam;
Bassani, F. (2003). *Polaritons*. In: *Electronic Excitations in Organic Based Nanostructures*. Edited by Agranovich, V. M. and Bassani, F. Elsevier, Amsterdam, pp. 129–183;
Weisbuch, C. and Benisty, H. (2006). *Solid State Commun.* **135**, 627.
9. Agranovich, V. M., Benisty, H., and Weisbuch, C. (1997). *Solid State Commun.* **102**, 631.
10. Lidzey, D. G., Bradley, D. D. C., Skolnick, M. S., Virgili, T., Walker, S., and Whittaker, D. M. (1998). *Nature* **395**, 53;
Lidzey, D. G., Bradley, D. D. C., Virgili, T., Armitage, A., and Skolnick, M. S. (1999). *Phys. Rev. Lett.* **82**, 3316;
Tartakovskii, A. I., Emam-Ismail, M., Lidzey, D. G., Skolnick, M. S., Bradley, D. D. C., Walker, S., and Agranovich, V. M. (2001). *Phys. Rev. B* **63**, 121302;
Hobson P. A. , Barnes W. L. , Lidzey, D. G., Gehring G. .A., Whittaker, D. M.,

Skolnick, M. S., and Walker, S. (2002). *Appl. Phys. Lett.* **81**, 3519;

Schouwink, P., Lupton, J., Von Berlepsch, H., Dahne L., and Mahr, R. F. (2002). *Phys. Rev. B*, **66** 081203(R);

Lidzey, D. G., Fox A. M., Rahn, M. D., Skolnick, M. S., Agranovich, V. M., and Walker, S. (2002). *Phys. Rev. B*, **65**, 195312;

Takada N., Kamata, T., and Bradley, D. D. C. (2003). *Appl. Phys. Lett.* **82**, 1812;

Lidzey, D. G. (2002). In: V. M. Agranovich, G. C. La Rocca (Eds.). *Organic Nanostructures: Science and Applications*. IOS Press, Amsterdam, p. 405;

Lidzey, D. G. (2003). In: V. M. Agranovich and F. Bassani (Eds.). *Electronic Excitations in Organic Based Nanostructures*. Elsevier, Amsterdam, Chapter 8.

11. Oulton, R. F., Takada, N., Koe J., Stavrinou, P. N. and Bradley, D. D. C. (2003). *Semicond. Sci. Technol.* **18**, 419.
12. Holmes, R. J. and Forrest, S. R. (2004). *Phys. Rev. Lett.* **93**, 186404 .
13. Holmes, R. J. and Forrest, S. R. (2005). *Phys. Rev. B* **71**, 235203.
14. Holmes, R. J. and Forrest, S. R. (2007). *Organic Electronics* **8**, 33.
15. Agranovich, V. M., Litinskaia, M., and Lidzey, D. G. (2003). *Phys. Rev. B* **67**, 85311.
16. Song, J.-H., He, Y., Nurmikko, A. V., Tischler, J., and Bulovic, V. (2004). *Phys. Rev. B.* **69**, 235330.
17. Litinskaia, M., Reineker, P., and Agranovich, V. M. (2004). *J. of Luminescence* **110**, 364.
18. Litinskaia, M. L., Reineker, P., and Agranovich, V. M. (2004). *Phys. Status Solidi A* **201**, 646.
19. Zoubi, H. and La Rocca, G. C. (2005). *Phys. Rev. B* **71**, 235316; (2005). *Phys. Rev. B* **72**, 125306.
20. Zoubi, H. (2006). *Phys. Rev. B* **74**, 045317.
21. Kèna-Cohen, S., and Forrest, S. R. (2008). *Phys. Rev. B*, **77**, 073205.
22. Savvidis, P. G., Baumberg, J. J., Stevenson, R. M., Skolnick, M. S., Whittaker, D. M., and Roberts, J. S. (2000). *Phys. Rev. Lett.* **84**, 1547.
23. Stevenson, R. M., Astratov, V. N., Skolnick, M. S., Whittaker, D. M., Emam-Ismail, M., Tartakovskii, A. I., Savvidis, P. G., Baumberg, J. J., and Roberts, J. S. (2000). *Phys. Rev. Lett.* **85**, 3680.
24. Savvidis, P. G., Baumberg, J. J., Stevenson, R. M., Skolnick, M. S., Whittaker, D. M., and Roberts, J. S. (2000). *Phys. Rev. B* **62**, R13278.
25. Alexandrou, A., Bianchi, G. Péronne, E., Hallé, B., Boeuf, F., André, R., Romestein, R., and Le Si Dang. (2001). *Phys. Rev. B* **64**, 233318.
26. Deng, H., Weihs, G., and Santori, C. (2002). *Science* **298**, 199.
27. Imamoglu, A., Ram, R. J. Pau, S., and Yamamoto, Y. (1996). *Phys. Rev. A* **53**, 4250;
    Yamamoto, Y. (2000). *Nature* **405**, 629.
28. Tassone, F., Bassani, F., and Andreani, L. C. (1992). *Phys. Rev. B*, **45**, 6023;

Andreani, L. C. (1995). In: E. Burstein and C. Weisbuch (Eds), *Confined Electrons and Photons*. Plenum Press, New York, p. 57.
29. Davydov, A. S. (1968). *Theory of molecular excitons*. Nauka, Moscow; English translation: (1971). Plenum, New York.
30. Born, M. and Huang, K. (1954). *Dynamical Theory of Crystal Lattices*. Clarendon Press, Oxford.
31. Landau L. D. and Lifshitz E. M. (1960). *Electrodynamics of Continuous Media*. Pergamon, Oxford.
32. Michetti, P. and La Rocca, G. C. (2005). *Phys. Rev. B* **71**, 115320.
33. Agranovich, V. M. and La Rocca, G. C. (2005). *Solid State Commun.* **135**, 544.
34. Bonnand, C., Bellessa, J., Symonds, C., and Plenet, J. C. (2006). *Appl. Phys. Lett.* **89**, 231119.
35. Citrin, D. S. and Back, T. D. (2006). *Phys. Stat. Solidi B* **243**, 2349.
36. Peierls, R. E. (1955). *Quantum Theory of Solids*. Clarendon Press, Oxford, (Chapter 2).
37. Knapp, E. W. (1984). *Chem. Phys.* **85**, 73.
38. Agranovich, V. M. and Gartstein, Yu. N. (2007). *Phys. Rev. B* **75**, 075302 (7 pages).
39. Lee, P. A. and Ramakrishnan, T. V. (1985). *Rev. Mod. Phys.* **57**, 287.

Chapter 11

1. Agranovich, V. M. and Galanin, M. D. (1982). *Electronic excitation transfer in condensed matter*. North–Holand, Amsterdam.
2. Silinsh, E. A. (1980). *Molecular Crystals: Their Electronic States*. Springer, Berlin, Heidelberg.
3. Pope, M. and Swenberg, C. E. (1999). *Electronic Processes in Organic Crystals and Polymers*. Oxford University Press, Oxford.
4. Agranovich, V. M. and Zakhidov, A. A. (1977). *Chem. Phys. Lett.* **50**, 278.
5. Hoffmann, M. (2003). *Mixing of Frenkel and Charge-Transfer Excitons and their Quantum Confinement in thin Films*. In: Agranovich, V. M. and Bassani, G. F. (Eds.), *Electronic Excitations in Organic Based Nanostructures. Thin Films and Nanostructures* **31**. Elsevier Academic Press, Amsterdam, pp. 221–292.
6. Zyss, J. (Ed.) (1994). *Molecular Nonlinear Optics, Materials, Physics, and Devices*. Academic Press, New York.
7. Haarer, D. and Philpott, M. R. (1983). In: *Spectroscopy and Excitation Dynamics of Condensed Molecular Systems*, edited by V. M. Agranovich and R. M. Hochstrasser. North–Holland, Amsterdam, pp. 27–82.
8. Haarer, D. (1974). *Chem. Phys. Lett.* **27**, 91.
9. Petelenz, P. (2002). *Charge transfer excitons in organics*. In: V. M. Agranovich and G. C. La Rocca. (Eds.) *Organic Nanostructures: Science and Ap-*

plications. IOS Press, Amsterdam, Oxford, Tokyo, Washington DC, p. 1.
10. Haug, H. and Koch, S. W. (1994). *Quantum Theory of the Optical and Electronic Properties of Semiconductors.* 3rd edn. World Scientific, Singapore.
11. Chemla, D. S. and Zyss, J. (1987). *Nonlinear Optical Properties of Organic Molecules and Crystals.* Academic Press, Orlando.
12. Rashba, E. I. and Sturge, M. D. (Eds.). (1982). *Excitons.* North Holland, Amsterdam.
13. Forrest, S. R. (1997). *Chemical Rev.* **97**, 1793.
14. Knoester, J. and Agranovich, V. M. (2003). *Frenkel and charge–transfer excitons in organic solids.* In: Agranovich, V. M. and Bassani, G. F. (Eds.), *Electronic Excitations in Organic Based Nanostructures. Thin Films and Nanostructures* **31**. Elsevier Academic Press, Amsterdam, pp. 1–96.
15. Sebastian, L., Weiser, G., and Bässler, H. (1981). *Chem. Phys.* **61**, 125.
16. Sebastian, L., Weiser, G., Peter, G., and Bässler, H. (1983). *Chem. Phys.* **75**, 103.
17. Weiser, G. (1992). *Phys. Rev. B* **45**, 14076.
18. Gelinck, G. H., Piet, J. J., and Warman, J. (1999). *Synth. Met.* **101**, 553.
19. Agranovich, V. M. and Ilinskii, K. N. (1994). *Phys. Lett. A* **191**, 309.
20. Kiselev, S. A., Hartung, E., Soos, Z. G., Forrest, S. R., and Agranovich, V. M. (1998). *Chemical Physics* **238**, 365.
21. So, F. F., Forrest, S. R., Shi, Y. Q., and Steier, W. H. (1990). *Appl. Phys. Lett.* **56**, 674;
Imanishi, Y., Hattori, S., Kakuta, A. and Numata, S. (1993). *Phys. Rev. Lett.* **71**, 2098;
Nanaka, T., Mori, Y., Nagai, N., Nakagawa, Y., Saeda, N., Nakahagi, T. and Ishitani, A. (1994). *Thin Solid Films* **239**, 214.
22. Umbach, E., Seidel, C., Taborski, J., Li, R., and Soukopp, A. (1995). *Phys. Stat. Sol. (B)* **192**, 389;
Umbach, E. and Fink, R. (2002). In: *Organic Nanostructures,* edited by V. M. Agranovich and G. C. La Rocca. SIF, IOS Press, p. 233.
23. Petty, M. C. (1996). *Langmuir–Blodgett Films.* Cambridge University Press, Cambridge, p. 134.
24. Bjornholm, T., Geiser, T., Larsen, J., Jorgensen, M., Brunfeldt, K., Schaumburg, K., and Bechgaard, K. (1993). *Synthetic Metals* **55–57**, 3813.
25. Okada–Shudo, Y., Kajzar, F., Meritt, C., and Kafafi, Z. (1998). *Synthetic Metals* **94**, 91.
26. Zhu, K.–D. and Kabayashi, T. (1995). *Phys. Lett. A.* **201**, 439.
27. Fahrenbruch, A. L. and Bube, A. H. (1983). *Fundamentals of Solar Cells. Photovoltaic Solar Energy Conversion.* Academic Press, New York.
28. Belinicher, V. I. and Sturman, B. I. (1980). *Sov. Phys.–Uspekhy* **130**, 415.
29. Agranovich, V. M., La Rocca, G. C., and Bassani, F. (1995). *Pis'ma Zh.Èksper. Theoret. Fiz.* **62**, 407; English translation: (1995). *JETP Lett.* **62**, 418.
30. Cherrof, G. and Keller, S. P. (1958). *Phys. Rev.* **111**, 98.
31. Pensak, L. (1958). *Phys. Rev.* **109**, 601.

32. Goldstein, B. (1958). *Phys. Rev.* **109**, 601.
33. Swanson, J. A. (1961). *IBM Journal of Research and Development* **5**, 210.
34. Savenije, T. J., Koehorst, R. B. M., and Schaafsma, T. J. (1995). *Chem. Phys. Lett.* **244**, 363.
35. Forrest, S. R. and Thompson, M. E., Eds. (2007). *Chemical Review* **107**, Number 4.
36. Woehrle, D., Tennigkeit, B., Elbej, J., Kreienhoop, L., and Schnupfeil, G. (1993). *Mol. Cryst. Liquid Cryst.* **230**, 221;
Gunster, S., Siebentritt, S., and Messner, D. (1993). *Mol. Cryst. Liquid Cryst.* **230**, 351;
Takahashi, K., Nakatani, S. I., Matsuda, T., Nanbu, H., Kamura, T., and Murata, K. (1994). *Chem. Lett.* **11**, 2001.
37. Gregg, B. A. and Kim, Y. I. (1994). *J. Phys. Chem.* **98**, 2412.
38. Kay, N., Humphrybaker, R., and Graetzel, M. (1994). *J. Phys. Chem.* **98**, 952.
39. Karl, N., Bauer, A., Holzaepfel, J., Marktanner, J., Moebius, M., and Stoelzle, F. (1994). *Mol. Cryst. Liquid Cryst.* **252**, 243.
40. Agranovich, V. M., La Rocca, G. C., and Bassani, F. (1995). *Chem. Phys. Lett.* **247**, 525.
41. Zakhidov, A. A. and Yoshino, K. (1994). *Synthetic Metals* **64**, 155.
42. Baltscheffsky, M. (Ed.) (1990). *Current Research in Photosynthesis*, Vol. I. Kluwer Academic, Dordrecht.
43. Deisenhofer, J. and Norrsi, J. R. (Eds.) (1993). *The Photosynthetic Reaction Center*. Academic Press, New York.
44. Pieranski, P. (1980). *Phys. Rev. Lett.* **43**, 569;
Kalia, R. K. and Vashishta, P. (1981). *J. Phys. C* **14**, L643.
45. Wigner, E. and Bardeen, J. (1935). *Phys. Rev.* **48**, 84.
46. Bardeen, J. (1936). *Phys. Rev.* **49**, 653.
47. Schmitt–Rink, S., Chemla, D. S., and Miller, D. A. B. (1985). *Phys. Rev. B* **32**, 6601.
48. Schmitt–Rink, S., Chemla, D. S., and Miller, D. A. B. (1989). *Adv. in Phys.* **38**, 89.
49. Green, B. I., Orenstein, J., and Schmitt–Rink, S. (1990). *Science* **247**, 679.
50. La Rocca, G. C. (2003). *Wannier-Mott Excitons in Semiconductors*. In: *Electronic Excitations in Organic Based Nanostructures*, edited by V. M. Agranovich and F. Bassani. *Thin Films and Nanostructures* **31**. Elsevier Academic Press, Amsterdam, pp. 97–128.
51. Bloembergen, N. (1965). *Nonlinear Optics*. Benjamin, New York.
52. Flytzanis, C. (1975). In: *Quantum Electronics*, edited by H. Rabin and C. L. Tang, Vol. **I**. Academic Press, New York, p. 1.
53. Shen, Y. R. (1984). *The Principles of Nonlinear Optics*. Wiley, New York.
54. Reineker, P., Agranovich, V. M., and Yudson, V. I. (1996). *Chem. Phys. Lett.* **260**, 621.
55. Agranovich, V. M. and Galanin, M. D. (1982). In: Agranovich, V. M. and

Maradudin, A. A. (Eds.) *Electronic Excitation Energy Transfer in Condensed Matter*. North–Holand, Amsterdam.
56. Agranovich, V. M., La Rocca, G. C., and Bassani, F.(1995). *Pis'ma Zh.Èksper. Theoret. Fiz.* **62**, 407; English translation: (1995). *JETP Lett.* **62**, 418.
57. Agranovich, V. M., La Rocca, G. C., and Bassani, F. (1995). *Chem. Phys. Lett.* **247**, 525.

Chapter 12

1. Otto, A. (1968). *Z. Phys.* **216**, 398.
2. Otto, A. (1975). *Spectroscopy of Surface Polaritons by Attenuated Total Reflection*. In: Seraphin, B. O (Ed.), *Optical Properties of Solids, New Developments*. North-Holland, Amsterdam, p. 677.
3. Kretschmann, E. (1971). *Z. Phys.* **241**, 313;
   (1972). *Opt. Commun.* **5**, 331;
   (1978). *Opt. Commun.* **26**, 41.
4. Agranovich, V. M. and Ginzburg, V. L. (1984). *Crystal Optics with Spatial Dispersion, and Excitons*. Springer, Berlin, Heidelberg.
5. Agranovich, V. M. and Mills, D. L. (Eds.) (1982). *Surface Polaritons – Electromagnetic Waves at Surfaces and Interfaces*. North–Holland, Amsterdam.
6. Schoenwald, J., Burstein, E., and Elson, J. M. (1973). *Solid State Commun.* **12**, 185.
7. Suslov, I. M. (1983). *Zh. Èksper. Theoret. Fiz.* **83**, 1079.
8. Suslov, I. M. (1983). *Zh. Èksper. Theoret. Fiz.* **84**, 1792.
9. Frenkel, Ya. J. (1931), *Phys. Rev.* **37**, 17.
10. Frenkel, Ya. J. (1931), *Phys. Rev.* **37**, 1276.
11. Anex, B. G and Simpson, W .T. (1960). *Rev. Mod. Phys.* **32**, 466.
12. Pockrand, I., Brillante, A., Philpott, M. R., and Swalen, J. D. (1978). *Optics Commun.* **27**, 91.
13. Brillante, A., Pockrand, I., Philpott, M. R., and Swalen, J. D. (1978). *Chem. Phys. Lett.* **57**, 395.
14. Philpott, M. R. and Swalen, J. D. (1978). *J. Chem. Phys.* **69**, 2912.
15. Agranovich, V. M. (1975). *Usp. Fiz. Nauk* **115**, 199.
16. Agranovich, V. M. (1982). In: Agranovich, V. M. and Mills, D. L. (eds.) (1982). *Surface Polaritons–Electromagnetic Waves at Surfaces and Interfaces*. North–Holland, Amsterdam, p. 187.
17. Maradudin, A. A. (1982). In: Agranovich, V. M. and Mills, D. L. (eds.) (1982). *Surface Polaritons – Electromagnetic Waves at Surfaces and Interfaces*. North–Holland, Amsterdam, p. 405.
18. Maradudin, A. A. (1981). *Festkörperprobleme* **XXII**, 25.
19. Tomlinson, W. J. (1980). *Optics Lett.* **5**, 323.
20. Agranovich, V. M., Babichenko, V. S., and Chernyak, V. Ya. (1980). *Pis'ma v Zh. Èksper. Theoret. Fiz.* **32**, 532.
21. Maradudin, A. A. (1981). *Z. Phys. B–Condensed Matter* **41**, 341.

22. Agranovich, V. M. (1980). In *Proc. VII Intern. Conf. on Raman Spectroscopy*, edited by W. F. Murphy. North Holland, Amsterdam, p. 111.
23. Chernyak, V. Ya. (1983). *Fiz. Tverd. Tela* **25**, 614.
24. Chen, Y. J. and Carter, G. M. (1982). *Appl. Phys. Lett.* **41**, 307.
25. Agranovich, V. M., Rupasov, V. I., and Chernyak, V. Ya. (1981). *Pis'ma v Zh. Èksper. Theoret. Fiz.* **33**, 196.
26. Sivukhin, D. V. (1956). *Zh. Èksper. Theoret. Fiz.* **21**, 94.
27. Sivukhin, D. V. (1956). *Zh. Èksper. Theoret. Fiz.* **21**, 367.
28. Philpott, M. R. (1973). *J. Chem. Phys.* **58**, 588.
29. Sugakov, V. I. (1969). *Ukrain. Fiz. Zh.* **14**, 1425.
30. Sugakov, V. I. (1972). *Fiz. Tverd. Tela* **14**, 1977.
31. Sugakov, V. I. (1969). *Ukrain. Fiz. Zh.* **18**, 1495.
32. Turlet, J. M. (1979). *These d'Etat*. Bordeaux.
33. Turlet, J. M. and Philpott, M. R. (1975). *J. Chem. Phys.* **62**, 2777.
34. Orrit, M., Bernard, G., Gernet, J., Turlet, J. M., and Kottis, P. (1981). In: *Recent Developments in Condensed Matter Physics*, Vol. **2**, edited by J. T. Devreese. Plenum, p. 459.
35. Philpott, M. R. and Turlet, J. M. (1976). *J. Chem. Phys.* **64**, 3852.
36. Turlet, J. M., Kottis, Ph., and Philpott, M. R. (1983). *Adv. Chem. Phys.* **54**, 303.
37. Agranovich, V. M. and Dubovsky, O. A. (1966). *Pis'ma v Zh. Èksper. Theoret. Fiz.* **3**, 345.
38. Tovstenko, V. I. (1979). *Chem. Phys. Lett.* **68**, 483.
39. Forrest, S. R. (1997). *Chem. Rev.* **97**, 1793.
40. Hennessy, M. H., Soos, Z. G., Pascal, R. A. Jr, and Girlando, A. (1999). *Chem. Phys.* **245**, 199.
41. Hoffmann, M., Schmidt, K., Agranovich, V. M., and Leo, K. (2000). *Chem. Phys.* **258**, 73.
42. Fidder, H., Knoester, J., and Wiersma, D. A. (1991). *J. Chem. Phys.* **95**, 7880.
43. Agranovich, V. M., Schmidt, K., and Leo, K. (2000). *Chem. Phys. Lett.* **325**, 308;
(2001). *Chem. Phys. Lett.* **3336**, 536.
44. Shen, Z. and Forrest, S. R. (1997). *Phys. Rev. B.* **55**, 10578.
45. Hädicke, E. and Graser, F. (1986). *Acta Cryst. C* **42**, 189.
46. Lovinger, A. J., Forrest, S. R., Kaplan, M. L., Schmidt, P. H., and Venkatesan, T. (1984). *J. Appl. Phys.* **55**, 476.
47. Lovinger, A. J., Forrest, S. R., Kaplan, M. L., Schmidt, P. H., and Venkatesan, T. (1983). *Bull. Am. Phys. Soc.* **28**, 363;
(1983). *Bull. Am. Phys. Soc.* **28**, 476.
48. Kaplan, M. L., Day, S. C., Lovinger, A. J., Schmidt, P. H., and Forrest, S. R. (1994). Full set of crystal structure data, private communication.
49. So, F. F., Forrest, S. R., Shi, Y. Q., and Steier, W. H. (1990). *Appl. Phys. Lett.* **56**, 674.

50. Leonhardt, M., Mager, O., and Port, H. (1999). *Chem. Phys. Lett.* **313**, 24.
51. Agranovich, V. M., Atanasov, R. D., and Bassani, G. F. (1992). *Chem. Phys. Lett.* **199**, 621.
52. Kravtsov, V. E. and Lerner, I. V. (1984). *Zh. Èksper. Theoret. Fiz.* **86**, 1332.
53. Agranovich, V. M., Lozovik, Yu. E., and Mekhtiev, M. A. (1970). *Zh. Èksper. Theoret. Fiz.* **59**, 246.

Chapter 13

1. Agranovich, V. M., Atanasov, R. D., and Bassani, G. F. (1994). *Solid State Commun.* **92**, 295.
2. La Rocca, G. C. (1996). *Physica Scripta T* **66**, 142.
3. Yudson, V. I., Reineker, P., and Agranovich, V. M. (1995). *Phys. Rev. B* **52**, R5543.
4. Engelmann, A., Yudson, V. I., and Reineker, P. (1998). *Phys. Rev. B* **57**, 1784.
5. Agranovich, V. M., Benisty, H., and Weisbuch, C. (1997). *Solid State Commun.* **102**, 631;
Benisty, H., Weisbuch, C., and Agranovich, V. M. (1998). *Physica E* **2**, 909.
6. Agranovich, V. M., La Rocca, G. C., and Bassani, F. (1997). *JETP Lett.* **66**, 748.
7. Basko, D. M., La Rocca, G. C., Bassani, F., and Agranovich, V. M. (1999). *Eur. Phys. J. B* **8**, 353.
8. Agranovich, V. M. and Basko, D. M. (1999). *JETP Lett.* **69**, 250.
9. Basko, D. M., Agranovich, V. M., Bassani, F., and La Rocca, G. C. (2000). *Eur. Phys. J. B* **13**, 653.
10. Agranovich, V. M., Basko D. M., La Rocca, G. C., and Bassani, F. (2001). *Synthetic Metals* **116**, 349.
11. Agranovich, V. M., Basko D. M. La Rocca, G. C., and Bassani, F. (1998). *J. Phys.: Condens. Matter* **10**, 9369.
12. Agranovich, V. M. and Bassani, F. (2003). (Eds.) *Electronic Excitations in Organic Based Nanostructures.* Elsevier, Amsterdam.
13. Agranovich, V. M. (2007). *Hybrid organic–inorganic nanostructures and light–matter interaction.* In: *Problems of Condensed Matter*, edited by A. L. Ivanov and S. G. Tikhodeev. Clarendon Press, Oxford, pp. 24–43.
14. Rashba, E. I. and Sturge, M. D. (Eds.). (1982). *Excitons.* North Holland, Amsterdam.
15. Green, B. I., Orenstein, J., and Schmitt-Rink, S. (1990). *Science*, 679.
16. Turlet, J. M., Kottis, Ph., and Philpott, M. R. (1983). *Adv. Chem. Phys.* **54**, 303.
17. Forrest, S. R. (1997). *Chemical Rev.* **97**, 1793; (1994). *Chemical Rev.* **92**, 295.
18. Bastard, G. (1988). *Wave Mechanics Applied to Semiconductor Heterostructures.* CNRS, Paris.

19. Atanasov, R., Bassani, F., and Agranovich, V. M. (1994). *Phys. Rev. B* **49**, 2658.
20. Cingolani, R. (1997). In: *Semiconductors and Semimetals*, Vol. **44**. Academic Press, New York, pp. 163–226.
21. Yamada, T., Hoshi, H., Manaka, T., Ishikawa, K., Takezoe, H., and Fukuda, A. (1996). *Phys. Rev. B* **53**, R13314.
22. Hebard, A. F., Haddon, R. C., Fleming, R. M., and Kortan, A. R. (1991). *Appl. Phys. Lett.* **59**, 2109.
23. Kohl, M., Heitmann, D., Grambow, P., and Ploog, K. (1990). *Phys. Rev. B* **42**, 2941.
24. Madelung, O. (Ed.) (1987). *Landolt–Börnstein*, Vol. **III/22a**. Springer, Berlin.
25. Schmitt–Rink, S., Chemla, D. S., and Miller, D. A. B. (1985). *Phys. Rev. B* **32**, 6601; (1989). *Adv. Phys.* **38**, 89.
26. Jaziri, S., Romdhane, S., Bouchriha, H., and Bennaceur, R. (1997). *Phys. Lett. A* **234**, 141.
27. Schultheis, L., Honold, A., Kuhl, J., Köhler, K., and Tu, C. W. (1986). *Phys. Rev. B* **34**, 9027.
28. Vinattieri, A. (1993). *Solid State Commun.* **88**, 189.
29. D'Andrea, A. and Muzi, R. (1995). *Solid State Commun.* **95**, 493.
30. Gradshteyn, I. S. and Ryzhik, I. M. (1994). *Table of Integrals, Series, and Products*. Fifth ed., edited by A. Jeffrey. Academic Press, Boston, San Diego, New York.
31. Loudon, R. (1959). *Am. J. Phys.* **27**, 649.
32. Bánayai L., Gailbraith, I., Ell, C., and Haug, H. (1987). *Phys. Rev. B* **36**, 6099.
33. Huong, N. Q. and Birman, J. L. (2000). *Phys. Rev. B* **61**, 13131.
34. Huong, N. Q. and Birman, J. L. (2003). *Phys. Rev. B* **67**, 075313.
35. Gao, Y., Huong, N. Q., Birman, J. L., and Potasek, M. J. (2004). *J. Appl. Phys.* **96**, 4839.
36. Gao, Y., Huong, N. Q., Birman, J. L., and Potasek, M. J. (2005). *Proc. SPIE Int. Soc. Opt. Engn.* **5592**, 272.
37. Birman, J. L. and Huong, N. Q. (2007). *J. Lumin.* **125**, 196.
38. La Rocca, G. C., Bassani, F., and Agranovich, V. M. (1996). In: *Notions and Perspectives of Nonlinear Optics*. Edited by O. Keller. World Scientific, Singapore;
(1995). *Nuovo Cimento* **17**, 1555.
39. Haug, H. and Koch, S. W. (1994). *Quantum Theory of the Optical and Electronic Properties of Semiconductors*. 3rd edn. World Scientific, Singapore.
40. Honold, A., Schultheis, L., Kuhl, J., and Tu, C. W. (1989). *Phys. Rev. B* **40**, 6442;
Deveaud, B., Clérot, F., Roy, N., Satzke, K., Sermage, B., and Katzer, D. S. (1991). *Phys. Rev. Lett.* **67**, 2355.
41. La Rocca, G. C. and Bassani, F. (1988). *Phys. Lett. A* **247**, 365.

42. Atanasov, R., Bassani, F., and Agranovich, V. M. (1994). *Phys. Rev. B* **50**, 7809.
43. Bloembergen, N., Chang, R. K., Jha, S. S., and Lee, C. H. (1968). *Phys. Rev.* **174**, 813.
44. Agranovich, V. M. and Ginzburg, V. L. (1984). *Crystal Optics with Spatial Dispersion, and Excitons*. Springer, Berlin, Heidelberg.
45. Roslyak, O. and Birman J. I. (2007). *Los Alamos National Laboratory, Preprint Archive, Condensed Matter* 1-10, arXiv:0704.1923v1 [cond-mat.mtrl-sci].
46. Agranovich, V. M. and Galanin, M. D. (1982). *Electronic excitation transfer in condensed matter*. North–Holand, Amsterdam.
47. Landau L. D. and Lifshitz E. M. (1960). *Electrodynamics of Continuous Media*. Pergamon, Oxford.
48. Basko, D. M. (2003). *Energy Transfer from a Semiconductor Nanostructure to an Organic Material and a New Concept for Light-Emitting Devices*. In: *Thin Films and Nanostructures. Electronic Excitations in Organic Based Nanostructures* **31**, edited by V. M. Agranovich and G. F. Bassani. Elsevier, Amsterdam. pp. 447–486.
49. Kos, S., Achermann, M., Klimov, V. I., and Smith, D. L. (2005). *Phys. Rev. B* **71**, 205309.
50. Deveaud, B., Kappei, L., Berney, J., Morier–Genoud F., Portella–Oberli, M. T., Szczytko, J., and Piermarocchi, C. (2005). *Chemical Physics* **318**, 104.
51. Pope, M., Kalmann, H., and Magnante, P. (1963). *J. Chem. Phys.* **38**, 2042.
52. Weisbuch, C. and Vinter, B. (1991). *Quantum Semiconductor Structures. Fundamentals and Applications*. Academic Press, Boston.
53. Achermann, M., Petrushka, M. A., Kos, S., Smith, D. L., Koleske, D. D., and Klimov, V. I. (2004). *Nature* **429**, 642.
54. Klimov, V. I. and Achermann, M. (2005). *Non–contact pumping of light emitters via non-radiative energy transfer*. U.S. Pat. Appl. Publ., 14 pp. CODEN: USXXCO US 2005253152 A1 20051117.
55. Achermann, M., Petrushka, M. A., Koleske, D. D., Crawford, M. H., and Klimov V. I. (2006). *Nano Letters* **6**, 1396.
56. Heliotis, G., Itskos, G., Murray, R., Dawson, M. D., Watson, I. M., and Bradley, D. D. C. (2006). *Adv. Mater.* **18**, 334.
57. Blumstengel, S., Sadofev, S., Xu, C., Puls, J., and Henneberger, F. (2006). *Phys. Rev. Lett.* **97**, 237401.
58. Okuno, T., Lipskii, A. A., Ogava, T., Amagai, I., and Masumoto, Y. (2000). *J. of Luminescence* **87**, 491.
59. Hong, S.-K., Yeon, K. H., and Nam, S. W. (2006). *Physica E* **31**, 48.
60. Basko, D. M., La Rocca, G. C., Bassani, F., and Agranovich, V. M. (2005). *Phys. Rev. B* **71**, 165330.
61. Nozik, A. J. (2002). *Physica E* **73**, 253318.
62. Beard, M. C., Knutsen, K. P., Yu, P., Luther, J. M., Song, Q., Metzger, W. K., Ellington, R. J., and Nozik, A. J. (2007). *Nano Letters* **7**, 2506.

63. Klimov, V. I. (2007). *Ann. Rev. Phys. Chem.* **53**, 635.
64. Zhang Q., Atay T., Tischler, J. R., Bradley, M. S., Bulovic, V., and Nurmikko, A. V. (2007). *Nature nanotechnology* **2**, 555.
65. Agranovich, V. M., and Czajkowski, G. (2008). Los Alamos Natl. Laboratory, Preprint Archive, Condensed Matter 1–4.arXiv: 0801.3794v1.
66. Agranovich, V. M., Benisty, H., and Weisbuch, C. (1997). *Solid State Commun.* **102**, 631.
67. Sumi, H. (1976). *J. Phys. Soc. Japan* **41**, 526;
 Stanley, R. P., Houdré, R., Weisbuch, C., Oesterle, U., and Ilegems, M. (1996). *Phys. Rev. B* **53**, 10995.
68. Lidzey, D. G., Bradley, D. D. C., Skolnick, M. S., Virgili, T., Walker, S., and Whittaker, D. M. (1998). *Nature* **395**, 53.
69. Lidzey, D. G., Bradley, D. D. C., Virgili, T., Armitage, A., and Skolnick, M. S. (1999). *Phys. Rev. Lett.* **82**, 3316.
70. Tartakovskii, A. I., Emam-Ismail, M., Lidzey, D. G., Skolnick, M. S., Bradley, D. D. C., Walker, S., and Agranovich, V. M. (2001). *Phys. Rev. B* **63**, 121302.
71. Lidzey, D. G. Bradley, D. D. C., Skolnick, M. S., and Walker, S. (2001). *Synthetic Metals* **124**, 37,
 Schouwink, P., Lupton, J. M., Belepsch, H. von, Daehne, L., and Mahr, R. F. (2002). *Phys. Rev. B* **66**, 081203(R);
 Takada, N., Kamuta, T., and Bradley, D. D. C. (2003). *Appl. Phys. Lett.* **82**, 1812.
72. Agranovich, V. M. and Ginzburg, V. L. (1984). *Crystal Optics with Spatial Dispersion, and Excitons.* Springer, Berlin, Heidelberg.
73. Vaubel, G. and Baessler, H. (1970). *Mol. Cryst. Liq. Cryst.* **12**, 39;
 Turlet, J.-M. and Philpott, M. R. (1975). *J. Chem. Phys.* **62**, 4260.
74. Tassone, F., Piermarocchi, C., Savona, V., Quattropani, A., and Schwendimann, P. (1997). *Phys. Rev. B* **56**, 7554.
75. Co–Jen Ho, Jefferson Rabbit, R., and Topp, M. R. (1987). *J. Phys. Chem.* **91**, 5599.
76. Fidder, H. (1993). *Collective Optical Response of Molecular Aggregates.* Thesis. University of Groningen.
77. Bonvalet, A., Nagle, J., Berger, V., Migus, A., Martin, J. L., and Joffre, M. (1996). *Phys. Rev. Lett.* **76**, 4392.
78. Holmes, R. J., Kena–Cohen, S., Menon, V. M., and Forrest, S. R. (2006). *Phys. Rev. B* **74**, 235211.
79. Wenus, J., Parashkov, R., Ceccarelli, S., Brehier, A., Lauret, J.–S., Skolnick, M. S., Deleporte, E., and Lidzey, D. G. (2006). *Phys. Rev. B* **74**, 235212.
80. Oulton, R. F., Takada, N., Koe J., Stavrinou, P. N., and Bradley, D. D. C. (2003). *Semicond. Sci. Technol.* **18**, 419.

Chapter 14

REFERENCES

1. Amerongen. H. van, Valkunas, L., and Grondelle, R. van. (2000). *Photosynthetic Excitons*. World Scientific, Singapore.
2. Förster, Th. (1948). *Ann. Phys.* **2**, 55.
3. Reineker, P. In: Höhler, G. (Ed.) (1982). *Exciton Dynamics in Molecular Crystals and Aggregates*. Springer, Berlin, p. 1.
4. Kenkre, V. M. In: Höhler, G. (Ed.) (1982). *Exciton Dynamics in Molecular Crystals and Aggregates*. Springer, Berlin, p. 55.
5. Silbey, R. (1983). In: *Spectroscopy and Excitation Dynamics of Condensed Molecular Systems*. Edited by V. M. Agranovich and R. M. Hochstrasser. North–Holland, Amsterdam, pp. 1–26.
6. May, V. and Kühn, O. (2000). *Charge and Energy Transfer Dynamics in Molecular Systems*. Wiley - VCH, Berlin.
7. Redfield, A. (1965). *Adv. Magn. Res.* **1**, 1.
8. Agranovich, V. M. and Galanin, M. D. (1982). *Electronic Excitation Energy Transfer in Condensed Matter*. North–Holand, Amsterdam.
9. Choi, S-I. and Rice, S. A. (1963). *J. Chem. Phys.* **38**, 366.
10. Kubo, R. (1957). *J. Phys. Soc. Jpn.* **12**, 570.
11. Zubarev, D. N. (1960). *Uspekhi Fiz. Nauk* **71**, 71; English translation: (1960). *Sov. Phys.–Usp.* **3**, 320.
12. Peierls, R. E. (1955). *Quantum Theory of Solids*. Clarendon Press, Oxford.
13. Frölich, H. (1937). *Proc. Roy. Soc.* **160**, 230.
14. Agranovich, V. M. and Konobeev, Yu. V. (1959). *Optika i spektroskopija* **6**, 242; English translation: (1959). *Opt. Spectrosc.* **6**, 115;
    (1963). *Fiz. Tverd. Tela* **5**, 1373; English translation: (1963). *Sov. Phys. - Solid State* **5**, 999;
    (1968). *Phys. Stat. Sol.* **27**, 435.
15. Katal'nikov, V. V. (1968). *Fiz. Tekh. Poluprowodn.* **2**, 1392. English translation: (1969). *Sov. Phys. Semicond.* **2**, 1165;
    (1974). *Fiz. Tverd. Tela* **16**, 3498. English translation: (1974). *Sov. Phys. - Solid State* **16**, 2271;
    Katal'nikov, V. V. and Rudenko, O. S. (1977). *Ukrain. Fiz. Zh.* **22**, 1362.
16. Agranovich, V. M. (1968). *Theory of Excitons*. Nauka, Moscow (in Russian).
17. Kenkre, V. M. and Knox, R. S. (1974). *Phys. Rev. B.* **9**, 5279;
    Kenkre, V. M. (1975). *Phys. Rev. B.* **11**, 1741.
18. Haken, H. and Strobl, G. (1973). *Z. Phys.* **262**, 135.
19. Trlifai, M. (1963). *Czech. J. Phys.* **13**, 631.
20. Kopelman, R. (1983). In: *Spectroscopy and Excitation Dynamics of Condensed Molecular Systems*. Edited by V. M. Agranovich and R. M. Hochstrasser. North–Holland, Amsterdam, pp. 139–184.
21. Pope, M. and Swenberg, C. E. (1999). *Electronic Processes in Organic Crystals and Polymers*. Oxford University Press, Oxford.
22. Fayer, M. D. (1983). In: *Spectroscopy and Excitation Dynamics of Condensed Molecular Systems*. Edited by V. M. Agranovich and R. M. Hochstrasser. North–Holland, Amsterdam, pp. 185–248.

23. Rose, T. S., Rigini, R., and Fayer, M. D. (1984). *Chem. Phys. Lett.* **106**, 13.
24. Rose, T. S., Newel, V. J., Meth, J. S., and Fayer, M. D. (1988). *Chem. Phys. Lett.* **145**, 475.
25. Agranovich, V. M., Ratner, A. M., and Salieva, M. Kh. (1986). *Solid State Commun.* **63**, 329;
Agranovich, V. M. and Leskova, T. A. (1988). *Solid State Commun.* **68**, 1029.
26. Agranovich, V. M., Ratner, A. M., and Salieva, M. Kh. (1988). *Chem. Phys.* **128**, 23.
27. Knoester, J. and Mukamel, S. (1991). *Phys. Rep.* **205**, 1.

Chapter 15

1. Bloch, F. (1930). *Z. Phys.* **61**, 206.
2. Bloch, F. (1932). *Z. Phys.* **74**, 295.
3. Bogoljubov, N. N. (1949). *Lectures on Quantum Statistics*. Kiev.
4. Bogoljubov, N. N. and Tyablikov, S. W. (1959). *Zh. Èksper. Theoret. Fiz.* **19**, 256.
5. Agranovich, V. M. and Toshich, V. S. (1967) *Zh. Èksper. Theoret. Fiz.* **53**, 149; (1968) *Sov. Phys. JETP* **26**, 104.
6. Dyson, F. J. (1956). *Phys. Rev.* **102**, 1230.
7. Kranendonk, J. Van. (1955). *Physica* **21**, 81, 749, 925.
8. Dubovsky, O. A. and Konobeev, Yu. V. (1970). *Fiz. Tverd. Tela* **12**, 406.
9. Landau, L. D. and Lifshitz, E. M. (1963). *Quantum Mechanics*. Pergamon Press, Oxford.
10. Moskalenko, S. A. (1958). *Optics and Spectroscopy* **5**, 147; Lampert, R. A. (1958). *Phys. Rev. Lett.* **1**, 450.
11. Haynes, J. R. (1966). *Phys. Rev. Lett.* **17**, 860.
12. Bogoljubov, N. N. (1947). *Izv. AN SSSR*, Ser. Phys. **11**, 77.
13. Abrikosov, A. A., Gorkov, L. P., and Dzyaloshinsky, I. E. (1962). *Methods of Quantum Field Theory in Statistical Physics*. Fizmatgiz, Moscow (in Russian); English translation: (1975) *Methods of Field Theory in Statistical Physics*. Dover.
14. Bocchieri, P. and Seneci, F. (1965). *Nuovo Cimento* **18 B**, 392.
15. Ilinskaia, A. V., and Ilinski, K. N. (1996). *J. Phys. A: Math. Gen.* **29**, L23.
16. Trlifai, M. (1963). *Czech. J. Phys.* **13**, 631.
17. Chesnut, D. B. (1964). *J. Chem. Phys.* **41**, 472.
18. Witkowski, A. (1966). *Acta Phys. Polon.* **30**, 431; (1967), *Acta Phys. Polon.* **31**, 1.
19. Simpson, O. (1957). *Proc. Roy. Soc.* **A 238**, 402.
20. Gergel, V. A., Kazarinov, R. F., and Suris, R. A. (1968). *Zh. Èksper. Theoret. Fiz.* **54**, 298.

21. Chandrasekhar, S. (1943). Stochastic Processes in Physics and Astronomy, *Rev. Mod. Phys.* **15**, 1–89.
22. Pope, M. and Burgos, J. (1967). *Molecular Crystals* **3**, 215.
23. Spano, F. C., Agranovich, V. M, and Mukamel, S. (1991). *J. Chem. Phys.* **95**, 1400.
24. Spano, F. C. (1995). *Chem. Phys. Lett.* **234**, 29.
25. Chakrabarti, A., Schmidt, A., Valencia, V., Fluegel, B., Masumdar, S., Armstrong, N., and Peyghambarian, N. (1998). *Phys. Rev. B* **57**, R4206.
26. Zoubi, H. and La Rocca, G. C. (2005). *Phys. Rev. B* **72**, 125306.
27. Lifshitz, I. M. (1948). *Zh. Èksper. Theoret. Fiz.* **18**, 293; Lifshitz, I. M. and Peresada, V. I. (1955). *Uchenye zapiski Kharkovskogo Universiteta* **64**; *Trudy Fiz. otd. fiz-mat f-ta* **6**, p. 37. These two papers are included into the volume of selected papers of I. M. Lifshitz (1987): *Physics of real crystals and disordered media*. Moscow, Izd. Nauka. A summary of these works one also can find in: Lifshitz, I. M., Azbel, M. Ya., and Kaganov, M. I. (1971). *Electron theory of metals*. Moscow, Izd. Nauka; English translation: (1973). Consultants Bureau, New York.
28. Litinskaia, M. (2008). *Phys. Rev. B.*, **77**, 155325.
29. Zoubi, H. (2006). *Phys. Rev. B* **74**, 045317.
30. Chesnut, D. B. and Suna, A. (1963). *J. Chem. Phys.* **39**, 2146.
31. Spano, F. S. and Knoester, J. (1994) *Advan. Magn. Opt. Reson.* **18**, 117.
32. Knoester, J. (2002). *Optical properties of molecular aggregates*. In: *Organic Nanostructures: Science and Applications*. Proceedings of the International School of Physics Enrico Fermi, Course CXLIX. Edited by V. M. Agranovich and G. C. La Rocca, Bologna, pp. 149–186.
33. Knoester, J. and Agranovich, V. M. (2003). *Frenkel and charge-transfer excitons in organic solids*. In: Agranovich, V. M. and Bassani, G. F. (Eds.), *Electronic Excitations in Organic Based Nanostructures. Thin Films and Nanostructures* **31**. Elsevier Academic Press, Amsterdam, pp. 1–96.
34. Egorov, V. V. and Alfimov, M. V. (2007). *Physics - Uspekhi* **50** (10), pp. 985–1029.
35. Knapp, E. W. (1984). *Chem. Phys.* **85**, 73.
36. Malyshev, V. and Moreno, P. (1995). *Phys. Rev. B* **51**, 14587.
37. Arkhipov, A. S., Astrakharchik, G. E., Belikov, A. V., and Lozovik, Yu. E. (2005). *JETP Lett.* **82**, 39.

Appendix

1. Landau, L. D. and Lifshitz, E. M. (1960). *Electrodynamics of Continuous Media*. Pergamon, Oxford.
2. Yu, P. Y. and Cardona, M. (1996). *Fundamentals of Semiconductors: Physics and Materials Properties*. Springer, Berlin.
3. Bastard, G. (1988). *Wave Mechanics Applied to Semiconductor Heterostructures*. Les Editions de Physique, Les Ulis, Paris.

4. Bassani, F. and Pastori–Parravicini, G. (1975). *Electronic States and Optical Transitions in Solids.* Pergamon, Oxford.
5. Landau, L. D. and Lifshitz, E. M. (1963). *Quantum Mechanics.* Pergamon Press, Oxford.
6. Agranovich, V. M. and Ginzburg, V. L. (1984). *Crystal Optics with Spatial Dispersion, and Excitons.* Springer, Berlin, Heidelberg.
7. Agranovich, V. M., La Rocca, G. C., and Bassani, F. (1997). *JETP Lett.* **66**, 748.
8. Basko D. M., La Rocca, G. C., Bassani, F., and Agranovich, V. M. (1999). *Eur. Phys. J. B* **8**, 353.
9. Shinada, M. and Sugano, S. (1966). *J. Phys. Soc. Jpn.* **21**, 1936.
10. Agranovich, V. M., Atanasov, R. D., and Bassani, G. F. (1994). *Solid State Commun.* **92**, 295.
11. Cingolani, R. (1997). In *Semiconductors and Semimetals* **44**. Academic Press, pp. 163–226.
12. Hoffmann, M. *private communication.*
13. Umlauff, M., Hoffmann, J., Kalt, H., Langbein, W., Hvam, J. M., Scholl, M., Söllner, J., Heuken, M., Jobst, B., and Hommel, D. (1998). *Phys. Rev. B* **57**, 1390.
14. Andreani, L. C., Tassone, F., and Bassani, F. (1991). *Solid State Commun.* **77**, 641.
15. Deveaud, B., Clérot, F., Roy, N., Satzke, K., Sermage, B., and Katzer, D. S. (1991). *Phys. Rev. Lett.* **67**, 2355.
16. Permogorov, S. (1975). *Phys. Stat. Solidi B* **68**, 9.
17. Permogorov, S. and Travnikov, V. (1979). *Solid State Commun.* **29**, 615.
18. Pelekanos, N., Ding, J., Fu, Q., Nurmikko, A. V., Durbin, S. M., Kobayashi, M., and Gunshor, R. L. (1991). *Phys. Rev. B* **43**, 9354.
19. Kohl, M., Heitmann, D., Grambow, P., and Ploog, K. (1990). *Phys. Rev. B* **42**, 2941.
20. Shah, J. (1996). *Ultrafast Spectroscopy of Semiconductors and Semiconductor Nanostructures.* Springer, Berlin.
21. Zimmermann, R. (1997). *Pure and Applied Chemistry* **69**, 1179.

TABLES

Table 2.1	Characters of the irreducible representations of group C_{2h}	26
Table 2.2	Characters of the irreducible representations of the group C_i.	27
Table 2.3	Character table for the irreducible representations of group D_{2h}.	27
Table 3.1	Crystallographic data	87
Table 3.2	Summary of free molecule data	88
Table 3.3	The angles formed by the mean and the longitudinal axis with the primitive cell vectors for the case of anthracene molecule	89
Table 3.4	Polarization ratio $P(b/a)$ and the Davydov splitting Δ_D in anthracene	91
Table 3.5	Exciton energies and oscillator strengths of anthracene	92
Table 3.6	Exciton energies and oscillator strengths of tetracene	95
Table 12.1	Fitting parameters for MePTCDI and PTCDA	355
Table 12.2	Energies of the lowest bulk and surface states in MePTCDI and PTCDA	356

FIGURES

Fig. 3.1	The hypochromatic effect for the lower excited state	68
Fig. 3.2	The same as in Figure 3.1, for a different set of parameters	68
Fig. 3.3	Ground state potential and asymmetric double-well potential associated with the phenomenon of exciton self-trapping	73
Fig. 3.4	Reflection spectra of the (001) face	86
Fig. 3.5	Derived absorption spectra of the (001) face	87
Fig. 3.6	The structure of the primitive cell in the anthracene crystal.	89
Fig. 3.7	Energy of the two exciton branches of the 0-0 component of the 3800 Å transition	93
Fig. 3.8	Spectrum of excitonic states in 1D molecular crystal with two molecules per unit cell	101
Fig. 4.1	Dispersion relation for Coulomb excitons and polaritons	114
Fig. 4.2	Polariton dispersion	116
Fig. 4.3	The dispersion of polariton in cubic crystals	120
Fig. 4.4	The dispersion relation $\mathcal{E}'(k)$ for angle $\theta = \pi/2$	133
Fig. 4.5	Excitonic dispersion in two-dimensional crystals with retardation	135
Fig. 6.1	C-phonon energy	186
Fig. 6.2	Functions Re $\{\Phi_2(E)\}$ and $\Phi_1(E) = -1$	186
Fig. 6.3	The functions $\Phi_1(E)$ and $\Phi_2(E)$ in the presence of a Fermi resonance	189
Fig. 6.4	Same as Fig. 6.3, with $\varepsilon_{\min} < \tilde{\varepsilon}_2 < \varepsilon_{\max}$	189
Fig. 6.5	Same as Fig. 6.3, with $\tilde{\varepsilon}_2 > \varepsilon_{\max}$	190
Fig. 6.6	Polariton dispersion in the Fermi resonance region, neglecting anharmonicity	191
Fig. 6.7	Same as Fig. 6.6, but with anharmonicity taken into account	193
Fig. 6.8	Dependence of the dielectric function on the frequency in the region of overtone frequencies	195
Fig. 6.9	The $\nu_1 + \nu_3$ band of CO_2	202
Fig. 6.10	Transmission spectrum of the $\nu_2 + \nu_3$ band of N_2O	203

Fig. 6.11	Polariton dispersion in the NH_4Cl crystal	204
Fig. 6.12	Fragments of phonon spectra in scattering by polaritons in the HIO_3 crystal	206
Fig. 6.13	The polariton wavepacket, created by coherent scattering of picosecond laser and Stokes pulses	207
Fig. 6.14	The polariton dispersion curve of ammonium chloride in the vicinity of the $2\nu_4$ two-phonon quasicontinuum	208
Fig. 6.15	Measured polariton dephasing rate $2/T_2$ ps^{-1} vs. polariton frequency	208
Fig. 6.16	Raman scattering spectra for $^{15}N_x{}^{14}N_{1-x}H_4Br$ crystals	212
Fig. 9.1	The levels near the boundary of a molecular crystal	246
Fig. 9.2	The dependence $E(z)$ at the boundary with the vacuum	247
Fig. 9.3	The dependence $E(z)$ in the case $S_B > S_A$	249
Fig. 9.4	The dependence $E(z)$ for the A-layer in the case $S_A > S_B$	249
Fig. 9.5	The dependence $E(z)$ for the A-layer in the case $S_B > S_A$	250
Fig. 9.6	Two-layer one-dimensional model of a crystal with one molecule in an elementary cell	257
Fig. 9.7	Sketch of a 1D interface structure under the influence of an electromagnetic field	262
Fig. 9.8	Dependence of the intensity I of a vibrations on the pumping intensity	264
Fig. 10.1	Dispersion relation for an organic microcavity	267
Fig. 10.2	Examples of the dispersion curves for microcavity polaritons	277
Fig. 10.3	A microcavity with J-aggregates	280
Fig. 10.4	An illustration of the wavevector and energy broadenings of polariton branches	283
Fig. 10.5	The dispersion curves of the coherent polaritonic states and of uncoupled cavity photon and the molecular excitation	287
Fig. 10.6	The energy spectrum of the polaritonic eigenstates in a 1D model microcavity	294
Fig. 10.7	Examples of the spatial structure of the the photon part of the polariton eigenstates in a 1D microcavity with disorder	295
Fig. 10.8	Comparison of the spatial structure of the photon and exciton components of one of the localized polariton eigenstates	296

REFERENCES

Fig. 10.9	Examples of the time evolution of spatially identical wavepackets built out of the polariton eigenstates of a perfect 1D microcavity	298
Fig. 11.1	Absorption (right side) and emission spectrum (left side) of a 170 micron thick crystal at 2 K	302
Fig. 11.2	The number of the dissociated pairs as a function of the total number of excitations at the donor–acceptor interface	308
Fig. 11.3	The steady state number of CTEs occupying the donor–acceptor interface	311
Fig. 11.4	Steady state distribution of the CTE electrostatic energies for different pumping intensities	312
Fig. 11.5	Position of the energy distribution peak as a function of the number of CT excitons	313
Fig. 11.6	Distribution of the nearest-neighbor distances between the charge–transfer excitons	314
Fig. 11.7	Normalized potential profile across a D–A interface	319
Fig. 11.8	Electrostatic potential along a D–A–N superlattice in the presence of 2D CTEs at the D–A interfaces	320
Fig. 12.1	Otto configuration	326
Fig. 12.2	Kretschmann configuration	326
Fig. 12.3	Experimental dispersion curve as a function of internal and external angles of incidence	333
Fig. 12.4	The molecular and exciton levels in a crystal with (a) one molecule and (b) two molecules in a crystal unit cell	338
Fig. 12.5	The bulk exciton and the site shift surface exciton levels in anthracene-type crystals	340
Fig. 12.6	b-polarized reflection spectrum of anthracene in the region of 0-0 transition	341
Fig. 12.7	The effect of a single SSSE transition on the bulk crystal reflection spectrum	342
Fig. 12.8	Lowest bulk and edge state energies for resonance ($E_\mathrm{F} = E_\mathrm{CT} = 0$)	354
Fig. 13.1	The physical configuration under study (a) and the Frenkel and Wannier–Mott excitons in nanostructure (b)	363
Fig. 13.2	The interaction parameter $\Gamma(k)$	367
Fig. 13.3	The dispersion $E_{u,\ell}(k)$ of the upper and lower hybrid exciton branches	369
Fig. 13.4	The same as on Fig. 13.3, but $\delta = 0$	369
Fig. 13.5	The same as in Fig. 13.3, but $\delta = -10$ meV	370

Fig. 13.6	Real and imaginary parts of χ in the linear regime and high excitation	383
Fig. 13.7	The sketch of the planar structure under discussion	387
Fig. 13.8	Schematic and optical properties of the hybrid quantum-well/nanocrystal structure	389
Fig. 13.9	Schematic of the hybrid inorganic–organic semiconductor heterostructures and a simplified energy-level diagram	391
Fig. 13.10	Sketch of the hybrid nanostructure	392
Fig. 13.11	Hybrid organic–inorganic multilayer film	397
Fig. 13.12	Schematics of the microcavity structure	399
Fig. 13.13	Scheme of hybrid structure with two cavities	399
Fig. 13.14	Bare reduced dispersion curves of cavity photon, W–M exciton and Frenkel exciton	405
Fig. 13.15	Dispersion curve for cavity polariton and coupling coefficients	406
Fig. 13.16	Energy relaxation of Wannier–Mott excitons when coupled by the cavity to Frenkel excitons	407
Fig. 14.1	The decay rate of the transient grating signal versus θ^2	422
Fig. D.1	Free L-exciton and Z-exciton lifetime	448
Fig. D.2	Free L-exciton lifetime τ (ns) versus the in-plane wavevector k (cm^{-1}) for three well widths and versus the barrier width L_b	449
Fig. D.3	The same as in Fig. D.2, but for the III–V semiconductor compounds	450
Fig. D.4	Localized X-exciton lifetime τ versus the localization length L	453
Fig. D.5	The same as in Fig. D.4, but for the III–V semiconductor compounds	454

INDEX

Aaviksoo, Ya., 138, 247
absorption
  band, 146
  coefficient, 382
absorption coefficient, 9, 146, 154, 198
absorption index, 144
acceptor, 384, 388
acoustic phonons, 450
activation energy, 420
additional boundary conditions (ABC), 236, 239
additive refraction approximation, 152
admolecules, 340
AFM, 396
Agranovich, V. M., vi, 35, 73, 101, 140, 247–249, 257, 262, 264, 280, 283, 287, 294–296, 298, 305, 308, 399, 405, 407, 417, 439, 446
Ajbinder, H. E., 34
AL, 358
Alfimov, M. V., 435
Anderson localization, 358, 420
Anex, B. G, 332
anharmonicity, 166
  electrooptic, 192
annihilation operator, 37, 373, 377
anthracene, 7, 10, 15, 34, 59, 85, 88, 91, 138, 143, 149, 246, 248, 250, 269, 302, 362, 418, 426
anticommutation rules, 37
anticommutator, 200
Antonyuk, B. P., vii
aromatic molecular crystals, 30
aromatic series, 149
Arrhenius law, 72
ATR, 325
attenuated total reflection (ATR), 196
Auger recombination, 397
azulene, 165

background dielectric constant ϵ_b, 147, 281
Bakalis, L. D., 49, 52
Bakhshiev, N., 148
Basko, D. M., vii, 386, 439, 446
Bassani, F., vii, 73, 109, 138, 439, 446
Bellessa, J., 266
Belousov, M. V., 205, 210, 211
Benisty, H., vii, 407

benzene, 10, 28, 143, 425
Berk, N. F., 34
Bessel function
  modified of zeroth order, 375
Bethe splitting, 15, 55
Bethe, H., 15
biexciton, 426
biexcitons, 98, 146, 199
  in CuCl crystal, 201
biphenyl, 143
biphonon, 167
  binding energy, 170
  energy levels, 175
  oscillator strength, 194
  transverse, 194
biphonons
  local, 168
  longitudinal, 194, 195, 203, 205
  quasilocal, 168
  surface, 194
  transverse, 195, 203
Birman, J. L., vii, 168, 377, 384
bistable energy transmission, 262
Bloch functions, 366, 441
Bloch states, 75
Bloch, F., 424
Bloembergen, N., 199
blue-shift, 381
Blume, M., 34
Blumstengel, S., 392
Bocchieri, P., 429
Bogani, F., 168, 172, 202
Bogoljubov, N. N., 424, 437
Boltzmann constant, 137
Boltzmann equation, 413
Boltzmann statistics, 416
Born approximation, 426
Born, M., 19, 105, 142, 144, 163
Born–Oppenheimer approximation, 85
Bose operators, 41, 103, 170
Bose–Einstein condensation, 431
  of cavity polaritons, 284
boson operators
  for cavity excitation, 402
  for Frenkel excitation, 402
  for Wannier excitation, 402
bosons, 269
bottle neck, 119
Bradley, D. D. C., 267, 390
Brillouin surfrace, 30

Brillouin zone, 2, 14, 29, 74, 86, 175, 203
  two-dimensional, 327
Brodin, M. S., 90
Broude, V. L., v, 6, 10, 83, 159

Califano, S., 168
CARS, 166, 205, 325
Carter, G. M., 336
cavity exciton–polaritons, 279
cavity polaritons, 267
cavity-polaritons, 400
CdS, 6
CdSe, 389, 396
CdSe/ZnS, 389
CdTe
  photovoltage, 316
Chakrabarti, A., 433
charge transfer (CT) excitons, 327, 345
charge transfer exciton, 301
charge-transfer excitons, v
charged
  excitons, 35
  Frenkel excitons, 35
  triplet excitons, 35
Chen, Y. J., 336
Cherrof, G., 316
Chesnut, D. B., 434
chlorophyll, 410
Choi, S-I., 32, 412
Clark, L. B., 85–87
Claus, R., 204
Claxton, T. A., 89
CO_2, 202
coherent active Raman spectroscopy (CARS), 166
coherent anti-Stokes Raman spectroscopy (CARS), 325
coherent potential approximation, 213
coherent quasiparticles, 282
coherent scattering, 206
cold photoconductivity, 306
commutation relations, 255
commutation rules, 41
compulsory degeneracy, 29
conduction band, 366
Cooper pairs, 430
coronene, 306
Coulomb excitons, 270
Coulomb gauge, 103
Coulomb interaction, 11, 173, 196, 327, 378
Coulomb surface excitons, 327, 330
  dispersion law, 332
Craig, D. P., v, 53, 61, 89
creation operator, 37, 377
  of the Frenkel exciton, 373

crystal
  uniaxial, 234
crystal Hamiltonian, 11
crystal space group, 24
crystalline hydrogen, 167
CT excitons, 327, 345
CTE, 301
CTIP, 332
Cummins, P. G., 143
CuPc, 372
CW, 323
cyanine dye, 396
cyclic boundary conditions, 14, 327
Czajkowski, G., vii

damping
  of biphonon states, 186
dark conductivity, 317
Davydov soliton transfer, 213
Davydov splitting, v, 7, 15, 16, 21, 32, 53, 55, 62, 72, 85, 87, 150, 268, 277
  anthracene, 91
  SSSE, 339
Davydov, A. S., v, 10, 15, 23, 69, 75, 89, 151
Debye temperature, 407, 420
Deigen, M. F., 33, 74
Demidenko, A. A., 228
density matrix, 379
depletion, 317
Deutch, J. M., 160
Deveaud, B., 138, 247, 387
Dexter, D. L., 160
diagonal disorder, 358
diamond, RSL spectra, 168
dielectric constant tensor, 142
  uniaxial crystal, 149
dielectric function, 194
dielectric tensor, 19, 140, 196
  transverse, 196
diffusion
  constant, 411
  equation, 411
  length, 411
diffusion coefficient, 34, 35
diffusive regime, 411
dipole moment, 142
dipole-allowed transitions, 144
dipole-forbidden transitions, 144
dispersion relation
  hybrid excitons, 368
donor, 384, 388
donor–acceptor (D-A) complex, 301
donor–acceptor interface, 305
Dows, D. A., 202
Dubovsky, O. A., vii, 156, 210, 426

Dunmur, D. A., 143, 165
durene, 33
Dyson equation, 82
Dyson, F. J., 424
Dzyaloshinsky, I. E., 215

EELS, 325
effective mass approximation, 4
Egorov, V. V., 435
Einstein coefficients, 8
electron, 366
electron energy loss spectroscopy (EELS), 325
electron–exciton complexes, 199
electronic paramagnetic resonance, 31
envelope functions, 366
EPR, 31, 32
ϵ_b, 281
Ewald method, 85
Ewald's method, 19
Ewald, P. P., 18, 105
excimer, 74
excitation transfer, 13
exciton, v, 1
  Bohr radius, 234
  de Broglie wavelength, 417
  dead layer, 247
  diffusion coefficient, 411
  diffusion constant, 420
  effective mass, 235
  group velocity, 414
  lifetime, 7, 421
  localized, 6, 70, 146, 329
  longitudinal, 235, 331
  mean free path, 410
  mean free time, 329
  mechanical, 20, 21, 150
  self-trapped, 418
  self-trapped (ST), 75
  transverse, 235
exciton surface polaritons, 332
exciton–phonon interaction, 69
excitonic band, 13
excitonic bands, 15
excitonic states, 36
excitons
  Coulomb, 11, 19, 140
  diffusion, 358
  in CuCl crystal, 201
  in layered structures, 246
  large radius, 2
  mechanical, 19, 140
  mobility, 431
  nonradiative decay, 7, 426
  singlet, 30
  singlet Frenkel, 31

  small radius, 2, 12, 143
  surface, 325
  triplet, 30
  triplets, 12
  two-dimensional, 238, 247
  Wannier–Mott, 2, 21

factor group, 24
Fano antiresonance, 204, 205
Fano, U., 105
Faraday effect, 165
Fayer, M. D., 421
FE, 1, 361
Fermi operators, 435
Fermi resonance, 167, 169, 180, 188, 251
  with plasmons, 196
Fermi resonance bands, 257
Fermi's golden rule, 119, 386, 406, 415, 444, 449
Fermi, E., 252, 257
Flytzanis, C., 201, 205
Forrest, S. R., vii, 268, 408
four-photon process, 201
Fourier representation, 183
Fox, D., 89
Förster energy transfer, 384, 410, 418
Förster mechanism, 159, 385
Förster radius, 384
Förster transfer (FRET), 391
Förster transfer rate, 444
Franck-Condon energy, 84
Frank–Condon principle, 71
Frenkel biexcitons, 99
Frenkel exciton, v, 1, 2, 36, 77, 177, 266, 301, 345, 424
  radius, 366
Frenkel excitons, 247, 269, 361, 373, 376, 381, 382
  surface, 327
  two-dimensional, 247
Frenkel, Ya. J., v, 1, 10, 15, 69, 75, 140, 151, 328
FRET, 391, 397
Frölich, H., 414
Fulton, R. L., 143

GaAlAs, 372
GaAs, 336, 400, 440
Gale, G. M., 206
Gardini, G., 168
Gartstein, Yu. N., vii, 295
gas-condensed matter shift, 41
Gergel, R. F., 431
Gergel, W. A., 430
germanium, 168
Giacintov, N. E., 34

Ginzburg, V. L., vi, 140
Goldhirsh, I., 94
Goldstein, B., 316
Gorelik, V. S., 205
Gotschev, I., 211
Green's function
  biphonon theory, 180
  many times, 200
  of self–adjoint operator, 175
  retarded, 182
  two-particle, 175
Grescishkin, W. S, 34
Gross, E. F., 5
group velocity, 5, 331, 336
growth direction, 362
Grun, J. B., 201
gyration tensor, 242
gyration vector, 242
gyrotropy
  superlattice linear response, 238
gyrotropy constant, 239

Haarer, D., 301, 302
H-aggregate, 345, 433
Haken, H., 3, 410
Haken–Strobl–Reineker model, 34, 410, 419
Hanamura, E., 138, 201
hard spheres approximation, 425
harmonic approximation, 253
Hartree–Fock approximation, 379
Haynes, J. R., 427
HCl, 168
He, 10
Heaviside' function, 218
heavy hole band, 440
Heisenberg equation, 182
Heisenberg Hamiltonian, 424
Heisenberg operators, 379
Heisenberg representation, 412
Heitler–London approximation, 10, 17, 36, 42, 49, 54, 346
Heliotis, G., 390, 391
Henneberger, F., 391
Herring, C., 29
HIO_3, 204
HLA, 49
Hochstrasser, R. M., v, 34, 165
Hoffmann, M., vii, 83, 302
Hoffmann, R., 67, 68
hole, 366
holes
  heavy, 366
  light, 366
Holmes, R. J., 267
Holstein

Hamiltonian, 84
  model, 84
Holstein, T., 84
Holstein–Primakoff representation, 97
Hong, H. K., 156
Hong, S.-K., 395
Hopfield, J. J., 105, 106
hopping integral, 346
Hornig, D. F., 168
Hoschen, J., 156
Huang, K., 19, 142, 144, 163
Hutchison, C. A., 32
hybridization parameter, 373
hyperpolarizabilities, 321, 323

impurities
  interstitial, 145
  substitutional, 145
induction (displacement) vector, 18
inelastic light scattering, 325
InGaAs, 387
InGaN, 390
InGaN/GaN QW, 389
interband polarization, 377
interband transitions, 440
internal field coefficients, 142
internal field correction, 143
intraband transitions, 440
inversion centre, 164
iodoform, 143
Ioffe–Regel criterion, 420
IQW, 362
irreducible representation, 25
isotopic shift, 209
isotopic substitution impurities, 208

J-aggregate, 52, 76, 284, 287, 345, 396, 407, 435
J-aggregated cyanine dye, 267, 279
J-band, 435
Jordan–Wigner transformation, 49, 52, 435
Jortner, J., 156
Joule losses, 392

Kamchatnov, A. M., vii, 251
Kane's energy, 366, 442
Karl, G., 180
Katalnikov, V. V., 418
Kazarinov, R. F., 430
Keller, S. P., 316
Kenkre, V. M., 419
Kerr nonlinearity, 321, 384
Kerr susceptibility, 245
Ketskemety, I., 9
Kèna-Cohen, S., 268

Khokhlov, Yu. K., 143, 163
kinetic coefficients, 239
Kirchhoff law, 9
Kiselev, S. A., vii, 305, 311–314
Klimov, V. I., 388, 389
Knapp, E. W., 435
Knoester, J., 49, 52, 73, 76
Knox, R. S., 3, 5, 229, 419
Konobeev, Yu. V., vii, 156, 228, 417, 426
Kopelman, R., 420
$\mathbf{k} \cdot \mathbf{p}$ perturbation theory, 441
Kr, 10
Krauzman, M., 168, 205
Kravets integral, 146
Kravtsov, V. E., vii
Kretschmann configuration, 325
Kretschmann, E., 325
Krishnan, R. S., 168
Krivenko, T. A., 211
Kronenberger, A., v
Kubo, R., 412
Kühn, O., 419

La Rocca, G. C., vii, 439, 446
Lagrange equations, 255
Lagrangian, 252, 255
Lalov, I. I., vii, 173, 211
Landau superfluidity conditions, 430
Landau, L. D., 75
Landauer, R., 316
Langmuir–Blodgett bimonolayer, 266
Langmuir–Blodgett films, 306, 314
Langmuir–Blodgett technique, 388
lattice vibrations, 77
LED, 360
LED devices, 320
LEDs, 387, 390
Leo, K., vii
Leonhardt, M., 356
Leskova, T. A., vii
Lidzey, D. G., 399, 408
Lifshitz, E. M., 209
Lifshitz, I. M., 433
light hole band, 440
light-emitting devices, 361
light-emitting devices (LEDs), 387
$LiNbO_3$, 204
Liouville equation, 410
Litinskaia, M., vii, 277, 433
local field method, 141
local states, 209
long-wavelength field, 18
longitudinal–transverse splitting
  of biphonons, 173
  of excitons, 266
Lorentz factor, 146, 148

Lorentz local field correction, 231
Lorentz, H. A., 230
Lorenz–Lorentz formula, 141, 144, 151, 356
Lozovik, Yu. E., vii
Lyddane–Sachs–Teller formula, 195
Lyons, L. E., 31
Lyubarsky, G. Ya., 2

magnons
  bound states, 167
Mahan, G. D., 52, 143, 160, 162
Malshukov, A. G., vii
Mangum B. W., 32
Maradudin, A. A., 335
Marisova, S. V., 90
Markov process, 418
master equation, 410
material equations, 122
Matsui, A., 72
Mavrin, B. N., 204
Maxwell equations, 280
Maxwell's equations, 141, 327, 343
May, V., 419
Mazo, R. M., 160
MC (microcavity), 400
McClure, D. S., 10, 78
McConnell, H. M., 30, 31, 33, 34
mean polarizability method, 356
MePTCDI, 345
Merrifield, R. E., 31, 82
Mitin, G. G., 208
molecular crystals, 1, 140
molecular level shift, 248
molecular liquids, 140
molecular orbital (MO) method, 85
Mössbauer effect, 151
Mukamel, S., vii
multiexcitons, 397
Munn, R. W., 165

^{15}N, 210
naphthalene, 10, 15, 26, 30, 32, 33, 59, 85, 91, 143, 149, 418, 425
Ne, 10
neutrons
  scattering, 167
$^{14}NH_4Br$, 210
NH_4Cl, 204, 207
Nienhuis, G., 160
noble gases, 10
Noe, L. J., 165
nongyrotropic crystals, 144
nonlinear polarizabilities, 199
nonlinear polarizability
  biphonons, 168

nonlinear polarizability tensor, 230
N$_2$O, 203
normal coordinates, 252
normal vibrations, 166
NTDCA, 372

Obreimov, I. V., v, 157
occupation numbers, 36
octupole, 143, 162
Okuno, T., 395
OMBD, 305
Onodera, Y., 156
optical phonon, 171
OQW, 362
organic dyes, 315
organic molecular beam deposition, 305
organic quantum well (OQW), 362
oriented gas, 10
Orrit, M., 136, 340, 341
Otto configuration, 326
Otto, A., 325
Ovander, L. N., 98

Pauli blocking factor, 380
paulions, 269
PbS, 398
PbSe, 395, 397, 398
Peierls, R. E., 1, 75, 151
Pekar, S. I., 33, 74, 75, 106, 125, 216
Pensak, L., 316
pentacene, 85, 93, 371
percolation model, 420
permutation rules, 40
perylene, 306, 345
Petelenz, P., 302
Peterson, D. L., 78, 83, 84
phase space filling, 322
phase space filling (PSF), 362
phenanthrene, 143
Philpott, M. R., vii, 83, 85, 93, 143, 301, 338, 340, 341
phonons, 166
  complexes, 167
  interaction, 166
  optical, bound states, 167
photogeneration, 397
photovoltaic effect, 315
picosecond transient grating (TG) method, 421
Pitaevsky, L. P., 170, 215
Pockrand, I., 266
Pogarev, D. E., 205
point group, 24
Poisson equation, 439
polariton, 105
  creation and annihilation operators, 192
  decay into two phonons, 204
  dispersion at the Fermi resonance, 193
  dispersion in Fermi resonance region, 190
  dispersion law, 204
  dispersion, biphonons, 168
  Fermi resonance, 190, 207
  sliding, 119
polaritons, 106, 167, 169, 190, 431
  dispersion law, 167, 194
  Raman-active, 206
  transverse, 194
polarizability, 141
  mean, approximation, 151
polarization, 141
  functions, 379
  intraband, 379
polarization ratio, 90, 355
polaron, 75, 328
polarons, 33
Polivanov, Yu. N., 205
polyfluorene, 390
polymethine dyes, 435
Pope, M., v, vii, 387, 420
POPOP, 391
populations, 379
porphyrin dyes, 267
Prevot, B., 168
Price, W. C., 211
Prikhotko, A. F., v, 10
Pringsheim, P., v
Proville, L., 172
PSF, 362
PTCDA, 302, 345, 372, 448
pumping intensity, 263
pyrene, 34, 74, 418

QD, 361
quadrupole, 143, 162
quantum confined Stark effect, 372
quantum confinement, 345
quantum dot, 361
quantum well, 361
  inorganic, 362
  organic, 362
quantum wire, 361
quantum wires, 372
quantum yield, 247
quasibiphonon energy, 198
quasibiphonons, 186, 199, 204, 205
Quattropani, A., 109
QW, 361
QWW, 361

Rabi splitting, 267, 268, 278, 286
Raman scattering, 179, 207

Raman scattering (RSL), 166
Rashba, E. I., v, 6, 10, 71, 74, 76, 82, 146, 154, 155, 159
Rayleigh surface waves, 325
reciprocal lattice, 14
Redfield theory, 410
Redfield, A., 410
reflection coefficient, 402
refraction index, 8, 198
refractive index, 144, 154
Reineker, P., vii, 34, 410
relaxation time, 336
retarded interaction, 103
Rice, S. A., 412
Robinson, G. W., 156
Ron, A., 168
Rose, T. S., 302
Roslyak, O., 384
rotating wave approximation, 254
rotation elements, 24
rotatory power, 165
RSL, 166
    by polaritons, 167, 190
    second order, 191
Rudoy, Yu, 94
Rupasov, V. I., vii
Ruvalds, I., 168
Rydberg formula, 5

Salvi, P. R., 202
saturation density, 382
Schettino, V., 202
Schlosser, D. W., 85
Schrödinger equation, 1, 14
    hydrogen-like atom, 3
Schrödinger picture, 36
Schrödinger representation, 38, 174
second harmonic, 201
second quantization, 36
secular equation, 14
self-induced transparency, 336
self-trapping, 71
Seneci, F., 429
sensitized fluorescence, 420
Sheka, E. F., v, 6, 89, 168
short-wavelength field, 18
Si, 336
σ-conjugated silane polymers, 267
Silbey, R., 89, 419
silicon, 168
Simpson, O., 430
Simpson, W. T., 78, 83, 84
Simpson, W .T., 332
singlet excitons, 7, 30
singlet state, 30
site group, 24, 25

site shift surface excitons (SSSE), 337
Smith, D. Y., 160
So, F. F., 356
solar cells, 315, 360
soliton, 259
solitons, 260
Soos, Z. G., 83
Spano, F. C., 432
spatial dispersion, 20, 142, 144, 155, 194
spin degeneracy, 378
Spin-Hamiltonian, 32
SSSE, 337
ST, 71, 75
Stark effect, 165
Stepanov, B. I., 9
Sterin, Kh. E., 204
Sternlicht, H., 30, 31, 33, 34
Strobl, G., 410
strong coupling, 361
Sugakov, V. I., 340, 341
sum rules, 62
Suna, A., 82, 434
superlattice
    organic, 233
    period, 233
superradiance, 247
    in two–dimensional excitons, 247
superriadance, 138
superrradiant radiative decay, 247
surface biphonons, 213
    dispersion law, 195
surface excitons, 325
    autolocalization, 359
    coherent states, 328
    dispersion law, 328
    group velocity, 334
surface polariton, 213, 327, 332
surface polaritons, 341
    nonlinear, 335
    self-focusing, 335
surface polarization current, 343
surface soliton, 336
surface state, 213
Suris, R. A., 430
susceptibility
    of the hybrid structure, 382
Suslov, I. M., 328
Swanson, J. A., 316
Swenberg, C. E., v, 34, 420

Talanina, I. B., vii
Tamm states, 314
Tamm surface levels, 325
Tamm surface states, 345
TC (thiacarbocyanine), 407
TDBC, 279, 396

terrylene, 400
tetracene, 85, 86, 93, 268, 269, 345, 371, 400
TG method, 421
three-photon absorption, 201
time-reversal operation, 29
Tomlinson, W. J., 335
Toshich, V. S., vii
Toshich, W. S., 98
total reflection, 331
Tovstenko, V. I., 344
Toyozawa, Y., 119, 156, 229
translation operator, 13
translational degeneracy, 12
transverse photons, 191
trioctylphosphine (TOP), 388
trioctylphosphine oxide (TOPO), 388
trions, 35
triplet excitons, 30
triplet state, 30
triplet-excitons, 7
Trlifaj, M., 419
Tserkovnikov, Yu., 94
Tubino, R., 168
Turlet, J. M., 340, 341
two-photon absorption, 201
Tyablikov, S. V., 94, 424, 437
Tyablikov–Bogoljubov method, 437

Umklapp processes, 104
uncertainty relation, 414

valence band, 366
van der Waals forces, 140, 257, 306
Van Hove singularities, 177
Van Kranendonk model, 167, 178, 202, 203
Van Kranendonk, J., 167, 179, 180, 424
vector potential, 103
vibronic states, 78
VOPc, 372

W–M, 361
Walmsley, S. M., v
Wannier equation, 379
Wannier excitons, 368
Wannier, G. H., 1
Wannier–Mott excitons, 2, 21, 270, 322, 361, 373, 374, 376, 406
Wannier-Mott exciton, v, 301
Warman, J., 305
wave
  extraordinary, 234
  ordinary, 234
WE, 368
weak coupling, 361

Weisbuch, C., vii, 267, 407
Weiss molecular field approximation, 308
Whiteman, J .D., 34
Wildenthal, H., vii
Winston, H., 24, 28
Winter, F. X., 204
Wolf, H. C., 10, 91
Wünsche, A., 143

Xe, 10

Yakhot, V., 94
Yatsiv, Sh., 89
Yudson, V. I., vii

Zakhidov, A. A., vii
Zawadowski, A., 168
zinc–blende structure, 366
ZnCdSe, 372
ZnO, 391
ZnS, 396
  photovoltage, 316
ZnSe, 395, 440
ZnSe/ZnCdSe, 368